sociology 8
the essentials

Margaret L. Andersen
University of Delaware

Howard F. Taylor
Princeton University

With
Kim A. Logio
Saint Joseph's University

CENGAGE
Learning®

Australia • Brazil • Mexico • Singapore • Spain • United Kingdom • United States

Sociology: The Essentials, **Eighth Edition**

Margaret L. Andersen, Howard F. Taylor, and Kim A. Logio

Vice President, General Manager: Erin Joyner

Product Manager: Seth Dobrin

Content Developer: Robert Jucha

Content Coordinator: Naomi Dreyer

Product Assistant: Coco Bator

Media Developer: John Chell

Market Development Manager: Michelle Williams

Content Project Manager: Cheri Palmer

Art Director: Caryl Gorska

Manufacturing Planner: Judy Inouye

Rights Acquisitions Specialist: Thomas McDonough

Production Service: Jill Traut, MPS Limited

Photo Researcher: Reba Frederics, PreMedia Global

Text Researcher: Pinky Subi, PreMedia Global

Copy Editor: Heather McElwain

Illustration and Composition: MPS Limited

Text Designer: Jeanne Calabrese

Cover Designer: Lee Friedman

Cover Image: Anthony Marsland/Stone/ Getty Images

For product information and technology assistance, contact us at
Cengage Learning Customer & Sales Support, 1-800-354-9706

For permission to use material from this text or product, submit all requests online at **www.cengage.com/permissions**
Further permissions questions can be emailed to
permissionrequest@cengage.com

Library of Congress Control Number: 2013941761

Student Edition:

ISBN-13: 978-1-285-43132-1

ISBN-10: 1-285-43132-4

Cengage Learning
200 First Stamford Place, 4th Floor
Stamford, CT 06902
USA

Cengage Learning is a leading provider of customized learning solutions with office locations around the globe, including Singapore, the United Kingdom, Australia, Mexico, Brazil, and Japan. Locate your local office at **www.cengage.com/global**

Cengage Learning products are represented in Canada by Nelson Education, Ltd.

To learn more about Cengage Learning Solutions, visit **www.cengage.com**

Purchase any of our products at your local college store or at our preferred online store **www.cengagebrain.com**

Printed in the United States of America
1 2 3 4 5 6 7 17 16 15 14 13

Brief Contents

Contents

6 Groups and Organizations 122

7 Deviance and Crime 144

PART III SOCIAL INEQUALITIES

8 Social Class and Social Stratification 170

9 Global Stratification 200

10 Race and Ethnicity 226

11 Gender 254

PART IV SOCIAL INSTITUTIONS

14 | Education and Health Care 336

15 | Economy and Politics 360

PART V SOCIAL CHANGE

16 Environment, Population, and Social Change 390

Boxes

Maps

Preface

Sociology: The Essentials is a book that teaches students the basic concepts, theories, and insights of the sociological perspective. With each new edition come new challenges—challenges that stem from new generations of students with different learning styles; challenges that stem from the diversity among students who will study this book; and challenges that stem from the changes that are taking place in society. One of the most important changes taking place today is how students learn and how they are engaged with their course material, often in the form of online learning resources. With that in mind, *Sociology: The Essentials*, eighth edition, takes full advantage of this revolutionary change by having a fully electronic version of the book available, which allows for personalized, fully online digital learning—a platform of content, assignments, and learning resources that will engage students in an interactive mode, while also offering instructors the opportunity to make individualized configurations of course work. Some will want to continue using the printed version of the book, still enhanced with various pedagogical features. But those who want to enhance their curriculum through online resources will be able to utilize the new MindTap Sociology in the way that best suits your course.

However the book is used, we have updated it to reflect the latest social changes and developments in sociological scholarship. Because we revise this book regularly, we are somewhat amazed, even as sociologists, to see how much can change even in the relatively short period of time between editions. Our book constantly adapts, not just to accommodate new scholarship that appears at an amazing pace, but also to recognize significant changes in society that occur.

In this edition, we have maintained the themes that have been the book's hallmark from the start: a focus on diversity in society, attention to society as both enduring and changing, the significance of social context in explaining human behavior, the increasing impact of globalization on all aspects of society, and a focus on the critical thinking and analysis of society that is fostered through sociological research and theory. We know that studying sociology opens new ways of looking at the world. As we teach our students, sociology is grounded in careful observation of social facts, as well as analyses of how society operates. For students and faculty alike, studying sociology can be exciting, interesting, and downright fun, even though it also deals with sobering social issues, such as the growing inequality that marks our time, as just one example.

In this book, we try to capture the excitement of the sociological perspective, while also introducing students to how sociologists do research and how they theoretically approach their subject matter. We know that most students in an introductory course will not become sociology majors, though we hope, of course, that our book and their teacher encourages them to do so. We want to give students, no matter their area of study, a way of thinking about the world around them that is not immediately apparent. This is especially reflected in a new feature of this edition—a short, boxed insert in every chapter entitled, "What Would a Sociologist Say?" Here, we take a common topic and, with informal writing, briefly discuss how a sociological perspective would approach understanding on that particular issue. We think this new feature will help students see the unique ways that sociologists view everyday topics—things as commonplace as the funeral of a superstar, finding a job, or sports in popular culture.

And, importantly, we want our book to be engaging and accessible to undergraduate readers, while also preserving the integrity of sociological research and theory. Our experience in teaching introductory students shows us that students can appreciate the revelations of sociological research and theory if presented in a way that engages them and connects to their lives. We have kept this in mind throughout this revision and have focused on material that students can understand and apply to their own social worlds.

CRITICAL THINKING AND DEBUNKING

We use the theme of *debunking* in the manner first developed by Peter Berger (1963) to look behind the facades of everyday life, challenging the ready-made assumptions that permeate commonsense thinking. Debunking is a way for students to develop their critical thinking, and we use the debunking theme to help students understand how society is constructed and sustained. This theme is highlighted in the **Debunking Society's Myths** feature found throughout each chapter.

In this edition, we also include a feature to help students see the relevance of sociology in their everyday lives. The box feature **See for Yourself** allows students to apply a sociological concept to observations from their own lives, thus helping them develop their critical abilities and understand the importance of the sociological perspective.

Critical thinking is a term widely used but often vaguely defined. We use it to describe the process by which students learn to apply sociological concepts to observable events in society. Throughout the book, we ask students to use sociological concepts to analyze and interpret the world they inhabit. This is reflected in the **Thinking Sociologically** feature that is also present in most chapters.

Because contemporary students are so strongly influenced by the media, we also encourage their critical thinking through the box feature called **A Sociological Eye on the Media.** These boxes examine sociological research that challenges some of the ideas and images portrayed in the media. This not only improves students' critical thinking skills but also shows them how research can debunk these ideas and images.

A FOCUS ON DIVERSITY

When we first wrote this book, we did so because we wanted to integrate the then new scholarship on race, gender, and class into the core of the sociological field. We continue to see race, class, and gender—or, more broadly, the study of inequality—as one of the core insights of sociological research and theory. With that in mind, diversity, and the inequality that sometimes results, is a central theme throughout this book. A boxed theme, **Understanding Diversity**, highlights this feature, but you will find that analysis of inequality, especially by race, gender, and class, is woven throughout the book.

SOCIAL CHANGE

The sociological perspective helps students see society as characterized both by constant change and social stability. How society changes and the events—both dramatic and subtle—that influence change are analyzed throughout this book. New material is added throughout the text that comments on the impact of the economic recession that began in 2008 and shows students how their lives—seemingly individual—are greatly influenced by social structures beyond their control.

GLOBAL PERSPECTIVE

One of the main things we hope students learn in an introductory course is how broad-scale conditions influence things within their everyday lives. Understanding this idea is a cornerstone of the sociological perspective and one of the main lessons learned in introductory courses. One way to see this is to help students understand how the increasingly global character of society affects day-to-day realities. Thus, we use a global perspective to examine how global changes are affecting all parts of life within the United States, as well as other parts of the world. This means more than including cross-cultural examples. It means, for example, examining phenomena such as migration and immigration or helping students understand that

their own consumption habits are profoundly shaped by global interconnections. The availability of jobs, too, is another way students can learn about the impact of an international division of labor on work within the United States. Our global perspective is found in the research and examples cited throughout the book, as well as in various chapters that directly focus on the influence of globalization on particular topics, such as work, culture, and crime. The map feature **Viewing Society in Global Perspective** also brings a global perspective to the subject matter.

NEW TO THE EIGHTH EDITION

We have made various changes to the eighth edition to make it current and to reflect new developments in sociological research. Taken together, these changes should make the eighth edition easier for instructors to teach and even more accessible and interesting for students.

As in the previous edition, we include a separate chapter on sociological research methods (Chapter Three, "Doing Sociological Research"), but we place it after the chapter on culture as a way of capturing student interest early. *Sociology: The Essentials* is organized into five major parts: "Introducing the Sociological Imagination" (Chapter 1); "Studying Society and Social Structure" (Chapters 2 through 7); "Social Inequalities" (Chapters 8 through 12); "Social Institutions" (Chapters 13 through 15); and "Social Change" (Chapter 16).

Part I, "Introducing the Sociological Imagination," introduces students to the unique perspective of sociology, differentiating it from other ways of studying society, particularly the individualistic framework students tend to assume. Within this section, **Chapter 1, "The Sociological Perspective,"** introduces students to the sociological perspective. The theme of debunking is introduced, as is the sociological imagination, as developed by C. Wright Mills. This chapter briefly reviews the development of sociology as a discipline, with a focus on the classical frameworks of sociological theory, as well as contemporary theories, such as feminist theory and postmodernism. The eighth edition adds examples from current events to capture student interest, including the impact of the recent recession, the high rate of suicide among veterans, the influence of social media, and the rise of the so-called "boomerang generation."

In **Part II, "Individuals and Society,"** students learn some of the core concepts of sociology. It begins with the study of culture in **Chapter 2, "Culture and the Media,"** reflecting the significance of the media in the lives of our students. There is a new section on how the widespread availability of Internet-based blogs, chat groups, and social networks is changing how people communicate, including about current events. We include a discussion of social media as a force shaping contemporary culture. The new box, "What Would a Sociologist Say?" asks students to think sociologically about why funerals and deaths of media superstars so captivate the public. We also include new data on television and computer habits and have added a discussion of blogging as an example of cultural change.

Chapter 3, "Doing Sociological Research," contains a discussion of the research process and the tools of sociological research—the survey, participant observation, controlled experiments, content analysis, historical research, and evaluation research. New to this chapter are recent studies, such as Alice Goffman's participant observation study of people on the run from the law; techniques new to this text such as the Solomon Four-Group Experimental Design; and concepts such as unobtrusive measurement.

Chapter 4, "Socialization and the Life Course," contains material on socialization theory and research, including agents of socialization such as the media, family, and peers. Research on how families teach about race is presented, and theories of socialization are discussed. The chapter also includes information about aging and the life course. In addition to updated statistics about childhood, adolescence, and aging, there is added discussion about how socialization has changed. The "What Would a Sociologist Say?" box, for example, looks at online interaction and how the socialization process takes place in cyberspace.

Chapter 5, "Social Interaction and Social Structure," emphasizes how changes in the macrostructure of society influence the microlevel of social interaction. We do this by focusing on technological changes that are now part of students' everyday lives and making the connection between changes at the societal level in the everyday realities of people's lives. New material is included on game theory, on interpersonal attraction, and on the demographic composition of Internet users. The material in this chapter gives attention to the influence of cyberspace on social interaction. Also new are a "Doing Sociological Research" box on the "Prisoner's Dilemma" game interaction, and a "What Would a Sociologist Say?" box on congressional debates.

In **Chapter 6, "Groups and Organizations,"** we study social groups and bureaucratic organizations, using sociology to understand the complex processes of group influence, organizational dynamics, and the bureaucratization of society. In this edition, we have added a discussion of organizational culture, using the scandal at Penn State as an example. We have also added a new discussion of diversity in organizations, based on new scholarship on that topic.

Chapter 7, "Deviance and Crime," includes the study of sociological theories and research on deviance with attention to labeling theory; modern-day corporate crime and deviance; and the effects of race, class, and gender on arrest rates. Deviance is seen as caused by the combination or *intersection* of personality variables and social–structural variables. The core material is illustrated with contemporary events, such as the rampage shooting in Arizona and the horrific mass murders of first-grade children in Newtown, Connecticut. Included in this chapter is recent research on opinions on gun ownership, the mass racialized incarceration of Blacks and Hispanics in U.S. prisons, and a discussion of what it means to be "made."

In **Part III, "Social Inequalities,"** each chapter explores a particular dimension of stratification in society. Beginning with the significance of class, **Chapter 8, "Social Class and Social Stratification,"** provides an overview of basic concepts central to the study of class and social stratification. The chapter has a substantial emphasis on the recent economic recession, including new material on the Occupy America movement. There is also a new discussion of the student debt crisis and new research on a wide range of topics, including the rise of the superrich, wealth differences by race, and concentrated poverty. Throughout, there is updated data on income, wealth, and poverty.

Chapter 9, "Global Stratification," follows with a particular emphasis on understanding the significance of global stratification, the inequality that has developed among, as well as within, various nations. We have added a discussion of the influence of *global outsourcing*. In this edition, we connect inequality within the United States to worldwide inequality. And we offer new information on child labor, sex trafficking as part of global inequality, and how the Gini coefficient can be used to compare inequality across nations.

Chapter 10, "Race and Ethnicity," is a comprehensive review of the significance of race and ethnicity in society, plus discussion of very recent studies of effects of Latino immigration, of skin color gradation in both Black and Hispanic communities, of different types of racism, and the relevance of net worth as opposed to annual income in Black communities. Also discussed is the new north-to-south "reverse" migration of some Blacks. We have added a section on multiracial identities, including a pro-and-con discussion of what has come to be called "multiracialism" and the 2010 census on multiracial identification. As well, we include new discussions of "whiteness" and of the effects of race versus social class. The chapter includes discussions of topics such as the new housing segregation, the disproportionate effects of the foreclosure crisis on minorities, the disproportionate exclusion of Blacks and Latinos from juries.

Chapter 11, "Gender," focuses on gender as a central concept in sociology closely linked to systems of stratification in society. The chapter links the social construction of gender to homophobia, and then is followed by a separate chapter on sexuality. This edition adds a discussion of the so-called postfeminist movement and discusses the controversial book, *The End of Men*. We revised the discussion of sociological and feminist theory. Throughout, we offered updated data on earnings, employment, and gender-based attitudes.

Chapter 12, "Sexuality," treats sexuality as a social construction and a dimension of social stratification and inequality. We have put more emphasis on the influence of feminist theory on the study of sexuality. The chapter also includes new research on transgender people, as well as updated data on attitudes about sexuality, including same-sex marriage. The chapter has been reorganized to strengthen the discussion of power as well as to emphasize race/class/gender analyses of sexual stereotypes. We also provide a new box on sexuality and disability, and updated data throughout.

Part IV, "Social Institutions," includes three chapters, each focusing on basic institutions within society. **Chapter 13, "Families and Religion,"** maintains its inclusion of important topics in the study of families, such as interracial dating, debates about same-sex marriage, fatherhood, gender roles within families, and family violence. But we have added material on important topics in family studies, including "boomerang families," the "third shift" of women's family care work, and child care. The section on religion has a new box on the rise of religious fundamentalism.

Chapter 14, "Education and Health Care," has been substantially reorganized and updated to reflect these two important topics of public policy and public debate. The section on education includes new information that considers school tracking and individualized education plans (IEP), exploring how the education system attempts to meet the needs of all students. We have added material on the current policy debates about No Child Left Behind and the Race to the Top education initiatives. In the section on health, we have new material on the Affordable Care Act, with a discussion of the debates around health care reform. We also provide new research on obesity and the health consequences of poor nutrition.

Chapter 15, "Economy and Politics," analyzes the state, power, and authority and bureaucratic government. It also contains a detailed discussion of theories of power in addition to coverage of the economy seen globally and characteristics of the labor force. For the eighth edition, we reorganized the chapter to put economy before politics, because the economy is driving so many contemporary issues. We then reorganized the material within the section on the economy, especially to emphasize diversity and the social organization of work. We provide new material on outsourcing, a new research box on precarious work, and more emphasis on the current economic crisis. We also include new research on myths about immigration and its effects on native-born workers. The section on politics includes a discussion of the influence of the Tea Party, as well as data on the 2012 elections. Also in this section is new material on democracy, authoritarianism, and totalitarianism. In addition, we provide a new section on the military as a social institution.

Part V, "Social Change," includes **Chapter 16, "Environment, Population, and Social Change."** This chapter has been substantially revised for this edition so that a sociological analysis of environmental issues frames the chapters. Thus, the chapter focuses on sustainability and climate change. We also provide a new section on social dimensions of disasters. In the discussion of population processes, we include much more on the changes bringing more diversity into the U.S. population. To illustrate sociological theory, we've provided a new section on "Globalization and Modernization" to emphasize modernization as a social process.

MindTap Sociology: The Personal Learning Experience

MindTap Sociology for *Sociology: The Essentials*, eighth edition, from Cengage Learning represents a new approach to a highly personalized, online learning platform. A fully online learning solution, MindTap Sociology combines all of a student's learning tools—readings, multimedia, activities, and assessments—into a singular learning path that guides students through an introduction to sociology course. Instructors personalize the experience by customizing the presentation of these learning tools for their students, even seamlessly introducing their own content into the learning path via "apps" that integrate into the MindTap platform. Learn more at **www.cengage.com/mindtap**.

MindTap Sociology for *Sociology: The Essentials*, eighth edition, is easy to use and saves instructors time by allowing them to:

- Seamlessly deliver appropriate content and technology assets from a number of providers to students, as they need them.
- Break course content down into movable objects to promote personalization, encourage interactivity, and ensure student engagement.
- Customize the course—from tools to text—and make adjustments "on the fly," making it possible to intertwine breaking news into their lessons and incorporate today's teachable moments.
- Bring interactivity into learning through the integration of multimedia assets (apps from Cengage Learning and other providers) and numerous in-context exercises and supplements; student engagement will increase, leading to better student outcomes.
- Track students' use, activities, and comprehension in real time, which provides opportunities for early intervention to influence progress and outcomes. Grades are visible and archived so students and instructors always have access to current standings in the class.
- Assess knowledge throughout each section: after readings, in activities, homework, and quizzes.
- Automatically grade all homework and quizzes.

- MindTap Sociology for *Sociology: The Essentials*, eighth edition features Aplia assignments, which help students learn to use their sociological imagination through compelling content and thought-provoking questions. Students complete interactive activities that encourage them to think critically in order to practice and apply course concepts. These valuable critical thinking skills help students become thoughtful and engaged members of society. Aplia for *Sociology: The Essentials*, eighth edition is also available as a standalone product. Login to CengageBrain.com for access.

Aplia

Aplia™ is now a part of MindTap Sociology and available separately. Aplia™ is an online interactive learning solution that improves comprehension and outcomes by increasing student effort and engagement. Founded by a professor to enhance his own courses, Aplia provides automatically graded assignments that were written to make the most of the web medium and contain detailed, immediate explanations on every question. Our easy-to-use system has been used by more than 2,000,000 students at over 1,800 institutions.

CourseReader for Sociology

CourseReader for Sociology, first edition, allows you to create a fully customized online reader in minutes. Access a rich collection of thousands of primary and secondary sources, readings, and audio and video selections from multiple disciplines. Each selection includes a descriptive introduction that puts concepts into context, and every selection is further supported by both critical thinking and multiple-choice questions designed to reinforce key points. This easy-to-use solution allows you to select exactly the content you need for your courses and is loaded with convenient pedagogical features like highlighting, printing, note taking, and downloadable MP3 audio files for each reading.

FEATURES AND PEDAGOGICAL AIDS

The special features of this book flow from its major themes: diversity, current theory and research, debunking and critical thinking, social change, and a global perspective. The features are also designed to help students develop critical thinking skills so that they can apply abstract concepts to observed experiences in their everyday life and learn how to interpret different theoretical paradigms and approaches to sociological research questions.

Critical Thinking Features

The feature **Thinking Sociologically** takes concepts from each chapter and asks students to think about these concepts in relationship to something they can easily observe in an exercise or class discussion. The feature **Debunking Society's Myths** takes certain common assumptions and shows students how the sociological perspective would inform such assumptions and beliefs.

See for Yourself

The feature **See for Yourself** provides students with the chance to apply sociological concepts and ideas to their own observations. This feature can also be used as the basis for writing exercises, helping students improve both their analytic skills and their writing skills.

An Extensive and Content-Rich Map Feature

We use the map feature that appears throughout the book to help students visualize some of the ideas presented, as well as to learn more about regional and international diversity. One map theme is **Mapping America's Diversity** and the other is **Viewing Society in Global Perspective.** These maps have multiple uses for instructional value, beyond instructing students about world and national geography. The maps have been designed primarily to show the differentiation by county, state, and/or country on key social facts.

High-Interest Theme Boxes

We use high-interest themes for the box features that embellish our focus on diversity and sociological research throughout the text. **Understanding Diversity** boxes further explore the approach to diversity taken throughout the book. In most cases, these box features provide personal narratives or other information designed to teach students about the experiences of different groups in society.

Because many are written as first-person narratives, they can invoke students' empathy toward groups other than those to which they belong—something we think is critical to teaching about diversity. We hope to show students the connections between race, class, and other social groups that they otherwise find difficult to grasp.

The box feature **Doing Sociological Research** is intended to show students the diversity of research questions that form the basis of sociological knowledge and, equally important, how the questions researchers ask influence the methods used to investigate the questions.

We see this as an important part of sociological research—that how one investigates a question is determined as much by the nature of the question as by allegiance to a particular research method. Some questions require a more qualitative approach; others, a more quantitative approach. In developing these box features, we ask: What is the central question sociologists are asking? How did they explore this question using sociological research methods? What did they find? What are the implications of this research? We deliberately selected questions that show the full and diverse range of sociological theories and research methods, as well as the diversity of sociologists. Each box feature ends with **Questions to Consider** to encourage students to think further about the implications and applications of the research.

What Would a Sociologist Say? boxes take a topic of interest and examine how a sociologist would likely interpret this subject. The topics are selected to capture student interest, such as a discussion of veteran suicides, hip-hop culture, and sex and popular culture. We think this box brings a sociological perspective to commonplace events.

The feature **A Sociological Eye on the Media,** found in several chapters, examines some aspect of how the media influence public understanding of some of the subjects in this book. We think this is important because sociological research often debunks taken-for-granted points of view presented in the media, and we want students to be able to look at the media with a more critical eye. Because of the enormous influence of the media, we think this is increasingly important in educating students about sociology. In addition to the features just described, we offer an entire set of learning aids within each chapter that promotes student mastery of the sociological concepts.

In-Text Learning Aids

Learning Objectives. We have added learning objectives to this edition, which appear near the beginning of every chapter. Matched to the major chapter headings, these objectives identify what we expect students to learn from the chapter. Faculty may choose to use these learning objectives to assess how well students comprehend the material. We tried to develop the learning objectives based on different levels of understanding and analysis, recognizing the various paths that students take in how they learn material.

Chapter Outlines. A concise chapter outline at the beginning of each chapter provides students with an overview of the major topics to be covered.

Key Terms. Key terms and major concepts appear in bold when first introduced in the chapter. A list of the key terms is found at the end of the chapter, which makes study more effective. Definitions for the key terms are found in the glossary.

Theory Tables. Each chapter includes a table that summarizes different theoretical perspectives by comparing and contrasting how these theories illuminate different aspects of different subjects.

Chapter Summary in Question-and-Answer Format. Questions and answers highlight the major points in each chapter and provide a quick review of major concepts and themes covered in the chapter.

A **Glossary** and complete **References** for the whole text are found at the back of the book.

SOCIOLOGY: THE ESSENTIALS, EIGHTH EDITION SUPPLEMENTS

Sociology: The Essentials, eighth edition, is accompanied by a wide array of supplements prepared to create the best learning environment inside as well as outside the classroom for both instructors and students. All the continuing supplements for *Sociology: The Essentials,* eighth edition, have been thoroughly revised and updated. We invite you to take full advantage of the teaching and learning tools available to you.

For Instructors

Instructor's Resource Manual. This supplement offers instructors brief chapter outlines, student learning objectives, American Sociological Association recommendations, key terms and people, detailed chapter lecture outlines, lecture/discussion suggestions, student activities, chapter worksheets, video suggestions, video activities, and Internet exercises. The eighth edition also includes a syllabus to help instructors easily organize learning tools such as Aplia and create lesson plans.

Test Bank. This instructor-reviewed test bank consists of a myriad of multiple-choice, true/false, short-answer, and essay questions for each chapter, all with page references to the text. Each multiple-choice item has the question type (factual, applied, or conceptual) indicated, and all test questions will be mapped to a learning objective for the chapter. All questions are also labeled as new, modified, or pickup so instructors know if the question is new to this edition of the test bank, modified but picked up from the previous edition of the test bank, or picked up straight from the previous edition of the test bank.

Cengage Learning Testing Powered by Cognero. This flexible, online system allows teachers to author, edit, and manage test bank content from multiple Cengage Learning solutions, create multiple test versions in an instant, and deliver tests from your LMS, your classroom, or wherever you want.

PowerPoint Slides. Preassembled Microsoft® PowerPoint® lecture slides with graphics from the text make it easy for you to assemble, edit, publish, and present custom lectures for your course.

The Sociology Video Library Vol. I–IV. These DVDs drive home the relevance of course topics through short, provocative clips of current and historical events. Perfect for enriching lectures and engaging students in discussion, many of the segments on this volume have been gathered from BBC Motion Gallery. Ask your Cengage Learning representative for a list of contents.

Acknowledgments

We relied on the comments of many reviewers to improve the book, and we thank them for the time they gave in developing very thoughtful commentaries on the different chapters. Thanks to Thea Alvarado, College of the Canyons; Maria Bryant, College of Southern Maryland; Kenneth Colburn, Butler University; Craig Cook, Crown College; Jason Crockett, Kutztown University; Keri Diggins, Scottsdale Community College; Lori Guasta, University of Colorado at Colorado Springs; Jamie Gusrang, Community College of Philadelphia; Kenneth Melichar, Piedmont College; Rachael Neal, Coe College; Robert H. Oxley, Schoolcraft College; Nancy Reeves, Gloucester County College; and Victor Thompson, Rider University.

We appreciate the efforts of many people who make this project possible. Margaret Andersen especially thanks Dana Brittingham whose extraordinary organizational skills make it possible to keep up with the many daily demands of this book and her other work. All three of us are fortunate to be working with a publishing team with great enthusiasm for this project. We thank all of the people at Cengage Learning who have worked with us on this and other projects. We especially thank Bob Jucha for shepherding this revision through, as he has done many times before. And we welcome our new editor Seth Dobrin to this project and thank him for his commitment to this work. We were also fortunate to work with Mark Kerr, executive editor during a transition to a new editor; we hope he sees some of his ideas reflected in the content of the book. Cheri Palmer is expert at overseeing the many aspects of production that are critical to the book's success. We especially thank Jill Traut of MPS Limited for her attention to the many aspects of production. We are appreciative of the fine eye of Heather McElwain for her careful copyediting of the manuscript, and Reba Frederics from Pre-Media Global for photographic research. Finally, our special thanks also go to our spouses Richard Morris Rosenfeld, Patricia Epps Taylor, and Jim Rau for their ongoing love and willingness to put up with us when we are frazzled by the project details! Finally, a special dedication goes to Olivia "Bunny" Pla, granddaughter of Howard F. and Patricia Epps Taylor.

About the Authors

Margaret L. Andersen is the Edward F. and Elizabeth Goodman Rosenberg Professor of Sociology at the University of Delaware where she also holds joint appointments in women's studies and Black American studies and currently serves as Associate Provost for Academic Affairs. She is the author of *On Land and On Sea: A Century of Women in the Rosenfeld Collection; Living Art: The Life of Paul R. Jones, African American Art Collector; Race and Ethnicity in Society: The Changing Landscape* (with Elizabeth Higginbotham); *Thinking about Women: Sociological Perspectives on Sex and Gender;* and *Race, Class and Gender* (with Patricia Hill Collins). She is a recipient of the American Sociological Association's Jessie Bernard Award and has received the Sociologists for Women in Society's Feminist Lecturer Award. She is the former vice president of the American Sociological Association, former president of the Eastern Sociological Society, and a recipient of the University of Delaware's Excellence in Teaching Award and the College of Arts and Sciences Award for Outstanding Teaching.

Howard F. Taylor was raised in Cleveland, Ohio. He graduated Phi Beta Kappa from Hiram College and has a Ph.D. in sociology from Yale University. He has taught at the Illinois Institute of Technology, Syracuse University, and Princeton University, where he is presently professor of sociology and former director of the Center for African American Studies. He has published over fifty articles in sociology, education, social psychology, and race relations. His books include *The IQ Game* (Rutgers University Press), a critique of hereditarian accounts of intelligence; *Balance in Small Groups* (Van Nostrand Reinhold), translated into Japanese; and the forthcoming *The SAT Triple Whammy: Race, Gender, and Social Class Bias.* He has appeared widely before college, radio, and TV audiences, including ABC's *Nightline.* He is past president of the Eastern Sociological Society, and a member of the American Sociological Association and the Sociological Research Association, an honorary society for distinguished research. He is a winner of the DuBois-Johnson-Frazier Award, given by the American Sociological Association for distinguished research in race and ethnic relations, and the President's Award for Distinguished Teaching at Princeton University. He lives in Pennington, New Jersey, with his wife, a corporate lawyer.

Kim A. Logio received her Ph.D. in sociology from the University of Delaware and is currently associate professor and chair of sociology at Saint Joseph's University in Philadelphia, Pennsylvania. She has been interviewed for local television and National Public Radio for her work on body image and race, class, and gender differences in nutrition and weight control behavior. She is a member of the American Sociological Association and the Eastern Sociological Society. She often teaches research methods and guides students through the completion of their undergraduate thesis projects. She has been awarded a teaching award at Saint Joseph's University. She lives in Delaware County, Pennsylvania, with her husband and three children.

sociology

the essentials

8

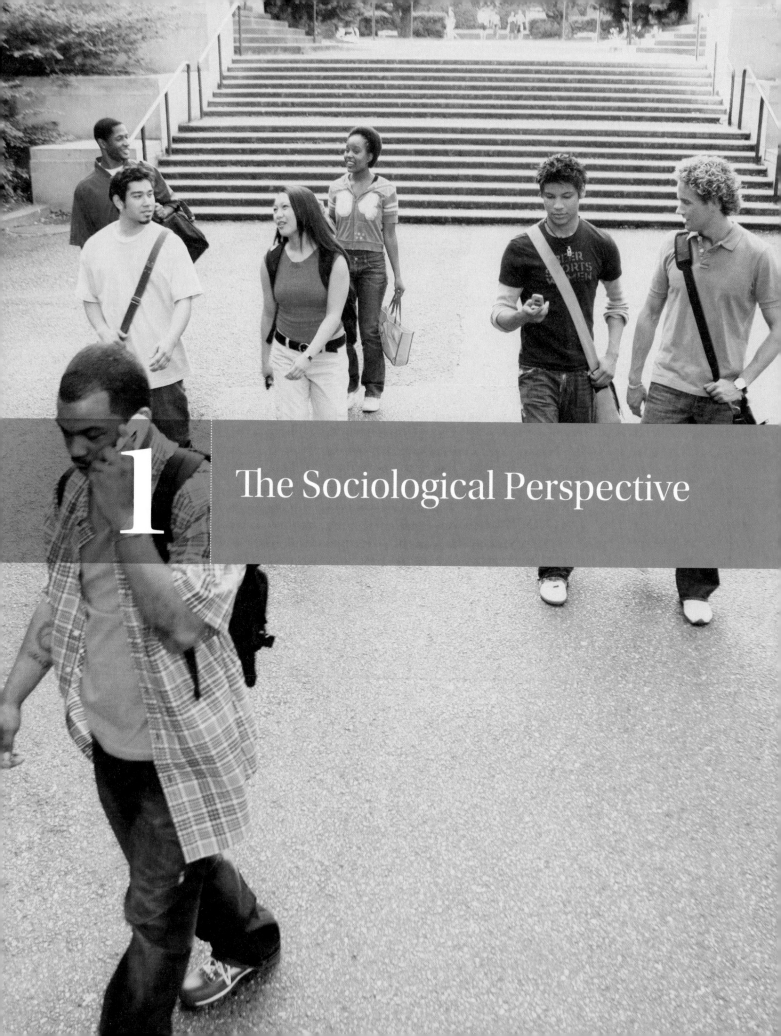

1 The Sociological Perspective

Andersen Ross/Blend Images/Corbis

imagine that you had been switched with another infant at birth. How different would your life be? What if your accidental family was very poor . . . or very rich? How might this have affected the schools you attended, the health care you received, and the possibilities for your future career? If you had been raised in a different religion, would this have affected your beliefs, values, and attitudes? Taking a greater leap, what if you had been born another sex or a different race? What would you be like now?

We are talking about changing the basic facts of your life—your family, social class, education, religion, sex, and race. Each has major consequences for who you are and how you will fare in life. These factors play a major part in writing your life script. Social location (meaning a person's place in society) establishes the limits and possibilities of a life.

Consider this:
- During economic recessions, families where the couple is less rigidly tied to traditional breadwinner/homemaker roles are less likely to experience family strain than is true for families with more traditional gender roles (Sherman 2009).
- The people least likely to attend college are those most likely to benefit from it (Brand and Xie 2010).
- During the housing foreclosure crisis in the recent recession, women of color were five times more likely than men of any color in the same income brackets to hold subprime mortgages—that is, mortgages with interest rates higher than the prime rate (Fishbein and Woodall 2006).
- Gender and racial diversity in for-profit business organizations is associated with increased sales revenues, more customers, and higher profits (Herring 2009).

These conclusions, drawn from current sociological research, describe some consequences of particular social locations in society. Although we may take our place in society for granted, our social location has a profound effect on our chances in life. The power of sociology is that it teaches us to see how society influences our lives and the lives of others, and it helps us explain the consequences of different social arrangements.

Sociology also has the power to help us understand the influence of major changes on people. Currently, rapidly developing technologies, increasing globalization, a more diverse population in the United States, and changes in women's roles are affecting everyone in society, although in different ways. How are these changes affecting your life? Perhaps you rely on social media to keep in touch with friends. Maybe your community is witnessing an increase in immigrants from other places. Or, maybe you see women and men trying hard to balance the needs of both work and family life. All of these are issues that guide sociological questions. Sociology explains some of the causes and consequences of these changes.

Although society is always changing, it is also remarkably stable. People generally follow established patterns of human behavior, and you can generally anticipate how people will behave in certain situations. You can even anticipate how different social conditions will affect different groups of people in society. This is what sociologists find so interesting: Society is marked by both change and stability. Societies continually evolve, creating the need for people to adapt to change while still following generally established patterns of behavior.

learning objectives

- Illustrate what is meant by saying that human behavior is shaped by social structure
- Question individualistic explanations of human behavior
- Describe the significance of diversity in studying contemporary society
- Explain the origins of sociological theory
- Compare and contrast major frameworks of sociological theory

WHAT IS SOCIOLOGY?

Sociology is the study of human behavior in society. Sociologists are interested in the study of people and have learned a fundamental lesson: Human behavior, even when seemingly "natural" or taken for granted, is shaped by social structures—structures that have their origins beyond the immediately visible behaviors of everyday life. In other words, *all human*

David Grossman/Alamy Limited

Sociology is the study of human behavior. What social behaviors do you see here?

behavior occurs in a social context. That context—the institutions and culture that surround us—shapes what people do and think. In this book, we will examine the dimensions of society and analyze the elements of social context that influence human behavior.

Sociology is a scientific way of thinking about society and its influence on human groups. Observation, reasoning, and logical analysis are the tools of sociologists, coupled with knowledge of the large body of theoretical and analytical work. Sociology is inspired by the fascination people have for observing people, but it goes far beyond casual observations. It builds from objective and accurate analyses that others can validate as reliable.

Every day, the media in their various forms (television, film, video, digital, and print) bombard us with social commentary. Media commentators provide endless opinion about the various and sometimes bizarre forms of behavior in our society. Sociology is different. Sociologists often appear in the media, and they study some of the same subjects that the media examine, such as crime, violence, or income inequality, but sociologists use specific research techniques and well-tested theories to explain social issues. Indeed, sociology can provide the tools for testing whether the things we hear about society are actually true. Much of what we hear in the media and elsewhere about society, although delivered with perfect earnestness, is misstated and sometimes completely wrong, as you will see in some of the "Debunking Society's Myths" examples featured throughout this book.

►key sociological concepts

As you build your sociological perspective, you must learn certain key concepts to begin understanding how sociologists view human behavior. Social structure, social institutions, social change, and social interaction are not the only sociological concepts, but they are fundamental to grasping the sociological perspective.

Social Interaction. Sociologists see **social interaction** as behavior between two or more people that is given meaning. Through social interaction, people react and change, depending on the actions and reactions of others. Because society changes as new forms of human behavior emerge, change is always in the works.

Social Structure. We define **social structure** as the organized pattern of social relationships and social institutions that together constitute society. Social structure is not a "thing," but refers to the fact that social forces not always visible to the human eye guide and shape human behavior. Acknowledging that social structure exists does not mean that humans have no choice in how they behave, only that those choices are largely conditioned by one's location in society.

Social Institutions. In this book, you will also learn about the significance of **social institutions**, defined as established and organized systems of social behavior with a particular and recognized purpose. The family, religion, marriage, government, and the economy are examples of major social institutions. Social institutions confront individuals at birth and transcend individual experience, but they still influence individual behavior.

Social Change. As you can tell, sociologists are also interested in the process of **social change**, the alteration of society over time. As much as sociologists see society as producing certain outcomes, they do not see society as fixed, nor do they see humans as passive recipients of social expectations. Sociologists view society as stable but constantly changing.

As you read this book, you will see that these key concepts—social interaction, social structure, social institutions, and social change—are central to the sociological imagination.

---

thinking SOCIOLOGICALLY

Q: What do the following people have in common?

First Lady Michelle Obama
Robin Williams (actor, comedian)
Ronald Reagan (former president)
Reverend Martin Luther King, Jr.
Regis Philbin (TV personality)
Reverend Jesse Jackson
Saul Bellow (novelist; Nobel Prize recipient)
Joe Theismann (former football player and
TV personality)
Congresswoman Maxine Waters (from California)
Senator Barbara Mikulski (from Maryland)

A: They were all sociology majors!

Source: Compiled by Peter Dreier, Occidental College. ●

The subject matter of sociology is everywhere. This is why people sometimes wrongly believe that sociology just explains the obvious. But sociologists bring a unique perspective to understanding social behavior and social change. Even though sociologists often do research on familiar topics, such as youth cultures or relations between women and men, they do so using particular research tools and specific frames of analysis (known as sociological theory). Psychologists, anthropologists, political scientists, economists, social workers, and others also study social behavior, although each has a different perspective or "angle" on people in society. Together, these fields of study (also called disciplines) make up what are called the social sciences.

THE SOCIOLOGICAL PERSPECTIVE

Think back to the opening of this chapter where you were asked to imagine yourself growing up under completely different circumstances. Our goal in that passage was to make you feel the stirring of the *sociological perspective*—the ability to see the societal patterns that influence individual and group life. The beginnings of the sociological perspective can be as simple as the pleasures of watching people or wondering how society influences people's lives. Indeed, many students begin their study of sociology because they are "interested in people." Sociologists convert this curiosity into the systematic study of how society influences different people's experiences within it.

C. Wright Mills (1916–1962) was one of the first to write about the sociological perspective in his classic book, *The Sociological Imagination* (1959). He wrote that the task of sociology was to understand the relationship between individuals and the society in which they live. He defined the **sociological imagination** as the ability to see the societal patterns that influence the individual as well as groups of individuals. Sociology should be used, Mills argued, to reveal how the context of society shapes our lives. He thought that to understand the experience of a given person or group of people, one had to have knowledge of the social and historical context in which people lived.

Think, for example, about the time and effort that many people put into their appearance. You might ordinarily think of this as merely personal grooming or an individual attempt to "look good," but there are significant social origins of this behavior. When you stand in front of a mirror, you are probably not thinking about how society is present in your reflection. But as you look in the mirror, you are seeing how others see you and are very likely adjusting your appearance with that in mind, even if not consciously. Therefore, this seemingly individual behavior is actually a very social act. If you are trying to achieve a particular look, you are likely doing so because of social forces that establish particular ideals, which are produced by industries that profit enormously from the products and services that people buy, even when they do so believing this is an individual choice.

Some industries suggest that you should be thinner or curvier, your pants should be baggy or straight, your breasts should be minimized or maximized—either way you need more products. Maybe you should have a complete makeover! Many people go to great lengths to try to achieve a constantly changing beauty ideal, one that is probably not even attainable (such as flawless skin, hair always in place, perfectly proportioned body parts). Sometimes trying to meet these ideals can even be hazardous to your physical and mental health.

The point is that the alleged standards of beauty are produced by social factors that extend far beyond an individual's concerns with personal appearance. Beauty ideals, like other socially established beliefs and practices, are produced in particular social and historical contexts. People may come up with all kinds of personal strategies for achieving these ideals: They may buy more products, try to lose more weight, get a Botox treatment, or even become extremely depressed and anxious if they think their efforts are failing. These personal behaviors may seem to be only individual issues, but they have basic social causes. That is, the origins of these behaviors exist beyond personal lives. The sociological imagination permits us to see that something as seemingly personal as how you look arises from a social context, not just individual behavior.

Sociologists are certainly concerned about individuals, but they are attuned to the social and historical context that shapes the experiences of individuals and groups. A distinction made by the sociological imagination is that made between *troubles* and *issues*. **Troubles** are privately felt problems that spring from events or feelings in a person's life. **Issues** affect large numbers of people and have their origins in the institutional arrangements and history of a society (Mills 1959). This distinction is the crux of the difference between individual experience and social structure, defined as

Personal troubles are felt by individuals who are experiencing problems; *social issues* arise when large numbers of people experience problems that are rooted in the social structure of society.

Spencer Platt/Getty Images

the organized pattern of social relationships and social institutions that together constitute society. Issues shape the context within which troubles arise. Sociologists employ the sociological perspective to understand how issues are shaped by social structures.

Mills used the example of unemployment to explain the meaning of troubles versus issues—an example that has particular resonance now, given the economic recession the United States has experienced and the personal troubles (including unemployment) that this has generated. When an individual person becomes unemployed—or cannot find work—he or she has a personal trouble. Think of the worry that many college graduates have experienced in trying to find work during the recession. In addition to financial problems that unemployment brings, a person may feel a loss of identity, may become depressed, may have to uproot a family and move, or—in the case of college students—may have to move back home with parents after graduation.

The problem of unemployment, however, is deeper than the experience of one person. Unemployment is rooted in the structure of society; this is what interests sociologists. What societal forces cause unemployment? Who is most likely to become unemployed at different times? How does unemployment affect an entire community (for instance, when a large plant shuts down) or an entire nation (such as during the economic downturn of recent years)? Sociologists know that unemployment causes personal troubles, but understanding unemployment is more than understanding one person's experience. It requires understanding the social structural conditions that influence people's lives.

The specific task of sociology, according to Mills, is to comprehend the whole of human society—its personal and public dimensions, historical and contemporary—and its influence on the lives of human beings. Mills had an important point: People often feel that things are beyond their control, meaning that they are being shaped by social forces larger than their own individual lives. Social forces influence our lives in profound ways, even though we may not always know how. Consider this: Sociologists have noted a current trend, popularly labeled "the boomerang generation" or "accordion families" (Newman 2012). This refers to the pattern whereby many young people, after having left their family home to attend college, are returning home after graduation. Although this may seem like an individual decision to save money on housing or live "free" while paying off student loans, when a whole generation experiences this living arrangement, there are social forces at work that extend beyond individual decisions. In other words, people feel the impact of social forces in their personal lives, even though they may not always know the full dimensions of those forces. This is where sociology comes into play—revealing the social forces that shape the different dimensions of our day-to-day lives.

Sociology is an **empirical** discipline. This means that sociological conclusions are based on careful and systematic observations, as we will see in Chapter 3 on sociological research methods. In this way, sociology is very different from ordinary common sense. For empirical observations to be useful to other observers, they must be gathered and recorded rigorously. Sociologists are also obliged to

reexamine their assumptions and conclusions constantly. Although the specific methods that sociologists use to examine different problems vary, as we will see, the empirical basis of sociology is what distinguishes it from mere opinion or other forms of social commentary.

Discovering Unsettling Facts

In studying sociology, it is crucial to examine the most controversial topics and to do so with an open mind, even when you see the most disquieting facts. The facts we learn through sociological research can be "inconvenient" because the data can challenge familiar ways of thinking. Consider the following:

- Even though many think of the Internet as promoting more impersonal social interaction, sociological research finds that people with Internet access are actually more likely to have romantic partners because of meeting people online (Rosenfeld and Thomas 2012).
- Despite the widespread idea promoted in the media that well-educated women are opting out of professional careers to become "stay-at-home moms," the proportion of college-educated White women who stay home with children has actually declined; those who opt out of work do so more typically because of frustration with how they are treated at work (Stone 2007).
- The number of women prisoners has increased at almost twice the rate of increase for men; two-thirds of women and half of men in prison are parents (Glaze and Maruschak 2008; Sabol and Couture 2008).

These facts provide unsettling evidence of persistent problems in the United States, *problems that are embedded in society, not just in individual behavior.* Sociologists try to reveal the social factors that shape society and determine the chances of success for different groups. Some never get the chance to go to college; others are unlikely to ever go to jail. These divisions persist because of people's placement within society.

Sociologists study not just the disquieting side of society. Sociologists may study questions that affect everyday life, such as how young boys and men are affected by changing gender roles (Kimmel 2008), worker–customer dynamics in nail salons (Kang 2010), or the expectations that young women and men have for combining work and family life (Gerson 2010). There are also many intriguing studies of unusual groups, such as cyberspace users (Kendall 2002), strip clubs and dancers (Price-Glynn 2010; Barton 2006), or heavily tattooed people, known as collectors (Irwin 2001). The subject matter of sociology is vast. Some research illuminates odd corners of society; other studies address urgent problems of society that may affect the lives of millions.

Debunking in Sociology

The power of sociological thinking is that it helps us see everyday life in new ways. Sociologists question actions and ideas that are usually taken for granted. Peter Berger (1963) calls this process "debunking." **Debunking** refers to looking behind the facades of everyday life—what Berger called the "unmasking tendency" of sociology (1963: 38). In other words, sociologists look at the behind-the-scenes patterns and processes that shape the behavior they observe in the social world.

Take schooling, for example: We can see how the sociological perspective debunks common assumptions about education. Most people think that education is primarily a way to learn and get ahead. Although this is true, a sociological perspective on education reveals something more. Sociologists have concluded that more than learning takes place in schools; other social processes are at work. Social cliques are formed where some students are "insiders" and others are excluded "outsiders." Young schoolchildren acquire not just formal knowledge but also the expectations of society and people's place within it. Race and class conflicts are often played out in schools (Lewis 2003). Poor children seldom have the same resources in schools as middle-class or elite children, and they are often assumed to be incapable of doing schoolwork and are treated accordingly. The somber reality is that schools may actually stifle the opportunities of some children rather than launch all children toward success (Kozol 2006).

Debunking is sometimes easier to do when looking at a culture or society different from one's own. Consider how behaviors that are unquestioned in one society may seem positively bizarre to an outsider. For a thousand years in China, it was usual for the elite classes to bind the feet of young girls to keep the feet from growing bigger—a practice allegedly derived from a mistress of the emperor. Bound feet were a sign of delicacy and vulnerability. A woman with large feet (defined as more than 4 inches long!) was thought to bring shame to her husband's household. The practice was supported by the belief that men were highly aroused by small feet, even though men never actually saw the naked foot. If they had, they might have been repulsed, because a woman's actual foot was U-shaped and often rotten and covered with dead skin (Blake 1994). Outside the social, cultural, and historical context in which it was practiced, footbinding seems bizarre, even dangerous. Feminists have pointed out that Chinese women were crippled by this practice, making them unable to move about freely and more dependent on men (Chang 1991).

This is an example of outsiders debunking a practice that was taken for granted by those within

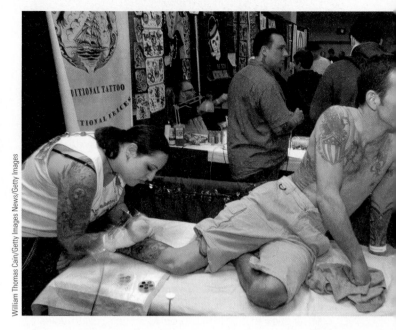

Cultural practices that seem bizarre to outsiders may be taken for granted or defined as appropriate by insiders.

the culture. Debunking can also call into question practices in one's own culture that may normally go unexamined. Strange as the practice of Chinese footbinding may seem to you, how might someone from another culture view wearing shoes that make it difficult to walk? Or piercing one's tongue or eyebrow? Many take these practices of contemporary U.S. culture for granted, just as they do Chinese footbinding. Until these cultural processes are debunked, seen as if for the first time, they might seem normal.

DOING **sociological research**

Debunking the Myths of Black Teenage Motherhood

Research Question: Sociologist Elaine Bell Kaplan knew that there was a stereotypical view of Black teen mothers that they had grown up in fatherless households where their mothers had no moral values and no control over their children. The myth of Black teenage motherhood also depicts teen mothers as unable to control their sexuality, as having children to collect welfare checks, and as having families who condone their behavior. Is this true?

Research Method: Kaplan did extensive research in two communities in the San Francisco Bay area—East Oakland and Richmond—both communities with a large African American population and typical of many inner-city, poor neighborhoods. Once thriving Black communities, East Oakland and Richmond are now characterized by high rates of unemployment, poverty, inadequate schools, crime, drug-related violence, and high numbers of single-parent households. Having grown up herself in Harlem, Kaplan knew that communities like those she studied have not always had these problems, nor have they condoned teen pregnancy. She spent several months in these communities, working as a volunteer in a community teen center that provided educational programs, day care, and counseling to

teen parents, and "hanging out" with a core group of teen mothers. She did extensive interviews with thirty-two teen mothers, supplementing them when she could with interviews with their mothers and, sometimes, the fathers of their children.

Research Results: Kaplan found that teen mothers adopt strategies for survival that help them cope with their environment, even though these same strategies do not help them overcome the problems they face. Unlike what the popular stereotype suggests, she did not find that the Black community condones teen pregnancy; quite the contrary, the teens felt embarrassed and stigmatized by being pregnant and experienced tension and conflict with their mothers, who saw their pregnancy as disrupting the hopes they had for their daughters' success. These conclusions run directly counter to the public image that such women do not value success and live in a culture that promotes welfare dependency.

Conclusions and Implications: Instead of simply stereotyping these teens as young and tough, Kaplan sees them as struggling to develop their own gender and sexual identity. Like other teens, they are highly vulnerable, searching for love and aspiring to create a meaningful

and positive identity for themselves. But failed by the educational system and locked out of the job market, the young women's struggle to develop an identity is compounded by the disruptive social and economic conditions in which they live.

Kaplan's research is a fine example of how sociologists debunk some of the commonly shared myths that surround contemporary issues. Carefully placing her analysis in the context of the social structural changes that affect these young women's lives, Kaplan provides an excellent example of how sociological research can shed new light on some of our most pressing social problems.

Questions to Consider

1. Suppose that Kaplan had studied middle-class teen mothers. What similarities and differences would you predict in the experiences of middle-class and poor teen mothers? Does race matter? In what ways does your answer debunk myths about teen pregnancy?

2. Make a list of the challenges you would face were you to be a teen parent. Having done so, indicate those that would be considered personal troubles and those that are social issues. How are the two related?

Source: Kaplan, Elaine Bell. 1996. *Not Our Kind of Girl: Unraveling the Myths of Black Teenage Motherhood*. Berkeley, CA: University of California Press.

---

debunking SOCIETY'S MYTHS

MYTH: Email scams promising to deliver a large sum of cash from some African bank if you contact the email deliverer prey on people who are just stupid or old.
SOCIOLOGICAL RESEARCH: Studies of such email scams indicate that Americans and Brits are especially susceptible to such scams because they play on widely held cultural stereotypes about Africa (that these are economically unsophisticated nations in which people are unable to manage money). These scams also exploit the American cultural belief that it is possible to "get rich quick"—reflecting a belief in individualism and the belief that anyone who tries hard enough can get ahead (Smith 2009).

Establishing Critical Distance

Debunking requires critical distance—that is, being able to detach from the situation at hand and view things with a critical mind. The role of critical distance in developing a sociological imagination is well explained by the early sociologist **Georg Simmel** (1858–1918). Simmel was especially interested in the role of *strangers* in social groups. Strangers have a position both inside and outside social groups; they are part of a group without necessarily sharing the group's assumptions and points of view. Because of this, the stranger can sometimes see the social structure of a group more readily than can people who are thoroughly imbued with the group's worldview. Simmel suggests that the sociological perspective requires a combination of nearness and distance. One must have enough critical distance to avoid being taken in by the group's definition of the situation, but be near enough to understand the group's experience.

Sociologists are not typically strangers to the society they study. You can acquire critical distance through a willingness to question the forces that shape social behavior. Often, sociologists become interested in things because of their own experiences. The biographies of sociologists are rich with examples of how their personal lives informed the questions they asked. Among sociologists are former ministers and nuns now studying the sociology of religion, women who have encountered sexism who now study the significance of gender in society, rock-and-roll fans studying music in popular culture, and sons and daughters of immigrants now analyzing race and ethnic relations (see the box "Understanding Diversity: Becoming a Sociologist").

THE SIGNIFICANCE OF DIVERSITY

The analysis of diversity is one central theme of sociology. Differences among groups, especially differences in the treatment of groups, are significant in any society, but they are particularly compelling in a society as diverse as that in the United States.

Defining Diversity

Today, the United States includes people from all nations and races. In 1900, one in eight Americans was not White; today, racial and ethnic minority groups (including African Americans, Hispanics, American Indians, Native Hawaiians, Asian Americans, and people of more than one race) represent 27 percent of Americans, and that proportion is growing (see Table 1.1 and Map 1.1, p. 12). These broad categories themselves are

In an increasingly diverse society, valuing and understanding diversity is a part of fully understanding society.

internally diverse, including, for example, those with long-term roots in the United States, as well as Cuban Americans, Salvadorans, Cape Verdeans, Filipinos, and many others.

Perhaps the most basic lesson of sociology is that people are shaped by the social context around them. In the United States, with so much cultural diversity, people will share some experiences, but not all. Experiences not held in common can include some of the most important influences on social development, such as language, religion, and the traditions of family and community. Understanding diversity means recognizing this diversity and making it central to sociological analyses.

In this book, we use the term *diversity* to refer to the variety of group experiences that result from the social structure of society. **Diversity** is a broad concept that includes studying group differences in society's opportunities, the shaping of social institutions by different social factors, the formation of group and individual identity, and the process of social change. Diversity includes the

table 1.1 U.S. Population Projections, 2010–2050

	2010	2020	2030	2040	2050
White	79.5%	78.0%	76.6%	75.3%	74.0%
Black	12.9%	13.0%	13.1%	13.0%	13.0%
American Indian and Alaskan Native	1.0%	1.1%	1.2%	1.2%	1.2%
Asian	4.6%	5.5%	6.3%	7.1%	7.8%
Native Hawaiian and Other Pacific Islander	0.2%	0.2%	0.2%	0.3%	0.3%
Two or more races	1.8%	2.1%	2.7%	3.2%	3.7%

Note: The U.S. census counts race and Hispanic ethnicity separately. Thus Hispanics may fall into any of the race categories. Those who identified themselves as Hispanic were 16 percent of the total U.S. population in the 2010 census.

Source: U.S. Census Bureau, 2012. *National Population Projections: Summary Table.* Washington, DC: U.S. Department of Commerce, **www.census.gov**

Becoming a Sociologist

Individual biographies often have a great influence on the subjects sociologists choose to study. The authors of this book are no exception. Margaret Andersen, a White woman, now studies the sociology of race and women's studies. Howard Taylor, an African American man, studies race, social psychology, and especially race and intelligence testing. Here, each of them writes about the influence of their early experiences on becoming a sociologist.

Margaret Andersen As I was growing up in the 1950s and 1960s, my family moved from California to Georgia, then to Massachusetts, and then back to Georgia. Moving as we did from urban to small-town environments and in and out of regions of the country that were very different in their racial character, I probably could not help becoming fascinated by the sociology of race. Oakland, California, where I was born, was highly diverse; my neighborhood was mostly White and Asian American. When I moved to a small town in Georgia in the 1950s, I was ten years old, but I was shocked by the racial norms I encountered. I had always loved riding in the back of the bus—our major mode of transportation in Oakland—and could not understand why this was no longer allowed. Labeled by my peers as an outsider because I was not southern, I painfully learned what it meant to feel excluded just because of "where you are from."

When I moved again to suburban Boston in the 1960s, I was defined by Bostonians as a southerner and ridiculed. Nicknamed "Dixie," I was teased for how I talked. Unlike in the South, where Black people were part of White people's daily lives despite strict racial segregation, Black people in Boston were even less visible. In my high school of 2500 or so students, Black students were rare. To me, the school seemed not much different from the strictly segregated schools I had attended in Georgia. My family soon returned to Georgia, where I was an outsider again; when I later returned to Massachusetts for graduate school in the 1970s, I worried about how a southerner would be accepted in this "Yankee" environment. Because I had acquired a southern accent, I think many of my teachers stereotyped me and thought I was not as smart as the students from other places.

These early lessons, which I may have been unaware of at the time, must have kindled my interest in the sociology of race relations. As I explored sociology, I wondered how the concepts and theories of race relations applied to women's lives. So much of what I had experienced growing up as a woman in this society was completely unexamined in what I studied in school. As the women's movement developed in the 1970s, I found sociology to be the framework that helped me understand the significance of gender and race in people's lives. To this day, I write and teach about race and gender, using sociology to help students understand their significance in society.

Howard Taylor I grew up in Cleveland, Ohio, the son of African American professional parents. My mother, Murtis Taylor, was a social worker and the founder and then president of a social work agency called the Murtis H. Taylor Human Services Center in Cleveland, Ohio. She is well known for her contributions to the city of Cleveland and was an early "superwoman," working days and nights, cooking, caring for her two sons, and being active in many professional and civic activities. I think this gave me an early appreciation for the roles of women and the place of gender in society, although I surely would not have articulated it as such at the time.

Courtesy of Howard Taylor

My father was a businessman in a then all-Black life insurance company. He was also a "closet scientist," always doing physics experiments, talking about scientific studies, and bringing home scientific gadgets. He encouraged my brother and me to engage in science, so we were always experimenting with scientific studies in the basement of our house. In the summers, I worked for my mother in the social service agency where she worked, as a camp counselor, and in other jobs. Early on, I contemplated becoming a social worker, but I was also excited by science. As a young child, I acquired my father's love of science and my mother's interest in society. In college, the one field that would gratify both sides of me, science and social work, was sociology. I wanted to study human interaction, but I also wanted to be a scientist, so the appeal of sociology was clear.

At the same time, growing up African American meant that I faced the consequences of race every day. It was always there, and like other young African American children, I spent much of my childhood confronting racism and prejudice. When I discovered sociology, in addition to bridging the scientific and humanistic parts of my interests, I found a field that provided a framework for studying race and ethnic relations. The merging of two ways of thinking, coupled with the analysis of race that sociology has long provided, made sociology fascinating to me.

Today, my research on race, class, gender, and intelligence testing seems rooted in these early experiences. I do quantitative research in sociology and see sociology as a science that reveals the workings of race, class, and gender in society.

Amber Alexander, University of Delaware

MAP 1.1

Mapping America's Diversity: A Changing Population

The nation is becoming increasingly diverse, but the distribution of minority groups differs in various regions of the country. Looking at this map, what factors do you think influence the distribution of the population?

Data: U.S. Census Bureau 2010. **www .census.gov**

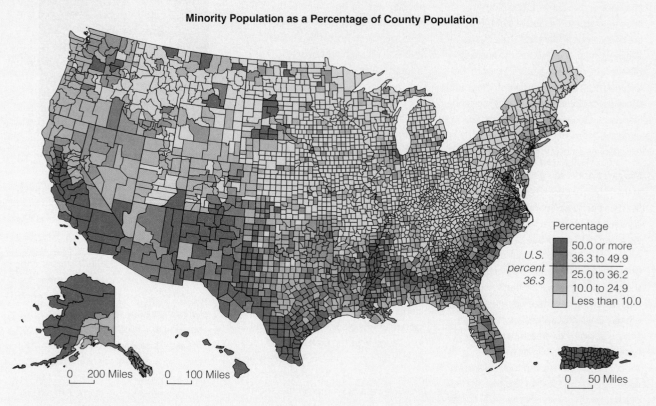

Minority Population as a Percentage of County Population

Percentage

U.S. percent 36.3

	50.0 or more
	36.3 to 49.9
	25.0 to 36.2
	10.0 to 24.9
	Less than 10.0

0   200 Miles

0   100 Miles

0   50 Miles

study of different cultural orientations, although diversity is not exclusively about culture.

Understanding diversity is crucial to understanding society because fundamental patterns of social change and social structure are increasingly patterned by diverse group experiences. There are numerous sources of diversity, including race, class, gender, and others as well. Age, nationality, sexual orientation, and region of residence, among other factors, also differentiate the experience of diverse groups in the United States. And as the world is increasingly interconnected through global communication and a global economy, the study of diversity also encompasses a global perspective—that is, an understanding of the international connections existing across national borders and the impact of such connections on life throughout the world.

thinking SOCIOLOGICALLY

What are some of the sources of diversity on your campus? How does this diversity affect social relations on campus? ●

Society in Global Perspective

No society can be understood apart from the global context that now influences the development of all societies. The social and economic system of any one society is increasingly intertwined with those of other

"*Actually, Lou, I think it was more than just my being in the right place at the right time. I think it was my being the right race, the right religion, the right sex, the right socioeconomic group, having the right accent, the right clothes, going to the right schools . . .*"

Warren Miller/The New Yorker Collection/Cartoonbank.com

nations. Coupled with the increasing ease of travel and telecommunication, this means that a global perspective is necessary to understand change both in the United States and in other parts of the world.

To understand globalization, you must look beyond the boundaries of your own society to see how patterns in any given society are increasingly being shaped by the connections between societies. Comparing and contrasting societies across different cultures is valuable. It helps you see patterns in your own society that you might otherwise take for granted, and it enriches your appreciation of the diverse patterns of culture that mark human society and human history. A global perspective, however, goes beyond just comparing different cultures; it also helps you see how events in one society or community may be linked to events occurring on the other side of the globe.

Globalization brings diverse cultures together, but it is also a process by which Western markets have penetrated much of the world.

For instance, return to the example of unemployment that C. Wright Mills used to distinguish between troubles and issues. One man may lose his job in Peoria, Illinois, and a woman in Los Angeles may employ a Latina domestic worker to take care of her child while she pursues a career. On the one hand, these are individual experiences for all three people, but they are linked in a pattern of globalization that shapes the lives of all three. The Latina domestic may have a family whom she has left in a different nation so that she can afford to support them. The corporation for which the Los Angeles woman works may have invested in a new plant overseas that employs cheap labor, resulting in the unemployment of the man in Peoria. The man in Peoria may have seen immigrant workers moving into his community, and one of his children may have made a friend at school who speaks a language other than English.

Such processes are increasingly shaping many of the subjects examined in this book—work, family, education, politics, just to name a few. Without a global perspective, you would not be able to fully understand the experience of any one of the people just mentioned, much less how these processes of change and global context shape society. Throughout this book, we will use a global perspective to understand some of the developments shaping contemporary life in the United States.

THE DEVELOPMENT OF SOCIOLOGICAL THEORY

Like the subjects it studies, sociology is itself a social product. Sociology first emerged in western Europe during the eighteenth and nineteenth centuries. In this period, the political and economic systems of Europe were rapidly changing. Monarchy, the rule of society by kings and queens, was disappearing, and new ways of thinking were emerging. Religion as the system of authority and law was giving way to scientific authority. At the same time, capitalism grew. Contact between different societies increased, and worldwide economic markets developed. The traditional ways of the past were giving way to a new social order. The time was ripe for a new understanding.

The Influence of the Enlightenment

The **Enlightenment** in eighteenth- and nineteenth-century Europe had an enormous influence on the development of modern sociology. Also known as the Age of Reason, the Enlightenment was characterized by faith in the ability of human reason to solve society's problems. Intellectuals believed that there were natural laws and processes in society to be discovered and used for the general good. Modern science was gradually supplanting traditional and religious explanations for natural phenomena with theories confirmed by experiments.

The earliest sociologists promoted a vision of sociology grounded in careful observation. **Auguste Comte** (1798–1857), a French philosopher who coined the term *sociology*, believed that just as science had discovered the laws of nature, sociology could discover the laws of human social behavior and thus help solve society's problems. This approach is called **positivism**, a system of thought, still prominent today, in which scientific observation and description is considered the highest form of knowledge, as opposed to, say, religious dogma or poetic inspiration. The modern scientific method, which guides sociological research, grew out of positivism.

Alexis de Tocqueville (1805–1859), a French citizen, traveled to the United States as an observer beginning in 1831. Tocqueville thought that democratic values and the belief in human equality positively influenced American social institutions and transformed personal relationships. Less admiringly, he felt that in the United States the tyranny of kings had been replaced by the "tyranny of the majority." He was referring to the ability of a majority to impose its will on everyone else in a democracy. Tocqueville also felt that despite the emphasis on individualism in American culture, Americans had little independence of mind, making them self-centered and anxious about their social class position (Collins and Makowsky 1972).

Another early sociologist is **Harriet Martineau** (1802–1876). Like Tocqueville, Martineau, a British citizen, embarked on a long tour of the United States in 1834. She was fascinated by the newly emerging culture in the United States. Her book *Society in America* (1837) is an analysis of the social customs that she observed. This important work was overlooked for many years, probably because the author was a woman. It is now recognized as a classic. Martineau also wrote the first sociological methods book, *How to Observe Morals and Manners* (1838), in which she discussed how to observe behavior when one is a participant in the situation being studied.

As one of the earliest observers of American culture, Harriet Martineau used the powers of social observation to record and analyze the social structure of American society. Long ignored for her contributions to sociology, she is now seen as one of the founders of early sociological thought.

Spencer Arnold/Stringer/Hulton Archive/Getty Images

Classical Sociological Theory

Of all the contributors to the development of sociology, the giants of the European tradition were Emile Durkheim, Karl Marx, and Max Weber. They are classical thinkers because the ideas they offered more than 150 years ago continue to influence our understanding of society, not just in sociology but in other fields as well (such as political science and history).

Emile Durkheim. During the early academic career of the Frenchman **Emile Durkheim** (1858–1917), France was in the throes of great political and religious upheaval. Anti-Semitism (hatred of Jews) was being expressed, along with ill feeling among other religions, as well. Durkheim, himself Jewish, was fascinated by how the public degradation of Jews by non-Jews seemed to calm and unify a large segment of the divided French public. Durkheim later wrote that public rituals have a special purpose in society, creating social solidarity, referring to the bonds that link the members of a group. Some of Durkheim's most significant works explore the question of what forces hold society together and make it stable.

According to Durkheim, people in society are glued together by belief systems (Durkheim 1947/1912). The rituals of religion and other institutions symbolize and reinforce the sense of belonging. Public ceremonies create a bond between people in a social unit. Durkheim thought that by publicly punishing people, such rituals sustain moral cohesion in society. Durkheim's views on this are further examined in Chapter 7, which discusses deviant behavior.

Durkheim also viewed society as an entity larger than the sum of its parts. He described this as society *sui generis* (which translates as "thing in itself"), meaning that society is a subject to be studied separately from the sum of the individuals who compose it. Society is external to individuals, yet its existence is internalized in people's minds—that is, people come to believe what society expects them to believe. Durkheim conceived of society as an integrated whole—each part contributing to the overall stability of the system. His work is the basis for *functionalism*, an important theoretical perspective that we will return to later in this chapter.

One contribution from Durkheim was his conceptualization of the *social*. Durkheim created the term **social facts** to indicate those social patterns that are *external* to individuals. Things such as customs and social values exist outside individuals, whereas psychological drives and motivation exist inside people. Social facts, therefore, are not to be explained by biology or psychology but are the proper subject of sociology; they are its reason for being.

A striking illustration of this principle was Durkheim's study of suicide (Durkheim 1951/1897). He analyzed rates of suicide in a society, as opposed to looking at individual (psychological) causes of suicide. He showed that suicide rates varied according to how

Suicide among Veterans

Currently, 6500 veterans commit suicide each year—more than the total number of soldiers killed in Afghanistan and Iraq since the start of those wars (Williams 2012). Veterans are, in fact, twice as likely as nonveterans in the general population to commit suicide, even though "natural" causes of death do not differ between these two groups (Kaplan et al. 2007). How would a sociologist explain this?

Certainly, there are psychological factors at work—post-traumatic stress, depression, and, sometimes, substance abuse—but sociological factors are at work, too. Durkheim would argue that this is a good example of anomic suicide. A soldier returning home is likely to encounter a far less structured environment than when in service where military life is highly structured. This can be a suicide-prone environment, especially if combined with unemployment, homelessness, or a disability. If you add to that a lack of social support services or benefits specifically to address the risk of suicide, you can have a potentially lethal social context.

Although sociologists do not ignore the psychological dimensions of behavior such as suicide, society involves other important social factors that produce this tragic behavior.

clear the norms and customs of the society were, whether the norms and customs were consistent with each other and not contradictory. *Anomie* (the breakdown of social norms) exists where norms were either grossly unclear or contradictory; the suicide rates were higher in such societies or such parts of a society. It is important to note that this condition is in society—external to individuals, but felt by them (Puffer 2009). In this sense, such a condition is truly societal.

Durkheim held that social facts, though they exist outside individuals, nonetheless pose constraints on individual behavior. Durkheim's major contribution was the discovery of the social basis of human behavior. He proposed that society could be known through the discovery and analysis of social facts. This is the central task of sociology (Coser 1977; Bellah 1973; Durkheim 1950/1938).

Durkheim thought that symbols and rituals were an important in producing social cohesion in society. You can witness this when shrines are spontaneously created in the aftermath of tragedies, such as this outpouring of solidarity following the mass shootings in an elementary school in Newtown, Connecticut.

EMMANUEL DUNAND/AFP/Newscom

Karl Marx. It is hard to imagine another scholar who has had as much influence on intellectual history as has **Karl Marx** (1818–1883). Along with his collaborator, Friedrich Engels, Marx not only changed intellectual history but also world history.

Marx's work was devoted to explaining how capitalism shaped society. He argued that capitalism is an economic system based on the pursuit of profit and the sanctity of private property. Marx used a class analysis to explain capitalism, describing capitalism as a system of relationships among different classes, including capitalists (also known as the bourgeois class), the proletariat (or working class), the petty bourgeoisie (small business owners and managers), and the *lumpenproletariat* (those "discarded" by the capitalist system, such as the homeless). In Marx's view, profit, the goal of capitalist endeavors, is produced through the exploitation of the working class. Workers sell their labor in exchange for wages, and capitalists make certain that wages are worth less than the goods the workers produce. The difference in value is the profit of the capitalist. In the Marxist view, the capitalist class system is inherently unfair because the entire system rests on workers getting less than they give.

Marx thought that the economic organization of society was the most important influence on what humans think and how they behave. He found that the beliefs of the common people tended to support the interests of the capitalist system, not the interests of the workers themselves. Why? Because the capitalist class controls the production of goods and the production of ideas. It owns the publishing companies, endows the universities where knowledge is produced, and controls information industries—thus shaping what people think.

Marx considered all of society to be shaped by economic forces. Laws, family structures, schools, and other institutions all develop, according to Marx, to suit economic needs under capitalism. Like other early

The United States is experiencing growing class inequality, symbolized by demonstrations during the Occupy America movement.

David Grossman/Alamy Limited

all opinions, including unpopular ones, and use the tools of rigorous sociological inquiry to understand why people believe and behave as they do.

An important concept in Weber's sociology is *verstehen* (meaning "understanding" and pronounced "ver-shtay-en"). **Verstehen**, a German word, refers to understanding social behavior from the point of view of those engaged in it. Weber believed that to understand social behavior, one had to understand the meaning that a behavior had for people. He did not believe sociologists had to be born into a group to understand it (in other words, he didn't believe "it takes one to know one"), but he did think sociologists had to develop some subjective understanding of how other people experience their world. One major contribution from Weber was the definition of *social action* as a behavior to which people give meaning (Weber 1962/1913; Parsons 1951b; Gerth and Mills 1946), such as placing a bumper sticker on your car that states pride in U.S. military troops.

Sociology in the United States

American sociology was built on the earlier work of Europeans, but unique features of U.S. culture contribute to its distinctive flavor. Less theoretical and more practical than their European counterparts, early American sociologists believed that if they exposed the causes of social problems, they could alleviate some of the consequences, which are measured in human suffering.

Early sociologists in both Europe and the United States conceived of society as an organism, a system of interrelated functions and parts that work together to create the whole. This perspective is called the **organic metaphor**. Sociologists saw society as constantly evolving, like an organism. The question many early sociologists asked was to what extent humans could shape the evolution of society.

Many were influenced in this question by the work of British scholar **Charles Darwin** (1809–1882), who revolutionized biology when he identified the process termed *evolution*, a process by which new species are created through the survival of the fittest. **Social Darwinism** was the application of Darwinian thought to society. According to the social Darwinists, the "survival of the fittest" is the driving force of social evolution as well. They conceived of society as an organism that evolved from simple to complex in a process of adaptation to the environment. They theorized that society was best left alone to follow its natural evolutionary course. Because social Darwinists believed that evolution always took a course toward perfection, they advocated a *laissez-faire* (that is, "hands-off") approach to social change. Social Darwinism was thus a conservative mode of thought; it assumed that the current arrangements in society were natural and inevitable (Hofstadter 1944).

Most other early sociologists in the United States took a more reform-based approach. Nowhere was the emphasis on application more evident than at the

sociologists, Marx took social structure as his subject rather than the actions of individuals. It was the *system* of capitalism that dictated people's behavior. Marx saw social change as arising from tensions inherent in a capitalist system—the conflict between the capitalist and working classes. Marx's ideas are often misperceived by U.S. students because communist revolutionaries throughout the world have claimed Marx as their guiding spirit. It would be naive to reject his ideas solely on political grounds. Much that Marx predicted has not occurred—for instance, he claimed that the "laws" of history made a worldwide revolution of workers inevitable, and this has not happened. Still, he left us an important body of sociological thought springing from his insight that society is systematic and structural and that class is a fundamental dimension of society that shapes social behavior.

Max Weber. **Max Weber** (1864–1920; pronounced "Vay-ber") was greatly influenced by and built upon Marx's work. But, whereas Marx saw economics as the basic organizing element of society, Weber theorized that society had three basic dimensions: political, economic, and cultural. According to Weber, a complete sociological analysis must recognize the interplay between economic, political, and cultural institutions (Parsons 1947). Weber is credited with developing a *multidimensional* analysis of society that goes beyond Marx's more one-dimensional focus on economics.

Weber also theorized extensively about the relationship of sociology to social and political values. He did not believe there could be a value-free sociology because values would always influence what sociologists considered worthy of study. Weber thought sociologists should acknowledge the influence of values so that ingrained beliefs would not interfere with objectivity. Weber professed that the task of sociologists is to teach students the uncomfortable truth about the world. Faculty should not use their positions to promote their political opinions, he felt; rather, they have a responsibility to examine

University of Chicago, where a style of sociological thinking known as the Chicago School developed. The Chicago School is characterized by thinkers who were interested in how society shaped the mind and identity of people. We study some of these thinkers, such as George Herbert Mead and Charles Horton Cooley, in Chapter 4. They thought of society as a human laboratory where they could observe and understand human behavior to be better able to address human needs, and they used the city in which they lived as a living laboratory.

Robert Park (1864–1944), from the University of Chicago, was a key founder of sociology. Originally a journalist who worked in several midwestern cities, Park was interested in urban problems and how different racial groups interacted with each other. He was also fascinated by the sociological design of cities, noting that cities were typically sets of concentric circles. At the time, the very rich and the very poor lived in the middle, ringed by slums and low-income neighborhoods (Coser 1977; Collins and Makowsky 1972; Park and Burgess 1921). Park would still be intrigued by how boundaries are defined and maintained in urban neighborhoods. You might notice this yourself. A single street crossing might delineate a Vietnamese neighborhood from an Italian one, an affluent White neighborhood from a barrio. The social structure of cities continues to be a subject of sociological research.

Many early sociologists of the Chicago School were women whose work is only now being rediscovered. **Jane Addams** (1860–1935) was one of the most renowned sociologists of her day. But, because she was a woman, she was never given the jobs or prestige that men in her time received. She was the only practicing sociologist ever to win a Nobel Peace Prize (in 1931), and she never had a regular teaching job. Instead, she used her skills as a research sociologist to develop community projects that assisted people in need (Deegan 1988). She was a leader in the settlement house movement providing services and doing research to improve the lives of slum dwellers, immigrants, and other poor people.

Another early sociologist, widely noted for her work in the anti-lynching movement, was **Ida B. Wells-Barnett** (1862–1931). Born a slave, Ida B. Wells-Barnett learned to read and write at Rust College, a school established for freed slaves, later receiving her teaching

Ida B. Wells-Barnett is now well known for her brave campaign against the lynching of African American people. Less known are her early contributions to sociological thought.

Bettmann/Corbis

credentials at Fisk University. She wrote numerous essays on the status of African Americans in the United States and was an active crusader against lynching and for women's rights, including the right to vote. Because she was so violently attacked—in writing and in actual threats—and because of her passionate work, she often had to write under an assumed name. Until recently, her contributions to the field of sociology have been largely unexamined. Interestingly, her grandson, Troy Duster (b. 1936), now a faculty member at New York University and the University of California, Berkeley, became the president of the American Sociological Association in 2004 (Giddings 2008; Henry 2008; Lengermann and Niebrugge-Brantley 1998).

W. E. B. DuBois (1868–1963; pronounced "due boys") was one of the most important early sociological thinkers in the United States. DuBois was a prominent Black scholar, a cofounder of the NAACP (National Association for the Advancement of Colored People) in 1909, a prolific writer, and one of the best American minds. He received the first Ph.D. ever awarded to a Black person in any field (from Harvard University), and he studied for a time in Germany, hearing several lectures by Max Weber.

DuBois was deeply troubled by the racial divisiveness in society, writing in a classic essay published in 1901 that "the problem of the twentieth century is the problem of the color line" (DuBois 1901: 354). Like many of his women colleagues, he envisioned a community-based, activist profession committed to social justice (Deegan 1988); he was a friend and collaborator with Jane Addams. He believed in the importance of a scientific approach to sociological questions, but he also thought that convictions always directed one's studies. Were he alive today, he might no doubt note that the problem of the color line still persists well into this, the twenty-first century.

Much of DuBois's work focused on the social structure of Black communities, one of his classic studies being of the city of Philadelphia. His book, *The Philadelphia Negro*, published in 1899, remains a classic study of African American urban life and its social institutions. One of the most lasting ideas from DuBois is his concept of "dual (or double) consciousness." DuBois saw African Americans as always seeing themselves in the

Jane Addams, the only sociologist to win the Nobel Peace Prize, used her sociological skills to try to improve people's lives. The settlement house movement provided social services to groups in need, while also providing a social laboratory in which to observe the sociological dimensions of problems such as poverty.

Bettmann/Corbis

eyes of others, a response that would be typical among any group oppressed by others. For DuBois, this dual consciousness led African Americans to always be alert to how others see them, but also to develop a strong collective identity of themselves as "Black" or, as we would say now, African American (DuBois 1903).

THEORETICAL FRAMEWORKS IN SOCIOLOGY

The founders of sociology have established theoretical traditions that ask basic questions about society and inform sociological research. The idea of theory may seem dry to you because it connotes something that is only hypothetical and divorced from "real life"; however, sociological theory is one of the tools that sociologists use to interpret real life. Sociologists use theory to organize their observations and apply them to the broad questions sociologists ask, such as: How are individuals related to society? How is social order maintained? Why is there inequality in society? How does social change occur?

Different theoretical frameworks within sociology make different assumptions and provide different insights about the nature of society. In the realm of *macrosociology* are theories that strive to understand society as a whole. Durkheim, Marx, and Weber were macrosociological theorists. Theoretical frameworks that center on face-to-face social interaction are known as *microsociology*. Some of the work derived from the Chicago School—research that studies individuals and group processes in society—is microsociological. Although sociologists draw from diverse theoretical perspectives to understand society, three broad traditions form the major theoretical perspectives that they use: functionalism, conflict theory, and symbolic interaction.

Functionalism

Functionalism has its origins in the work of Durkheim, who you will recall was especially interested in how social order is possible and how society remains relatively stable. **Functionalism** interprets each part of society in terms of how it contributes to the stability of the whole. As Durkheim suggested, functionalism conceptualizes society as more than the sum of its component parts. Each part is "functional" for society—that is, contributes to the stability of the whole. The different parts are primarily the institutions of society, each of which is organized to fill different needs and each of which has particular consequences for the form and shape of society. The parts each then depend on one another.

The family as an institution, for example, serves multiple functions. At its most basic level, the family has a reproductive role. Within the family, infants receive protection and sustenance. As they grow older, they are exposed to the patterns and expectations of

their culture. Across generations, the family supplies a broad unit of support and enriches individual experience with a sense of continuity with the past and future. All these aspects of family can be assessed by how they contribute to the stability and prosperity of society. The same is true for other institutions.

The functionalist framework emphasizes the consensus and order that exist in society, focusing on social stability and shared public values. From a functionalist perspective, disorganization in the system, such as deviant behavior and so forth, leads to change because societal components must adjust to achieve stability. This is a key part of functionalist theory—that when one part of society is not working (or is *dysfunctional*, as they would say), it affects all the other parts and creates social problems. Change may be for better or worse; changes for the worse stem from instability in the social system, such as a breakdown in shared values or a social institution no longer meeting people's needs (Eitzen and Baca Zinn 2012; Collins, R. 1974; Turner 1974; Merton 1968).

Functionalism was a dominant theoretical perspective in sociology for many years, and one of its major theorists was **Talcott Parsons** (1902–1979). In Parsons's view, all parts of a social system are interrelated, with different parts of society having different basic functions. Functionalism was further developed by **Robert Merton** (1910–2003). Merton saw that social practices often have consequences for society that are not immediately apparent, not necessarily the same as the stated purpose. He suggested that human behavior has both manifest and latent functions. *Manifest functions* are the stated and intended goals of social behavior. *Latent functions* are neither stated nor intended.

---

thinking SOCIOLOGICALLY

What are the *manifest functions* of grades in college?

What are *the latent functions*? ●

Critics of functionalism argue that its emphasis on social stability is inherently conservative and that it understates the roles of power and conflict in society. Critics also disagree with the explanation of inequality offered by functionalism—that it persists because social inequality creates a system for the fair and equitable distribution of societal resources. Functionalists would argue that it is fair and equitable that the higher social classes earn more money because they are more important (functional) to society. Critics of functionalism argue that functionalism is too accepting of the status quo. Functionalists would counter this argument by saying that, regardless of the injustices that inequality produces, inequality serves a purpose in society: It provides an incentive system for people to work and promotes solidarity among groups linked by common social standing.

Conflict Theory

Conflict theory emphasizes the role of coercion and power, a person's or group's ability to exercise influence and control over others, in producing social order. Whereas functionalism emphasizes cohesion within society, conflict theory emphasizes strife and friction. Derived from the work of Karl Marx, conflict theory pictures society as fragmented into groups that compete for social and economic resources. Social order is maintained not by consensus but by domination, with power in the hands of those with the greatest political, economic, and social resources. When consensus exists, according to conflict theorists, it is attributable to people being united around common interests, often in opposition to other groups (Dahrendorf 1959; Mills 1956).

According to conflict theory, inequality exists because those in control of a disproportionate share of society's resources actively defend their advantages. The masses are not bound to society by their shared values but by coercion at the hands of the powerful. In conflict theory, the emphasis is on social control, not consensus and conformity. Groups and individuals advance their own interests, struggling over control of societal resources. Those with the most resources exercise power over others; inequality and power struggles are the result. Conflict theory gives great attention to class, race, and gender in society because these are seen as the grounds of the most pertinent and enduring struggles in society.

Whereas functionalists find some benefit to society in the unequal distribution of resources, conflict theorists see inequality as inherently unfair, persisting only because groups who are economically advantaged use their social position to their own betterment. Their dominance even extends to the point of shaping the beliefs of other members of the society by controlling public information and having major influence over institutions such as education and religion. From the conflict perspective, power struggles between conflicting groups are the source of social change. Typically, those with the greatest power are able to maintain their advantage at the expense of other groups.

Conflict theory has been criticized for neglecting the importance of shared values and public consensus in society while overemphasizing inequality. Like functionalist theory, conflict theory finds the origins of social behavior in the structure of society, but it differs from functionalism in emphasizing the importance of power.

Symbolic Interaction

The third major framework of sociological theory is **symbolic interaction theory**. Instead of thinking of society in terms of abstract institutions, symbolic interactionists consider immediate social interaction to be the place where "society" exists. Because of the human capacity for reflection, people give meaning to their behavior, and this is how they interpret the different behaviors, events, or things that are significant for sociological study.

Because of this, symbolic interaction, as its name implies, relies extensively on the symbolic meaning that people develop and rely on in the process of social interaction. Symbolic interaction theory emphasizes face-to-face interaction and thus is a form of microsociology, whereas functionalism and conflict theory are more macrosociological.

Derived from the work of the Chicago School, symbolic interaction theory analyzes society by addressing the subjective meanings that people impose on objects, events, and behaviors. Subjective meanings are given primacy because, according to symbolic interactionists, people behave based on what they *believe*, not just on what is objectively true. Thus society is considered to be socially constructed through human interpretation (Blumer 1969; Berger and Luckmann 1967; Shibutani 1961). Symbolic interactionists see meaning as constantly modified through social interaction. People interpret one another's behavior, and it is these interpretations that form the social bond. These interpretations are called the "definition of the situation." For example, why would young people smoke cigarettes even though all objective medical evidence points to the danger of doing so? The answer is in the definition of the situation that people create. Studies find that teenagers are well informed about the risks of tobacco, but they also think that "smoking is cool," that they themselves will be safe from harm, and that

Symbolic interaction theory can help explain why people might do things that otherwise seem contrary to what one might expect.

table 1.2 Classical Theorists Reflect on the Economic Recession

	Major Concepts	What's the Big Idea?	An Applied Example: The Economic Recession
EMILE DURKHEIM 1858–1917 	Society *sui generis* Social solidarity Social facts	Social structures produce social forces that impinge on individuals even when they are not immediately visible; social solidarity is produced through identifying some as "other" or not belonging.	In times of economic crises, people may blame others, such as immigrants or "foreigners" for taking jobs from those perceived as citizens.
KARL MARX (1818–1883) 	Capitalism .Class conflict	Capitalism is built on the exploitation of laboring groups for the profit of others. Class conflict is embedded in the system of capitalism that then shapes other social institutions.	It is no surprise that inequality is growing; the forces of capitalism mean that the rich will amass the most resources, with everyone else becoming worse off.
MAX WEBER (1864–1920) 	Multidimensional analysis *Verstehen*	Cultural values interact with economic and political systems to produce society; no one factor determines the character of society.	Even when the economy is stagnant, cultural beliefs in hard work and the Protestant ethic mean that people will blame individuals, not the system, for failure.
W. E. B. DUBOIS (1868–1963) 	Color line Double consciousness	Racial inequality structures social institutions in the United States. Those who are oppressed by race develop a dual consciousness, ever aware of their status in the eyes of others but also having a collective identity as African American.	The "problem of the color line" extends into the twenty-first century, as African American people and other people of color are those most likely to be disadvantaged by economic stress.

Bettmann/Corbis

Bettmann/Corbis

Akg-images/Newscom

Special Collections and Archives, W. E. B. Du Bois Library, University of Massachusetts at Amherst

smoking projects an image—a positive identity for boys as a "tough guy" and for girls as fun-loving, mature, and glamorous. Smoking is also defined by young women as keeping you thin—an ideal constructed through dominant images of beauty. In other words, the symbolic meaning of smoking overrides the actual facts regarding smoking and risk.

thinking SOCIOLOGICALLY

Think about the example given about smoking, and using a symbolic interaction framework, how would you explain other risky behaviors, such as steroid use among athletes or eating disorders among young women? ●

Symbolic interaction interprets social order as constantly negotiated and created through the interpretations people give to their behavior. In observing society, symbolic interactionists see not simply facts but "social constructions," the meanings attached to things, whether those are concrete symbols (like a certain way of dress or a tattoo) or nonverbal behaviors. To a symbolic interactionist, society is highly subjective—existing in the minds of people, even though its effects are very real.

Functionalism, conflict theory, and symbolic interaction theory are by no means the only theoretical frameworks in sociology. For some time, however, they have provided the most prominent general explanations of society. Each has a unique view of the social realm. None is a perfect explanation of society, yet each has something to contribute. Functionalism gives special weight to the order and cohesion that usually characterizes society. Conflict theory emphasizes the inequalities and power imbalances in society. Symbolic interaction emphasizes the meanings that humans give to their behavior. Together, these frameworks provide a rich, comprehensive perspective on society, individuals within society, and social change, and they can shed light on existing social problems (see Table 1.2).

Feminist Theory

Contemporary sociological theory has been greatly influenced by the development of **feminist theory**. Prior to the emergence of second-wave feminism (the feminist movement emerging in the 1960s and 1970s), women were largely absent and invisible within most sociological work—indeed, within most academic work. When seen, they were strongly stereotyped in traditional roles as wives and mothers. Feminist theory developed to understand the status of women in society and with the purpose of using that knowledge to better women's lives.

Feminist theory has created vital new knowledge about women and has also transformed what

▶ careers in **sociology**

Now that you understand a bit more what sociology is about, you may ask, "What can I do with a degree in sociology?" This is a question we often hear from students. There is no single job called "sociologist" like there is "engineer" or "nurse" or "teacher," but sociology prepares you well for many different kinds of jobs, whether with a bachelor's degree or a postgraduate education. The skills you acquire from your sociological education are useful for jobs in business, health care, criminal justice, government agencies, various nonprofit organizations, and other job venues.

For example, the research skills one gains through sociology can be important in analyzing business data or organizing information for a food bank or homeless shelter. Students in sociology also gain experience working with and understanding those with different cultural and social backgrounds; this is an important and valued skill that employers seek. Also, the ability to dissect the different causes of a social problem can be an asset for jobs in various social service organizations.

Some sociologists have worked in their communities to deliver more effective social services. Some are employed in business organizations and social services where they use their sociological training to address issues such as poverty, crime and delinquency, population studies, substance abuse, violence against women, family social services, immigration policy, and any number of other important issues. Sociologists also work in the offices of U.S. representatives and senators, doing background research on the various issues addressed in the political process.

These are just a few examples of how sociology can prepare you for various careers. A good way to learn more about how sociology prepares you for work is to consider doing an internship while you are still in college.

For more information about careers in sociology, see the booklet, "21st Century Careers with an Undergraduate Degree in Sociology," available through the American Sociological Association (www.asanet.org).

Critical Thinking Exercise

1. Read a national newspaper over a period of one week and identify any experts who use a sociological perspective in their commentary. What does this suggest to you as a possible career in sociology? What are some of the different subjects about which sociologists provide expert information?

2. Identify some of the students from your college who have finished degrees in sociology. What different ways have they used their sociological knowledge?

is understood about men. Feminist scholarship in sociology, by focusing on the experiences of women, provides new ways of seeing the world and contributes to a more complete view of society. Feminist theory is a now vibrant and rich perspective in sociology, and

table 1.3 Three Classical Sociological Frameworks

Basic Questions	Functionalism	Conflict Theory	Symbolic Interaction
What is the relationship of individuals to society?	Individuals occupy fixed social roles.	Individuals are subordinated to society.	Individuals and society are interdependent.
Why is there inequality?	Inequality is inevitable and functional for society.	Inequality results from a struggle over scarce resources.	Inequality is demonstrated through the importance of symbols.
How is social order possible?	Social order stems from consensus on public values.	Social order is maintained through power and coercion.	Social order is sustained through social interaction and adherence to social norms.
What is the source of social change?	Society seeks equilibrium when there is social disorganization.	Change comes through the mobilization of people struggling for resources.	Change evolves from an ever-evolving set of social relationships and the creation of new meaning systems.
Major Criticisms			
	This is a conservative view of society that underplays power differences among and between groups.	The theory understates the degree of cohesion and stability in society.	There is little analysis of inequality, and it overstates the subjective basis of society.

© Cengage Learning

it has added much to how people understand the sociology of gender—and its connection to other social factors, such as race and class. Along with the classical traditions of sociology, feminist theory is included throughout this book in the context of particular topics.

Whatever the theoretical framework used, theory is evaluated in terms of its ability to explain observed social facts (see Table 1.3). The sociological imagination is not a single-minded way of looking at the world. It is the ability to observe social behavior and interpret that behavior in light of societal influences.

chapter summary

What is sociology?
Sociology is the study of human behavior in society. The *sociological imagination* is the ability to see societal patterns that influence individuals. Sociology is an *empirical* discipline, relying on careful observations as the basis for its knowledge.

What is debunking?
Debunking in sociology refers to the ability to look behind things taken for granted, looking instead to the origins of social behavior.

Why is diversity central to the study of sociology?
One of the central insights of sociology is its analysis of social diversity and inequality. Understanding *diversity* is critical to sociology because it is necessary to analyze *social institutions* and because diversity shapes most of our social and cultural institutions.

When and how did sociology emerge as a field of study?
Sociology emerged in western Europe during the *Enlightenment* and was influenced by the values of critical reason, humanitarianism, and positivism. *Auguste Comte*, one of the earliest sociologists, emphasized sociology as a positivist discipline. *Alexis de Tocqueville* and *Harriet Martineau* developed early and insightful analyses of American culture.

What are some of the basic insights of classical sociological theory?
Emile Durkheim is credited with conceptualizing society as a social system and with identifying *social facts* as patterns of behavior that are external to the individual. *Karl Marx* showed how capitalism shaped the development of society. *Max Weber* sought to explain society through cultural, political, and economic factors.

What are the major theoretical frameworks in sociology?

Functionalism emphasizes the stability and integration in society. *Conflict theory* sees society as organized around the unequal distribution of resources and held together through power and coercion. *Symbolic interaction* theory emphasizes the role of individuals in giving meaning to social behavior, thereby creating society. *Feminist theory* is the analysis of women and men in society and is intended to improve women's lives.

Key Terms

conflict theory 19
debunking 8
diversity 10
empirical 7
Enlightenment 13
feminist theory 21

functionalism 18
issues 6
organic metaphor 16
positivism 14
social change 5
social Darwinism 16

social facts 14
social institution 5
social interaction 5
social structure 5
sociological imagination 5
sociology 4

symbolic interaction
   theory 19
troubles 6
verstehen 16

2 Culture and the Media

Defining Culture

The Elements of Culture

Cultural Diversity

The Mass Media and Popular Culture

Theoretical Perspectives on Culture
    and the Media

Cultural Change

Chapter Summary

in one contemporary society known for its technological sophistication, people—especially the young—walk around with plugs in their ears. The plugs are connected to small wires that are themselves coated with a plastic film. These little plastic-covered wires are then connected to small devices made of metal, plastic, silicon, and other modern components, although most of the people who use them have no idea how they are made. When turned on, the device puts music into people's ears or, in some cases, shows pictures and movies on a screen about the size of a postage stamp. Some of the people who use these devices wouldn't even consider walking around without them; it is as if the device shields them from some of the other elements of their culture.

The same people who carry these devices around have other habits that, when seen from the perspective of someone unfamiliar with this culture, might seem peculiar and certainly highly ritualized. Apparently, when the young people in this society go away to school, most take a large number of various technological devices along with them. Many of them sleep with one of these devices turned on all night. It looks like a large box—some square, others flat—and it projects pictures and sound when the user clicks buttons on another small device that, though detached from the bigger box, can be placed anywhere in the room. If you click the buttons on this portable device, the pictures and sound coming forth from the larger box will change possibly hundreds of times, revealing a huge assortment of images that seem to influence what people in this culture believe and, in many cases, how they behave. They say that in over 40 percent of the households in this culture, this device is turned on 24 hours a day (Gitlin 2002)!

The young people in this culture seem to get up every day and immediately go to another device where they do things with unusual names, such as to "text" or "tweet" their friends (who, by the way, may be nowhere near them), pushing buttons with their thumbs on a small device with a tiny screen. Indeed, it seems that everything these young people do involves looking at some kind of screen, enough so that one of the authors of this book has labeled their generation "screenagers."

Not everyone in this culture has access to all of these devices, although many want them. Indeed, having more of the devices seems to be a mark of one's social status, that is, how you are regarded in this culture. But very few people know where the devices are made, what they are made of, or how they work, even though the young often ridicule older people for not understanding how the devices work or why they are so important to them.[1]

From outside the culture, these practices seem strange, yet few within the culture think the behaviors associated with these devices are anything but perfectly ordinary. Most of the time, people do not spend much time thinking about the meaning of the behaviors associated with these devices unless, for some reason, they suddenly do not work.

You have surely guessed that the practices described here are taken from U.S. culture: iPods/iPads, smartphones, television/video viewing. These are such daily practices that they practically define modern American culture. Unless they are somehow interrupted, most people do not think much about their influence on society, on people's relationships, or on people's definitions of themselves.

When viewed from the outside, cultural habits that seem perfectly normal often seem strange. Take an example from a different culture. The Tchikrin people—a remote culture of the central Brazilian rain forest—paint their bodies in elaborate designs. Painted bodies communicate to others the relationship of the person to his or her body, to society, and to the spiritual world. The designs and colors symbolize the balance the Tchikrin people think exists between biological powers and the integration of people into the social group. The Tchikrin also associate hair with sexual powers; lovers get a special thrill from using their teeth to pluck an eyebrow or eyelash from their partner's face (Turner 1969). To the Tchikrin people, these practices are no more unusual or exotic than the daily habits we practice in the United States.

To study culture, to analyze it and measure its significance in society, we must separate ourselves from judgments such as "strange" or "normal." We must see a culture as it is seen by insiders, but we cannot be completely taken in by that view. We should know the culture as insiders and understand it as outsiders.

[1]This introduction is inspired by a classic article on the "Nacirema"—*American*, backwards—by Horace Miner (1956). But it is also written based on essays students at the University of Delaware wrote regarding the media blackout exercise described on page 43. Students have written that, without access to their usual media devices, they felt they "had no personality!" and that the period of the blackout was "the worst forty-eight hours of my life!"

The Tchikrin people of the Brazilian rain forest paint elaborate and beautiful designs on their bodies that define the relationship of people to social groups. Are there ways that cultural practices in the United States also define social relationships?

learning objectives

- Define culture
- Recall the elements of culture
- Explain the significance of cultural diversity
- Relate the influence of the mass media and popular culture
- Compare and contrast theoretical explanations of culture and the media
- Discuss the components of cultural change

DEFINING CULTURE

Culture is the complex system of meaning and behavior that defines the way of life for a given group or society. It includes beliefs, values, knowledge, art, morals, laws, customs, habits, language, and dress, among other things. Culture includes ways of thinking as well as patterns of behavior. Observing culture involves studying what people think, how they interact, and the objects they use.

In any society, culture defines what is perceived as beautiful and ugly, right and wrong, good and bad. Culture helps hold society together, giving people a sense of belonging, instructing them on how to behave, and telling them what to think in particular situations. Culture gives meaning to society.

Culture is both material and nonmaterial. **Material culture** consists of the objects created in a given society—its buildings, art, tools, toys, print and broadcast media, and other tangible objects, such as those discussed in the chapter opener. In the popular mind, material artifacts constitute culture because they can be collected in museums or archives and analyzed for what they represent. These objects are significant because of the meaning they are given. A temple, for example, is not merely a building, nor is it only a place of worship. Its form and presentation signify the religious meaning system of the faithful.

Nonmaterial culture includes the norms, laws, customs, ideas, and beliefs of a group of people. Non-material culture is less tangible than material culture, but it has a strong presence in social behavior. Examples of nonmaterial culture are numerous and found in the patterns of everyday life. In some cultures, people eat with utensils; in others, people do not. The eating utensils are part of material culture, but the belief about whether to use them is nonmaterial culture.

It is cultural patterns that make humans so interesting. Is it culture that distinguishes human beings from animals? Some animal species develop what we might call culture. Chimpanzees, for example, learn behavior through observing and imitating others, a point proved by observing the different eating practices among chimpanzees in the same species but raised in different groups (Whiten et al. 1999). Others have observed elephants picking up the dead bones of other elephants and fondling them, perhaps evidence of grieving behavior (Meredith 2003). Dolphins are known to have a complex auditory language. And most people think that their pets communicate with them. Apparently, humans are not unique in their ability to develop systems of communication. But some scientists generally conclude that animals lack the elaborate symbol-based cultures common in human societies. Perhaps, as Charles Darwin wrote, "The difference in mind between man and the higher animals, great as it is, certainly is one of degree and not of kind" (Darwin, cited in Gould 1999).

Studying animal groups reminds us of the interplay between biology and culture. Human biology sets limits and provides certain capacities for human life and the development of culture. Similarly, the environment in which humans live establishes the possibilities and limitations for human society. Nutrition, for instance, is greatly influenced by environment, thereby affecting human body height and weight. Not everyone can shoot baskets like LeBron James or lob a tennis ball like Venus and Serena Williams, but with training and conditioning, people can enhance their physical abilities. Biological limits exist, but cultural factors have an enormous influence on the development of human life.

Morgan Lane Photography/Shutterstock.com

and maybe there is even a drinking ritual associated with turning a particular age. Or, if you are older, say turning forty or fifty, perhaps people kid you about "being over the hill" and decorate your office in black crepe paper. Such are the cultural rituals associated with birthdays in the United States.

But what if you had been born in another culture? Traditionally, in Vietnam, everyone's birthday is celebrated on the first day of the year and few really acknowledge the day they were born. In Russia, you might get a birthday pie, not a cake, with a birthday message carved into the crust. In Newfoundland, you might get ambushed and have butter rubbed on your nose for good luck—the butter on your nose being considered too greasy for bad luck to catch you. Many of these cultural practices are being changed by the infusion of Western culture, but they show how something as seemingly "normal" as celebrating your birthday has strong cultural roots.

What are the norms associated with birthday parties that you have attended? What social factors influence these parties? How do these reflect the values in U.S. culture? ●

Characteristics of Culture

Across societies, there are common characteristics of culture, even when the particulars vary. These different characteristics are as follows:

1. **Culture is shared**. Culture would have no significance if people did not hold it in common. Culture is collectively experienced and collectively agreed upon. The shared nature of culture is what makes human society possible. The shared basis of culture may be difficult to see in complex societies where groups have different traditions, perspectives, and ways of thinking and behaving. In the United States, for example, different racial and ethnic groups have unique histories, languages, and beliefs—that is, different cultures. Even within these groups, there are diverse cultural traditions. Latinos, for example,

thinking SOCIOLOGICALLY

Celebrating Your Birthday!

Birthday cake, candles, friends singing "Happy birthday to you!" Once a year, you feel like the day is yours, and your friends and family gather to celebrate with you. Some people give you presents, send cards,

Cultural shapes many things, including how people dress. The U.S. students in the upper left may not even think of themselves as displaying culture, but their manner of dress as students is a reflection of their culture. Compare this cultural display to others: In the upper right, students from Mali (in western Africa) are dressed in their traditional style. In the lower left, students are wearing the shalwar kameez, the traditional dress in south and central Asian countries. And, in the lower right, a woman student is wearing a traditional kimono at an archery shoot in Kyoto.

comprise many groups with distinct origins and cultures. Still, there are features of Latino culture, such as the Spanish language and some values and traditions, that are shared. Latinos also share a culture that is shaped by their common experiences as minorities in the United States. Similarly, African Americans have created a rich and distinct culture that is the result of their unique experience within the United States. What identifies African American culture are the practices and traditions that have evolved from both the U.S. experience and African and Caribbean traditions. Placed in another country, such as an African nation, African Americans would likely recognize elements of their culture, but they would also feel culturally distinct as Americans.

Within the United States, culture varies by age, race, region, gender, ethnicity, religion, class, and other social factors. A person growing up in the South is likely to develop different tastes, modes of speech, and cultural interests from a person raised in the West. Despite these differences, there is a common cultural basis to life in the United States. Certain symbols, language patterns, belief systems, and ways of thinking are distinctively American and form a common culture, even though great cultural diversity exists.

2. **Culture is learned**. Cultural beliefs and practices are usually so well learned that they seem perfectly natural, but they are learned nonetheless. How do people come to prefer some foods to others? How is musical taste acquired? Culture may be

taught through direct instruction, such as a parent teaching a child how to use silverware or teachers instructing children in songs, myths, and other traditions in school.

Culture is also learned indirectly through observation and imitation. Think of how a person learns what it means to be a man or a woman. Although the "proper" roles for men and women may never be explicitly taught, one learns what is expected from observing others. A person becomes a member of a culture through both formal and informal transmission of culture. Until the culture is learned, the person will feel like an outsider. The process of learning culture is referred to by sociologists as *socialization*, discussed in Chapter 4.

3. **Culture is taken for granted**. Because culture is learned, members of a given society seldom question the culture of which they are a part, unless for some reason they become outsiders or establish some critical distance from the usual cultural expectations. People engage unthinkingly in hundreds of specifically cultural practices every day; culture makes these practices seem "normal." If you suddenly stopped participating in your culture and questioned each belief and every behavior, you would soon find yourself feeling detached and perhaps a little disoriented; you might even become ineffective at functioning within your group.

You can see this if you travel outside of your culture, such as visiting a foreign country. Even the simplest things, such as how you eat or even use the toilet, may seem strange and have to be learned. As a result, tourists tend to stand out when in a foreign culture. They rarely have much knowledge of the culture they are visiting and, even when they are well informed, typically approach the society from their own cultural orientation.

But you do not have to leave your home country to observe this. Cultural differences within a society also shape social relations. For example, students who have been raised in a cultural group that teaches them to be quiet and not outspoken might be perceived as stupid or "slow" if in a classroom where they are expected to assert themselves and be aggressive in debate. Native American students, for example, may experience this, and if a teacher is not aware of these cultural differences, such students may be penalized simply for observing their cultural traditions. You can probably think of many other examples in which cultural misunderstanding can lead to isolation of those perceived as different or even to overt conflict. Culture binds us together, but lack of communication across cultures can have negative consequences.

4. **Culture is symbolic**. The significance of culture lies in the meaning it holds for people. **Symbols** are things or behaviors to which people give meaning; the meaning is not inherent in a symbol but is bestowed by the meaning people give it. The U.S. flag, for example, is literally a decorated piece of cloth. Its cultural significance derives not from the cloth of which it is made but from its meaning as a symbol of freedom and democracy, as was witnessed by the widespread flying of the flags after the terrorist attacks on the United States on September 11, 2001.

That something has symbolic meaning does not make it any less important or influential than objective facts. Symbols are powerful expressions of human culture. Think of the Confederate flag. Those who object to the Confederate flag being displayed on public buildings see it as a symbol of racism and the legacy of slavery. Those who defend it see it as representing southern heritage, a symbol of group pride and regional loyalty. Similarly, the use of Native American mascots to name and represent sports teams is symbolic of the exploitation of Native Americans. Native American activists and their supporters see the use of Native American mascots as derogatory and extremely insulting, representing gross caricatures of Native American traditions. (Think of the Washington Redskins, the Cleveland Indians, or the Atlanta Braves' "tomahawk chop.") The protests that have developed over controversial symbols are indicative of the enormous influence of cultural symbols.

Symbolic attachments can guide human behavior. For example, people stand when the national anthem is sung and may feel emotional from displays of the cross or the Star of David. Under some conditions, people organize mass movements to protest what they see as the defamation of important symbols, such as the burning of a flag or the burning of a cross. The significance of the symbolic value of culture can hardly be overestimated. Learning a culture means not just engaging in particular behaviors but also learning their symbolic meanings within the culture.

5. **Culture varies across time and place**. Culture develops as humans adapt to the physical and social environment around them. Culture is not fixed from one place to another. In the United States, for example, there is a strong cultural belief in

DOING **sociological research**

Tattoos: Status Risk or Status Symbol?

Research Question: Not so long ago, tattoos were considered a mark of social outcasts. They were associated with gang members, sailors, and juvenile delinquents. But now tattoos are in vogue—a symbol of who's trendy and hip. How did this happen that a once stigmatized activity associated with the working class became a statement of middle-class fashion?

Research Method: This is what sociologist Katherine Irwin wanted to know when she first noticed the increase in tattooing among the middle class. Irwin first encountered the culture of tattooing when she accompanied a friend getting a tattoo in a shop she calls Blue Mosque. She started hanging out in the shop and began a four-year study using participant observation in the shop, along with interviews of people getting their first tattoos. Irwin also interviewed some of the parents of tattooees and potential tattooees.

Research Results: Irwin found that middle-class tattoo patrons were initially fearful that their desire for a tattoo would associate them with low-status groups, but they reconciled this by adopting attitudes that associated tattooing with middle-class values and norms. Thus, they defined tattooing as symbolic of independence, liberation, and freedom from social constraints. Many of the women defined tattooing as symbolizing toughness and strength—values they thought rejected more conventional ideals of femininity.

Some saw tattoos as a way of increasing their attachment to alternative social groups or to gain entrée into "fringe" social worlds. Although tattoos held different cultural meanings to different groups, people getting tattooed used various techniques (what Irwin calls "legitimation techniques") to counter the negative stereotypes associated with tattooing.

Conclusions and Implications: Irwin concludes that people try to align their behavior with legitimate cultural values and norms even when that behavior seemingly falls outside of prevailing standards.

Questions to Consider

1. Do you think of tattoos as fashionable or deviant? What do you think influences your judgment about this, and how might your judgment be different were you in a different culture, age group, or historical moment?

2. Are there fashion adornments that you associate with different social classes? What are they, and what kinds of judgment (positive and negative) do people make about them? Where do these judgments come from, and why are they associated with social class?

Source: Irwin, Katherine. 2001. "Legitimating the First Tattoo: Moral Passage through Informal Interaction." *Symbolic Interaction* 24 (March): 49–73.

---

scientific solutions to human problems; consequently, many think that problems of food supply and environmental deterioration can be addressed by scientific breakthroughs, such as genetic engineering to create high-yield tomatoes or cloning

Tattooing, once considered a working-class symbol, has now become stylish and common, both among celebrities and in the general public.

cows to eliminate mad cow disease. In other cultural settings, different solutions may seem preferable. Indeed, some religions think of genetic engineering as trespassing on divine territory.

Culture also varies over time. As people encounter new situations, the culture that emerges is a mix of the past and present. Second-generation immigrants to the United States are raised in the traditions of their culture of origin, and children of immigrants typically grow up with both the traditional cultural expectations of their parents' homeland and the cultural expectations of a new society. Adapting to the new society can create conflict between generations, especially if the older generation is intent on passing along their cultural traditions. The children may be more influenced by their peers and may choose to dress, speak, and behave in ways that are characteristic of their new society but unacceptable to their parents.

To sum up, culture is concrete because we can observe the cultural objects and practices that define human experience. Culture is abstract because it is a way of thinking, feeling, believing, and behaving. Culture links the past and the present because it is the knowledge that makes us part of human groups. Culture gives shape to human experience.

THE ELEMENTS OF CULTURE

Culture is multifaceted, consisting of both material and nonmaterial things, some parts of culture being abstract, others more concrete. The different elements of culture include language, norms, beliefs, and values (see Table 2.1).

Language

Language is a set of symbols and rules that, combined in a meaningful way, provides a complex communication system. The formation of culture among humans is made possible by language. Learning the language of a culture is essential to becoming part of a society, and it is one of the first things children learn. Indeed, until children acquire at least a rudimentary command of language, they seem unable to acquire other social skills. Language is so important to human interaction that it is difficult to think of life without it; indeed, as one commentator on language has said, "Life is lived as a series of conversations" (Tannen 1990: 13).

Think about the experience of becoming part of a social group. When you enter a new society or a different social group, you have to learn its language to become a member of the group. This includes any special terms of reference used by the group. Lawyers, for example, have their own vocabulary and their own way of constructing sentences called, not always kindly, "legalese." Becoming a part of any social group—a friendship circle, fraternity or sorority, or any other group—involves learning the language they use. Those who do not share the language of a group cannot participate fully in its culture.

Language is fluid and dynamic and evolves in response to social change. Think, for example, of how the introduction of computers has affected the English language. People now talk about "downloading apps" and providing "input." Only a few years ago, had you said you were going to "text" your friends, no one would have known what you were talking about. Text messaging has also introduced its own language: BFF (best friends forever), LOL (laughing out loud), and GTG (got to go)—a new language shared among those in the text-messaging culture. Unheard of not that longer ago, these expressions are now commonplace—in other words, a new form of culture.

Does Language Shape Culture? Language is clearly a big part of culture. Edward Sapir (writing in the 1920s) and his student Benjamin Whorf (writing in the 1950s) thought that language was central in determining social thought. Their theory, the **Sapir–Whorf hypothesis**, asserts that language determines other aspects of culture because language provides the categories through which social reality is defined. Sapir and Whorf thought that language determines what people think because language forces people to perceive the world in certain terms (Whorf 1956; Sapir 1921).

If the Sapir–Whorf hypothesis is correct, then speakers of different languages have different perceptions of reality. Whorf used the example of the social meaning of time to illustrate cultural differences in how language shapes perceptions of reality. He noted that the Hopi Indians conceptualize time as a slowly turning cylinder, whereas English-speaking people conceive of time as running forward in one direction at a uniform pace. Linguistic constructions of time shape how the two different cultures think about time and therefore how they think about reality. In Hopi culture, events are located not in specific moments of time but in "categories of being," as if everything is in a state of becoming, not fixed in a particular time and place (Carroll 1956). In contrast, the English language locates things in a definite time and place, placing great importance on verb tense, with things located precisely in the past, present, or future.

Recent critics do not think that language determines culture to the extent that Sapir and Whorf proposed. Language does not single-handedly dictate the perception of reality—but, no doubt, language has a strong influence on culture. Most scholars now see two-way causality between language and culture. Asking whether language determines culture or vice versa is like asking which came first, the chicken or the egg. Language and culture are inextricable, each shaping the other.

table 2.1 Elements of Culture

	Definition	Examples
Language	A set of symbols and rules that, put together in a meaningful way, provides a complex communication system	English; Spanish; hieroglyphics
Norms	The specific cultural expectations for how to behave in a given situation	Behavior involving use of personal space; manners
Folkways	General standards of behavior adhered to by a group	Cultural forms of dress; food habits
Mores	Strict norms that control moral and ethical behavior	Religious doctrines; formal law
Values	Abstract standards in a society or group that define ideal principles	Liberty; freedom
Beliefs	Shared ideas about what is true held collectively by people within a given culture	Belief in a higher being

© Cengage Learning

Consider again the example of time. Contemporary Americans think of the week as divided into two parts: *weekdays* and *weekends*, words that reflect how we think about time. When does a week end? Having language that defines the weekend encourages us to think about the weekend in specific ways. It is a time for rest, play, chores, and family. In this sense, language shapes how we think about the passage of time—we look forward to the weekend, we prepare ourselves for the workweek—but the language itself (the very concept of the weekend) stems from patterns in the culture—specifically, the work patterns of advanced capitalism. The capitalist work ethic makes it morally offensive to merely "pass the time"; instead, time is to be managed. Concepts of time in preindustrial, agricultural societies follow a different pattern. In agricultural societies, time and calendars are based on agricultural and seasonal patterns; the year proceeds according to this rhythm, not the arbitrary units of time of weeks and months. This shows how language and culture shape each other.

Social Inequality in Language. The language of any culture reflects the nature of that society. Thus, in a society where there is inequality, language is likely to communicate assumptions and stereotypes about different social groups. What people say—including what people are called—reinforces patterns of inequality in society (Moore 1992). We see this in what different groups in the United States are called (see also the box on page 34, "Understanding Diversity: The Social Meaning of Language"). What someone is called can be significant because it imposes an identity on that person. This is why the names for various racial and ethnic groups have been so heavily debated. Thus, for years, many Native Americans objected to being called "Indian," because it was a term White conquerors created about them. To emphasize their native roots in the Americas, the term *Native American* was adopted. Now, though many prefer to be called by their actual origin, *Native American* and *American Indian* are also used interchangeably. Likewise, Asian Americans tend to be offended by being called "Oriental," an expression that stemmed from Western (that is, European and American) views of Asian nations.

Language reflects the social value placed on different groups, and it reflects power relationships, depending on who gets to name whom. Derogatory terms such as *redneck, white trash*, or *trailer park trash* stigmatize people based on regional identity and social class. This is also why it is so demeaning when derogatory terms are used to describe racial–ethnic groups. For example, throughout the period of Jim Crow segregation in the American South, Black men, regardless of their age, were routinely called "boy" by Whites. Calling a grown man a "boy" is an insult; it diminishes his status by defining him as childlike. Referring to a woman as a "girl" has the same effect. Why are young women, even well into their twenties, routinely referred to as "girls"? Just as does calling a man "boy," this diminishes women's status.

Living in a multicultural society often juxtaposes diverse cultures, even in public places.

AP Photo/Paul Sakuma

---

debunking SOCIETY'S MYTHS

MYTH: Bilingual education discourages immigrant children from learning English and thus blocks their assimilation into American culture and reduces their chances for a good education.

SOCIOLOGICAL PERSPECTIVE: Studies of students who are fluent bilinguals show that they outperform both English-only students and students with limited bilingualism. Moreover, preserving the use of native languages can better meet the need for skilled bilingual workers in the labor market (Portes 2002). ●

Note, however, that terms such as *girl* and *boy* are pejorative only in the context of dominant and subordinate group relationships. African American women, as an example, often refer to each other as "girl" in informal conversation. The term *girl* used between those of similar status is not perceived as derogatory, but when used by someone in a position of dominance, such as when a male boss calls his secretary a "girl," it is demeaning. Likewise, terms such as *dyke, fag*, and *queer* are terms lesbians and gay men sometimes use without offense in referring to each other, even though the same terms are offensive to lesbians and gays when others use them. By reclaiming these terms as positive within their own culture, lesbians and gays build cohesiveness and solidarity (Due 1995). These examples show that power relationships between groups supply the social context for the connotations of language.

In sum, language can reproduce the inequalities that exist in society. At the same time, changing the language that people use can, to some extent, alter social stereotypes and thereby change the way people think.

Norms

Social norms are another component of culture. **Norms** are the specific cultural expectations for how to behave in a given situation. Society without norms would be chaos; with norms in place, people know how to act, and social interactions are consistent, predictable, and learnable. There are norms governing every situation. Sometimes they are implicit; that is, they need not be spelled out for people to understand them. For example, when joining a line, there is an implicit norm that you should stand behind the last person, not barge in front of those ahead of you. Implicit norms may not be formal rules, but violation of these norms may nonetheless produce a harsh response. Implicit norms may be learned through specific instruction or by observation of the culture; they are part of a society's or group's customs. Norms are explicit when the rules governing behavior are written down or formally communicated. Typically, specific sanctions are imposed for violating explicit norms.

see FOR YOURSELF

Identify a *norm* that you commonly observe. Construct an experiment in which you, perhaps with the assistance of others, violate the norm. Record how others react and note the sanctions engaged through this norm violation exercise. Note: Be careful not to do anything that puts you in danger or causes serious problems for others. ●

In the early years of sociology, William Graham Sumner (1906) identified two types of norms: folkways and mores. **Folkways** are the general standards of behavior adhered to by a group. You might think of folkways as the ordinary customs of different group cultures. How you dress is an example of a cultural folkway. Other examples are the ways that people greet each other, decorate their homes, and prepare their food. Folkways may be loosely defined and loosely adhered to, but they nevertheless structure group customs and implicitly govern much social behavior.

Mores (pronounced "more-ays") are strict norms that control moral and ethical behavior. Mores provide strict codes of behavior, such as the injunctions, legal and religious, against killing others and committing adultery. Mores are often upheld through rules or **laws**, the written set of guidelines that define right and wrong in society. Basically, laws are formalized mores. Violating mores can bring serious repercussions. When any social norm is violated, the violator is typically punished.

Social sanctions are mechanisms of social control that enforce folkways, norms, and mores. The seriousness of a social sanction depends on how strictly the norms or mores are held. **Taboos** are those behaviors that bring the most serious sanctions. Dressing in an unusual way that violates the folkways of dress may bring ridicule but is usually not seriously punished. In some cultures, the rules of dress are strictly interpreted, such as the requirement by Islamic fundamentalists that women who appear in public have their bodies cloaked and faces veiled. It would be considered a taboo for a woman in this culture to appear in public without being veiled. The sanctions for doing so can be as severe as whipping, branding, banishment, even death.

Sanctions can be positive or negative, that is, based on rewards or punishment. When children learn social norms, for example, correct behavior may elicit positive sanctions; the behavior is reinforced through praise, approval, or an explicit reward. Early on, for example, parents might praise children for learning to put on their own clothes; later, children might get an allowance if they keep their rooms clean. Bad behavior earns negative sanctions, such as getting spanked or grounded. In society, negative sanctions may be mild or severe, ranging from subtle mechanisms of control, such as ridicule, to overt forms of punishment, such as imprisonment, physical coercion, or death.

One way to study social norms is to observe what happens when they are violated. Once you become aware of how social situations are controlled by norms, you can see how easy it is to disrupt situations where adherence to the norms produces social order. **Ethnomethodology** is a theoretical approach in sociology based on the idea that you can discover the normal social order through disrupting it. As a technique of study, ethnomethodologists often deliberately disrupt social norms to see how people respond and try to reinstate social order (Garfinkel 1967).

In a famous series of ethnomethodological experiments, college students were asked to pretend they were boarders in their own homes for a period of fifteen minutes to one hour. They did not tell their families what they were doing. The students were instructed to be polite, circumspect, and impersonal; to use terms of formal address; and to speak only when spoken to. After the experiment, two of the participating students reported

Our reliance on technology as a form of culture was vividly demonstrated during Hurricane Sandy when millions were without the power needed to communicate on their various electronic devices. How has this technology changed cultural practices?

The Social Meaning of Language

Language reflects the assumptions of a culture. This can be seen and exemplified in several ways:

- **Language affects people's perception of reality.**
  Example: Researchers have found that using male pronouns, even when intended to be gender-neutral, produces male-centered imagery and ideas. Studies also find that when college students look at job descriptions that use masculine pronouns, they assume that women are not qualified for the job (Gastil 1990; Switzer 1990; Hamilton 1988).

- **Language reflects the social and political status of different groups in society.**
  Example: A term such as *woman doctor* suggests that men are the standard and women the exception. The term *workingwomen* (used to refer to women who are employed) also suggests that women who do not work for wages are not working. Ask yourself what the term *working man* connotes and how this differs from *working woman*.

- **Groups may advocate changing language referring to them as a way of asserting a positive group identity.**
  Example: Advocates for the disabled challenge the term *handicapped*, arguing that it stigmatizes people who may have many abilities, even if they are physically distinctive. Also, although someone may have one disabling condition, she or he may otherwise be perfectly able.

- **The implications of language emerge from specific historical and cultural contexts.**
  Example: The naming of so-called races comes from the social and historical processes that define different groups as inferior or superior. Racial labels do not come just from physical, national, or cultural differences. The term *Caucasian*, for example, was coined in the seventeenth century when racist thinkers developed alleged scientific classification systems to rank different societal groups. Alfred Blumenbach used the label *Caucasian* to refer to people from the Caucasus of Russia whom he thought were more beautiful and intelligent than any group in the world.

- **Language can distort actual group experience.**
  Example: Terms used to describe different racial and ethnic groups homogenize experiences that may be unique. Thus the terms *Hispanic* and *Latino* lump together Mexican Americans, island Puerto Ricans, U.S.-born Puerto Ricans, as well as people from Honduras, Panama, El Salvador, and other Central and South American countries. *Hispanic* and *Latino* point to the shared experience of those from Latin cultures, but like the terms *Native American* and *American Indian*, they obscure the experiences of unique groups, such as the Lakota, Nanticoke, Cherokee, Yavapai, or Navajo.

- **Language shapes people's perceptions of groups and events in society.**
  Example: Following Hurricane Katrina in New Orleans, African American people taking food from abandoned stores were described as "looting" and White people as "finding food." Also, Native American victories during the nineteenth century are typically described as "massacres"; comparable victories by White settlers are described in heroic terms (Moore 1992).

- **Terms used to define different groups change over time and can originate in movements to assert a positive identity.**
  Example: In the 1960s, *Black American* replaced the term *Negro* because the civil rights and Black Power movements inspired Black pride and the importance of self-naming (Smith et al. 1992). Earlier, *Negro* and *colored* were used to define African Americans. Currently, it is popular to refer to all so-called racial groups as "people of color." This phrase was derived from the phrase "women of color," created by feminist African American, Latina,* Asian American, and Native American women to emphasize their common experiences. Some people find the use of "color" in this label offensive because it harkens back to the phrase "colored people," a phrase generally seen as paternalistic and racist because it was a label used by dominant groups to refer to African Americans prior to the civil rights movement. The phrase "women of color" now has a more positive meaning than the earlier term *colored women* because it is meant to recognize common experiences, not just label people because of their presumed skin color.

In this book, we have tried to be sensitive to the language used to describe different groups. We recognize that the language we use is fraught with cultural and political assumptions and that what seems acceptable now may be offensive later. Perhaps the best way to solve this problem is for different groups to learn as much as they can about one another, becoming more aware of the meaning and nuances of naming and language and more conscious of the racial assumptions embedded in the language. Greater sensitivity to the language used in describing different group experiences is an important step in promoting better intergroup relationships.

Latina is the feminine form in Spanish and refers to women; *Latino*, to men.

that their families treated the experiment as a joke; another's family thought the daughter was being extra nice because she wanted something. One family believed that the student was hiding some serious problem. In all the other cases, parents reacted with shock, bewilderment, and anger. Students were accused of being mean, nasty, impolite, and inconsiderate; the parents demanded explanations for their sons' and daughters' behavior (Garfinkel 1967). Through this experiment, the student researchers were able to see that even the informal norms governing behavior in one's home are carefully structured. By violating the norms of the household, the norms were revealed (Garfinkel 1967).

Ethnomethodological research teaches us that society proceeds on an "as if" basis. That is, society exists because people behave as if there were no other way to do so. Usually, people go along with what is expected of them. Culture is actually "enforced" through the social sanctions applied to those who violate social norms. Usually, specific sanctions are unnecessary because people have learned the normative expectations. When the norms are violated, their existence becomes apparent (see also Chapter 5).

Beliefs

As important as social norms are the beliefs of people in society. **Beliefs** are shared ideas held collectively by people within a given culture about what is true. Shared beliefs are part of what binds people together in society. Beliefs are also the basis for many norms and values of a given culture. In the United States, beliefs that are widely held and cherished are the belief in God and the belief in democracy.

Some beliefs are so strongly held that people find it difficult to cope with ideas or experiences that contradict them. Someone who devoutly believes in God may find atheism intolerable; those who believe in reincarnation may seem irrational to those who think life ends at death. Similarly, those who believe in magic may seem merely superstitious to those with a more scientific and rational view of the world.

Whatever beliefs people hold, they orient us to the world. They provide answers to otherwise imponderable questions about the meaning of life. Beliefs provide a meaning system around which culture is organized. Whether belief stems from religion, myth, folklore, or science, it shapes what people take to be possible and true. Although a given belief may be logically impossible, it nonetheless guides people through their lives.

Values

Deeply intertwined with beliefs are the values of a culture. **Values** are the abstract standards in a society or group that define ideal principles. Values define what is desirable and morally correct; thus values determine what is considered right and wrong, beautiful and ugly, good and bad. Although values are abstract, they provide a general outline for behavior. Freedom, for example, is a value held to be important in U.S. culture, as is equality. Values are ideals forming the abstract standards for group behavior, but they are also ideals that may not be realized in every situation.

Values can be a basis for cultural cohesion, but they can also be a source of conflict. Some of our most contested issues can often be traced to value conflicts. Should government play a role in the provision of health care? Should sex education be taught in schools?

Cultural values can clash when groups have strongly held, but clashing, value systems. Values can be a source of cultural cohesion, but also of cultural conflict. What are some of the different values that are being debated in society?

Should public schools allow school prayer? Should women have the right to choose to terminate a pregnancy? These and numerous other examples you can likely identify are matters of great debate—debates made more heated by the value conflicts that lie at the core of these public issues.

Values guide the behavior of people in society; they also shape the social norms in a given culture. An example of the impact that values have on people's behavior comes from an American Indian society known as the Kwakiutl (pronounced "kwa-kee-YOO-tal"), a group from the coastal region of southern Alaska, Washington State, and British Columbia. The Kwakiutl developed a practice known as *potlatch*, in which wealthy chiefs would periodically pile up their possessions and give them away to their followers and rivals (Wolcott 1996; Harris 1974; Benedict 1934). The object of potlatch was to give away or destroy more of one's goods than did one's rivals. The potlatch reflected Kwakiutl values of reciprocity, the full use of food and goods, and the social status of the wealthiest chiefs in Kwakiutl society. (By the way, chiefs did not lose their status by giving away their goods because the goods were eventually returned in the course of other potlatches. They would even burn large piles of goods, knowing that others would soon replace their wealth through other potlatches.)

Compare this practice with the patterns of consumption in the United States. Imagine the CEOs of major corporations regularly gathering up their wealth and giving it away to their workers and rival CEOs! In the contemporary United States, *conspicuous consumption* (consuming for the sake of displaying one's wealth) celebrates values similar to those of the potlatch: High-status people demonstrate their position by accumulating more material possessions than those around them (Veblen 1953/1899).

Together, norms, beliefs, and values guide the behavior of people in society. It is necessary to understand how they operate in a situation to understand why people behave as they do.

CULTURAL DIVERSITY

It is rare for a society to be culturally uniform. As societies develop and become more complex, different cultural traditions appear. Or, diversity may be part of a complex past wherein different groups have long and engaged histories with each other, sometimes fraught with conflict. In the United States, diversity stems from religious, ethnic, and racial differences, as well as regional, age, gender, and class differences. Currently, 13 percent of people in the United States are foreign born. In a single year, immigrants from more than 100 countries come to the United States (U.S. Census Bureau 2012a). Whereas earlier immigrants were predominantly from Europe, Latin America and Asia are now the greatest sources of new immigrants. One result is a large increase in the number of U.S. residents for whom English is the second

language. Cultural diversity is clearly a characteristic of contemporary American society.

The richness of American culture stems from the many traditions that different groups have brought with them to this society, as well as from the cultural forms that have emerged through their experience within the United States. Jazz, for example, is one of the few musical forms indigenous to the United States. An indigenous art form refers to something that originated in a particular region or culture. However, jazz also has roots in the musical traditions of slave communities and African cultures. Since the birth of jazz, cultural greats such as Ella Fitzgerald, Count Basie, Duke Ellington, Billie Holiday, and numerous others have not only enriched the jazz tradition but have also influenced other forms of music, including rock and roll.

Native American cultures have likewise enriched the culture of our society, as have the cultures that various immigrant groups have brought with them to the United States. With such great variety, how can the United States be called one culture? The culture of the United States, including its languages, arts, food customs, religious practices, and dress, can be seen as the sum of the diverse cultures that constitute this society.

Dominant Culture

Two concepts from sociology help us understand the complexity of culture in a given society: dominant culture and subculture. The **dominant culture** is the culture of the most powerful group in a society. It is the cultural form that receives the most support from major institutions and that constitutes the major belief system. Although the dominant culture is not the only culture in a society, it is commonly believed to be "the" culture of a society despite the other cultures present. Social institutions in the society perpetuate the dominant culture and give it a degree of legitimacy that other cultures do not share. Quite often, the dominant culture is the standard by which other cultures in the society are judged. The dominant culture is the standard by which other cultures in the society are judged.

A dominant culture need not be the culture of the majority of people; rather, it is simply the culture of that group in society with enough power to define the cultural framework. As an example, think of a college or university that has a strong system of fraternities and sororities. On campus, the number of students belonging to fraternities and sororities is probably a numerical minority of the total student body, but the cultural system established by the Greeks may dominate campus life nonetheless. In a society as complex as the United States, it is hard to isolate a single dominant culture, although there is a widely acknowledged "American" culture that is considered to be the dominant one. Stemming from middle-class values, habits, and economic resources, this culture is strongly influenced by instruments of culture such as television, the fashion industry, and Anglo-European traditions, and includes diverse elements such as fast food, Christmas shopping, and professional

MAP 2.1

Mapping America's Diversity: English Language Not Spoken at Home

With increased immigration and greater diversity in the U.S. population, evidence of cultural diversity can be seen in many homes—language being one type of evidence. This map shows the regional differences in the percentage of the population over age 5 who speak a language other than English at home. For the United States as a whole, 17.9 percent of the population—almost one-fifth—fit into this category. Eight percent of the population say they speak English less than very well. What implications does this have for the regions most affected? How might it influence relations between different generations within households?

Source: U.S. Census Bureau. 2010. "American FactFinder." **www.census.gov**

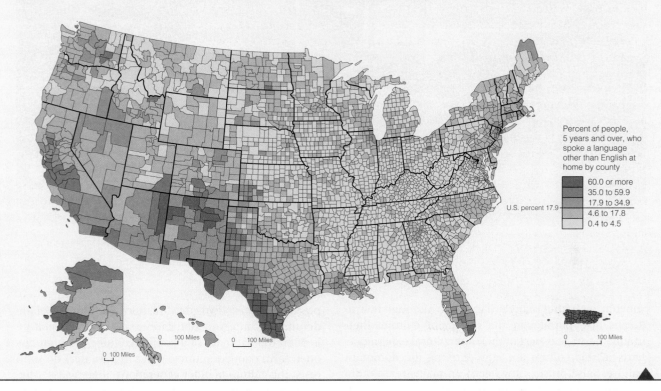

Percent of people, 5 years and over, who spoke a language other than English at home by county

- 60.0 or more
- 35.0 to 59.9
- 17.9 to 34.9
- 4.6 to 17.8
- 0.4 to 4.5

U.S. percent 17.9

sports. It is also a culture that emphasizes achievement and individual effort—a cultural tradition that we will later see has a tremendous impact on how many in the United States view inequality (see Chapter 8).

Subcultures

Subcultures are the cultures of groups whose values and norms of behavior differ to some degree from those of the dominant culture. Members of subcultures tend to interact frequently with one another and share a common worldview. They may be identifiable by their appearance (style of clothing or adornments) or perhaps by language, dialect, or other cultural markers. You can view subcultures along a continuum of how well they are integrated into the dominant culture. Subcultures typically share some elements of the dominant culture and coexist within it, although some subcultures may be quite separated from the dominant one. This separation occurs because they are either unwilling or unable to assimilate into the dominant culture, that is, share its values, norms, and beliefs (Dowd and Dowd 2003).

Rap and hip-hop music first emerged as a subculture where young African Americans developed their own style of dress and music to articulate their resistance to the dominant White culture. Now, rap and hip-hop have been incorporated into mainstream youth culture. Indeed, they are now global phenomena, as cultural industries have turned hip-hop and rap into a profitable commodity. Even so, rap still expresses an oppositional identity for Black and White youth and other groups who feel marginalized by the dominant culture (Watkins 1999).

Some subcultures retreat from the dominant culture, as do the Amish, some religious cults, and some communal groups. In these cases, the subculture is actually a separate community that lives as independently from the dominant culture as possible. Other subcultures may coexist with the dominant society, and members of the subculture may participate in both the subculture and the dominant culture.

Subcultures also develop when new groups enter a society. Puerto Rican immigration to the U.S. mainland, for example, has generated distinct Puerto Rican

The Amish people form a subculture in the United States, although preserving their traditional way of life can be a challenge in the context of contemporary society.

subcultures within many urban areas. Although Puerto Ricans also partake in the dominant culture, their unique heritage is part of their subcultural experience. Parts of this culture are now entering the dominant culture. Salsa music, now heard on mainstream radio stations, was created in the late 1960s by Puerto Rican musicians who were expressing the contours of their working-class culture (Sanchez 1999; Boggs 1992). The themes in salsa reflect the experience of barrio people and mix the musical traditions of other Latin music, including rumba, mambo, and cha-cha. As with other subcultures, the boundaries between the dominant culture and the subculture are permeable, resulting in cultural change as new groups enter society.

thinking SOCIOLOGICALLY

Identify a group on your campus that you would call a *subculture*. What are the distinctive norms of this group? Based on your observations of this group, how would you describe its relationship to the dominant culture on campus? ●

Countercultures

Countercultures are subcultures created as a reaction against the values of the dominant culture. Members of the counterculture reject the dominant cultural values, often for political or moral reasons, and develop cultural

practices that explicitly defy the norms and values of the dominant group. Nonconformity to the dominant culture is often the hallmark of a counterculture. Youth groups often form countercultures. Why? In part, they do so to resist the culture of older generations, thereby asserting their independence and identity. Some also argue that young people establish countercultures because they have so little power in society that they have to construct their own cultures to have some sort of status, or social standing, at least among their peers (Milner 2004). Thus, countercultures among youth, like other countercultures, usually have a unique way of dress, their own special language, perhaps even different values and rituals.

Some countercultures directly challenge the dominant society. The white supremacist movement is an example. People affiliated with this movement have an extreme worldview, one that is in direct opposition to dominant values. White supremacist groups have developed a shared worldview, one based on extreme hostility to racial minorities, gays, lesbians, and feminists. Because of their self-contained culture—one focused on hate—they can be very dangerous (Ferber 1998; Stern 1996).

Countercultures may also develop in situations where there is political repression and some groups are forced "underground." Under a dictatorship, for example, some groups may be forbidden to practice their religion or speak their own language. In Spain, under the dictator Francisco Franco, people were forbidden to speak Catalan—the language of the region around

Cultural diffusion is occurring as U.S. culture is being exported to other nations, as well as the other way around. This photo shows the Old Navy store that opened in Tokyo, Japan.

Barcelona. When Franco died in 1975 and Spain became more democratic, the Catalan language flourished—both in public speaking and in the press.

Ethnocentrism and Cultural Relativism

Because culture tends to be taken for granted, it can be difficult for people within a culture to see their culture as anything but "the way things are." It can thus be difficult to view other cultures without making judgments based on one's own cultural views. **Ethnocentrism** is the habit of seeing things only from the point of view of one's own group. Judging one culture by the standards of another culture is ethnocentric. An ethnocentric perspective prevents you from understanding the world as others experience it, and it can lead to narrow-minded conclusions about the worth of diverse cultures.

Any group can be ethnocentric. Also, ethnocentrism can be extreme or subtle—as in the example of social groups who think their way of life is better than that of any other group. Is there such a ranking among groups in your community? Fraternities and sororities often build group rituals around such claims, youth groups see their way of life as superior to adults, and urbanites may think their cultural habits are more sophisticated than those of groups labeled "country hicks." Ethnocentrism is a powerful force because it combines a strong sense of group solidarity with the idea of group superiority.

Ethnocentrism can build group solidarity, but it also discourages intergroup understanding. Understanding ethnocentrism is critical to understanding some of the major conflicts that are shaping current history. Taken to extremes, ethnocentrism can lead to overt political conflict, war, terrorism, even *genocide*, which is the mass killing of people based on their membership in a particular group. Understanding ethnocentrism can help explain the belief of groups such as al Qaeda that terrorism is justified as a religious *jihad* (defined as a religious

struggle to defend Islamic faith). You might wonder how people could believe so much in the righteousness of their religious faith that they would murder people. Ethnocentrism is a key part of the answer. Understanding ethnocentrism does not excuse such behavior, but it helps you understand how such murderous behavior can occur, though it would be overly simple to explain current political conflicts only in terms of ethnocentrism.

Ethnocentrism can also help you to understand the view that many nations now have of the United States (see Figure 2.1)—a fact that people within the United States have difficulty understanding because we hold ethnocentric views of our own culture, as if it is superior to all others. Many other nations do not see U.S. culture in the positive light that U.S. citizens might expect. As this figure shows, cultural values in the Islamic world can clash with those of the West and are part of the complexity of U.S. relations with those cultures.

Contrasting with ethnocentrism is cultural relativism. **Cultural relativism** is the idea that something can be understood and judged only in relation to the cultural context in which it appears. This does not make

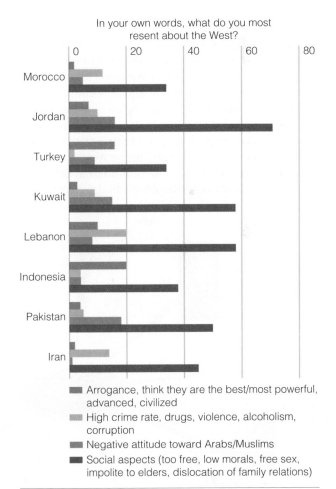

In your own words, what do you most resent about the West?

Morocco
Jordan
Turkey
Kuwait
Lebanon
Indonesia
Pakistan
Iran

■ Arrogance, think they are the best/most powerful, advanced, civilized
■ High crime rate, drugs, violence, alcoholism, corruption
■ Negative attitude toward Arabs/Muslims
■ Social aspects (too free, low morals, free sex, impolite to elders, dislocation of family relations)

FIGURE 2.1 Islamic Views of the West

Data: Burkholder, Richard. 2003. "Iraq and the West: How Wide Is the Morality Gap?" *The Gallup Poll*, Princeton, NJ. **www.gallup.com.** Copyright © 2003 Gallup, Inc. All rights reserved. The content is used with permission; however, Gallup retains all rights of republication.

every cultural practice morally acceptable, but it suggests that without knowing the cultural context, it is impossible to understand why people behave as they do. For example, in the United States, burying or cremating the dead is the cultural practice. It may be difficult for someone from this culture to understand that in parts of Tibet, with a ruggedly cold climate and the inability to dig the soil, the dead are cut into pieces and left for vultures to eat. Although this would be repulsive (and illegal) in the United States, within Tibetan culture, this practice is understandable.

Understanding cultural relativism gives insight into some controversies, such as the international debate about the practice of clitoridectomy—a form of genital mutilation. In a clitoridectomy (sometimes called female circumcision), all or part of a young woman's clitoris is removed, usually not by medical personnel, often in very unsanitary conditions, and without any painkillers. Sometimes, the lips of the vagina may be sewn together. Human rights and feminist organizations have documented this practice in some countries on the African continent, in some Middle Eastern nations, and in some parts of Southeast Asia; estimates are that around two million girls per year worldwide are at risk. This practice is most frequent in cultures where women's virginity is highly prized and where marriage dowries depend on some accepted proof of virginity.

From the point of view of Western cultures, clitoridectomy is genital mutilation and an example of violence against women. Many have called for international intervention to eliminate the practice, but there is also a debate about whether disgust at this practice should be balanced by a reluctance to impose Western cultural values on other societies. Should cultures have the right of self-determination or should cultural practices that maim people be treated as violations of human rights? This controversy is unresolved. The point is to see that understanding a cultural practice requires knowing the cultural values on which it is based.

The Globalization of Culture

The infusion of Western culture throughout the world seems to be accelerating as the commercialized culture of the United States is marketed worldwide. One can go to quite distant places in the world and see familiar elements of U.S. culture, whether it is McDonald's in Hong Kong, Old Navy in Japan, or Disney products in western Europe. From films to fast food, the United States dominates international mass culture, largely through the influence of capitalist markets, as conflict theorists would argue. The diffusion of a single culture throughout the world is referred to as **global culture**. Despite the enormous diversity of cultures worldwide, U.S. markets increasingly dominate fashion, food, entertainment, and other cultural values, thereby creating a more homogenous world culture. Global culture is increasingly marked by capitalist interests, squeezing out the more diverse folk cultures that have been common throughout the world (Steger 2009).

Does increasing globalization of culture change traditional cultural values? Some worry that globalization imposes Western values on non-Western cultures, thus eroding long-held cultural traditions. But global economic change can also introduce more tolerant values to cultures that might have had a narrower worldview previously. As globalization occurs, *both* economic changes *and* traditional cultural values shape the emerging national culture of different societies (Inglehart and Baker 2000).

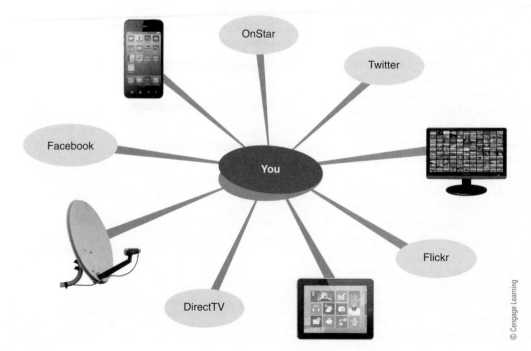

You can see how strong cultural monopolies have become if you just imagine how surrounded you are, even as an individual, by various devices (many of them owned by the same company) that deliver culture to you.

Photos: Phone, Maxx-Studio/Shutterstock.com; Satellite Dish, Roobcio/Shutterstock.com; LCD, Oleksiy Mark/Shutterstock; Tablet, Telnov Oleksii/Shutterstock.com

© Cengage Learning

The conflict between traditional and more commercial values is now being played out in world affairs, with some arguing that the conflicts we see in international relations are partially rooted in a struggle between the values of a consumer-based, capitalist Western culture and the traditional values of local communities. Benjamin Barber (1995) expresses this as the struggle between "McWorld" and "Jihad"—the tension between global commerce and parochial values. As some people resist the influence of market-driven values, movements to reclaim or maintain ethnic and cultural identity can intensify. Thus you can witness a proliferation of culturally based movements, including strong feelings of nationalism, such as among extremist groups in the Middle East, but also pro-democratic movements in parts of the Middle East.

THE MASS MEDIA AND POPULAR CULTURE

Increasingly, culture in the United States—as well as in many parts of the world—is dominated and shaped by the mass media. Indeed, the culture of the United States is so infused by the media that, when people think of U.S. culture, they are likely thinking of something connected to the media—television, film, and so forth. The term **mass media** refers to the channels of communication that are available to wide segments of the population—the print, film, and electronic media (radio and television), as well as the Internet, including Facebook.

The mass media have extraordinary power to shape culture, including what people believe and the information available to them. If you doubt this, observe how much the mass media affect your everyday life. A YouTube video "goes viral" and everyone seems to be talking about it. Or, friends may talk about last night's episode of a particular show or laugh about the antics of their favorite sitcom character. You may have even met your partner or spouse via the electronic media. Your way of dressing, talking, and even thinking has likely been shaped by the media, despite the fact that most people deny this, claiming "they are just individuals." You can find the mass media everywhere—in airports, elevators, classrooms, bars and restaurants, and hospital waiting rooms. Enter an elevator in a major hotel and you might find CNN or the Weather Channel on twenty-four hours a day. You may even be born to the sounds and images of television, since they are turned on in many hospital delivery rooms. Television is now so ever-present in our lives that 42 percent of all U.S. households are now called "constant television households"—that is, those households where television is on most of the time (Gitlin 2002). For many families, TV is the "babysitter." Ninety-eight percent of all homes in the United States have at least one television. The average person consumes some form of media sixty-eight hours a week—more time than they likely spend in school or at work; thirty-one of these hours are spent watching television (U.S. Census Bureau 2012a). More than half (59 percent) of young Americans (those aged 18 to 29) even report that they spend too much time on their cell phones and on the Internet (Newport 2012).

For most Americans, television consumes half of all leisure time (U.S. Bureau of Labor Statistics 2012a). Television is a powerful transmitter of culture. Even with all of the channels and choices available, television portrays a very homogeneous view of culture because in seeking the widest possible audience, networks and sponsors find the most common ground and take few risks. The mass media also shape our understanding of social problems by determining the range of opinion or information that is defined as legitimate and by deciding which experts will be called on to elaborate an issue (Gitlin 2002). Turn on a news talk show, for example, and ask yourself who gets to lead the public discussion of current events. Are the diverse groups in society represented at the table? Do some perspectives seem off limits or outside the boundaries of the media discourse? What age, race, gender, and social class are those who seem to get the most time on air?

Now, however, the widespread availability of Internet-based blogs, chat groups, and social networks is radically changing how people communicate, including about current events. Young people, especially, spend more time using computers for games and other leisure activities than they use for reading (U.S. Bureau of Labor Statistics 2012a). Facebook, Twitter, Foursquare, and other electronic networks have become such a common form of interaction that they are now referred to as **social media**—the term used to refer to the vast networks of social interaction that new media have inspired. Such usage increases the possibility of democratic participation by allowing the open discussion and transmittal of information (Ferdinand 2000). At the same time, however, these forms of communication can mean increased surveillance, both by governments and by hackers. As with other forms of culture, how these networks are used and controlled is a social process.

Despite the vast reach of the mass media, many—including you, perhaps—believe that it has little effect on their beliefs and values, no matter how much they enjoy it. The influence of the mass media is made apparent by trying to do without it—even for a brief period of time. Simply getting away from all of the forms of media that permeate daily life may be extremely difficult to do, as you will see if you try the experiment in the "See for Yourself" box on page 43. Turn it all off for a short period of time, and see if you feel suddenly "left out" of society. Then ask yourself how the mass media influence your life, your opinions, your values, even how you look!

The Organization of Mass Media

Mass media are not only a pervasive part of daily life, but they are also a huge business. On average, consumers spend $900 per year on media consumption, most of which is for television. That may not seem like much until you realize that the television industry (including cable) is a multibillion dollar industry that is organized by powerful economic interests (U.S. Census Bureau 2012a)!

Increasingly, the media are owned by a small number of companies—companies that form huge media monopolies. This means that a few very powerful groups—media conglomerates—are the major producers and distributors of culture. A single corporation can control a huge share of television, radio, newspapers, music, publishing, film, and the Internet. As the production of popular culture becomes concentrated in the hands of just a few, there may be less diversity in the content.

The organization of the mass media as a system of economic interests means that there is enormous power in the hands of a few to shape the culture of the whole society. Sociologists refer to the concentration of cultural power as **cultural hegemony** (pronounced "heh-JEM-o-nee"), defined as the pervasive and excessive influence of one culture throughout society. Cultural hegemony means that people may conform to cultural patterns and interests that benefit powerful elites, even without those elites overtly forcing people into conformity. Thus, on the one hand, although there seems to be enormous choice in what media forms people consume, the cultural messages are largely homogenous (meaning "same"). Cultural hegemony produces a homogeneous mass culture, even when it appears that there is vast choice. Thus cultural monopolies are a means through which powerful groups gain the assent of those they rule. The concept of cultural hegemony implies that culture is highly politicized, even if it does not appear so. Through cultural hegemony, those who control cultural institutions can also control people's political awareness because they create cultural beliefs that make the rule of those in power seem inevitable and right. As a result, political resistance to the dominant culture is blunted (Gramsci 1971). We explore this idea further in the following discussion on sociological theories of culture.

The Media and Popular Culture

Because the mass media pervade the whole society, the media influence such things as popular styles, language, and value systems. Together, these create **popular culture**, meaning the beliefs, practices, and objects that are part of everyday traditions. In a society so dominated by the mass media, popular culture, including such things as music and films, mass-marketed books and magazines, large-circulation newspapers, and Internet websites, are mass-produced. Popular culture is distinct from elite culture, which is shared by only a select few but is highly valued. Unlike elite culture (sometimes referred to as "high culture"), popular culture is mass-consumed and has enormous significance in the formation of public attitudes and values. Popular culture is also supported by patterns of mass consumption, as the many objects associated with popular culture are promoted and sold to a consuming public.

The distinction between popular and elite culture means that various segments of the population consume culture in different ways. This, too, is affected by patterns of social class, race, and gender in the society. Although popular culture may be widely available and relatively cheap for consumers, some groups derive their cultural experiences from expensive theater shows or opera performances where tickets may cost hundreds of dollars. Meanwhile, millions of "ordinary" citizens get their primary cultural experience from television, movie rentals, and increasingly, the Internet. Even something as seemingly common as Internet usage reflects patterns of social class differences in society, as you can see in Figure 2.2. The **digital divide** is a term used to refer to persistence

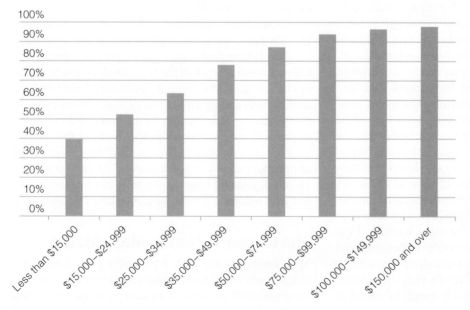

Percent using the Internet

FIGURE 2.2 Social Class and the Internet Even with the widespread availability of the Internet, there are still significant social class differences in who has household access. What difference do you think this makes in the daily lives of those in different income brackets? What social policies might you suggest for remedying this so that there is less of a *"digital divide"*?

Data: U.S. Census Bureau. 2012. *Statistical Abstract 2011.* **www.census.gov**

of inequality in people's access to electronic information. This inequality has led many to advocate for free wireless service in some cities as a way of trying to make Internet access more democratic.

Race, Gender, and Class in the Media

Many sociologists argue that the mass media can promote narrow definitions of who people are and what they can be. What is considered beautiful, for example? Is there a universal idea of beauty, or do the media promote different ideals for different groups?

The mass media have a huge impact on how we see beauty and who or what is defined as beautiful; moreover, these beauty ideals change over time. Youth is defined as beautiful, whereas aging is not. Light skin is promoted as more beautiful than dark skin, regardless of race, although being tan is seen as more beautiful than being pale. In African American women's magazines, the models typified as most beautiful are often those with the clearly Anglo features of light skin, blue eyes, and straight or wavy hair. European facial features are also pervasive in the images of Asian and Latino women appearing in U.S. magazines. The point is that the media communicate that only certain forms of beauty are culturally valued. These ideals are not "natural" somehow. They are constructed by those who control cultural and economic institutions (Craig 2002; Gimlin 2002).

You can learn a lot about how the media shape popular culture by looking carefully and systematically at the images produced and disseminated via the media. *Content analyses* of the media (a research method discussed in the following chapter) show distinct patterns of how race, gender, and class are depicted in various media forms. On prime-time television, men are still a large majority of the characters shown. Over the years, there has been an increase in the extent to which women are depicted in professional jobs, but such images usually depict professional women as young (suggesting that career success comes early), thin, and beautiful. In music videos, women are more present, but typically wearing sexy and skimpy clothing and more often the object of another's gaze than is true for their male counterparts; Black women especially are represented in sexualized ways (Collins 2004; Emerson 2002).

---

see **FOR YOURSELF**

Two Days Without the Media

Suppose that you lived for a few days without use of the mass media that permeate our lives. How would this affect you? In an intriguing experiment, Charles Gallagher (a sociologist at La Salle University) has developed a research project for students in which he asks them to stage a media blackout in their lives for just forty-eight hours. You can try this yourself.

Begin by keeping a written log for forty-eight hours of exactly how much time you spend with some form of media. Include all time spent watching television, on the Internet, reading books and magazines, listening to music, viewing films, even using cell phones—any activity that can be construed as part of the media monopoly on people's time.

Next, eliminate all use of the media, except for that required for work, school, and emergencies for a forty-eight-hour period, keeping a journal as you go of what happens, what you are thinking, what others say, and how people interact with you. *Warning: If you try the media blackout, be sure to have some plan in place for having your family and/or friends contact you in case of an emergency!* When one of the authors of this book (Andersen) had her students do this experiment, they complained even before starting that they wouldn't be able to do it! But they had to try. What happened?

First, Andersen's students had help: The week of the assignment came during a hurricane on the East Coast when many were without power for several days. This did not deter the students from thinking they just *had to have* their DVD players, music, TV, and cell phones! Many of the students said they could not stand being without access to the media—even for a few hours. Most could not go the full two days without using the media.

Most reported that they felt isolated during the media exercise, not just from information, but also mostly from other people. They were excluded from conversations with friends about what happened on a given television episode or about film characters or movie stars profiled in magazines and from playing computer games. One even wrote that without the media, she felt that she had no personality! Without their connection to the media, students felt alienated, isolated, and detached, although most also reported that they studied more without the distraction of the media. A most interesting finding was that several reported that they were much more reflective during this time and had more meaningful conversations with friends.

After trying this experiment, think about the enormous influence that the mass media have in shaping everyday life, including your self-concept and your relationship with other people. What does this exercise teach you about *cultural hegemony*? The role of the mass media in shaping society? How would each of the following theoretical frameworks explain what happened during your media blackout: functionalism, conflict theory, feminist theory, or symbolic interaction?

Source: Personal correspondence, Charles Gallagher, La Salle University. ●

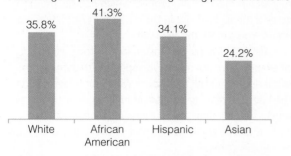

Percentage of population watching during prime-time hours

FIGURE 2.3 Prime-Time Television Usage by Race and Ethnicity

Source: The Nielsen Company. 2009. *Ethnic Trends in Media.*
www.blog.nielsen.com

man, women lounging around on plush furniture, assembled to resemble a stereotypical harem—with plush, overstuffed cushions, lush gardens, and often Middle Eastern tapestries on the walls, thereby producing stereotypes about the supposed sexual excess and availability of Middle Eastern women. Research documents numerous examples of stereotyped portrayals in the media—stereotypes you will see for yourself if you step outside of the taken-for-granted views with which you ordinarily observe the media.

see FOR YOURSELF

Watch a particular kind of television show (situation comedy, sports broadcast, children's cartoon, or news program, for example) and make careful written notes on the depiction of different groups in this show. How often are women and men or boys and girls shown?

How are they depicted? You could also observe the portrayal of Asian Americans, Native Americans, African Americans, or Latinos. What do your observations tell you about the cultural ideals that are communicated through *popular culture*? ●

Even though African Americans watch more television than White people do (see Figure 2.3), they are generally confined to a narrow variety of character types, depicted in stereotypical ways. In recent years, the number of African American characters shown in television has come to match their proportion in the population, but largely because of their casting in situation comedies and in programs that are mostly minority. Latinos and Asians are vastly underrepresented in the media, relative to their proportion in the population, and Native Americans are nonexistent (Signorielli 2009). Latinos are often stereotyped as criminals, passionate lovers, or comic figures. African American men are most often seen as athletes and sports commentators, criminals, or entertainers. Women who work as football sports commentators are typically on the sidelines, reporting not so much on the play of the game as on human interest stories or injury reports—suggesting that women's role in sports is limited to that of nurturer. It is difficult to find a single show where Asians are the principal characters—usually they are depicted in silent roles, as sidekicks, domestic workers, or behind-the-scenes characters. Native Americans make occasional appearances, where they usually are depicted as mystics or warriors. Jewish women are generally invisible on popular TV programming, except when they are ridiculed in stereotypical roles. Arab Americans are likewise stereotyped, depicted as terrorists, rich oil magnates, or in the case of women, as perpetually veiled and secluded (Read 2003; Mandel 2001).

In a good example of how race and gender stereotypes merge in the mass media, one analyst carefully analyzed multiple episodes of the popular show, *The Bachelor* (Dubrofsky 2006). Supposedly, the women all have an equal chance at being selected as the bachelor's mate, the basic concept of the show asserting heterosexual relationships as the most appropriate romance. But, as Dubrofsky shows, women of color are never chosen as the bachelor's mate; they are, in fact, eliminated early from the competition. Equally revealing, Dubrofsky shows how the show's set suggests a harem-like quality—multiple women available to one

Class stereotypes abound in the media and popular culture as well, with working-class men typically portrayed as being ineffectual, even buffoonish (Dines and Humez 2002; Butsch 1992). This has been demonstrated in research by sociologist Laura Grindstaff, who spent six months working on two popular talk shows. For her research, she did careful participant observation and interviewed the production staff and talk show guests. She found that to get airtime, guests had to enact social class stereotypes, acting vulgar and loud. She concluded that, although these popular talk shows give ordinary people a place to air their problems and be heard, the shows exploit the working class, making a spectacle of their troubles (Grindstaff 2002; Press 2002).

Recently, there has been increased representation of gays and lesbians in the media, after years of being virtually invisible or only the subject of ridicule. As advertisers have sought to expand their commercial markets, they are showing more gay and lesbian characters on television. This makes gays and lesbians more visible, although critics point out that they are still cast in narrow and stereotypical terms, or in comical roles (such as in *Modern Family*). Cultural visibility for any group is important because it validates people and can influence the public's acceptance of and generate support for equal rights protection (Gamson 1998).

Television is not the only form of popular culture that influences public consciousness, class, gender, and race. Music, film, books, and other industries play a significant role in molding public consciousness. What images do these cultural forms produce? You can look for yourself. Try to buy a birthday card that contains neither an age or gender stereotype, or watch TV or a movie and see how different gender and race groups are portrayed. You will likely find that women are depicted as trying to get the attention of men; African Americans are more likely than Whites to be seen singing and dancing.

Do these images matter? Studies find that exposure to traditional sexualized imagery in music videos has a negative effect on college students' attitudes, for example, holding more adversarial attitudes about sexual relationships (Kalof 1999). Other studies find that even when viewers see media images as unrealistic, they think that others find the images important and will evaluate them accordingly; this has been found to be especially true for young White girls who think boys will judge them by how well they match the media ideal (Milkie 1999). Although people do not just passively internalize media images and do distinguish between fantasy and reality (Hollander 2002; Currie 1997), such images form cultural ideals that have a huge impact on people's behavior, values, and self-image.

THEORETICAL PERSPECTIVES ON CULTURE AND THE MEDIA

Sociologists study culture and the media in a variety of ways, asking a variety of questions about the relationship of culture to other social institutions and the role of culture in modern life (See Table 2.2 on p. 46). One important question for sociologists studying the mass media is whether these images have any effect on those who see them. Do the media create popular values or reflect them? The **reflection hypothesis** contends that the mass media reflect the values of the general population (Tuchman 1979). The media try to appeal to the most broad-based audience, so they aim for the middle ground in depicting images and ideas. Maximizing popular appeal is central to television program development; media organizations spend huge amounts on market research to uncover what people think and believe and what they will like. Characters are then created with whom people will identify. Interestingly, the images in the media with which we identify are distorted versions of reality. Real people seldom live like the characters on television, although part of the appeal of these shows is how they build upon, but then mystify, the actual experiences of people.

The reflection hypothesis assumes that images and values portrayed in the media reflect the values existing in the public, but the reverse can also be true—that is, the ideals portrayed in the media also influence the attitudes and values of those who see them. This has been illustrated in research on music videos. In a controlled experiment, the researchers exposed college men and women to hip-hop videos with high sexual content. Following their viewing, men in the sample expressed greater sexual objectification of women, more sexual permissiveness, stereotypical gender attitudes, and acceptance of rape myths; the findings did not hold for women in the sample (Kistler and Lee 2010). Although there is not a simple and direct relationship between the content of mass media images and what people think of themselves, clearly these mass-produced images can have a significant impact on who we are and what we think.

Culture and Group Solidarity

Many sociologists have studied particular forms of culture and have provided detailed analyses of the content of cultural artifacts, such as images in certain television programs or genres of popular music. Other sociologists take a broader view by analyzing the relationship of culture to other forms of social organization. Beginning with some of the classical sociological theorists (see Chapter 1), sociologists have studied the relationship of culture to other social institutions. Max Weber looked at the impact of culture on the formation of social and economic institutions. In his classic analysis of the Protestant work ethic and capitalism, Weber argued that the Protestant faith rested on cultural beliefs that were highly compatible with the development of modern capitalism. By promoting a strong work ethic and a need to display material success as a sign of religious salvation, the Protestant work ethic indirectly but effectively promoted the interests of an emerging capitalist economy. (We revisit this issue in Chapter 13.) In other words, culture influences other social institutions.

Many sociologists have also examined how culture integrates members into society and social groups. Functionalist theorists, for example, believe that norms and values create social bonds that attach people to society. Culture therefore provides coherence and stability in society. Robert Putnam examines this idea in his book *Bowling Alone* (2000), in which he argues that there has been a decline in *civic engagement*—defined as participation in voluntary organizations, religious activities, and other forms of public life—in recent years. As people become less engaged in such activities, there is a decline in the shared values and norms of the society so that social disorder results. Sociologists are debating the extent to which there has been such a decline in public life, but from a functionalist perspective, the point is that participation in a common culture is an important social bond—one that unites society (Etzioni et al. 2001).

Classical theoretical analyses of culture have placed special emphasis on nonmaterial culture—the values, norms, and belief systems of society. Sociologists who use this perspective emphasize the integrative function of culture, that is, its ability to give people a sense of belonging in an otherwise complex social system (Smelser 1992). In the broadest sense, they see culture as a major integrative force in society, providing societies with a sense of collective identity and commonly shared worldviews.

Culture, Power, and Social Conflict

Whereas the emphasis on shared values and group solidarity drives one sociological analysis of culture, conflicting values drives another. Conflict theorists (see Chapter 1) have analyzed culture as a source of power in society. You can find numerous examples throughout human history where conflict between different cultures has actually shaped the course of world affairs. One such example comes from the Middle East and the situation for the Kurdish people. The Kurds are an ethnic group (see Chapter 10) who speak their own language and inhabit an area in the Middle East that includes parts of Iraq, Iran, Turkey, and Syria, although they mostly live in northern Iraq. Most are Sunni Muslims, and they have experienced years of political and economic repression and, under Saddam Hussein, mass murder. Attempting to eliminate Kurds altogether, Saddam Hussein ordered the execution of over 180,000 people in Kurdish villages, often using chemical and biological weapons (O'Leary 2002). This and other examples of so-called *ethnic cleansing* show how cultural conflict can be driven by intense group hatred and powerful forms of domination.

Conflict theorists see contemporary culture as produced within institutions that are based on inequality and capitalist principles. As a result, the cultural values and products that are produced and sold promote the economic and political interests of the few—those who own or benefit from these cultural industries. As we have seen, this is especially evident in the study of the mass media and popular culture marketed to the masses by entities with a vast economic stake in distributing their products. Conflict theorists conclude that the cultural products most likely to be produced are consistent with the values, needs, and interests of the most powerful groups in society. The evening news, for example, is typically sponsored by major financial institutions and oil companies. Conflict theorists then ask how this commercial sponsorship influences the content of the news. If the news were sponsored by labor unions, would conflicts between management and workers always be defined as "labor troubles," or might newscasters refer instead to "capitalist troubles"?

Conflict theorists see culture as increasingly controlled by economic monopolies. Whether it is books, music, films, news, or other cultural forms, monopolies in the communications industry (where culture is increasingly located) have a strong interest in protecting the status quo. As media conglomerates swallow up smaller companies and drive out smaller, less-efficient competitors, the control that economic monopolies have over the production and distribution of culture becomes enormous. Mega-communications companies then influence everything—from the movies and television shows you see to the books you read in school.

However, culture can also be a source of political resistance and social change. Reclaiming an indigenous culture that had been denied or repressed is one way that groups mobilize to assert their independence. An example from within the United States is the *repatriation movement* among American Indians who have argued for the return of both cultural artifacts and human remains held in museum collections. Many American Indians believe that, despite the public good that is derived from studying such remains and objects, cultural independence and spiritual respect outweigh such scientific arguments (Thornton 2001). Other social movements, such as the gay and lesbian movement, have also used cultural performance as a means of political and social protest. Cross-dressing, drag shows, and other forms of "gender play" can be seen as cultural performances that challenge homophobia and traditional sexual and gender roles (Rupp and Taylor 2003).

A final point of focus for sociologists studying culture from a conflict perspective lies in the concept of cultural capital. **Cultural capital** refers to the cultural resources that are deemed worthy (such as knowledge of elite culture) and that give advantages to groups possessing such capital. This idea has been most developed by the French sociologist Pierre Bourdieu (1984), who

---

table 2.2 Theoretical Perspectives on Culture

	According to:		
Functionalism	Conflict Theory	Symbolic Interaction	New Cultural Studies
Culture...			
Integrates people into groups	Serves the interests of powerful groups	Creates group identity from diverse cultural meanings	Is ephemeral, unpredictable, and constantly changing
Provides coherence and stability in society	Can be a source of political resistance	Changes as people produce new cultural meanings	Is a material manifestation of a consumer-oriented society
Creates norms and values that integrate people in society	Is increasingly controlled by economic monopolies	Is socially constructed through the activities of social groups	Is best understood by analyzing its artifacts—books, films, and television images

© Cengage Learning

Classical Theorists on Hip-Hop!

Perhaps you are a fan of hip-hop. You love the beat, the style, and it might even influence how you dress. Fans of different forms of popular culture typically just "like" it—but sociology also provides a way to think about popular culture—where it originates, who and what it influences, and how it is organized in social institutions. This gives you a different way of thinking about popular culture. Suppose some of the classical theorists of sociology were asked to comment on the popularity of hip-hop. What might they say? Here is an imagined conversation among them.

Emile Durkheim: I notice that young people can name hip-hop musicians that others in the society do not recognize. This commonly happens because different generations tend to grow up within a shared music culture. Whether it's hip-hop, country, or pop, music cultures bind groups together by creating a sense of shared and collective identity.

For young people, this makes them feel like part of a generation instead of being completely alienated from an otherwise adult-dominated culture.

Karl Marx: It is interesting that White youth are now the major consumers of hip-hop. Hip-hop originated from young, Black youth who are disadvantaged by the economic system of society. Now capitalism has appropriated this creative work and turned it into a highly profitable commodity that benefits dominant groups who control the music industry. As this has happened, the critical perspective originated by young, Black urban men has been supplanted by race and gender stereotypes that support the interests of the powerful.

Max Weber: Emile and Karl, you just see it one way. It's not that you are wrong, but you have to take a multidimensional view. Yes, hip-hop is an economic and

a cultural phenomenon, but it is also linked to power in society. Haven't you noticed how political candidates try to use popular music to appeal to different political constituencies? Don't be surprised to find hip-hop artists performing at political conventions! That's what I find so intriguing: Hip-hop is an economic, cultural, *and* political phenomenon.

W. E. B. DuBois: I've said that Black people have a "double consciousness"—one where they always have to see themselves through the eyes of a world that devalues them—American and "Black" at the same time. But, concurrently, the "twoness" that Black people experience generates wonderful cultural forms such as hip-hop that reflect the unique spirit of African Americans. I once wrote that "there is no true American music but the wild sweet melodies of the Negro slave" (DuBois 1903: 14), but I wish I had lived to see this new spirited and soulful form of musical expression!

---

sees the appropriation of culture as one way that groups maintain their social status.

Bourdieu argues that members of the dominant class have distinctive lifestyles that mark their status in society. Their ability to display this cultural lifestyle signals their importance to others; that is, they possess cultural capital. From this point of view, culture has a role in reproducing inequality among groups. Those with cultural capital use it to improve their social and economic position in society. Sociologists have found a significant relationship, for example, between cultural capital and grades in school; that is, those from the more well-to-do classes (those with more cultural capital) are able to parlay their knowledge into higher grades, thereby reproducing their social position by being more competitive in school admissions and, eventually, in the labor market (Hill 2001; Treiman 2001).

Symbolic Interaction and the Study of Culture

Especially productive when applied to the study of culture has been *symbolic interaction theory*—a perspective that analyzes behavior in terms of the meaning

people give it. (See Chapter 1.) The concept of culture is central to this orientation. Symbolic interaction emphasizes the interpretive basis of social behavior, and culture provides the interpretive framework through which behavior is understood.

Symbolic interaction also emphasizes that culture, like all other forms of social behavior, is socially constructed; that is, culture is produced through social relationships and in social groups, such as the media organizations that produce and distribute culture. People do not just passively submit to cultural norms. Rather, they actively make, interpret, and respond to the culture around them. Culture is not one-dimensional; it contains diverse elements and provides people with a wide range of choices from which to select how they will behave (Swidler 1986). Culture, in fact, represents the creative dimension of human life.

In recent years, a new interdisciplinary field known as *cultural studies* has emerged that builds on the insights of the symbolic interaction perspective in sociology. Sociologists who work in cultural studies are often critical of classical sociological approaches to studying culture, arguing that the classical approach has overemphasized nonmaterial culture, that is, ideas, beliefs, values, and norms. The new scholars of cultural studies find that material culture has increasing importance in modern society (Walters 1999; Crane 1994). This includes cultural forms that are recorded through print, film, artifacts, or the electronic media. Postmodernist theory has greatly influenced new cultural studies (see Chapter 1). *Postmodernism* is based on the idea that society is not an objective thing; rather, it is found in the words and images that people use to represent behavior and ideas. Given this orientation, postmodernism often analyzes common images and cultural products found in everyday life.

Classical theorists have tended to study the unifying features of culture; cultural studies researchers tend to see culture as more fragmented and unpredictable. To them, culture is a series of images—images that can be interpreted in multiple ways, depending on the viewpoint of the observer. From the perspective of new cultural studies theorists, the ephemeral and rapidly changing quality of contemporary cultural forms is reflective of the highly technological and consumer-based culture on which the modern economy rests. Modern culture, for example, is increasingly dominated by the ever-changing, but ever-present, images that the media bombard us with in everyday life. The fascination that cultural studies theorists have for these images is partially founded in illusions that such a dynamic and rapidly changing culture produces.

CULTURAL CHANGE

In one sense, culture is a conservative force in society; it tends to be based on tradition and is passed on through generations, conserving and regenerating the values and beliefs of society. Culture is also increasingly based on institutions that have an economic interest in maintaining the status quo. People are also often resistant to cultural change because familiar ways and established patterns of doing things are hard to give up. But in other ways, culture is completely taken for granted, and it may be hard to imagine a society different from that which is familiar.

Imagine, for example, the United States without fast food. Can you do so? Probably not. Fast food is so much a part of contemporary culture that it is hard to imagine life without it. Consider these facts about fast-food culture:

- The average person in the United States consumes three hamburgers and four orders of French fries per week.
- People in the United States spend more money on fast food than on movies, books, magazines, newspapers, videos, music, computers, and higher education combined.
- One in eight workers has at some point been employed by McDonald's.

Percent using online news

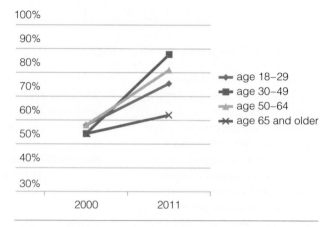

FIGURE 2.4 Who Gets News How? For many, reading the daily newspaper has been a cultural tradition, but this is changing. Many wonder if print newspapers will actually vanish over time. How do you get your news? Looking at age and online news consumption in this figure, what does the data suggest?

Source: U.S. Census Bureau. 2012a. *Statistical Abstract 2011*. Washington, DC: U.S. Department of Commerce. **www.census.gov**

- McDonald's is the largest private operator of playgrounds in the United States and the single largest purchaser of beef, pork, and potatoes.
- Ninety-six percent of American schoolchildren can identify Ronald McDonald—only exceeded by the number who can identify Santa Claus (Schlosser 2001).

Eric Schlosser, who has written about the permeation of society by fast-food culture, has written that "a nation's diet can be more revealing than its art or literature" (2001: 3). He relates the growth of the fast-food industry to other fundamental changes in American society, including the vast numbers of women entering the paid labor market, the development of an automobile culture, the increased reliance on low-wage service jobs, the decline of family farming, and the growth of agribusiness. One result is a cultural emphasis on uniformity, not to mention increased fat and calories in people's diets.

This example shows how cultures can change over time, sometimes in ways that are hardly visible to us unless we take a longer-range view or, as sociologists would do, question that which surrounds us. Culture is a dynamic, not static, force in society, and it develops as people respond to various changes in their physical and social environments. (see, for example, Figure 2.4).

Culture Lag

Sometimes cultures adjust slowly to changing cultural conditions, and the result can be **culture lag** (Ogburn 1922). Some parts of culture may change more rapidly than others; thus one aspect of culture may "lag" behind another. Rapid technological change is often

Death of a Superstar

When Whitney Houston, fabulous superstar, extraordinary singer, and beautiful woman, tragically died in February 2012, millions of people grieved her passing. Her funeral was broadcast live on several major national television networks, with over 14 million people tuning in, far exceeding the usual number of television viewers during that time of day (*The New York Times*, February 21, 2012). How can people be so moved by someone's death, even when they do not know her personally?

We live in a celebrity culture, one in which the public seems endlessly fascinated by the lives of stars, especially those from the world of entertainment and popular culture. If you look at media coverage of the deaths of superstars, you will likely see a common tale told through the media coverage: The tragic and premature loss of someone with enormous talent who rose from common origins to soaring heights of wealth, popularity, and power. The very lyrics in

one of Whitney Houston's songs, "Didn't we almost have it all?" reverberate in the cultural tale relayed through the media—that is, the American dream that one can rise from humble beginnings to "having it all." As sociologist Karen Sternheimer writes, "Celebrity and fame are unique manifestations of our sense of American social mobility; they provide the illusion that material wealth is possible for anyone" (2011: xiii).

Emile Durkheim would say (as would functionalist theorists) that celebrity funerals have a sociological dimension. That is, they produce the *collective consciousness*, thus binding us together in a cultural system and reaffirming our collective beliefs and values.

Whitney Houston's death is not the first time that the public has grieved over a superstar (think of Michael Jackson, Elvis Presley, Marilyn Monroe, James Dean), nor will it be the last. But you don't have to wait for a tragic death to see the cultural ideal of the American dream retold through the media. Observe celebrity culture with a sociological perspective and ask yourself where, when, and how you see the American dream replayed through various media reports.

don Emmert/AFP/Getty Images/Newscom

attended by culture lag because some elements of the culture do not keep pace with technological innovation. In today's world, we have the technological ability to develop efficient, less-polluting rapid transit, but changing people's transportation habits is difficult.

When culture changes rapidly or someone is suddenly thrust into a new cultural situation, the result can be **culture shock**, the feeling of disorientation when one encounters a new or rapidly changed cultural situation. Even moving from one cultural environment to another within one's own society can make a person feel out of place. The greater the difference between cultural settings, the greater the culture shock. International students who come to study in the United States experience this routinely; just imagine how you might feel were you to pick up and move to a foreign country to enroll in college.

Sources of Cultural Change

There are several causes of cultural change, including (1) a change in the societal conditions, (2) cultural diffusion, (3) innovation, and (4) the imposition of cultural change by an outside agency. Let us examine each.

1. **Cultures change in response to changed conditions in the society.** Economic changes, population changes, and other social transformations all influence the development of culture. A change in the makeup of a society's population may be enough by itself to cause a cultural transformation. The high rate of immigration in recent years has brought many cultural changes to the United States. Many major cities, such as Miami and Los Angeles, have a Latin feel because of the large Hispanic population. But cultural change from immigration is now apparent in locations throughout

the United States. Markets selling Asian, Mexican, and Middle Eastern foods are increasingly common; school districts include students who speak a huge variety of languages; popular music bears the imprint of different world cultures. This is not the first time U.S. culture has changed because of immigration. Many national traditions stem from the patterns of immigration that marked the earlier part of the twentieth century—think of St. Patrick's Day parades, Italian markets, and Chinatowns.

2. **Cultures change through cultural diffusion.** **Cultural diffusion** is the transmission of cultural elements from one society or cultural group to another. In our world of instantaneous communication, cultural diffusion is swift and widespread. This is evident in the degree to which worldwide cultures have been Westernized. Cultural diffusion also occurs when subcultural influences enter the dominant group. Dominant cultures are regularly enriched by minority cultures. An example is the influence of Black and Latino music on other musical forms. Rap music, for example, emerged within inner-city African American neighborhoods, describing and analyzing in its own form the economic and political conditions of the urban ghetto. Now, rap music is listened to by White as well as Black audiences and is part of youth culture in general. Cultural diffusion is one thing that drives cultural evolution, especially in a society such as ours that is lush with diversity.

3. **Cultures change as the result of innovation, including inventions and technological developments.** Cultural innovations can create dramatic changes in society. Think, for example, of how the invention of trolleys, subways, and automobiles changed the character of cities. People no longer walked to work; instead, cities expanded outward to include suburbs. Furthermore, the invention of the elevator let cities expand not just out, but also up.

Now, the development of computer technology infiltrates every dimension of life. It is hard to overestimate the effect of innovation on contemporary cultural change. Technological innovation is so rapid and dynamic that one generation can barely maintain competence with the hardware of the next. The smallest laptop or handheld computer today weighs hardly more than a few ounces, and its capabilities rival that of computers that filled entire buildings only twenty years ago.

What are some of the social changes that technology change is creating? People can now work and be miles—even nations—away from their places of employment. Families can communicate from multiple sites; children can be paged; grandparents can receive live photos of a family event; criminals are tracked via cellular technology; music can be

stolen without even going into a music store. Conveniences multiply with the growth of such technology, but so do the invasions of privacy and, perhaps, identity theft. In such a rapidly changing technological world, it is hard to imagine what will be common in just a few years.

One new form of culture that has emerged from technological change is the *blog*. A blog (short for "web log") is a chronological display of entries that people make on the Internet on topics that can be about anything, but most often involve politics or popular culture. A blog is an active diary of sorts in which multiple people participate, either as those who make the entries ("bloggers") or those who simply read them.

This new cultural phenomenon raises interesting questions for sociological research. Studies of blogs to date find, for example, that women are a small proportion of bloggers—only 10 percent of the bloggers on the most widely used political sites. Some use blogs as support systems—for example, a gay person in a very traditional and isolated community may participate in a blog that provides a national community of support. One study in China found that many women are using blogs to subvert traditional concepts of womanhood (Schaffer and Xianlin 2007; Dolan 2006; Harp and Tremayne 2006).

The use of blogs is a good example of how technological innovation can create new forms of culture. Unlike traditional communities, blogging communities can cross vast geographic distances that connect people who might not ever meet face to face. Just as town meetings might have created a sense of community in the past, cyberspace communities can now create new "imagined communities." Some suggest that blogs can actually create a more democratic society by directly engaging more people in political discussion and activity (Perlmutter 2008).

4. **Cultural change can be imposed.** Change can occur when a powerful group takes over a society and imposes a new culture. The dominating group may arise internally, as in a political revolution, or it may appear from outside, perhaps as an invasion. When an external group takes over the society of a "native," or indigenous, group—as White settlers did with Native American societies—they typically impose their own culture while prohibiting the indigenous group from expressing its original cultural ways. Manipulating the culture of a group is a way of exerting social control. Many have argued that public education in the United States, which developed during a period of mass immigration, was designed to force White, northern European, middle-class values onto a diverse immigrant population that was perceived to be potentially unruly and politically disruptive. Likewise, the schools run

by the Bureau of Indian Affairs have been used to impose dominant group values on Native American children (Snipp 1996).

Resistance to political oppression often takes the form of a cultural movement that asserts or revives the culture of an oppressed group; thus cultural expression can be a form of political protest. Identification with a common culture can be the basis for group solidarity, as found in the example of the "Black pride" movement in the 1970s, whose influence is still felt today by having encouraged Black Americans to celebrate their African heritage with Afro hairstyles, African dress, and African awareness. Cultural solidarity has also been encouraged among Latinos through La Raza Unida (meaning "the race," or "the people, united"). Cultural change can promote social change, just as social change can transform culture.

chapter summary

What is culture?
Culture is the complex and elaborate system of meaning and behavior that defines the way of life for a group or society. It is shared, learned, taken for granted, symbolic, and emergent and varies from one society to another.

How do sociologists define norms, beliefs, and values?
Norms are rules of social behavior that guide every situation and may be formal or informal. When norms are violated, *social sanctions* are applied. *Beliefs* are strongly shared ideas about the nature of social reality. *Values* are the abstract concepts in a society that define the worth of different things and ideas.

What is the significance of diversity in human cultures?
As societies develop and become more complex, cultural diversity can appear. The United States is highly diverse culturally, with many of its traditions influenced by immigrant cultures and the cultures of African Americans, Latinos, and Native Americans. The *dominant culture* is the culture of the most powerful group in society. *Subcultures* are groups whose *values* and cultural patterns depart significantly from the *dominant culture*.

What is the sociological significance of the mass media and popular culture?
Elements of popular culture, such as the *mass media*, have an enormous influence on groups' beliefs and values, including images associated with racism and sexism. *Popular culture* includes the beliefs, practices, and objects of everyday traditions.

What do different sociological theories reveal about culture?
Sociological theory provides different perspectives on the significance of culture. *Functionalist theory* emphasizes the influence of values, norms, and beliefs on the whole society. *Conflict theorists* see culture as influenced by economic interests and power relations in society. *Symbolic interactionists* emphasize that culture is socially constructed. This has influenced new cultural studies, which interpret culture as a series of images that can be analyzed from the viewpoint of different observers.

How do cultures change?
There are several sources of cultural change, including change in societal conditions, *cultural diffusion*, innovation, and the imposition of change by dominant groups. As cultures change, *culture lag* can result, meaning that sometimes cultural adjustments are out of sync with each other. People who experience new cultural situations may experience *culture shock*.

Key Terms

beliefs 35
counterculture 38
cultural capital 46
cultural diffusion 50
cultural hegemony 42
cultural relativism 39
culture 26
culture lag 48

culture shock 49
digital divide 42
dominant culture 36
ethnocentrism 39
ethnomethodology 33
folkways 33
global culture 40
language 31

laws 33
mass media 41
material culture 26
mores 33
nonmaterial culture 27
norms 33
popular culture 42
reflection hypothesis 45

Sapir–Whorf hypothesis 31
social media 41
social sanctions 33
subculture 37
symbols 29
taboo 33
values 35

3 Doing Sociological Research

Hill Street Studios/Getty Images

The Research Process

The Tools of Sociological Research

Research Ethics: Is Sociology Value Free?

Chapter Summary

you have now seen some of the interesting things sociologists study through a glimpse into the sociology of culture. You also have a basic foundation in sociological perspective and the major concepts in the field. We turn now to the tools sociologists use to study social phenomena: the tools of sociological research methods. These tools are varied, and the best tool to use depends on the sociological question that is being asked. Let us start with this example.

Suppose you wanted to do some sociological research on how homeless people lived. What is life like for them? How dangerous is it? Where are the homeless to be found? Do they interact and associate with each other? Do they work at all, and if so, doing what? Do they feel rejected by society? Do they really sleep on park benches at night? In his study entitled *Sidewalk*, sociologist Mitch Duneier (1999) wanted to know all these things, plus more. So he decided to study a group of homeless people by living with them. And that is exactly what he did. He lived with them on park benches and in doorways on New York City's lower East Side. He spent four years with them. He interacted with them. He worked with them—a group consisting largely of African American men who sold books and magazines on the street. Duneier himself is White: He tells how becoming accepted into this society of African American men was itself an interesting and challenging process.

Contrary to popular belief, he discovered that these men make up a rather well-organized minisociety, with a social status structure, rules, norms, and a culture. He discovered many unknown elements of this "sidewalk society." Duneier used a method of sociological research called *participant observation*.

Did you ever wonder what happens to people, both women and men, who are on the lam from the law—perhaps for committing some sort of crime, such as armed robbery, burglary, assault, or even something minor such as breaking a curfew? Sociologist Alice Goffman (2009) actually lived in secret with several such people who were fleeing from legal prosecution in her "On the Run" study. Like Duneier's study, Goffman's was a participant observation study. She and the others lived together in a Philadelphia ghetto. They were not simply an unorganized bunch in a street gang, but instead an organized group of fifteen people who evolved a distinct subculture that, for example, informed them about strategies for evading the police and the courts, and that contained norms pertaining to interpersonal behavior—such as specifying punishment for anyone who "ratted" on another group member. The point is that the group was in effect an organized small society with its own membership, social structure, and culture.

In this chapter, we examine the participant observation method plus other methods of sociological research. Each method is different from the others, but they all share a common goal: a deeper understanding of how society operates.

learning objectives

- Learn how and in what ways sociology is a science
- Find out about the techniques of sociological research, and their advantages and disadvantages
- Understand that ethical values have a role in science

THE RESEARCH PROCESS

Sociological research is the tool sociologists use to answer questions. There are various methods that sociologists use to do research, all of which involve rigorous observation and careful analysis.

As we saw in the chapter opener, sociologists Mitch Duneier (1999) as well as Alice Goffman (2009) examined several questions about a group of people by living with them. They were engaged in what is called **participant observation**—a sociological research technique in which a researcher actually becomes simultaneously both participant in and observer of that which she or he studies.

In another example of participant observation, sociologist Peter Moskos (2008), as research for his doctoral dissertation, actually went through a police academy and spent two years as a beat policeman in a major American city, thus subjecting himself to both the rigid discipline of the police force as well as the dangers of the street in this role (see the "Doing Sociological Research: A Cop in the Hood," at the end of this chapter).

Sociologists do other kinds of sociological research as well. Some approaches are more structured and focused than participant observation, such as survey research. Other methods may involve the use of official records or interviews. The different approaches used reflect the different questions asked in the first place. Other methods may require statistical analysis of a large

set of quantitative information. Either way, the chosen research method must be appropriate to the sociological question being asked. (In the Doing Sociological Research boxes throughout this book, we explore different research projects that sociologists have done, showing what question they started with, how they did their research, and what they found.)

However it is done, research is an engaging and demanding process. It requires skill, careful observation, and the ability to think logically about the things that spark your sociological curiosity.

Sociology and the Scientific Method

Sociological research derives from what is called the *scientific method*, originally defined and elaborated by the British philosopher **Sir Francis Bacon** (1561–1626). The **scientific method** involves several steps in a research process, including observation, hypothesis testing, analysis of data, and drawing conclusions. Since its beginnings, sociology has attempted to adhere to the scientific method. To the degree that it has succeeded, sociology is a science. Yet, there is also an art to developing sociological knowledge. Sociology aspires to be both scientific and humanistic, but sociological research varies in how strictly it adheres to the scientific method. Some sociologists test hypotheses (discussed later); others use more open-ended methods, such as the studies by Duneier and by Goffman.

Science is *empirical*, meaning it is based on careful and systematic observation, not just on conjecture. Although some sociological studies are highly *quantitative* and statistically sophisticated, others are *qualitatively* based, that is, based on more interpretive observations, not statistical analysis. Both quantitative and qualitative studies are empirical. Sociological studies may be based on surveys, observations, and many other forms of analysis, but they always depend on an empirical underpinning.

Sociological knowledge is not the same as philosophy or personal belief. Philosophy, theology, and personal experience can deliver insights into human behavior, but at the heart of the scientific method is the notion that a theory must be *testable*. This requirement distinguishes science from purely humanistic pursuits such as theology and literature.

One wellspring of sociological insight is **deductive reasoning**. When a sociologist uses deductive reasoning, he or she creates a specific research question about a focused point that is based on a more general or universal principle (see Figure 3.1). Here is an example of deductive reasoning: One might reason that because Catholic doctrine forbids abortion, Catholics would then be less likely than other religious groups to support abortion rights. This notion is "deduced" from a general principle (Catholic doctrine). You could test this notion (the research question) via a survey. As it turns out, the testing of this hypothesis shows that it is

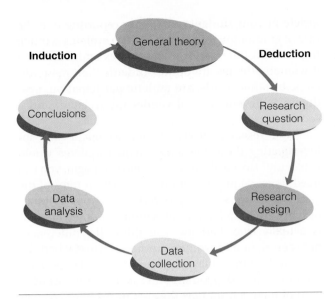

FIGURE 3.1 The Research Process Research can begin by asking a research question derived from general theory or earlier studies, but it can also begin with an observation or even from the conclusion of prior research. One's research question is the basis for a research design and the subsequent collection of data. As this figure shows, the steps in the research process flow logically from what is being asked (figure based on Babbie 2013; Wallace 1983 and 1971).

© Cengage Learning

incorrect: Surveys show that Catholics as a group are on average *more* likely to support abortion rights than are other religious groups. That may come to you as a bit of a surprise! That is why we do research.

Inductive reasoning—another source of sociological insight—reverses this logic: That is, it arrives at general conclusions from specific observations. For example, if you observe that most of the demonstrators protesting abortion in front of a family planning clinic are evangelical Christians, you might infer that strongly held religious beliefs are important in determining human behavior. Again referring to Figure 3.1, inductive reasoning would begin with one's observations. Either way—deductively or inductively—you are engaged in research.

Research Design

When sociologists do research, they engage in a process of discovery. They organize their research questions and procedures systematically—their research site being the social world. Through research, sociologists organize their observations and interpret them.

Developing a Research Question. Sociological research is an organized practice that can be described in a series of steps (see Figure 3.1). The first step in sociological research is to develop a research question. One source of research questions is past research. For any number of reasons, a sociologist might disagree with a research finding and decide to carry out further

research or develop a detailed criticism of previous research. A research question can also begin from an observation that you make in everyday life, such as wondering about the lives of homeless people.

Developing a sociological research question typically involves reviewing existing studies on the subject, such as past research reports or articles. This process is often called a *literature review*. Digital technology has vastly simplified the task of reviewing past studies, that is, the "literature." Researchers who once had to burrow through paper indexes and card catalogs to find material relevant to their studies can now scan much larger swaths of material in far less time using online databases. The catalogs of most major libraries in the world are accessible on the Internet, as are specialized indexes, professional research journals, discussion groups, and other research tools developed to assist sociological researchers.

Increasingly, many journals that report new sociological research are now available online in full-text format, such as on JSTOR (for "journal storage"), or *Sociological Abstracts*. You must be careful using the Internet for research, however. How do you know when something found on the web is valid or true? A lot of what is found on the web is of questionable accuracy, that is, unsubstantiated by accurate research or empirical study. Pay attention, for example, to what person or group has posted the website. Is it a political organization? An organization promoting a cause? A person expressing an opinion? See the box "A Sociological Eye on the Media: Research and the Media" on pages 58–59 for some guidelines about interpreting what you see on the web and in the media.

When you review prior research, you may wonder if the same results would be found if the study were repeated, perhaps examining a different group or studying the phenomenon at a different time. Research that is repeated exactly, but on a different group of people or in a different time or place, is called a **replication study**. Suppose earlier research

The research process involves several operations that can be performed on the computer, such as entering data in numerical form and writing findings in a research report.

found that women managers have fewer opportunities for promotion than do men. You might want to know if this still holds true. You would then replicate the original study, probably using a different group of women and men managers, but asking the same questions that were asked earlier. A replication study can tell you what changes have occurred since the original study and may also refine the results of the earlier work. Research findings should be reproducible: If the research is sound, other researchers who repeat a study should get the same results, unless, of course, some identifiable change, or no identifiable change, has occurred in the interim.

Sociological research questions can also come from casual observation of human behavior. Perhaps you have observed the seating patterns in your college dining hall at lunch and wondered why people sit with the same group day after day. Does the answer point to similarity among the people on the basis of race, gender, age, or perhaps political views? Answering this question would be an example of inductive reasoning: going from a specific observation (such as seating patterns at lunch) to a generalization (a theory about the effects of race and gender). Researcher Beverly Tatum (1997) found that seating patterns in a college dining room depended heavily upon race and also gender.

Creating a Research Design. A **research design** is the overall logic and strategy underlying a research project. Sociologists engaged in research may distribute questionnaires, interview people, or make direct observations in a social setting or laboratory. They might analyze cultural artifacts, such as magazines, newspapers, television shows, or other media. Some do research using historical records. Others base their work on the analysis of social policy. All these are forms of sociological observation. Research design consists of choosing the observational technique best suited to a particular research question.

thinking SOCIOLOGICALLY

If you wanted to conduct research that would examine the relationship between student alcohol use and family background, what measures, or indicators, would you use to get at the two variables: alcohol use and family background? How might you design your study? ●

Suppose you wanted to study the career goals of student athletes. In reviewing earlier studies, perhaps you found research discussing how athletics is related to academic achievement (Messner, S. 2011; Schacht 1996; Messner, M. 1992). You might also have read an article in your student newspaper reporting that the rate of graduation for women college athletes is much higher than the rate for men athletes and wondered if women athletes are better students than men athletes. In other words, are athletic participation, academic achievement, and gender interrelated, and if so, how?

Your research design would lay out a plan for investigating these questions. Which athletes would you study? How will you study them? To begin, you will need to get sound data on the graduation rates of the groups you are studying to verify that your assumption of better graduation rates among women athletes is actually true. Perhaps, you think, the differences between men and women are not so great when the men and women play the same sports. Or perhaps the differences depend on other factors, such as what kind of financial support they get or whether coaches encourage academic success. To study the influence of coaches, you might observe interactions between coaches and student athletes, recording what coaches say about class work. As you proceed, you would probably refine your research design and even your research question. Do coaches encourage different traits in men and women athletes? To answer this question, you have to build into your research design a comparison of coaches interacting with men and with women. Perhaps you even want to compare female and male coaches and how they interact with women and men. *The details of your research design flow from the specific questions you ask.*

Quantitative versus Qualitative Research. The research design often involves deciding whether the research will be qualitative or quantitative or perhaps some combination of both. **Quantitative research** is that which uses numerical analysis. In essence, this approach reduces the data into numbers, for example, the percentage of teenage mothers in California. **Qualitative research** is somewhat less structured than quantitative research, yet still focuses on a central research question. Qualitative research allows for more interpretation and nuance in what people say and do and thus can provide an in-depth look at a particular social behavior. Both forms of research are useful, and both are used extensively in sociology.

Some research designs involve the testing of **hypotheses**. A hypothesis (pronounced "hy-POTH-uh-sis") is a prediction or a hunch, a tentative assumption that one intends to test. If you have a research design that calls for the investigation of a very specific hunch, you might formulate a hypothesis. Hypotheses are often formulated as if–then statements. For example:

Hypothesis: If a person's parents are racially prejudiced, then that person will, on average, be more prejudiced than a person whose parents are relatively free of prejudice.

Some research is done by analyzing the content of various cultural artifacts. Content analysis is one tool of sociological research.

This is merely a hypothesis or expectation, not a demonstration of fact. Having phrased a hypothesis, the sociologist must then determine if it is true or false. To test the preceding example, one might take a large sample of people and determine their prejudice level by interviews or some other mechanism. One would then determine the prejudice level of their parents, perhaps by interviewing their parents. (One would of course have to develop beforehand a questionnaire that accurately measures "prejudice.") According to the hypothesis, one would expect to find more prejudiced children from prejudiced parents and more nonprejudiced children from nonprejudiced parents. If this association is found, the hypothesis is supported. If it is not found, then the hypothesis would be rejected.

Not all sociological research follows the model of hypothesis testing, but all research does include a plan for how **data** will be gathered. (Note that *data* is the plural form; one says, "data are used . . .," not "data is used. . . .") Data can be qualitative or quantitative; either way, they are still data. Sociologists often try to convert their observations into a quantitative form (see the Statistics in Sociology box on pages 64–65).

Sociologists frequently design research to test the influence of one variable on another. A **variable** is a characteristic of a person or group that can have more than one value or score. The notion of "a variable" is very central to sociological research. A variable can be relatively straightforward, such as age or income, or a variable may be more abstract, such as social class or degree of prejudice. In much sociological research, variables are analyzed to understand how they influence each other. With proper measurement techniques and a good research design, the relationships between different variables can be discerned. In the example of student athletes given previously, the variables you use would likely be student graduation rates, gender, and perhaps the sport played. In the hypothesis about race prejudice, parental prejudice and their child's prejudice would be the two variables you would study.

An **independent variable** (see Figure 3.2) is one that a researcher wants to test as the presumed cause of something else. The **dependent variable** is one on which there is a presumed effect. That is, if *X* is the independent variable, then *X* leads to *Y*, the dependent variable. In the previous example of the hypothesis, the amount of prejudice of a parent is the independent variable and the amount of prejudice of a child is the dependent variable. In some sociological research, *intervening variables*—variables that fall between the independent and dependent variables (see Figure 3.2)—are also studied.

Sociological research proceeds through the study of concepts. A **concept** is any abstract characteristic or attribute that can potentially be measured. Social class and social power are concepts. These are not things that can be seen directly, although they are key concepts in the field of sociology. When sociologists want to study concepts, they must develop ways of "seeing" them.

Variables are sometimes used to show more abstract concepts that cannot be directly measured, such as the concept of social class. In such cases, the variables studied are **indicators**—something that points to or reflects an abstract concept. An indicator is a way of "seeing" a concept. An example is shown in Map 3.1 (page 60) using the United Nations' human development index. Here, the human development index is composed of several *indicators*, including life expectancy and educational attainment, combined to show levels of well-being. "Level of well-being" is the *concept*.

The **validity** of a measurement (an indicator) is the degree to which it accurately measures or reflects a concept. To ensure the validity of their findings,

FIGURE 3.2 The Analysis of Variables Much sociological research seeks to find out whether some independent variable (*X*) affects an intervening variable (*Z*), which in turn affects a dependent variable (*Y*).

© Cengage Learning

Research and the Media

On any given day, if you watch the news, read a newspaper, or search the web, you are likely to learn about various new research studies purporting some new finding. How do you know if the research results reported in the media are accurate?

Most people are not likely to check the details of the study or have the research skills to verify the study's claims. But one benefit of learning the basic concepts and tools of sociological research is to be able to critically assess and judge the research frequently reported in the media. The following questions will help:

1. **What are the major variables in the study? Are the researchers claiming a causal connection between two or more variables?** For example, the press reported that one way parents can reduce the chances of their children becoming sexually active at an early age is to quit smoking (O'Neil 2002). The reseasrcher who conducted this study actually claimed there was no direct link between parental smoking and teen sex, although she did find a correlation between parents' risky behaviors—smoking, heavy drinking, and not using seat belts—and children's sexual activity. She argued that parents who engage in unsafe activities provide a model for their children's own risky behavior (Wilder and Watt 2002).

Just because there is a link, or "correlation," between two variables does not necessarily mean one caused the other. Seeing parental behavior as a model for what children do is hardly the same thing as seeing parents' smoking as the cause of early sexual activity!

2. **How have researchers defined and measured the major topics of their study?** For example, if someone claims that 10 percent of all people are gay, how is "being gay" defined? Does it mean having had only one such experience over one's entire lifetime or does it mean actually having a gay identity? Does the definition include both gay and lesbian behavior? Does it also include bisexual behavior? The difference matters because a particular definition may inflate the number reported. Sometimes you must look up the original study, which may be online, to learn how things are defined or how they are measured. Ask yourself if the same conclusions would be reached had the researchers used different definitions and measurements.

3. **Is the research based on a truly representative scientific sample, or is it a biased sample?** You might have to go to the original source of the study to learn

this, but often the sample will be reported in the press (even if in nonscientific language). For example, a study widely reported in the media had headlines exclaiming "Study Links Working Mothers to Slower Learning" (Lewin 2002). But if you read the news report closely, you will learn that this study included only White, non-Hispanic families. Black and Hispanic children were dropped from some of the published results because there were too few cases in the sample to make meaningful statistical comparisons, thus resulting in a *biased sample* (see page 61) (Brooks-Gunn et al. 2002). Another study by the same research team found that there were no significant effects of mother's employment on children's intellectual development among African American or Hispanic children (Waldfogel et al. 2002). The point is not that the study is invalid, but that its results have more limited implications than the headlines suggest.

4. **Is there false generalization in the media report?** Often a study has more limited claims in the scientific version than what is reported in the media. Using the example just given about the connection between maternal employment and children's learning, it would be a big mistake to generalize from the study's

researchers usually use more than one indicator for a particular concept. If two or more chosen measures of a concept give similar results, it is likely that the measurements are giving an accurate—that is, valid—depiction of the concept. For example, using a person's occupation, years of formal education, and annual earnings—namely, using three indicators of her or his social class—would likely be more valid than using only one indicator.

Sociologists also must be concerned with the **reliability** of their research results. A measurement is reliable if repeating the measurement under the same circumstances gives the same result. If a person is given a survey or test two or three times and every time the test gives different results, then the reliability of the test is poor. One way to ensure that sociological measurements are reliable is to use measures that have proved sound in past studies. Another technique

results to all children and families. Remember that some groups were not included.

5. **Can the study be replicated?** *Replication* means *accurately repeated.* Unless there is full disclosure of the research methodology (that is, how the study was conducted), this will not be possible. But you can ask yourself how the study was conducted, whether the procedures used were reasonable and logical, and whether the researchers made good decisions in constructing their research question and research design. If possible, you might be able to obtain the original study upon which the media coverage was based.

6. **Who sponsored the study and do they have a vested interest in the study's results?** Find out if a group or organization with a particular vested interest in the outcome sponsors the research. For example, would you give as much validity to a study of environmental pollution that was funded and secretly conducted by a chemical company as you would a study on the same topic conducted by independent scientists who openly report their research methods and results and who had no connection with the chemical company? Research sponsored by interested parties does not necessarily negate research findings, but it can raise questions about the researchers' objectivity and the standards of inquiry they used.

7. **Who benefits from the study's conclusions?** Although this question does not necessarily challenge the study's findings, it can help you think about whom the findings are likely to help.

8. **What assumptions did the researchers have to make to ask the question they did?** For example, if you started from the assumption that poverty is not the individual's fault but is the result of how society is structured, would you study the values of the poor or perhaps the values of policymakers? When research studies explore matters where social values influence people's opinions, it is especially important to identify the assumptions made by certain questions.

9. **What are the implications of the study's claims?** Thinking through the policy implications of a given result can often help you see things in a new light, particularly given how the media tend to sensationalize much of what is reported.

   Consider the study of maternal employment and children's intellectual development examined in question 3. If you take the media headlines at face value, you might leap to the conclusion that working mothers hurt their children's intellectual development, and you might then think it would be best if mothers quit their jobs and stayed at home. But is this a reasonable implication of this study? Does the study not have just as many implications for day-care policies as it does for encouraging stay-at-home mothers? Especially when reported research studies involve politically charged topics (such as issues of "family values"; or even something like "gun control"), it is important to ask questions that explore various implications of social policies.

10. **Do these questions mean you should never believe anything you hear in the media?** Of course not. Thinking critically about research does not mean being negative or cynical about everything you hear or read. The point is not to reject all media claims out of hand, but instead to be able to evaluate good versus bad research. All research has limitations. Learning the basic tools of research, even if you never conduct research yourself or pursue a career where you would use such skills, can make you a better-informed citizen and prevent you from being duped by claims that are neither scientifically nor sociologically valid.

is to have a variety of people gather the data to make certain the results are not skewed by the tester's appearance, personality, and so forth. The researcher must be sensitive to all factors that affect the reliability of a study.

Sometimes sociologists want to gather data that would almost certainly be unreliable if the subjects (the people in the study) knew they were being studied. Knowing that they are being studied might cause people to change their behavior, a phenomenon in research known as the **Hawthorne effect**, an effect first discovered while observing work groups at a Western Electric plant in Hawthorne, Illinois. The work groups mysteriously increased their productivity right after they were observed by the researchers—an effect the researchers themselves did not notice at first. An example of this effect would be a professor who wants to measure student attentiveness by observing

MAP 3.1

Viewing Society in Global Perspective: Human Development Index

The human development index is a series of indicators developed by the United Nations and used to show the differing levels of well-being in nations around the world. The index is calculated using a number of indicators, including life expectancy, educational attainment, and standard of living. (Are these reasonable indicators of well-being? What else might you use?)

Data: United Nations Development Program. 2013. Human Development Report: International Human Development Indicators. http://hdr.undp.org/en/data/map/

HDI: Human Development Index (HDI) Value (2011)

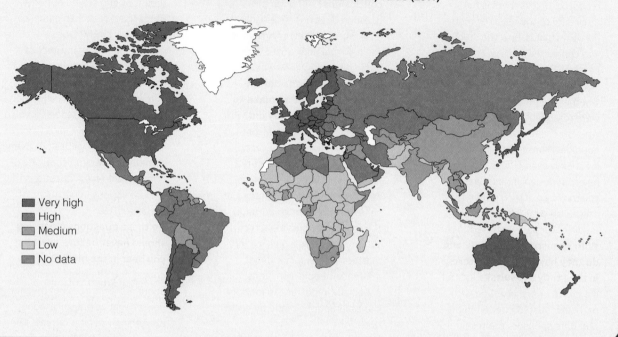

- Very high
- High
- Medium
- Low
- No data

how many notes are taken during class. Students who know they are being scrutinized will magically become more diligent! (In the natural sciences, such as physics, the effect of studying or observing something upon that which is being studied is called the *Heisenberg Principle of Indeterminacy*, named after the German physicist Werner Heisenberg, who first noted it: By studying an object, you change it and thus cannot know its exact state before it was studied. Note that in sociology, participant observation of the covert type, for example, is designed to get around this problem.)

Gathering Data. After research design comes data collection. During this stage of the research process, researchers interview people, observe behaviors, or collect facts that throw light on the research question. When sociologists gather original material, the product is known as *primary*. Examples include the answers to questionnaires or notes made while observing group behavior. Sociologists often rely on *secondary data*, namely data that some other party has already gathered and organized. This can include national opinion polls, census data, national crime statistics, or data from an earlier study made available by the original researcher. Secondary data may also come from official sources, such as university records, city or county records, national health statistics, or historical records.

When gathering data, the groups that sociologists often want to study are so large or so dispersed that research on the whole group is impossible. To construct a picture of the entire group, sociologists take data from a subset of the group and extrapolate to get a picture of the whole. A **sample** is any subset of people (or groups or categories) of a population. A **population** is a relatively large collection of people (or groups or categories) that a researcher studies and about which generalizations are made. Suppose a sociologist wants to study the students at your school. All the students together constitute the population being studied. A survey could be done that reached every student, but conducting a detailed interview with every student would be highly impractical. If the sociologist wants the sort of information that can be gathered only during a personal interview, she would study only a portion, or sample, of all the students at your school.

How is it possible to draw accurate conclusions about a population by studying only part of it? The secret lies in making sure that the sample is *representative* of the population as a whole. The sample should have the same mix of people as the larger population and in the same proportions. If the sample is representative, then the researcher can *generalize* what she finds from the sample to the entire population. For example, if she interviews a sample of 100 students and finds that 10 percent of them are in favor of a tuition increase, and if the sample is representative of the population, then she can conclude that about 10 percent of *all* the students at your school are in favor of a tuition increase. Note that a sample of 5 or 6 students would probably result in generalizations of poor quality, because the sample is not large enough to be representative. A *biased* (nonrepresentative) sample can lead to grossly inaccurate conclusions.

The best way to ensure a representative sample is to make certain that the sample population is selected randomly. A scientific **random sample** gives everyone in the population an equal chance of being selected. Quite often, striking and controversial research findings prove to be distorted by inadequate sampling. The man-on-the-street survey, much favored by TV and radio news reports, and certain other media as well, is the least scientific type of sample and the least representative. (The person-on-the-street sample includes only those who were available at that particular time and place and thus ignores those who were not there.)

Analyzing the Data. After the data have been collected, whether primary or secondary data, they must be analyzed. **Data analysis** is the process by which sociologists organize collected data to discover the patterns and uniformities that the data reveal. The analysis may be statistical or qualitative. When the data analysis is completed, conclusions and generalizations can be made.

Data analysis is labor intensive, but it is also an exciting phase of research. Here is where research discoveries are made. Sometimes while pursuing one question, a researcher will stumble across an unexpected finding, referred to by researchers as **serendipity**. A serendipitous finding is something that emerges from a study that was not anticipated, perhaps the discovery of an association between two variables that the researcher was not looking for or some pattern of behavior that was outside the scope of the research design. Such findings can be minor sidelines to the researcher's major conclusions or, in some cases, lead

A census taker interviews a man in his home.

Rhoda Sidney/PhotoEdit

to major new discoveries. They are part of the excitement of doing sociological research.

A good example of serendipity involved the "Baby Einstein" early intervention program (see the box "Doing Sociological Research: The 'Baby Einstein' Program: A Farce?" on page 67). This program promised to dramatically increase early infant language development by means of exposing very young children (two years old or younger) to various videos (DVDs). It was later discovered that the videos actually *inhibited* language development in children of this age!

Reaching Conclusions and Reporting Results. The final stage in research is developing conclusions, relating findings to sociological theory and past research, and reporting the findings. An important question researchers will ask at this stage is whether their findings can be generalized. **Generalization** is the ability to draw conclusions from specific data and to apply them to a broader population. Researchers ask: Do my results apply only to those people who were studied, or do they also apply to the broader population beyond? Assuming that the results have wide application, a researcher can then ask if the findings refine or refute existing theories and whether the research has direct application to practical social issues. Using the earlier example of the relationship between parent and offspring prejudice, if you found that racially prejudiced people did tend to have racially prejudiced parents (thus supporting your hypothesis), then you might report these results in a paper or research report. You might also ask: What kinds of programs for reducing prejudice do the results of your study suggest?

The Person-on-the-Street Interview

One way research sociologists assure that a sample from some population is *representative* of that population is to take what is called a *scientific random sample*, as explained on page 61. But to get a scientific random sample, each individual in a designated population must be somehow identified or enumerated. Each such enumerated individual must be given a chance of selection that is equal to that of every other enumerated individual in the designated, known population. If, say, the population were all registered full- and part-time undergraduate students in your college or university, the registrar's list of students would be sufficient. The researcher would then draw out a specified number (N) of students at random, by shuffling the list and then taking the top N number—say, 100 students. (Alternatively, each student could be assigned a number and then 100 of these numbers would be selected using a computerized "random numbers table.") The resulting 100 students would then be a true random sample from a known population (all registered students). The sample of 100 could thus be assumed to be representative of the population of all students at the school. The sociologist would then interview or study this N of students.

In a person-on-the-street survey, the population is not known; no population is specified. One does not know whether or not a given individual is in or not in some (unknown) population. Hence one definitely cannot regard such samples as truly random (representative) from a population.

THE TOOLS OF SOCIOLOGICAL RESEARCH

There are several tools or techniques sociologists use to gather data. Among the most widely used are survey research, participant observation, controlled experiments, content analysis, historical research, and evaluation research.

The Survey: Polls, Questionnaires, and Interviews

Whether in the form of a questionnaire, interview, or telephone poll, surveys are among the most commonly used tools of sociological research. Questionnaires are typically distributed to large groups of people. The *return rate* is the percentage of questionnaires returned out of all those distributed or initially requested. A low return rate introduces possible bias because the small number of responses may not be representative of the whole group.

Like questionnaires, interviews provide a structured way to ask people questions. They may be conducted face-to-face, by phone, or electronically, as by mail (email) or even Facebook. Interview questions may be open-ended or closed-ended, though the open-ended form is particularly accommodating if respondents wish to elaborate.

Typically, a survey questionnaire will solicit data about the respondent (the person you are studying), such as income, occupation or employment status (employed or unemployed), years of formal education, yearly income, age, race, and gender, coupled with additional questions that throw light upon a particular research question. For *closed-ended* questions, people must reply from a list of possible answers, like a multiple-choice test. For *open-ended* questions, the respondent is allowed to elaborate on her or his answer. Closed-ended questions are generally (though not always) analyzed quantitatively, and open-ended questions are generally (though not always) analyzed qualitatively. Thus a survey can involve both qualitative as well as quantitative research.

As a research tool, surveys make it possible to ask specific questions about a large number of topics and then to perform sophisticated analyses to find patterns and relationships among variables. The disadvantages of surveys arise from their rigidity (see Table 3.1, page 66). Responses may not accurately capture the opinions of the respondent or may fail to capture nuances in people's behavior and attitudes. Also, what people say and what they do are not always the same. Survey researchers must be persistent to get answers that are truthful, one reason for allowing respondents to be anonymous. Survey researchers sometimes get at this problem by asking essentially the same question in different ways. In this way, *validity* is increased.

Participant Observation

A unique and interesting way for sociologists to collect data and study society is to actually become part of the group they are studying. This is the method of *participant observation*. (*Non-participant observation* is also used as a technique for research. For example, one may wish to study a work group in a factory without actually participating in the group itself.) Two roles are played at the same time: subjective participant and objective observer. Usually the group is aware that the sociologist is studying them, but not always. Participant observation is sometimes called *field research*, a term borrowed from anthropology.

Participant observation combines subjective knowledge gained through personal involvement and objective knowledge acquired by disciplined recording of what one has seen. The subjective component supplies a dimension of information that is lacking in survey data.

---

debunking SOCIETY'S MYTHS

MYTH: People who are just hanging out together and relaxing don't care much about social differences between them.

SOCIOLOGICAL RESEARCH: Even casual groups have organized social hierarchies. That is, they make distinctions within the group that give some people higher status than others. This has been shown in participant observation studies such as Duneier's (1999) study of the homeless on New York City's lower East Side; in Anderson's (1976) study of the people just hanging out in "Jelly's Bar"; in Anderson's (1999) *Code of the Street* study, which showed a rigid hierarchy among those engaged in street crime; in Alice Goffman's (2009) study of people "on the run" from the law; and in the study of "Pam's Place" (Gimlin 2002, 1996), which showed the class-like distinctions hairdressers and their customers made between each other. ●

Street Corner Society (1943), a classic work by sociologist William Foote Whyte, documents one of the first qualitative participant observation studies ever done. Whyte studied the "Cornerville Gang," a group of Italian American men whose territory was a street corner in Boston in the late 1930s and early 1940s. Although not Italian, Whyte learned to speak the language, lived with an Italian family, and then infiltrated the gang by befriending the gang's leader, whose pseudonym was "Doc." Doc was Whyte's **informant**, a person with whom the participant observer works closely to learn about the group. For the duration of the study, Doc was the only gang member who knew that Whyte was doing research on his gang. This represents what is called **covert participant observation**, in which the members of the group being studied do not know that they are being researched. This is one means of trying to reduce the Hawthorne effect. (If the group is told that they are being studied and that they are the research subjects, then it is called **overt participant observation**. Sometimes the group members inadvertently find out that they are research subjects and may become angry because of the discovery. In this case, covert

participant observation is by accident transformed into overt participant observation.)

Most social scientists of the 1940s and 1950s thought gangs were socially disorganized, random deviant groups, but Whyte's study showed otherwise—as have participant observation studies since then, notably those of Anderson (1999, 1990, 1976) and Goffman (2009), as examples. He found that the Cornerville Gang, and by implication other urban street corner gangs as well, was a highly organized minisociety with its own social hierarchy (social stratification), morals, practices, and punishments (sanctions) for deviating from the norms of the gang.

There are a few built-in weaknesses to participant observation as a research technique. We already mentioned that it is very time-consuming. Participant observers have to cull data from vast amounts of notes. Such studies usually focus on fairly small groups, posing problems of generalization. Participant observation can also pose real physical dangers to the researcher, such as being "found out" or "outed" if one is studying a street gang using covert participant observation (Sanchez-Jankowski 1991). Observers may also lose their objectivity by becoming too much a part of what they study. If this happens—the observer becomes so much a part of the group that he or she is no longer a scientific observer but rather a participant—it is called "going native" and is seen as one of the disadvantages of participant observation research. These limitations aside, participant observation has been the source of some of the most arresting and valuable studies in sociology (see "A Cop in the Hood" on page 69).

Controlled Experiments

Controlled experiments are highly focused ways of collecting data and are especially useful for determining a pattern of cause and effect. To conduct a controlled experiment, two groups are created, an *experimental group*, which is exposed to the factor or variable one is examining, and the *control group*, which is not. In a controlled experiment, external influences are either eliminated or equalized, that is, held constant, between the experimental and the control group. This is necessary in order to establish cause and effect.

Suppose you wanted to study whether violent television programming causes aggressive behavior in children. You could conduct a controlled experiment to investigate this question. The behavior of children would be the dependent variable (variable Y); the independent variable (variable X) is whether or not the children are exposed to violent programming. To investigate your question, you would expose an experimental group of children (under monitored conditions) to a movie containing lots of violence (ultimate fighting, for example, or gunfighting). The control group would watch a movie

statistics in sociology

Certain fundamental statistical concepts are basic to sociological research. Although not all sociologists do quantitative research, basic statistics are important to carrying out and interpreting sociological studies.

A **percentage** is the same as parts per hundred. To say that 22 percent of U.S. children are poor tells you that for every 100 children randomly selected from the whole population, approximately 22 will be poor. A **rate** is the same as parts per some number, such as per 10,000 or 100,000. The homicide rate in 2009 was about 7.2, meaning that for every 100,000 people in the population, approximately 7 were murdered. A rate is meaningless without knowing the numeric base on which it is founded; it is always the number per some other number.

A **mean** is the same as an average. Adding a list of fifteen numbers and dividing by fifteen gives the mean. The **median** is often confused with the mean but is actually quite different. The median is the midpoint in a series of values arranged in numeric order. In a list of fifteen numbers arrayed in numeric order, the eighth number (the middle number) is the median. In some cases, the median is a better measure than the mean because the mean can be skewed ("pulled" up or down) by extremes at either end. Another often-used measure is the **mode**, which is simply the value (or score) that appears most frequently in a set of data.

Let's illustrate the difference between mean and median using national income distribution as an example. Suppose that you have a group of ten people. Two make $10,000 per year, seven make $40,000 per year, and one makes $1 million per year. If you calculate the mean (the average), it comes to $130,000. The median, on the other hand, is $40,000, a figure that more accurately suggests the income profile of the group. That single million-a-year earner dramatically distorts, or skews, the picture of the group's income. If we want information about how the group lives in general, we are wiser to use the median income figure as a rough guide, not the mean. Note also that the mode in this example is the same as the median: $40,000.

Sociologists frequently examine the relationship between two variables. **Correlation** is a widely used technique for analyzing the patterns of association, or correlation, between pairs of variables such as income and education. We might begin with a questionnaire that asks for annual earnings (which we designate as the dependent variable, Y) and level of education (the independent variable, X). Correlation analysis delivers two types of information: It tells us the "direction" of the relationship between X and Y and also the "strength" of that relationship. The direction of a relationship is positive (that is, a positive correlation exists) if X is low when Y is low and if X is high when Y is high. But there is also a correlation if Y is low when X is high (or vice versa); this is a negative, or inverse, correlation. The strength of a correlation is simply how closely or tightly the variables are associated, regardless of the direction of correlation. With this example, you might well find a positive correlation between education (X) and annual earnings (Y), and we would also be interested in the strength of this correlation.

A correlation does not necessarily imply cause and effect. A correlation is simply an association, one whose cause must be explained by means other than simple correlation analysis. A **spurious correlation** exists when there is no meaningful causal connection between apparently associated variables.

Another widely used method of analyzing sociological data is **cross-tabulation**, a way of seeing if two variables are related by breaking them down into categories for comparison. Take the following example. In a Gallup Poll (2012), the following question was asked: "Do you feel that the laws covering the sale of firearms should be made more strict, or less strict?" The following results, a cross-tabulation of answers

---

that is free of violence. Beforehand, the children would be assigned randomly to the experimental group or the control group (this is called *experimental randomization*) in order to make the composition of the two groups as much alike as possible. Aggressiveness in the children (the dependent variable) would be measured twice: a *pretest* measurement made before the movies are shown and a *posttest* measurement made afterward. You would take pretest and posttest measures on both the control and the experimental groups. Studies of this sort actually find that the children who watched the violent movie are indeed more violent and aggressive afterward than those who watched a movie containing no violence (Taylor et al. 2012; Worchel et al. 2000; Bushman 1998).

An extension of such a study would be the *Solomon Four-Group Experimental Design*. In this kind of study, *only* a posttest measure of aggressiveness is taken on experimental and control groups. In other words, there would be four separate groups or experimental conditions, as follows:

- One experimental group gets X (the violent movie) plus a pretest plus a posttest.
- Another experimental group gets X plus the posttest only.
- One control group gets both the pretest plus the posttest, but no X.
- Finally, one control group gets only the posttest, that is, no X and no pretest.

That way, the effect of taking a pretest measure in the first place can be determined: By comparing the control group having only the posttest measure of aggressiveness to the control group having both the pretest and the posttest measures, one can assess the effect of having taken a posttest measure versus not taking it. In other words, taking a pretest measure might

to the question (the dependent variable) by gender (the independent variable), were obtained:

	More Strict:	Less Strict:
Women:	74%	26%
Men:	53%	47%

Source: **www.gallup.com**

As you can see from this cross-tabulation, women and men differed on the question. In general, women wanted more strict laws than did men. This means that the two variables—gender and the answer to the question—are related.

The now infamous massacre of twenty first-grade children in a Connecticut school with an assault rifle and automatic pistols had just taken place has influenced public attitudes toward gun control. This horrific incident is discussed in Chapter 7. Preliminary evidence suggests that the percentage of both women as well as men wanting more strict gun control will increase.

Statistical information is notoriously easy to misinterpret, willfully or accidentally. Examples of some statistical mistakes include the following:

- **Citing a correlation as a cause.** A correlation reveals an association between things (variables). Correlations do not necessarily indicate that one causes the other. Sociologists often say: "Correlation is not proof of causation."

- **Overgeneralizing.** Statistical findings are limited by the extent to which the sample group actually reflects or represents the population from which the sample was obtained. Generalizing beyond the population is a misuse of statistics. Studying only men and then generalizing conclusions to both men and women would be an example of overgeneralizing. This kind of mistake is fairly common in the media and also in some sociological research.

- **Interpreting probability as certainty.** Probability is a statement about chance or likelihood only. For example, in the cross-tabulation given previously, women are more likely than men to favor strict gun control. This means that women have a higher probability (a greater chance) of favoring strict gun control than men; it does not mean that all women favor strict gun control or that all men do not.

- **Building in bias.** In a famous advertising campaign, public taste tests were offered between two soft drinks. A wily journalist verified that in at least one site, the brand sold by the sponsor of the test was a few degrees colder (thus presumably better tasting) than its competitor when it was given to the people being tested, which biased the results. Bias can also be built into studies by careless wording on questionnaires.

- **Faking data.** Perhaps one of the worse misuses of statistics is actually making up, or faking, data. A famous instance of this occurred in a study of identical twins who were separated early in life and raised apart (Burt 1966). The researcher wished to show that despite their separation, the twins remained highly similar in certain traits, such as measured intelligence (IQ), thus suggesting that their (identical) genes caused their striking similarity in intelligence. It was later shown that the data were fabricated (Mackintosh 1995; Taylor 1980; Hearnshaw 1979; Kamin 1974).

- **Using data selectively.** Sometimes a survey includes many questions, but the researcher reports on only a few of the answers. Doing so makes it quite easy to misstate the findings. Researchers often do not report findings that show no association between variables, but these can be just as telling as associations that do exist. For example, researchers on gender differences typically report the differences they find between men and women, but seldom publish their findings when the results for men and women are identical. This tends to exaggerate the differences between women and men and falsely confirms certain social stereotypes about gender differences.

Mark Richards/PhotoEdit

The men in this bar, as shown by participant observation as in Anderson's (1976) classic study of "Jelly's Bar" in *A Place on the Corner*, reveal status differences among themselves that they create, such as (in descending status order) "regulars," "hoodlums," and "winos."

itself affect the people you are studying and thus affect the amount of their aggressiveness.

Among its advantages, a controlled experiment can establish causation, and it can zero in on a single independent variable. On the downside, controlled experiments can be artificial. They are for the most part performed in a contrived laboratory setting (unless it is what is called a *field experiment*), and they tend to eliminate many real-life effects. Analysis of controlled experiments includes making judgments about how much the artificial setting has affected the results (see Table 3.1).

Content Analysis

Researchers can learn a vast amount about a society by analyzing *cultural artifacts* such as newspapers, magazines, TV programs, or popular music. **Content analysis** is a way of measuring by examining the

table 3.1 Comparison of Six Research Techniques

Technique (Tool)	Qualitative Analysis or Quantitative Analysis	Advantages	Disadvantages
The survey (polls, questionnaires, interviews)	Usually quantitative, often qualitative	Permits the study of a large number of variables; results can be generalized to a larger population if sampling is accurate	Difficult to focus in great depth on a few variables; difficult to measure subtle nuances in people's attitudes
Participant observation	Usually qualitative	Studies actual behavior in its home setting; affords great depth of inquiry	Is very time-consuming; difficult to generalize beyond the research setting
Controlled experiment	Usually quantitative	Focuses on only two or three variables; able to study cause and effect	Difficult or impossible to measure large number of variables; may have an artificial quality
Content analysis	Can be either qualitative or quantitative	A way of measuring culture	Limited by studying only cultural products or artifacts (music, TV programs, stories, other), rather than people's actual attitudes
Historical research	Usually qualitative	Saves time and expense in data collection; takes differences over time into account	Data often reflect biases of the original researcher and reflect cultural norms that were in effect when the data were collected
Evaluation research	Can be either qualitative or quantitative	Evaluates the actual outcomes of a program or strategy; often direct policy application	Limited in the number of variables that can be measured; maintaining objectivity is problematic if research is done or commissioned by administrators of the program being evaluated

© Cengage Learning

cultural artifacts of what people write, say, see, and hear. The researcher studies not people but the communications the people produce as a way of creating a picture of their society.

Content analysis is frequently used to measure cultural change and to study different aspects of culture (Lamont 1992). Sociologists also use content analysis as an indirect way to determine how social groups are perceived—they might examine, for example, how Asian Americans are depicted in television dramas or how women are depicted in advertisements.

Children's books have been the subject of many content analyses. In acknowledgment of their impact on the development of youngsters, a team of sociologists compared images of Black Americans in children's books from the 1930s to the present (Pescosolido et al. 1997). They obtained three important findings: First, they found a declining representation of African Americans from the 1930s through the 1950s, with practically no representation from 1950 through 1964. Beginning in 1964, an increase in representation lasted until the mid-1970s, when the appearance of African American characters leveled off. Second, they found that the symbolic images of African Americans did change significantly over time. In the 1960s—a period of much racial unrest—African Americans were mostly

Controlled experimentation shows that some media violence tends to desensitize children to the effects of violence, including engendering less sympathy for victims of violence (Baumeister and Bushman 2008; Huesmann et al. 2003; Cantor 2000). Many also think that violent video games (another form of media) may be a cause of school shootings, where youth go on a rampage of gunfire against fellow students. Perhaps there is some link here; it is too simplistic to see a direct causal connection between viewing violence and actually engaging in it. For one thing, such an argument ignores the broader social context of violent behavior (including such things as the availability of guns, family characteristics, youth alienation from school, to name a few) (Sternheimer 2007).

Dwayne Newton/PhotoEdit

The "Baby Einstein" Program: A Farce?

Research Methodology: Several years back (starting in 1997), the Baby Einstein program, acquired by Disney Productions in 2001, advertised that it could greatly increase language development and other skills (the dependent variables) by subjecting very young children (two years old or younger) to the various toys, flash cards, DVDs, and books they marketed.

Research Question and Hypothesis: The claim was that such exposure would produce earlier and better language development, comparing a sample of infants before exposure to several months after exposure. The implication was that exposure to this Baby Einstein program would result in faster and better language development than that in infants not exposed to the program. *After all, what could be more obvious?* Of course early exposure to the program would result in better language development than would nonexposure to the program. How could the results possibly be otherwise? Thousands upon thousands of parents with high hopes purchased the Baby Einstein products.

Research Results: This example shows precisely why doing research is so important! Several years after the Baby Einstein program was begun, anxious parents began contacting the founders of the program and telling them that their infants were not responding well to the Baby Einstein DVDs, flash cards, and so on. Furthermore, two University of Washington professors discovered that, in fact, children exposed to the program and its gadgets had actually *slowed down* their language development relative to children who had not been in the program. In addition, children exposed to the program revealed greater attention deficit afterward.

Retest and Conclusions: As a test for these observations, experimental *as well as* control conditions for children, matched on age, were created, and measures of (dependent variables) language development, reading speed, attention span, and other dependent variables were carried out over time. As it turns out, the findings disproved what "obvious common sense" told us; namely, the experimental group babies (the Baby Einstein conditions) performed less well on these dependent variables than did babies in the control conditions (those not exposed to the program)!

Implications: These results were upsetting to the original founders of the program, including Disney Productions. Presently, there are class-action lawsuits being brought against Disney Productions on the grounds that the "Baby Einstein" materials were fraudulent and not educational, as they were initially advertised. Furthermore, the American Academy of Pediatrics has since recommended that children younger than two years of age not be exposed to the kinds of DVDs and videos marketed by Baby Einstein.

Questions to Consider

1. Do you think that the lack of predicted effect of the original Baby Einstein program might have been produced by cultural differences among the studied infants? How would you measure culture and cultural differences? Would any cultural effects on language development among infants depend upon the infant's race? Gender? Family's annual income?

2. Do you think that the failure of the original Baby Einstein program might have been in part because of the extremely young age of the tested infants? How might one correct for this in a new study?

Source: Lewin, Tamar. 2010. "Baby Einstein Founder Goes to Court." *The New York Times* (January 13): A15. **www.nytimes.com**

---

portrayed in "safe," distant images, such as in secondary and nearly invisible occupational roles. Third, they found few portrayals of Black adults in intimate, egalitarian, or interracial relationships. Recent research on stereotyping generally confirms these three findings (Baumeister and Bushman 2008: 419–421).

Content analysis has the advantage of being *unobtrusive*, or "nonreactive." The research can have no effect at all on the person being studied because the cultural artifact has already been produced. Hence content analysis will reveal very little if any Hawthorne effect. Content analysis is limited in what it can study, however, because it is based on mass communication— either visual, oral, or written. It cannot tell us what people think about these images or whether they affect people's behavior. Other methods of research, such as interviewing or participant observation, would be used to answer these questions. Nonetheless, content analysis can be very insightful.

Historical Research

Historical research examines sociological themes over time. It is commonly done in historical archives, such as official records, church records, town archives, private diaries, or oral histories. The sources of this sort of material are critical to its quality and applicability. Oral histories have been especially illuminating, most dramatically in revealing the unknown histories of groups that have been ignored or misrepresented in other historical accounts. For example, when developing an account of the spirituality of Native Americans, one would be misguided to rely solely on the records left by Christian missionaries or

U.S. Army officials. These records would give a useful picture of how Whites perceived Native American religion, but they would be a very poor source for discovering how Native Americans understood their own spirituality.

In a similar vein, the writings of a slave owner can deliver fascinating insights into slavery, but a slave owner's diary will certainly present a different picture of slavery as a social institution than will the written or oral histories of former slaves themselves.

Handled properly, comparative and historical research is rich with the ability to capture long-term social changes and is the perfect tool for sociologists who want to ground their studies in historical or comparative perspectives.

Evaluation Research

Evaluation research assesses the effect of policies and programs on people in society. If the research is intended to produce policy recommendations, then it is called *policy research.*

Suppose you want to know if an educational program is actually improving student performance. You could design a study that measured the academic performance of two groups of students, one that participates in the program and one that does not (a "control" group). If the academic performance of students in the program is better than that of those not in the program, and if the groups are alike in other ways (they are often matched to accomplish this), you would conclude that the program was effective. If the academic performance of the students in the program ended up being the same (or even worse) as those not in the program, then you would conclude that the program was not effective. (See the box "The 'Baby Einstein Program': A Farce?" on page 67.) If you use this research to recommend social policy, you would be doing policy research.

RESEARCH ETHICS: IS SOCIOLOGY VALUE FREE?

The topics dealt with by sociology are often controversial. People have strong opinions about social questions, and in some cases, the settings for sociological work are highly politicized. Imagine spending time in an urban precinct house to do research on police brutality or doing research on acquired immunodeficiency syndrome (AIDS) and sex education in a conservative public school system. Under these conditions, can sociology be scientifically objective? How do researchers balance their own political and moral commitments against the need to be objective and open-minded? Sociological knowledge has an intimate connection to political values and social views. Often the very purpose of sociological research is to gather data as a step in creating social policy. Can sociology be value free? Should it be?

This is an important question without a simple answer. Most sociologists do not claim to be value free, but they do try as best they can to produce objective research. It must be acknowledged that researchers make choices throughout their research that can influence their results. The problems sociologists choose to study, the people they decide to observe, the research design they select, and the type of media they use to distribute their research can all be influenced by the personal values of the researcher.

Sociological research often raises ethical questions. In fact, ethical considerations of one sort or another exist with any type of research. In a survey, the person being questioned is often not told the purpose of the survey or who is funding the study. Is it ethical to conceal this type of information?

In controlled experiments, *deception* is often employed, as in the now-famous studies by Stanley Milgram, to be reviewed later in Chapter 6, where people were led to believe that they were causing harm to another, when in fact they were not. Researchers often reveal the true purpose of an experiment only after it is completed. This is called **debriefing**. The deception is therefore temporary. But does that lessen the potential ethical violation? Maybe the effects of deception become longer lasting. Does deception lessen any potential damage to the self-concept of the subject or respondent, or does it actually increase this damage?

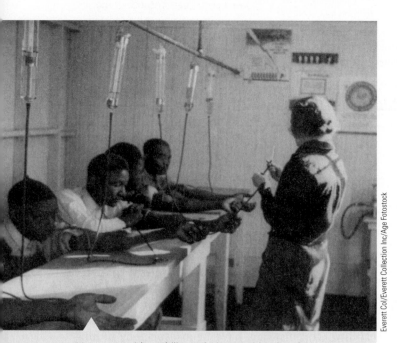

Here men with syphilis are being examined to determine the "progress" of syphilis. These unfortunate men were experimental subjects in the U.S. government's infamous Tuskegee Syphilis Study, one of the clearest ethical violations in all the history of science.

Everett Col/Everett Collection Inc/Age Fotostock

DOING **sociological research**

A Cop in the Hood: Participant Observation

Research Methodology: An excellent recent example of participant observation research of the overt type is sociologist Peter Moskos's twenty months as a bona fide police officer in Baltimore, Maryland. For his doctoral dissertation research, Moskos, who is White, underwent the standard six months of training in the police academy and was then assigned to Baltimore's Eastern District, a heavily African American and depressed ghetto with a heavy drug trade. A true participant observer, he became a police officer. He got to know and trust the other officers with whom he worked, and he became familiar with the social life of the homeless individuals, drug dealers, and neighborhood residents in East Baltimore. He lived minute by minute and day by day with the ever-present extreme dangers of police work, carried a Glock semiautomatic pistol with a seventeen-shot clip (which he never had to fire but had to "show" on occasion), and discovered that "danger creates a bond" among police officers. He wrote his field notes each day after work—numbering 350 typed, single-spaced pages overall. His study ranks with other classic participant observation studies in sociology, such as Whyte's *Street Corner Society* (1943), Anderson's *A Place on the Corner* (1976) and *Streetwise* (1990), and Duneier's *Sidewalk* (1999).

Research Results: Moskos's study is important because, among other things, it dispels a number of myths that the public has about police officers and

police work. For example, many think that summoning the police by calling 911 will get a quick solution to the problem—whether it be a drug deal taking place, an incident of domestic violence, or gunfire on the street. Although police are indeed generally quick to respond, in reality, the drug deal or the domestic violence reconvenes immediately after the police leave the scene. Moskos even concludes that, unfortunately, 911 is "a joke."

Many assume that if a suspected drug dealer is standing close to a vial of cocaine in the street, the observing police officer will report that he or she "saw" the dealer throw the vial into the street. Moskos found, however, that this was rarely the case: The vast majority of officers over the vast majority of such incidents reported "seeing" the dealer toss the vial only if they indeed saw the dealer do so and were able to verify this act by another officer witnessing it. A veteran officer warned Moskos that "if you don't see him drop it, then just kick it or crush it."

In his further demystification of the police and police culture, Moskos describes his fellow police officers not as power-hungry, thrill-seeking bullies, but as hard-working people who marshal their own weaknesses and strengths to cope with unique job conditions.

Also of importance is Moskos's discovery of certain elements of *social structure* characterizing street drug trade. For example, virtually each and every illicit drug transaction on the street corner involves five social roles

in addition to the person who actually purchases a drug or drugs: lookouts (who watch for police cars, the lowest-status role in the street transaction); steerers (who "hawk" or advertise their drug to passersby); money men (who collect the money paid for the drug); slingers (who actually give the drugs to the purchaser); and gunmen (who stand ready in the shadows in case they feel needed). Engaging in such roles serves the function of limiting the legal liability of each individual in the event of arrests.

Implications: Such insights into the social structure and culture of street activity (in this case, street drug trade) ranks Moskos's work with other participant observation studies that reveal structure and culture—for example, those of Whyte, Anderson, and Duneier.

Questions to Consider

1. In studies that employ participant observation, there is always a danger of becoming too involved with one's research subjects, thus causing the researcher to lose a certain amount of objectivity. This is called "going native." Do you think that this might have happened, or did not happen, to Moskos? Elaborate.

2. Would you like to do a study of a police force, much as did Moskos? Using what you have learned, briefly describe how you might go about starting such a participant observation study.

Source: Moskos, Peter. 2008. *Cop in the Hood: My Year Policing Baltimore's Eastern District*. Princeton, NJ: Princeton University Press.

(Some damage to the self-concept of subjects in the Milgram experiments was indeed found when the subjects realized that they were easily duped into causing what they thought was serious harm to another human. See Chapter 6.)

One of the clearest ethical violations in all of the history of science has come to be known as the Tuskegee Syphilis Study. The study was conducted at

the Tuskegee Institute in Macon County, Alabama, a historically Black college. For this study, begun in 1932 by the government's United States Public Health Service, a sample of about 400 Black males who were infected with the sexually transmitted disease syphilis (this was the "experimental" group) were allowed to go untreated medically for over forty years. Another 200 Black males who had not contracted syphilis were used as a control group. The purpose of the study was to examine the effects of "untreated syphilis in the male negro." The study was not unlike similar "studies" carried out against Jews by Hitler's Nazi regime in Germany at the same time—just before and during World War II in the 1930s and early 1940s. Jews who were injected with debilitating illnesses remained medically untreated. Untreated syphilis causes blindness, mental retardation, and death, and this is how many of the untreated Black men in the Tuskegee study fared over the forty-plus-year period.

In the 1950s, penicillin was discovered as an effective treatment for infectious diseases, including syphilis, and was widely available. Nonetheless, the scientists conducting the study decided *not* to give penicillin to the infected men in the study on the grounds that it would "interfere" with the study of the physical and mental harm caused by untreated syphilis! The U.S. government itself authorized the study to be continued until the early 1970s—that is, until quite recently. By the mid-1970s, pressure from the public and the press caused the federal government to terminate the study, but by then it was too late to save approximately 100 men who had already died of the ravages of untreated syphilis, plus many others who were forced to live with major mental and physical damage.

Following the ethical horrors of studies such as the Tuskegee study, the American Sociological Association (ASA) has since developed a professional code of ethics (see the 2012–13 ASA website for the full code of ethics). The federal government also has many regulations about the protection of human subjects. Ethical researchers adhere to these guidelines and must ensure that research subjects are not subjected to physical, mental, or legal harm. Research subjects must also be informed of the rights and responsibilities of both researcher and subject. Sociologists, like other scientists, also should not involve people in research without what is called **informed consent**—that is, getting agreement to participate from the respondents or subjects after the purposes of the study are explained in detail to them. There may be exceptions to the need for informed consent, such as when observing people in public places. Sociologists also take measures to avoid identifying their respondents and to assure confidentiality through the use of pseudonyms or by not using names at all and by assigning random ID (code) numbers to all respondents during data analysis.

chapter summary

What is sociological research?
Sociological research is used by sociologists to answer questions and, in many cases, to test *hypotheses*. The research method one uses depends upon the question that is asked.

Is sociological research scientific?
Sociological research is derived from the *scientific method*, meaning that it relies on empirical observation and, sometimes, the testing of *hypotheses*. The research process involves several steps: developing a research question, designing the research, collecting data, analyzing data, and developing conclusions. Different research designs are appropriate to different research questions, but sociologists have to be concerned with the *validity*, the *reliability*, and the *generalization* of their results. Applying one's results obtained from a sample to a broader population is an example of generalization.

What is the difference between qualitative research and quantitative research?
Qualitative research is research that is relatively unstructured, does not rely heavily upon statistics, and is closely focused on a question being asked. *Quantitative research* is research that uses statistical methods. Both kinds of research are used in sociology.

What are some of the statistical concepts in sociology?
Through research, sociologists are able to make statements of *probability*, or likelihood. Sociologists use *percentages* and *rates*. The *mean* is the same as an average. The *median* represents the midpoint in an array of values or scores. The *mode* is the most common value or score. *Correlation* and *cross-tabulation* are statistical procedures that allow sociologists to see how two (or more) different variables are associated. There have been instances of misuse of statistics in the behavioral and social sciences, including sociology, and these have resulted in incorrect conclusions.

What different tools of research do sociologists use?
The most common tools of sociological research are surveys and interviews, *participant observation*, controlled experiments, *content analysis*, comparative and historical research, and *evaluation research*. Each method has its own strengths and weaknesses. You can better generalize from surveys, for example, than *participant observation*, but *participant observation* is better for capturing subtle nuances and depth in social behavior.

Can sociology be value free?

Although no research in any field can always be value free, sociological research nonetheless strives for objectivity while recognizing that the values of the researcher may have some influence on the work. One of the worst cases of ethical violation in scientific research was the Tuskegee Syphilis Study. There are ethical dilemmas in doing sociological research, such as whether one should attempt to avoid the Hawthorne effect by collecting data without letting research subjects (people) know they are being observed.

Key Terms

concept 57
content analysis 65
controlled
   experiment 63
correlation 64
covert participant
   observation 63
cross-tabulation 64
data 57
data analysis 61
debriefing 68

deductive reasoning 54
dependent variable 57
evaluation research 68
generalization 61
Hawthorne effect 59
hypothesis 56
independent variable 57
indicator 57
inductive reasoning 55
informant 63
informed consent 70

mean 64
median 64
mode 64
overt participant
   observation 63
participant observation 54
percentage 64
population 60
qualitative research 56
quantitative research 56
random sample 61

rate 64
reliability 58
replication study 55
research design 56
sample 60
scientific method 54
serendipity 61
spurious correlation 64
validity 57
variable 57

4

Socialization and the Life Course

The Socialization Process

Agents of Socialization

Theories of Socialization

Growing Up in a Diverse Society

Aging and the Life Course

Resocialization

Chapter Summary

do you think you could define what it means to be human? Biologists, geneticists, and many other natural scientists have attempted to identify the specific makeup of humans. In the summer of 2000, scientists working on the human genome project announced that they had deciphered the human genetic code. By mapping the complex structure of DNA (deoxyribonucleic acid) on high-speed computers, scientists identified the 3.12 billion chemical base pairs in human DNA and put them in proper sequence, unlocking the genetic code of human life. Scientists likened this to assembling "the book of life," that is, having the knowledge to make and maintain human beings. The stated purpose of the human genome project is to see how genetics influences the development of disease, but it raises numerous ethical questions about human cloning and the possibility of creating human life in the laboratory. Is our genetic constitution what makes us human? Suppose you created a human being in the laboratory but left that creature without social contact. Would the "person" be human?

Knowing the sequence of the human genome may raise the specter of making human beings in the laboratory, but without society, what would humans be like? What is the distinction between being "biologically" human and being "socially" human? Sociologists argue that we need both to be fully part of society. Thanks to the human genome project, we now have an understanding of what it means to be "biologically" human. But is that being fully human? Sociologist **Robert Merton** (1957) noted that we are "not born human." When we are born with our particular genetic makeup, we are simply biologically human. As we go from infancy, through the toddler years, from childhood, through adulthood and old age, our genetic DNA is unchanging. But the ongoing interaction we have with other human beings during our lifetime provides the guidelines, or the blueprint, for what it truly means to be part of human society. From the moment of birth, we are being "taught" how to behave and what is expected of us to be "normal" or well-socialized people.

Consider how the rare cases of *feral children*, who have been raised in the absence of human contact, provide some clues as to what happens during human development when a person has little or no social contact. One such case, discovered in 1970, involved a young girl given the pseudonym of Genie. When her blind mother appeared in the Los Angeles County welfare office seeking assistance for herself, caseworkers first thought the girl was six years old. In fact, she was thirteen, although she weighed only 59 pounds and was 4 feet, 6 inches tall. She was small and withered, unable to stand up straight, incontinent, and severely malnourished. Her eyes did not focus, and she had two nearly complete sets of teeth. A strange ring of calluses circled her buttocks. She could not talk. As the case unfolded, it was discovered that the girl had been kept in nearly total isolation for most of her life, never learning verbal language or any form of social interaction.

Interacting with other humans is a critical part of becoming human. Another example is the story of Shin Dong-hyuk who escaped from a North Korean prison known as Camp 14. His story explains how he was born in the prison camp and was socialized entirely into the rules and expectations of prison life. As a result, he never knew family love, loyalty, or how to interact with others. Not until he met a new prisoner who described the world beyond the prison boundaries did Shin Dong-hyuk know that he could hope for something more. After his escape at age 23, his story has been told and offers insight into how socialization shapes beliefs, behaviors, and expectations for the future (Harden 2012).

Genes may confer skin and bone and brain, but only by learning the values, norms, and roles that culture bestows on people do we become social beings—literally, human beings. Sociologists refer to this process as socialization—the subject of this chapter.

learning objectives

- Explain the socialization process
- Identify the different agents responsible for socialization over the life course
- Compare theories of socialization
- Explore how socialization differs across cultures
- Identify the stages of life from childhood to old age
- Discuss the possibilities for resocialization throughout the life course

THE SOCIALIZATION PROCESS

Socialization is the process through which people learn the expectations of society. To be a fully socialized member of society means to have internalized the expected *norms* of that society. **Internalization** occurs when behaviors and assumptions are learned so thoroughly that people no longer question them, but simply accept them as correct. The lessons that are internalized can have a powerful influence on behavior and attitudes.

Let's start with behavior. Not all human beings *act* the same way. In fact, within any one culture, not all people act the same. Not all Americans act the same; not all college students act the same; not even all of your friends within your social circle act the same. Yet, the socialization process guides each of us in how to behave within our given roles. **Roles** are the expected behavior associated with a given status in society. When you occupy a social role, you tend to take on the expectations of others. For example, when you enter a new group of friends, you probably observe their behavior, their language, their dress, perhaps even their opinions of others, and often modify your own behavior accordingly. Before you know it, you are a member of the group, perhaps socializing others into the same set of expectations. This can happen throughout your life course, as you join new groups, socially, academically, or within a work organization. You are socialized to the group's norms, taking on your specific role and interacting with the group in socially acceptable ways.

Formal and informal learning, through schools and other socialization agents, are important elements of the socialization process. In this photo, hearing-impaired children are learning sign language.

Richard T. Nowitz/Encyclopedia/Corbis

International Adoption and Interracial Families

In the United States, interracial families are on the rise. Interracial families occur because of interracial marriages and because of interracial adoption. Interracial marriages produce multiracial children who have one parent who shares part of their racial–ethnic identity to teach them about cultural expectations and traditions. But consider an internationally adopted child. What happens, for example, when two White Americans are raising a child born in China? How is cultural socialization different for these families? When the parents are racially and ethnically different from an adopted child, the process of socialization may be very different. Since 1999, nearly 67,000 U.S. adoptions of children from China have occurred. In 2011 alone, 2587 children were adopted from China. This number represents almost 28 percent of all inter-country adoptions for that year (U.S. Department of State 2011). The parents of these children have a choice about how to socialize their Chinese

child into American society. For some, a "color-blind" approach seems best. This involves passing on the parental cultural norms and values with little regard for the racial and ethnic difference between them and their children. Research shows the color-blind approach is less popular today than in years past, and that parents who adopted from eastern Europe or Russia are more likely to employ the color-blind approach to cultural socialization (Lee et al. 2006).

A second, and increasingly more common, approach to cultural socialization of Chinese adopted children results from the parental beliefs in enculturation and racialization. This means that the parents balance socializing their children into American culture while also providing opportunities for them to learn about and participate in Chinese cultural activities (Lee et al. 2006). Chinese schools provide weekend classes for students to learn Chinese language and culture. Chinese New Year celebrations

and trips to urban Chinatowns become part of the family traditions.

Another integral part of raising children who are racially and ethnically different from their parents is ongoing discussions about race, discrimination, and difference. Parents will discuss potential experiences of racial discrimination at school and other places. Some parents even discuss the adoption with school officials and teachers (Lee et al. 2006).

Socialization starts within families at infancy and continues throughout the life course. Internationally adopted children face a unique set of challenges at all stages of socialization. As babies, they are taken from homes or orphanages and immersed into American families and American homes. In addition to learning about the new people in their life, they face adjusting to new food, new language, new styles of dress, and all the other "newness" of a foreign culture. From infancy to adolescence to adulthood, adoptees face challenges in balancing their place in a new, and possibly interracial, family.

What about attitudes? By means of socialization, people absorb their culture—customs, habits, laws, practices, and means of expression. Socialization is the basis for **identity**: how one defines oneself. Identity is both personal and social. To a great extent, it is bestowed by others, because we come to see ourselves as others see us. Socialization also establishes **personality**, defined as a person's relatively consistent pattern of behavior, feelings, predispositions, and beliefs. For example, maybe your college campus has taken great steps toward environmental sustainability. Going green is a movement that so many of your friends, classmates, and instructors talk about and feel strongly about. You are probably already socialized into reducing, reusing, and recycling. Your attitude likely matches your behavior. Maybe you are not the most "green" person you know and even occasionally throw a plastic bottle in the regular trash. But you do not deliberately thwart efforts of environmental sustainability. You see the importance in recycling and generally believe in the movement. As a member of the college or university community, you have taken on the attitudes and behaviors

consistent with that community with regard to environmental conservation.

The socialization experience differs for individuals, depending on factors such as age, race, gender, and class, as well as more subtle aspects of personality. Women and men encounter different socialization patterns as they grow up because each gender brings with it different social expectations (see Chapter 11). Likewise, growing up Jewish, Asian, Latino, or African American involves different socialization experiences. In the "Understanding Diversity: International Adoption and Interracial Families" box, we discuss issues that Chinese children face growing up in White families. Conflicts between adopted culture and biological race are evident for many interracially adopted children. Such conflicts can be particularly acute when a person grows up within different, even overlapping, cultures.

The Nature–Nurture Controversy

Examining the socialization process helps reveal the degree to which our lives are *socially constructed*, meaning that the organization of society and the life outcomes of people within it are the result of social definitions and processes. Is it "nature" (what is natural) or is it "nurture" (what is social)—or both—that makes us human? This question has been the basis for debate for many years.

From a sociological perspective, what a person becomes results more from social experiences than from *innate* (inborn or natural) traits, although innate traits do have some influence on culture, as we saw in Chapter 2. For example, people may be born with a great capacity for knowledge, but without a good education, those people are unlikely to achieve their full potential and may not be recognized as intellectually gifted.

From a sociological perspective, nature (genetics) provides a certain stage for what is possible, but society provides the full drama of what we become. Our values and social attitudes are not inborn; they emerge through the social relations we have with others and our social position in society. Such factors as your family environment, how people of your social group are treated, and the historic influences of the time all shape how we are nurtured by society.

Perhaps the best way to understand the nature-nurture controversy is not that one or the other fully controls who we become, but that life involves a complex interplay, or *interaction*, between genetic and social influences on human beings. The emphasis in sociology, however, is to see the social realities of our lives as extremely important in shaping human experience (Joseph 2010; Ledger 2009).

Socialization as Social Control

Sociologist Peter Berger pointed out that not only do people live in society but society also lives in people (Berger 1963). Socialization is, therefore, a mode of social control. **Social control** is the process by which groups and individuals within those groups are brought into conformity with dominant social expectations. Sometimes an individual rebels and attempts to resist this conformity, but because people generally conform to cultural expectations, socialization gives society a certain degree of predictability. Patterns are established that become the basis for social order.

To understand how socialization is a form of social control, imagine the individual in society as surrounded by a series of concentric circles (see Figure 4.1). Each circle is a layer of social controls, ranging from the most subtle, such as the expectations of others, to the most overt, such as physical coercion and violence. Coercion and violence are usually not necessary to extract conformity because learned beliefs and the expectations of others are enough to keep people in line. These socializing forces can be subtle because even when a person disagrees with others, he or she can feel pressure to conform and may experience stress and discomfort in choosing not to conform. People learn through a lifetime of experience that deviating

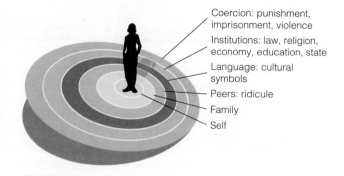

Coercion: punishment, imprisonment, violence

Institutions: law, religion, economy, education, state

Language: cultural symbols

Peers: ridicule

Family

Self

FIGURE 4.1 Socialization as Social Control Though we are all individuals, the process of socialization also keeps us in line with society's expectations. This may occur subtly through peer pressure or, in some circumstances, through coercion and/or violence.

© Cengage Learning

from the expectations of others invites peer pressure, ridicule, and other social judgments that remind one of what is expected.

Conformity and Individuality

Saying that people conform to social expectations does not eliminate individuality. We are all unique to some degree. Our uniqueness arises from different experiences, different patterns of socialization, the choices we make, and the imperfect ways we learn our roles; furthermore, people resist some of society's expectations. Sociologists warn against seeing human beings as totally passive creatures because people interact with their environment in creative ways. Yet, most people conform, although to differing degrees. Socialization is profoundly significant, but this does not mean that people are robots. Instead, socialization emphasizes the adaptations people make as they learn to live in society.

Some people conform too much, for which they pay a price. Socialization into men's roles can encourage aggression and a zeal for risk-taking. Men have a lower life expectancy and higher rate of accidental death than do women, probably because of the risky behaviors associated with men's roles, that is, simply "being a man" (Kimmel and Messner 2004). Women's gender roles carry their own risks. Striving excessively to meet the beauty ideals of the dominant culture can result in feelings of low self-worth and may encourage harmful behaviors, such as smoking or severely restricting eating to keep one's weight down. Being a man or woman is not inherently bad for your health, but conforming to gender roles to an extreme can compromise your physical and mental health (Jones 2001). Women and girls are more likely than men and boys, for example, to suffer from eating disorders or to have an unhealthy self-image (Algars et al. 2010; Neighbors and Sobal 2007).

The Consequences of Socialization

Socialization is a lifelong process with consequences that affect how we behave toward others and what we think of ourselves. First, *socialization establishes self-concepts*. **Self-concept** is how we think of ourselves as the result of the socialization experiences we have over a lifetime. Socialization is also influenced by various social factors, as shown in Figure 4.2, which describes how students' self-concepts are shaped by gender. Notice those traits where there is a big difference between how men and women self-evaluate.

Second, *socialization creates the capacity for role-taking*. As we learn societal expectation, we create the ability to see ourselves through the perspective of another. Socialization is fundamentally reflective; that is, it involves self-conscious human beings seeing and reacting to the expectations of others. The capacity for reflection and the development of identity are ongoing. This is how we establish our roles in society. As we encounter new situations in life, such as going away to college or getting a new job, we are able to see what is expected and to adapt to the situation accordingly, becoming a young adult student or an employee of a company. Of course, not all people do so successfully.

This can become the basis for social deviance (explored in Chapter 7) or for many common problems in social and psychological adjustment.

Third, *socialization creates the tendency for people to act in socially acceptable ways*. Through socialization, people learn the normative expectations attached to social situations and the expectations of society in general. As a result, socialization creates some predictability in human behavior and brings some order to what might otherwise be social chaos.

Finally, *socialization makes people bearers of culture*. Socialization is the process by which people learn and internalize the attitudes, beliefs, and behaviors of their culture. At the same time, socialization is a two-way process—that is, a person is not only the recipient of culture but also is the creator of culture, passing cultural expectations on to others. The main product of socialization, then, is society itself.

AGENTS OF SOCIALIZATION

Socialization agents are people, or sources, or structures that pass on social expectations. Everyone is a socializing agent because social expectations are

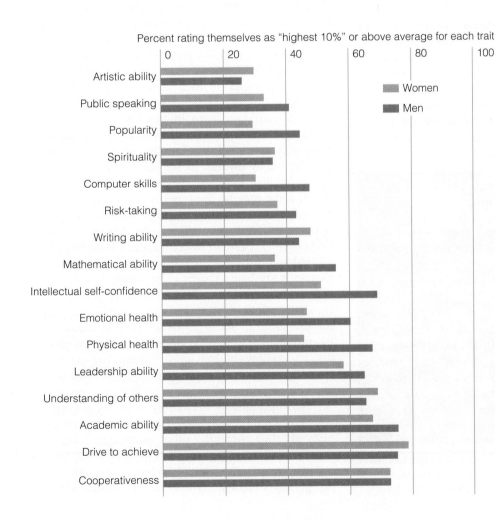

Percent rating themselves as "highest 10%" or above average for each trait

Women
Men

FIGURE 4.2 Student Self-Concepts: The Difference Gender Makes Men and women differ in how they identify certain characteristics of their personality. What patterns can you see in these data? What characteristics do women identify compared to the ones men choose?

Data: Based on national sample of first-year college students, fall 2011.

Source: Pryor, John H., et al. 2011. *The American Freshman: National Norms Fall 2011.* Higher Education Research Institute. Los Angeles, CA: University of California, Los Angeles.

communicated in countless ways and in every interaction people have, whether or not intentionally. When people are simply doing what they consider "normal," they are communicating social expectations to others. When you dress a particular way, you may not feel you are telling others they must dress that way. Yet, when everyone in the same environment dresses similarly, some expectation about appropriate dress is clearly being conveyed. People feel pressure to become what society expects of them even though the pressure may be subtle and unrecognized.

---

thinking SOCIOLOGICALLY

Think about the first week that you were attending college. What expectations were communicated to you and by whom? Who were the most significant *socialization agents* during this period? Which expectations were communicated formally and which informally? If you were analyzing this experience sociologically, what would be some of the most important concepts to help you understand how one "becomes a college student"? Were there expectations for dress? For transportation around campus? ●

Socialization does not occur simply between individual people; it occurs in the context of social institutions. Recall from Chapter 1 that institutions are established patterns of social behavior that persist over time. Institutions are a level of society above individuals. Many social institutions shape the process of socialization, including, as we will see, the family, the media, peers, religion, sports, and schools.

Conformity at school socializes people into the expectations of being a student.

Thomas Imo/Alamy Limited

The Family

For most people, the family is the first source of socialization. Through families, children are introduced to the expectations of society. Children learn to see themselves through their parents' eyes. Thus, how parents define and treat a child is crucial to the development of the child's sense of self.

An interesting example of the subtlety in familial socialization comes from a study comparing how U.S. and Japanese mothers talk to their children (Fernald and Morikawa 1993). Observers watched mothers from both cultures speak to their infant children, ranging in age from six to nineteen months. Both Japanese and U.S. mothers simplified and repeated words for their children, but strong cultural differences were evident in their context. U.S. mothers focused on naming objects for their babies: "Is that a *car*?" "Kiss the *doggy*." Japanese mothers were more likely to use their verbal interactions with children as an opportunity to practice social routines such as "give me" and "thank you." The behavior of the Japanese mothers implied that the name of the object was less important than the polite exchange. Japanese mothers also were more likely to use sounds to represent the objects, such as "oink-oink" for a pig, or "vroom-vroom" for a vehicle. U.S. mothers were more likely to use the actual names for objects. The researchers interpreted these interactions as reflecting the beliefs and practices of each culture. Japanese mothers used objects as part of a ritual of social exchange, emphasizing polite routines, whereas U.S. mothers focused on labeling things. In each case, the child receives a message about what is most significant in the culture.

What children learn in families is certainly not uniform. Even though families pass on the expectations of a given culture, families within that culture may be highly diverse, as we will see in Chapter 13. Some families may emphasize educational achievement; some may be more permissive, whereas others emphasize strict obedience and discipline. Even within families, children may experience different expectations based on gender or birth order (being born first, second, or third). Researchers have found, for example, that sons are being taught to be tough, but that both sons and daughters are being taught egalitarian roles (Epstein and Ward 2011). Living in a family experiencing the strain of social problems such as alcoholism, unemployment, domestic violence, or teen pregnancy also affects how children are socialized. The specific effects of different family structures and processes are the basis for ongoing and extensive sociological research.

As important as the family is in socializing the young, it is not the only socialization agent. As children grow up, they encounter other socializing influences, sometimes in ways that might contradict family expectations. Parents who want to socialize their children in less gender-stereotyped ways might be frustrated by the

Television shows such as "The New Normal" (cast shown here) now portray a different family structures, such as depicted here with a gay couple using a surrogate to have a baby.

ALI ADLER IS HERE PRODUCTIONS/RYAN MURPHY PRODUCTIONS/20TH CENTURY FOX TV/THE KOBAL COLLECTION/Picture Desk

influence of the media, which promotes highly gender-typed toys and activities to boys and girls. These multiple influences on the socialization process create more complex attitudes and behaviors among people.

The Media

As we saw in Chapter 2, the mass media increasingly are important agents of socialization. Television alone has a huge impact on what we are socialized to believe and become. Add to that the print messages received in books, comics, newspapers, and now blogs and the Internet, plus images from film, music, video games, and radio, and you begin to see the enormous influence the media have on the values we form, our images of society, our desires for ourselves, and our relationships with others. These images are powerful throughout our lifetimes, but many worry that their effect during childhood may be particularly deleterious.

The high degree of violence in the media has led to the development of a rating system for televised programming. There is no doubt that violence is extensive in the media. Analysts estimate that by age 18, the average child will have witnessed at least 18,000 simulated murders on television (Wilson et al. 2002). Research continues to examine the relationship between exposure to media violence and different types of aggressive behaviors (Gentile, Mathieson, and Crick 2011).

Media violence also tends to desensitize children to the effects of violence, including engendering less sympathy for victims of violence (Baumeister and Bushman 2008; Huesmann et al. 2003). Many also think that violent video games (another form of media) may contribute to school shootings, where an armed individual—often a student at the particular school—randomly shoots and wounds or kills one or more individuals, usually other students but also teachers (see Chapter 7; Newman 2010). Including television, video games, music lyrics, and movies, media portrayals of violence desensitize viewers regarding the danger of weapons. Guns such as the ones used in the school shooting in Newtown, Connecticut, have deadly consequences; yet, media images of them pervade society.

Perhaps there is some link here, but it is too simplistic to see a direct causal connection between viewing violence and actually engaging in it. For one thing, such an argument ignores the broader social context of violent behavior, including such things as the availability of guns, family characteristics, parental control, youth alienation from school, to name a few (Newman 2010).

Violence in the media is not solely to blame for violent behavior in society. Children do not watch television in a vacuum; they live in families where they learn different values and attitudes about violent behavior, and they observe the society around them, not just the images they see in fictional representations. Most likely, children are influenced not only by the images of televised and filmed violence but also by the social context in which they live. The images of violence in the media in some ways only reflect the violence in society. The sociological question is whether or not media reflects societal reality, or if reality is influenced by the images presented in the media.

The media expose us to numerous images that shape our definitions of ourselves and the world around us. What we think of as beautiful, sexy, politically acceptable, and materially necessary is strongly influenced by the media. If every week, as you read a newsmagazine, someone shows you the new car that will give you status and distinction, the message is clear and we begin to think that our self-worth can be measured by the car we drive. If every weekend, as we watch televised sports, someone tells us that to have fun we should drink the right beer, we come to believe that parties are perceived as better when everyone is drinking. The values represented in the media, whether they are about violence, racist and sexist stereotypes, or any number of other social images, have a great effect on what we think and who we come to be.

Peers

Peers are those with whom you interact on equal terms, such as friends, fellow students, and coworkers. Among peers, there are no formally defined superior and subordinate roles, although status distinctions commonly arise in peer group interactions. Without peer approval, most people find it hard to feel socially accepted.

Interaction in Cyberspace

As a student at a college or university, you likely do not remember a world without the Internet. Use of the Internet started out small but has grown to include an online version of almost all activities that can be done face-to-face. Everything from watching the news, researching the latest statistics, staying in touch with friends, and participating in a class discussion are now possible in front of a computer screen instead of in front of another person.

Socialization involves the ongoing process of learning how to interact with others and what the social norms are for communicating. George Herbert Mead and Charles Horton Cooley, sociologists you will read about a bit later in the chapter, explain how a sense of self develops through the expectations and judgments of others in a social environment. Cooley outlined the importance of the *looking-glass self*. Identity is developed by balancing how we think we appear to others and how those others judge us. Mead emphasized our different roles that we take on as a result of our relationship to others. Both Cooley and

Mead provided the groundwork for symbolic interaction theory to explain how socialization happens within a social environment.

When communication happens in an online community, how are norms for interaction altered? In what ways have you been socialized into the expected behavior of online communication?

Research details how participants in an online discussion group navigate self-presentation (Lee 2006). Private information is often concealed with techniques such as using a false name or a "user" name that is unidentifiable. Public information, however, is carefully revealed through personal narratives that others in the discussion group can use to learn about the person. Cooley's concept that we develop identity by how others perceive us takes on new meaning for online communication. If the user limits how much private information is made public, then the user has greater control over the perception by others. Consider your online interactions. What do your privacy settings on social networking

sites say about what you are trying to conceal? How much can you influence the way you are perceived online?

Additionally, the roles assumed by participants in an online discussion group are different than in-class course discussions. Mead argues that there is distinction between the part of our personality that is self-defining (the "I") and the part of our personality that is conforming to what others expect of us (the "me"). If online discussions allow participants to be unseen, will this change what contribution will be made to the discussion? Are you more comfortable saying something contrary in a discussion in person or online?

Symbolic-interactionist theory provides a good starting point for how to think about socialization in online communities. Social interaction is still crucial to understanding the development of self. Online communities, through e-courses, social networking sites, or digital chat rooms, provide a new forum for social interaction. Sociologists utilize core theoretical ideas like the looking-glass self and role-taking to explain this type of interaction.

Peers are an important agent of socialization. Young girls and boys learn society's images of what they are supposed to be through the socialization process.

Jaimie Duplass/iStockphoto.com

Peers are enormously important in the socialization process. Peer cultures for young people often take the form of *cliques*—friendship circles where members identify with each other and hold a sense of common identity. You probably had cliques in your high school and may even be able to name them. Did your school have "jocks," "goths," "nerds," "freaks," "stoners," and so forth? Sociologists studying cliques have found that they are formed based on a sense of exclusive membership, like the in-groups and out-groups we will examine in Chapter 6. Cliques are cohesive but also have an internal hierarchy, with certain group leaders having more power and status than other members. Interaction techniques, like making fun of people, produce group boundaries, defining who's in and who's out. The influence of peers is strong in childhood and adolescence, but it also persists into adulthood.

A relatively recent phenomenon on high school grounds and on college campuses is bullying—the systematic, consistent long-time beating or verbally berating of a single student, who is chosen to be the victim by

This is a memorial to Tyler Clementi, the 18-year-old first-year student at Rutgers University who took his own life after one of his peers posted a video of him suggesting a romantic relationship between him and another man.

Emmanual Dunand/AFP/Getty Images

a clique. School bullying is serious business and nothing to be ignored, because it often has dire consequences. There are instances in which bullying has resulted in the suicide of the victim. In 2010, a gay student at Rutgers University took his own life after his college roommate posted a video linking him romantically to another male student. The incident received national media attention, highlighting the bullying of gay students. The roommate who filmed the victim was convicted on 15 criminal counts, including invasion of privacy and bias intimidation.

As agents of socialization, peers are important sources of social approval, disapproval, and support. This is one reason groups without peers of similar status are often at a disadvantage in various settings, such as women in male-dominated professions or minority students on predominantly White campuses. Being a "token" or an "only," as it has come to be called, places unique stresses on those in settings with relatively few peers from whom to draw support (Thoits 2009). This is one reason those who are minorities in a dominant group context often form same-sex or same-race groups for support, social activities, and the sharing of information about how to succeed in their environment.

Religion

Religion is another powerful agent of socialization, and religious instruction contributes greatly to the identities children construct for themselves. Children tend to develop the same religious beliefs as their parents. Even those who renounce the religion of their youth are deeply affected by the attitudes, images, and beliefs instilled by early religious training. Very often, those who disavow religion return to their original faith at some point in their life, especially if they have strong ties to their family of origin and if they form families of their own (Wuthnow 2010).

Religious socialization influences a large number of beliefs that guide adults in how they organize their lives,

including beliefs about moral development and behavior, the roles of men and women, and sexuality, to name a few. Higher religiosity is connected to sexist views, especially among men (Maltby et al. 2010). Religious socialization also influences beliefs about sexuality, including the likelihood of tolerance for gay and lesbian sexuality (Whitehead and Baker 2012). Religion can even influence child-rearing practices, including the use of physical nurturing and strict discipline.

Sports

Most people perhaps think of sports as something that is just for fun and relaxation—or perhaps to provide opportunities for college scholarships and athletic careers—but sports are also an agent of socialization. Through sports, men and women learn concepts of self that stay with them in their later lives.

Sports are also where many ideas about gender differences are formed and reinforced (Eitzen 2012; Messner 2009). For men, success or failure as an athlete can be a major part of a man's identity. Even for men who have not been athletes, knowing about and participating in sports is an important source of men's gender socialization. Men learn that being competitive in sports is considered a part of manhood. Indeed, the attitude that "sports builds character" runs deep in the culture. Sports are supposed to pass on values such as competitiveness, the work ethic, fair play, and a winning attitude. Sports are considered to be where one learns to be a man.

---

debunking SOCIETY'S MYTHS

MYTH: Many people feel that sports are for the most part played just for the fun of it.
SOCIOLOGICAL PERSPECTIVE: Although sports are a form of entertainment, playing sports is also a source for socialization into roles, such as gender roles. ●

Michael Messner's research on men and sports reveals the extent to which sports shape masculine identity. Messner interviewed thirty former athletes: Latino, Black, and White men from poor, working-class, and middle-class backgrounds. All of them spoke of the extraordinary influence of sports on them as they grew up. Not only are sports a major source of gender socialization, but many working-class, African American, and Latino men often see sports as their only possibility for a good career, even though the number of men who succeed in athletic careers is a minuscule percentage of those who hold such hopes.

Messner's research shows that, for most men, playing or watching sports is often the context for developing relationships with fathers, even when the father is absent or emotionally distant in other areas of life. Older brothers and other male relatives also socialize young men into sports. For many of the men in Messner's study, the athletic accomplishments of other family members created uncomfortable pressure to perform and compete, although on the whole, they recalled their early sporting years with positive emotions. It was through sports relationships with male peers, more than anyone else, however, that the men's identity was shaped. As boys, the men could form "safe" bonds with other men; still, through sports activity, men learned homophobic attitudes (that is, fear and hatred of homosexuals) and rarely developed intimate, emotional relationships with each other (Messner 2002).

Sports historically have been less significant in the formation of women's identity, although this has changed, largely as the result of Title IX. Title IX (1972) opened more opportunities in athletics to girls and women by legally defining the exclusion of women from school sports as sex discrimination. Women who participate in sports typically develop a strong sense of bodily competence—something usually denied to them by the prevailing, unattainable cultural images of women's bodies. Sports also give women a strong sense of self-confidence and encourage them to seek challenges, take risks, and set goals (Eitzen 2012; Blinde et al. 1994).

Still, athletic prowess, highly esteemed in men, is not tied to cultural images of womanliness. Quite the contrary, women who excel at sports are sometimes stereotyped as lesbians, or "butches," and may be ridiculed for not being womanly enough. These stereotypes reinforce traditional gender roles for women, as do media images of women athletes that emphasize family images and the personality of women athletes (Eitzen 2012; Cavalier 2003). Research in the sociology of sports shows how activities as ordinary as shooting baskets on a city lot, playing on the soccer team for one's high school, or playing touch football on a Saturday afternoon can convey powerful cultural messages about our identity and our place in the world. Sports are a good example of the power of socialization in our everyday lives.

Schools

Once young people enter kindergarten (or, even earlier, day care), another process of socialization begins. At home, parents are the overwhelmingly dominant source of socialization cues. In school, teachers and other students are the source of expectations that encourage children to think and behave in particular ways. The expectations encountered in schools vary for different groups of students. These differences are shaped by a number of factors, including teachers' expectations for different groups and the resources that different parents can bring to bear on the educational process. The parents of children attending elite, private schools, for example, often have more influence on school policies and classroom activities than do parents in low-income communities. In any context, studying socialization in the schools is an excellent way to see the influence of gender, class, and race in shaping the socialization process.

---

debunking SOCIETY'S MYTHS

MYTH: Schools are primarily places where young people learn skills and other knowledge.

SOCIOLOGICAL PERSPECTIVE: There is a *hidden curriculum* in schools where students learn expectations associated with race, class, and gender relations in society as influenced by the socialization process. ●

For example, research finds that teachers respond differently to boys and girls in school. Boys receive more attention from teachers than do girls. Even when teachers respond negatively to boys who are misbehaving, they are paying more attention to the boys (American Association of University Women 2010; Sadker and Sadker 1994). Social class stereotypes also affect teachers' interactions with students. Teachers are likely to perceive working-class children and poor children as less bright and less motivated than middle-class children; teachers are also more likely to define working-class students as troublemakers (Dunne and Gazely 2008; Oakes et al. 2000). These negative appraisals are *self-fulfilling prophecies*, meaning that the expectations they create often become the cause of actual behavior in the children; thus they affect the odds of success for children. (We will return to a discussion of self-fulfilling prophecies in Chapter 14.)

Boys also receive more attention in the curriculum than girls. The characters in texts are more frequently boys; the accomplishments of boys are more likely portrayed in classroom materials; and boys and men are more typically depicted as active players in history, society, and culture (American Association of University Women 1998). This is called the *hidden curriculum* in the schools—the informal and often subtle messages about social roles that are conveyed through classroom interaction and classroom materials—roles that are clearly linked to gender, race, and class.

In schools, boys and girls are quite often segregated into different groups, with significant sociological consequences. Differences between boys and girls become exaggerated when they are defined as distinct groups. Seating boys and girls in separate groups or sorting them into separate play groups heightens gender differences and greatly increases the significance of gender in the children's interactions with each other. Equally important is that gender becomes less relevant in the interactions between boys and girls when they are grouped together in common working groups, although gender does not disappear altogether as an influence. Barrie Thorne (1993), who has observed gender interaction in schools, concludes from her observations that gender has a "fluid" character and that gender relations between boys and girls can be improved through conscious changes that discourage gender separation.

While in school, young people acquire identities and learn patterns of behavior that are congruent with the needs of other social institutions. Sociologists using conflict theory to understand schools would say that U.S. schools reflect the needs of a capitalist society. School is typically the place where children are first exposed to a hierarchical, bureaucratic environment. Not only does school teach them the skills of reading, writing, and other subject areas, but it is also where children are trained to respect authority, be punctual, and follow rules—thereby preparing them for their future lives as workers in organizations that value these traits.

---

see FOR YOURSELF

Visit a local day-care center, preschool, or elementary school and observe children at play. Record the activities they are involved in, and note what both girls and boys are doing. Do you observe any differences between boys' and girls' play? What do your observations tell you about *socialization* patterns for boys and girls? ●

THEORIES OF SOCIALIZATION

Knowing that people become socialized does not explain how it happens. Different theoretical perspectives explain socialization, including psychoanalytic theory, social learning theory, and symbolic interaction theory. Each perspective, including functionalism as well as conflict theory, carries a unique set of assumptions about socialization and its effect on the development of the self (see Table 4.1).

Psychoanalytic Theory

Psychoanalytic theory originates in the work of **Sigmund Freud** (1856–1939). Perhaps Freud's greatest contribution was the idea that the unconscious mind shapes human behavior. Freud is also known for developing the technique of *psychoanalysis* to help discover the causes of psychological problems in the recesses of troubled patients' minds. Freud's approach depicts the human psyche in three parts: the id is about impulses; the superego is about the standards of society and morality; and the ego is about reason and common sense.

The psychoanalytic perspective interprets human identity as relatively fixed at an early age in a process greatly influenced by one's family.

Social Learning Theory

Whereas psychoanalytic theory places great importance on the *internal* unconscious processes of the human mind, **social learning theory** considers the formation of identity to be a learned response to *external* social stimuli (Bandura and Walters 1963). Social learning theory emphasizes the societal context of socialization. Identity is regarded not as the product of the unconscious but as the result of modeling oneself (called role modeling) in response to the expectations of others. According to social learning theory, behaviors and attitudes develop in response to *reinforcement* and encouragement from those around us. Reinforcement comes to us as *positive reinforcement* (reward) or *negative reinforcement* (punishment). Behavior that is positively reinforced is more likely to be repeated, whereas behavior that is negatively reinforced is not. A major tenant of social learning theory is the principle that positive

---

table 4.1 Theories of Socialization

	Social Learning Theory	Functional Theory	Conflict Theory	Symbolic Interaction Theory
How each theory views:				
Individual learning process	People respond to social stimuli in their environment.	People internalize the role expectations that are present in society.	Individual and group aspirations are shaped by the opportunities available to different groups.	Children learn through taking the role of significant others.
Formation of self	Identity is created through the interaction of mental and social worlds.	Internalizing the values of society reinforces social consensus.	Group consciousness is formed in the context of a system of inequality.	Identity emerges as the creative self interacts with the social expectations of others.
Influence of society	Young children learn the principles that shape the external world.	Society relies upon conformity to maintain stability and social equilibrium.	Social control agents exert pressure to conform.	Expectations of others form the social context for learning social roles.

Social learning theory emphasizes how people model their behaviors and attitudes on those of others.

reinforcement *plus* the presence of an admired role model makes the particular behavior highly likely.

Functionalism and Conflict Theory

Sociologists use a variety of theoretical perspectives to understand the socialization process, including those just described. They can also draw from the major theoretical frameworks we have introduced to understand socialization. From the vantage point of functionalist theory, socialization integrates people into society because it is the mechanism through which they internalize social roles and the values of society. This reinforces social consensus because it encourages at least some degree of conformity. Thus socialization is one way that society maintains its stability.

Conflict theorists would see this differently. Because of the emphasis in conflict theory on the role of power and coercion in society, conflict theorists thinking about socialization would be most interested in how group identity is shaped by patterns of inequality in society. A person's or group's identity always emerges in a context, and if that context is one marked by different opportunities for different groups, then one's identity will be shaped by that fact. This may help you understand why, for example, women are more likely to choose college majors in areas of study that have traditionally been associated with women's work opportunities (that is, in the so-called helping professions and in the arts and humanities and less frequently in math and sciences). Furthermore, though social control agents pressure people to conform, people also resist oppression. Thus the identities of people oppressed in society often include some form of resistance to oppression. This can help you understand why members of racial groups who identify with their own group, not the dominant White group, tend to have higher self-esteem (that is, a stronger valuing of self). In other words, resisting the expectations of a dominant group (such as being subservient or internalizing a feeling of inferiority) can actually heighten one's perceived self-worth.

Symbolic Interaction Theory

Recall that *symbolic interaction theory* centers on the idea that human actions are based on the meanings people attribute to behavior; these meanings emerge through social interaction (Blumer 1969). Symbolic interaction has been especially important in developing an understanding of socialization. People learn identities and values through socialization. For example, learning to become a good student means taking on the characteristics associated with that role. Because roles are socially defined, they are not real, like objects or things, but are real because of the meanings people give them.

For symbolic interactionists, meaning is constantly reconstructed as people act within their social environments. The **self** is what we imagine we are; it is not only an interior bundle of drives, instincts, and motives. Because of the importance attributed to reflection in symbolic interaction theory, symbolic interactionists use the term *self*, rather than the term *personality*, to refer to a person's identity. Symbolic interaction theory emphasizes that human beings make conscious and meaningful adaptations to their social environment. From a symbolic-interactionist perspective, identity is not something that is unconscious and hidden from view, but is socially bestowed and socially sustained (Berger 1963).

Two theorists have greatly influenced the development of symbolic-interactionist theory in sociology. **Charles Horton Cooley** (1864–1929) and **George Herbert Mead** (1863–1931) were both sociologists at the University of Chicago in the early 1900s (see Chapter 1). Cooley and Mead saw the self developing in response to the expectations and judgments of others in their social environment.

Charles Horton Cooley postulated the **looking-glass self** to explain how our conception of self arises through considering our relationships to others (Cooley 1967/1909, 1902). The development of the looking-glass self emerges from (1) how we think we appear to others; (2) how we think others judge us; and (3) how the first two make us feel—proud, embarrassed, or other feelings. The looking-glass self involves perception and effect, the perception of how others see us and the effect of others' judgment on us (see Figure 4.3).

How others see us is fundamental to the idea of the looking-glass self. In seeing ourselves as others do, we respond to the expectations others have of us. This means that the formation of the self is fundamentally a social process—one based in the interaction people have with each other, as well as the human capacity for self-examination. One unique feature of human life is the ability to see ourselves through others' eyes. People can imagine themselves in relationship to others and develop a definition of themselves accordingly. From a symbolic-interactionist perspective, the reflective process is key to the development of the self. If you

Stephen Simpson/Taxi/Getty Images

FIGURE 4.3 The Looking-Glass Self The looking-glass self refers to the process by which we attempt to see ourselves as others see us. This also helps us identify what roles we play in society and in relation to others. *Drawing conceptualized by Norman Andersen.*

the behavior. Although children in the imitation stage have little understanding of the behavior being copied, they are learning to become social beings. For example, think of young children who simply mimic the behavior of people around them (such as pretending to read a book, but doing so with the book upside down).

In the second stage, the **play stage**, children begin to take on the roles of significant people in their environment, not just imitating but incorporating their relationship to the other. Especially meaningful is when children take on the role of **significant others**, those with whom they have a close affiliation. A child pretending to be his mother may talk to himself as the mother would. The child begins to develop self-awareness, seeing himself or herself as others do.

In the third stage of socialization, the **game stage**, children become capable of taking on multiple roles at the same time. These roles are organized in a complex system that gives the children a more general or comprehensive view of the self. In this stage, children begin to comprehend the system of social relationships in which they are located. The children not only see themselves from the perspective of a significant other, but also understand how people are related to each other and how others are related to them. This is the phase where children internalize (incorporate into the self) an abstract understanding of how society sees them.

Mead compared the lessons of the game stage to a baseball game. In baseball, all roles together make the game. The pitcher does not just throw the ball past the batter as if they were the only two people on the field; rather, each player has a specific role, and each role intersects with the others. The network of social roles and the division of labor in the baseball game is a social system, like the social systems children must learn as they develop a concept of themselves in society.

grow up with others who think you are smart and sharp witted, chances are you will develop this definition of yourself. If others see you as dull witted and withdrawn, chances are good that you will see yourself this way. George Herbert Mead agreed with Cooley that children are socialized by responding to others' attitudes toward them. According to Mead, social roles are the basis of all social interaction.

Taking the role of the other is the process of putting oneself into the point of view of another. To Mead, role-taking is a source of self-awareness. As people take on new roles, their awareness of self changes. According to Mead, identity emerges from the roles one plays. He explained this process in detail by examining childhood socialization, which he saw as occurring in three stages: the imitation stage, the play stage, and the game stage (Mead 1934). In each phase of development, the child becomes more proficient at taking the role of the other. In the first stage, the **imitation stage**, children merely copy the behavior of those around them. Role-taking in this phase is nonexistent because the child simply mimics the behavior of those in the surrounding environment without much understanding of the social meaning of

---

see **FOR YOURSELF**

Childhood Play and Socialization

The purpose of this exercise is to explain how *childhood socialization* is a mechanism for passing on social *norms* and *values*. Begin by identifying a form of play that you engaged in as a young child. What did you play? Who did you play with? Was it structured or unstructured play? What were the rules? Were they formal or informal, and who controlled whether they were observed?

1. Now think about what *norms* and *values* were being taught to you by way of this play. Do they still affect you today? If so, how?
2. How does your experience compare to those of students in your class who differ from you in terms of gender, race, ethnicity, regional origin, and so forth. Are there differences in learned norms and values that can be attributed to these different social characteristics? ●

In the game stage, children learn more than just the roles of significant others in their environment. They also acquire a concept of the **generalized other**—the abstract composite of social roles and social expectations. In the generalized other, they have an example of community values and general social expectations that adds to their understanding of self; however, children do not all learn the same generalized other. Depending on one's social position (that is, race, class, gender, region, or religion), one learns a particular set of social and cultural expectations.

If the self is socially constructed through the expectations of others, how do people become individuals? Mead answered this by saying that the self has two dimensions: the "I" and the "me." The "I" is the unique part of individual personality, the active, creative, self-defining part. The "me" is the passive, conforming self, the part that reacts to others. In each person, there is a balance between the I and the me, similar to the tension Freud proposed between the id and the superego. Mead differed from Freud, however, in his judgment about when identity is formed. Freud felt that identity was fixed in childhood and henceforth driven by internal, not external, forces. In Mead's version, social identity is always in flux, constantly emerging (or "becoming") and dependent on social situations. Over time, identity stabilizes as one learns to respond consistently to common situations.

Social expectations associated with given roles change as people redefine situations and as social and historical conditions change; thus the social expectations learned through the socialization process are not permanently fixed. For example, as more women enter the paid labor force and as men take on additional responsibilities in the home, the expectations associated with motherhood and fatherhood are changing. Men now experience some of the role conflicts that women have faced in balancing work and family. As the roles of mother and father are redefined, children are learning new socialization patterns; however, traditional gender expectations maintain a remarkable grip. Despite many changes in family life and organization, young girls are still socialized for motherhood and young boys are still socialized for greater independence and autonomy.

GROWING UP IN A DIVERSE SOCIETY

Understanding the institutional context of socialization is important for understanding how socialization affects different groups in society. Socialization makes us members of our society. It instills in us the values of the culture and brings society into our self-definition, our perceptions of others, and our understanding of the world around us. Socialization is not, however, a uniform process, as the different examples developed in this chapter show. In a society as complex and diverse as the United States, no two people will have exactly the same experiences. We can find similarities between us, often across vast social and cultural differences, but variation in social contexts creates vastly different socialization experiences.

Furthermore, current changes in the U.S. population are creating new multiracial and multicultural environments in which young people grow up. Schools, as an example, are in many places being transformed by the large number of immigrant groups entering the school system. In such places, children come into contact with other children from a variety of different groups. This creates a new context in which children form their social values and learn their social identities (see the box "Doing Sociological Research: Race Socialization among Young Adults" on page 89).

One task of the sociological imagination is to examine the influence of different contexts on socialization. Where you grow up; how your family is structured; what resources you have at your disposal; your racial–ethnic identity, gender, and nationality—all shape the socialization experience. Socialization experiences for all groups are shaped by many factors that intermingle and intersect to form the context for socialization.

One way that this has been demonstrated is in research by sociologist Annette Lareau (2003). Over an extended period of time, Lareau and her research assistants carefully observed White and Black families from middle-class, working-class, and poor backgrounds. The researchers spent many hours in the homes of the families studied, including following the children and parents as they went about their daily routines. Based on these detailed observations, they observed important class differences in how families—both Black and White—socialize their children.

The middle-class children were highly programmed in their activities, their lives filled with various organized activities—music lessons, sports, school groups, and so forth. In contrast, the working-class and poor children, regardless of race, were less structured in their activities, and economic constraints were a constant theme in their daily lives. But the pace of life for working-class and poor children was slower and more relaxed. These children had more unstructured playtime, whereas middle-class children's lives were a constant barrage of highly structured activities with

The family serves as a major agent of socialization, especially of the young. Different racial–ethnic groups have different family traditions.

sisters greet the infant with different expectations, depending on whether it is a boy or a girl. Socialization does not come to an end as we reach adulthood; rather, it continues through our lifetime. As we enter new situations, and even as we interact in familiar ones, we learn new roles and undergo changes in identity.

Sociologists use the term **life course** perspective to describe and analyze the connection between people's personal attributes, the roles they occupy, the life events they experience, and the social and historical aspects of these events. The life course perspective underscores the point made by C. Wright Mills (introduced in Chapter 1) that personal biographies are linked to specific social–historical periods. Thus different generations are strongly influenced by large-scale events (such as war, immigration, economic prosperity, or depression, for example).

The phases of the life course are familiar: childhood, youth and adolescence, adulthood, and old age. These phases of the life course bind different generations and define some of life's most significant events, such as birth, marriage, retirement, and death.

Childhood

During childhood, socialization establishes one's initial identity and values. In this period, the family is an extremely influential source of socialization, but experiences in school, peer relationships, sports, religion, and the media also have a profound effect. Children acquire knowledge of their culture through countless subtle cues that provide them with an understanding of what it means to live in society.

Socializing cues begin as early as infancy, when parents and others begin to describe their children based on their perceptions. Frequently, these perceptions are derived from the cultural expectations parents have for children. Parents of girls may describe their babies as "sweet" and "cuddly," whereas boys are described as "strong" and "alert." Even though it is difficult to physically identify baby boys and girls when they are infants, parents in this culture dress even their tiny infants in colors and styles that typically distinguish one gender from the other.

The lessons of childhood socialization come in myriad ways, some more subtle than others. Much socialization in early childhood takes place through play and games. Games using traditional gender roles will likely lead to girls growing up to fulfill feminine roles and boys growing up to fulfill masculine roles. Research that compared children raised in gay households with children raised in heterosexual households revealed some important differences (Goldberg et al. 2012). Generally, the research finds that children of lesbian or gay parents play games that are less gender-stereotyped and that children of heterosexual couples played more divergent, gender-specific games (Goldberg et al. 2012).

Beyond understanding how children are socialized into adult roles, we should consider the importance of

intense time demands. Lareau argues that middle-class families engaged in *concerted cultivation* of childhood, meaning they made "deliberate and sustained effort to stimulate children's development and to cultivate their cognitive and social skills" (Lareau 2003: 238). Working-class and poor children experienced more "natural growth," that is, childhood experiences that allow them to develop in a less-structured environment with more time for creative play.

As a result, middle-class children tend to learn a more individualized self-concept and a sense of entitlement, but the price is an overly programmed daily life. Working-class and poor children experience obvious costs in that they have more financial constraints, but even more fundamentally, Lareau argues, they are left unable to negotiate their way through various social institutions as effectively as the middle class. In a sense, these childhood socialization patterns are also reshaping the class system in which children will likely find themselves as adults. Middle-class children are being prepared, even if inadvertently, for lives with a sense of privilege and entitlement; working-class children, for responding to the directives of others. In this way, patterns of socialization occurring because of social class origins are training children to take their place in the class system that will likely mark their adult lives. Thus social class is an important—although often invisible—force shaping the socialization of young people.

AGING AND THE LIFE COURSE

Socialization begins the moment a person is born. As soon as the sex of a child is known (which now can be even before birth), parents, grandparents, brothers, and

MAP 4.1

Viewing Society in Global Perspective: Children as a Percentage of the Population

Throughout the world, the proportion of children as a percentage of the population of a given country tends to be higher in those countries that are most economically disadvantaged and most overpopulated. In such countries, children contribute to household income by working, often in circumstances that are hazardous to their health. What consequences do you think the proportion of children in a given society has for the society as a whole?

Source: U.S Census Bureau. 2012.
www.census.gov/prod/2011pubs /12statab/intlstat.pdf

Children as a Percentage of Population*

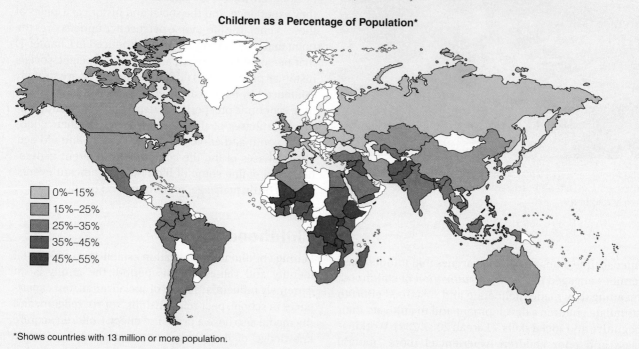

- 0%–15%
- 15%–25%
- 25%–35%
- 35%–45%
- 45%–55%

*Shows countries with 13 million or more population.

children and childhood to society. In the United States, we value children as our future leaders, inventors, and activists. We also recognize that children will likely become parents. How we treat our children now will influence future generations. We have strict guidelines for protecting children, including laws against child labor. Some cultures, however, rely more heavily on children for labor or household income contributions. Consider how much of the population in various countries is made up of children. (See Map 4.1 for the percentage of populations that are children under age 15 across the world.)

Adolescence

Only recently has adolescence been thought of as a separate phase in the life cycle. Until the early twentieth century, children moved directly from childhood roles to adult roles. It was only when formal education was extended to all classes that adolescence emerged as a particular phase in life when young people are regarded as no longer children, but not yet adults. There are no clear boundaries to adolescence, although it generally lasts from junior high school until the time one takes on adult roles by getting a job, marrying, and so forth. Adolescence can include the period through high school and extend right up through college graduation.

Erik Erikson (1980), the noted psychologist, stated that the central task of adolescence is the formation of a consistent identity. Adolescents are trying to become independent of their families, but they have not yet moved into adult roles. Conflict and confusion can arise as the adolescent swings between childhood and adult maturity. Some argue that adolescence is a period of delayed maturity. Although society expects adolescents to behave like adults, they are denied many privileges associated with adult life. Until age 18, they cannot vote or marry without permission, and sexual activity is condemned. In addition, until age 21, they cannot legally drink alcohol. The tensions of adolescence have been blamed for numerous social problems, such as drug and alcohol abuse, youth violence, and the school dropout rate.

The issues that young people face are a good barometer of social change across generations. Today's young people face an uncertain world where adult roles are less predictable than in the past. Marriage later in life, high divorce rates, frequent technological change, and economic recession all create a confusing environment for young people. Studies of adolescents find that, in this context, young people understand the need

DOING **sociological research**

Race Socialization among Young Adults

Many parents teach their children about their cultural heritage and encourage cultural pride. In a diverse American society, parents are also charged with socializing their children to respond to potential discrimination. A research study by Deborah Rivas-Drake shows that Latinos develop ethnic identities directly from family socialization and messages about future discrimination.

Research Questions: How do the warnings issued by parents about racial discrimination influence the expectations of young adults? Rivas-Drake researched Latino families to try to understand how young adults perceive the likelihood of anti-Latino bias and how that influences their ethnic identity. She also examined how cultural socialization regarding ethnic identity influenced psychological adjustments as an adult, such as depression and self-esteem. Rivas-Drake developed a research project to examine Latino young adults who are pursuing higher education and how their racial–ethnic socialization influences their development.

Research Method: Rivas-Drake sampled 227 Latino students from one university who were eighteen years of age or older. The mean age among the sample was

about nineteen years old, and 65 percent of the sample were women. Seventy-two percent of the sample had one parent who was born in another country. She administered an online survey in 2008 that asked a series of questions that measured how their parents socialized them with regard to ethnicity and race, and questions about perceived barriers to opportunities and resources because of racial bias.

Research Results: Overall, students had more socialization from parents regarding culture than they did about expectations for future bias. But most students in the sample did agree that discrimination did exist and that Latinos had fewer resources and opportunities than White students. Latino students who reported greater parental cultural socialization had stronger ethnic identity. Those students who reported receiving greater preparation for racial barriers to opportunity reported less ethnic identity but greater understanding of the status of their ethnic group. Latino students who were prepared by their parents to expect racial bias were more aware of barriers to opportunity, and consequently, the self-esteem and depressive symptoms of these students were more affected.

Conclusions and Implications: This research highlights how parents socialize their children and the consequences of that socialization for Latino young adults. The college students in this sample revealed that, if they were socialized to have strong ethnic identities, their Latino status was more central to their identity. They were then better adjusted as young adults in college. Latinos who reported having parents who prepared them for racial bias and barriers to opportunity were more likely to understand the status of Latino groups relative to other American groups. These students faced greater challenges in their overall well-being as young adults.

Questions to Consider

1. What is your earliest memory of a cultural lesson from your family? Do you remember a time when you thought your family traditions were different from other traditions?

2. How did your upbringing prepare you for college life? Did your parents have specific expectations for you regarding education? What about other expectations?

Source: Rivas-Drake, Deborah. 2011. "Ethnic-Racial Socialization and Adjustment among Latino College Students: The Mediating Roles of Ethnic Centrality, Public Regard, and Perceived Barriers to Opportunity." *Journal of Youth and Adolescence* 40: 606–619.

---

for flexibility, specialization, and, likely, frequent job change. Although the media stereotype adolescents as slackers, most teens are willing to work hard, do not engage in criminal or violent activity, and have high expectations for an education that will lead to a good job. Many, however, find that their expectations are out of alignment with the opportunities that are actually available, particularly during periods of economic downturn, such as the recent recession.

Patterns of adolescent socialization vary significantly by race, social class, and gender. National surveys find some intriguing class and race differences in how young people think about work and play in their lives. In general, the most economically privileged young people see their activities as more like play than work, whereas those less privileged are more likely to define their activities as work. Additional research finds that adolescents learn from parents about gender-stereotyped courses

and job options, and often choose a course of study that will lead to gender-specific careers (Tenenbaum 2009).

Adulthood

Socialization does not end when one becomes an adult. Building on the identity formed in childhood and adolescence, adult socialization is the process of learning new roles and expectations in adult life. More than at earlier stages in life, *adult socialization* involves learning behaviors and attitudes appropriate to specific situations and roles.

Youths entering college, to take an example from young adulthood, are newly independent and have new responsibilities. In college, one acquires not just an education but also a new identity. Those who enter college directly from high school may encounter conflicts with their family over their newfound status. Older students who work and attend college may experience difficulties (defined as *role conflict;* see Chapter 5) trying to meet dual responsibilities, especially if their family is not supportive. Meeting multiple and conflicting demands may require a returning student to develop different expectations about how much she can accomplish or to establish different priorities about what she will attempt.

Adult life is peppered with events that may require adults to adapt to new roles. Marriage, a new career, starting a family, entering the military, getting a divorce, or dealing with a death in the family all transform an individual's previous social identity. In today's world, these transitions through the life course are not as orderly as they were in the past. Where there was once a sequential and predictable trajectory of schooling, work, and family roles through one's twenties and thirties, that is no longer the case. Younger generations now experience diverse patterns in the sequencing of work, schooling, and family formation—even returning home—than was true in the past. These changes complicate the life course, and people have to make different adaptations to these changing roles (Cooksey and Rindfuss 2001; Rindfuss et al. 1999).

Becoming a full adult is thus taking longer than before. This has led some (such as social psychologist J. Arnett [2010] and Arnett and Tanner [2010]) to coin the term "emerging adulthood" to describe the path of today's twenty-somethings to age thirty. This is simply another way of saying "extended adolescence." As seen in Table 4.2, today's young people stay in school longer, marry later in life if they marry at all, and delay participation in the labor force. This certainly does not indicate "slacking," however. On the contrary, it simply means that progression into adulthood is more complex and involves more tasks and effort than in the past; it therefore takes longer. Hence, the transition into adulthood has been slowed down somewhat. The period of

twenty-something is now the upper end of adolescence rather than the lower end of adulthood. With these social changes, people have to be inventive in the roles they occupy because some of the old normative expectations no longer apply.

Another part of learning a new role is **anticipatory socialization**, the learning of expectations associated with a role a person expects to enter in the future. Anticipatory socialization allows a person to foresee the expectations associated with a new role and to learn what is expected in that role in advance.

In the transition from an old role to a new one, individuals often vacillate between their old and new identities as they adjust to fresh settings and expectations. An interesting example is *coming out*, the process of identifying oneself as gay or lesbian. This can be either a public coming out or a private acknowledgment of sexual orientation. The process can take years and generally means coming out to a few people, at first selective family members or friends who are likely to have the most positive reaction. Coming out is rarely a single event, but occurs in stages on the way to developing a new identity.

Age and Aging

Passage through adulthood involves many transitions. In our society, one of the most difficult transitions is the passage to old age. We are taught to fear aging in this society, and many people spend a lot of time and money trying to keep looking young. Unlike many other societies, ours does not revere the elderly, but instead devalues them, making the aging process even more difficult.

It is easy to think that aging is just a natural fact. Despite desperate attempts to hide gray hair, eliminate wrinkles, and reduce middle-aged bulge, aging is inevitable. The skin creases and sags, the hair thins, metabolism slows, and bones become less dense and more brittle by losing bone mass. Although aging is a physical process, the social dimensions of aging are just as important, if not more important, in determining the aging process. Just think about how some people appear to age much more rapidly than others. Some sixty-year-olds look only forty, and some forty-year-olds look sixty. These differences result from combinations of biological and social factors, such as genetics, eating and exercise habits, stress, smoking habits, pollution in the physical environment, and many other factors. The social dimensions of aging are what interest sociologists.

Although the physiology of aging proceeds according to biological processes, what it *means* to grow older is a social phenomenon. **Age stereotypes** are preconceived judgments about what different age groups are like. Stereotypes abound for both old

table 4.2 Showing the Transition to Adulthood

	1990	2010
Percentage aged 20–21 in school	39.7 %	52.4 %
Median age at first marriage		
Women	23.9 years	26.1 years
Men	26.1 years	28.2 years
Fertility rate, women aged 15–19	59.9 (per 1000 women)	29.3 (per 1000 women)
Percentage aged 16–19 in labor force		
Women	51.6%	35.0 %
Men	55.7 %	34.9 %

Source: U.S. Census Bureau. 2011. *Current Population Survey March and Annual Social and Economic Supplements, 2011 and earlier; Current Population Survey June 2010,* October 2010; Bureau of Labor Statistics (2012a).

The stresses of life that accompany age can change a person in many ways, as is evident in these "before" and "after" photographs of President Barack Obama.

and young people. Young people, especially teenagers, are stereotyped as irresponsible, addicted to loud music, lazy ("slackers"), sloppy, and so on; the elderly are stereotyped as forgetful, set in their ways, mentally dim, and unproductive. Though like any stereotype, these stereotypes are largely myths, they are widely believed. Age stereotypes also differ for different groups. Older women are stereotyped as having lost their sexual appeal, contrary to the stereotype of older men as handsome or "distinguished" and desirable. Gender is, in fact, one of the most significant factors in age stereotypes. Age stereotypes are also reinforced through popular culture. Advertisements depict women as needing creams and lotions to hide "the telltale signs of aging." Men are admonished to cover the patches of gray hair that appear or to use other products to prevent baldness. Entire industries are constructed on the fear of aging that popular culture promotes. Facelifts, tummy tucks, and vitamin advertisements all claim to "reverse the process of aging," even though the aging process is a fact of life.

Age Prejudice and Discrimination. **Age prejudice** refers to a negative attitude about an age group that is generalized to all people in that group. Prejudice against the elderly is prominent. The elderly are often thought of as childlike and thus incapable of adult responsibility. Prejudice relegates people to a perceived lower status in society and stems from the stereotypes associated with different age groups.

Age discrimination is the different and unequal treatment of people based solely on their age. Whereas age prejudice is an attitude, age discrimination involves actual behavior. As an example, people may talk "baby talk" to the elderly. This reinforces the stereotype of the elderly as childlike and incompetent. Some forms of age discrimination are illegal. The Age Discrimination in Employment Act, first passed in 1967 but amended several times since, protects people from age discrimination in employment. It states that age discrimination is a violation of the individual's civil rights. An employer can neither hire nor fire someone based solely on age, nor segregate or classify workers based on age. Age discrimination cases have become one of the most frequently filed cases through the Equal Employment Opportunity Commission (EEOC), the federal agency set up to monitor violations of civil rights in employment.

Ageism is a term sociologists use to describe the institutionalized practice of age prejudice and discrimination. More than a single attitude or an explicit act of discrimination, ageism is structured into the institutional fabric of society. Like racism and sexism, ageism encompasses both prejudice and discrimination, but it is also manifested in the structure of institutions. As such, it does not have to be intentional or overt to affect how age groups are treated. Ageism in society means that, regardless of laws that prohibit age discrimination, a person's age is a significant predictor of his or her life chances. Resources are distributed in society in ways that advantage some age groups and disadvantage others; cultural belief systems devalue the elderly; society's systems of care are often inadequate to meet people's needs as they grow old—these are the manifestations of ageism, a persistent and institutionalized feature of society.

Age Stratification.

Most societies produce age hierarchies—systems in which some age groups have more power and better life chances than others. **Age stratification** refers to the hierarchical ranking of different age groups in society. Age stratification exists because processes in society ensure that people of different ages differ in their access to society's rewards, power, and privileges. As we will see, in the United States and elsewhere, age is a major source of inequality (see Figure 4.4).

Age is an *ascribed status;* that is, age is determined by when you were born. Different from other ascribed statuses, which remain relatively constant over the duration of a person's life, age changes steadily throughout your life. Still, you remain part of a particular generation—something sociologists call an **age cohort**—an aggregate group of people born during the same period.

People in the same age cohort share the same historical experiences—wars, technological developments, and economic fluctuations—although they might do so in different ways, depending on other life factors. Living through the Great Depression, for example, shaped an entire generation's attitudes and behaviors, as did growing up in the 1960s, as will being a member of the contemporary youth generation. Depending on how the major economic recession of 2008–2013 shakes out, it too may significantly shape the current youth generation. Recall from Chapter 1 that C. Wright Mills saw the task of the sociological imagination as analyzing the relationship between biography and history. Understanding the experiences of different age cohorts is one way you can do this. People who live through the same historic period experience a similar impact of that period in their personal lives. The troubles and triumphs they experience and the societal issues they face are rooted in the commonality established by their age cohort. There is variation in how the old are treated. In many societies, older people are given enormous respect. There may be traditions to honor the elders, and they may be given authority over decisions in society, as they are perceived as most wise. On the other hand, among some cultures, adults who can no longer contribute to the society because of old age or illness may be perceived as extreme burdens and thus may be banished from the society altogether.

Why does society stratify people on the basis of age? Once again, we find that the three main theoretical perspectives of sociological analysis—functionalism, conflict theory, and symbolic interaction—offer different explanations (see Table 4.3). Functionalist sociologists ask whether the grouping of individuals contributes in some way to the common good of society. From this perspective, adulthood is functional to society because adults are seen as the group contributing most fully to it; the elderly are not. Functionalists argue that older people are seen as less useful and are therefore granted lower status in society. Youth are in between. The constraints and expectations placed on youth—they are prohibited from engaging in a variety of "adult" activities, expected to go to school, not expected to support themselves—are seen to free them from the cares of adulthood and give them time and opportunity to learn an occupation and prepare to contribute to society.

According to the functionalist argument, the elderly voluntarily withdraw from society by retiring and lessening their participation in social activities such as church, civic affairs, and family. **Disengagement theory**, drawn from functionalism, predicts that as people age, they gradually withdraw from participation in society and are simultaneously relieved of responsibilities. This withdrawal is functional to society because it provides for an orderly transition from one generation to the next. The young presumably infuse the roles they take over from the elderly with youthful energy and stamina. According to the functionalist argument, the diminished usefulness of the elderly justifies their depressed earning power and their relative neglect in social support networks.

Conflict theory focuses on the competition over scarce resources between age groups. Among the most important scarce resources are jobs. Unlike functionalist theory, conflict theory offers an explanation of why both youth and the elderly are assigned lower status in society and are most likely to be poor. Barring youth and the elderly from the labor market eliminates these groups from competition, improving the prospects for middle-aged workers. Removed from competition, both

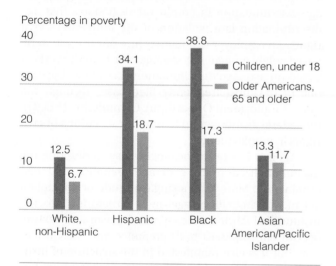

Percentage in poverty

FIGURE 4.4 Poverty by Age and Race (Percentage)
Poverty in the United States is clearly divided by race. Consider the number of Black children living in poverty compared to Whites. What are some reasons older Americans have a lower percentage living in poverty?

Source: DeNavas-Walt, Carmen, Bernadette D. Proctor, and Jessica C. Smith. 2012. *Income, Poverty, and Health Insurance Coverage in the United States: 2011.* Current Population Reports, P60–243. Washington, DC: U.S. Census Bureau.

table 4.3 Sociological Theories of Aging

	Functional Theory	Conflict Theory	Symbolic Interaction
Age differentiation	Contributes to the common good of society because each group has varying levels of utility in society	Results from the different economic status and power of age cohorts	Occurs in most societies, but the social value placed on different age groups varies across diverse cultures
Age groups	Are valued according to their usefulness in society	Compete for resources in society, resulting in generational inequities and thus potential conflict	Are stereotyped according to the perceived value of different groups
Age stratification	Results from the functional value of different age cohorts	Intertwines with inequalities of class, race, and gender	Promotes ageism, which is institutionalized prejudice and discrimination against old people

© Cengage Learning

the young and the old have very little power, and like other minorities, they are denied access to the resources they need to change their situation. Conflict theory also helps explain that competition can emerge between age groups, such as deciding whether to limit Social Security payments to save for future generations.

Symbolic interaction theory analyzes the different meanings attributed to social entities. Symbolic interactionists ask what meanings become attached to different age groups and to what extent these meanings explain how society ranks such groups. Definitions of aging are socially constructed, as we saw in our discussion of age stereotypes. Moreover, in some societies, the elderly may be perceived as having higher status than in other societies. Symbolic interaction considers the role of social perception in understanding the sociology of age. Age clearly takes on significant social meaning—meaning that varies from society to society for a given age group and that varies within a society for different age groups.

Growing old in a society such as the United States with such a strong emphasis on youth means encountering social stereotypes about the old, adjusting to diminished social and financial resources, and sometimes living in the absence of social supports, even when facing some of life's most difficult transitions, such as declining health and the loss of loved ones. Still, many people experience old age as a time of great satisfaction and enjoy a sense of accomplishment connected to work, family, and friends. The degree of satisfaction during old age depends to a great extent on the social support networks established earlier in life—evidence of the continuing influence of socialization.

For many, old age is a time for new accomplishments and achievements, such as for this marathon runner. As the population ages better, our stereotypes about age are changing.

Piluhin/Alamy Limited

Rites of Passage

A **rite of passage** is a ceremony or ritual that marks the transition of an individual from one role to another. Rites of passage define and legitimize abrupt role changes that begin or end each stage of life. The ceremonies surrounding rites of passage are often dramatic and infused with awe and solemnity. Examples include graduation ceremonies; weddings; and religious affirmations, such as the Jewish ceremony of the bar mitzvah for boys or the bat mitzvah for girls, confirmation for Catholics, and adult baptism for many Christian denominations.

Formal promotions or entry into some new careers may also include rites of passage. Completing police academy training or being handed one's diploma are examples. Such rites usually include family and friends, who watch the ceremony with pride; people frequently keep mementos of these rites as markers of the transition through life's major stages. Bridal showers and baby showers have been analyzed as rites of passage. At a shower, the person who is being honored is about to assume a new role and identity—from young woman to wife or mother. Rites of passage entail public announcement of the new status for the benefit of both the individual and those with whom the newly anointed person will interact. In the absence of such rituals, the transformation of identity would not be formally recognized, perhaps leaving uncertainty in the youngster or the community about the individual's worthiness, preparedness, or community acceptance.

Sociologists have noted that in the U.S. population as whole, there is no standard and formalized rite of passage marking the transition from childhood to adulthood. As a consequence, the period of adolescence is attended by ambivalence and uncertainty. As adolescents hover between adult and child status, they may not have the clear sense of identity that a rite of passage can provide. However, although there is no universal ceremony in our culture by which young people

are noted as moving from child to adult, some social class and ethnic subcultures do mark the occasion. Among the wealthy, the debutante's coming-out celebration is a traditional introduction of a young woman to adult society. Latinos may celebrate the *quinceañera* (fifteenth birthday) of young girls. A tradition of the Catholic Church, this rite recognizes the girl's coming of age, while also keeping faith with an ethnic heritage. Dressed in white, she is introduced by her parents to the larger community. Formerly associated mostly with working-class families and other Latinos, the quinceañera has also become popular among affluent Mexican Americans, who may match New York debutante society by spending as much as $50,000 to $100,000 on the event.

RESOCIALIZATION

Most transitions people experience in their lifetimes involve continuity with the former self as it undergoes gradual redefinition. Sometimes, however, adults are forced to undergo a radical shift of identity. **Resocialization** is the process by which existing social roles are radically altered or replaced (Fein 1988). Resocialization is especially likely when people enter institutional settings

Empire/Universal Images Group/Getty Images

Judy Griesedieck/Encyclopedia/Corbis

Ernst Heiniger/Science Source/Photo Researchers

AP Photo/Boris Grdanoski

Every culture has important *rites of passage* that mark the transition from one phase in the life course to another. Here, different cultural traditions distinguish the rites of passage associated with marriage: a traditional Nigerian wedding (upper left); a young American couple (upper right); a Shinto (Japanese) bride taking a marital pledge by drinking sake (lower left); and a newlywed orthodox Christian couple in Macedonia (lower right).

where the institution claims enormous control over the individual. Examples include the military, prisons, monastic orders, and some cults (see also Chapter 6 for a discussion of total institutions). When military recruits enter boot camp, they are stripped of personal belongings, their heads are shaved, and they are issued identical uniforms. Although military recruits do not discard their former identities, the changes brought about by becoming a soldier can be dramatic and are meant to make the military one's *primary group*, not one's family, friends, or personal history. The military represents an extreme form of resocialization in which individuals are expected to subordinate their identity to that of the group. In such organizations, individuals are interchangeable, and group consensus (meaning, in the military, unanimous, unquestioned subordination to higher ranks) is an essential component of group cohesion and effectiveness. Military personnel are expected to act as soldiers, not as individuals.

Resocialization often occurs when people enter hierarchical organizations that require them to respond to authority on principle, not out of individual loyalty. The resocialization process promotes group solidarity and generates a feeling of belonging. Participants in these settings are expected to honor the symbols and objectives of the organization; disloyalty is seen as a threat to the entire group.

Hazings are good examples of rites of passage that often accompany induction into a group.

thinking SOCIOLOGICALLY

Find three to five adults (young or old) who have just entered a new stage of life (getting a new or first job, getting married, becoming a grandparent, retiring, entering a nursing home, and so forth), and ask them to describe this new experience. Ask questions such as what others expect of them in this new role, how these expectations are communicated to them, what changes they see in their own behavior, and what expectations they have of their new situation. What do your observations tell you about *adult socialization*? ●

Resocialization may involve degrading initiates physically and psychologically with the aim of breaking down or redefining their old identity. They may be given menial and humiliating tasks and be expected to act in a subservient manner. Social control in such a setting may be exerted by peer ridicule or actual punishment. Fraternities and sororities offer an interesting everyday example of this pattern of resocialization through initiation rituals.

The Process of Conversion

Resocialization also occurs during what people popularly think of as conversion. A conversion is a far-reaching transformation of identity, often related to religious or political beliefs. People usually think of conversion in the context of cults, but it happens in other settings as well.

John Walker Lindh was a U.S. citizen when the United States entered the Iraq war in 2000. He joined the Taliban in Afghanistan and was later charged with conspiring to kill Americans abroad and supporting terrorist organizations. Lindh is an example of an *extreme conversion*. He was raised Catholic in an affluent family, but he converted to Islam as a teenager, changing not just his ideas, but also his dress. Neighbors described him as being transformed from "a boy who wore blue jeans and T-shirts to an imposing figure in flowing Muslim garb" (Robertson and Burke 2001). As a young man, he traveled to Yemen and Pakistan to study language and the Koran and was introduced there to the Taliban. News sources have since named him the "American Taliban."

As when people join religious cults, this is extreme conversion, but conversion happens in less extreme situations, too. People may convert to a different religion, thereby undergoing resocialization by changing beliefs and religious practices. Or someone may become strongly influenced by the beliefs of a *social movement*, such as the Tea Party political movement, and abruptly or gradually change beliefs—even identity—as a result.

The Brainwashing Debate

Extreme examples of resocialization are seen in the phenomenon popularly called "brainwashing." In the popular view of brainwashing, converts have their previous identities totally stripped; the transformation is seen as so complete that only deprogramming can restore the former self. You can picture the fictional character of Jason Bourne from the *Bourne Identity* books and movies. Potential candidates of brainwashing include people who enter religious cults, prisoners of war, and hostages. Sociologists have examined brainwashing to illustrate the process of resocialization. As the result of their research, sociologists have cautioned against using the word *brainwashing* when referring to this form of conversion. The term implies that humans are mere puppets or passive victims whose free will can be taken away during these conversions (Robbins 1988). In religious cults, however, converts do not necessarily drop their former identity. Many scholars reject the brainwashing term to describe a resocialization into a new religion or a cult following of some kind.

Sociological research has found that the people most susceptible to cult influence are the most suggestible, primarily young adults who are socially isolated, drifting, and having difficulty performing in other areas (such as in their jobs or in school). Such people may choose to affiliate with cults voluntarily. Despite the widespread belief that people have to be deprogrammed to be freed from the influence of cults, many people are able to leave on their own (Robbins 1988). So-called brainwashing is simply a manifestation of the social influence people experience through interaction with others. Even in cult settings, socialization is an interactive process, not just a transfer of group expectations to passive victims.

Forcible confinement and physical torture can be instruments of extreme resocialization. Under severe captivity and deprivation, a captured person may come to identify with the captor; this is known as the **Stockholm syndrome**. In traditional psychology, this same phenomenon was called "identification with the aggressor." In such instances, the captured person has become *dependent* on the captor. On release, the captive frequently needs debriefing, or deprogramming. Prisoners of war and hostages may not lose free will altogether, but they do lose freedom of movement and association, which makes prisoners intensely dependent on their captors and therefore vulnerable to the captor's influence.

The Stockholm syndrome can help explain why some battered women do not leave their abusing spouses or boyfriends. Dependent on their abuser both financially and emotionally, battered women often develop identities that keep them attached to men who abuse them, a clear example of identification with an aggressor. In these cases, outsiders often think the women should leave instantly, whereas the women themselves may find leaving difficult, even in the most abusive situations.

chapter summary

What is socialization, and why is It significant for society?
Socialization is the process by which human beings learn the social expectations of society. Socialization creates the expectations that are the basis for people's attitudes and behaviors. Through socialization, people conform to social expectations, although people still express themselves as individuals.

What are the agents of socialization?
Socialization agents are those who pass on social expectations. They include the family, the media, peers, sports, religious institutions, and schools, among others. The family is usually the first source of socialization. The media also influence people's values and behaviors. *Peers* are an important source of individual identity; without peer approval, most people find it hard to be socially accepted. Schools also pass on expectations that are influenced by gender, race, and other social characteristics of people and groups.

What theoretical perspectives do sociologists use to explain socialization?
Psychoanalytic theory sees the self as driven by unconscious drives and forces that interact with the expectations of society. *Social learning theory* sees identity as a learned response to social stimuli such as reward–punishment and role models. *Functionalism* interprets

socialization as key to social stability because socialization establishes shared roles and values. *Conflict theory* interprets socialization in the context of inequality and power relations. *Symbolic interaction theory* sees people as "constructing" the self as they interact with the environment and give meaning to their experience. Charles Horton Cooley described this process as the *looking-glass self*. Another sociologist, George Herbert Mead, described childhood socialization as occurring in three stages: imitation, play, and games.

Does socialization mean that everyone grows up the same?
Socialization is not a uniform process. Growing up in different environments and in such a diverse society means that different people and different groups are exposed to different expectations. Factors such as family structure, social class, regional differences, and many others influence how one is socialized.

Does socialization end during childhood?
Socialization continues through a lifetime, although childhood is an especially significant time for the formation of *identity*. Adolescence is also a period when peer cultures have an enormous influence on the formation of people's self-concepts. *Adult socialization* involves the learning of specific expectations associated with new roles.

What are the social dimensions of the aging process?

Although aging is a physiological process, its significance stems from social meanings attached to aging. *Age prejudice* and *age discrimination* result in the devaluation of older people. *Age stratification*—referring to the inequality that occurs among different age groups—is the result.

What does resocialization mean?

Resocialization is the process by which existing social roles are radically altered or replaced. It can take place in an organization that maintains strict social control and demands that the individual conform to the needs of the group or organization. Examples are religious conversion, excessive influence via social interaction ("brainwashing"), and the Stockholm syndrome.

Key Terms

age cohort 92
age discrimination 91
age prejudice 91
age stereotype 90
age stratification 92
ageism 91
anticipatory socialization 90
disengagement theory 92

game stage 85
generalized other 86
identity 75
imitation stage 85
internalization 74
life course 87
looking-glass self 84
peers 79

personality 75
play stage 85
psychoanalytic theory 83
resocialization 94
rite of passage 93
roles 74
self 84
self-concept 77

significant others 85
social control 76
social learning theory 83
socialization 74
socialization agents 77
Stockholm syndrome 96
taking the role
   of the other 85

5 Social Structure and Social Interaction

John Henley/Getty Images

What Is Society?

What Holds Society Together?

Types of Societies

Social Interaction and Society

Theories About Analyzing
   Social Interaction

Interaction in Cyberspace

Chapter Summary

picture a college classroom on your campus. Students sit, and some are taking notes; others, listening; a few, perhaps, sleeping. The class period ends and students stand, gathering their books, backpacks, bags, and other gear. As they stand, many whip out their cell phones, place them to their ears, and quickly push buttons that connect them to a friend. As the students exit the room, many are engaged in social interaction—chatting with their friends: some by phone, others by text messaging ("texting"), some by talking face-to-face. Few, if any, of them realize that their behavior is at that moment influenced by society—a society whose influence extends into their immediate social relationships, even when the contours of that society—its social structure—are likely invisible to them.

These same students might plug a music player into their ears as they move on to their next class, possibly tuning in to the latest sounds while tuning out the sounds of the environment around them. Some will return to their residences and perhaps text message friends, download some music, or connect with "friends" on Facebook. Surrounding all of this behavior are social changes that are taking place in society, including changes in technology, in global communication, and in how people now interact with each other. How we make sense of these changes requires an understanding of the connection between society and social interaction. In this way, a sociological perspective can help you see the relationship between individuals and the larger society of which they are a part.

- Define society and identify ways in which society is held together
- Identify the types of societies
- Understand social interaction as a "game" within society
- Learn what theories are used to analyze social interaction and know how they differ
- Discover in what ways cyberspace interaction has changed society

WHAT IS SOCIETY?

In Chapter 2, we studied culture as one force that holds society together. *Culture* is the general way of life, including norms, customs, beliefs, and language. Human **society** is a system of social interaction that includes both culture and social organization. Within a society, members have a common culture, even though there may also be great diversity within it. Members of a society think of themselves as distinct from other societies, maintain ties of social interaction, and have a high degree of interdependence. The interaction they have, whether based on harmony or conflict, is one element of society. Within society, **social interaction** is behavior between two or more people that is given meaning by them. Social interaction is how people relate to each other and form a social bond.

Social interaction is the foundation of society, but society is more than a collection of individual social actions. Emile Durkheim, the classical sociological theorist, described society as *sui generis*—a Latin phrase meaning "a thing in itself, of its own particular kind." To sociologists, seeing society *sui generis* means that society is more than just the sum of its parts. Durkheim saw society as an organism, something comprising different parts that work together to create a unique whole. Just as a human body is not just a collection of organs but is alive as a whole organism with relationships between its organs, society is not only a simple collection of individuals, groups, or institutions but is a whole entity that consists of all these elements and their interrelationships.

Durkheim's point—central to sociological analysis—is that *society is much more than the sum of the individuals in it*. Society takes on a life of its own. It is patterned by humans and their interactions, but it is something that endures and takes on shape and structure beyond the immediacy of any given group of people. This is a basic idea that guides sociological thinking.

You can think of it this way: Imagine how a photographer views a landscape. The landscape is not just the sum of its individual parts—mountains, pastures, trees, or clouds—although each part contributes to the whole. The power and beauty of the landscape is that all its parts *relate* to each other, some in harmony and some in contrast, to create a panoramic view. The photographer who tries to capture this landscape will likely use a wide-angle lens. This method of photography captures the breadth and comprehensive scope of what the photographer sees. Similarly, sociologists try to picture society as a whole, not only by seeing its individual parts but also by recognizing the relatedness of these parts and their vast complexity.

Macro- and Microanalysis

Sociologists use different lenses to see the different parts of society. Some views are more macroscopic—that is, sociologists try to comprehend the whole of society, how it is organized, and how it changes. This is called **macroanalysis**, a sociological approach that takes the broadest view of society by studying large patterns of social interaction that are vast, complex, and highly differentiated. You might do this by looking at a whole society or comparing different total societies to each other. For example, as we opened this chapter, you saw that large-scale changes in technology influence even the most immediate social interaction that we have with other people. Thus, whereas only a few years ago it would not have been imaginable to create a network of friends in cyberspace, today it is a common practice, especially for young people.

Other views are more microscopic—that is, the focus is on the smallest, most immediately visible parts of social life, such as specific people interacting with each other. This is called **microanalysis**. In this approach, sociologists study patterns of social interactions that are relatively small, less complex, and less differentiated—the microlevel of society. Again, thinking of how this

Lain Masterton/Alamy Limited

The introduction of new technologies is transforming the nature of human communication. As more young people become adept with these tools, what will the future bring?

chapter opened, you might want to study how people engage in "texting" each other on a one-to-one basis. How are they similar or different, on the basis of age, or gender, or social class, or race? For example, do people text (that is, interact) with each other within racial groups more than between racial groups? Observing this would be an example of microanalysis.

Thus a sociologist who studies social interaction via texting or on the Internet would be engaging in microanalysis but might interpret what is found in the context of macrolevel processes (such as race relations in society). Just as a photographer might use a wide-angle lens to photograph a landscape or a telephoto lens for a closer view, sociologists use both macro- and microanalyses to reveal different dimensions of society.

In this chapter, we continue our study of sociology by starting with the macrolevel of social life (by studying total social structures), then continuing through the microlevel (by studying groups and face-to-face interaction). The idea is to help you see how large-scale dimensions of society shape even the most immediate forms of social interaction.

Sociologists use the term **social organization** to describe the order established in social groups at any level. Specifically, social organization brings regularity and predictability to human behavior; social organization is present at every level of interaction, from the whole society to the smallest groups.

Social Institutions

Societies are identified by their cultural characteristics and the social institutions that compose each society. A **social institution** (or simply an institution) is an established and organized system of social behavior with a recognized purpose. The term refers to the broad systems that organize specific functions in society. Unlike individual behavior, social institutions cannot be directly observed, but their impact and structure can still be seen. For example, the family is an institution that provides for the care of the young and the transmission of culture. Religion is an institution that organizes sacred beliefs. Education is the institution through which people learn the information and skills needed to live in the society.

The concept of the social institution is important to sociological thinking. You can think of social institutions as the enduring consequences of social behavior, but what fascinates sociologists is how social institutions take on a life of their own. For example, you were likely born in a hospital, which itself is part of the health care institution. The simple act of birth, which you might think of as an individual experience, is shaped by the structure of this social institution. Thus, you were likely delivered by a doctor, accompanied by nurses and, perhaps, a midwife—each of whom exists in a specific social relationship to the health care institution. Each of these people is in an *institutional* role. Moreover, this social institution also shaped the practices surrounding your birth. Thus, you might have been initially removed from your mother and examined by a doctor, which is very different from the institutional practices in other societies.

The major institutions in society include the family, education, work and the economy, the political institution (or state), religion, and health care, as well as the mass media, organized sports, and the military. These are all complex structures that exist to meet certain needs that are necessary for society to exist. *Functionalist theorists* have traditionally identified these needs (functions) as follows (Parsons 1951a; Aberle et al. 1950).

1. *The socialization of new members of the society.* This is primarily accomplished by the family, but involves other institutions as well, such as education.
2. *The production and distribution of goods and services.* The economy is generally the institution that performs this set of tasks, but this may also involve the family as an institution—especially in societies where production takes place within households.
3. *Replacement of society's members.* All societies must have a means of replacing members who die, move or migrate away, or otherwise leave the society. Families are typically organized to do this.
4. *The maintenance of stability and existence.* Certain institutions within a society (such as the government, the police force, and the military) contribute toward the stability and continuance of the society.
5. *Providing the members with an ultimate sense of purpose.* Societies accomplish this task by creating national anthems, for instance, and by encouraging patriotism in addition to providing basic values and moral codes through institutions such as religion, the family, and education.

In contrast to functionalist theory, *conflict theory* further notes that because conflict is inherent in most societies, the social institutions of society do not provide for all its members equally. Some members are provided for better than others, thus demonstrating that institutions affect people by granting more power to some social groups than to others. Using the example of the health care institution given previously, some groups have considerably less power within the institution than do others. Thus nurses are generally subordinate to doctors and doctors to hospital administrators. And beyond these specific actors within the health care institutions, different social groups in society have more or less power within social institutions. Therefore, racial and ethnic minorities in general have poorer access to health care than others; the poor have less access, as do those of lower social class status. (For more information, see Chapter 14 on health care.)

Social Structure

Sociologists use the term **social structure** to refer to the organized pattern of social relationships and social institutions that together compose society. Social structures

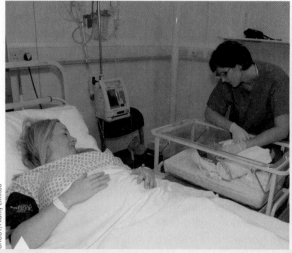

Birth, though a natural process, occurs within social institutions—institutions that vary in different societies, depending on the social organization of society. Here you see how birth in the United States, which is mainly defined as a medical event, contrasts with a health assistant attending a birth in rural Mexico.

are not immediately visible to untrained observers; nevertheless, they are present, and they affect all dimensions of human experience in society. Social structural analysis is a way of looking at society in which the sociologist analyzes the patterns in social life that reflect and produce social behavior.

Social class distinctions are an example of a social structure. Class shapes the access that different groups have to the resources of society, and it shapes many interactions people have with each other. People may form cliques with those who share similar class standing, or they may identify with certain values associated with a given class. Class then forms a social structure—one that shapes and guides human behavior at all levels, no matter how overtly visible or invisible this structure is to someone at a given time.

The philosopher Marilyn Frye aptly uses the metaphor of a birdcage to describe the concept of social structure (Frye 1983). She notes that if you look closely at only one wire in a cage, you cannot see the other wires. You might then wonder why the bird within does not fly away. Only when you step back and see the whole cage instead of a single wire do you understand why the bird does not escape. Social structure, like the birdcage, confines people; their motion and mobility are restricted; their lives are shaped by social structure. Just as the birdcage is a network of wires, so is society a network of social structures, both micro and macro.

WHAT HOLDS SOCIETY TOGETHER?

What holds societies together? We ask this question throughout this chapter. This central question in sociology was first addressed by Emile Durkheim, the French sociologist writing in the late 1800s and early 1900s. He argued that people in society had a **collective consciousness**, defined as the body of beliefs common to a community or society that give people a sense of belonging and a feeling of moral obligation to its demands and values. According to Durkheim, collective consciousness gives groups social solidarity because members of a group feel they are part of one society.

Where does the collective consciousness come from? Durkheim argued that it stems from people's participation in common activities, such as work, family, education, and religion—in short, society's institutions.

Mechanical and Organic Solidarity

According to Durkheim, there are two types of social solidarity: mechanical and organic. **Mechanical solidarity** arises when individuals play similar—rather than different—roles within the society. Individuals in societies marked by mechanical solidarity share the same values and hold the same things sacred. This particular kind of cohesiveness is weakened when a society becomes more complex. Contemporary examples of mechanical solidarity are rare because most societies of the world have been absorbed in the global trend for greater complexity and interrelatedness. Before European conquest, Native American groups were bound together by at least a partial mechanical solidarity. Indeed, many Native American groups are now trying to regain the vestige of mechanical solidarity on which their cultural heritage rests, but they are finding that the superimposition of White institutions on Native American life interferes with the adoption of traditional ways of thinking and being, which prevents mechanical solidarity from gaining its original strength even though this view is not intended to treat all Native American tribes or groups the same.

In contrast, **organic** (or contractual) **solidarity** occurs when people play a great variety of roles, and unity is based on role differentiation, not similarity. The United States and other industrial societies are built on organic solidarity, and each is cohesive because of the differentiation within each. Roles are no longer necessarily similar, but they are necessarily interlinked—the performance of multiple roles is necessary for the execution of society's complex and integrated functions.

Durkheim described this state as the **division of labor**, defined as the relatedness of *different* tasks that develop in complex societies. The labor force within the contemporary U.S. economy, for example, is divided according to the kinds of work people do. Within any division of labor, tasks become distinct from one another, but they are still woven into a whole.

The division of labor is a central concept in sociology because it represents how the different pieces of society fit together. The division of labor in most contemporary societies is often marked by distinctions such as age, gender, race, and social class. In other words, if you look at who does what in society, you will see that women and men tend to do different things; this is the gender division of labor. Similarly, old and young to some extent do different things; this is a division of labor by age. This is crosscut by the racial division of labor, the pattern whereby those in different racial–ethnic groups tend to do different work—or are often forced to do different work—in society. At the same time, the division of labor is also marked by class distinctions, with some groups providing work that is highly valued and rewarded and others doing work that is devalued and poorly rewarded. As you will see throughout this book, gender, race, and class intersect and overlap in the division of labor in society.

Gemeinschaft and Gesellschaft

Different societies are held together by different forms of solidarity. Some societies are characterized by what the German sociologist Ferdinand Tönnies called **gemeinschaft**, a German word that means "community"; other societies are characterized as **gesellschaft**, which literally means "society" (Tönnies 1963/1887). Each involves a type of solidarity or cohesiveness. Those societies that are *gemeinschafts* (communities) are characterized by a sense of "we" feeling, a very moderate division of labor, strong personal ties, strong family relationships, and a sense of personal loyalty. The sense of solidarity between members of the gemeinschaft society arises from personal ties; small, relatively simple social institutions; and a collective sense of loyalty to the whole society. People tend to be well integrated into the whole, and social cohesion comes from deeply shared values and beliefs (often, sacred values). Social control need not be imposed externally because control comes from

the internal sense of belonging that members share. You might think of a small community church as an example.

In contrast, in societies marked by *gesellschaft*, an increasing importance is placed on the secondary relationships people have—that is, less intimate and more instrumental relationships such as work roles instead of family or community roles. Gesellschaft is characterized by less prominence of personal ties, a somewhat diminished role of the nuclear family, and a lessened sense of personal loyalty to the total society. The solidarity and cohesion remain, and it can be very cohesive, but the cohesion comes from an elaborated *division of labor* (thus, *organic* solidarity), greater flexibility in social roles, and the instrumental ties that people have to one another.

Social solidarity under gesellschaft is weaker than in the gemeinschaft society, however. Gesellschaft is more likely than gemeinschaft to be torn by class conflict because class distinctions are less prominent, though still present, in the gemeinschaft. Racial–ethnic conflict is more likely within gesellschaft societies because the gemeinschaft tends to be ethnically and racially very homogeneous; it is often characterized by only one racial or ethnic group. This means that conflict between gemeinschaft societies, such as ethnically based wars, can be very high because both groups have a strong internal sense of group identity that may be intolerant of others.

In sum, complexity and differentiation are what make the gesellschaft cohesive, whereas similarity and unity cohere the gemeinschaft society. In a single society, such as the United States, you can conceptualize the whole society as gesellschaft, with some internal groups marked by gemeinschaft. Our national motto seems to embody this idea: *e pluribus unum* (unity within diversity), although clearly this idealistic motto has only been partly realized.

TYPES OF SOCIETIES

In addition to comparing how different societies are bound together, sociologists are interested in how social organization evolves in different societies. Simple things such as the size of a society can also shape its social organization, as do the different roles that men and women engage in as they produce goods, care for the old and young, and pass on societal traditions. Societies also differ according to their resource base—whether they are predominantly agricultural or industrial, for example, and whether they are sparsely or densely populated.

Thousands of years ago, societies were small, sparsely populated, and technologically limited. In the competition for scarce resources, larger and more technologically advanced societies dominated smaller ones. Today, we have arrived at a global society with highly evolved degrees of social differentiation and inequality, notably along class, gender, racial, and ethnic lines (Nolan and Lenski 2008).

Sociologists distinguish six types of societies based on the complexity of their social structure, the amount of overall cultural accumulation, and the level of their technology. They are *foraging, pastoral, horticultural, agricultural* (these four are called *preindustrial* societies), and then *industrial* and *postindustrial* societies (see Table 5.1). Each type of society can still be found on Earth, although all but the most isolated societies are rapidly moving toward the industrial and postindustrial stages of development.

These different societies vary in the basis for their organization and the complexity of their division of labor. Some, such as foraging societies, are subsistence economies, where men and women hunt and gather food but accumulate very little. Others, such as pastoral societies and horticultural societies, develop a more elaborate division of labor as the social roles that are needed for raising livestock and farming become more numerous. With the development of agricultural societies, production becomes more large scale and strong patterns of social differentiation develop, sometimes taking the form of a caste system or even slavery.

The key driving force that distinguishes these different societies from each other is the development of technology. All societies use technology to help fill human needs, and the form of technology differs for the different types of society.

Preindustrial Societies

A **preindustrial society** is one that directly uses, modifies, and/or tills the land as a major means of survival. There are four kinds of preindustrial societies, listed here by degree of technological development: foraging (or hunting–gathering) societies, pastoral societies, horticultural societies, and agricultural societies (see Table 5.1).

In *foraging (hunting–gathering) societies*, the technology enables the hunting of animals and gathering of vegetation. The technology does not permit the refrigeration or processing of food, hence these individuals must search continuously for plants and game. Because hunting and gathering are activities that require large amounts of land, most foraging societies are nomadic; that is, they constantly travel as they deplete the plant supply or follow the migrations of animals. The central

table 5.1 Types of Societies

		Economic Base	Social Organization	Examples
Preindustrial Societies	*Foraging societies*	Economic sustenance dependent on hunting and foraging	Gender is important basis for social organization, although division of labor is not rigid; little accumulation of wealth	Pygmies of Central Africa
	Pastoral societies	Nomadic societies, with substantial dependence on domesticated animals for economic production	Complex social system with an elite upper class and greater gender role differentiation than in foraging societies	Bedouins of Africa and Middle East
	Horticultural societies	Society marked by relatively permanent settlement and production of domesticated crops	Accumulation of wealth and elaboration of the division of labor, with different occupational roles (farmers, traders, craftspeople, and so on)	Ancient Aztecs of Mexico; Inca Empire of Peru
	Agricultural societies	Livelihood dependent on elaborate and large-scale patterns of agriculture and increased use of technology in agricultural production	Caste system develops that differentiates the elite and agricultural laborers; may include system of slavery	American South, pre–Civil War
Industrial Societies		Economic system based on the development of elaborate machinery and a factory system; economy based on cash and wages	Highly differentiated labor force with a complex division of labor and large formal organizations	Nineteenth and most of twentieth-century United States and Western Europe
Postindustrial Societies		Information-based societies in which technology plays a vital role in social organization	Education increasingly important to the division of labor	Contemporary United States, Japan, and others

© Cengage Learning

institution is the family, which serves as the means of distributing food, training children, and protecting its members. There is usually role differentiation on the basis of gender, although the specific form of the gender division of labor varies in different societies. The Pygmies of Central Africa are an example of a foraging society.

In *pastoral societies*, technology is based on the domestication of animals. Such societies tend to develop in desert areas that are too arid to provide rich vegetation. The pastoral society is nomadic, necessitated by the endless search for fresh grazing grounds for the herds of their domesticated animals. The animals are used as sources of hard work that enable the creation of a material surplus. Unlike a foraging society, this surplus frees some individuals from the tasks of hunting and gathering and allows them to create crafts, make pottery, cut hair, build tents, and apply tattoos. The surplus generates a more complex and differentiated social system with an elite or upper class and more role differentiation on the basis of gender. The nomadic Bedouins of Africa and the Middle East are pastoral societies.

In *horticultural societies*, hand tools are used to cultivate the land, such as the hoe and the digging stick. The individuals in horticultural societies practice ancestor worship and conceive of a deity or deities (God or gods) as a creator. This distinguishes them from foraging societies that generally employ the notion of numerous spirits to explain the unknowable. Horticultural societies recultivate the land each year and tend to establish relatively permanent settlements and villages. Role differentiation is extensive, resulting in different and interdependent occupational roles such as farmer, trader, and craftsperson. The ancient Aztecs of Mexico and the Incas of Peru represent examples of horticultural societies.

The *agricultural society* is exemplified by the pre–Civil War American South, a society of slavery. Such societies have a large and complex economic system that is based on large-scale farming. Such societies rely on technologies such as use of the wheel and use of metals. Farms tend to be considerably larger than the cultivated land in horticultural societies. Large and permanent settlements characterize agricultural societies, which also exhibit dramatic social inequalities. A rigid caste system develops, separating the peasants, or slaves, from the controlling elite caste, which is then freed from manual work allowing time for art, literature, and philosophy, activities of which they can then claim the lower castes are incapable. The American pre-Civil War South and its system of slavery is a good example of an agricultural society. In fact, some argue that the present surviving system of sharecropping in the American South and Southwest is a slave-like agricultural society (Bell 1992).

Different types of societies produce different kinds of social relationships. Some may involve more direct and personal relationships (called gemeinschafts), whereas others produce more fragmented and impersonal relationships (called gesellschafts).

Industrial Societies

An *industrial society* is one that uses machines and other advanced technologies to produce and distribute goods and services. The Industrial Revolution began over 250 years ago when the steam engine was invented in England, delivering previously unattainable amounts of mechanical power for the performance of work. Steam engines powered locomotives, factories, and dynamos and transformed societies as the Industrial Revolution spread. The growth of science led to advances in farming techniques such as crop rotation, harvesting, and ginning cotton, as well as industrial-scale projects such as dams for generating hydroelectric power. Joining these advances were developments in medicine, new

MAP 5.1

Mapping America's Diversity: Population Density

As this map shows, population density (measured as the number of people per square mile) varies enormously in different regions and areas of the country. In what ways do you think the density of a given area might affect people's social interaction?

Source: U.S. Census Bureau. 2009. *American FactFinder.* **www.census.gov**

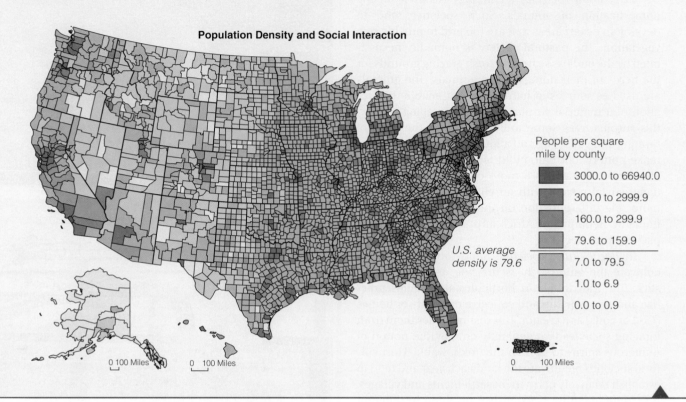

Population Density and Social Interaction

People per square mile by county

- 3000.0 to 66940.0
- 300.0 to 2999.9
- 160.0 to 299.9
- 79.6 to 159.9

U.S. average density is 79.6

- 7.0 to 79.5
- 1.0 to 6.9
- 0.0 to 0.9

0 100 Miles   0 100 Miles   0   100 Miles

techniques to prolong and improve life, and the emergence of birth control to limit population growth.

Unlike agricultural societies, industrial societies rely on a highly differentiated labor force and the intensive use of capital and technology. Large formal organizations are common. The task of holding society together, falling on institutions such as religion in preindustrial societies, now falls more on the institutions that have a high division of labor, such as the economy and work, government, politics, and large bureaucracies.

Within industrial societies, the forms of gender inequality that we see in contemporary U.S. society tend to develop. With the advent of industrialization, societies move to a cash-based economy, with labor performed in factories and mills paid on a wage basis and household labor remaining unpaid. This introduced what is known as the *family-wage economy*, in which families become dependent on wages to support themselves, but work within the family (housework, child care, and other forms of household work) is unpaid and therefore increasingly devalued (Tilly and Scott 1978). In addition, even though women (and young children) worked in factories and mills from the first inception of industrialization, the family-wage economy is based on the idea that men are the primary breadwinners. A system of inequality in men's and women's wages was introduced—an economic system that even today continues to produce a wage gap between men and women.

Industrial societies tend to be highly productive economically, with a large working class of industrial laborers. People become increasingly urbanized as they move from farmlands to urban centers or other areas where factories are located. Immigration is common in industrial societies, particularly because industries are forming where there is a high demand for more, cheap labor.

Industrialization has brought many benefits to U.S. society—a highly productive and efficient economic system, expansion of international markets, extraordinary availability of consumer products, and for many, a good working wage. Industrialization has, at the same time, also produced some of the most serious social problems that our nation faces: industrial pollution, an overdependence on consumer goods, wage inequality and job dislocation for millions, and problems of crime and crowding in urban areas (see Map 5.1 on population density).

Postindustrial Societies

In the contemporary era, a new type of society is emerging. Whereas most twentieth-century societies can be characterized in terms of their generation of material goods, **postindustrial society** depends economically on the production and distribution of services, information, and knowledge. Postindustrial societies are information-based societies in which technology plays a vital role in the social organization. The United States is fast becoming a postindustrial society, and Japan may be even further along. Many of the workers provide services such as administration, education, legal services, scientific research, and banking, or they engage in the development, management, and distribution of information, particularly in the areas of computer use and design. Central to the economy of the postindustrial society are the highly advanced technologies of computers, robotics, and genetic engineering. Multinational corporations globally link the economies of postindustrial societies.

The transition to a postindustrial society has a strong influence on the character of social institutions. Educational institutions acquire paramount importance in the postindustrial society, and science takes an especially prominent place. For some, the transition to a postindustrial society means more discretionary income for leisure activities—tourism, entertainment, and relaxation industries (spas, massage centers, and exercise) become more prominent—at least for people in certain classes and in the absence of severe economic recession, which has recently plagued not only the United States but Japan, Germany, France, Greece, and other technologically advanced countries as well. As with the United States in the last several years, the transition to postindustrialism has meant permanent joblessness for many. For others, it has meant the need to hold down more than one job simply to make ends meet.

SOCIAL INTERACTION AND SOCIETY

You can see by now that society is an entity that exists above and beyond individuals. Also, different societies are marked by different forms of *social organization*. Although societies differ, emerge, and change, they are also highly predictable. Your society shapes virtually every aspect of your life from the structure of its social institutions to the more immediate ways that you interact with people. It is to that level—the microlevel of society—that we now turn.

Groups

At the microlevel, society is made up of many different social groups. At any given moment, each of us is a member of many groups simultaneously, and we are subject to their influence: family, friendship groups, athletic teams, work groups, racial and ethnic groups, and so on. Groups impinge on every aspect of our lives and are a major determinant of our attitudes and values regarding everything from personal issues such as sexual attitudes and family values to major social issues such as the death penalty and physician-assisted suicide.

To sociologists, a **group** is a collection of individuals who

- interact and communicate with each other;
- share goals and norms; and,
- have a subjective awareness of themselves as "we," that is, as a distinct social unit.

To be a group, the social unit in question must possess all three of these characteristics. We will examine the nature and behavior of groups in greater detail in Chapter 6.

In sociological terms, not all collections of people are groups. People may be lumped together into *social categories* based on one or more shared characteristics, such as teenagers (an age category), truck drivers (an occupational category), and even those who have lost their life savings and pensions as a result of criminal Ponzi investment schemes, such as occurred in the fall of 2008, when many unknowingly invested money with the now-infamous criminal Bernard Madoff (more about him, similar others, and Ponzi investment schemes in Chapter 7).

Social categories can become social groups, depending on the amount of "we" feeling the group has. Only when there is this sense of common identity, as defined in the previous characteristics of groups, is a collection of people an actual group. For example, all people nationwide watching TV programs at 8 o'clock Wednesday evening form a distinct social unit, an *audience*. But they are not a group because they do not interact with one another, nor do they possess an awareness of themselves as "we." However, if many viewers were to come together for a convention where they could interact and develop a "we" feeling, such as do fans of the long-running book and movie series about Harry Potter, they would constitute a group.

We now know that people do not need to be face-to-face to constitute a group. Online communities, for example, are people who interact with each other regularly, share a common identity, and think of themselves as being a distinct social unit. On the Internet community Facebook, for example, you may have a group of "friends," some of whom you know personally and others whom you only know online. But these *friends*, as they are known on Facebook, make up a social group that might interact on a regular, indeed, daily basis—possibly even across great distances.

Groups also need not be small or "close-up" and personal. *Formal organizations* are highly structured social groupings that form to pursue a set of goals. Bureaucracies such as business corporations or municipal governments or associations such as the Parent-Teacher Association (PTA) are examples of formal organizations. A deeper analysis of bureaucracies and formal organizations appears in Chapter 6.

Status

Within groups, people occupy different statuses. **Status** is an established position in a social structure that carries with it a degree of social rank or value. A status is a rank in society. For example, the position "vice president of the United States" is a status, one that carries relatively high prestige. "High school teacher" is another status; it carries less prestige than "vice president of the United States," but more prestige than, say, "cabdriver." Statuses occur within institutions and also within groups. "High school teacher" is a status within the education institution. Other statuses in the same institution are "student," "principal," and "school superintendent." Within a given group, people may occupy different statuses that can be dependent on a variety of factors, such as age or seniority within the group.

Typically, a person occupies many statuses simultaneously. The combination of statuses composes a **status set**, which is the complete set of statuses occupied by a person at a given time (a term originally introduced by sociological theorist Robert Merton [1968]). A person may occupy different statuses in different institutions. Simultaneously, a person may be a bank president (in the economic institution), voter (in the political institution), church member (in the religious institution), and treasurer of the PTA (in the education institution). Each status may be associated with a different level of prestige.

Sometimes the multiple statuses of an individual conflict with one another. **Status inconsistency** exists where the statuses occupied by a person bring with them significantly different amounts of prestige and thus differing expectations. For example, someone trained as a lawyer, but working as a cabdriver, experiences status inconsistency. Some recent immigrants from Vietnam and Korea have experienced status inconsistency. Many refugees who had been in high-status occupations in their home country, such as teachers, doctors, and lawyers, could find work in the United States only as grocers or technicians—jobs of relatively lower status than the jobs they left behind. A relatively large body of research in sociology has demonstrated that status inconsistency—in addition to low status itself—can lead to stress and depression (Taylor et al. 2013; Thoits 2009; Taylor and Hornung 1979; Lenski 1954).

Achieved statuses are those attained by virtue of individual effort. Most occupational statuses—police officer, pharmacist, or boat builder—are achieved statuses. In contrast, **ascribed statuses** are those occupied from the moment a person is born. Your biological sex is an ascribed status. Yet, even ascribed statuses are not exempt from the process of social construction.

For most individuals, race is an ascribed status fixed at birth, although an individual with one light-skinned African American parent and one White parent may appear to be White and may go through life as a White person. Within the African American community, this is called *passing*, although this term is used somewhat less often now than it was several years ago. Ascribed status may not be rigidly defined, as for individuals who define themselves as *biracial* or *multiracial* (see also Chapter 10). Finally, ascribed statuses can arise through means beyond an individual's control, such as severe disability or chronic illness.

Some seemingly ascribed statuses, such as gender, can become achieved statuses. Gender, typically thought of as fixed at birth, is a social construct. You can be born female or male (this is your sex), but becoming a woman or a man is the result of social behaviors associated with your ascribed status. In other words, gender is also achieved. People who cross-dress, have a sex change, or develop some characteristics associated with the other sex are good examples of how gender is achieved, but you do not have to see these exceptional behaviors to observe that. People "do" gender in everyday life. They put on appearances and behaviors that are associated with their presumed gender (Andersen 2011; West and Fenstermaker 1995; West and Zimmerman 1987). If you doubt this, ask yourself what you did today to "achieve" your gender status. Did you dress a certain way? Wear "manly" cologne or deodorant? Splash on a "feminine" fragrance? And so on. These behaviors—all performed at the microlevel—reflect the macrolevel of your gender status.

---

debunking SOCIETY'S MYTHS

MYTH: Gender is an *ascribed status* where one's gender identity is established at birth.

SOCIOLOGICAL PERSPECTIVE: Although one's biological sex identity is an ascribed status, gender is a social construct and thus is also an *achieved status*—that is, accomplished through routine, everyday behavior, including patterns of dress, speech, touch, and other social behaviors. Sex is not the same as gender (Andersen 2011). ●

The line between achieved and ascribed status can be hard to draw. Social class, for example, is determined by occupation, education, and annual income—all of which are achieved statuses—yet one's job, education, and income are known to correlate strongly with the social class of one's parents. Hence, one's social class status is at least partly—though not perfectly—determined at birth. It is an achieved status that includes an inseparable component of ascribed status as well.

Although people occupy many statuses at one time, it is usually the case that one status is dominant, called the **master status**, overriding all other features of the person's identity. The master status may be imposed by others, or a person may define his or her own master status. A woman judge, for example, may carry the master status "woman" in the eyes of many. She is seen not just as a judge, but also as a woman judge, thus making gender a master status (Webster and Hysom 1998). A master status can completely supplant all other statuses in someone's status set. Being in a wheelchair is another example of a master status. Consider, for example, the case of a person in a wheelchair who is at the same time a medical doctor, an author, and a painter. People will see the wheelchair, at least at first, as the most important, or salient, part of identity, ignoring other statuses that define someone as a person. For a time, that person will be known as "that wheelchair guy" or "that wheelchair doctor."

In role modeling, a person imitates the behavior of an admired other.

thinking SOCIOLOGICALLY

Make a list of terms that describe who you are. Which of these are *ascribed statuses* and which are *achieved statuses*? What do you think your *master status* is in the eyes of others? Does one's *master status* depend on who is defining you? What does this tell you about the significance of social judgments in determining who you are? ●

Roles

A **role** is the behavior others expect from a person associated with a particular status. Statuses are occupied; roles are acted or "played." The status of police officer carries with it many expectations; this is the role of police officer. Police officers are expected to uphold the law, pursue suspected criminals, assist victims of crimes, fill out forms for reports, and so on. Usually, people behave in their roles as others expect them to, but not always. When a police officer commits a crime, such as physically brutalizing someone, he or she has violated the role expectations. Role expectations may vary according to the role of the observer—whether the person observing the police officer is a member of a minority group, for example.

As we saw in Chapter 4, social learning theory predicts that we learn attitudes and behaviors in response to the positive reinforcement and encouragement received from those around us. This is important in the formation of our own identity in society. "I am Linda, the skater," or "I am John, the guitarist." These identities are often obtained through **role modeling**, a process by which we imitate the behavior of another person we admire who is in a particular role. A ten-year-old girl or boy who greatly admires the teenage expert skateboarder next door will attempt, through role modeling, to closely imitate the tricks that neighbor performs on the skateboard. As a result, the formation of the child's self-identity is significantly influenced.

A person may occupy several statuses and roles at one time. A person's **role set** includes all the roles occupied by the person at a given time. Thus a person may be not only Linda the skater but also Linda the student, the daughter, and the lover. Roles may clash with each other, a situation called **role conflict**, wherein two or more roles are associated with contradictory expectations. Notice that in Figure 5.1 some of the roles diagrammed for this college student may conflict with others. Can you speculate about which might and which might not?

In U.S. society, some of the most common forms of role conflict arise from the dual responsibilities of job

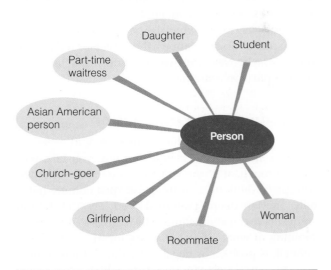

FIGURE 5.1 Roles in a College Student's Role Set
Identify the different roles that you occupy and draw a similar diagram of your own role set. Then identify which roles are consistent with each other and which might produce *role conflict* and *role strain*.

© Cengage Learning

and family. The parental role demands extensive time and commitment, and so does the role of worker. Time given to one role is time taken away from the other. Although the norms pertaining to workingwomen and workingmen are rapidly changing, it is still true that women are more often expected to uphold traditional role expectations associated with their gender role and are more likely held responsible for minding the family when job and family conflict. The sociologist Arlie Hochschild captured the predicament of today's women when she described the "second shift": an employed mother spends time and energy all day on the job, only to come home to the "second shift" of family and home responsibilities (Hochschild 2003, 1997, and Hochschild and Machung 1989).

Hochschild has found that some companies have instituted "family-friendly" policies, designed to reduce the conflicts generated by the "second shift." Ironically, however, in her study she found that few workers take advantage of programs such as more flexible hours, paid maternity leave, and job sharing—except for the on-site child care that actually allowed parents to work more!

Hochschild's studies point to the conflict between two social roles: family roles and work roles. Her research is also illustrative of a different sociological concept: **role strain**, a condition wherein a single role brings conflicting expectations. Different from role conflict, which involves tensions *between* two roles, role strain involves conflicts within a single role. In Hochschild's study, the work role has not only the expectations traditionally associated with work but also the expectation that one "love" one's work and be as devoted to it as to one's family. The result is role strain. The role of student also often involves role strain. For example, students are expected to be independent thinkers, yet they feel—quite correctly—that they are often required to simply repeat on an exam what a professor tells them. The tension between the two competing expectations is an example of role strain.

Everyday Social Interaction

You can also see the influence of society in everyday behavior, including such basics as how you talk, patterns of touch, and who you are attracted to. Although you might think of such things as "just coming naturally," they are deeply patterned by society. The cultural context of social interaction really matters in our understanding of what given behaviors mean. An action is defined as positive or negative by the cultural context because social behavior is that to which people give meaning. An action that is positive in one culture can be negative in another. For example, shaking the right hand in greeting is a positive action in the United States, but the same action in East India or certain Arab countries might be an insult. Social and cultural context matter. A kiss on the lips is a positive act in most cultures, yet

if a stranger kissed you on the lips, you would probably consider it a negative act, perhaps even repulsive.

Verbal and Nonverbal Communication. We saw in the culture chapter (Chapter 2) how patterns of social interaction are embedded in the language we use, and language is deeply influenced by culture and society. Furthermore, communication is not just what you say, but also how you say it and to whom. You can see the influence of society on *how* people speak, especially in different contexts. Under some circumstances, a pause in speaking may communicate emphasis; for others, it may indicate uncertainty. Cultural differences across society make this obvious. Thus, during interactions between Japanese businessmen, long periods of silence often occur. Unlike U.S. citizens, who are experts in "small talk" and who try at all costs to avoid periods of silence in conversation, Japanese people do not need to talk all the time and regard periods of silence as desirable opportunities for collecting their thoughts (Worchel et al. 2000; Fukuda 1994). American businesspeople in their first meetings with Japanese executives often think, erroneously, that these silent interludes mean the Japanese are responding negatively to a presentation. Even though some find the Japanese mode of conversation highly uncomfortable, getting used to it is a key tool in successful negotiations.

Nonverbal communication is also a form of social interaction and can be seen in various social patterns. A surprisingly large portion of our everyday communication with others is nonverbal, although we are generally only conscious of a small fraction of the nonverbal "conversations" in which we take part. Consider all the nonverbal signals exchanged in a casual chat: body position, head nods, eye contact, facial expressions, touching, and so on. As noted just previously, the length of a silence period during a conversation is itself a type of nonverbal communication. Studies of nonverbal communication, like those of verbal communication, show that it is much influenced by social forces, including the relationships between diverse groups of people. The meanings of nonverbal communications depend heavily on race, ethnicity, social class, and gender, as we shall see.

For example, patterns of touch (called **tactile communication**) are strongly influenced by gender. Parents vary their touching behavior depending on whether the child is a boy or a girl. Boys tend to be touched more roughly; girls, more tenderly and protectively. Such patterns continue into adulthood, where women touch each other more often in everyday conversation than do men. Women are on the average more likely to touch and hug as an expression of emotional support, whereas men touch and hug more often to assert power or to express sexual interest (Baumeister and Bushman 2008; Worchel et al. 2000). Clearly, there are also instances where women touch to express sexual interest and/or dominance, but research shows that, in general, touch is a supportive activity for women. For men, touch is often a dominance-asserting activity, except in athletic contexts where hugging and patting among men is a supportive activity (Worchel et al. 2000).

In observing patterns of touch, you can see where social status influences the meaning of nonverbal behaviors. Professors, male or female, may pat a man or woman student on the back as a gesture of approval; students will rarely do this to a professor. Male professors touch students more often than do female professors, showing the additional effect of gender. Because such patterns of touching reflect power relationships between women and men, they can also be offensive and may even involve sexual harassment (see Chapters 11 and 15).

You can also see the social meaning of interaction by observing how people use personal space. **Proxemic communication** refers to the amount of space between interacting individuals. Although people are generally unaware of how they use personal space, usually the more friendly people feel toward each other, the closer they will stand. In casual conversation, friends stand closer to each other than do strangers. People who are sexually attracted to each other stand especially close, whether the sexual attraction is gay, lesbian, or heterosexual. According to anthropologist E. T. Hall (1966; Hall and Hall 1987), we all carry around us a *proxemic bubble* that represents our personal, three-dimensional space. When people we do not know enter our proxemic bubble, we feel threatened and may take evasive action. Friends stand close; enemies tend to avoid interaction and keep far apart. According to Hall's theory, we attempt to exclude from our private space those whom we do not know or do not like, even though we may not be fully aware that we are doing so.

Patterns of touch reflect differences in the power that is part of many social relationships.

The proxemic bubbles of different ethnic groups on average have different sizes. Hispanic people tend to stand much closer to each other than do White, middle-class Americans; their proxemic bubble is, on average, smaller. Similarly, African Americans also tend to stand close to each other while conversing. Interaction distance is quite large between White, middle-class, British males—their average interaction distances can be as much as several feet.

Proxemic interactions also differ between men and women (Taylor et al. 2013; Romain 1999; Tannen 1990). Women of the same race and culture tend to stand closer to each other in casual conversation than do men of the same race and culture. When a Middle Eastern man (who has a relatively small proxemic bubble) engages in conversation with a White, middle-class, U.S. man (who has a larger proxemic bubble), the Middle Eastern man tends to move toward the White American, who tends to back away. You can observe the negotiations of proxemic space at cocktail parties or any other setting that involves casual social interaction.

In a society as diverse as the United States, understanding how diversity shapes social interaction is an essential part of understanding human behavior. Ignorance of the meanings that gestures have in a society can get you in trouble. For example, some Mexicans and Mexican Americans may display the right hand held up, palm inward, all fingers extended, as an obscene gesture meaning "screw you many times over." This provocative gesture has no meaning at all in Anglo (White) society. (However, giving someone "the finger"— middle finger extended ("flipping the bird")—certainly does carry meaning in not only Anglo society but also in Latin society and many other societies as well! It is a bit of nonverbal communication that is nearly universally recognized.)

see FOR YOURSELF

Riding in Elevators

1. Try a simple experiment. Ride in an elevator and closely observe the behavior of everyone in the elevator with you. Write down in a notebook such things as how far away people stand from each other. Note the differences carefully, even in estimated inches. What do they look at? Do they tend to stand in the corners? Do they converse with strangers or the people they are with? If so, what do they talk about?

2. Now return to the same elevator and do something that breaks the usual norms of elevator behavior, such as standing too close to someone. (You will have to get up a lot of nerve to do this!) How did people react? What did they do? How did you feel?

How does this experiment show how social norms are maintained through informal norms of social control? ●

Likewise, people who grow up in urban environments learn to avoid eye contact on the streets. Staring at someone for only two or three seconds can be interpreted as a hostile act, if done man to man (Anderson 1999, 1990). If a woman maintains mutual eye contact with a male stranger for more than two or three seconds, she may be assumed by the man to be sexually interested in him. In contrast, during sustained conversation with acquaintances, women maintain mutual eye contact longer than do men (Romain 1999).

Interpersonal Attraction

We have already asked, "What holds society together?" This was asked at the macroanalysis level—that is, the level of society. But what holds relationships together—or, for that matter, makes them fall apart? You will not be surprised to learn that formation of relationships has a strong social structural component—that is, it is patterned by social forces and can to a great extent be predicted.

Humans have a powerful desire to be with other human beings; in other words, they have a strong need for *affiliation*. We tend to spend about 75 percent of our time with other people when doing all sorts of activities—eating, watching television, studying, doing hobbies, working, and so on (Cassidy and Shaver 1999). People who lack all forms of human contact are very rare in the general population, and their isolation is usually rooted in psychotic or schizophrenic disorders. Extreme social isolation at an early age causes severe disruption of mental, emotional, and language development, as we saw in Chapter 4.

Konrad Lorenz, the animal behaviorist, shows that adult Graylag geese that have *imprinted* on him the moment they were hatched will follow him anywhere, as though he were their mother goose (from Tweed Roosevelt, personal communication)!

The affiliation tendency has been likened to **imprinting**, a phenomenon seen in newborn or newly hatched animals who attach themselves to the first living creature they encounter, even if it is of another species (Lorenz 1966). Studies of geese and squirrels show that once the young animal attaches itself to a human experimenter, the process is irreversible. The young animal prefers the company of the human to the company of its own species! A degree of imprinting may be discernible in human infant attachment, but researchers note that the process is more complex, more changeable, and more influenced by social factors in infants.

Somewhat similar to affiliation is *interpersonal attraction*, a nonspecific positive response toward another person. Attraction occurs in ordinary day-to-day interaction and varies from mild attraction (such as thinking your grocer is a "nice person") all the way to deep feelings of love. According to one view, attractions fall on a continuum ranging from hate to strong dislike to mild dislike to mild liking to strong liking to love. Another view is that attraction and love are two different continua, able to exist separately. In this view, you can actually like someone a whole lot, but not be in love. Conversely, you can feel passionate love for someone, including strong sexual feelings and intense emotion, yet not really "like" the person. Have you ever been in love with someone you did not particularly like?

Can attraction be scientifically predicted? Can you identify with whom you are most likely to fall in love? The surprising answer to these questions is a loud, although somewhat qualified, "yes." Most of us have been raised to believe that love is impossible to measure and certainly impossible to predict scientifically. We think of love, especially romantic love, as quick and mysterious—a lightning bolt. Couples report falling in love at first sight, thinking that they were "meant for each other" (McCollum 2002). Countless novels and stories support this view, but extensive research in sociology and social psychology suggests otherwise: In a probabilistic sense, love can be predicted beyond the level of pure chance. Let us take a look at some of these intriguing findings.

A strong determinant of your attraction to others is simply whether you live near them, work next to them, or have frequent contact with them. (This is a *proxemic* determinant.) You are more likely to form friendships with people from your own city than with people from a thousand miles away. One classic study even showed that you are more likely to be attracted to someone on your floor, your residence hall, or your apartment building than to someone even two floors down or two streets over (Festinger et al. 1950). Subsequent studies continue to show this effect (Baumeister and Bushman 2008). Such is the effect of proximity in the formation of human friendships.

Now, though the general principle still holds, many people form relationships without being in close

Romantic love is idealized in this society as something that "just happens," but research shows that interpersonal attraction follows predictable patterns.

Jacqueline Veissid/Photodisc/Jupiter Images

We hear that "beauty is only skin deep." Apparently, that is deep enough. To a surprisingly large degree, the attractions we feel toward people of either gender are based on our perception of their physical attractiveness (Baumeister and Bushman 2008). A vast amount of research over the years has consistently shown the importance of attractiveness in human interactions: Adults react more leniently to the bad behavior of an attractive child than to the same behavior of an unattractive child (Berscheid and Reis 1998). Teachers evaluate cute children of either gender as "smarter" than unattractive children with identical academic records (Worchel et al. 2000). In studies of mock jury trials, attractive defendants, male or female, receive lighter jury-recommended sentences on average than do unattractive defendants convicted of the same crime (Gilbert et al. 1998).

Of course, standards of attractiveness vary between cultures and between subcultures within the same society. What is highly attractive in one culture may be repulsive in another. In the United States, there is a maxim that you can never be too thin—a major cause of eating disorders such as *anorexia* and *bulimia*, especially among White women (Hesse-Biber 2007). The maxim is oppressive for women in U.S. society, yet it is clearly highly culturally relative, even within U.S. culture. What is considered "overweight" or "fat" is indeed a social construction (Atkins 2011). Among many African Americans, "chubbiness" (itself a social construction) in women is considered attractive. Such women are called "healthy" and "phatt" or "thick," which means the same as "stacked" or curvaceous.

Similar cultural norms often apply in certain U.S. Hispanic populations. The skinny woman is not considered attractive. Nonetheless, studies show that anorexia and bulimia are now increasing among women of color, showing how cultural norms can change—even though Black women, in general, are more satisfied with their body image than White women (Atkins 2011; Lovejoy 2001; Fitzgibbon and Stolley 2002; see also Chapter 14).

Studies of dating patterns among college students show that the more attractive one is, the more likely one will be asked on a date. This applies to gay and lesbian dating as well as to heterosexual dating (Berscheid and Reiss 1998). However, one very important exception can be added to this finding: Physical attractiveness predicts only the early stages of a relationship. When one measures relationships that last a while, other factors come into play, principally religion, political attitudes, social class background, educational aspirations, and race. Perceived physical attractiveness may predict who is attracted to whom initially, but other variables are better predictors of how long a relationship will last.

So, do "opposites attract"? Not according to the research. We have all heard that people are attracted to

proximity, such as in online dating. Studies of Internet dating show that, even in this cyberworld, social norms still apply. Studies of Internet dating find, for example, that unlike other dating behavior, there is pressure to disclose more secrets about oneself in a shorter period of time on the Internet (Lawson and Leck 2006).

Our attraction to another person is also greatly affected simply by how frequently we see that person or even his or her photograph. When watching a movie, have you ever noticed that the central character seems more attractive at the end of the movie than at the beginning? This is particularly true if you already find the person very attractive when the movie begins. Have you ever noticed that the fabulous-looking person sitting next to you in class looks better every day? You may be experiencing *mere exposure effect:* The more you see someone in person—or even in a photograph—the more you like him or her. In studies where people are repeatedly shown photographs of the same face, the more often a person sees a particular face, then other things being constant, the more he or she likes that person (Moreland and Beach 1992).

There are two qualifications to the effect. First, overexposure can result when a photograph is seen too often. The viewer becomes saturated and ceases to like the pictured person more with each exposure. Second, the initial response of the viewer can determine how much liking will increase. If someone starts out liking a particular person, seeing that person frequently will increase the liking for that person; however, if one starts out disliking the pictured person, the amount of dislike tends to remain about the same, regardless of how often one sees the person (Taylor et al. 2013).

their "opposite" in personality, social status, background, and other characteristics. Many of us grow up believing this to be true. However, if the research tells us one thing about interpersonal attraction, it is that with only a few exceptions we are attracted to people who are *similar* or *even identical* to us in socioeconomic status, race, ethnicity, religion, perceived personality traits, and general attitudes and opinions (Taylor et al. 2013; Baumeister and Bushman 2008; Brehm et al. 2002). "Dominant" people tend to be attracted to other dominant people, not to "submissive" people. "Verbally aggressive" people tend to be attracted to others who are also verbally aggressive and not to someone who is verbally withdrawn or verbally shy. Couples tend to have similar opinions about political issues of great importance to them, such as attitudes about abortion, crime, animal rights, gun violence, and whom to vote for as president. Overall, couples tend to exhibit strong cultural or subcultural similarity, not difference.

There are exceptions, of course. We sometimes fall in love with the *exotic*—the culturally or socially different. Novels and movies return endlessly to the story of the rich young woman who falls in love with a rough-and-ready biker, but such a pairing is by far the exception and not the rule. That rich young woman is far more likely to fall in love with a rich young man. When it comes to long-term relationships, including both friends and lovers (whether heterosexual, lesbian, gay, or bisexual), humans vastly prefer a great degree of similarity, even though, if asked, they might deny it. In fact, the less similar a heterosexual relationship is with respect to race, social class, age, and educational aspirations (how far in school the person wants to go), then the quicker the relationship is likely to break up (Silverthorne and Quinsey 2000; Worchel et al. 2000; Berscheid and Reis 1998).

Many young romantic relationships, regrettably, come to an end. On campus, relationships tend to break up most often during gaps in the school calendar, such as winter and spring breaks. Summers are especially brutal on relationships formed during the academic year. Breakups are seldom mutual. Almost always, only one member of the pair wants to break off the relationship, whereas the other wants to keep it going. The sad truth means that the next time you hear that a breakup was "mutual," you will know this is probably a lie or self-deception.

THEORIES ABOUT ANALYZING SOCIAL INTERACTION

Groups, statuses, and roles form a web of social interaction. Sociologists have developed different ways of conceptualizing and understanding social interaction. Functionalist theory, discussed in Chapter 1, is one such concept. Here we detail four others: the social construction of reality, ethnomethodology, impression management, social exchange and game theory (refer to Table 5.2). The first three theories come directly from the symbolic interaction perspective.

table 5.2 Theories of Social Interaction

	The Social Construction of Reality	Ethnomethodology	Dramaturgy	Social Exchange Theory	Game Theory
Interprets society as:	Organized around the subjective meaning that people give to social behavior	Held together through the consensus that people share around social norms; you can discover these norms by violating them	A stage on which actors play their social roles and give impression to those in their "audience"	A series of interactions that are based on estimates of rewards and punishments	A system in which people strategize "winning" and "losing" in their interactions with each other
Analyzes social interaction as:	Based on the meaning people give to, or attribute to, actions in society	A series of encounters in which people manage their impressions in front of others	Enactment of social roles played before a social audience	A rational balancing act involving perceived costs and benefits of a given behavior	Calculated risks to balance rewards and punishments

© Cengage Learning

The Social Construction of Reality

What holds society together? This is a basic question for sociologists, one that, as we have seen, has long guided sociological thinking. Sociologists note that society cannot hold together without something that is shared—a shared social reality.

Some sociological theorists have argued convincingly that there is little actual reality beyond that produced by the process of social interaction itself. This is the principle of the *social construction of reality*, the idea that our perception of what is real is determined by the subjective meaning that we attribute to an experience, a principle central to symbolic interaction theory (Blumer 1969; Berger and Luckmann 1967). Hence, there is no objective "reality" in itself. Things do not have their own intrinsic meaning. We subjectively *impose* meaning on things.

Children do this routinely—impose inherent meaning on things. Upon seeing a marble roll off a table, the child attributes causation (meaning) to the marble: The marble rolled off the table "because it wanted to." Such perceptions carry into adulthood: The man walking down the street who accidentally walks smack into a telephone pole, at first thought glares at the pole, as though the pole somehow caused the accident! He inadvertently attributes causation and meaning to an inanimate object—the telephone pole (Heider 1958).

Considerable evidence exists that people do just that; they force meaning on something when doing so allows them to see or perceive what they want to perceive—even if that perception seems to someone else to be contrary to actual fact. They then come to believe that what they perceived is indeed "fact." A classic and convincing study of this is Hastorf and Cantril's (1954) study of Princeton and Dartmouth students who watched a film of a game of basketball between the two schools. Both sets of students watched the same film. The students were instructed to watch carefully for rule infractions by each team. The results were that the Princeton students reported twice as many rule infractions involving the Dartmouth team as the Dartmouth students saw. The Dartmouth students saw about twice as many rule infractions by Princeton as the Princeton students saw! Remember that they all saw exactly the same game—the same "facts." We see the "facts" we want to see, as a result of the social construction of reality. Subsequent research has strongly supported the Hastorf and Cantril findings (Taylor et al. 2013; Baumeister and Bushman 2008; Ross 1977; Jones and Nisbett 1972).

As we saw in Chapter 1, our perceptions of reality are determined by what is called the *definition of the situation*: We observe the context in which we find ourselves and then adjust our attitudes and perceptions accordingly. Sociological theorist W. I. Thomas embodies this idea in his well-known dictum that *situations defined as real are real in their consequences* (Thomas 1966/1931). The Princeton and Dartmouth students saw different "realities" depending on what college they were attending, and the consequences (the perceived rule infractions) were very real to them.

The definition of the situation is a principle that can also affect a "factual" event such as whether an emergency room patient is perceived to be dead by the doctors. In his insightful research in the emergency room of a hospital, Sudnow (1967) found that patients who arrived at the emergency room with no discernible heartbeat or breathing were treated differently by the attending physician depending on the patient's age. A person in his or her early twenties or younger was not immediately pronounced "dead on arrival" (DOA). Instead, the physicians spent a lot of time listening for and testing for a heartbeat, stimulating the heart, examining the patient's eyes, giving oxygen, and administering other stimulation to revive the patient. If the doctor obtained no lifelike responses, the patient was pronounced dead. Older patients, however, were on the average less likely to receive such extensive procedures. The older person was examined less thoroughly and often was pronounced dead on the spot with only a brief stethoscopic examination of the heart. In such instances, how the physicians defined the situation—how they socially constructed the reality of death—was certainly real in its consequence for the patient!

Ethnomethodology

Our interactions are guided by rules that we follow. Sometimes these rules are nonobvious and subtle. These rules are the *norms* of social interaction. Again, what holds society together? Society cannot hold together without norms, but what rules do we follow? How do we know what these rules or norms are? An approach in sociology called *ethnomethodology* is a clever technique for finding out.

Ethnomethodology (Garfinkel 1967), after *ethno* for "people" and *methodology* for "mode of study," is a technique for studying human interaction *by deliberately disrupting social norms and observing how individuals attempt to restore normalcy*. The idea is that to study such norms, one must first break them, because the subsequent behavior of the people involved will reveal just what the norms were in the first place. In the "See For Yourself" elevator example you were asked to perform previously (see page 111), an application of ethnomethodology would be standing too close to someone on the elevator (this is the norm violation) and observing what that person does as a result (which would be the norm-restoration behavior).

Ethnomethodology is based on the premise that human interaction takes place within a consensus, and interaction is not possible without this consensus. The consensus is part of what holds society together. According to Garfinkel, this consensus will be revealed by people's *background expectancies*, namely, the norms for behavior that they carry with them into situations of interaction. It is presumed that these expectancies are to a great degree shared, and thus studying norms by deliberately violating them will reveal the norms that most people bring with them into interaction. The ethnomethodologist argues that you cannot simply walk up to someone and ask what norms the person has and uses, because most people will not be able to articulate them. We are not wholly conscious of what norms we use even though they are shared. Ethnomethodology is designed to "uncover" those norms.

The recently aired CNN TV program called "What Would You Do?" employs what is in effect ethnomethodology, though in a nonsystematic and relatively uncontrolled way. For example, in one episode, a father is seen in a restaurant very loudly scolding his own small child for accidentally dropping a few crumbs on the floor. The extremely loud scolding represents a norm violation in this context. The father is in alliance with the TV producers. The point is to see what the observing people in the restaurant do, namely, engage in what the ethnomethodologist would call norm-restoration behavior. They found that many people looked but did not intervene. A few did intervene, such as by asking the father why he was so loud, saying that his punishment was too severe.

Ethnomethodologists in actual research often use ingenious procedures for uncovering norms by thinking up clever ways to interrupt "normal" interaction. In a clever study, sociology professor William Gamson had one of his students go into a grocery store where jelly beans, normally priced at that time at 49 cents per pound, were on sale for 35 cents. The student engaged the saleswoman in conversation about the various candies and then asked for a pound of jelly beans. The saleswoman then wrapped them and asked for 35 cents. The rest of the conversation went like this:

> **Student:** Oh, only 35 cents for all those nice jelly beans? There are so many of them. I think I will pay $1 for them.
>
> **Saleswoman:** Yes, there are a lot, and today they are on sale for only 35 cents.
>
> **Student:** I know they are on sale, but I want to pay $1 for them. I just love jelly beans, and they are worth a lot to me.
>
> **Saleswoman:** Well, uh, no, you see, they are selling for 35 cents today, and you wanted a pound, and they are 35 cents a pound.
>
> **Student:** (voice rising) I am perfectly capable of seeing that they are on sale at 35 cents a pound. That has nothing to do with it. It is just that I personally feel that they are worth more, and I want to pay more for them.
>
> **Saleswoman:** (becoming quite angry) What is the matter with you? Are you crazy or something? Everything in this store is priced more than what it is worth. Those jelly beans probably cost the store only a nickel. Now do you want them or should I put them back?

At this point, the student became quite embarrassed, paid the 35 cents, and hurriedly left (Gamson and Modigliani 1974).

The point here is that the saleswoman approached the situation with a presumed normative consensus, a consensus that became revealed by its deliberate violation by the student. The puzzled saleswoman took measures to attempt to normalize the interaction, even to *force* it to be normal (see Table 5.2).

Impression Management and Dramaturgy

Another way of analyzing social interaction is to study *impression management*, a term coined by symbolic interaction theorist Erving Goffman (1959). **Impression management** is a process by which people control how others perceive them. A student handing in a term paper late may wish to give the instructor the impression that it was not the student's fault but was because of uncontrollable circumstances ("my computer hard drive crashed," "my dog ate the last hard copy," and so on). The impression that one wishes to "give off" (to use Goffman's phrase) is that "I am usually a very diligent person, but today—just today—I have been betrayed by circumstances."

Impression management can be seen as a type of con game. *We willfully attempt to manipulate others' impressions of us.* Goffman regarded everyday interaction as a series of attempts to con the other. In fact, trying in various ways to con the other is, according to Goffman, at the very center of much social interaction and social organization in society: Social interaction is just a big con game!

Perhaps this cynical view is not true of all social interaction, but we do present different "selves" to others in different settings. The settings are, in effect, different stages on which we act as we relate to others. For this reason, Goffman's theory is sometimes called the *dramaturgy model* of social interaction, a way of analyzing interaction that assumes the participants are actors on a stage in the drama of everyday social life. People present different faces (give off different impressions) on different stages (in different situations or different roles) with different others. To your mother, you may present yourself as the dutiful, obedient daughter, which may not be how you present yourself to a friend. Perhaps you think acting like a diligent student makes you seem a jerk, so you hide from your friends that you are really interested in a class or enjoy your homework. Analyzing

DOING sociological research

Doing Hair, Doing Class

Research Question: Sociologist Debra Gimlin was curious about a common site for social interaction—hair salons. She noticed that the interaction that occurs in hair salons is often marked by differences in the social class status of clients and stylists. Her research question was: How do women attempt to cultivate the cultural ideals of beauty, and in particular, how is this achieved through the interaction between hair stylists and their clients?

Research Method: She did her research by spending more than 200 hours observing social interaction in a hair salon. She watched the interaction between clients and stylists and conducted interviews with the owner, the staff, and twenty women customers. During the course of her fieldwork, she recorded her observations of the conversations and interaction in the salon, frequently asking questions of patrons and staff. The patrons were mostly middle and upper-middle class; the stylists, working class. All the stylists were White, as were most of the clients.

Research Results: "Beauty work" as Gimlin calls it, involves the stylist bridging the gap between those who seek beauty and those who define it; her (or his) role is to be the expert in beauty culture,

bringing the latest fashion and technique to clients. Beauticians are also expected to engage in some "emotion work"—that is, they are expected to nurture clients and be interested in their lives; often they are put in the position of sacrificing their professional expertise to meet clients' wishes.

According to Gimlin, because stylists typically have lower class status than their clients, this introduces an element into the relationship that stylists negotiate carefully in their routine social interaction. Hairdressers emphasize their special knowledge of beauty and taste as a way of reducing the status differences between themselves and their clients. In this way, they manage the impressions their clients are thought to have. They also try to nullify the existing class hierarchy by conceiving an alternative hierarchy, not one based on education, income, or occupation but only on the ability to style hair competently. Thus stylists describe clients as perhaps "having a ton of money," but unable to do their hair or know what looks best on them. Stylists become confidantes with clients, who often tell them highly personal information about their lives—another attempt at impression management. Appearing to create personal relationships with their clients, even

though they never see them outside the salon, also reduces status differences.

Conclusions and Implications: Gimlin concludes that beauty ideals are shaped in this society by an awareness of social location and cultural distinctions. As she says, "Beauty is ... one tool women use as they make claims to particular social statuses" (1996: 525).

Questions to Consider
The next time you get your hair cut, you might observe the social interaction around you and ask how class, gender, and race shape interaction in the salon or barbershop that you use. Try to get someone in class to collaborate with you so that you can compare observations in different salon settings. In doing so, you will be studying how gender, race, and class shape social interaction in everyday life.

1. Would you expect the same dynamic in a salon where men are the stylists?
2. Do Gimlin's findings hold in settings where the customers and stylists are not White or where they are all working class?
3. In your opinion, would Gimlin's findings hold in an African American men's barbershop?

Source: Gimlin, Debra. 1996. "Pamela's Place: Power and Negotiation in the Hair Salon." *Gender & Society* 10 (October): 505–526.

---

impression management reveals that we try to con the other into perceiving us as we want to be perceived. The box "Doing Sociological Research: Doing Hair, Doing Class" shows how impression management can be involved in many settings, including the everyday world of the hair salon.

A study by Albas and Albas (1988) demonstrates just how pervasive impression management is in social interaction. The Albases studied how college students interacted with one another when the instructor returned graded papers during class. Some students got good grades ("aces"), others got poor grades ("bombers"), but both employed a variety of devices (cons) to maintain or give off a favorable impression. For example, the aces wanted to

show off their grades, but they did not want to appear to be braggarts, so they casually or "accidentally" let others see their papers, such as by dropping them on the floor, face up. In contrast, the bombers hid or covered their papers to hide their poor grades, said they "didn't care" what they got, or simply lied about their grades.

One thing that Goffman's theory makes clear is that social interaction is a very perilous undertaking. Have you ever been *embarrassed*? Of course you have; we all have. Think of a really big embarrassment that you experienced. Goffman defines embarrassment as a spontaneous reaction to a sudden or transitory challenge to our identity: We attempt to restore a prior perception of our "self" by others. Perhaps you were

The Congress and Game Theory

Members of the U. S. Congress often bitterly debate issues—such as the intense budget debates in both the House and the Senate immediately following the election of President Obama at the end of 2012. Sociologists have noted that such debates inevitably involve game-like trade-offs between members of Congress. A sociologist might thus use some form of social exchange theory, or game theory, to analyze such acrimonious verbal exchanges. Does such a given interchange constitute a zero-sum game, or a non-zero sum game? A zero-sum game would be exemplified by, say, the Republican Speaker of the House promising to deliver 15 votes "for" issue X if his "opponent" (say, the Democratic Minority Leader) agreed to "give up" 15 votes.

giving a talk before a class and then suddenly forgot the rest of the talk. Or perhaps you recently bent over and split your pants. Or perhaps you are a man and barged accidentally into a women's bathroom. All these actions will result in embarrassment, causing you to "lose face."

You will then attempt to *restore face* ("save face"), that is, eliminate the conditions causing the embarrassment. You thus will attempt to con others into perceiving you as they might have before the embarrassing incident. One way to do this is to shift blame from the self to some other, for example, claiming in the first example that the teacher did not give you time to adequately memorize the talk; or in the second example, claiming that you will never buy that particular, obviously inferior brand of pants again; or in the third example, claiming that the sign saying "Women's Room" was not clearly visible. All these represent deliberate manipulations (cons) to save face on your part—to restore the other's prior perception of you.

Social Exchange and Game Theory

Another way of analyzing social interaction is through the social exchange model (see Table 5.2). The *social exchange model* of social interaction holds that our interactions are determined by the rewards or punishments that we receive from others (Cook and Gervasi 2006; Wright 2000). A fundamental principle of exchange theory is that an interaction that elicits approval from another (a type of reward) is more likely to be repeated than an interaction that incites disapproval (a type of punishment). According to the exchange principle, one can predict whether a given interaction is likely to be repeated or continued by calculating the degree of reward or punishment inspired by the interaction.

Rewards can take many forms. They can include tangible gains such as gifts, recognition, and money, or subtle everyday rewards such as smiles, nods, and pats on the back. Similarly, punishments come in many varieties, from extremes such as public humiliation, beating, banishment, or execution, to gestures as subtle as a raised eyebrow or a frown. For example, if you ask someone out for a date and the person says yes, you have gained a reward, and you are likely to repeat the interaction. You are likely to ask the person out again, or to ask someone else out. If you ask someone out, and he or she glares at you and says, "No *kind* of way!," then you have elicited a punishment that will probably cause you to shy away from repeating this type of interaction with that person.

Social exchange theory has grown partly out of **game theory**, a mathematic and economic theory that predicts that human interaction has the characteristics of a "game," namely, strategies, winners and losers, rewards and punishments, and profits and costs (Stevens 2011; Kuhn and Nasar 2002; Wright 2000). Simply asking someone out for a date indeed has a gamelike aspect to it, and you will probably use some kind of strategy to "win" (have the other agree to go out with you) and "get rewarded" (have a pleasant or fun time) at minimal "cost" to you (you don't want to spend a large amount of money on the date or you do not want to get into an unpleasant argument on the date). The interesting thing about game theory is that it sees human interaction as just that: a game.

If in a given interchange between persons A and B the amount of reward to person A is exactly equal to the amount of loss to person B, then it is called a **zero-sum game** (reward plus loss will equal zero). A simple example would be person A receiving a $1,000 gift from person B—the reward to person A is the same amount as the loss to person B. To take another example, the game of poker is a zero-sum game: Person A's winnings exactly equal B's losses. This applies even if there are more than one "person Bs."

If on the other hand the amounts of reward and punishment for persons A and B are unequal, then it is a **non-zero sum game** (amount of reward plus amount of loss ≠ zero). If you, a male, ask a woman out for a date and she accepts, this is reward for you. But if she rejects your offer, this is punishment for you and either a neutral or even a reward for her! Hence what you get (punishment) and what she gets (neutrality or reward) do not sum to zero—unless of course she attains a hefty amount of glee from rejecting you, in which case it would then indeed be zero-sum! Otherwise, it is

DOING **sociological research**

The Prisoner's Dilemma Game

Research Method: The "prisoner's dilemma" is a classic "trade-off" game in the study of social interaction. Different varieties of it are often used in sociological studies of social interaction. The dilemma arises in a story about two criminals, whom we will call Bart and Mack. They are arrested on suspicion of having committed armed robbery. They are found by the police to be carrying concealed weapons, but they do not have enough evidence to link them to the robbery. Accordingly, the police question them *separately*. Both men are invited to confess to the crime and hence betray each other.

Research Question: What happens to either of them depends upon how each reacts. How do real people react if put in this prisoner's dilemma situation?

Research Results: This is a hypothetical exercise with different potential outcomes. One possibility is that neither confesses to the crime. This represents a *cooperative alternative* for each person. They cooperate with each other and reject the deals offered by the police. If this happens, both will get only a light prison sentence.

Another possibility is that one person will confess (the competitive alternative) while the other will not (the cooperative alternative). If Bart confesses and Mack does not, then the police will let Bart go free as a reward for testimony against Mack, who will get a long prison sentence. The outcomes are reversed if Bart does not confess and Mack does.

The last possible result is that both confess. In this case, both receive moderate prison terms. The "dilemma" is thus whether to confess and betray your partner, or hold out (not confess) and cooperate with him. How might you run an experiment to examine these different potential outcomes?

Conclusions and Implications: What real-life situations represent situations like the prisoner's dilemma. What about arguments with your brother or sister when you were growing up? Here is a good one: What if both you and a friend have cheated on a final exam? Should you "tell on" him or her to the instructor? Should your friend tell on you?

Questions to Consider

1. Do you think that the results would be the same, or different, if both of the subjects in the prisoner's dilemma experiment were of opposite gender instead of the same gender? Speculate about it.

2. What if they were of the same gender but of different races—one Black, one White? Speculate.

Source: Baumeister, Roy F., and Brad J. Bushman. 2008. *Social Psychology and Human Nature*. Belmont, CA: Thompson/Wadsworth.

---

non-zero sum. We thus see that the "game of love" is indeed a game, whether zero-sum or non-zero sum.

INTERACTION IN CYBERSPACE

When people interact and communicate with one another by means of personal computers—through some virtual community such as email, Twitter, Facebook, LinkedIn, and the like or other computer-to-computer interactions—then they are engaging in **cyberspace interaction** (or virtual interaction).

The character of cyberspace interaction is changing rapidly as new technologies emerge. Not long ago, nonverbal interaction was absent in cyberspace as people could not "see" what others were like. But with the introduction of video-based cyberspace, such as photos on Facebook and MySpace, and Skype, people can now display still and moving images of themselves. These images provide new opportunities, as we noted previously, for what sociologists would call the presentation of self and impression management. Sometimes this comes with embarrassing consequences. The young college student who displays a seminude or nude photo of herself or himself, projecting a sexual presentation of self, may be horrified if one of the parents or a potential employer visits the Facebook site! Furthermore, the photo could be intercepted by a disgruntled boyfriend, reproduced, and made to "go viral" (seen by hundreds or thousands of people). Cyberspace interaction is becoming increasingly common among all age, gender, and race groups, although clear patterns are also present in who is engaged in this form of social interaction and how people use it (Hargittai 2008; see Table 5.3). Women, for example, used to lag behind men in Internet usage but have now caught up. Internet usage is also related to race (Whites have the most usage but not by a large margin); age (youngest use it most); annual earnings (those with highest earnings use it most); education (more education means more usage); and location of residence being urban, suburban, or rural (rural residents use it the least).

Although women and men are roughly the same overall in Internet usage (see Table 5.3), gender differences can still be found in the *type* of usage. Women are more likely to use email to write to friends and family, share news, plan events, and forward jokes. And women are more likely to report that email nurtures their relationships. Men, on the other hand, use the Internet more to transact business, and they look for a wider array of information than women do. Men are also more likely to use the Internet for hobbies, including such things as sports fantasy leagues, downloading music, and listening to radio. Men more than women in the age 18 to 24 range tend to use social networking sites to make new friends and to flirt, whereas women in this age range are somewhat more likely to use it to stay in touch with friends that they already have (Pew Internet and American Life Project 2012).

table 5.3 Demographics of Internet Users

Following is the percentage of each group who use the Internet according to a 2012 survey.

Total Adults	
Men	81%
Women	81%
Race/Ethnicity	
White, non-Hispanic	83%
Black, non-Hispanic	74%
Hispanic (English-speaking)	73%
Age	
18–29	95%
30–49	89%
50–64	77%
65+	52%
Household Income	
Less than $30,000 per year	68%
$30,000 to $49,999	86%
$50,000 to $74,999	95%
$75,000+	97%
Educational Attainment	
Less than high school	47%
High school	72%
Some college	90%
College+	96%

Source: The Pew Internet and American Life Project. August–September, 2012. http://pewinternet.com/

It is too early to know the implications of these cyberspace interactions. Some think it will make social life more alienating, with people developing weaker social skills and less ability for successful face-to-face interaction. Some studies have noted that people can develop extremely close and in-depth relationships as a result of their interaction in cyberspace (Hargittai 2008). The Internet also creates more opportunity for people to misrepresent themselves or even create completely false—or even stolen—identities. But studies find that computer-mediated interactions also follow some of the same patterns that are found in face-to-face interaction. People still "manage" identities in front of a presumed audience; they project images of self to others that are consistent with the identity they have created for themselves, and they form social networks that become the source for evolving identities, just as people do in traditional forms of social interaction.

In this respect, cyberspace interaction is an application of Goffman's principle of *impression management*. The person can put forward a totally different and wholly created self, or identity. One can "give off," in Goffman's terms, any impression one wishes and, at the same time, know that one's true self is protected by anonymity. This gives the individual quite a large and free range of roles and identities from which to choose. As predicted by symbolic interaction theory, of which Goffman's is one variety, *the reality of the situation grows out of the interaction process itself.* This is a central point of symbolic interaction theory and is central to sociological analysis generally: Interaction creates reality.

Cyberspace interaction has thus resulted in new forms of social interaction in society—in fact, a new social order containing both deviants and conformists. These new forms of social interaction have their own rules and norms, their own language, their own sets of beliefs, and practices or rituals—in short, all the elements of culture, as defined in Chapter 2. For sociologists, cyberspace also provides an intriguing new venue in which to study the connection between society and social interaction.

chapter summary

What is society?
Society is a system of social interaction that includes both culture and *social organization*. Society includes *social institutions*, or established organized social behavior, and exists for a recognized purpose; *social structure* is the patterned relationships within a society.

What holds society together?
According to theorist Emile Durkheim, society with all its complex social organization and culture, is held together, depending on overall type, by *mechanical solidarity* (based on individual similarity) and *organic solidarity* (based on a *division of labor* among dissimilar individuals). Two other forms of social organization also contribute to the cohesion of a society: *gemeinschaft* ("community," characterized by cohesion based on friendships and loyalties) and *gesellschaft* ("society," characterized by cohesion based on complexity and differentiation).

What are the types of societies?
Societies across the globe vary in type, as determined mainly by the complexity of their social structures, their division of labor, and their technologies. From least to most complex, they are *foraging, pastoral, horticultural, agricultural* (these four constitute *preindustrial* societies), *industrial*, and *postindustrial* societies.

What are the forms of social interaction in society?
All forms of social interaction in society are shaped by the structure of its social institutions. A *group* is a collection of individuals who interact and communicate with each other, share goals and norms, and have a subjective awareness of themselves as a distinct social unit. *Status* is a hierarchical position in a structure; a *role* is the expected behavior associated with a particular status. A *role* is the behavior others expect from a person associated with a particular status. Patterns of social interaction influence nonverbal interaction as well as patterns of attraction and affiliation.

What theories are there about social interaction?
Social interaction takes place in society within the context of social structure and social institutions. Social interaction is analyzed in several ways, including the *social construction of reality* (we impose meaning and reality on our interactions with others); *ethnomethodology* (deliberate interruption of interaction to observe how a return to "normal" interaction is accomplished); *impression management* (a person "gives off" a particular impression to "con" the other and achieve certain goals, as in *cyberspace interaction*); and *social exchange* and *game theory* (one engages in gamelike reward and punishment interactions to achieve one's goals).

How is technology changing social interaction?
Increasingly, people engage with each other through *cyberspace interaction*. Social norms develop in cyberspace as they do in face-to-face interaction, but a person in cyberspace can also manipulate the impression that he or she gives off, thus creating a new "virtual" self.

Key Terms

achieved status **108**
ascribed status **108**
collective consciousness **102**
cyberspace interaction **119**
division of labor **103**
ethnomethodology **115**
game theory **118**
gemeinschaft **103**
gesellschaft **103**

group **107**
impression management **116**
imprinting **112**
macroanalysis **100**
master status **108**
mechanical solidarity **102**
microanalysis **100**
nonverbal communication **110**
non-zero sum game **118**

organic solidarity **103**
postindustrial society **107**
preindustrial society **104**
proxemic communication **111**
role **109**
role conflict **109**
role modeling **109**
role set **109**
role strain **110**
social institution **101**

social interaction **100**
social organization **101**
social structure **101**
society **100**
status **108**
status inconsistency **108**
status set **108**
tactile communication **110**
zero-sum game **118**

6

Groups and Organizations

D. Trozzo/Alamy

Types of Groups

Social Influence in Groups

Formal Organizations and Bureaucracies

Functionalism, Conflict Theory, and
    Symbolic Interaction: Theoretical
    Perspectives

Chapter Summary

it's saturday night. You feel like staying in, perhaps to read a book, play video games, watch a movie. You're just not "up" for going out as you often do. Just as you are settling it, you get a text from a friend saying, "Hey, let's party; great band at our favorite place. Let's go." Very soon, another friend texts, "I'm in! See you there." And another, "Me too." Before you know it, you are in the club, enjoying yourself but perhaps wishing you had just had a quiet night at home. The next morning, as you nurse your headache, you wonder why you went. You had really wanted a quiet night alone, but you soon found yourself surrounded by others, doing what they were doing, even though it wasn't how you had planned to spend your evening. What happened?

The answer is that you were subjected to group behavior—one of the most interesting phenomena in the social world. We like to think of ourselves as individuals and, of course, we are. But even as individuals, our behavior is strongly influenced by the groups to which we belong. And at any given moment, we belong to multiple groups, some with more influence than others. Understanding group behavior is critical to understanding people's behavior.

Consider this: If someone told you that you could catch a spaceship to a next level of existence, beyond anything you had ever known on Earth, would you take a lethal combination of drugs and alcohol to get you there? Surely not, you must be thinking! But that is precisely what thirty-nine members of the Heaven's Gate cult did in 1997 in a mansion in Rancho Santa Fe, California. They were told by the group leader that a spaceship was coming, following the tail of Comet Hale–Bopp and that they would be transported to a better place. Although seemingly completely irrational, this behavior can only be understood by analyzing how these thirty-nine individuals became subject to the control of such an extremist group—in other words, succumbing to group pressure.

Eighteen men and twenty-one women committed mass suicide as part of the Heaven's Gate cult in 1997, all of them dressed alike in dark clothes and Nike sneakers. This is an extreme example of group conformity.

Group pressure also escalates violent behaviors. You might recall a horrific rape that occurred in New Delhi, India in 2012, when seven men gang-raped a twenty-three-year-old medical school student on a public bus. The young woman died two weeks later from the severe injuries. Rape is a violent act even when committed by one person, but research finds that rape involving more than one perpetrator—that is, group rape—is usually far more violent and involves more severe forms of violation than rape by a single perpetrator. Scholars conclude that the group behavior involved in a gang rape intensifies violence as the group members succumb to the power of a group leader and/or feel they must participate or they will be ostracized by the group (Woodhams et al. 2012).

In less dramatic and disturbing examples, group influence also shapes all kinds of ordinary behavior. Juries are groups and jury decision making is clearly influenced by group processes. As the jury deliberates, a consensus is formed as more members of the group (that is, the jury) move to a particular verdict. Moreover, the larger the faction, the less willing an individual juror will be to defy the weight of group opinion. As we shall see, this is *group size effect:* an effect of sheer numbers in the group independent of the effects of individual actions and thoughts (Vidmar and Hans 2007).

You can probably think of examples in your own experience when you succumbed to group pressure, even when your individual judgment told you to do or think something different. Perhaps you smoke cigarettes, knowing full well that they are very harmful to your health. Or maybe you have purchased something from the latest fashion trend, even though you really could not spare the money. Or maybe you have gone out drinking with friends because "everybody was doing it."

People are highly subject to the social influences of groups. Whether a relatively small group—such as a jury or your friendship circle—or a large bureaucratic organization, such as a the government or a work organization, people are influenced by the sociological forces of group behavior.

- Understand the sociological concept of groups and the different forms groups take
- Analyze the social processes that produce conformity in groups
- Explain the social structures that characterize formal organizations and bureaucracies
- Compare and contrast the major sociological theories in terms of how they understand groups and organizations

TYPES OF GROUPS

Each of us is a member of many groups simultaneously. We have relationships in groups with family, friends, team members, and professional colleagues. Within these groups are gradations in relationships: We are generally closer to our siblings (our sisters and brothers) than to our cousins; we are intimate with some friends, merely sociable with others. If we count all our group associations, ranging from the powerful associations that define our daily lives to the thinnest connections with little feeling (other pet lovers, other company employees), we will uncover connections to literally hundreds of groups.

What is a group? Recall from Chapter 5, a **group** is two or more individuals who interact, share goals and norms, and have a subjective awareness as "we," that is, as a distinct social unit. To be considered a group, a social unit must have all three characteristics, although some groups are more bound together than others. Consider two superficially similar examples: The individuals in a line waiting to board a train are unlikely to have a sense of themselves as one group. A line of prisoners chained together and waiting to board a bus to the penitentiary is more likely to have a stronger sense of common feeling.

As you remember from the previous chapter, certain gatherings are not groups in the strict sense, but may be *social categories* (for example, teenagers, truck drivers) or *audiences* (everyone watching a movie). The importance of defining a group is not to perfectly decide if a social unit is a group—an unnecessary endeavor—but to help us understand the behavior of people in society. As we inspect groups, we can identify characteristics that reliably predict trends in the behavior of the group and even the behavior of individuals in the group.

The study of groups has application at all levels of society, from the attraction between people who fall in love to the characteristics that make some corporations drastically outperform their competitors—or that lead them into bankruptcy. The aggregation of individuals into groups has a transforming power, and sociologists understand the social forces that make these transformations possible. In this chapter, we move from the *microlevel* of analysis (the analysis of groups and

The annual running of the bulls in Pamplona, Spain is a death-defying exercise in group behavior.

face-to-face social influence) to the relatively more *macrolevel* of analysis (the analysis of formal organizations and bureaucracies).

Dyads and Triads: Group Size Effects

Even the smallest groups are of acute sociological interest. A **dyad** is a group consisting of exactly two people. A **triad** consists of three people. This seemingly minor distinction, first scrutinized by the German sociologist **Georg Simmel** (1858–1918), can have critical consequences for group behavior (Simmel 1902). Simmel was interested in discovering the effects of size on groups, and he found that the mere difference between two and three people spawned entirely different group dynamics (the behavior of a group over time).

Imagine two people standing in line for lunch. First one talks, then the other, then the first again. The interaction proceeds in this way for several minutes. Now a third person enters the interaction. The character of the interaction suddenly changes: At any given moment, two people are interacting more with each other than either is with the third. When the third person wins the attention of the other two, a new dyad is formed, supplanting the previous pairing. The group, a triad, then consists of a dyad (the pair that is interacting) plus an *isolate*.

Triadic segregation is what Simmel called the tendency for triads to segregate into a pair and an isolate (a single person). A triad tends to segregate into a **coalition** of the dyad against the isolate. The isolate then has the option of initiating a coalition with either member of the dyad. This choice is a type of social advantage, leading Simmel to coin the principle of *tertius gaudens*,

a Latin term meaning "the third one gains." Simmel's reasoning has led to numerous contemporary studies of coalition formation in groups (for example, Konishi and Ray 2003).

For example, interactions in a triad often end up as "two against one." You may have noticed this principle of coalition formation in your own conversations. Perhaps two friends want to go to a movie you do not want to see. You appeal to one of them to go instead to a minor league baseball game. She wavers and comes over to your point of view. Now you have formed a coalition of two against one. The friend who wants to go to the movies is now the isolate. He may recover lost social ground by trying to form a new coalition by suggesting a new alternative (going bowling or to a different movie). This flip-flop interaction may continue for some time, demonstrating another observation by Simmel: A triad is a decidedly unstable social grouping, whereas dyads are relatively stable. The minor distinction between dyads and triads is one person but has important consequences because it changes the character of the interaction within the group. Simmel is known as the discoverer of **group size effect**—the effects of group number on group behavior *independent of the personality characteristics of the members themselves.*

Primary and Secondary Groups

Charles Horton Cooley (1864–1929), a famous sociologist of the Chicago School of sociology, introduced the concept of the **primary group**, defined as a group consisting of intimate, face-to-face interaction

One of the best examples of the primary group is that consisting of parent and child.

Sharing the Journey

Modern society is often characterized as remote, alienating, and without much feeling of community or belonging to a group. This image of society has been carefully studied by sociologist Robert Wuthnow, who noticed that people in the United States are increasingly looking to small groups as places where they can find emotional and spiritual support and where they find meaning and commitment, despite the image of society as an increasingly impersonal force.

Research Question: Wuthnow began his research by noting that, even with the individualistic culture of U.S. society, small groups play a major role in this society. He saw the increasing tendency of people to join recovery groups, reading groups, spiritual groups, and myriad other support groups. Wuthnow began his research by asking these specific questions: What motivates people to join support groups? How do these groups function? What do members like most and least about such groups? His broadest question, however, was to wonder how the wider society is influenced by the proliferation of small support groups.

Research Methods: To answer these questions, a large research team of fifteen scholars designed a study that included both a quantitative and a qualitative dimension. They distributed a survey to a representative sample of more than 1000 people in the United States.

Supplementing the survey were interviews with more than 100 support group members, group leaders, and clergy. The researchers chose twelve groups for extensive study; researchers spent six months to three years tracing the history of these groups, meeting with members and attending group sessions.

Research Results: Based on this research, Wuthnow concludes that the small group movement is fundamentally altering U.S. society. Forty percent of all Americans belong to some kind of small support group. As the result of people's participation in these groups, social values of community and spirituality are undergoing major transformation. People say they are seeking community when they join small groups, whether the group is a recovery group, a religious group, a civic association, or some other small group. People turn to these small groups for emotional support more than for physical or monetary support.

Conclusions and Implications: Wuthnow argues that large-scale participation in small groups has arisen in a social context in which the traditional support structures in U.S. society, such as the family, no longer provide the sense of belonging and social integration that they provided in the past. Geographic mobility, mass society, and the erosion of local ties all contribute to this trend. People still seek a sense of community, but they create it

in groups that also allow them to maintain their individuality. In voluntary small groups, you are free to leave the group if it no longer meets your needs.

Wuthnow also concludes that these groups represent a quest for spirituality in a society when, for many, traditional religious values have declined. As a consequence, support groups are redefining what is sacred. They also replace explicit religious tenets imposed from the outside with internal norms that are implicit and devised by individual groups. At the same time, these groups reflect the pluralism and diversity that characterize society. In the end, they buffer the trend toward disintegration and isolation that people often feel in mass societies.

Questions to Consider

1. Are you a member of a voluntary small group? If so, what sense of community does the group provide for you? How do you maintain your sense of individuality at the same time?

2. What social changes do you observe in the world around you that might encourage people to join various support groups?

3. Some people join support groups in the aftermath of major life transition—a death, recovery from addiction, the desire to lose weight, and so on. What does group membership in such a situation provide for individuals?

Source: Wuthnow, Robert. 1994. *Sharing the Journey: Support Groups and America's New Quest for Community*. New York: Free Press.

---

and relatively long-lasting relationships. Cooley had in mind the family and the early peer group. In his original formulation, primary was used in the sense of "first," the intimate group of the formative years (Cooley 1967/1909). The insight that there was an important distinction between intimate groups and other groups proved extremely fruitful. Cooley's somewhat narrow concept of family and childhood peers has been elaborated upon over the years to include a variety of intimate relations as examples of primary groups.

Primary groups have a powerful influence on an individual's personality or self-identity. The effect of

family on an individual can hardly be overstated. The weight of peer pressure on school children is particularly notorious. Street gangs are a primary group, and their influence on the individual is significant; in fact, gang members frequently think of themselves as a family. Inmates in prison very frequently become members of a gang—primary groups perhaps based mainly upon race or ethnicity—as a matter of their own personal survival. The intense camaraderie formed among Marine Corps units in boot camp and in war, such as the war in Afghanistan, is another classic example of primary group formation and the resulting intense effect on individuals and upon their survival.

Support groups, such as this group therapy session, often provide people a feeling of community, even when faced with individual troubles.

In contrast to primary groups are **secondary groups**, those that are larger in membership, less intimate, and less long lasting. Secondary groups tend to be less significant in the emotional lives of people. Secondary groups include all the students at a college or university, all the people in your neighborhood, and all the people in a bureaucracy or corporation.

---

thinking SOCIOLOGICALLY

Identify a *group* of which you are a part. How does one become a member of this group? Who gets included and who is excluded? Does the group share any unique language or other cultural characteristics (such as dress, jargon, or other group identifiers)? Does anyone ever leave the group, and if so, why?

1. Would you describe this group mainly as a *primary* or a *secondary group*? Why?

2. Now think about this group from the perspective of functionalist theory, conflict theory, and symbolic interaction theory. Is there a hierarchy within the group? Is there competition between group members? What social meanings do members of the group share? ●

Primary and secondary groups serve different needs. Primary groups give people intimacy, companionship, and emotional support. These human desires are termed **expressive needs** (also called socioemotional needs). Family and friends share and amplify your good fortune, rescue you when you misbehave, and cheer you up when life looks grim. Many studies have shown the overwhelming influence of family and friendship groups on religious and political affiliation, as shown in the box "Doing Sociological Research: Sharing the Journey" (Wuthnow 1994).

Secondary groups serve **instrumental needs** (also called task-oriented needs). Athletic teams form to have fun and win games. Political groups form to raise funds and influence the government. Corporations form to make profits, and employees join corporations to earn a living. The true distinction between primary and secondary groups is in how intimate the group members feel about one another and how dependent they are on the group for sustenance and identity.

Secondary groups occasionally take on the characteristics of primary groups, even if temporarily. This is precisely what happened to a group of miners who became trapped for nearly three months a half mile below the surface in Chile's Atacama Desert. When the thirty-three miners were eventually rescued, an event that was covered live on the international news, we learned that this was a very striking example of the transition from a largely secondary group to an exceptionally close-knit primary group. A strong leader (foreman Luis Urzúa) insisted that, "It was one for all and all for one down there." As the men reported it, the experience transformed all thirty-three men into a large and very close family (primary group) even as they later coped with their newfound fame, celebrity, and requests to endorse products (Padgett et al. 2011).

Reference Groups

Primary and secondary groups are groups to which members belong. Both are called *membership groups*. In contrast, **reference groups** are those to which you may or may not belong but use as a standard for evaluating your values, attitudes, and behaviors (Merton and Rossi 1950). Reference groups are generalized versions of role models. They are not "groups" in the sense that the individual interacts within (or in) them. Do you pattern your behavior on that of sports stars, musicians, military officers, or business executives? If so, those models are reference groups for you.

Imitation of reference groups can have both positive and negative effects. Members of a Little League baseball team may revere major league baseball players and attempt to imitate laudable behaviors such as tenacity and sportsmanship. But young baseball fans are also liable to be exposed to tantrums, fights, and tobacco chewing and spitting. This illustrates that the influence of a reference group can be both positive and negative.

Reference groups do not have to be actual people. As we have been seeing throughout this book, the media can serve as a powerful reference group, influencing how people perceive themselves. Positive representations of one's reference group can promote strong self-esteem; negative representations and negative

David Mercado/Reuters

Initiation into a group can mean losing one's individual identity, especially when strict conformity to the group is enforced. New initiates into military life routinely have their hair cut, symbolic of the dominance of group identity over individual identity.

stereotypes, such as of racial–ethnic groups, can produce diminished self-esteem. The representation of racial and ethnic groups in a society can have a striking positive effect perhaps even among children acquiring their lifetime set of group affiliations (Harris 2006; Zhou and Bankston 2000).

In-Groups and Out-Groups

When groups have a sense of themselves as "us," they will also have a complementary sense of other groups as "them." The distinction is commonly characterized as *in-groups* versus *out-groups*. The concept was originally elaborated by the early sociological theorist **W. I. Thomas** (Thomas 1931). College fraternities and sororities certainly exemplify "in" versus "out." So do families. So do gangs—especially so. The same can be true of the members of your high school class, your sports team, your racial group, your gender, and your social class.

Attribution theory is the principle that we all make inferences about the personalities of others, such as concluding what the other is "really like." These attributions depend on whether you are in the in-group or the out-group. Thomas F. Pettigrew has summarized the

research on attribution theory, showing that individuals commonly generate a significantly distorted perception of the motives and capabilities of other people's acts based on whether that person is an in-group or out-group member (Baumeister and Bushman 2008; Gilbert and Malone 1995; Pettigrew 1992).

Pettigrew describes the misperception as **attribution error**, meaning errors made in attributing causes for people's behavior to their membership in a particular group, such as a racial group. Attribution error has several dimensions, all tending to favor the in-group over the out-group. All else being assumed equal, we tend to perceive people in our in-group positively and those in out-groups negatively, regardless of their actual personal characteristics:

1. When onlookers observe improper behavior by an out-group member, onlookers are likely to attribute the deviance to the disposition (the personality) of the wrongdoer. Disposition refers to the perceived "true nature" or "inherent nature" of the person, often considered to be genetically determined. For example, a White person may see a Hispanic person carrying a knife and, without further information, attribute this behavior to the presumed "inherent tendency" for Hispanics to be violent. The same would be true if a Hispanic person, without additional information, assumed that all Whites have the same "inherent tendency" to be racist.

2. When the *same* behavior is exhibited by an in-group member, the perception is commonly held that the act is due to the *situation* of the wrongdoer, not to the in-group member's inherent disposition or personality. For example, a White person may see another White person carrying a knife and

Bonnie Jo Mount/The Washington Post/Getty Images

As with the African American women's sorority, Delta Sigma Theta, groups often use clothing styles and colors to signify group belonging.

conclude, without further information, that the weapon must be carried for protection in a dangerous area.

3. If an out-group member is seen to perform in some laudable way, the behavior is often attributed to a variety of special circumstances, and the out-group member is seen as "the exception."

4. An in-group member who performs in the same laudable way is given credit for a worthy personality disposition.

Typical attribution errors include misperceptions between racial groups and between men and women. If a White police officer shoots a Black or Latino, a White individual, given no additional information, is likely to assume that the victim instigated the shooting and thus "deserved" to be shot. On the other hand, a Black person is more likely to assume that the police officer fired unnecessarily, perhaps because the officer is dispositionally assumed to be a racist (Taylor et al. 2013; Kluegel and Bobo 1993).

A related phenomenon has been seen in men's perceptions of women coworkers. Meticulous behavior in a man is perceived positively and is seen by other men as "thorough"; in a woman, the *exact same* behavior is perceived negatively and is considered "picky." Behavior applauded in a man as "aggressive" is condemned in a woman exhibiting the same behavior as "pushy" or "bitchy" (Uleman et al. 1996).

Social Networks

As already noted, no individual is a member of only one group. Social life is far richer than that. A **social network** is a set of links between individuals, between groups, or between other social units, such as bureaucratic organizations or even entire nations (Aldrich and Ruef 2006; Centeno and Hargittai 2003; Mizruchi 1992). One could say that any given person belongs simultaneously to several networks (Wasserman and Faust 1994). With the development of *social media* (see Chapter 2), networks that may have once been face-to-face have now developed through electronic media, such as Facebook and Twitter. The development of social media brings a new dimension to the study and analysis of networks because you may be in a network with people you do not even know. Nonetheless, your group of friends, or all the people on an electronic mailing list to which you subscribe, or all of your Facebook subscribers are social networks, some human, some electronic.

Networks can be critical to your success in life. Numerous research studies indicate that people get jobs via their personal networks more often than through formal job listings, want ads, or placement agencies (Ruef et al. 2003; Petersen et al. 2000; Granovetter 1995, 1974). Getting a job is more often a matter of whom you know than what you know. Who you know, and whom they know in turn, is a social network that may have a marked effect on your life and career.

This job candidate does a last-minute check of his resume just before being interviewed by a company representative who contacted the job candidate through a social network. Like many others during the recent recession, he was let go from his previous job.

Jim West/Alamy Limited

Networks form with all the spontaneity of other forms of human interaction (Wasserman and Faust 1994; Knoke 1992; Mintz and Schwartz 1985). Networks evolve, such as social ties within neighborhoods, professional contacts, and associations formed in fraternal, religious, occupational, and volunteer groups. Networks to which you are only *weakly* tied (you may know only one person in your neighborhood) provide you with access to that entire network, hence the sociological principle that there is "strength in weak ties" (Petersen et al. 2000; Montgomery 1992; Granovetter 1973).

Networks based on race, class, and gender form with particular readiness. This has been especially true of job networks. The person who leads you to a job is likely to have a similar social background. Research indicates that the "old boy network"—any network of White, male corporate executives—is less important than it used to be, although it is certainly not by any means gone. The diminished importance of the old boy network is because of the increasing prominence of women and minorities in business organizations. In fact, among African American and Latino individuals, one's family can provide network contacts that can lead to jobs and upward mobility (Dominguez and Watkins 2003). Still, as we will see later in this chapter, women and minorities are considerably underrepresented in corporate life, especially in high-status jobs, and since 2004, the presence of women and racial minorities on corporate boards has actually declined (Alliance for Board Diversity 2011). Some recent research shows that despite recent gains on the part of minorities (relative to Whites), Blacks and Latinos

Finding a Job: The Invisible Hand

Is getting a job simply a matter of getting the right credentials and training? Hardly, according to sociologists. Even in a good job market, one needs the help of social networks to find a job. This has been clearly demonstrated by sociologist Deirdre Royster, who compared the experiences of two groups of men: One group was White, the other Black. Both had graduated from vocational school. The Black and White men had comparable educational credentials, the same values, and the same work ethic; yet, the White men were far more likely to gain employment than were the Black men. Why? Royster's research revealed that the most significant difference between the two groups was access to *job networks*, just as Granovetter's work on job networks would predict (Royster 2003; Granovetter 1995).

---

are still disproportionately harmed by lack of network contacts (Smith 2007).

Networks can reach around the world, but how big is the world? How many of us, when we discover someone we just met is a friend of a friend, have remarked, "My, it's a small world"? Research into what has come to be known as the *small world problem* has shown that networks make the world a lot smaller than you might think.

Original small world researchers Travers and Milgram wanted to test whether a document could be routed via the U.S. postal system to a complete stranger more than 1000 miles away using only a chain of acquaintances (Watts 1999; Watts and Strogatz 1998; Kochen 1989; Lin 1989; Travers and Milgram 1969). If so, how many steps would be required? The researchers organized an experiment in which approximately 300 senders were all charged with getting a document to one receiver, a complete stranger. (Remember that all this was well before the advent of the desktop computer in the 1980s.) The receiver was a male Boston stockbroker. The senders were one group of Nebraskans and one group of Bostonians chosen completely at random. Every sender in the study was given the receiver's name, address, occupation, alma mater, year of graduation, wife's maiden name, and hometown. They were asked to send the document directly to the stockbroker if they knew him on a first-name basis. Otherwise, they were asked to send the folder to a friend, relative, or acquaintance known on a first-name basis who might be more likely than the sender to know the stockbroker.

How many intermediaries do you think it took, on average, for the document to get through? (Most people estimate from twenty to hundreds.) *The average number of intermediate contacts was only 6.2!* However, only about one-third of the documents actually arrived at the target. This was quite impressive, considering that the senders did not know the target person—hence the current expression that any given person in the country is on average only about "six degrees of separation" from any other person. In this sense, the world is indeed "small."

This original small world research has recently been criticized on two grounds: First, only one-third of the documents actually reached the target person. The 6.2 average intermediaries applied only to these completed chains. Thus two-thirds of the initial documents never reached the target person. For these people, the world was certainly not "small." Second, the sending chains tended to closely follow occupational, social class, and ethnic lines, just as general network theory would predict (Kleinfeld 1999; Wasserman and Faust 1994). Thus the world may indeed be "small," but only for people in your immediate social network (Ruef et al. 2003; Watts 1999).

A study of Black national leaders by Taylor and associates shows that Black leaders form a very closely knit network, one considerably more closely knit than longer-established White leadership networks (Domhoff 2002; Jackson 2000; Jackson et al. 1995, 1994; Taylor 1992; Alba and Moore 1982; Moore 1979; Kadushin 1974; Mills 1956). The world is indeed quite "small" for America's Black leadership. Included in the study were Black members of Congress, mayors, business executives, military officers (generals and full colonels), religious leaders, civil rights leaders, media personalities, entertainment and sports figures, and others. The study found that when considering only direct personal acquaintances—not indirect links involving intermediaries—one-fifth of the entire national Black leadership network know each other directly as a friend or close acquaintance. The Black leadership network is considerably more closely connected than White leadership networks. The Black network had greater *density*. Add only one intermediary, the friend of a friend, and the study estimated that almost *three-quarters* of the entire Black leadership network are included. Therefore, any given Black leader can generally get in touch with three-quarters of all other Black leaders in the country either by knowing them personally (a "friend") or via only one common acquaintance (a "friend of a friend"). That's pretty amazing when one realizes that the study is considering the population of Black leaders in the entire country.

Streaking, or running nude in a public place—relatively popular among college students in the 1970s and early 1980s and still popular on some campuses—is more common as a group activity than as a strictly individual one. This illustrates how the group can provide the people in it with *deindividuation*, or diffusion of responsibility among group members—a type of merging of self with group. This allows individuals to feel less responsibility or blame for their actions, thus convincing them that the group must share the blame.

SOCIAL INFLUENCE IN GROUPS

The groups in which we participate exert tremendous influence on us. We often fail to appreciate how powerful these influences are. For example, who decides what you should wear? Do you decide for yourself each morning, or is the decision already made for you by fashion designers, role models, and your peers? Consider how closely your hair length, hair styling, and choice of jewelry have been influenced by your peers. Did you invent your skinny jeans, your dreadlocks, or your blue blazer? People who label themselves as nonconformists often conform rigidly to the dress code and other norms of their in-group.

A group such as one's family even influences your adult life long after children have grown up and formed households of their own. The choices of political party among adults (Republican, Democratic, or Independent) correlate strongly with the party of one's parents, again demonstrating the power of the primary group. Seven out of ten people vote with the political party of their parents, even though these same people insist that they think for themselves when voting (Worchel et al. 2000; Jennings and Niemi 1974). Furthermore, most people share the religious affiliation of their parents, although they will insist that they chose their own religion, free of any influence by either parent.

We all like to think we stand on our own two feet, immune to a phenomenon as superficial as group pressure. The conviction that one is impervious to social influence results in what social psychologist Philip Zimbardo calls the *not-me syndrome:* When confronted with a description of group behavior that is disappointingly conforming and not individualistic, most individuals counter that some people may conform to social pressure, "but not me"; or "some people yield quickly to styles of dress, but not me"; or "some people yield to autocratic authority figures, but not me" (Taylor et al. 2013; Zimbardo et al. 1977). But sociological experiments often reveal a dramatic gulf between what people *think* they will do and what they *actually do.* The conformity study by Solomon Asch discussed next is a case in point.

The Asch Conformity Experiment

We learned in the previous section that social influences are evidently quite strong. Are they strong enough to make us disbelieve our own senses? Are they strong enough to make us misperceive what is objective, actual fact? In a classic piece of work known as the Asch conformity experiment, Solomon Asch showed that even simple objective facts cannot withstand the distorting pressure of group influence (Asch 1955, 1951).

Examine the two illustrations in Figure 6.1. Which line on the right is more nearly equal in length to the line on the left (line S)? Line B, obviously. Could anyone fail to answer correctly?

In fact, Solomon Asch discovered that social pressure of a rather gentle sort was sufficient to cause an astonishing rise in the number of wrong answers. Asch lined up five students at a table and asked which line in the illustration on the right is the same length as the line on the left. Unknown to the fifth student, the first four were *confederates*—collaborators with the experimenter who only pretended to be participants. For several rounds, the confederates gave correct answers to Asch's tests. The fifth student also answered correctly, suspecting nothing. Then on subsequent trials the first student

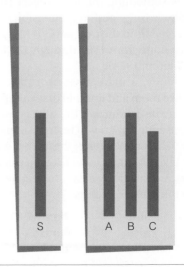

FIGURE 6.1 Lines from Asch Experiment

Source: Asch, Solomon. 1956. "Opinion and Social Pressure." *Scientific American* 19 (July): 31–36.

gave a wrong answer. The second student gave the same wrong answer. Third, wrong. Fourth, wrong. Then came the fifth student's turn.

In Asch's experiment, fully *one-third* of all students in the fifth position gave the same wrong answer as the confederates at least half the time. Forty percent gave "some" wrong answers. Only one-fourth of the students consistently gave correct answers in defiance of the invisible pressure to conform.

Line length is not a vague or ambiguous stimulus. It is clear and objective. Wrong answers from one-third of all subjects is a very high proportion. The subjects fidgeted and stammered while doing it, but they did it nonetheless. Those who did not yield to group pressure showed even more stress and discomfort than those who yielded to the (apparent) opinion of the group.

Would you have gone along with the group? Perhaps, perhaps not. Sociological insight grows when we acknowledge the fact that fully a third of all participants will yield to the group. The Asch experiment has been repeated many times over the years, with students and nonstudents, old and young, in groups of different sizes, and in different settings (Baumeister and Bushman 2008; Worchel et al. 2000). The results remain essentially the same. A third to a half of the participants make a judgment contrary to fact, yet in conformity with the group. Finally, the Asch findings have consistently revealed a *group size effect:* The greater the number of individuals (confederates) giving an incorrect answer (from five up to fifteen confederates), the greater the number of subjects per group giving an incorrect answer.

The Milgram Obedience Studies

What are the limits of social pressure? In terms of moral and psychological issues, judging the length of a line is a small matter. What happens if an authority figure demands obedience—a type of conformity—even if the task is something the test subject (the person) finds morally wrong and reprehensible? A chilling answer emerged from the now famous Milgram Obedience Studies done from 1960 through 1973 by Stanley Milgram (Milgram 1974).

In this study, a naive research subject entered a laboratory-like room and was told that an experiment on learning was to be conducted. The subject was to act as a "teacher," presenting a series of test questions to another person, the "learner." Whenever the learner gave a wrong answer, the teacher would administer an electric shock.

The test was relatively easy. The teacher read pairs of words to the learner, such as

blue	box
nice	house
wild	duck

The teacher then tested the learner by reading a multiple-choice answer, such as

blue   **sky**   **ink**   **box**   **lamp**

The learner had to recall which term completed the pair of terms given originally, in this case, "blue box."

If the learner answered incorrectly, the teacher was to press a switch on the shock machine, a formidable-looking device that emitted an ominous hum when activated (see Figure 6.2). For each successive wrong answer, the teacher was to increase the intensity of the shock by 15 volts.

The machine bore labels clearly visible to the teacher: Slight Shock, Moderate Shock, Strong Shock, Very Strong Shock, Intense Shock, Extreme Intensity Shock, Danger: Severe Shock, and lastly, *XXX* at 450 volts. As the voltage rose, the learner responded with squirming, groans, then screams.

The experiment was rigged. The learner was a confederate. No shocks were actually delivered. The true purpose of the experiment was to see if any "teacher" would go all the way to 450 volts. If the subject (teacher) tried to quit, the experimenter responded with a sequence of prods:

"Please continue."

"The experiment requires that you continue."

"It is absolutely essential that you continue."

"You have no other choice, you must go on."

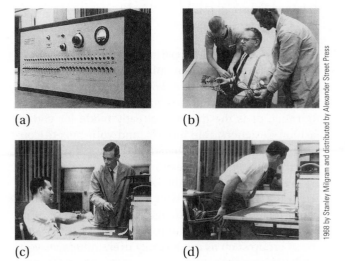

(a)　　　　　　　　　(b)

(c)　　　　　　　　　(d)

1968 by Stanley Milgram and distributed by Alexander Street Press

FIGURE 6.2 Milgram's Setup These photographs show how intimidating and authoritative the Milgram experiment must have been. The first picture (a) shows the formidable-looking shock generator. The second (b) shows the role player, who pretends to be getting the electric shock, being hooked up. The third (c) shows an experimental subject (seated) and the experimenter (in lab coat, standing). The fourth picture (d) shows a subject terminating the experiment prematurely, that is, before giving the maximum shock level (voltage) of 450 volts. A large majority (65 percent) of subjects did not do this and actually went all the way to the maximum shock level.

Source: Milgram, Stanley. 1974. *Obedience to Authority: An Experimental View.* New York: Harper & Row, p. 26.

In the first experiment, fully 65 percent of the volunteer subjects ("teachers") went *all the way* to 450 volts on the shock machine!

Milgram himself was astonished. Before carrying out the experiment, he had asked a variety of psychologists, sociologists, psychiatrists, and philosophers to guess how many subjects would actually go all the way to 450 volts. The opinion of these consultants was that only one-tenth of one percent (one in one thousand) would actually do it!

What would you have done? Remember the "not-me" syndrome. Think about the experimenter saying, "You have no other choice, you must go on." Most people claim they would refuse to continue as the voltage escalated. The importance of this experiment derives in part from how starkly it highlights the difference once again between what people think they will do and what they actually do.

Milgram devised a series of additional experiments in which he varied the conditions to find out what would cause subjects *not* to go all the way to 450 volts. He moved the experiment from an impressive university laboratory to a dingy basement to counteract some of the tendency for people to defer to a scientist conducting a scientific study. One learner was then instructed to complain of a heart condition. Still, well over half of the subjects delivered the maximum shock level. Speculating that women might be more humane than men (all prior experiments used only male subjects), Milgram did the experiment again using only women subjects. The results? Exactly the same. Social class background made no difference. Racial and ethnic differences had no detectable effect on compliance rate.

At the time that the Milgram experiments were conceived, the world was watching the trial in Jerusalem of World War II Nazi Adolf Eichmann. Millions of Jews, Gypsies, homosexuals, and communists were murdered between 1939 and 1945 by the Nazi party, led by Adolf Hitler. As head of the Gestapo's "Jewish section," Eichmann oversaw the deportation of Jews to concentration camps and the mass executions that followed. Eichmann disappeared after the war, was abducted in Argentina by Israeli agents in 1961, and transported to Israel, where he was tried and ultimately hanged for crimes against humanity.

The world wanted to see what sort of monster could have committed the crimes of the Holocaust, but a jarring picture of Eichmann emerged. He was slight and mild mannered, not the raging ghoul that everyone expected. He insisted that although he had indeed been a chief administrator in an organization whose product was mass murder, he was guilty only of doing what he was told to do by his superiors. He did not hate Jews, he said. In fact, he had a Jewish half-cousin whom he hid and protected. He claimed, "I was just following orders."

How different was Adolph Eichmann from the rest of us? The political theorist Hannah Arendt dared to suggest in her book *Eichmann in Jerusalem* (1963) that evil on a giant scale is banal. It is not the work of monsters, but an accident of civilization. Arendt argued that we need only look into ourselves to find the villain.

The Iraqi Prisoners at Abu Ghraib: Research Predicts Reality?

We have just learned that ordinary people will do horrible things to other humans simply because of the influence of the group, because of an authority figure, or because of a combination of both. This has been the lesson of the Asch studies and the Milgram studies. Recent events in the world have once again shown vividly and clearly how accurate such sociological and psychological experiments are in the prediction of actual human behavior.

In the spring of 2004, it was revealed that American soldiers who were military police guards at a prison in Iraq (the prison was named Abu Ghraib) had engaged in severe torture of Iraqi prisoners of war. The torture included sexual abuse of the prisoners—having male prisoners simulate sex with other male prisoners, positioning their mouths next to the genitals of another male prisoner, and other such acts. Still other acts of torture involved physical abuse such as beatings, stomping on the fingers of prisoners (thus fracturing them), and a large number of other physical acts of torture, including bludgeoning, some allegedly resulting in deaths of prisoners. Such tortures are clearly outlawed by the Geneva Conventions and by clearly stated U.S. principles of war. Both male and female guards participated in these acts of torture. The guards later claimed that they were simply following orders, either orders directly given or indirectly assumed. At the time, President George W. Bush and then Secretary of Defense Donald H. Rumsfeld both claimed that the acts of torture were merely the acts of a "corrupt few" and that the vast majority of American soldiers would never engage in such horrible acts. Since then, it has come to light via CIA memoranda that certain kinds of torture were indeed formal U.S. policy.

Now consider what we know from research. The Milgram studies strongly suggest that many ordinary soldiers who were not at all "corrupt," at least not more than average, would indeed engage in these acts of torture, particularly if they believed that they were under orders to do so, or if they believed that they would not be punished in any way if they did. The American soldiers must bear a significant portion of the responsibility for their own behavior. Nonetheless, the causes of the soldiers' behaviors lie not in the personalities of a "corrupt few" (their "natures") but in the social structure and group pressures of the situation.

The soldiers (guards) in the Abu Ghraib prison may not have received direct orders to torture prisoners, but they did so nonetheless. A now classic study of a simulated prison by Haney, Banks, and Zimbardo (1973) shows this effect quite clearly. In this study, Stanford University students were told by an experimenter to enter a dungeon-like basement. Half were told to pretend to be guards (to role-play being a guard) and half were told that they were prisoners (to role-play being a prisoner). Which students were told what was randomly determined.

After two or three days, the guards, completely on their own, began to act very sadistically and brutally toward the prisoners—having them strip naked, simulate sex, act subservient, and so on. Interestingly, the prisoners for the most part did just what the guards wanted them to do, no matter how unpleasant the requested act! The experiment was so scary that the researchers terminated the experiment after six days—more than a week early.

Remember that this study was conducted in 1973—thirty-one years *before* Abu Ghraib. Yet, this simulated prison study (as well as the Asch and Milgram studies) predicted quite precisely how both "guards" and "prisoners" would act in a real prison situation. Group influence effects uncovered by the Asch as well as the Milgram studies ruled in both the simulated prison of 1973 as well as the only too real Iraq prison of 2004.

debunking SOCIETY'S MYTHS

MYTH: People in groups are just individuals who make up their own minds about how to think and behave.
SOCIOLOGICAL PERSPECTIVE: The Asch, Milgram, and simulated prison experiments conclusively show that people are profoundly influenced by group pressure, often causing them to make up their minds contrary to objective fact and even to deliberately cause harm to another person. ●

Groupthink

Wealth, power, and experience are apparently not enough to save us from social influences. **Groupthink**, as described by I. L. Janis, is the tendency for group members to reach a consensus opinion, even if that decision is downright stupid (Janis 1982).

Janis reasoned that because major government policies are often the result of group decisions, it would be fruitful to analyze group dynamics that operate at the highest level of government—for instance, in the office of the president of the United States. The president makes decisions based upon group discussions with his advisers. The president is human and thus susceptible to group influence. Janis discovered a common pattern of misguided thinking in his investigations of presidential decisions. He surmised that outbreaks of groupthink had several things in common:

1. An illusion of invulnerability
2. A falsely negative impression of those who are antagonists to the group's plans
3. Discouragement of dissenting opinion
4. An illusion of unanimity

Groupthink influences many important decisions, such as going to war, in close-knit presidential administrations. The U.S. Senate Select Committee on Intelligence fingered groupthink as leading to the intelligence failures that drove us into war in Iraq and that led government leaders to underestimate the terrorist threats to the United States prior to 9/11. Groupthink is not inevitable when a team gathers to make a decision, but it is common and appears in all sorts of groups, from student discussion groups to the highest councils of power (Paulus et al. 2001).

debunking SOCIETY'S MYTHS

MYTH: A group of experts brought together in a small group will solve a problem according to their collective expertise.
SOCIOLOGICAL PERSPECTIVE: Groupthink can lead even the most qualified people to make disastrous decisions because people in groups in the United States tend to seek consensus at all costs. ●

Risky Shift

The term *groupthink* is commonly associated with group decision making with consequences that are not merely unexpected but disastrous. Another group phenomenon, **risky shift**, may help explain why the products of groupthink are frequently calamities. Have you ever found yourself in a group engaged in a high-risk activity that you would not do alone? When you created mischief as a child, were you not usually part of a group? If so, you were probably in the thrall of risky shift—the general tendency for groups to be more risky than individuals taken singly.

Risky shift was first observed by James Stoner (1961). Stoner gave study participants descriptions of a situation involving risk, such as one in which people seeking a job must choose between job security and a potentially lucrative but risky advancement. The participants were then asked to decide how much risk the person should take. Before performing his study, Stoner believed that individuals in a group would take less risk than individuals alone, but he found the opposite: After his groups had engaged in open discussions, they favored greater risk than they would have before discussion.

See FOR YOURSELF

Think of a time when you engaged in some risky behavior. What group were you part of, and how did the group influence your behavior? How does this illustrate the concept of *risky shift*? Is there more risky shift with more people in the group? If so, this would illustrate a group size effect. ●

Stoner's research has stimulated literally hundreds of studies using males and females, different nationalities, different tasks, and other variables (Taylor et al. 2013; Worchel et al. 2000). The results are complex. Much, but not all, group discussion leads to greater risk-taking. In subcultures that value caution above daring, as in some work groups of Japanese and Chinese firms, group decisions are less risky after discussion than before. The shift can occur in either direction, driven by the influence of group discussion, but there is generally some kind of shift in one direction or the other rather than no shift at all (Kerr 1992). This is called **polarization shift**.

What causes risky shifts? The most convincing explanation is that deindividuation occurs. **Deindividuation** is the sense that one's self has merged with a group. In terms of risk-taking, one feels that responsibility (and possibly blame) is borne not only by oneself but also by the group. This seems to have happened among the American prison guards who tortured prisoners at Abu Ghraib prison: Each guard could convince himself or herself that responsibility, hence blame, was to be borne by the group as a whole. The greater the number of people in a group, the greater the tendency toward deindividuation. In other words, deindividuation is a *group size effect*. As groups get larger, trends in risk-taking are amplified.

FORMAL ORGANIZATIONS AND BUREAUCRACIES

Groups, as we have seen, are capable of greatly influencing individuals. The study of groups and their effects on the individual represent an example of *microanalysis*, to use a concept introduced in Chapter 5. In contrast, the study of formal organizations and bureaucracies, a subject to which we now turn, represents an example of *macroanalysis*. The focus on groups drew our attention to the relatively small and less complex, whereas the focus on organizations draws our attention to the relatively large and structurally more complex.

A **formal organization** is a large secondary group, highly organized to accomplish a complex task or tasks and to achieve goals efficiently. Many of us belong to various formal organizations: work organizations, schools, and political parties, to name a few. Organizations are formed to accomplish particular tasks and are characterized by their relatively large size, compared with a small group such as a family or a friendship circle. Often organizations consist of an array of other organizations. The federal government is a huge organization comprising numerous other organizations, most of which are also vast. Each organization within the federal government is also designed to accomplish specific tasks, be it collecting your taxes, educating the nation's children, or regulating the nation's transportation system and national parks.

Organizations develop routine practices that result in the production of an organizational culture. **Organizational culture** refers to the collective norms and values that shape the behavior of people within an organization; in other words, it is the environment of the organization. Organizational culture is present in any organization and involves both formal and informal norms. The culture of an organization may be reflected in certain symbols and rituals, perhaps even a certain style of dress. Organizational culture guides the behaviors of people within the organization, shaping what is perceived to be appropriate and inappropriate. Indeed, organizations appear very different depending upon their culture. Corporate organizational culture has tended to be somewhat formal and restrictive, although new and innovative companies, such as Google and Facebook have transformed these traditional organizational cultures. But, even when it is informal, organizational culture guides behavior within the organization. And it does not take explicit rules to regulate organizational behavior; comments from coworkers or bosses may be enough to enforce such organizational norms.

Organizational culture can also produce problems for organizations, as was recently found in the scandal at Penn State University involving the sexual abuse of young boys by assistant football coach Jerry Sandusky. The special investigative report that analyzed these incidents and made recommendations to the university specifically pointed at organizational culture as a contributing factor in the failure of Penn State University to stop Sandusky's behavior. Certainly, the individual behavior of Sandusky, now serving at least thirty years in prison, is to blame, but the Special Counsel's report also blames the organizational culture of the university for failing to investigate when repeated reports of Sandusky's behavior came forward. The report cited the repeated failure of four university leaders (the president, the famed football coach Joe Paterno, the athletic director, and the university vice president) for participating in an organizational culture that resisted outside perspectives, had the pervasive goal of protecting the university's reputation, and had an excessive reverence for football, placing football above the protection of children.

Organizations tend to be persistent, although they are also responsive to the broader social environment where they are located (DiMaggio and Powell 1991). Organizations are frequently under pressure to respond to changes in the society by incorporating new practices and beliefs into their structure. Business corporations, as an example, have had to respond to increasing global competition; they do so by expanding into new international markets, developing a globally focused workforce, and trimming costs by eliminating workers and various layers of management.

Organizations can be tools for innovation, depending on the organization's values and purpose. Rape crisis centers are examples of organizations that originally emerged from the women's movement because of the perceived need for services for rape victims. Rape crisis centers have, in many cases, changed how police departments and hospital emergency personnel respond to rape victims. By advocating changes in rape law and services for rape victims, rape crisis centers have generated change in other organizations as well (Schmitt and Martin 1999; Fried 1994).

Types of Organizations

Sociologists Blau and Scott (1974) and Etzioni (1975) classify formal organizations into three categories distinguished by their types of membership affiliation: normative, coercive, and utilitarian.

Normative Organizations. People join **normative organizations** to pursue goals that they consider worthwhile. They obtain personal satisfaction, but no monetary reward for membership in such an organization. In many instances, people join the normative organization for the social prestige that it offers. Many are service and charitable organizations and are often called *voluntary organizations*. They include organizations such as Kiwanis clubs, political parties, religious organizations, the National Association for the Advancement of Colored People (NAACP), B'nai B'rith, La Raza, and other similar voluntary organizations that are concerned with specific issues. Such groups have been created to meet particular needs, sometimes ones that members see as unmet by other organizations.

Gender, class, race, and ethnicity all play a role in who joins what voluntary organization. Social class is reflected in the fact that many people do not join certain organizations simply because they cannot afford to join. Membership in a professional organization, as one example, can cost hundreds of dollars each year. Those who feel disenfranchised, however, may join grassroots organizations—voluntary organizations that spring from specific local needs that people think are unmet. Tenants may form an organization to protest rent increases or lack of services, or a new political party may emerge from people's sense of alienation from existing party organizations. African Americans, Latinos, and Native Americans have formed many of their own voluntary organizations in part because of their historical exclusion from traditional White voluntary organizations. Some of these are vibrant, ongoing organizations in their own right (such as the African American organizations Delta Sigma Theta and

Alpha Kappa Alpha sororities and the fraternities Alpha Phi Alpha, Kappa Alpha Psi, and Omega Psi Phi; see Giddings 1994).

Coercive Organizations. **Coercive organizations** are characterized by membership that is largely involuntary. Prisons are an example of organizations that people are coerced to "join" by virtue of punishment for their crime. Similarly, mental hospitals are coercive organizations: People are placed in them, often involuntarily, for some form of psychiatric treatment. In many respects, prisons and mental hospitals are similar in their treatment of inmates or patients. They both have strong security measures such as guards, locked and barred windows, and high walls (Rosenhan 1973; Goffman 1961).

The sociologist Erving Goffman has described coercive organizations as total institutions. A **total institution** is an organization that is cut off from the rest of society and one in which resident individuals are subject to strict social control (Goffman 1961). Total institutions include two populations: the "inmates" and the staff. Within total institutions, the staff exercises complete power over inmates, for example, nurses over mental patients and guards over prisoners. The staff administers all the affairs of everyday life, including basic human functions such as eating and sleeping. Rigid routines are characteristic of total institutions, thus explaining the common complaint by those in hospitals that they cannot sleep because nurses repeatedly enter their rooms at night, regardless of whether the patient needs medication or treatment. However, the problem of such rigid routines has eased somewhat in some institutions.

Utilitarian Organizations. The third type of organization named is **utilitarian**. These are large organizations, either for-profit or nonprofit, that individuals join for specific purposes, such as monetary reward. Large business organizations that generate profits (in the case of for-profit organizations) and salaries and wages for the organization's employees (as with either for-profit or nonprofit organizations) are utilitarian organizations. Examples of large, for-profit organizations include Microsoft, Amazon.com, and Google. Examples of large nonprofit organizations that pay salaries to employees are colleges and universities, the Educational Testing Service (ETS), churches, and organizations such as the National Collegiate Athletic Association (NCAA).

Bureaucracy

As a formal organization develops, it is likely to become a **bureaucracy**, a type of formal organization characterized by an authority hierarchy, a clear division of labor, explicit rules, and impersonality. Bureaucracies are notorious for their unwieldy size and complexity as well as their reputation for being remote and cumbersome organizations that are highly impersonal and machinelike in their operation. The federal government is a good example of a cumbersome bureaucracy that many believe is ineffective because of its sheer size. Numerous other formal organizations have developed into huge bureaucracies: Microsoft, Disney, many universities, hospitals, state motor vehicle registration systems, and some law firms.

The early sociological theorist **Max Weber** (1947/1925) analyzed the classic characteristics of a bureaucracy. These characteristics represent what he called an **ideal type**—a model rarely seen in reality but that defines the principal characteristics of a social form. The characteristics of bureaucracies described as an ideal type are:

1. *High degree of division of labor and specialization.* The notion of the specialist embodies this criterion. Bureaucracies ideally employ specialists in the various positions and occupations, and these specialists are responsible for a specific set of duties. Sociologist Charles Perrow (2007, 1994, 1986) notes that many modern bureaucracies have hierarchical authority structures and an elaborate division of labor.

2. *Hierarchy of authority.* In bureaucracies, positions are arranged in a hierarchy so that each is under the supervision of a higher position. Such hierarchies are often represented in an *organization chart*, a diagram in the shape of a pyramid that shows the relative rank of each position plus the lines of authority between each. These lines of authority are often called the "chain of command," and they show not only who has authority, but also who is responsible to whom and how many positions are responsible to a given position.

3. *Rules and regulations.* All the activities in a bureaucracy are governed by a set of detailed rules and procedures. These rules are designed, ideally, to cover almost every possible situation and problem that might arise, including hiring, firing, salary scales, and rules for sick pay and absences.

4. *Impersonal relationships.* Social interaction in the (ideal) bureaucracy is supposed to be guided by *instrumental* criteria, such as the organization's rules, rather than by *expressive needs*, such as personal attractions or likes and dislikes. The ideal is that the objective application of rules will minimize matters such as personal favoritism—giving someone a promotion simply because you like him or her or firing someone because you do not like him or her. Of course, as we will

see, sociologists have pointed out that bureaucracy has "another face"—the *informal* social interaction that keeps the bureaucracy working and often involves interpersonal friendships and social ties, typically among people taken for granted in these organizations, such as the support staff.

5. *Career ladders.* Candidates for the various positions in the bureaucracy are supposed to be selected on the basis of specific criteria, such as education, experience, and standardized examinations. The idea is that advancement through the organization becomes a career for the individual. Some organizations, such as some universities and some law firms, have a policy of *tenure*—a guarantee of continued employment until one's retirement from the organization.

6. *Efficiency.* Bureaucracies are designed to coordinate the activities of many people in pursuit of organizational goals. Ideally, all activities have been designed to maximize this efficiency. The whole system is intended to keep social–emotional relations and interactions at a minimum and instrumental interaction at a maximum.

Bureaucracy's "Other Face"

All the characteristics of Weber's "ideal type" are general defining characteristics. Rarely do actual bureaucracies meet this exact description. A bureaucracy has, in addition to the ideal characteristics of structure, an *informal structure*. This includes social interactions, even network connections, in bureaucratic settings that ignore, change, or otherwise bypass the formal structure and rules of the organization. This informal structure often develops among those who are taken for granted in organizations, such as secretaries and administrative assistants—who are most often women. Sociologist Charles Page (1946) coined the phrase "bureaucracy's other face" to describe this condition.

This other face is informal culture. It has evolved over time as a reaction to the formality and impersonality of the bureaucracy. Thus administrative assistants and secretaries will sometimes "bend the rules a bit" when asked to do something more quickly than usual for a boss they like and bend the rules in another direction for a boss they do not like by slowing down or otherwise sabotaging the boss's work. Researchers have noted, for example, that secretaries and assistants may well have more authority than their job titles and salaries suggest. As a way around the cumbersome formal communication channels within the organization, the informal network, or "grapevine," often works better, faster, and sometimes even more accurately than the formal channels. As with any culture, the informal culture in the bureaucracy has its own norms or rules. One is not supposed to "stab friends in the back," such as by "ratting on" them to a boss or spreading a rumor about them that is intended to get them fired. Yet, just as with any norms, there is deviation from the norms, and "backstabbing" and "ratting" does happen.

Bureaucracy's other face can also be seen in the workplace subcultures that develop, even in the largest bureaucracies. Some sociologists interpret the subcultures that develop within bureaucracies as people's attempts to humanize an otherwise impersonal organization. Keeping photographs of family and loved ones in the office, placing personal decorations on one's desk (if allowed), and organizing office parties are some ways people resist the impersonal culture of bureaucracies. Of course, this informal culture can also become exclusionary, increasing the isolation that some workers feel at work. Gay and lesbian workers may feel left out when other workers gossip about people's heterosexual dates; minority workers may be excluded from the casual conversations in the workplace that connect nonminority people to one another.

The informal norms that develop within the modern-day bureaucracy often cause worker productivity to go up or down, depending on the norms and how they are informally enforced. The classic 1930s Hawthorne studies, so named because they were carried out at the Western Electric telephone plant in Hawthorne, Illinois (Roethlisberger and Dickson 1939), discovered that small groups of workers developed their own ideas—their own norms—about how much work they should produce each day. If someone produced too many completed tasks in a day, he would make the rest of the workers "look bad" and run the risk of having the organization raise its expectations of how much work the group might be expected to produce. Because of this, anyone producing too much was informally labeled a "rate buster," and that person was punished by some act, such as punches on the shoulder (called "binging") or by group ridicule (called "razzing"). By the same token, one could be accused of producing too little, in which case he was labeled a "chiseler" and punished in the same way by either binging or razzing. This informal culture of bureaucracy's other face continues today in a manner similar to the culture initially discovered in the early Hawthorne studies (Ritzer 2010; Perrow 2007, 1986).

Problems of Bureaucracies

In contemporary times, problems have developed that grow out of the nature of the complex bureaucracy. Two problem areas already discussed are the occurrence of risky shift in work groups and the development of groupthink. Additional problems include a tendency to *ritualism* and the potential for *alienation* on the part of those within the organization.

Ritualism. Rigid adherence to rules can produce a slavish following of them, regardless of whether it accomplishes the purpose for which the rule was originally designed. The rules become ends in themselves rather than means to an end: This is **organizational ritualism**.

A classic example of the consequences of *organizational ritualism* was the tragedy involving the space shuttle *Challenger* in 1986. Only seconds after liftoff, as hundreds watched live, the *Challenger* exploded, killing schoolteacher Christa McAuliffe, and six other crew members. Many still remember where they were and exactly what they were doing when they heard about the tragedy. The failure of the essential O-ring gaskets on the solid fuel booster rockets of the *Challenger* shuttle caused the catastrophic explosion. It was revealed later that the O-rings were known to become brittle at below-freezing temperatures, as was the temperature at the launch pad the evening before the *Challenger* lifted off.

Why did the managers and engineers at NASA (National Aeronautics and Space Administration) allow the shuttle to lift off given these conditions and their prior knowledge? The managers had all the information about the O-rings before the launch. Furthermore, engineers had warned them against the danger. In a detailed analysis of the decision to launch, sociologist Diane Vaughan (1996) uncovered both risky shift and organizational ritualism within the organization. The NASA insiders, confronted with signals of danger, proceeded as if nothing was wrong when they were repeatedly faced with the evidence that something was indeed *very* wrong. They in effect *normalized* their own behavior so that their actions became acceptable to them, representing nothing out of the ordinary. This is an example of organizational ritualism, as well as what Vaughan calls the "normalization of deviance."

Unfortunately, history repeated itself on February 1, 2003, when the space shuttle *Columbia*, upon its return from space, broke up in a fiery descent into the atmosphere above Texas, killing all who were aboard. The evidence shows that a piece of hard insulating foam separated from an external fuel tank during launch and struck the shuttle's left wing, damaging it and dislodging its heat-resistant tiles that are necessary for reentry. The absence of these tiles caused a burn-up upon reentry into the atmosphere. With eerie similarity to the earlier 1986 *Challenger* accident, subsequent analysis concluded that a "flawed institutional culture" and—citing sociologist Diane

Vaughan—a normalization of deviance accompanying a gradual erosion of safety margins were among the causes of the *Columbia* accident (Schwartz and Wald 2003).

No single individual was at fault in either accident. The story is not one of evil but rather of the ritualism of organizational life in one of the most powerful bureaucracies in the United States. It is a story of rigid group conformity within an organizational setting and of how deviant behavior is redefined, that is, socially constructed, just as also happened in the Penn State Sandusky scandal. Organizational culture overshadows individual good judgment, creating a decrease in safety and increased risk. This is one of the hazards of organizational behavior.

Alienation. The stresses on rules and procedures within bureaucracies can result in a decrease in the overall cohesion of the organization. This often psychologically separates a person from the organization and its goals. This state of *alienation* results in increased turnover, tardiness, absenteeism, and overall dissatisfaction with the organization.

Alienation can be widespread in organizations where workers have little control over what they do or where workers themselves are treated like machines employed on an assembly line, doing the same repetitive action for an entire work shift. Alienation is not restricted to manual labor, however. In organizations where workers are isolated from others, where they are expected only to implement rules, or where they think they have little chance of advancement, alienation can be common. As we will see, some organizations have developed new patterns of work to try to minimize worker alienation and thus enhance their productivity.

The McDonaldization of Society

Sometimes the problems and peculiarities of bureaucracy can have effects on the total society. This has been the case with what George Ritzer (2010) has called **McDonaldization**, a term coined from the well-known fast-food chain. In fact, 90 percent of U.S. children between ages 3 and 9 visit McDonald's each month! Ritzer noticed that the principles that characterize fast-food organizations are increasingly dominating more aspects of U.S. society, indeed, of societies around the world. McDonaldization refers to the increasing and ubiquitous presence of the fast-food *model* in most organizations that shape daily life. Work, travel, leisure, shopping, health care, politics, and even education have all become subject to McDonaldization. Each industry is based on a principle of high and efficient productivity, which translates into a highly rational social

Evidence of the "McDonaldization of society" can be seen everywhere, perhaps including on your own campus. Shopping malls, food courts, sports stadiums, even cruise ships reflect this trend toward standardization.

Bloomberg/Getty Images

organization, with workers employed at low pay but with customers experiencing ease, convenience, and familiarity.

Ritzer argues that McDonald's has been such a successful model of business organization that other industries have adopted the same organizational characteristics, so much so that their nicknames associate them with the McDonald's chain: McPaper for *USA Today*, McChild for child-care chains like KinderCare, and McDoctor for the drive-in clinics that deal quickly and efficiently with minor health and dental problems. Finally, the efficiency and predictability characteristic of the various Starbuck coffee shops represent what Ritzer calls "Starbuckization"—a type of McDonaldization.

Based in part upon Max Weber's concept of the ideal bureaucracy mentioned earlier, Ritzer identifies four dimensions of the McDonaldization process: efficiency, calculability, predictability, and control:

1. *Efficiency* means that things move from start to finish in a streamlined path. Steps in the production of a hamburger are regulated so that each hamburger is made exactly the same way—hardly characteristic of a home-cooked meal. Business

can be even more efficient if the customer does the work once done by an employee. In fast-food restaurants, the claim that you can "have it your way" really means that you assemble your own sandwich or salad.

2. *Calculability* means there is an emphasis on the quantitative aspects of products sold: size, cost, and the time it takes to get the product. At McDonald's, branch managers must account for the number of cubic inches of ketchup used per day; likewise, ice cream scoopers in chain stores measure out predetermined and exact amounts of ice cream.

3. *Predictability* is the assurance that products will be exactly the same, no matter when or where they are purchased. Eat an Egg McMuffin in New York, and it will likely taste just the same as an Egg McMuffin in Los Angeles or Paris! Ditto for a tall decaf cappuccino from Starbucks.

4. *Control* is the primary organizational principle that lies behind McDonaldization. Behavior of the customers and workers is reduced to a series of machinelike actions. Ultimately, efficient technologies replace much of the work that humans once performed.

McDonaldization clearly brings many benefits. There is a greater availability of goods and services to a wide proportion of the population; instantaneous service and convenience to a public with less free time; predictability and familiarity in the goods bought and sold; and standardization of pricing and uniform quality of goods sold, to name a few benefits. However, this increasingly rational system of goods and services also spawns irrationalities. For example, the majority of workers at McDonald's lack full-time employment, have no worker benefits, have no control over their workplace, and quit on average after only four or five months.

Diversity in Organizations

The hierarchical structuring of positions within organizations results in the concentration of power and influence with a few individuals at the top. Because organizations tend to reflect patterns within the broader society, this hierarchy, like that of society, is marked by inequality in race, gender, and class relations. Although the concentration of power in organizations is incompatible with the principles of a democratic society, organizations are structured by hierarchies and discrimination is still quite pervasive. Especially among the power elite, White men still predominate. And studies find that, even though the presence of women and people of color is growing in positions of organizational leadership, they tend to take on the same values as the dominant group, evidence once again of group conformity (Zweigenhaft and Domhoff 2006).

A classic study by Rosabeth Moss Kanter (1977) shows how the structure of organizations leads to obstacles in the advancement of groups who are underrepresented in the organization. People who are underrepresented in the organization become tokens; they feel put "out front" and under the all-too-watchful eyes of their superiors as well as peers. As a result—as research since Kanter's has shown—they often suffer severe stress (Smith 2007; Jackson 2000). They may be assumed to be incompetent, getting their position simply because they are women, minorities, or both—even in instances where the person has had superior admissions qualifications. This is stressful for a person and shows that tokenism can have very negative consequences (Guttierez y Muhs et al. 2012).

Social class, in addition to race and gender, plays a part in determining people's place within formal organizations. Employees of middle- and upper-class origins in organizations make higher salaries and wages and are more likely to get promoted than are people

Few organizational boards and executive committees contain minorities and women: When present, they are often tokens.

of lower social class origins, even for individuals who are of the same race or ethnicity. This even holds for people coming from families of lower social class status who are as well educated as their middle- and upper-class coworkers. Thus their lower salaries and lack of promotion cannot necessarily be attributed to a lack of education. In this respect, their treatment in the bureaucracy only perpetuates rather than lessens the negative effects of the social class system in the United States.

The social class stratification system in the United States produces major differences in the opportunities and life chances of individuals, and the bureaucracy simply carries these differences forward. Class stereotypes also influence hiring practices in organizations. Personnel officers look for people with "certain demeanors," a code phrase for those who convey middle-class or upper-middle-class standards of dress, language, manners, and so on, which some people may be unable to afford or may not possess.

Even as the structure of organizations reproduces the race, class, and gender inequalities that permeate society, ample research now finds that diversity within organizations has numerous benefits. Diverse groups—that is, diverse people—bring different experiences and perspectives to organizations and to organizational decision making. Of course, the problem is that there is still pressure on such people to conform to the dominant culture of the organization. That pressure to confirm can silence dissent, as groupthink would suggest, especially if those who bring diversity to the organization, such as women, gays, lesbians, transgender people, and people of color, are treated as tokens or silenced because they are different. When people are tokens in organizations, they are pressured not to stand out or, when they speak out, they may be ignored—or, worse, others take credit for their ideas.

Nonetheless, new research on diversity is consistently demonstrating the benefit of diversity for all kinds of organizations. In schools, all students learn more when in classrooms where there are people from different backgrounds (Gurin et al. 2002). In business organizations, racial diversity is associated with increased sales revenues, more customers, a stronger market share, and higher profits (Herring 2009). There is ample evidence that companies are now much more aware of the fact that innovation is more likely to occur in diverse work organizations (Page 2007). On college campuses, more cross-race interaction produces a more positive campus climate (Valentine et al. 2012).

debunking SOCIETY'S MYTHS

MYTH: Diversity is a real problem for organizations.
SOCIOLOGICAL PERSPECTIVE: Despite the challenges posed by trying to create more diverse work organizations, research shows that more diverse organizations have greater profits and are more innovative (Herring 2009; Page 2007). •

FUNCTIONALISM, CONFLICT THEORY, AND SYMBOLIC INTERACTION: THEORETICAL PERSPECTIVES

All three major sociological perspectives—functionalism, conflict theory, and symbolic interaction—are exhibited in the analysis of formal organizations and bureaucracies (see Table 6.1). The functional perspective, based

table 6.1 Theoretical Perspective on Organizations

	Functionalist Theory	Conflict Theory	Symbolic Interaction Theory
Central Focus	Positive functions (such as efficiency) contribute to unity and stability of the organization.	Hierarchical nature of bureaucracy encourages conflict between superior and subordinate, men and women, and people of different racial or class backgrounds.	Stresses the role of self in the bureaucracy and how the self develops and changes.
Relationship of Individual to the Organization	Individuals, like parts of a machine, are only partly relevant to the operation of the organization.	Individuals are subordinated to systems of power and experience stress and alienation as a result.	Interaction between superiors and subordinates forms the structure of the organization.
Criticism	Hierarchy can result in dysfunctions such as ritualism and alienation.	De-emphasizes the positive ways that organizations work.	Tends to downplay overall social organization.

© Cengage Learning

in this case on the early writing of Max Weber, argues that certain functions, called *eufunctions* (that is, positive functions), characterize bureaucracies and contribute to their overall unity. The bureaucracy exists to accomplish these eufunctions, such as efficiency, control, impersonal relations, and a chance for individuals to develop a career within the organization. As we have seen, however, bureaucracies develop the "other face" (informal interaction and culture, as opposed to formal or bureaucratic interaction and culture) as well as problems of ritualism and alienation of people from the organization. These latter problems are called *dysfunctions* (negative functions), which have the consequence of contributing to disunity, lack of harmony, and less efficiency in the bureaucracy.

The conflict perspective argues that the hierarchical or stratified nature of the bureaucracy in effect encourages rather than inhibits conflict among individuals within it. These conflicts are between superior and subordinate, as well as between racial and ethnic groups, men and women, and people of different social class backgrounds, hampering smooth and efficient running of the bureaucracy. Furthermore, conflict theory helps us understand the power structures that exist in organizations—both the formal ones that come from the organizational hierarchy and the less formal ways that power is exercised between people and among groups within the organization.

Symbolic interaction theory stresses the role of the self in any group and especially how the self develops as a product of social interaction. Within organizations, people may feel that their "self" becomes subordinated to the larger structure of the organization. This is especially true in bureaucratic organizations where individuals often feel overwhelmed by the sheer complexity of working through bureaucratic structures. But symbolic interaction also emphasizes the creativity of human beings as social actors and thus would be a good perspective to use if analyzing how people change organizational structures and cultures.

chapter summary

What are the types of groups?
Groups are a fact of human existence and permeate virtually every facet of our lives. Group size is important, as is the otherwise simple distinction between dyads and triads. *Primary groups* form the basic building blocks of social interaction in society. *Reference groups* play a major role in forming our attitudes and life goals, as do our relationships with in-groups and out-groups. *Social networks* partly determine things such as who we know and the kinds of jobs we get. Networks based on race—ethnicity, social class, and other social factors are extremely closely connected—are very dense.

How strong is social influence?
The social influence groups exert on us is tremendous, as seen by the Asch conformity experiments. The Milgram experiments demonstrated that the interpersonal influence of an authority figure can cause an individual to act against his or her deep convictions. The torture and abuse of Iraqi prisoners of war by American soldiers/prison guards serves as testimony to the powerful effects of both social influence and authority structures. The Iraqi tortures were in effect experimentally predicted by a simulated prison study done in the United States over thirty years earlier.

What is the importance of groupthink and risky shift?
Groupthink can be so pervasive that it adversely affects group decision making and often results in group decisions that by any measure are simply stupid. *Risky shift* (and *polarization shift*) similarly often compel individuals to reach decisions that are at odds with their better judgment.

What are the types of formal organizations and bureaucracies, and what are some of their problems?
There are several types of *formal organizations*, such as *normative, coercive,* or *utilitarian*. Weber typified *bureaucracies* as organizations with an efficient division of labor, an authority hierarchy, rules, impersonal relationships, and career ladders. Bureaucratic rigidities often result in organizational problems such as ritualism and resulting "normalization of deviance." The *McDonaldization* of society has resulted in greater efficiency, calculability, and control in many industries, probably at the expense of some individual creativity. Formal organizations perpetuate society's inequalities on the basis of race–ethnicity, gender, and social class. Current research finds however that innovation in organizations is more likely if there is greater diversity—and thus a variety of perspectives—within the organization.

What do functional, conflict, and symbolic interaction theories say about organizations? Functional, conflict, and symbolic interaction theories highlight and clarify the analysis of organizations by specifying both organizational functions and dysfunctions (*functional theory*); by analyzing the consequences of hierarchical, gender, race, and social class conflict in organizations (*conflict theory*); and, finally, by studying the importance of social interaction and integration of the self into the organization (*symbolic interaction theory*).

Key Terms

attribution error 128
attribution theory 128
bureaucracy 136
coalition 125
coercive organization 136
deindividuation 135
dyad 125
expressive needs 127

formal organization 135
group 124
group size effect 125
groupthink 134
ideal type 137
instrumental needs 127
McDonaldization 139

normative
  organization 136
organizational culture 135
organizational
  ritualism 138
polarization shift 135
primary group 125

reference group 127
risky shift 134
secondary group 127
social network 129
total institution 136
triad 125
utilitarian organization 136

7 Deviance and Crime

Defining Deviance

Sociological Theories of Deviance

Forms of Deviance

Crime and Criminal Justice

Chapter Summary

in the early 1970s, an airplane carrying forty members of an amateur rugby team crashed in the Andes Mountains in South America. The twenty-seven survivors were marooned at 12,000 feet in freezing weather and deep snow. There was no food except for a small amount of chocolate and some wine. A few days after the crash, the group heard on a small transistor radio that the search for them had been called off.

Scattered in the snow were the frozen bodies of dead passengers. Preserved by the freezing weather, these bodies became, after a time, sources of food. At first, the survivors were repulsed by the idea of eating human flesh, but as the days wore on, they agonized over the decision about whether to eat the dead crash victims, eventually concluding that they had to eat if they were to live.

In the beginning, only a few ate the human meat, but soon the others began to eat too. The group experimented with preparations as they tried different parts of the body. They developed elaborate rules (social norms) about how, what, and whom they would eat. Some could not bring themselves to cut the meat from the human body, but would slice it once someone else had cut off large chunks. They all refused to eat certain parts—the lungs, skin, head, and genitals.

After two months, the group sent out an expedition of three survivors to find help. The group was rescued, and the world learned of their ordeal. Their cannibalism (the eating of other human beings) generally came to be accepted as something they had to do to survive. Although people might have been repulsed by the story, the survivors' behavior was understood as a necessary adaptation to their life-threatening circumstances. The survivors also maintained a sense of themselves as good people even though what they did profoundly violated ordinary standards of socially acceptable behavior in most cultures in the world (Henslin 1993; Miller 1991; Read 1974).

Was the behavior of the Andes crash survivors socially deviant? Were the people made crazy by their experience, or was this a normal response to extreme circumstances?

Compare the Andes crash to another case of human cannibalism. In 1991, in Milwaukee, Wisconsin, Jeffrey Dahmer pled guilty to charges of murdering at least fifteen men in his home. Dahmer lured the men—eight of them African American, two White, and one a fourteen-year-old Laotian (Asian) boy—to his apartment, where he murdered and dismembered them, then cooked and ate some of their body parts. For those he considered most handsome, he boiled the flesh from their heads so that he could save and admire their skulls. Dahmer was seen as a total social deviant, someone who violated every principle of human decency. Even hardened criminals were disgusted by Dahmer. In fact, he was killed in prison by another inmate in 1994.

Why was Dahmer's behavior considered so deviant when that of the Andes survivors was not? The answer can be found by looking at the situation in which these behaviors occurred. For the Andes survivors, eating human flesh was essential for survival. For Dahmer, however, it was murder. From a sociological perspective, the deviance of cannibalism resides not just in the act itself but also in the social context in which it occurs. The exact same behavior—eating other human beings—is considered reprehensible in one context and acceptable in another. That is the essence of the sociological explanation: The nature of deviance is not only in the personality of the deviant person, nor is it inherently in the deviant act itself. Instead, it is a product of social structure.

learning objectives

- Using the different theories of deviance, understand that deviant behavior results from a combination of social-cultural and individual factors
- Present a sociological definition of deviance
- Describe different "forms" of deviance and be able to give examples
- Understand how crime is measured

DEFINING DEVIANCE

Sociologists define **deviance** as *behavior that is recognized as violating expected rules and norms*. Deviance is more than simple nonconformity; it is behavior that departs significantly from social expectations. In the sociological perspective on deviance, there are four main identifying characteristics:

- Deviance emerges in a social context, not just the behavior of individuals; sociologists see deviance in terms of group processes and judgments.
- Not all behaviors are judged similarly by all groups; what is deviant to one group may be normative (not deviant) to another.

- Established rules and norms are socially created, not just morally decided or individually imposed.
- Deviance lies not just in behavior itself but also in the social responses of groups to behavior by others.

Sociological Perspectives on Deviance

Strange, unconventional, or nonconformist behavior is often understandable in its sociological context. Consider suicide. Are people who commit suicide mentally disturbed, or might their behavior be explained by social factors? Think about it. There are conditions under which suicide may well be acceptable behavior—for example, someone who commits suicide in the face of a terminal illness compared to a despondent person who jumps from a window.

Sociologists distinguish two types of deviance: formal and informal. *Formal deviance* is behavior that breaks laws or official rules. Crime is an example. There are formal sanctions against formal deviance, such as imprisonment and fines. *Informal deviance* is behavior that violates customary norms. Although such deviance may not be specified in law, it is judged to be deviant by those who uphold the society's norms.

The study of deviance can be divided into the study of why people violate laws or norms and the study of how society reacts. *Labeling theory* is discussed in detail later, but it recognizes that deviance is not just in the breaking of norms or rules but it includes how people react to those behaviors. Social groups are known to actually *create* deviance "by making the rules whose infraction constitutes deviance, and by applying those rules to particular people and labeling them as outsiders" (Erikson 1994, 1966; Becker 1963: 9).

The Context of Deviance. Even the most unconventional behavior can be understood if we know the context in which it occurs. Behavior that is deviant in one circumstance may be normal in another, or behavior may be ruled deviant only when performed by certain people. For example, people who break gender stereotypes may be judged as deviant even though their behavior is considered normal for the other sex. Heterosexual men and women who kiss in public are the image of romance; lesbians and gay men who even dare to hold hands in public are often seen as flaunting their sexual orientation and are thus regarded as "deviant."

The definition of deviance can also vary over time. Acquaintance rape (also called "date rape"), for example, was not considered social deviance until fairly recently. Women have been presumed to mean yes when they said no, and men were expected to "seduce" women through aggressive sexual behavior. Even now, women who are raped by someone they know may not think of it as rape. If they do, they may find that prosecuting the offender is difficult because others do not think of it as rape, especially under certain circumstances, such as the woman being drunk. Thus, what is in fact rape may not be seen as such by everyone.

The sociologist Emile Durkheim (covered in detail later in this chapter) argued that one reason acts of deviance are publicly punished is that the social order is threatened by deviance. Judging those behaviors as deviant and punishing them confirms general social standards. Therein lies the value of widely publicized trials, public executions, or the historical practice of displaying a wrongdoer in the stocks, which held one's feet fast, or the pillory, which held the hands and head. Passersby were permitted to hurl both stones and large rocks at those people so immobilized. The punishment affirms the collective beliefs of the society, reinforces social order, and inhibits future deviant behavior, especially as defined by those with the power to judge others.

---

debunking SOCIETY'S MYTHS

MYTH: Deviance is bad for society because it disrupts normal life.

SOCIOLOGICAL PERSPECTIVE: Deviance tends to stabilize society. By defining some forms of behavior as deviant, people are affirming the social norms of groups. In this sense, society actually *creates* deviance to some extent. ●

Durkheim argued that societies actually *need* deviance to know what presumably normal behavior is. In this sense, Durkheim considered deviance "functional" for society (Erikson 1994, 1966; Durkheim 1951/1897). You could observe Durkheim's point in the aftermath of the terrorist attacks on New York City's World Trade Center towers on September 11, 2001—known now as "9/11." Horrified by the sight of hijacked planes flying into the World Trade Center and the Pentagon and crashing in a Pennsylvania field, U.S. citizens responded through publicly demonstrating strong patriotism. Durkheim would interpret these terrorist acts as deviance producing strong social solidarity. This was one of Durkheim's most important insights: *Deviance produces social solidarity. Instead of breaking society up, deviance produces a pulling together, or social solidarity.*

The Influence of Social Movements. The perception of deviance may also be influenced by social movements, which are networks of groups that organize to support or resist changes in society (see Chapter 16). With a change in the social climate, formerly acceptable behaviors may be newly defined as deviant. Smoking, for instance, was once considered glamorous, sexy, and "cool." Now, smokers are widely scorned as polluters and, despite strong lobbying by the tobacco industry, regulations against smoking have proliferated.

Whereas only 17 percent of the public in 1987 thought that smoking should be banned in restaurants, over half (59 percent) thought so by 2010 (Gallup Organization 2010). Even the public's perception of danger from secondhand smoke has increased dramatically, from 36 percent in 1994 perceiving secondhand smoke to be dangerous to 55 percent in 2010 (Gallup Organization 2010). The increase in public disapproval of smoking results as much from social and political movements as it does from the known health risks. The success of the antismoking movement has come from the mobilization of constituencies able to articulate to the public that smoking is dangerous. Note that the key element here is the ability of people to mobilize— not just the evidence of risk. In other words, there has to be a social response for deviance to be defined as such; scientific evidence of harm in and of itself is not enough.

The Social Construction of Deviance. Perhaps because it violates social conventions or because it sometimes involves unusual behavior, deviance captures the public imagination. Commonly, however, the public understands deviance as the result of individualistic or personality factors. Many people see deviants as crazy, threatening, "sick," or in some other ways inferior, but sociologists see deviance as influenced by society— the same social processes and institutions that shape all social behavior.

Deviance, for example, is not necessarily irrational or "sick" and may be a positive and rational adaptation to a situation. Think of the Andes survivors discussed in this chapter's opener. Was their action (eating human flesh) irrational, or was it an inventive and rational response to a dreadful situation? To use another example, are gangs the result of the irrational behavior of maladjusted youth, or are they rational responses to social situations?

Sociological studies of gangs in the United States shed light on this question. The family situations of gang members are often problematic, although girls in gangs tend to be more isolated from their families than are boys in gangs (Fleisher 2000; Esbensen-Finn et al. 1999). Given the class, race, and gender inequality minority youth face, many turn to gangs for the social support they lack elsewhere (Walker-Barnes and Mason 2001; Moore and Hagedorn 1996). For example, some poor, young Puerto Rican girls live in relatively confined social environments with little opportunity for educational or occupational advancement. Their community expects them to be "good girls" and to remain close to their families. Joining a gang is one way to reject these restrictive roles (Messerschmidt 1997; Campbell 1987). Are these young women irrational or just doing the best they can to adapt to their situation? Sociologists interpret their behavior as an understandable adaptation to conditions of poverty, racism, and sexism.

Once considered "cool," smokers are now considered to be deviants, scorned as polluters, and often banished to outside office buildings, as here.

Jeff Greenberg/PhotoEdit

tattooing were associated with gangs and disrespectable people only a few years ago, it is now considered fashionable among young, middle-class people.

In sum, a sociological perspective on deviance asks: Why is deviance more common in some groups than others? Why are some more likely to be labeled deviant than others, even if they engage in the exact same behavior? How is deviance related to patterns of inequality in society? Sociologists do not ignore individual psychology but integrate it into an explanation of deviance that focuses on the social conditions surrounding the behavior, going beyond explanations of deviance that root it in the individual personality.

The Medicalization of Deviance. Commonly, people will say that someone who commits a very deviant act is "sick." This common explanation is what sociologists call the **medicalization of deviance** (Conrad and Schneider 1992). Medicalizing deviance attributes deviant behavior to a "sick" state of mind, where the solution is to "cure" the deviant through therapy or other psychological treatment.

An example is found in alcoholism. Some evidence indicates that alcoholism may have a genetic basis, and certainly alcoholism must be understood at least in part in medical terms, but viewing alcoholism *solely* from a medical perspective ignores the social causes that influence the development and persistence of this behavior. Practitioners know that medical treatment alone does

Also, in some subcultures or situations, deviant behavior is encouraged and praised. Have you ever been egged on by friends to do something that you thought was deviant, or have you done something you knew was wrong? Many argue that the reason so many college students drink excessively is that the student subculture encourages them to do so—even though students know it is harmful. Similarly, the juvenile delinquent regarded by school authorities as defiant and obnoxious is rewarded and praised by peers for the very behaviors that school authorities loathe. Much deviant behavior occurs, or escalates, because of the social support received from others. (Recall the discussion of risky shift in Chapter 6.)

Some behavior patterns defined as deviant are also surprisingly similar to so-called normal behavior. Is a heroin addict who buys drugs with whatever money he can find so different from a business executive who spends a large proportion of his discretionary income on alcohol? Each may establish a daily pattern that facilitates drug use; each may select friends based on shared interests in drinking or taking drugs; and each may become so physically, emotionally, and socially dependent on their "fix" that life seems unimaginable without it. Which of the two is more likely to be considered deviant?

The point is that deviance is both created and defined within a social context. It is not just weird, pathological, or irrational behavior. Sociologists who study deviance understand it in the context of social relationships and society. They define deviance in terms of existing social norms and the social judgments people make about one another. Indeed, deviant behavior can sometimes be indicative of changes that are taking place in the cultural folkways. Whereas body piercing and

Dieter Wertz/Alamy

What seems deviant to some observers may seem perfectly ordinary to others, illustrating the point that the meaning of deviance occurs in a social context.

not solve the problem. The social relationships, social conditions, and social habits of those with alcoholism must be altered, or the behavior is likely to recur.

thinking SOCIOLOGICALLY

Ask some of your friends to explain why rape occurs. What evidence of the *medicalization of deviance* exists in your friends' answers? ●

Sociologists criticize the medicalization of deviance for ignoring the effects of social structures on the development of deviant behavior. From a sociological perspective, deviance originates in society, not just in individuals. Changing the incidence of deviant behavior requires changes in society in addition to changes in individuals. Deviance, to most sociologists, is not a pathological state but an *adaptation to the social structures* in which people live. Factors such as family background, social class, racial inequality, and the social structure of gender relations in society produce deviance, and these factors must be considered to explain it.

SOCIOLOGICAL THEORIES OF DEVIANCE

Sociologists have drawn on several major theoretical traditions to explain deviant behavior, including functionalism, conflict theory, and symbolic interaction theory.

Functionalist Theories of Deviance

Recall that functionalism is a theoretical perspective that interprets all parts of society, even those that may seem dysfunctional, as instead contributing to the stability of the whole. At first glance, deviance seems to be dysfunctional for society. Functionalist theorists argue otherwise (see Table 7.1). They contend that deviance is functional because it creates social cohesion. Branding certain behaviors as deviant provides contrast with behaviors that are considered normal, giving people a heightened sense of social order. Norms are meaningless unless there is deviance from them; thus

deviance is necessary to clarify what society's norms are. Group coherence then comes from sharing a common definition of legitimate, as well as deviant, behavior. The collective identity of a group is affirmed when group members ridicule or condemn others they define as deviant. To give an example, think about how many people define gay men as deviant. Although lesbians and gay men have rejected this label, labeling homosexuality as deviant is one way of affirming the presumed normality of heterosexual behavior. Labeling someone else an "outsider" is, in other words, a way of affirming one's "insider" identity (Becker 1963).

Durkheim: The Study of Suicide. The functionalist perspective on deviance stems originally from the work of **Emile Durkheim**. Recall that one of Durkheim's central concerns was how society maintains its coherence (or social order). Durkheim saw deviance as functional for society because it produces solidarity among society's members. He developed his analysis of deviance in large part through his analysis of suicide. Through this work, he discovered a number of important sociological points. First, he criticized the usual psychological interpretations of why people commit suicide, turning instead to sociological explanations with data to back them up. Second, he emphasized the role of social structure in producing deviance. Third, he pointed to the importance of people's social attachments to society in understanding deviance. Finally, he elaborated the functionalist view that deviance provides the basis for social cohesion. His studies of suicide illustrate these points.

Durkheim was the first to argue that the causes of suicide were to be found in social factors, not individual personalities. Observing that the rate of suicide in a society varied with time and place, Durkheim looked for causes linked to these factors other than emotional stress. Durkheim argued that suicide rates are affected by the different social contexts in which they emerge. He looked at the degree to which people feel integrated into the structure of society and their social surroundings as social factors producing suicide.

Durkheim analyzed three types of suicide: anomic suicide, altruistic suicide, and egoistic suicide. **Anomie,**

table 7.1 Sociological Theories of Deviance

Functionalist Theory	Symbolic Interaction Theory	Conflict Theory
Deviance creates social cohesion.	Deviance is a learned behavior, reinforced through group membership.	Dominant classes control the definition of and sanctions attached to deviance.
Deviance results from structural strains in society.	Deviance results from the process of social labeling, regardless of the actual commission of deviance.	Deviance results from social inequality in society.
Deviance occurs when people's attachment to social bonds is diminished.	Those with the power to assign deviant labels themselves produce deviance.	Elite deviance and corporate deviance go largely unrecognized and unpunished.

© Cengage Learning

as defined by Durkheim, is the condition that exists when social regulations in a society break down: The controlling influences of society are no longer effective, and people exist in a state of relative normlessness. The term *anomie* refers not to an individual's state of mind, but instead to social conditions.

Anomic suicide occurs when the disintegrating forces in the society make individuals feel lost or alone. Teenage suicide is often cited as an example of anomic suicide. Studies of college campuses, for example, trace the cause of campus suicides to feelings of depression and hopelessness (Langhinrichsen-Rohling et al. 1998). As already noted in Chapter 1, the recent increase in suicide among returning veterans may well constitute anomic suicide, for example, if they return from war feeling as if no one understands them. Suicide is more likely committed by those who have been sexually abused as children or by those whose parents are alcoholics (Thakkar et al. 2000; Bryant and Range 1997).

Altruistic suicide occurs when there is excessive regulation of individuals by social forces. An example is someone who commits suicide for the sake of a religious or political cause. For example, after hijackers on September 11, 2001 ("9/11") took control of four airplanes—crashing two into the World Trade Center in New York, one into the Pentagon, and despite the intervention of passengers, one into a Pennsylvania farm field—many wondered how anyone could do such a thing, killing themselves in the process. Although sociology certainly does not excuse such behavior, it can help explain it. Terrorists and suicide bombers are so regulated by their extreme beliefs that they are willing to die and kill as many people as possible to achieve their goals. As Durkheim argued, altruistic suicide results when individuals are excessively dominated by the expectations of their social group. People who commit altruistic suicide subordinate themselves to collective expectations, even when death is the result.

Egoistic suicide occurs when people feel totally detached from society. This helps explain the high rate of suicide among the elderly in the United States. People over seventy-five years of age have one of the highest rates of suicide, presumably because the elderly lose many of their social ties to society (National Center for Health Statistics 2010a). Ordinarily, people are integrated into society by work roles, ties to family and community, and other social bonds. When these bonds are weakened through retirement or loss of family and friends, the likelihood of egoistic suicide increases.

Egoistic suicide is also more likely to occur among people who are not well integrated into social networks (Berkman et al. 2000). Thus it should not be surprising that women have lower suicide rates than men (National Center for Health Statistics 2010a). Sociologists explain this fact as a result of men being less embedded in social relationships of care and responsibility than women (Watt and Sharp 2001).

Durkheim's major point is that suicide is a social, not just an individual, phenomenon. Recall from Chapter 1 that Durkheim sees sociology as the discovery of the social forces that influence human behavior. One recent study shows that these social forces that affect behavior such as suicides among youths are *multilevel:* The higher the degree of integration of the individual into structural "levels," such as the family, the peer group, religion, the neighborhood, and the school, then the lower the risk of suicide (Maimon and Kuhl 2008). As individualistic as suicide might seem, Durkheim uncovered the influence of social structure even here. In fact, suicide also varies considerably by state (see Map 7.1)—another structural "level."

Rampage Shooting as Egoistic Suicide? Durkheim's principle of egoistic suicide can help you understand the horrific acts of mass murder rampages that have occurred in the United States—taking place at schools, movie theaters, political gatherings, shopping malls, and other such public and semipublic gatherings. Fully *two-thirds* of the mass murderers in these massacres either die by their own hand or are shot by the police at the scene—a phenomenon known to law enforcement as "suicide by cop" (Cloud 2012).

Among the most recent of rampage shootings is one now unfortunately firmly etched into contemporary American history: the mass killings of elementary schoolchildren by Adam Lanza. He killed twenty first-grade children and six teachers, totaling twenty-six dead at the Sandy Hook Elementary School in Newtown, Connecticut, on Friday morning, December 14, 2012. He employed four semiautomatic weapons with high-capacity clips. News of this unthinkable tragedy paralyzed the entire nation, and as was the case in many prior rampage shootings, Lanza shot and killed himself with a semiautomatic weapon immediately after killing the children and teachers. He did so after shooting his own mother four times in the head, killing her at her home. Shortly thereafter, it was revealed that the guns Lanza used were owned by his mother.

In ways that are far too haunting, the actions of Lanza in Connecticut mimicked those of now infamous Seung-Hui Cho, a college student at Virginia Tech University who in 2007 shot and killed thirty-two students, wounded fourteen others, and then killed himself, bringing the total killed to thirty-three—the highest ever among rampage shootings in recent U.S. history. Like Lanza of the Connecticut elementary school shootings, Cho was heavily armed with semiautomatic pistols. During their acts, both Cho and Lanza wore a mask, both wore military fatigues, and both carried several semiautomatic weapons simultaneously, and these weapons had high-capacity (extra-large) clips of bullets. Finally, and for our analysis more importantly, both were social isolates with very few or no close friends.

MAP 7.1

Mapping America's Diversity: Suicide Rates

Many factors can influence the suicide rate in different contexts. As discussed in the text, suicides can be caused by multiple structural and cultural factors, and sometimes these factors may be differently distributed by state or region. What are some of the social facts about the different states and regions that might affect the different rates of suicide you see in this map? What, in particular, might you guess about such social facts characterizing the states with the highest suicide rates?

Source: Data from U.S. Census Bureau, 2008. *The 2007 Statistical Abstract*: National Data Book. Washington, DC: U.S. Government Printing Office.

Suicide Rates by State

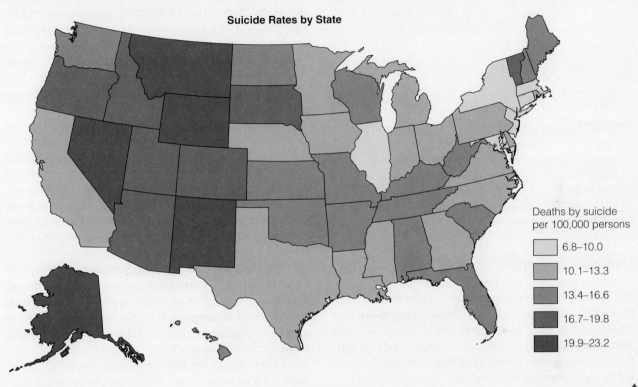

Deaths by suicide per 100,000 persons

- 6.8–10.0
- 10.1–13.3
- 13.4–16.6
- 16.7–19.8
- 19.9–23.2

There are social–structural elements that are common to the Newtown, Connecticut shootings, the Virginia Tech shootings, and other school shootings, such as the rampage killings at Columbine High School in Littleton, Colorado, in 1999. The acts in all three cases were committed by individuals who could be characterized as extremely socially isolated and utterly outside a network of peers. All of the perpetrators in these shootings were social isolates, and all four committed suicide immediately after their carnage. In Durkheim's sense, all of these instances represented examples of egoistic suicide, given the attributes of social isolation, lack of integration into society, troubled individual histories, and a desire to "make their mark" in history by killing the largest number of individuals possible in a single attack (Newman et al. 2006). The "egoistic" aspect of this last characteristic is certainly apparent.

In Tucson, Arizona, Jared Loughner on January 8, 2011, fired thirty-one bullets in fifteen seconds from a single oversize clip from his recently acquired Glock semiautomatic pistol, killing six people and seriously wounding thirteen. He thus hit nineteen people in fifteen seconds. Among the wounded were a district court judge (who died from wounds received) and popular Arizona Congresswoman Gabrielle Giffords, who survived a bullet that passed through the entire right side of her brain. (As of this writing, she has miraculously convalesced, and has given several important speeches—on gun control, no less.) Not only are there personality (psychological) similarities among these mass murderers, but all of these individuals also shared striking common *sociological* conditions:

1. All had become socially isolated, even from once close friends—especially Lanza.
2. Each had frequently delivered verbally disjointed and aggressive outbursts in the classroom, displaying troubled social interactions.
3. Each had previously vaguely hinted that they wanted to be famous for a single act of some sort.
4. All were men.
5. Each act took place in the social context of a culture that decidedly encourages gun ownership.

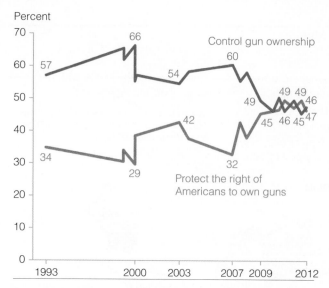

Percent

FIGURE 7.1 Views on Gun Control Public advocacy of gun ownership increased steadily from 2000, and support for gun ownership declined since then, but has recently increased, no doubt as result of the Lanza massacre in Connecticut.

Source: Pew Research Center Publications. July 2012.
http://pewresearch.org/pubs/

Look carefully at Figure 7.1. Advocacy of the right to own guns increased markedly since the early 2000s, but has recently leveled off. Proposals for gun control have increased—in quick sequence. This does not mean that anyone advocating gun ownership is going to shoot someone right away. What it does mean is that the cultural context within which all of the rampage shooters acted was sympathetic to gun use.

It might also be noted that each of the shooters had exhibited characteristics of a schizophrenic personality since their mid- to late teens—a classic pattern indicative of the onset of schizophrenia. Thus, in line with Durkheim's own original theoretical reasoning, egoistic suicide arises from the *combination* or *interaction* of both psychological and sociological factors.

Merton: Structural Strain Theory. The functionalist perspective on deviance has been further elaborated by the sociologist **Robert Merton** (1910–2003). Merton's **structural strain theory** traces the origins of deviance to the tensions caused by the gap between cultural goals and the means people have available to achieve those goals. Merton noted that societies are characterized by both culture and social structure. Culture establishes goals for people in society; social structure provides, or fails to provide, the means for people to achieve those goals. In a well-integrated society, according to Merton, people use accepted means to achieve the goals society establishes. In other words, the goals and means of the society are in balance. When the means

are out of balance with the goals, deviance is likely to occur. According to Merton, this imbalance, or disjunction, between cultural goals and structurally available means can actually *compel* the individual into deviant behavior (Merton 1968).

To explain further, a collective goal in U.S. society is the achievement of economic success. The legitimate means to achieve such success are education and jobs, but not all groups have equal access to those means. The result is structural strain that produces deviance. According to Merton, poor and working-class individuals are most likely to experience these strains because they internalize the same goals and values as the rest of society but have blocked opportunities for success. Structural strain theory therefore helps explain the high correlation that exists between unemployment and crime.

Figure 7.2 illustrates how strain between cultural goals and structurally available means can produce deviance. *Conformity* is likely to occur when the goals are accepted and the means for attaining the goals are made available to the individual by the social structure. If this does not occur, then cultural–structural strain exists, and at least one of four possible forms of deviance is likely to result: innovative deviance, ritualistic deviance, retreatism deviance, or rebellion.

Consider the case of female prostitution: The prostitute has accepted the cultural values of the dominant society—obtaining economic success and material wealth. Yet if she is poor, then the structural means to attain these goals are less available to her and turning to prostitution may result.

Other forms of deviance also represent strain between goals and means. *Retreatism deviance* becomes likely when neither the goals nor the means are available. Examples of retreatism are those with severe alcoholism or people who are homeless or reclusive. *Ritualistic deviance* is illustrated in the case of college women with eating disorders, such as *bulimia* (purging oneself after eating). The cultural goal of extreme

This is Adam Lanza, the rampage shooter who killed 20 small first-grade children and 6 adults at the Sandy Hook Elementary School in Newtown, Connecticut, in December of 2012. He used four semiautomatic large-clip weapons in this unspeakable massacre. These horrors were immediately followed by state and federal proposals for gun control legislation.

Rampage Shootings

Most laypeople will attribute violent deviant behavior, such as rampage shootings, to personality predispositions and personality flaws of offending individuals. Clearly, psychological predispositions do play a role in the actions of such individuals. Yet sociological research in general, and the study of deviant behavior in particular, has quite conclusively shown that such behavior is the product of the *combination* or *intersection* of personality variables with social structural variables. This was the case with four rampage shootings, including the Newtown, Connecticut, Sandy Hook Elementary School shootings by Adam Lanza; the Virginia Tech shootings by Seung-Hui Cho; the Columbine, Colorado, school shootings carried out by Klebold and Harris; and Jared Loughner's shooting of Congresswoman Gabrielle Giffords and eighteen others in Tucson, Arizona.

	Cultural goals accepted?	Institutionalized means toward goal available?
Conformity	Yes	Yes
Innovative deviance	Yes	No
Ritualistic deviance	No	Yes
Retreatism deviance	No	No
Rebellion	No (old goals) Yes (new goals)	No (old means) Yes (new means)

© Cengage Learning

FIGURE 7.2 Merton's Structural Strain Theory

thinness is perceived as unattainable, even though the means for trying to attain it are plentiful, for example, good eating habits and proper diet methods (Sharp et al. 2000). Finally, *rebellion* as a form of deviance is likely to occur when new goals are substituted for more traditional ones, and also new means are undertaken to replace older ones, as by force or armed combat. Many right-wing extremist groups, such as the American Nazi Party, "skinheads," and the Ku Klux Klan (KKK), are examples of this type of deviance. Finally, rampage shootings are examples of this kind of deviance.

Social Control Theory. Taking functionalist theory in another direction, Travis Hirschi has developed social control theory to explain deviance. **Social control theory**, a type of functionalist theory, suggests that deviance occurs when a person's (or group's) attachment to social bonds is weakened (Gottfredson and Hirschi 1995, 1990; Hirschi 1969). According to this view, people internalize social norms because of their attachments to others. People care what others think of them and therefore conform to social expectations because they accept what people expect. You can see here that social control theory, like the functionalist framework from which it stems, assumes the importance of the socialization process in producing conformity to social rules. When that conformity is broken, deviance occurs.

Social control theory assumes there is a common value system within society, and breaking allegiance to that value system is the source of social deviance. This theory focuses on how deviants are (or are not) attached to common value systems and what situations break people's commitment to these values. Social control theory suggests that most people probably feel some impulse toward deviance at times but that the attachment to social norms prevents them from actually participating in deviant behavior. Sociologists find that juveniles whose parents exercise little control over violent behavior and who learn violence from aggressive peers are most likely to engage in violent crimes (Heimer 1997), as was precisely the case with the two teenagers (Klebold and Harris) who killed twelve students and a teacher at Columbine High School (Newman et al. 2006). This also appears to have been the case with the Adam Lanza shootings in Connecticut: Lanza's mother had very limited contact with Lanza, who lived isolated in his mother's basement, and Lanza's father lived in another state.

Functionalism: Strengths and Weaknesses. Functionalism emphasizes that social structure, not just individual motivation, produces deviance. Functionalists argue that social conditions exert pressure on individuals to behave in conforming or nonconforming ways. Types of deviance are linked to one's place in the social structure; thus a poor person blocked from economic opportunities may use armed robbery to achieve economic goals, whereas a stockbroker may use insider trading to achieve the same. Functionalists acknowledge that people choose whether to behave in a deviant manner but believe that they make their choice from among socially prestructured options. The emphasis in functionalist theory is on social structure, not individual action. In this sense, functionalist theory is highly sociological.

Functionalists also point out that what appears to be dysfunctional behavior may actually be functional for the society. An example is the fact that most people consider prostitution to be dysfunctional behavior. From the point of view of an individual, that is true: It demeans the women who engage in it, puts them at physical risk, and subjects them to sexual exploitation. From the view of functionalist theory, however, prostitution supports and maintains a social system that links women's gender roles with sexuality, associates sex with commercial activity, and defines women as passive sexual objects and men as sexual aggressors. In other words, what appears to be deviant may actually serve various purposes for society.

Critics of the functionalist perspective argue that it does not explain how norms of deviance are first established. Despite its analysis of the ramifications of deviant behavior for society as a whole, functionalism does little to explain why some behaviors are defined as normative and others as illegitimate. Who determines social norms and on whom such judgments are most likely to be imposed are questions seldom asked by anyone using a functionalist perspective. Functionalists see deviance as having stabilizing consequences in society, but they tend to overlook the injustices that labeling someone deviant can produce. Others would say that the functionalist perspective too easily assumes that deviance has a positive role in society; thus functionalists rarely consider the differential effects that the administration of justice has on different groups. The tendency in functionalist theory to assume that the system works for the good of the whole too easily ignores the inequities in society and how these inequities are reflected in patterns of deviance. These issues are left for sociologists who work from the perspectives of conflict theory and symbolic interaction.

Conflict Theories of Deviance

Recall that conflict theory emphasizes the unequal distribution of power and resources in society. It links the study of deviance to social inequality. Based on the work of Karl Marx (see Chapter 1), conflict theory sees a dominant class as controlling the resources of society and using its power to create the institutional rules and belief systems that support its power. Like functionalist theory, conflict theory is a *macrostructural* approach; that is, both theories look at the structure of society as a whole in developing explanations of deviant behavior.

Because some groups of people have access to fewer resources in capitalist society, they are forced into crime to sustain themselves. Conflict theory posits that the economic organization of capitalist societies produces deviance and crime. The high rate of crime among the poorest groups, especially economic crimes such as theft, robbery, prostitution, and drug selling, are a result of the economic status of these groups. Rather than emphasizing values and conformity as a source of deviance as do functional analyses, conflict theorists see crime in terms of power relationships and economic inequality (Grant and Martínez 1997).

The upper classes, conflict theorists point out, can also better hide crimes they commit because affluent groups have the resources to mask their deviance and crime. As a result, a working-class man who beats his wife is more likely to be arrested and prosecuted than an upper-class man who engages in the same behavior. In addition, those with greater resources can afford to buy their way out of trouble by paying bail, hiring expensive attorneys, or even resorting to bribes.

Corporate crime is crime committed within the legitimate context of doing business. Conflict theorists expand our view of crime and deviance by revealing the significance of such crimes. They argue that appropriating profit based on exploitation of the poor and working class is inherent in the structure of capitalist society. **Elite deviance** refers to the wrongdoing of wealthy and powerful individuals and organizations (Simon 2007). Elite deviance includes what early conflict theorists called *white-collar crime* (Sutherland and Cressey 1978; Sutherland 1940). Elite deviance includes tax evasion; illegal campaign contributions; illegal investment schemes that steal money from innocent investors; corporate scandals, such as fraudulent accounting practices that endanger or deceive the public but profit the corporation or individuals within it; and even government actions that abuse the public trust. Several examples of elite deviance are covered in detail later in this chapter.

The ruling groups in society develop numerous mechanisms to protect their interests according to conflict theorists who argue that law, for example, is created by elites to protect the interests of the dominant class. Thus law, supposedly neutral and fair in its form and implementation, works in the interest of the most well-to-do (Weisburd et al. 2001, 1991; Spitzer 1975).

Conflict theory emphasizes the significance of social control in managing deviance and crime. **Social control** is the process by which groups and individuals within those groups are brought into conformity with dominant social expectations. Social control, as we saw in Chapter 4, can take place simply through socialization, but dominant groups can also control the behavior of others through marking them as deviant. An example is the historic persecution of witches during the Middle Ages in Europe and during the early colonial period in America (Ben-Yehuda 1986; Erikson 1966). Witches often were women who were healers

Social processes, as noted by conflict theories of deviance, strongly influence deviant behavior, as in this case of cocaine use.

they take it as partial evidence of the differential treatment of these groups by the criminal justice system.

Conflict Theory: Strengths and Weaknesses. The strength of conflict theory is its insight into the significance of power relationships in the definition, identification, and handling of deviance. It links the commission, perception, and treatment of crime to inequality in society and offers a powerful analysis of how the injustices of society produce crime and result in different systems of justice for disadvantaged and privileged groups. Not without its weaknesses, however, critics point out that laws protect most people, not just the affluent, as conflict theorists argue.

In addition, although conflict theory offers a powerful analysis of the origins of crime, it is less effective in explaining other forms of deviance. For example, how would conflict theorists explain the routine deviance of middle-class adolescents? They might point out that consumer marketing drives much of middle-class deviance. Profits are made from the accoutrements of deviance—rings in pierced eyebrows, "gangsta" rap music, and so on—but economic interests alone cannot explain all the deviance observed in society. As Durkheim argued, deviance is functional for the whole of society, not just those with a major stake in the economic system.

Symbolic Interaction Theories of Deviance

Whereas functionalist and conflict theories are *macrosociological* theories, certain *microsociological* theories of deviance look directly at the interactions people have with one another as the origin of social deviance. *Symbolic interaction theory* holds that people behave as they do because of the meanings people attribute to situations (see Chapter 1). This perspective emphasizes the meanings surrounding deviance, as well as how people respond to those meanings. Symbolic interaction emphasizes that deviance originates in the interaction between different groups and is defined by society's reaction to certain behaviors.

Symbolic interactionist theories of deviance originated in the perspective of the Chicago School of sociology. **W. I. Thomas** (1863–1947), one of the early sociologists from the University of Chicago, was among the first to develop a sociological perspective on social deviance. Thomas explained deviance as *a normal response to the social conditions in which people find themselves.* Thomas was one of the first to argue that delinquency was caused by the social disorganization brought on by slum life and urban industrialism; he saw deviance as a problem of social conditions, less so of individual character or individual personality.

Differential Association Theory. Thomas's work laid the foundation for a classic theory of deviance: differential association theory. **Differential association**

and midwives—those whose views were at odds with the authority of the exclusively patriarchal hierarchy of the church, then the ruling institution.

One implication of conflict theory, especially when linked with labeling theory, is that the power to define deviance confers an important degree of social control. **Social control agents** are those who regulate and administer the response to deviance, such as the police and mental health workers. Members of powerless groups may be defined as deviant for even the slightest infraction against social norms, whereas others may be free to behave in deviant ways without consequence. Oppressed groups may actually engage in more deviant behavior, but it is also true that they have a greater likelihood of being labeled deviant and incarcerated or institutionalized, whether or not they have actually committed an offense. This is evidence of the power wielded by social control agents.

When powerful groups hold stereotypes about other groups, the less powerful people are frequently assigned deviant labels. As a consequence, the least powerful groups in society are subject most often to social control. You can see this in the patterns of arrest data. All else being equal, poor people are more likely to be considered criminals and therefore more likely to be arrested, convicted, and imprisoned than middle- and upper-class people. The same is true of Latinos, Native Americans, and African Americans. Sociologists point out that this does not necessarily mean that these groups are somehow more criminally prone; rather,

theory, a type of symbolic interaction theory, interprets deviance, including criminal behavior, as behavior one learns through interaction with others (Sutherland and Cressey 1978; Sutherland 1940). Edwin Sutherland argued that becoming a criminal or a juvenile delinquent is a matter of learning criminal ways within the primary groups to which one belongs. To Sutherland, people become criminals when they are more strongly socialized to break the law than to obey it. Differential association theory emphasizes the interaction people have with their peers and others in their environment. Those who "differentially associate" with delinquents, deviants, or criminals learn to value deviance. The greater the frequency, duration, and intensity of their immersion in deviant environments, the more likely it is that they will become deviant.

Consider the career path of con artists and hustlers. Hustlers seldom work alone. Like any skilled worker, they have to learn the "tricks of the trade." A new recruit becomes part of a network of other hustlers who teach the recruit the norms of the deviant culture (Prus and Sharper 1991). Crime also tends to run in families. This does not necessarily mean that crime is passed on in genes from parent to child. It means that youths raised in deviant families are more likely socialized to become deviant themselves (Miller 1986). Differential association theory offers a compelling explanation for how deviance is culturally transmitted—that is, people pass on deviant expectations through the social groups in which they interact, of which the family is but one.

Critics of differential association theory have argued that this perspective tends to blame deviance on the values of particular groups. Differential association has been used, for instance, to explain the higher rate of crime among the poor and working class, arguing that this higher rate of crime occurs because they do not share the values of the middle class. Such an explanation, critics say, is class biased, because it overlooks the deviance that occurs in the middle-class culture and among elites. Disadvantaged groups may share the values of the middle class but cannot necessarily achieve them through legitimate means (a point, you will remember, made by Merton's structural strain theory).

Labeling Theory and Stigmatization. Labeling theory, a branch of symbolic interaction theory, interprets the responses of others as the most significant factor in understanding how deviant behavior is both created and sustained (Becker 1963). The work of contemporary labeling theorists such as Becker stems from the work of W. I. Thomas, who it will be recalled wrote, "If men define situations as real, they are real in their consequences" (Thomas and Thomas 1928: 572). A *label* is the assignment or attachment of a deviant identity to a person by others, including by agents of social institutions; therefore, people's reactions, not the action itself, produce deviance as a result of the labeling process.

Linked with conflict theory, labeling theory shows how those with the power to label an act or a person deviant and to impose sanctions—such as police, court officials, school authorities, experts, teachers, and official agents of social institutions—wield great power in determining societal understandings of deviance. Furthermore, because deviants are handled through bureaucratic organizations, bureaucratic workers "process" people according to rules and procedures, seldom questioning the basis for those rules or willing or able to challenge them (Montada and Lerner 1998; Margolin 1992; Cicourel 1968).

Once the label is applied, it sticks, and it is difficult for a person labeled deviant to shed the label—namely, to recover a nondeviant identity. To give an example, once a social worker or psychiatrist labels a client mentally ill, that person will be treated as mentally ill, regardless of his or her actual mental state. Pleas by the accused that he or she is mentally sound are typically taken as further evidence of mental illness. It is a kind of "catch-22"! Insistence by the labeled person that they are indeed "not mentally ill" is taken as evidence that they are in fact mentally ill!

A person need not have actually engaged in deviant behavior to be labeled deviant; yet, once applied, the label sticks. Labeling theory helps explain why convicts released from prison have such high rates of *recidivism* (return to criminal activities). Convicted criminals are formally and publicly labeled wrongdoers. They are treated with suspicion ever afterward and have great difficulty finding legitimate employment: The label "ex-con" defines their future options.

It is in fact exceedingly difficult for an ex-con to find employment after release from prison, even more so if the person is male and Black or Hispanic. In a clever study, Pager (2007) had pretrained role-players pose as ex-cons looking for a job. These role-players went into the job market and were interviewed for various jobs; all of them used the same preset script during the interview. The idea of the study was to see how many of them would be invited back for another interview. The results were staggering: Blacks who were *not* ex-cons were *less* likely to be invited back for a job interview than were Whites who *were* ex-cons, even though White ex-cons were not invited back in large numbers. All ex-cons had trouble being invited back, but even more so for Black and Hispanic ex-cons. So the effect of race alone exceeded the effect of incarceration alone. These upsetting differences could not be attributed to differences in interaction displayed during the interview, because everyone used the exact same prepared script.

Researchers Bruce Western (2007) and Jeffrey Reiman (2007) note that the prison system in the United States is in effect designed to *train* and *socialize* prisoners into a career of secondary deviance and to display to the public that crime is a threat primarily from the poor (see the box, "Doing Sociological Research: The Rich Get Richer and the Poor Get Prison," on page 165). Reiman sees that the goal of the prison system is not to reduce crime but to impress upon the public that crime is inevitable and that it originates only from the lower classes. Prisons accomplish this, even if unintentionally, by demeaning prisoners and stigmatizing them as different from "decent citizens," not training them in marketable skills. As a consequence, these people will never be able to pay their debt to society, and the prison system has created the very behavior it intended to eliminate.

Labeling theory suggests that deviance refers not just to something one does but to something one becomes. **Deviant identity** is the definition a person has of himself or herself as a deviant. Most often, deviant identities emerge over time (Simon 2007; Lemert 1972). A person addicted to drugs, for example, may not think of herself as a junkie until she realizes she no longer has nonusing friends. The formation of a deviant identity, like other identities, involves a process of social transformation in which a new self-image and new public definition of a person emerges. This is a process that involves how people view deviants and how deviants view themselves. Studies of tattoo "collectors" (that is, those who are very heavily tattooed) find, for example, that if collectors first learn to interpret tattooing as a desirable thing, they then begin to feel connected to a subculture of other collectors and eventually come to see their tattoos as part of themselves (Irwin 2001; Vail 1999; Montada and Lerner 1998).

A social **stigma** is an attribute that is socially devalued and discredited. Some stigmas result in people being labeled deviant. The experiences of people who are disabled, disfigured, or in some other way stigmatized are studied in much the same way as other forms of social deviance. Like other deviants, people with stigmas are stereotyped and defined only in terms of their presumed deviance.

Think, for example, of how people with disabilities are treated in society. Their disability can become a **master status** (see Chapter 5), a characteristic of a person that overrides all other features of the person's identity (Goffman 1963). Physical disability can become a master status when other people see the disability as the defining feature of the person; a person with a disability becomes "that blind woman" or "that paralyzed guy." People with a particular stigma are often all seen to be alike. This may explain why stigmatized individuals of high visibility are often expected to represent the whole group.

People who suddenly become disabled often have the alarming experience of their new master status rapidly erasing their former identity. People they know may treat and see them differently. A master status may also prevent people from seeing other parts of a person. A person with a disability may be assumed to have no meaningful sex life, even if the disability is unrelated to sexual ability or desire. Sociologists have argued that the negative judgments about people with stigmas tend to confirm the "usualness" of others (Goffman 1963). For example, when welfare recipients are stigmatized as lazy and undeserving of social support, others are indirectly promoted as industrious and deserving. Stigmatized individuals are thus measured against a presumed norm and may be labeled, stereotyped, and discriminated against.

Sometimes, people with stigmas bond with others, perhaps even strangers. This can involve an acknowledgment of "kinship" or affiliation that can be as subtle as an understanding look, a greeting that makes a connection between two people, or a favor extended to a stranger who the person sees as sharing the presumed stigma. Public exchanges are common between various groups that share certain forms of disadvantage, such as people with disabilities, lesbians and gays, or members of other minority groups.

see FOR YOURSELF

Perform an experiment by doing something mildly deviant for a period, such as carrying around a teddy bear doll and treating it as a live baby, or standing in the street and looking into the air, as though you are looking at something up there. Make a record of how others respond to you, and then ask yourself how labeling theory is important to the study of deviance. Then take your experiment a step further and ask yourself how people's reactions to you might have differed had you been of another race or gender. You might want to structure this question into your experiment by teaming up with a classmate of another race or gender. You could then compare each of your responses to the same behavior. A note of caution: Do not do anything illegal or dangerous; even the most seemingly harmless acts of deviance can generate strong (and sometimes hostile) reactions, so be careful in planning your experiment! ●

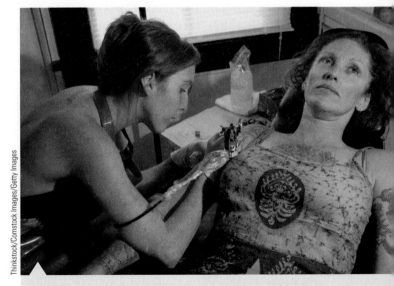

Thinkstock/Comstock Images/Getty Images

Extensive tattooing is regarded by many as deviant, although it may seem perfectly ordinary in the context of some peer groups.

Deviant Careers. In the ordinary context of work, a career is the sequence of movements a person makes through different positions in an occupational system (Becker 1963). A **deviant career**—a direct outgrowth of the labeling process—is the sequence of movements people make through a particular subculture of deviance. Deviant careers can be studied sociologically, like any other career. Within deviant careers, people are socialized into new "occupational" roles and encouraged, both materially and psychologically, to engage in deviant behavior. The concept of a deviant career emphasizes that there is a progression through deviance: Deviants are recruited, given or denied rewards, and promoted or demoted. As with legitimate careers, deviant careers involve an evolution in the person's identity, values, and commitment over time. Deviants, like other careerists, may have to demonstrate their commitment to the career to their superiors, perhaps by passing certain tests of their mettle, such as when a gang expects new members to commit a crime, perhaps even shoot someone.

Within deviant careers, rites of passage may bring increased social status among peers. Punishments administered by the authorities may even become badges of honor within a deviant community. Similarly, labeling a teenager "bad" for behavior that others think is immoral may actually encourage the behavior to continue because the juvenile may take this as a sign of success as a deviant.

Deviant Communities. The preceding discussion continues to indicate an important sociological point: Deviant behavior is not just the behavior of maladjusted individuals; it often takes place within a group context and involves group response. Some groups are actually organized around particular forms of social deviance; these are called **deviant communities** (Mizruchi 1983; Blumer 1969; Erikson 1966; Becker 1963).

Like subcultures and countercultures, deviant communities maintain their own values, norms, and rewards for deviant behavior. Joining a deviant community closes one off from conventional society and tends to solidify deviant careers because the deviant individual receives rewards and status from the in-group. Disapproval from the out-group may only enhance one's status within. Deviant communities also create a worldview that solidifies the deviant identity of their members. They may develop symbolic systems such as emblems, forms of dress, publications, and other symbols that promote their identity as a deviant group. Gangs wear their "colors," prostitutes have their own vocabulary of *tricks* and *johns*; skinheads have their insignia and music. All are examples of deviant communities. Ironically, subcultural norms and values reinforce the deviant label both inside and outside the deviant group, thereby reinforcing the deviant behavior.

Some deviant communities are organized specifically to provide support to those in presumed deviant

Some deviance develops in deviant communities, such as the neo-Nazis/"skinheads" shown marching here. Such right-wing extremist groups have remained relatively constant in numbers for the last several years.

categories. Groups such as Alcoholics Anonymous, Weight Watchers, and various twelve-step programs help those identified as deviant overcome their deviant behavior. These groups, which can be quite effective, accomplish their mission by encouraging members to accept their deviant identity as the first step to recovery.

A Problem with Official Statistics. Because labeling theorists see deviance as produced by those with the power to assign labels, they question the value of official statistics as indicators of the true extent of deviance. Reported rates of deviant behavior are themselves the product of socially determined behavior, specifically the behavior of identifying what is deviant. Official rates of deviance are produced by people in the social system who define, classify, and record certain behaviors as deviant and others as legitimate. Labeling theorists are more likely to ask how behavior becomes labeled deviant than they are to ask what motivates people to become deviant (Best 2011, 2007, 2001; Kitsuse and Cicourel 1963).

For example, in the aftermath of the terrorist attacks on the World Trade Center, officials debated whether to count the deaths of thousands as murder or as a separate category of terrorism. The decision would change the official rate of deviance by inflating or deflating the reported crime rate of murder in New York City in that year. In the end, these deaths were not counted in the murder rate. Labeling theorists think that official rates of deviance do not necessarily reflect only the actual commission of crimes or deviant acts; instead, the official rates reflect social judgments.

In another example, official rape rates are underestimates of the actual extent of rape, largely due to victims' reluctance to report. Also, police are less likely to "count" some rapes, such as those in which the victim is a prostitute, was drunk at the time of the assault, or had a previous relationship with the assailant. Moreover, rapes resulting in death are classified as homicides and therefore do not appear in the official statistics on rape (Babbie 2013).

Labeling Theory: Strengths and Weaknesses. The strength of labeling theory is its recognition that the judgments people make about presumably deviant behavior have powerful social effects. Labeling theory does not, however, explain why deviance occurs in the first place. It may illuminate the consequences of a young man's violent behavior, but it does not explain the actual origins of the behavior. Put bluntly, it does not explain why some people initially become deviant and others do not.

FORMS OF DEVIANCE

Although there are many forms of deviance, the sociology of deviant behavior has focused heavily on subjects such as mental illness, social stigmas, and crime. As we review each, you will also see how the different sociological theories about deviance contribute to understanding each subject. In addition, you will see how the social context of race, class, and gender relationships shape these different forms of deviance.

Mental Illness

Sociological explanations of mental illness look to the social systems in which mental illness is defined, identified, and treated, even though it is typical for many to think of mental illness only in psychological terms. This has several implications for understanding mental illness. Functionalist theory suggests, for example, that by recognizing mental illness, society also upholds normative values about more conforming behavior. Symbolic interaction theory tells us that mentally ill people are not necessarily "sick," but rather are the victims of societal reactions to their behavior. Some go so far as to say there is no such thing as mental illness, only people's reactions to unusual behavior (Szasz 1974). From this point of view, people learn faulty self-images and then are cast into the role of patient when treated by therapists.

Labeling theory, combined with conflict theory, suggests that those people with the fewest resources are most likely to be labeled mentally ill. Women, racial minorities, and the poor all suffer higher rates of reported mental illness and more serious disorders than do groups of higher social and economic status. Furthermore, research over the years has consistently shown that middle- and upper-class people are more likely to receive some type of psychotherapy for their illness. Poorer individuals and minorities are more likely to receive only physical rehabilitation and medication,

with no accompanying psychotherapy (Simon 2007; Hollingshead and Redlich 1958).

---

debunking SOCIETY'S MYTHS

MYTH: Mental illness is an abnormality best studied exclusively by psychologists and physicians.
SOCIOLOGICAL PERSPECTIVE: Mental illness follows patterns associated with race, class, and gender relations in society and is subject to a significant labeling effect. Those who study and treat mental illness benefit from combining a sociological perspective with both medical and psychological knowledge. ●

Sociologists give two explanations for the correlation between social status and mental illness. On the one hand, the stresses of being in a low-income group, being a racial minority, or being a woman in a sexist society all contribute to higher rates of mental illness; the harsher social environment is a threat to mental health. On the other hand, the same behavior that is labeled mentally ill for some groups may be tolerated and not so labeled in others. For example, behavior considered crazy in a homeless woman (who is likely to be seen as "deranged") may be seen as merely eccentric or charming when exhibited by a rich person.

Substance Abuse: Drugs and Alcohol

As with mental illness and stigmas, sociologists study the social factors that influence drug and alcohol use. Who uses what and why? How are users defined by others? These questions guide sociological research on substance abuse.

One of the first things to ask when thinking about drugs and alcohol is why using one substance is considered deviant and stigmatizing, and using another is not. How do such definitions of deviance change over time?

For example, alcohol is a legal drug. Whether one is labeled an alcoholic depends in large part on the social context in which one drinks, not solely on the amount of alcohol consumed. For years, the businessman's lunch where executives drank two or three martinis was viewed as normative. Drinking wine from a bottle in a brown bag on the street corner is considered highly deviant; having martinis in a posh bar is seen as cool—even though one martini contains considerably more alcoholic content than a swig of wine.

Sociological understandings challenge views of drug and alcohol use as stemming solely from inherent individual propensities that lead to substance abuse. Patterns of use vary by factors such as age, gender, and race (see Figure 7.3). Age is one significant predictor of illegal drug use. Young people are on average more likely to use marijuana and cocaine and binge drink than are people who are somewhat older, although there are, of course, exceptions.

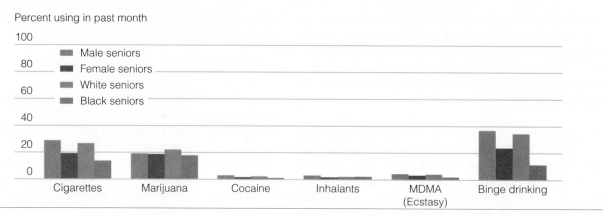

FIGURE 7.3 Use of Selected Substances by High School Seniors

Source: National Center for Health Statistics. 2011. *National Survey on Drug Use and Health.* Hyattsville, MD: U.S. Department of Health and Human Services.

CRIME AND CRIMINAL JUSTICE

The concept of deviance in sociology is a broad one, encompassing many forms of behavior—both legal and illegal, ordinary and unusual. **Crime** is one form of deviance, specifically, behavior that violates particular criminal laws. Not all deviance is crime. Deviance becomes crime when institutions of society designate it as violating a law or laws. *Deviance* is behavior that is recognized as violating rules and norms of society. Those rules may be formal laws, in which case the deviant behavior would be called *crime*, or informal customs or habits, in which case the deviant behavior would not be called crime.

Criminology is the study of crime from a scientific perspective. Criminologists include social scientists such as sociologists who stress the societal causes and treatment of crime. All the theoretical perspectives on deviance that we examined earlier contribute to our understanding of crime (see Table 7.2). According to the functionalist perspective, crime may be *necessary* to hold society together—a profound hypothesis. By singling out criminals as socially deviant, others are defined as good. The nightly reporting of crime on television is a demonstration of this sociological function of crime. Conflict theory suggests that disadvantaged groups are more likely to become criminal; it also sees the well-to-do as better able to hide their crimes and less likely to be punished. Symbolic interaction helps us understand how people learn to become criminals or come to be accused of criminality, even when they may be innocent. Each perspective traces criminal behavior to social conditions rather than only to the intrinsic tendencies or personalities of individuals.

Measuring Crime: How Much Is There?

Is crime increasing in the United States? One would certainly think so from watching the media. Images of violent crime abound and give the impression that crime is a constant threat and is on the rise. Data on crime actually show that violent crime peaked in 1990, but *decreased* through the 1990s and has continued to decline through 2011 (see Figure 7.4). Data about crime come from the Federal Bureau of Investigation (FBI) based on reports from police departments across the nation. The data are distributed annually in the *Uniform Crime Reports* and are the basis for official

table 7.2 Sociological Theories of Crime

Functionalist Theory	Symbolic Interaction Theory	Conflict Theory
Societies require a certain level of crime in order to clarify norms.	Crime is behavior that is learned through social interaction.	The lower the social class, the more the individual is *forced* into criminality.
Crime results from social structural strains (such as class inequality) within society.	Labeling criminals and stigmatizing them tends to reinforce rather than deter crime.	Inequalities in society by race, class, gender, and other forces tend to produce criminal activity.
Crime may be functional to society, thus difficult to eradicate.	Institutions with the power to label, such as prisons, actually produce rather than lessen crime.	Reducing social inequality in society is likely to reduce crime.

© Cengage Learning

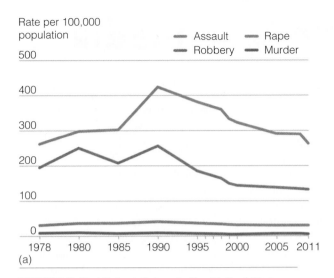

(a)

FIGURE 7.4A Violent Crime in the United States

Source: Federal Bureau of Investigation. 2012. *Uniform Crime Reports.* Washington, DC: U.S. Department of Justice. **www.fbi.gov**

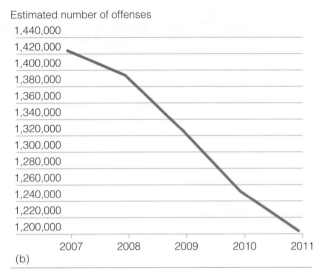

(b)

FIGURE 7.4B Violent Crime, Five-Year Trend 2007–2011

Source: Federal Bureau of Investigation. 2011. *Uniform Crime Reports.* Washington, DC: U.S. Department of Justice. **www.fbi.gov**

reports about the extent of crime and its rise and fall over time. These data show that although media coverage of crime—especially as reported in TV news—has remained high and about the same, the officially reported rate of assault and robbery has decreased, although rape and murder have remained roughly the same.

A second major source of crime data is the *National Crime Victimization Surveys* published by the Bureau of Justice Statistics in the U.S. Department of Justice. These data are based on surveys in which national samples of people are periodically asked if they have been the victims of one or more criminal acts (see Figures 7.5 and 7.6 on pages 165 and 166). These surveys clearly show that the likelihood of being a victim of crime is influenced by one's race, gender, and social class.

Both of these sources of data—the *Uniform Crime Reports* and the *National Crime Victimization Surveys*—are subject to the problem of underreporting. About half to two-thirds of all crimes may not be reported to police, meaning that much crime never shows up in the official statistics. Certain serious crimes, such as rape, are significantly underreported. Victims may be too upset to report a rape to the police, or they may believe that the police will not believe a rape has occurred. Equally significant, the victim may not want to undergo the continued emotional stress of an investigation and trial. Recall from earlier in this chapter that certain kinds of noncriminal deviance, such as suicide, are also underreported, particularly by upper-income families, because of embarrassment to the deceased person's family.

Another problem arises in the attempt to measure crime by means of official statistics. The FBI's *Uniform Crime Reports* stress what are called **index crimes**, which include the violent crimes of murder, manslaughter, rape, robbery, and aggravated assault,

plus property crimes of burglary, larceny-theft, and motor vehicle theft. These crimes are committed mostly by individuals who are disproportionately Black, Hispanic, and poor. Statistics based on these offenses do not reflect the crimes that tend to be committed by middle-class and upper-class people, such as tax violations, insider trading, fraudulent investment and accounting schemes, embezzlement, and other so-called elite crimes. The official statistics provide a relatively inflated picture for index crimes but an underreported picture of elite crimes, giving a biased picture of crime.

A final result is, unfortunately, that the public sees the stereotypic "criminal" as a poor or working-class person, most likely an African American or Latino male, not as a middle- or upper-class White person who has committed embezzlement. The official statistics give biased support to the stereotype. This in turn perpetuates the public belief that the "typical" criminal is lower class and minority instead of upper class and nonminority.

Personal and Property Crimes. The *Uniform Crime Reports* are subject to the same biases in official statistics mentioned earlier, but they are the major source of information on patterns of crime and arrest, with crimes classified into four categories. **Personal crimes** are violent or nonviolent crimes directed against people. Included in this category are murder, aggravated assault, forcible rape, and robbery. As Figure 7.4 shows, aggravated assault is the most frequently reported personal crime, although it is presently decreasing.

Hate crimes refer to assaults and other malicious acts (including crimes against property) motivated by various forms of social bias, including that based on race, religion, sexual orientation, ethnic/national origin, or disability. This form of crime has been increasing in recent years, especially against gays and lesbians.

a sociological eye ON THE media

Images of Violent Crime

The media routinely drive home two points to the consumer: Violent crime is always high and may be increasing over time, and there is much random violence constantly around us. The media bombard us with stories of "wilding," in which bands of youths kill random victims. Many of us think road rage is extensive (which it is not) and completely random. The media vividly and routinely report such occurrences as pointless, random, and probably increasing.

The evidence shows that although violent crime in the United States increased during the 1970s and 1980s, it nonetheless began to decrease in 1990 and continues to decrease nationally through the present. For example, both robbery and physical assaults have declined dramatically since 1990 (see Figure 7.4a). Yet, according to research (Best 2011, 2007, 1999; Glassner 1999), the media have consistently given a picture that violent crime has increased during this same period and, furthermore, that the violence is completely unpatterned and random.

No doubt there are occasions when victims are indeed picked at random. But the statistical rule of randomness could not possibly explain what has come to be called random violence, a vision of patternless chaos that is advanced by the media. If randomness truly ruled, then each of us would have an equal chance of being a victim—and of being a criminal. This is assuredly not the case. The notion of random violence, and the notion that it is increasing, ignores virtually everything that criminologists, psychologists, sociologists, and extensive research studies know about crime: It is highly patterned and significantly predictable, beyond sheer chance, by taking into account the social structure, social class, location, race–ethnicity, gender, labeling, age, whom one's family members are, and other such variables and forces in society that affect both criminal and victim.

The correct central picture, then, is clearly not conveyed in the media. Some have speculated that the picture maintained in the media of increasing crime is simply a tool to increase viewer ratings. But criminal violence is not increasing, but decreasing, and it is not random, but highly patterned and even predictable.

Property crimes involve theft of property without threat of bodily harm. These include burglary (breaking and entering), larceny (the unlawful taking of property, but without unlawful entry), auto theft, and arson. Property crimes are the most frequent criminal infractions.

Finally, so-called **victimless crimes** violate laws but are not listed in the FBI's serious crime index. These include illicit activities, such as gambling, illegal drug use, and prostitution, in which there is no complainant. Nonetheless, there is clearly at least some degree of victimization in such crimes: Some researchers see in many instances prostitution as containing at least one victim—the prostitute himself or herself. Enforcement of these crimes is typically not as rigorous as the enforcement of crimes against people or property, although periodic crackdowns occur, such as the current policy of mandatory sentencing for drug violations.

Elite and White-Collar Crime. Sociologists use the term *white-collar crime* to refer to criminal activities by people of high social status who commit their crimes in the context of their occupation (Sutherland and Cressey 1978). White-collar crime includes activities such as embezzlement (stealing funds from one's employer), involvement in illegal stock manipulations (insider trading), and a variety of violations of income tax law, including tax evasion. Also included are manipulations of accounting practices to make one's company appear profitable, thus artificially increasing the value of the company's stock.

Until very recently, white-collar crime seldom generated great concern in the public mind—far less than the concern directed at street crime. In terms of total dollars, however, white-collar crime is even more consequential for society. Scandals involving prominent white-collar criminals are coming to the public eye more frequently, such as the Ponzi scheme by white-collar criminal Bernard L. Madoff in 2008. Madoff ran a **Ponzi scheme**—*a con game* whereby a central person (Mr. Madoff) collects money from a large number of people, including friends and relatives, and then promises to invest their dollars with a high rate of interest for them. In Madoff's case, he promised a 10 percent rate or more of annual return, a very high rate even while the U.S. economy was souring. Actually, the money was never invested at all, but was used to pay off earlier investors. This is the key principle of a Ponzi scheme.

In the Madoff case, investors were led to believe that their money was under the competent control of Madoff himself, and because they trusted him, they never saw any records or stock certificates. For the scheme to work, Madoff had to convince new recruits that they could make a great deal of money if they just

Bernard Madoff, mastermind of a massive Ponzi investment scam that bilked over $55 billion from unsuspecting investors.

"left it to him." In the meantime, Madoff siphoned off a portion of the collected funds for himself and lived lavishly, with several large houses and estates in various parts of the world. When arrested in 2008, Madoff had processed $55 billion over the years—that is $55 billion, not $55 million! Madoff will spend the rest of his life in prison for his efforts, because his sentence was 150 years. Most of the duped individuals were close friends, family, and several celebrities. Many lost everything—their life savings, retirement pension accounts, and other investments. None of these people have been paid back to date, and probably never will be.

Such schemes are often called *pyramid schemes*, with large numbers of recent investors at the bottom of the pyramid and smaller numbers of the older (original) investors at the top. Such illegal schemes are named for one Carlo (or Charles) Ponzi, an immigrant who invented and perfected it in the 1920s in the United States.

Corporate Crime and Deviance: Doing Well, Doing Time

Corporations and even entire governments may engage in deviance—behavior that can be very costly to society. Sociologists estimate that the costs of corporate crime may be as high as $200 billion every year, dwarfing the take from street crime (roughly $15 billion), which most people incorrectly imagine is the bulk of criminal activity. Tax cheaters in business alone probably skim $50 billion a year from the IRS, three times the value of street crime. Taken as a whole, the cost of corporate crime is almost 6000 times the amount taken in bank robberies in a given year and 11 times the total amount for all theft in a year (Reiman 2007).

Corporate crime and deviance is wrongdoing that occurs within the context of a formal organization or bureaucracy and is actually sanctioned by the norms and operating principles of the bureaucracy (Simon 2007). This can occur within any kind of organization—corporate, educational, governmental, or religious. It exists once deviant behavior becomes institutionalized in the routine procedures of an organization. Sociological studies of corporate deviance show that this form of deviance is embedded in the ongoing and routine activities of organizations (Lee and Ermann 1999; Punch 1996). Individuals within the organization may participate in the deviant behavior with little awareness that their behavior is illegitimate. In fact, their actions are likely to be defined as in the best interests of the organization—business as usual. New members who enter the organization learn to comply with the organizational expectations or leave.

One of the most upsetting recent examples of massive corporate malfeasance involved a world-famous and time-honored American institution: the Johnson and Johnson Co., manufacturer of medical supplies from bandages (Band-Aid) to baby oil to artificial limbs. The company is the manufacturer of the now infamous DePuy artificial hip joint, adopted by thousands since 2005 to replace their own failing hip joints (Meier 2013).

It turns out that Johnson and Johnson executives knew *years before* they officially recalled the faulty DePuy artificial hip joint in 2010 that it had a deadly design flaw. In the interest of maintaining high profits, the company deliberately concealed evidence of the design flaw from physicians, patients, and their families. Evidently, the wish to maintain high profits exceeded the wish to make the patients well and to save their lives. The problem was that the DePuy artificial joint contained a metal ball fitted onto the top end of the thigh bone (the femur) that is inserted into a metal cup implanted into the hip bone. Hence, leg movements involved the motion of metal against metal, generating poisonous metal fragments into patients' bloodstreams. Consultants and medical researchers discovered the flaw several years before Johnson and Johnson recalled the DePuy joint, yet company executives totally ignored these research results. Presently, the company faces more than 10,000 lawsuits in connection with the device.

Another well-known case of accounting fraud concerned the Enron Corporation of Houston, Texas. When company executives found their own personal stock holdings in Enron declining in value in the early 2000s, they quickly and secretly sold their own stock, but did not allow their rank-and-file employees to sell their stock. The stock these unfortunate employees held declined rapidly over a several-month period, wiping out their retirement accounts, or "nest eggs." Yet the Enron executives hired the nationally known accounting firm of Arthur Andersen to submit "cooked" books, which concealed the stock loses. The responsible executives are presently serving lengthy prison terms.

Besides Jeffrey K. Skilling, chief executive of Enron, other offending corporate executives have recently been in the news and are currently serving prison sentences, including Lee Farkas of Taylor, Bean and Whitaker (securities fraud, conspiracy, and bank and wire fraud); Bernard J. Ebbers of WorldCom (nine counts of securities fraud and filing false claims); and high-living L. Dennis Kozlowski of Tyco International (twenty-two counts including securities fraud, grand larceny, and falsifying records) (*The New York Times*, October 14, 2011, p. A1).

Organized Crime

The structure of crime and criminal activity in the United States often takes on an organized, almost institutional character. This is crime in the form of mob activity and racketeering, known as organized crime.

Organized crime is crime committed by structured groups typically involving the provision of illegal goods and services to others. Organized crime syndicates are typically stereotyped as the Mafia, but the term can refer to any group that exercises control over large illegal enterprises, such as the drug trade, illegal gambling, prostitution, weapons smuggling, or money laundering. These organized crime syndicates are often based on racial, ethnic, or family ties, with different groups dominating and replacing each other in different criminal "industries," at different periods in U.S. history.

A key concept in sociological studies of organized crime is that these industries are organized along the same lines as legitimate businesses; indeed, organized crime has taken on a corporate form (Best 2008; Carter 1999). There are likely senior partners who control the profits of the business, workers who manage and provide the labor for the business, and clients who buy the services that organized crime provides, such as prostitution and drug dealing. In-depth studies of the organized crime underworld are difficult, owing to its secretive nature and dangers. As organized crime has moved into seemingly legitimate corporate organizations, it is even more difficult to trace, although some sociologists have penetrated underworld networks and provided fascinating accounts of how these crime networks are organized (Carter 1999).

In the Mafia, or *Cosa Nostra* (meaning "our thing"), considerable social status and power within the organization is achieved when one becomes a "made member." Prospective candidates start at the bottom of the organization's hierarchy and are observed by the bosses for a lengthy period of time, sometimes several years. Women are almost never picked. The prospective candidate is required to perform various jobs that require violence or intimidation, such as the collection of debts. As time goes on, recruits will be tested for their loyalty and competence by being asked to perform more daring criminal assignments. Most prospective members are expected to participate in a murder before being inducted into the Cosa Nostra. After their induction, the newly made members are expected to be ready to kill their own sons or brothers should it be learned that they have become informants against the organization (DeChamplain 2010).

Race, Class, Gender, and Crime

Arrest data show a very clear pattern of differential arrests along lines of race, gender, and class. To sociologists, the central question posed by such data is whether this reflects actual differences in the extent of crime among different groups or whether this reflects differential treatment by the criminal justice system. The answer is "both" (D'Alessio and Stolzenberg 2003).

Sociologists show that prosecution by the criminal justice system is significantly related to patterns of race, gender, and class inequality. We see this in the bias of official arrest statistics, in treatment by the police, in patterns of sentencing, and in studies of imprisonment.

Arrest statistics show a strong correlation between social class and crime, the poor being more likely than others to be arrested for crimes. Does this mean that the poor commit more crimes? To some extent, yes. Unemployment and poverty are related to crime (Reiman and Leighton 2009; Best 2007; Britt 1994). The reason is simple: Those who are economically deprived often see no alternative to crime, as Merton's structural strain theory would predict.

Moreover, law enforcement is concentrated in lower income and minority areas. People who are better off are further removed from police scrutiny and better able to hide their crimes. When and if white-collar criminals are prosecuted and convicted, they tend to receive somewhat lighter sentences. Middle- and upper-income people may be perceived as being less in need of imprisonment because they likely have a job and high-status people to testify for their good character. And white-collar crime is simply perceived as less threatening than crimes by the poor. Class also predicts who most likely will be victimized by crime, with those at the highest ends of the socioeconomic scale least likely to be victims of violent crime (Figure 7.5).

Bearing in mind the factors that affect the official rates of arrest and conviction—bias of official statistics, influence of powerful individuals, discrimination in patterns of arrest, differential policing—there remains

The Rich Get Richer and the Poor Get Prison

Research Question: Jeffrey Reiman and Paul Leighton (2009) have studied U.S. prisons by asking: (1) What goes on in prisons? and (2) What are the perceptions of prisons held by those in society?

Research Method: Reiman and Leighton used field research in prisons to answer these questions.

Research Results: The researchers found that the prison system in the United States, instead of serving as a way to rehabilitate criminals, is in effect designed to train and socialize inmates into a career of crime. It is also designed in such a way as to assure the public that crime is a threat primarily from the poor and that it originates at the lower rungs of society. Prisons contain elements that seem designed to accomplish this view.

Conclusions and Implications: One can "construct" a prison that ends up looking like a U.S. prison. First, continue to label as criminal those who engage in crimes that have no unwilling victim, such as prostitution or gambling. Second, give prosecutors and judges broad discretion to arrest, convict, and sentence based on appearance, dress, race, and apparent social class. Third, treat prisoners in a painful and demeaning manner, as one might treat children. Fourth, make certain that prisoners are not trained in a marketable skill that would be useful upon their release. And, finally, assure that prisoners will forever be labeled and stigmatized as different from "decent citizens," even after they have paid their debt to society. Once an ex-con, always an ex-con. One has thus just socially constructed a U.S. prison, an institution that will continue to generate the very thing that it claims to eliminate.

Questions to Consider

1. In your own opinion, how accurate is this "construction" of the U.S. prison? Do you know anyone who is currently in or recently in prison? Interview them and get their opinion.

2. How persistent in the coming years do you think this vision of the U.S. prison system will be?

Sources: Reiman, Jeffrey H., and Paul Leighton. 2009. *The Rich Get Richer and the Poor Get Prison: A Reader.* Upper Saddle River, NJ: Pearson.

---

evidence that the actual commission of crime varies by race. Why? Again, sociologists find a compelling explanation in social structural conditions. Racial minority groups are far more likely than Whites to be poor, unemployed, and living in single-parent families. These social facts are all predictors of a higher rate of crime. Note, too, as Figure 7.6 shows, that African Americans are generally more likely to be victimized by crime.

Recently women's participation in crime has been increasing, the result of several factors. Women are now more likely to be employed in jobs that present

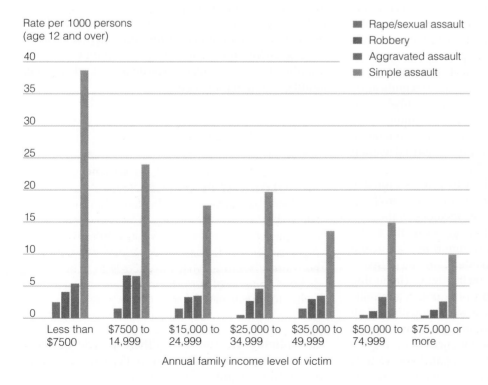

FIGURE 7.5 Victimization in Crime: A Class Phenomenon

Source: U.S Bureau of Justice Statistics. 2011. *Criminal Victimization.* Washington, DC: U.S. Department of Justice. **www.bjs.gov**

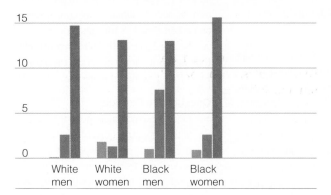

FIGURE 7.6 Crime Victimization by Race and Gender

Source: U.S. Bureau of Justice Statistics. 2011. *Criminal Victimization.* Washington, DC: U.S. Department of Justice. **www.bjs.gov**

opportunities for crimes, such as property theft, embezzlement, and fraud. Violent crime by women has also increased notably since the early 1980s, possibly because the images that women have of themselves are changing, making new behaviors possible. Most significant, crime by women is related to their continuing disadvantaged status in society. Just as crime is linked to socioeconomic status for men, so it is for women (Belknap 2001).

Despite recent achievements, many women remain in disadvantaged, low-wage positions in the labor market. At the same time, changes in the social structure of families mean that more women are economically responsible for their children without the economic support of men. Disadvantaged women may turn to illegitimate means of support, a trend that may be exacerbated by reductions in welfare support.

Women are somewhat less likely than men to be victimized by crime, although this varies significantly by race and age. Black women are more likely than White women to be victims of assault; young Black women are especially vulnerable. Divorced, separated, and single women are more likely than married women to be crime victims (see Figure 7.6).

For all women, victimization by rape is probably the greatest fear. Although rape is the most underreported crime, it has until recently been one of the fastest growing—something criminologists explain as the result of a greater willingness to report *and* an actual increase in the extent of rape (Federal Bureau of Investigation 2012). Approximately 90,000 rapes (including attempted rapes) are reported to the police annually. Officials estimate that this is a small fraction of all rapes committed. Many women are reluctant to report rape because they fear the consequences of having the criminal justice system question them. Rape victims are least likely to report the assault when the assailant is

someone known to them, even though a large number of rapes are committed by someone the victim knows.

A disturbingly frequent form of rape is *acquaintance rape*—rape committed by an acquaintance or someone the victim has just met. The extent of acquaintance rape is difficult to measure. The Bureau of Justice Statistics finds that 3 percent of college women experience rape or attempted rape in a given college year, and 13 percent report being stalked (Fisher et al. 2000). Research finds that acquaintance rape is linked to men's acceptance of various rape myths (such as believing that a woman's "no" means "yes"), the use of alcohol, and the peer support that men receive in some all-male groups and organizations, such as fraternities (Taylor et al. 2012; Belknap et al. 1999; Boeringer 1999; Ullman et al. 1999).

Sociologists have argued that the causes of rape lie in women's status in society—that women are treated as sexual objects for men's pleasure. The relationship between women's status and rape is also reflected in data revealing who is most likely to become a rape victim. African American women, Latinas, and poor women have the highest likelihood of being raped, as do women who are single, divorced, or separated. Young women are also more likely to be rape victims than older women (U.S. Bureau of Justice Statistics 2012). Sociologists interpret these patterns to mean that the most powerless women are also most subject to this form of violence.

The Criminal Justice System: Police, Courts, and the Law

Whether in the police station, the courts, or prison, the factors of race, class, and gender are highly influential in the administration of justice in this society. Those in the most disadvantaged groups are more likely to be defined and identified as deviant independently of their behavior and, having encountered these systems of authority, are more likely to be detained and arrested, found guilty, and punished.

---

debunking SOCIETY'S MYTHS

MYTH: The criminal justice system treats all people according to the neutral principles of law.

SOCIOLOGICAL PERSPECTIVE: Race, class, and gender continue to have an influential role in the administration of justice. For example, even when convicted of the same crime as Whites, African American and Latino male defendants with the same prior arrest record as Whites are more likely to be arrested, sentenced, and to be sentenced for longer terms than White defendants. ●

The Policing of Minorities. There is little question that minority communities are policed more heavily than White neighborhoods; moreover, policing in minority communities has a different effect from that in White, middle-class communities. To middle-class Whites, the presence of the police is generally reassuring, but for African Americans and Latinos, an encounter with a police officer can be terrifying. Regardless of what they are doing at the time, minority people, minority men in particular, are perceived as a threat, especially if they are observed in communities where they "don't belong."

Racial profiling has recently come to the public's attention, although it is a practice that has a long history. Often referred to half in jest by African Americans as the offense of "DWB," or "driving while Black," **racial profiling** on the part of a police officer is the use of race alone as the criterion for deciding whether to stop and detain someone on suspicion of having committed a crime. Police officers often argue that they "have no choice." The police argue that racial profiling is justified because a high proportion of Blacks and Hispanics commit crimes; but, although the crime rate for Blacks and Hispanics is higher than that of Whites, race is a particularly bad basis for suspicion because the vast majority of Blacks and Hispanics, like the vast majority of Whites, do not commit any crime at all. As evidence of this, studies have found that *eight out of every ten* automobile searches carried out by state troopers on the New Jersey Turnpike over ten years were conducted on vehicles driven by Blacks and Hispanics; the vast majority of these searches turned up no evidence of contraband or crimes of any sort (Kocieniewski and Hanley 2000; Cole 1999).

Racial minorities are also more likely than the rest of the population to be victims of excessive use of force by the police, also called police brutality. Most cases of police brutality involve minority citizens, and there is usually no penalty for the officers involved. Increasing the number of minority police officers has some small effect on how the police treat minorities.

Race and Sentencing. What happens once minority citizens are arrested for a crime? Bail is set higher for African Americans and Latinos than for Whites, and minorities have less success with plea bargains. Once on trial, minority defendants are found guilty more often than White defendants. At sentencing, African Americans and Hispanics are likely to get longer sentences than Whites, *even when they have the same number of prior arrests and socioeconomic background as Whites*. Young African American men, as well as Latino men, are sentenced more harshly than any other group, and once sentenced they are less likely to be released on probation (Doermer and Demuth 2010; Western 2007; Steffensmeier and Demuth 2000). In fact, Blacks and Hispanics who have *already received* the death penalty are more likely to be executed, rather than being pardoned or having the execution postponed, than are Whites who have committed the same crime (Jacobs et al. 2007). Any number of factors

influences judgments about sentencing, including race of the judge, severity of the crime, race of the victim, and the gender of the defendant, but throughout these studies, race is shown to consistently matter—and matter a lot.

Prisons: Rehabilitation or Mass Racialized Incarceration? Racial minorities account for *more than half* of the federal and state male prisoners in the United States. Blacks have the highest rates of imprisonment, followed by Hispanics, then Native Americans and Asians. (Native Americans and Asian Americans together are less than one percent of the total prison population.) Hispanics are the fastest-growing minority group in prison (West and Sabol 2010). Native Americans, though a small proportion of the prison population, are still overrepresented in prisons. In theory, the criminal justice system is supposed to be unbiased, able to objectively weigh guilt and innocence. The reality is that the criminal justice system reflects the racial and class stratification and biases in society.

The United States and Russia have the highest rate of incarceration in the world. Yet at the same time, while the proportions of individuals in the prison population have increased over time, there has been a recent leveling off. Although it is certainly true that, as we have already noted, Blacks and Hispanics themselves commit proportionately more crime than Asians and Native American Indians, it is nonetheless also demonstrably true that the structure of the U.S. criminal justice system disproportionately *propels* Blacks and Hispanics into prison at a greater rate than same-aged Whites who have the same criminal record. This is because unemployment is much higher for Blacks and Hispanics. Also, federal and state officials have traditionally been more hostile to people of color who run afoul of the criminal justice system than to Whites who do so. This disproportionately affects minority defendants who are more likely to have been convicted of earlier offenses. The situation is so severe and there are so many minority people in prison now that sociologist Bruce Western calls them a *new color caste* in U.S. society—in other words, a society unto itself (Western 2007).

Women in prison face unique problems, in part because they are in a system designed for men and run mostly by men, which tends to ignore the particular needs of women. For example, 25 percent of the women entering prison are pregnant or have just given birth, but they often get no prenatal or obstetric care. Male prisoners are trained for such jobs as auto mechanics, whereas women are more likely to be trained in relatively lower-status jobs such as beauticians and launderers. The result is that few women offenders are rehabilitated by their experience in prison.

The United States, then, is putting offenders in prison at a record pace. Is crime being deterred? Are prisoners being rehabilitated? Or are they simply being *warehoused*—put on a shelf? If the deterrence argument were correct, we would expect that increasing the risk of imprisonment would lower the rate of crime. For example, we would expect drug use to decline as enforcement of drug laws increases. In the past few years, there has been a marked increase in drug law enforcement but not the expected decrease in drug use. Using drugs as an example, then, it appears that the threat of imprisonment does not deter crime.

Terrorism as International Crime: A Global Perspective

Crime now crosses international borders and has become global, as we see with terrorism. The FBI includes terrorism in its definition of crime, seeing it as violent action to achieve political ends (White 2002). Thus terrorism is a crime that violates both international and domestic laws. It is a crime that crosses national borders, and to understand it requires a global perspective.

Terrorism is globally linked to other forms of international crime. It is suspected that profits from international drug trade of the terrorist organization al Qaeda, led by Osama bin Laden, helped finance the September 11, 2001, New York City terrorist attacks that led to the destruction of the World Trade Center Twin Towers and the deaths of almost 3000 people. Therefore, a global perspective on crime involves recognizing the global basis of some international crime networks that cross national borders (Binns 2003).

Many nations have long experienced terrorism in the form of bombings, hijackings, suicide attacks, and other terrorist crimes. But the attacks of September 11 focused the world's attention on the problem of terrorism in new ways, including increased fears of **bioterrorism**—the form of terrorism involving the dispersion of chemical or biological substances intended to cause widespread disease and death. Another form of terrorism, and thus cause for international concern, is **cyberterrorism**, the use of the computer to commit one or more terrorist acts. Terrorists may use computers in a number of ways. Data-destroying computer viruses may be implanted electronically in an enemy's computer. Another use would be to employ "logic bombs" that lie dormant for years until they are electronically instructed to overwhelm a computer system. The use of the Internet to serve the needs of international terrorists has already become a reality (*The New York Times*, January 2012; Jucha 2002).

On Tax Day, April 15, 2013, at the finish line of the annual Boston Marathon run, the air on Boston's Boylston Street was brutally shattered by two terrific explosions of terrorist bombs. The bombs were detonated at ground level, thus shredding legs and torsos of both runners and bystanders. Over 140 were severely injured and three were killed. Two suspects were immediately identified: Tamerlan Tsarnaev, killed by police in a shootout shortly after the bombing, and his brother, Dzhokhar, who was arrested within days of the bombing.

chapter summary

What is the difference between deviance and crime?
Deviance is behavior that violates norms and rules of society, and *crime* is a type of deviant behavior that violates the formal criminal law. *Criminology* is the study of crime from a scientific perspective.

How do sociologists conceptualize and explain deviance and crime?
Deviance is behavior that is recognized as violating expected rules and norms and that should be understood in the social context in which it occurs. Psychological explanations of deviance place the cause of deviance primarily within the individual. Sociologists emphasize the total social context in which deviance occurs. Sociologists see deviance more as the result of group and institutional, not individual, processes.

What does sociological theory contribute to the study of deviance and crime?
Functionalist theory sees both deviance and crime as functional for the society because it affirms what is

acceptable by defining what is not. *Structural strain theory*, a type of functionalist theory, predicts that societal inequalities actually force and compel the individual into deviant and criminal behavior. *Conflict theory* explains deviance and crime as a consequence of unequal power relationships and inequality in society. *Symbolic interaction theory* explains deviance and crime as the result of meanings people give to various behaviors. *Differential association theory*, a type of symbolic interaction theory, interprets deviance as behavior learned through social interaction with other deviants. *Labeling theory* (including the study of stigmatization), also a type of symbolic interaction theory, argues that societal reactions to behavior produce deviance, with some groups having more power than others to assign deviant labels to people.

What are the major forms of deviance?
Mental illness and substance abuse are major forms of deviance that sociologists study, although deviance comprises many different forms of behavior.

Sociological explanations of mental illness focus on the social context in which mental illness develops and is treated. Substance abuse includes alcohol and drug abuse but is not limited to these two forms.

What are the connections between inequality, deviance, and crime?

Sociological studies of crime analyze the various types of crimes, such as personal and property crimes, *elite* and *white-collar crime*, *corporate crime*, and *organized crime*. Many types of crimes are underreported, such as rape and certain elite and corporate crimes. Sociologists study the conditions, including race, class, and gender inequalities, that produce crime and shape how different groups are treated by the criminal justice system, such as showing group differences in sentencing.

How is crime related to race, class, and gender?

In general, crime rates for a variety of crimes are higher among minorities than among Whites; among poorer people than among middle- or upper-class people; and among men than among women. Women, especially minority women, are more likely to be victimized by serious crimes such as rape or violence from a spouse or boyfriend.

Key Terms

altruistic suicide 150
anomic suicide 150
anomie 149
bioterrorism 168
crime 160
criminology 160
cyberterrorism 168
deviance 146

deviant career 158
deviant community 158
deviant identity 157
differential association
    theory 155
egoistic suicide 150
elite deviance 154
hate crime 161

index crimes 161
labeling theory 156
master status 157
medicalization of
    deviance 148
organized crime 164
personal crimes 161
Ponzi scheme 162

property crimes 162
racial profiling 167
social control 154
social control agents 155
social control theory 153
stigma 157
structural strain theory 152
victimless crimes 162

8 Social Class and Social Stratification

Social Differentiation and Social Stratification

The Class Structure of the United States: Growing Inequality

The Distribution of Income and Wealth

Analyzing Social Class

Social Mobility: Myths and Realities

Why Is There Inequality?

Poverty

Chapter Summary

one afternoon in a major U.S. city, two women go shopping. They are friends—wealthy, suburban women who shop for leisure. They meet in a gourmet restaurant and eat imported foods while discussing their children's private schools. They also talk about the volunteer work they do in the local hospital, but they don't worry too much about their own health care, at least not financially, because both are fully covered through ample health insurance policies. After lunch, they spend the afternoon in exquisite stores—some of them large, elegant department stores; others, intimate boutiques where the staff know them by name. When one of the women stops to use the bathroom in one store, she enters a beautifully furnished room with an upholstered chair, a marble sink with brass faucets, fresh flowers on a wooden pedestal, shining mirrors, an ample supply of hand towels, and jars of lotion and soaps. The toilet is in a private stall with solid doors. In the stall, there is soft toilet paper and another small vase of flowers.

The same day, in a different part of town, another woman goes shopping. She lives on a marginal income earned as a stitcher in a textiles factory. Her daughter badly needs a new pair of shoes because she has outgrown last year's pair. The woman goes to a nearby discount store where she hopes to find a pair of shoes for under $15, but she dreads the experience. She knows her daughter would like other new things—a bathing suit for the summer, a pair of jeans, and a blouse. But this summer, the daughter will have to wear hand-me-downs because medical bills over the winter have depleted the little money left after food and rent. For the mother, shopping is not recreation but a bitter chore reminding her of the things she is unable to get for her daughter.

While this woman is shopping, she, too, stops to use the bathroom. She enters a vast space with sinks and mirrors lined up on one side of the room and several stalls on the other. The tile floor is gritty and gray. The locks on the stall doors are missing or broken. Some of the overhead lights are burned out, so the room has dark shadows. In the stall, the toilet paper is coarse. When the woman washes her hands, she discovers there is no soap in the metal dispensers. The mirror before her is cracked. She exits quickly, feeling as though she is being watched.

Two scenarios, one society. The difference is the mark of a society built upon class inequality. The signs are all around you. Think about the clothing you wear. Are some labels worth more than others? Do others in your group see the same marks of distinction and status in clothing labels? Do some people you know never seem to wear the "right" labels? Whether it is clothing, bathrooms, schools, homes, or access to health care, the effect of class inequality is enormous, giving privileges and resources to some and leaving others struggling to get by.

Great inequality divides society. Nevertheless, most people think that equal opportunity exists for all in the United States. The tendency is to blame individuals for their own failure or attribute success to individual achievement. Many people think the poor are lazy and do not value work. At the same time, the rich are often admired for their supposed initiative, drive, and motivation. Neither is an accurate portrayal. There are many hardworking individuals who are poor, and most rich people have inherited their wealth rather than earned it themselves.

Observing and analyzing class inequality is fundamental to sociological study. What features of society cause different groups to have different opportunities? Why is there such an unequal allocation of society's resources? Sociologists respect individual achievements but have found that the greatest cause for disparities in material success is the organization of society. Instead of understanding inequality as the result of individual effort, sociologists thus study the social structural origins of inequality.

learning objectives

- Explain how class is a social structure
- Describe the class structure of the United States
- Identify the different components of class inequality
- Analyze the extent of social mobility in the United States
- Compare and contrast theoretical models of class inequality
- Investigate the causes and consequences of U.S. poverty

SOCIAL DIFFERENTIATION AND SOCIAL STRATIFICATION

All social groups and societies exhibit social differentiation. **Status**, as we have seen earlier, is a socially defined position in a group or society. **Social differentiation** is the process by which different statuses develop in any group, organization, or society. Think of a sports organization. The players, the owners, the managers,

Social class differences make it seem as if some people are living in two different societies.

the fans, the cheerleaders, and the sponsors all have a different status within the organization. Together, they constitute a whole social system, one that is marked by social differentiation.

---

see FOR YOURSELF

Take a shopping trip to different stores and observe the appearance of stores serving different economic groups. What kinds of bathrooms are there in stores catering to middle-class clients? The rich? The working class? The poor? Which ones allow the most privacy or provide the nicest amenities? What fixtures are in the display areas? Are they simply utilitarian with minimal ornamentation, or are they opulent displays of consumption? Take detailed notes of your observations, and write an analysis of what this tells you about social class in the United States. ●

Status differences can become organized into a hierarchical social system. Social stratification is a relatively fixed, hierarchical arrangement in society by which groups have different access to resources, power, and perceived social worth. **Social stratification** is a system of structured social inequality. Using sports as an example again, you can see that many of the players

Social Class and Sports

Sports are a huge part of American culture. Whether you are an athlete, a fan, or just an observer, sports are a window into how social class shapes some of our most popular activities.

Start with the ideas of functionalism and the work of classical theorist Emile Durkheim. Durkheim was interested in the cultural symbols and events that bind people together. Think of how many sports symbols, such as jerseys, hats, and bumper stickers, are common sights in everyday life. These symbols project an identity to others that make a claim about being part of collective group. But,

sometimes, they reflect social class locations, too. Rich people, for example, are not likely to be wearing NASCAR caps, but may well have yacht club logos on their polo shirts and ties.

But, as Max Weber would point out, class, power, and prestige are all tangled up in sports. There are significant class differences associated with different sports, some having more prestige than others. Prestige is also interwoven with power, as you can see during political elections, where you see politicians all over the place—at tailgate parties and hanging out in the expensive box seats. But, beyond

the connection between power, politics, and prestige, sports is big business.

Corporate profits and sponsorship are very apparent, even in college sports. Look at the advertisements on most college scoreboards. Various plays in a football game might be featured as an "AT&T All-America Play of the Week!" Think of how much money is spent on commercials.

Social class in the world of sports is everywhere, even though the workers who help put on events are often invisible. Some of the athletes are very highly paid, but working-class people serve the food in stadiums, clean up after the fans leave, and take out all the trash. Sports are an amazing example of a capitalist social system.

earn extremely high salaries, although most do not. Those who do are among the elite in this system of inequality. But it is the owners who control the resources of the teams and hold the most power in this system. Sponsors (including major corporations and media networks) are the economic engines on which this system of stratification rests; fans are merely observers who pay to watch the teams play, but the revenue they generate is essential for keeping this system intact. Altogether, sports are systems of stratification because the groups that constitute the organization are arranged in a hierarchy where some have more resources and power than others. Some provide resources; others take them. And, even within the field of sports, there are

huge differences in which teams—and which sports— are among the elite.

All societies seem to have a system of social stratification, although they vary in the degree and complexity of stratification. Some societies stratify only along a single dimension, such as age, keeping the stratification system relatively simple. Most contemporary societies are more complex, with many factors interacting to create different social strata. In the United States, social stratification is strongly influenced by class, which is in turn influenced by matters such as one's occupation, income, and education, along with race, gender, and other influences such as age, region of residence, ethnicity, and national origin (see Table 8.1).

---

table 8.1 Inequality in the United States

- One in five (22 percent) children in the United States lives in poverty, including 38.8 percent of African American children, 34.1 percent of Hispanic children, 1.5 percent of White children, and 13.5 percent of Asian American children (DeNavas-Walt et al. 2012).

- The rate of poverty among people in the United States has been increasing since 2000 (DeNavas-Walt et al. 2012).

- Among women heading their own households, more than one-third (34 percent) lives below the poverty line (DeNavas-Walt et al. 2012).

- One percent of the U.S. population controls 35 percent of the total wealth in the nation; the bottom half only hold only one percent of all wealth.

- Between 2007 and 2010, American families saw their net worth decline, largely because of declines in the value of housing; those in the top 10 percent saw little or no decline (Bricker et al. 2012).

- The average CEO of a major company has a salary of $13 million per year; workers earning the minimum wage make $15,080 per year if they work 40 hours a week for 52 weeks and hold only one job (**www.aflcio.org**).

© Cengage Learning

Estate, Caste, and Class

Stratification systems can be broadly categorized into three types: estate systems, caste systems, and class systems. In an **estate system** of stratification, the ownership of property and the exercise of power are monopolized by an elite class who have total control over societal resources. Historically, such societies were feudal systems where classes were differentiated into three basic groups—the nobles, the priesthood, and the commoners. Commoners included peasants (usually the largest class group), small merchants, artisans, domestic workers, and traders. The nobles controlled the land and the resources used to cultivate the land, as well as all the resources resulting from peasant labor.

Estate systems of stratification are most common in agricultural societies. Although such societies have been largely supplanted by industrialization, there are still examples of societies that have a small but powerful landholding class ruling over a population that works mainly in agricultural production. Unlike the feudal societies of the European Middle Ages, however, contemporary estate systems of stratification display the influence of international capitalism. The "noble class" comprises not knights who conquered lands in war, but international capitalists or local elites who control the labor of a vast and impoverished group of people, such as in some South American societies where landholding elites maintain a dictatorship over peasants who labor in agricultural fields.

In a **caste system**, one's place in the stratification system is an *ascribed status* (see Chapter 5), meaning it is a quality given to an individual by circumstances of birth. The hierarchy of classes is rigid in caste systems and is often preserved through formal law and cultural practices that prevent free association and movement between classes. The system of apartheid in South Africa was a stark example of a caste system. Under apartheid, the travel, employment, associations, and place of residence of Black South Africans were severely restricted. Segregation was enforced using a pass system in which Black South Africans could not be in White areas unless for purposes of employment; those found without passes were arrested, often sent to prison without ever seeing their families again. Interracial marriage was illegal. Black South Africans were prohibited from voting; the system was one of total social control where anyone who protested was imprisoned. The apartheid system was overthrown in 1994 when Nelson Mandela, held prisoner for twenty-seven years of his life, was elected president of the new nation of South Africa; a new national constitution guaranteeing equal rights to all was ratified in 1996.

In **class systems**, stratification exists, but a person's placement in the class system can change according to personal achievements; that is, class depends to some degree on *achieved status*, defined as status that is earned by the acquisition of resources and power, regardless of one's origins. Class systems are more open than caste systems because position does not depend strictly on birth, and classes are less rigidly defined than castes because the divisions are blurred by those who move between one class and the next.

Despite the potential for movement from one class to another, in the class system found in the United States, class placement still depends heavily on one's social background. Although ascription (the designation of ascribed status according to birth) is not the basis for social stratification in the United States, the class a person is born into has major consequences for that person's life. Patterns of inheritance; access to exclusive educational resources; the financial, political, and social influence of one's family; and similar factors all shape one's likelihood of achievement. Although there is no formal obstacle to movement through the class system, individual achievement is very much shaped by an individual's class of origin.

In common terms, *class* refers to style or sophistication. In sociological use, **social class** (or *class*) is the social structural position that groups hold relative to the economic, social, political, and cultural resources of society. Class determines the access different people have to these resources and puts groups in different positions of privilege and disadvantage. Each class has members with similar opportunities who tend to share a common way of life. Class also includes a cultural component in that class shapes language, dress, mannerisms, taste, and other preferences. *Class is not just an attribute of individuals; it is a feature of society.*

The social theorist Max Weber described the consequences of stratification in terms of **life chances**, meaning the opportunities that people have in common by virtue of belonging to a particular class. Life chances include the opportunity for possessing goods, having an income, and having access to particular jobs. Life chances are also reflected in the quality of everyday life. Whether you dress in the latest style or wear another person's discarded clothes, have a vacation in an exclusive resort, take your family to the beach for a week, or have no vacation at all, these life chances are the result of being in a particular class.

Class is a structural phenomenon; it cannot be directly observed. Nonetheless, you can "see" class through various displays that people project, often unintentionally, about their class status. What clothing do you wear? Do some objects worn project higher class status than others? How about cars? What class status is displayed through the car you drive or, for that matter, whether you even have a car or use a bus to get to work? In these and myriad other ways, class is projected to others as a symbol of our presumed worth in society.

Social class can be observed in the everyday habits and presentations of self that people project. Common objects, such as clothing and cars, become symbols of

The vast inequality that exists in the United States became more evident during the economic recession of recent years—a recession that resulted in housing foreclosures for thousands of Americans.

see FOR YOURSELF

Status Symbols in Everyday Life

You can observe the everyday reality of social class by noting the status that different ordinary objects have within the context of a class system. Make a list of every car brand you can think of—or, if you prefer, every clothing label. Then rank your list with the highest status brand (or label) at the top of the list, going down to the lowest status. Then answer the following questions:

1. Where does the presumed value of this object come from? Does the value come from the actual cost of producing the object or something more subjective?
2. Do people make judgments about people wearing or driving the different brands you have noted? What judgments do they make? Why?
3. What consequences do you see (positive and negative) of the ranking you have observed? Who benefits from the ranking and who does not?

What does this exercise reveal about the influence of *status symbols* in society? ●

one's class status. As such, they can be ranked not only in terms of their economic value but also in terms of the status that various brands and labels carry. The interesting thing about social class is that a particular object may be quite ordinary, but with the right "label," it becomes a *status symbol* and thus becomes valuable. Take the example of Vera Bradley bags. These paisley bags are made of ordinary cotton with batting. Not long ago, such cloth was cheap and commonplace, associated with rural, working-class women. If such a bag were sewn and carried by a poor person living on a farm, the bag (and perhaps the person!) would be seen as ordinary, almost worthless. But, transformed by the right label (and some good marketing), Vera Bradley bags have become status symbols, selling for a high price (often a few hundred dollars—a price one would never pay for a simple cotton purse). Presumably, having such a bag denotes the status of the person carrying it. (See also the box "See for Yourself: Status Symbols in Everyday Life.")

The early sociologist Thorstein Veblen described the class habits of Americans as **conspicuous consumption**, meaning the ostentatious display of goods to define one's social status. Writing in 1899, Veblen said, "Conspicuous consumption of valuable goods is a means of respectability to the gentleman of leisure" (Veblen 1953/1899: 42). Although Veblen identified this behavior as characteristic of the well-to-do (the "leisure class," he called them), conspicuous consumption today marks the lifestyle of many. Indeed, mass consumerism is a hallmark of both the rich and the middle class, and even of many working-class people's lifestyles. What examples of this do you see among your associates?

Because sociologists cannot isolate and measure social class directly, they use other *indicators* to serve as measures of class. A prominent indicator of class is income; other common indicators are education, occupation, and place of residence. These indicators alone do not define class, but they are often accurate measures of the class standing of a person or group. We will see that these indicators tend to be linked. A good income, for example, makes it possible to afford a house in a prestigious neighborhood and an exclusive education for one's children. In the sociological study of class, indicators such as income and education have had enormous value in revealing the outlines and influences of the class system.

THE CLASS STRUCTURE OF THE UNITED STATES: GROWING INEQUALITY

People think of the United States as a land of opportunity where those who work hard can get ahead and anyone may become rich. But despite these beliefs, class divisions in the United States are real, and inequality is growing. Perhaps this has become more apparent to people in recent years as the nation experienced a recession and a very fragile economic situation. Millions lost their homes and retirement savings and other investments. Many in the middle and working class feel that their way of life is slipping away. For the first time in our nation's history, only 17 percent of the public thinks that children today will be better off than their parents; 62 percent no longer believe this (Rasmussen Reports 2009).

The class structure in the United States means very different living conditions for those of vast wealth and everyone else.

but the point here is that these structural changes are having a profound effect on the life chances of people in different social classes. Many in the working class, for example, once largely employed in relatively stable manufacturing jobs with decent wages and good benefits, now likely work, if they work at all, in lower-wage jobs with fewer benefits, such as health care and pensions. Middle-class families have amassed large sums of debt, sometimes to support a middle-class lifestyle, but also perhaps to pay for their children's education.

Thus economic problems that produce inequality are not purely economic: They are social, both in their origins and their consequences. Take home ownership as an example. Home ownership is one of the main components of the American dream. For most Americans, owning one's own home is the primary means of attaining economic security. It is also the key to other resources—good schools, cleaner neighborhoods, and an investment in the future. Likewise, losing your home is more than just a financial crisis—it reverberates through various aspects of your life. And the odds of having a home—indeed, the odds of losing your home—are profoundly connected to social factors, such as your race and your gender.

Housing foreclosure is a trauma for anyone who experiences it, but foreclosure has hit some groups especially hard. The racial segregation of Hispanic and, especially, African American neighborhoods is a major contributing cause to the high rate of mortgage foreclosures (Rugh and Massey 2010). Researchers find that African Americans are almost twice as likely to experience foreclosure as White Americans (Bocian et al. 2010; see also Figure 8.1). Moreover, women are 32 percent more likely than men to have *subprime mortgages* (that

Even aside from the economic recession, the gap between the rich and the poor in the United States is greater than in other industrialized nations, and it is larger than at any time in the nation's history. Many analysts argue that this gap is the central problem of the age—contributing to crime and violence, political division, threats to democracy, and increased anxiety and frustration felt by large segments of the population (Reich 2010).

Many factors have contributed to growing inequality in the United States, including the profound effects of national and global economic changes. Many think of the economic problems of the nation as stemming from individual greed on Wall Street, and this likely plays a role, but social inequality stems from systemic—that is, social structural—conditions, particularly what is called *economic restructuring*.

Economic restructuring refers to the decline of manufacturing jobs in the United States, the transformation of the economy by technological change, and the process of globalization. We examine economic restructuring more in Chapter 15 on the economy,

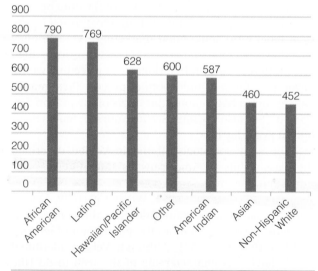

Number of foreclosures

FIGURE 8.1 Foreclosures by Race and Ethnicity, 2007–2009

Source: Bocian, Debbie, Wei Li, and Keith S. Ernst. 2010. *Foreclosures by Race and Ethnicity: The Demographics of a Crisis.* Center for Responsible Lending. **www.responsiblelending.org**

Note: Number is per 10,000 loans to owner/occupants of property.

is, mortgages with an interest rate *higher* than the prime lending rate). And Black women earning double the region's median income were nearly *five times more likely* to receive subprime mortgages than White men with similar incomes (Fishbein and Woodall 2006).

Some might argue that foreclosures occur because individual people have made bad decisions—buying homes beyond their means. But, institutional lending practices also target particular groups, making them more vulnerable to the economic forces that can shatter individual lives. Lenders may see African Americans as a greater credit risk, but they also know that the value of real estate is less in racially segregated neighborhoods. And discriminatory practices in the housing market have been well documented (Squires 2007; Oliver and Shapiro 2006).

The sociological point is that economic problems have a sociological dimension and cannot be explained by individual decisions alone. Economic policies also have different effects for different groups—sometimes intended, sometimes not. Wealthy people, as an example, typically pay a far lower tax rate than the middle class, because much of their money comes from investments, not income, and income is taxed at a much higher rate than investment income. Furthermore, various tax loopholes (such as home mortgage deductions, tax shelters on real estate investments, or even offshore banking deposits) can significantly reduce the tax burden by those with the most resources.

Corporations benefit the most from the tax structure; corporate taxes have fallen in recent years, while most individual Americans are paying more in federal tax than ever before (Johnston 2000). And the Congressional bailouts of recent years have had far more benefit for corporations than for the individuals suffering from the financial crisis.

THE DISTRIBUTION OF INCOME AND WEALTH

Understanding inequality requires knowing some basic economic and sociological terms. Inequality is often presented as a matter of differences in income, which is one important measure of class standing. But, in addition to income inequality, there are vast inequalities in who owns what—that is, the wealth of different groups.

Income is the amount of money brought into a household from various sources (wages, investment income, dividends, and so on) during a given period. In recent years, income growth has been greatest for those at the top of the population; for everyone else, income (controlling for the value of the dollar) has either been relatively flat or grown at a far lesser rate). This has contributed to growing inequality in the United States. But inequality becomes even more apparent when you consider both wealth and income.

Wealth is the monetary value of everything one actually owns. It is calculated by adding all financial assets (stocks, bonds, property, insurance, savings, value of investments, and so on) and subtracting debts, which gives a dollar amount that is one's **net worth**. Wealth allows you to accumulate assets over generations, giving advantages to subsequent generations that they might not have had on their own. Unlike income, wealth is cumulative—that is, its value tends to increase through investment; it can be passed on to the next generation, giving those who inherit wealth a considerable advantage in accumulating more resources.

To understand the significance of wealth compared to income in determining class location, imagine

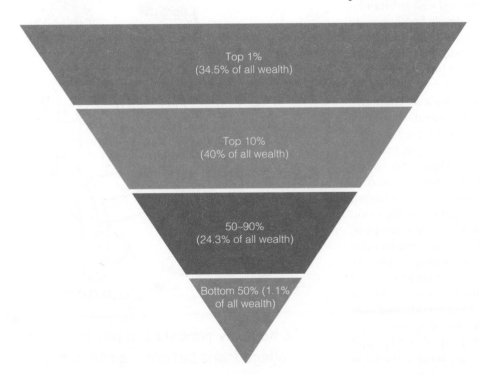

Top 1%
(34.5% of all wealth)

Top 10%
(40% of all wealth)

50–90%
(24.3% of all wealth)

Bottom 50% (1.1% of all wealth)

FIGURE 8.2 How Is Wealth in the United States Distributed?

Source: Levine, Linda. 2012. *An Analysis of the Distribution of Wealth across Households.* Congressional Research Service. **www.crs.gov**

Note: Percentages as a percent of all wealth.

"It's a dog's life," or so the saying goes. But even dogs have their experiences shaped by the realities of social class.

Radius Images/Alamy

AP Photo/Ric Feld

two college students graduating in the same year, from the same college, with the same major and same grade point average. Imagine further that both get jobs with the same salary in the same organization. Yet, in one case, parents paid all the student's college expenses and gave her a car upon graduation. The other student worked while in school and graduated with substantial debt from student loans. This student's family has no money with which to help support the new worker. Who is better off? Same salary, same credentials, but wealth (even if modest) matters. It gives one person an advantage that will be played out many times over as the young worker buys a home, finances her own children's education, and possibly inherits additional assets.

Where is all the wealth? The wealthiest 1 percent own 34.5 percent of all net worth; the bottom half hold only 1.1 percent of all wealth (see Figure 8.2; Levine 2012). Moreover, there has been an increase in the concentration of wealth since the 1980s, making the United States one of the most "unequal" nations in the world (Mishel et al. 2008). The growth of wealth by a select few, though long a feature of the U.S. class system, has also reached historic levels. As just one example, John D. Rockefeller is typically heralded as one of the wealthiest men in U.S. history. But comparing Rockefeller with Bill Gates, controlling for the value of today's dollars, Gates has far surpassed Rockefeller's riches (Myerson 1998).

In contrast to the vast amount of wealth and income controlled by elites, a very large proportion

a.bacall

"I hope my parents can pay off their college loans before I go to college."

www.CartoonStock.com

The Student/Debt Crisis

Numerous recent reports show that students are struggling over rising levels of debt from student loans. In 2010, a record one in five households in the United States had outstanding student debt, an increase from 9 percent as recently as 1989 (Fry 2012). Leaving college or graduate school with large amounts of debt impedes one's ability to get financially established.

All students are at risk of accruing debt, given the rising cost of education and the higher interest rates now associated with student loans. But data show that some groups are more vulnerable than others, adding to the inequalities that accrue across different groups. Among those in the bottom fifth of income earners, student debt, on average, takes up 24 percent of all income; for the top fifth of earners, only 9 percent of income. For those in the middle, student debt consumes 12 percent of income (Fry 2012).

The highest amount of student debt is also among those under 35, those who are just beginning careers and, possibly, families. Race also matters. Black students are more likely to borrow money for college than other groups—and to borrow more; 80 percent of Black students have outstanding student loans, compared to 65 percent of Whites, 67 percent of Hispanics, and 54 percent of Asian students. Moreover, levels of debt are highest among Blacks – an average of $28,692, compared to $24,772 for Whites, $22,886 for Hispanics, and $21,090 for Asians (Demos and Young Invincibles 2011).

How does this reality of student debt influence the experience of those you see in your own environment? What are the sociological causes of this significant social problem?

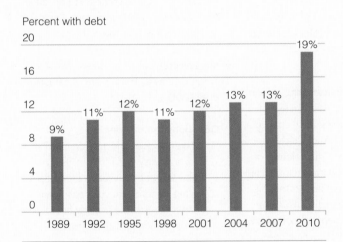

FIGURE 8.3 Households with Outstanding Student Debt

Source: Pew Research Center, Social and Demographic Trends Project. **www.pewsocialtrends.org/2012/09/26/a-record-one-in-five-households-now-owe-student-loan-debt.** Released September 26, 2012. A Record One-in-Five Households Now Owe Student Loan Debt Burden Greatest on Young, Poor by Richard Fry.

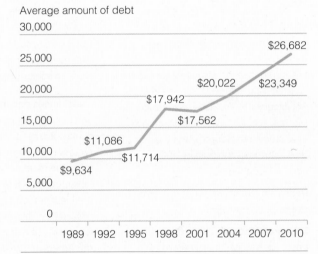

FIGURE 8.4 Average Amount of Household Student Debt

Source: Pew Research Center, Social and Demographic Trends Project. **www.pewsocialtrends.org/2012/09/26/a-record-one-in-five-households-now-owe-student-loan-debt.** Released September 26, 2012. A Record One-in-Five Households Now Owe Student Loan Debt Burden Greatest on Young, Poor by Richard Fry.

of Americans have hardly any financial assets once debt is subtracted. Figure 8.5 shows the net worth of different parts of the population, and you can see that most of the population has very low net worth. One-fifth of the population has zero or negative net worth, usually because their debt exceeds their assets. The American dream of owning a home, a new car, taking annual vacations, and sending one's children to good schools—not to mention saving for a comfortable retirement—is increasingly unattainable for many. When you see the amount of income and wealth a small segment of the population controls, a sobering picture of class inequality emerges. Students themselves may be experiencing this burden, as levels of debt from student loans have escalated in recent years.

Despite the prominence of rags-to-riches stories in American legend, most wealth in this society is inherited. A few individuals make their way into the elite class by virtue of their own success, but this is rare. The upper class is also overwhelmingly White and Protestant. The wealthy also exercise tremendous political power by funding lobbyists, exerting their social and personal influence on other elites, and contributing heavily to political campaigns. Studies of elites also find that they tend to be politically quite conservative (Zweigenhaft and Domhoff 2006; Burris 2000).

DOING **sociological research**

The Fragile Middle Class

The hallmark of the middle class in the United States is its presumed stability. Home ownership, a college education for children, and other accoutrements of middle-class status (nice cars, annual vacations, an array of consumer goods) are the symbols of middle-class prosperity. As has become clear in the recent economic recession, the rising rate of debt, foreclosure, and bankruptcy among the American middle class shows that the middle class is not as secure as it has been presumed to be.

Personal bankruptcy has risen dramatically with more than one and a half million nonbusiness filings for bankruptcy in 2010. How can this be happening in such a prosperous society? Two-thirds filed because of job loss; one-half because of health reasons. One-third are women filing alone (U.S. Courts 2010). Sociologists Teresa Sullivan, Elizabeth Warren, and Jay Lawrence Westbrook have studied bankruptcy, and their research shows the fragility of the middle class in recent times.

Research Question: What is causing the rise of bankruptcy?

Research Method: This study is based on an analysis of official records of bankruptcy in five states, as well as on detailed questionnaires given to individuals who filed for bankruptcy.

Research Results: The research findings of Sullivan and her colleagues debunk the idea that bankruptcy is most common among poor people. Instead, they found bankruptcy is mostly a middle-class phenomenon representing a cross-section of those in this class (meaning that those who are bankrupt are matched on the demographic characteristics of race, age, and gender with others in the middle class). They also debunk the notion that bankruptcy is rising because it is so easy to file. Rather, they found many people in the middle class so overwhelmed with debt that they cannot possibly pay it off. Most often people file for bankruptcy as a result of job loss and lost wages. But divorce, medical problems, housing expenses, and credit card debt also drive many to bankruptcy court.

Conclusions and Implications: Sullivan and her colleagues explain the rise of bankruptcy as stemming from structural factors in society that fracture the stability of the middle class. The volatility of jobs under modern capitalism is one of the biggest factors, but add to this the "thin safety net"—no health insurance for many, but rising medical costs. Also, the American dream of owning one's own home means many are "mortgage poor"—extended beyond their ability to keep up.

In addition, the United States is a credit-driven society. Credit cards are routinely mailed to people in the middle class, encouraging them to buy beyond their means. You can now buy virtually anything on credit: cars, clothes, doctor's bills, entertainment, groceries. You can even use one credit card to pay off other credit cards. Indeed, it is difficult to live in this society without credit cards. Increased debt is the result. Many are simply unable to keep up with compounding interest and penalty payments, and debt takes on a life of its own as consumers cannot keep up with even the interest payments on debt.

Sullivan, Warren, and Westbrook conclude that increases in debt and uncertainty of income combine to produce the fragility of the middle class. Their research shows that "even the most secure family may be only a job loss, a medical problem, or an out-of-control credit card away from financial catastrophe" (2000: 6).

Questions to Consider

1. Have you ever had a credit card? If so, how easy was it to get? Is it possible to get by without a credit card?

2. What evidence do you see in your community of the fragility or stability of different social class groups?

Source: T. A. Sullivan, E. Warren, and J. L. Westbrook, *The Fragile Middle Class: Americans in Debt.* Copyright © 2000 by Yale University Press.

They travel in exclusive social networks that tend to be open only to those in the upper class. They tend to intermarry, their children are likely to go to expensive schools, and they spend their leisure time in exclusive resorts.

Race also influences the pattern of wealth distribution in the United States; for every dollar of wealth White Americans hold, Black Americans have only 26 cents. The median net worth of White households is $110,729; Hispanics, only $7,424; and Blacks, $4,959. Moreover, although all households saw a drop in their net worth during the economic downturn, Whites saw a 16 percent drop—Hispanics, a 66 percent drop; and Blacks, a 53 percent drop (Kochhar et al. 2011). At all levels of income, occupation, and education, Black families have lower levels of wealth than similarly situated White families. Being able to draw on assets during times of economic stress means that families with some resources can better withstand difficult times than those without assets. Even small assets, such as home ownership or a savings account, provide protection from crises such as increased rent, a health emergency, or unemployment. Because the effects of wealth are intergenerational—that is, they accumulate over time—just providing equality of opportunity in the present does not address the differences in class status that Black and White Americans experience (Oliver and Shapiro 2006).

What explains the disparities in wealth by race? Wealth accumulates over time. Thus government policies in the past have prevented Black Americans from being able to accumulate wealth. Discriminatory housing policies, bank lending policies, tax codes, and so forth

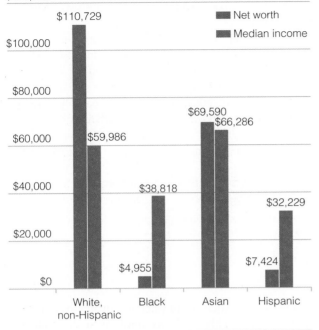

FIGURE 8.5 Median Income and Net Worth by Race, 2010

Source: U.S. Census Bureau. 2012c. *Historic Income Tables: Households, Table H-5; Net Worth and Asset Ownership of Households: 2010.* Washington, DC: U.S. Census Bureau. **www.census.gov**

have disadvantaged Black Americans, resulting in the differing assets Whites and Blacks in general hold now. Even though some of these discriminatory policies have ended, many continue. Either way, their effects persist, resulting in what sociologists Melvin Oliver and Thomas Shapiro call the *sedimentation of racial inequality.*

Understanding the significance of wealth in shaping life chances for different groups also challenges the view that all Hispanics have similar experiences and wealth. Cuban Americans and Spaniards are similar to Whites in their wealth holdings, whereas Mexicans, Puerto Ricans, Dominicans, and other Hispanic groups more closely resemble African Americans on the various indicators of wealth and social class. Likewise, one can better understand differences in class status among Asian American groups by carefully considering the importance not just of income, education, and occupation, but also patterns in the net assets of different groups (Oliver and Shapiro 2006). Without significant wealth holdings, families of any race are less able to transmit assets from previous generations to the next generation, one main support of social mobility.

ANALYZING SOCIAL CLASS

The class structure of the United States is elaborate, arising from the interactions of race and gender inequality with class, the presence of old mixed with new wealth, the income and wealth gap between the haves and have-nots, a culture of entrepreneurship and individualism,

and in recent times, accelerated globalization and high rates of immigration. Given this complexity, how do sociologists conceptualize social class?

Class as a Ladder. One way to conceptualize the class system is as a ladder, with different class groups arrayed up and down the rungs, each rung corresponding to a different level in the class system. Conceptualized this way, social class is the common position groups hold in a status hierarchy (Lucal 1994; Wright 1979); class is indicated by factors such as levels of income, occupational standing, and educational attainment. People are relatively high or low on the ladder depending on the resources they have and whether those resources are education, income, occupation, or any of the other factors known to influence people's placement (or ranking) in the stratification system. Indeed, an abundance of sociological research has stemmed from the concept of **status attainment**, the process by which people end up in a given position in the stratification system. Status attainment research describes how factors such as class origins, educational level, and occupation produce class location.

The laddered model of class suggests that stratification in the United States is hierarchical but somewhat fluid. That is, the assumption is that people can move up and down different "rungs" of the ladder—or class system. In a relatively *open class system* such as the United States, people's achievements do matter, although the extent to which people rise rapidly and dramatically through the stratification system is less than the popular imagination envisions. Some people do begin from modest origins and amass great wealth and influence (celebrities such as Bill Gates, Oprah Winfrey, and millionaire athletes), but these are the exceptions, not the rule. Some people move down in the class system, but as we will see, most people remain relatively close to their class of origin. When people rise or fall in the class system, the distance they travel is usually relatively short, as we will see further in the section on social mobility.

The image of stratification as a laddered system, with different gradients of social standing, emphasizes that one's **socioeconomic status (SES)** is derived from certain factors. Income, occupational prestige, and education are the three measures of socioeconomic status that have been found to be most significant in determining people's placement in the stratification system.

The **median income** for a society is the midpoint of all household incomes. In other words, half of all households earn more than the median income; half earn less. In 2011, median household income in the United States was $50,054 (DeNavas-Walt et al. 2012). Those bunched around the median income level are considered middle class, although sociologists debate which income brackets constitute middle-class standing because the range of what people think of as "middle class" is quite large. Nonetheless, income is a significant indicator of social class standing, although not the only one.

Income Distribution: Should Grades Be the Same?

Figure 8.6 shows the income distribution within the United States. Imagine that grades in your class were distributed based on the same curve. Let's suppose that after students arrived in class and sat down, different groups received their grades based on where they were sitting in the room and in the same proportion as the U.S. income distribution. Only students in the front receive As; the back, Ds and Fs. The middle of the room gets the Bs and Cs. Write a short essay answering the following questions based on this hypothetical scenario:

1. How many students would receive As, Bs, Cs, Ds, and Fs?

2. Would it be fair to distribute grades this way? Why or why not?

3. Which groups in the class might be more likely to support such a distribution? Who would think the system of grade distribution should be changed?

4. What might different groups do to preserve or change the system of grade distribution? What if you really needed an A, but got one of the Fs? What might you do?

5. Are there circumstances in actual life that are beyond the control of people and that shape the distribution of income?

6. How is social stratification maintained by the beliefs that people have about merit and fairness?

Adapted from: Brislen, William, and Clayton D. Peoples. 2005. "Using a Hypothetical Distribution of Grades to Introduce Social Stratification." *Teaching Sociology* 33 (January): 74–80. ●

Occupational prestige is a second important indicator of socioeconomic status. **Prestige is the value others assign to people and groups. Occupational prestige is the subjective evaluation people give to jobs.** To determine occupational prestige, sociological researchers typically ask nationwide samples of adults to rank the general standing of a series of jobs. These subjective ratings provide information about how people perceive the worth of different occupations. People tend to rank professionals, such as physicians, professors, judges, and lawyers highly, with occupations such as electrician, insurance agent, and police officer falling in the middle. Occupations with low occupational prestige are maids, garbage collectors, and shoe shiners (Nakao and Treas 1994; Davis and Smith 1984). These rankings do not reflect the worth of people within these positions but are indicative of the judgments people make about the worth of these jobs.

The final major indicator of socioeconomic status is **educational attainment**, typically measured as the total years of formal education. The more years of education attained, the more likely a person will have higher class status. The prestige attached to occupations is strongly tied to the amount of education the job requires; the more education people think is needed for a given occupation, the more occupational prestige people attribute to that job (Ollivier 2000; MacKinnon and Langford 1994; Blau and Duncan 1967).

Taken together, income, occupation, and education are good indicators of people's class standing. Using the laddered model of class, you can describe the class system in the United States as being divided into several classes: upper, upper middle, middle, lower middle, and lower class. The different classes are arrayed up and down, like a ladder, with those with the most money, education, and prestige on the top rungs and those with the least at the bottom.

In the United States, the *upper class* owns the major share of corporate and personal wealth (see Figure 8.7); it includes those who have held wealth for generations as well as those who have recently become rich. Only a very small proportion of people actually constitute the upper class, but they control vast amounts of wealth and power in the United States. Those in this class are elites who exercise enormous control throughout society. Some wealthy individuals can wield as much power as entire nations (Friedman 1999).

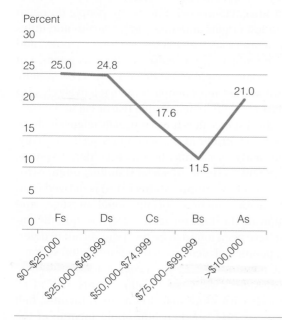

FIGURE 8.6 Income Distribution in the United States

This graph shows the percentage of the total population that falls into each of the five income groups. Would it be fair were course grades to be distributed by the same percentages?

Source: DeNavas-Walt, Carmen, Bernadette Proctor, and Jessica C. Smith. 2012. *Income, Poverty, and Health Insurance Coverage in the United States: 2011.* Washington, DC: U.S. Census Bureau.

Social class influences many things, including the leisure time people experience. Few can even imagine having something like the yacht pictured on the left, owned by Larry Ellison, founder of Oracle.

Even the term *upper class*, however, can mask the degree of inequality in the United States. You might consider those in the top 10 percent as upper class, but within this class are the super rich, or those popularly known as the "one percent," so labeled by the Occupy America movement, in contrast to the remaining 99 percent. Since about 1980, the share of income (not to mention wealth) going to the top one percent has increased to levels not seen in the United States since 1920, which was labeled at the time as the "Gilded Age" because of the concentration of wealth and income in the hands of a few. Income distribution now matches that of the Gilded Age—and, given the trends, may well exceed it. Sociological research finds that this new concentration of income among the superrich is the result of several trends, including the lowest tax rates for high incomes, a more conservative shift in Congress, diminishing union membership, and asset bubbles in the stock and housing markets (Volscho and Kelly 2012). How rich is rich? Each year, the business magazine *Forbes* publishes a list of the 400 wealthiest families and individuals in the country. By 2012, you had to have at least $1 billion to be on the list! Bill Gates and Warren Buffet are the two wealthiest people on the list—Gates with an estimated worth of $66 billion; Buffet, $46 billion. Even in the face of the massive economic downturn for so many in the United States, only two people in the top twenty of the group had less money than the year before. Although a substantial portion of those on the list describe themselves as "self-made," that is, living the American dream, most of these were still able to borrow from parents, in-laws, or spouses; though they may have built their fortunes, they did so with a head start on accumulation (Kroll and Dolan 2012). The best predictor of future wealth still remains the family into which you are born.

Those in the upper class with newly acquired wealth are known as the *nouveau riche*. Luxury vehicles, high-priced real estate, and exclusive vacations may mark the lifestyle of the newly rich. Larry Ellison, who made his fortune as the founder of the software company Oracle, is the third wealthiest person in the United States; he has a megayacht that is 482 feet long, five stories high, with 82 rooms inside. The megayacht also includes an indoor swimming pool, a cinema, a space for a private submarine, and a basketball court that doubles as a helicopter launch pad (see above).

FIGURE 8.7 Average Net Worth by Income Quintile

Recall that one's net worth is the value of everything owned minus one's debt. A quintile is one-fifth of a population, shown here for five different income brackets. You can see here the vast differences in wealth holdings by those in these different income brackets. How would one's wealth holdings affect your ability to withstand some sort of emergency—an illness, unemployment, a recession, and so forth?

Source: U.S. Census Bureau. 2012. *Historic Income Tables: Households, Table H-5; Net Worth and Asset Ownership of Households: 2010.* Washington, DC: U.S. Census Bureau. **www.census.gov**

Chart data — Average Net Worth by Income Quintile:
- Lowest quintile: $5,193
- Second lowest quintile: $33,333
- Third lowest quintile: $72,191
- Second highest quintile: $143,358
- Highest quintile: $333,168

Labor unions, traditionally dominated by White men in the skilled trades, are not only more diverse but also represent workers in occupations typically thought of as "white-collar" work.

The *upper-middle class* includes those with high incomes and high social prestige. They tend to be well-educated professionals or business executives. Their earnings can be quite high indeed, even millions of dollars a year. It is difficult to estimate exactly how many people fall into this group because of the difficulty of drawing lines between the upper, upper-middle, and middle classes. Indeed, the upper-middle class is often thought of as "middle class" because their lifestyle sets the standard to which many aspire, but this lifestyle is actually unattainable by most. A large home full of top-quality furniture and modern appliances, two or three relatively new cars, vacations every year (perhaps a vacation home), high-quality college education for one's children, and a fashionable wardrobe are simply beyond the means of a majority of people in the United States.

The *middle class* is hard to define in part because being "middle class" is more than just economic position. Half of all Americans identify themselves as middle class (Morin and Motel 2012), even though they vary widely in lifestyle and in resources at their disposal. But the idea that the United States is an open class system leads many to think that the majority have a middle-class lifestyle; thus the middle class becomes the ubiquitous norm even though many who consider themselves middle class have a tenuous hold on this class position.

The *lower-middle class* includes workers in the skilled trades and low-income bureaucratic workers, many of whom may actually think they are middle class. Also known as the *working class*, this class includes blue-collar workers (those in skilled trades who do manual labor) and many service workers, such as secretaries, hairstylists, food servers, police, and firefighters. A medium to low income, education, and occupational prestige define the lower-middle class relative to the class groups above it. The term *lower* in this class designation refers to the relative position of the group in the stratification system, but it has a pejorative sound to many people, especially to people who are members of this class, many of whom think of themselves as middle class.

The *lower class* is composed primarily of displaced and poor. People in this class have little formal education and are often unemployed or working in minimum-wage jobs. People of color and women make up a disproportionate part of this class. The poor include the *working poor*—those who work at least twenty-seven hours a week but whose wages fall below the federal poverty level. Four percent of all people working full time now live below the poverty line, a proportion that has generally increased over time. Although this may seem a small number, it includes 8.9 million adults. Black and Hispanic workers are twice as likely to be among the working poor as White or Asian workers, and women are more likely than men to be so (U.S. Bureau of Labor Statistics 2012a).

The concept of the **urban underclass** has been added to the lower class (Wilson 1987). The underclass includes those who are likely to be permanently unemployed and without much means of economic support. The underclass has little or no opportunity for movement out of the worst poverty. Rejected from the economic system, those in the underclass may become dependent on public assistance or illegal activities. Structural transformations in the economy have left large groups of people, especially urban minorities, in these highly vulnerable positions. The growth of the urban underclass has exacerbated the problems of urban poverty and related social problems (Wilson 2009, 1996, 1987).

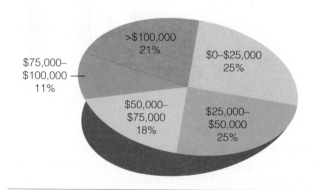

FIGURE 8.8 Percent of U.S. Population in Different Income Brackets This pie chart shows the percentage of the population that falls into each of the five income brackets.

Source: DeNavas-Walt, Carmen, Bernadette Proctor, and Jessica C. Smith. 2012. *Income, Poverty, and Health Insurance Coverage in the United States: 2011.* Washington, DC: U.S. Department of Commerce. **www.census.gov**

Class Conflict. A second way of conceptualizing the class system is conflict theory, derived from the work of Karl Marx. Conflict theory defines classes in terms of their structural relationship to other classes and their relationship to the economic system. The analysis of class from this sociological perspective interprets inequality as resulting from the unequal distribution of power and resources in society (see Chapter 1). Sociologists who work from a conflict perspective see classes as facing off against each other, with elites exploiting and dominating others. The key idea in this model is that class is not simply a matter of what individuals possess in terms of income and prestige; instead, class is defined by the relationship of the classes to the larger system of economic production (Vanneman and Cannon 1987; Wright 1985).

From a conflict perspective, the middle class, or the *professional-managerial class*, includes managers, supervisors, and professionals. Members of this group have substantial control over other people, primarily through their authority to direct the work of others, impose and enforce regulations in the workplace, and determine dominant social values. Although, as Marx argued, the middle class is controlled by the ruling class, members of this class tend to identify with the interests of the elite. The professional-managerial class, however, is caught in a contradictory position between elites and the working class. Like elites, those in this class have some control over others, but like the working class, they have minimal control over the economic system (Wright 1979). Karl Marx argued that as capitalism progresses, more and more of those in the middle class drop into the working class as they are pushed out of managerial jobs into working-class jobs or as professional jobs become organized more along the lines of traditional working-class employment.

Has this happened? Not to the extent Marx predicted. He thought that ultimately there would be only two classes—the capitalist and the proletariat. To some extent, however, this is occurring. Classes have become more polarized, with the well-off accumulating even more resources and the middle class seeing their income as either flat or falling, measured in constant dollars. Rising levels of debt among the middle class have contributed to this growing inequality. Many now have a fragile hold on being middle class: The loss of a job, a family emergency, such as the death of a working parent, divorce, disability, or a prolonged illness can quickly leave middle-and working-class families in a precarious financial state. At the same time, high salaries for CEOs, tax loopholes that favor the rich, and sheer greed are concentrating more wealth in the hands of a few.

Members of the working class have little control over their own work lives; instead, they generally have to take orders from others. This concept of the working class departs from traditional blue-collar definitions of working-class jobs because it includes many so-called white-collar workers (secretaries, salespeople, and nurses), any group working under the rules imposed by managers. The middle class may exercise some autonomy at work, but the working class has little power to challenge decisions of those who supervise them, except insofar as they can organize collectively, as in unions, strikes, or other collective work actions.

Whether you see the class system as a ladder or as a system of conflict, you can see that the class structure in the United States is hierarchical. Class position gives different people access to jobs, income, education, power, and social status, all of which bestow further opportunities on some and deprive others of success. People sometimes move from one class to another (although this is not the norm), but the class structure is a system with boundaries built into it, generating class conflict. The middle and working classes shoulder much of the tax burden for social programs, producing resentment by these groups toward the poor. At the same time, corporate taxes have declined and tax loopholes for the rich have increased, an indication of the privilege that is perpetuated by the class system. Whatever features of the class system different sociologists study, they see class stratification as a dynamic process—one involving the interplay of access to resources, judgments about different groups, and the exercise of power by a few.

Diverse Sources of Stratification

Class is only one basis for stratification in the United States. Factors such as age, ethnicity, and national origin have a tremendous influence on stratification. Race and gender are two primary influences in the stratification system in the United States. In fact, analyzing class without also analyzing race and gender can be misleading. Race, class, and gender, as we are seeing throughout this book, are overlapping systems of stratification that people experience simultaneously. A working-class Latina, for example, does not experience herself as working class at one moment, Hispanic at another moment, and a woman the next. At any given point in time, her position in society is the result of her race, class, *and* gender status. In other words, class position is manifested differently depending on one's race and gender, just as gender is experienced differently depending on one's race and class, and race is experienced differently depending on one's gender and class. Depending on one's circumstances, race, class, or gender may seem particularly salient at a given moment in a person's life. For example, a Black middle-class man stopped and interrogated by police when driving through a predominantly White middle-class neighborhood may at that moment feel his racial status as his single most outstanding characteristic, but at all times his race, class, and gender influence his life chances. As social categories, race, class, and gender shape all people's experience in this society, not just those who are disadvantaged (Andersen and Collins 2013).

Class also significantly differentiates group experience within given racial and gender groups. Latinos, for example, are broadly defined as those who trace

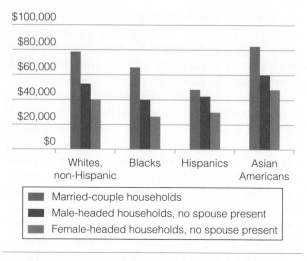

FIGURE 8.9 Median Annual Income by Race and Household Status As illustrated in this graph, married-couple households have the highest median income in all racial–ethnic groups; female-headed households, the least. Which groups reach median income status ($50,054 for households in 2011), and what does this tell you about the combined influence of family type, gender, and race/ethnicity?

Source: U.S. Census Bureau. 2012. *Current Population Survey*, Table HINC-02. *Age of Householder-Households, by Total Money Income in 2011, Type of Family, Race and Hispanic Origin of Householder.* Washington, DC: U.S. Census Bureau. **www.census.gov**

their origins to regions originally colonized by Spain. The ancestors of this group include both White Spanish colonists and the natives who were enslaved on Spanish plantations. Today, some Latinos identify as White, others as Black, and others by their specific national and cultural origins. The very different histories of those categorized as Latino are matched by significant differences in class. Some may have been schooled in the most affluent settings; others may be virtually unschooled. Those of upper-class standing may have had little experience with prejudice or discrimination; others may have been highly segregated into barrios and treated with extraordinary prejudice. Latinos who live near each other geographically in the United States and who are the same age and share similar ancestry may have substantially different experiences based on their class standing. Neither class, race, nor gender, taken alone, can be considered an adequate indicator of different group experiences. As you can see in Figure 8.9, even one's household status affects class standing.

The Race–Class Debate. The relationship between race and class is much debated among sociologists. The Black middle class goes all the way back to the small numbers of free Blacks in the eighteenth and nineteenth centuries (Frazier 1957), expanding in the twentieth century to include those who were able to obtain an education and become established in industry, business, or a profession. Although wages for Black middle-class and professional workers never

matched those of Whites in the same jobs, the Black middle class has had relatively high prestige within the Black community. Many sociologists conclude that the class structure among African Americans has existed alongside the White class structure, separate and different.

In recent years, both the African American and Latino middle classes have expanded, primarily as the result of increased access to education and middle-class occupations for people of color (Higginbotham 2001; Pattillo-McCoy 1999). Although middle-class Blacks and Latinos may have economic privileges that others in these groups do not have, their class standing does not make them immune to the negative effects of race. Asian Americans also have a significant middle class, but they have also been stereotyped as the most successful minority group because of their presumed educational achievement, hard work, and thrift. This stereotype is referred to as the *myth of the model minority* and includes the idea that a minority group must adopt alleged dominant group values to succeed. This myth about Asian Americans obscures the significant obstacles to success that Asian Americans encounter, and it ignores the hard work and educational achievements of other racial and ethnic groups. The idea that Asian Americans are the "model minority" also obscures the high rates of poverty among many Asian American groups (Lee 1996).

Despite recent successes, many in the Black middle class have a tenuous hold on this class status. The Black middle class remains as segregated from Whites as the Black poor, and continuing racial segregation in neighborhoods means that Black middle-class neighborhoods are typically closer to Black poor neighborhoods than the White middle-class neighborhoods are to White poor ones. This exposes many in the Black middle class to some of the same risks as those in poverty. This is not to say that the Black middle class has the same experience as the poor, but it challenges the view that the Black middle class "has it all" (Lacy 2007; Pattillo-McCoy 1999). Furthermore, Black Americans are still much more likely to be working class than middle class; they are also more likely to be working class than are Whites (Horton et al. 2000).

The Influence of Gender and Age. Despite decades of legislation in place to protect women from discrimination and to provide equal pay for equal work, women's median income still lags behind that of men. The median income for women, even among those employed full time, is far below the national median income level. In 2011, when median income for men working year-round and full-time was $48,202, the median income of women working year-round, full time was $37,118—77 percent of men's income (DeNavas-Walt et al. 2012). This is largely because most women work in gender-segregated jobs, a phenomenon we will explore further in the chapter on gender inequality.

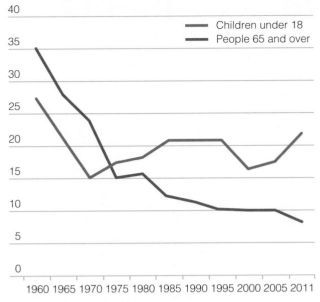

Percent in poverty

— Children under 18
— People 65 and over

FIGURE 8.10 Poverty among the Old and the Young, 1965–2011

Source: DeNavas-Walt, Carmen, Bernadette D. Proctor, and Jessica C. Smith. 2012. *Income, Poverty, and Health Insurance Coverage in the United States: 2011*. Washington, DC: U.S. Department of Commerce. **www.census.gov**

Age, too, is a significant source of stratification. The age group most likely to be poor are children, a whopping 22 percent of whom live in poverty in the United States. This represents a change from the recent past when the aged were most likely to be poor (see Figure 8.10). Although many elderly people are now poor (9 percent of those age 65 and over), far fewer in this age category are poor than was the case not many years ago (DeNavas-Walt et al. 2012). This shift reflects the greater affluence of the older segments of the population—a trend that is likely to continue as the current large cohort of middle-class baby boomers grow older.

SOCIAL MOBILITY: MYTHS AND REALITIES

Popular legends extol the possibility of anyone becoming rich in the United States. The well-to-do are admired not just for their style of life but also for their supposed drive and diligence. The admiration for those who rise to the top makes it seem like anyone who is clever enough and works hard can become fabulously rich. The assumption is that the United States class system is a **meritocracy**—that is, a system in which one's status is based on merit or accomplishments, not other social characteristics. As the word suggests, in a meritocracy people move up and down through the class system based on merit, not based on other characteristics. Is this the case in the United States?

Defining Social Mobility

Social mobility is a person's movement over time from one class to another. Social mobility can be up or down, although the American dream emphasizes upward movement. Mobility can be either *inter*generational, occurring between generations, as when a daughter rises above the class of her mother or father, or *intra*generational, occurring within a generation, as when a person's class status changes as the result of business success (or disaster).

Societies differ in the extent to which social mobility is permitted. Some societies are based on *closed class systems*, in which movement from one class to another is virtually impossible. In a caste system, for example, mobility is strictly limited by the circumstances of one's birth. At the other extreme are *open class systems*, in which placement in the class system is based on individual achievement, not ascription. In open class systems, there are relatively loose class boundaries, high rates of class mobility, and weak perceptions of class difference.

The Extent of Social Mobility

Does social mobility occur in the United States? Social mobility is much more limited than people believe. Success stories of social mobility do occur, but research finds that experiences of mobility over great distances are rare, certainly far less than believed. Most people remain in the same class as their parents. What mobility exists is typically short in distance, and some people actually drop to a lower status, referred to as *downward social mobility*. Research finds that rates of upward social mobility are highest among White men, followed by White women, then Black men, and finally, Black women (Mazumder 2008). Currently, half of all Americans say that they are not moving forward, and one-third say they have fallen further back (Acs and Zimmerman 2008). Indeed, one-third of the public now says they are lower class, compared to 25 percent only a few years ago (Morin and Motel 2012). Social mobility between generations is becoming even more rigid than in the past (Sawhill and McClanahan 2006).

Social mobility is influenced most by factors that affect the whole society, not just by individual characteristics. Just being born in a particular generation can have a significant influence on one's life chances. The fears of today's young, middle-class people that they will be unable to achieve the lifestyles of their parents show the effect that being in a particular generation can have on one's life chances. In other words, when mobility occurs, it is usually because of societal changes that create or restrict opportunities, including such changes as economic cycles, changes in the occupational structure, and demographic factors, such as the number of college graduates in the labor force (Beller and Hout 2006). But mobility in the United

Reproducing Class Stereotypes

The media have a substantial impact on how people view the social class system and different groups within it. Especially because people tend to live and associate with people in their own class, how they see others can be largely framed by the portrayal of different class groups in the media. Research has found this to be true and, in addition, has found that mass media have the power to shape public support for policies on public assistance.

To begin with, the media overrepresent the lifestyle of the most comfortable classes. It is the rare family that can afford the home decor and fashion depicted in soap operas, ironically most likely watched by those in the working class. Media portrayals, such as those found on television talk shows as well as sports, tend to emphasize stories of upward mobility. When the working class is depicted, it tends to be shown as deviant, reinforcing class antagonism

and giving viewers a sense of moral and "class superiority" (Gersch 1999).

Content analyses of the media also find that the poor are largely invisible in the media (Mantsios 2010). Those poor people who are depicted in television and magazines are more often portrayed as Black than is actually the case, leading people to overestimate the actual number of Black poor. The elderly and working poor are rarely seen (Clawson and Trice 2000; Gilens 1996). Representations of welfare overemphasize themes of dependency, especially when portraying African Americans. Women are also more likely than men to be represented as dependent (Misra et al. 2003). And rarely are welfare activists shown as experts; rather, public officials are typically given the voice of authority (Ryan 1996). One result is that the media end up framing the "field of thinkable solutions to public problems" (Sotirovic 2001, 2000), but do so within

a context that ignores the social structural context of social issues.

Source: Gersch, Beate. 1999. "Class in Daytime Talk Television." *Peace Review 11* (June): 275–281; Sotirovic, Mira. 2001. "Media Use and Perceptions of Welfare." *Journal of Communication* 51 (December): 750–774; Sotirovic, Mira. 2000. "Effects of Media Use on Audience Framing and Support for Welfare." *Mass Communication & Society* 2–3 (Spring–Summer): 269–296; Bullock, Heather E., Karen Fraser, and Wendy R. Williams. 2001. "Media Images of the Poor." *The Journal of Social Issues* 57 (Summer): 229–246; Clawson, Rosalee A., and Rakuya Trice. 2000. "Poverty as We Know It: Media Portrayals of the Poor." *The Public Opinion Quarterly* 64 (Spring): 53–64; Gilens, Martin. 1996. "Race and Poverty in America: Public Misperceptions and the American News Media." *Public Opinion Quarterly* 60 (Winter): 515–541; Misra, Joy, Stephanie Moller, and Marina Karides. 2003. "Envisioning Dependency: Changing Media Depictions of Welfare in the 20th Century." *Social Problems* 50 (November): 482–504; Ryan, Charlotte. 1996. "Battered in the Media: Mainstream News Coverage of Welfare Reform." *Radical America* 26 (August): 29–41. Further resources: See the film, *Class Dismissed: How TV Frames the Working Class*, Media Education Foundation. **www.mediaed.org**

States is not impossible. Indeed, many have immigrated to this nation with the knowledge that their life chances are better here than in their countries of origin. And the social mobility that does exist is greatly influenced by education. But, in sum, social mobility is much more limited than the American dream of mobility suggests.

Class Consciousness

Because of the widespread belief that mobility is possible, people in the United States, compared to many other societies, tend not to be very conscious of class. **Class consciousness** is the perception that a class structure exists along with a feeling of shared identification with others in one's class—that is, those with whom you share life chances (Centers 1949). Notice that there are two dimensions to the definition of class consciousness: the idea that a class structure exists and one's class identification.

There has been a long-standing argument that Americans are not very class conscious because of the belief that upward mobility is possible and because of the belief in individualism that is part of the culture. Images of opulence also saturate popular culture, making it seem that such material comforts are available to

anyone. The faith that upward mobility is possible ironically perpetuates inequality because, if people believe that everyone has the same chances of success, they are likely to think that whatever inequality exists must be fair or the result of individual success and failure.

Class inequality in any society is usually buttressed by ideas that support (or actively promote) inequality. Beliefs that people are biologically, culturally, or socially different can be used to justify the higher position of some groups. If people believe these ideas, the ideas provide legitimacy for the system. Karl Marx used the term **false consciousness** to describe the class consciousness of subordinate classes who had internalized the view of the dominant class. Marx argued that the ruling class controls subordinate classes by infiltrating their consciousness with belief systems that are consistent with the interests of the ruling class. If people accept these ideas, which justify inequality, they need not be overtly coerced into accepting the roles designated for them by the ruling class.

There have been times when class consciousness was higher, such as during the labor movement of the 1920s and 1930s. Then working-class people had a very high degree of class consciousness and mobilized on behalf of workers' rights. We see this happening again in the current economic downturn as more people identity themselves as lower class. But the formation of a relatively large middle class and a relatively high standard of living mitigates class discontent. Racial and ethnic divisions also make strong alliances within various classes less stable. Growing inequality could result in a higher degree of class consciousness, but this has not yet developed into a significant class-based movement for change.

WHY IS THERE INEQUALITY?

Stratification occurs in all societies. Why? This question originates in classical sociology in the works of Karl Marx and Max Weber, theorists whose work continues to inform the analysis of class inequality today.

Karl Marx: Class and Capitalism

Karl Marx (1818–1883) provided a complex and profound analysis of the class system under capitalism—an analysis that, although more than 100 years old, continues to inform sociological analyses and has been the basis for major world change. Marx defined classes in relationship to *the means of production*, defined as the system by which goods are produced and distributed. In Marx's analysis, two primary classes exist under capitalism: the *capitalist class*, those who own the means of production, and the *working class* (or proletariat), those who sell their labor for wages. There are further divisions within these two classes: the *petty bourgeoisie*, small business owners and managers (those whom you might think of as middle class) who identify with the interests of the capitalist class but do not own the means of production, and the *lumpenproletariat*, those who have become unnecessary as workers and are then discarded. (Today, these would be the underclass, the homeless, and the permanently poor.)

Marx thought that with the development of capitalism, the capitalist and working class would become increasingly antagonistic (something he referred to as class struggle). As class conflicts became more intense, the two classes would become more polarized, with the petty bourgeoisie becoming deprived of their property and dropping into the working class. This analysis is still reflected in contemporary questions about whether the classes are becoming more polarized, with the rich getting richer and everyone else worse off, as we have seen.

In addition to the class struggle that Marx thought would characterize the advancement of capitalism, he also thought that capitalism was the basis for other social institutions. Capitalism is the *infrastructure* of society, with other institutions (such as law, education, the family, and so forth) reflecting capitalist interests. Thus, according to Marx, the law supports the interests of capitalists; the family promotes values that socialize people into appropriate work roles; and education reflects the interests of the capitalist class. Over time, capitalism increasingly penetrates society, as we can

Frances Roberts/Alamy

In a society where there is excessive consumerism, shopping becomes a leisure activity, not just a necessity.

clearly see with the corporate mergers that characterize modern life and the predominance of capitalist values in society's institutions.

Why do people support such a system? Here is where ideology plays a role. **Ideology** refers to belief systems that support the status quo. According to Marx, the dominant ideas of a society are promoted by the ruling class. Through their control of the communications industries in modern society, the ruling class is able to produce ideas that buttress their interests.

Much of Marx's analysis boils down to the consequences of a system based on the pursuit of profit. If goods were exchanged at the cost of producing them, no profit would be produced. Capitalist owners want to sell commodities for more than their actual value—more than the cost of producing them, including materials and labor. Because workers contribute value to the system and capitalists extract value, Marx saw capitalist profit as the exploitation of labor. Marx believed that as profits became increasingly concentrated in the hands of a few capitalists, the working class would become increasingly dissatisfied. The basically exploitative character of capitalism, according to Marx, would ultimately lead to its destruction as workers organized to overthrow the rule of the capitalist class. *Class conflict* between workers and capitalists, he argued, was inescapable, with revolution being the inevitable result. Perhaps the class revolution that Marx predicted has not occurred, but the dynamics of capitalism that he analyzed are unfolding before us.

At the time Marx was writing, the middle class was small and consisted mostly of small business owners and managers. Marx saw the middle class as dependent on the capitalist class, but exploited by it, because the middle class did not own the means of production. He saw middle-class people as identifying with the interests of the capitalist class because of the similarity in their economic interests and their dependence on the capitalist system. Marx believed that the middle class failed to work in its own best interests because it falsely believed that it benefited from capitalist arrangements. Marx thought that in the long run the middle class would pay for their misplaced faith when profits became increasingly concentrated in the hands of a few and more and more of the middle class dropped into the working class. Because he did not foresee the emergence of the large and highly differentiated middle class we have today, not every part of Marx's theory has proved true. Still, his analysis provides a powerful portrayal of the forces of capitalism and the tendency for wealth to belong to a few, whereas the majority work only to make ends meet. He has also influenced the lives of billions of people under self-proclaimed Marxist systems that were created in an attempt, however unrealized, to overcome the pitfalls of capitalist society.

Max Weber: Class, Status, and Party

Max Weber (1864–1920) agreed with Marx that classes were formed around economic interests, and he agreed that material forces (that is, economic forces) have a powerful effect on people's lives. However, he disagreed with Marx that economic forces are the primary dimension of stratification. Weber saw three dimensions to stratification:

- *class* (the economic dimension);
- *status* (or prestige, the cultural and social dimension); and,
- *party* (or power, the political dimension).

Weber is thus responsible for a *multidimensional view* of social stratification because he analyzed the connections between economic, cultural, and political systems. Weber pointed out that, although the economic, social, and political dimensions of stratification are usually related, they are not always consistent. A person could be high on one or two dimensions, but low on another. A major drug dealer is an example: high wealth (economic dimension) and power (political dimension) but low prestige (social dimension), at least in the eyes of the mainstream society, even if not in other circles.

Weber defined *class* as the economic dimension of stratification—how much access to the material goods of society a group or individual has, as measured by income, property, and other financial assets. A family with an income of $200,000 per year clearly has more access to the resources of a society than a family living on an income of $50,000 per year. Weber understood that a class has common economic interests and that economic well-being was the basis for one's life chances. But, in addition, he thought that people were also stratified based on their status and power differences.

Status, to Weber, is the prestige dimension of stratification—the social judgment or recognition given to a person or group. Weber understood that class distinctions are linked to status distinctions—that is, those with the most economic resources tend to have the highest status in society, but not always. In a local community, for example, those with the most status may be those who have lived there the longest, even if newcomers arrive with more money. Although having power is typically related to also having high economic standing and high social status, this is not always the case, as you saw with the example of the drug dealer.

Finally, *party* (or what we would now call power) is the political dimension of stratification. It is the capacity to influence groups and individuals even in the face of opposition. Power is also reflected in the ability of a person or group to negotiate their way through social institutions. An unemployed Latino man wrongly accused of a crime, for instance, does not have much power to negotiate his way through the criminal justice system. By comparison, business executives accused of corporate crime can afford expensive lawyers and thus frequently go unpunished or, if they are found guilty, serve relatively light sentences in comparatively pleasant facilities. Again, Weber saw power as linked to economic standing, but he did not think that economic standing was always the determining cause of people's power.

Marx and Weber explain different features of stratification. Both understood the importance of the economic basis of stratification, and they knew the significance of class for determining the course of one's life. Marx saw people as acting primarily out of economic interests. Weber refined the sociological analyses of stratification to account for the subtleties that can be observed when you look beyond the sheer economic dimension to stratification, stratification being the result of economic, social, and political forces. Together, Marx and Weber provide compelling theoretical grounds for understanding the contemporary class structure.

Functionalism and Conflict Theory: The Continuing Debate

Marx and Weber were trying to understand why differences existed in the resources and power that different groups in society hold. The question persists of why there is inequality. Two major frameworks in sociological theory—functionalist and conflict theory—take quite different approaches to understanding inequality (see Table 8.2, p. 191).

The Functionalist Perspective on Inequality. Functionalist theory views society as a system of institutions organized to meet society's needs (see Chapter 1). The functionalist perspective emphasizes that the parts of society are in basic harmony with each other; society is characterized by cohesion, consensus, cooperation, stability, and persistence (Eitzen and Baca Zinn 2012;

table 8.2 Functionalist and Conflict Theories of Stratification

Interprets	Functionalism	Conflict Theory
Inequality	The purpose of inequality is to motivate people to fill needed positions in society.	Inequality results from a system where those with the most resources exploit and control others.
Reward system	Greater rewards are attached to higher positions to ensure that people will be motivated to train for functionally important roles in society.	Inequality prevents the talents of those at the bottom from being discovered and used.
Classes	Some groups are rewarded because their work requires the greatest degree of talent and training.	Classes conflict with each other as they vie for power and economic, social, and political resources.
Elites	The most talented are rewarded in proportion to their contribution to the social order.	The most powerful reproduce their advantage by distributing resources and controlling the dominant value system.
Class consciousness/ ideology	Beliefs about success and failure confirm the status of those who succeed.	Elites shape societal beliefs to make their unequal privilege appear to be legitimate and fair.
Poverty	Poverty serves economic and social functions in society.	Poverty is inevitable because of the exploitation built into the system.
Social policy	Because the system is basically fair, social policies should only reward merit.	Because the system is basically unfair, social policies should support disadvantaged groups.

© Cengage Learning

Merton 1957; Parsons 1951a). Different parts of the social system complement one another and are held together through social consensus and cooperation. To explain stratification, functionalists propose that the roles filled by the upper classes—such as governance, economic innovation, investment, and management—are essential for a cohesive and smoothly running society and hence are rewarded in proportion to their contribution to the social order (Davis and Moore 1945).

According to the functionalist perspective, social inequality serves an important purpose in society: It motivates people to fill the different positions in society that are needed for the survival of the whole. Functionalists think that some positions in society are more important than others and require the most talent and training. Rewards attached to those positions (such as higher income and prestige) ensure that people will make the sacrifices needed to acquire the training for functionally important positions (Davis and Moore 1945). Higher class status thus comes to those who acquire what is needed for success (such as education and job training). In other words, functionalist theorists see inequality as based on a reward system that motivates people to succeed.

The Conflict Perspective on Inequality. Conflict theory also sees society as a social system, but unlike functionalism, conflict theory interprets society as being held together through conflict and coercion. From a conflict-based perspective, society comprises competing interest groups, some with more power than others. Different groups struggle over societal resources and compete for social advantage. Conflict theorists argue that those who control society's resources also hold power over others. The powerful are also likely to act to reproduce their advantage and try to shape societal beliefs to make their privileges appear to be legitimate and fair. In sum, conflict theory emphasizes the friction in society rather than the coherence and sees society as dominated by elites.

From the perspective of conflict theory, derived largely from the work of Karl Marx, social stratification is based on class conflict and blocked opportunity. Conflict theorists see stratification as a system of domination and subordination in which those with the most resources exploit and control others. They also see the different classes as in conflict with each other, with the unequal distribution of rewards reflecting the class interests of the powerful, not the survival needs of the whole society (Eitzen and Baca Zinn 2012). According to the conflict perspective, inequality provides elites with the power to distribute resources, make and enforce laws, and control value systems; elites use these powers in ways that reproduce inequality. Others in the class structure, especially the working class and the poor, experience blocked mobility.

Conflict theorists argue that the consequences of inequality are negative. From a conflict point of view, the more stratified a society, the less likely that society will benefit from the talents of its citizens; inequality limits the life chances of those at the bottom, preventing their talents from being discovered and used. To the waste of talent is added the restriction of human creativity and productivity.

The Debate between Functionalist and Conflict Theories. Implicit in the argument of each perspective is criticism of the other perspective. Functionalism assumes that the most highly rewarded jobs are the most important for society, whereas conflict theorists argue that some of the most vital jobs in society—those that sustain life and the quality of life, such as farmers, mothers, trash collectors, and a wide range of other laborers—are usually the least rewarded. Conflict theorists also criticize functionalist theory for assuming that the most talented get the greatest rewards. They point out that systems of stratification tend to devalue the contributions of those left at the bottom and to under-utilize the diverse talents of all people (Tumin 1953). In contrast, functionalist theorists contend that the conflict view of how economic interests shape social organization is too simplistic. Conflict theorists respond by arguing that functionalists hold too conservative a view of society and overstate the degree of consensus and stability that exists.

The debate between functionalist and conflict theorists raises fundamental questions about how people view inequality. Is it inevitable? How is inequality maintained? Do people basically accept it? This debate is not just academic. The assumptions made from each perspective frame public policy debates. Whether the topic is taxation, poverty, or homelessness, if people believe that anyone can get ahead by ability alone, they will tend to see the system of inequality as fair and accept the idea that there should be a differential reward system. Those who tend toward the conflict view of the stratification system are more likely to advocate programs that emphasize public responsibility for the well-being of all groups and to support programs and policies that result in more of the income and wealth of society going toward the needy.

POVERTY

Despite the relatively high average standard of living in the United States, poverty afflicts millions of people. Poor health care, failures in the education system, and crime are all related to poverty. Who is poor, and why is there so much poverty in an otherwise relatively affluent society?

The federal government has established an official definition of poverty used to determine eligibility for government assistance and to measure the extent of poverty in the United States. The **poverty line** is the amount of money needed to support the basic needs of a household, as determined by government; below this line, one is considered officially poor. To determine the poverty line, the Social Security Administration takes a low-cost food budget (based on dietary information provided by the U.S. Department of Agriculture) and multiplies by a factor of three, assuming that a family spends approximately one-third of its budget on food. The resulting figure is the official poverty line, adjusted slightly each year for increases in the cost of living. In 2011, the official poverty line for a family of four was $23,021. Although a cutoff point is necessary to administer antipoverty programs, this definition of poverty can be misleading. A person or family earning $1 above the cutoff point would not be officially categorized as poor.

see FOR YOURSELF

Using the current federal *poverty line* ($23,021 for a family of four) as a guide, develop a monthly budget that does not exceed this income level and that accounts for all of your family's needs. For purposes of this exercise, assume that you head a family of four, the figure on which this poverty threshold is based. Base your budget on the actual costs of such things in your locale (rent, food, transportation, utilities, clothing, and so forth). Don't forget to account for taxes (state, federal, and local), health care expenses, your children's education, car repairs, and so on. What does this exercise teach you about those who live below the poverty line? ●

Who Are the Poor?

There are now more than 46 million poor people in the United States, representing 15 percent of the population. After the 1950s, poverty declined in the United States. Although the poverty rate has generally been increasing since 2000, it rose even more during the economic recession. Although the majority of the poor are White, disproportionately high rates of poverty are also found among Asian Americans, Native Americans, Black Americans, and Hispanics. Of those considered poor, 28 percent are Native Americans, 27.6 percent are African Americans, 25 percent are Hispanics, 12.3 percent are Asians, and 9.8 percent are non-Hispanic Whites (DeNavas-Walt et al. 2012; U.S. Census Bureau 2012c). Among Hispanics, there are further differences among groups. Puerto Ricans—the Hispanic group with the lowest median income—have been most likely to suffer increased poverty, probably because of their concentration in the poorest segments of the labor market and their high unemployment rates (Hauan et al. 2000; Tienda and Stier 1996). Asian American poverty has also increased substantially in recent years, particularly among the most recent immigrant groups, including Laotians, Cambodians, Vietnamese, Chinese, and Korean immigrants; Filipino, Japanese, and Asian Indian families have lower rates of poverty (White House 2012).

The vast majority of the poor have always been women and children, but the percentage of women and children considered to be poor has increased in recent years. The term **feginization of poverty** refers to the large proportion of the poor who are women and children. This trend results from several factors, including the dramatic growth of female-headed households, a decline in the proportion of the poor who are elderly (not matched by a decline in the poverty of women and children), and continuing wage inequality between women and men. The large number of poor women is associated with a commensurate large number of poor children. By 2011, 22 percent of all children in the United States (those under age 18) were poor, including 9 percent of non-Hispanic White children, 37 percent of Black children, 34 percent of Hispanic children, and 13 percent of Asian American children (DeNavas-Walt et al. 2012).

One-third of all families headed by women are poor (see Figure 8.11). In recent years, wages for young workers have declined; because most unmarried mothers are quite young, there is a strong likelihood that their children will be poor. Because of the divorce rate and generally little child support provided by men, women are also increasingly likely to be without the contributing income of a spouse and for longer periods of their lives. Women are more likely than men to live with children and to be financially responsible for them. However, women without children also suffer a high poverty rate, compounded in recent times by the fact that women now live longer than before and are less likely to be married than in previous periods.

The poor are not a one-dimensional group. They are racially diverse, including Whites, Blacks, Hispanics, Asian Americans, and Native Americans. They are diverse in age, including not just children and young mothers, but also men and women of all ages, and especially a substantial number of the elderly, many of whom live alone. They are also geographically diverse, to be found in areas east and west, south and north, urban and rural.

As Map 8.1 shows, poverty rates are generally higher in the South and Southwest. What the map cannot show, however, is concentrated poverty. **Concentrated poverty** means that there are areas of counties, cities, or states where larger percentages of people are poor. Such areas then have higher rates of crime, poor schools, few job opportunities, poor health and housing, and less access to services. Concentrated poverty is highest among African Americans and American Indians

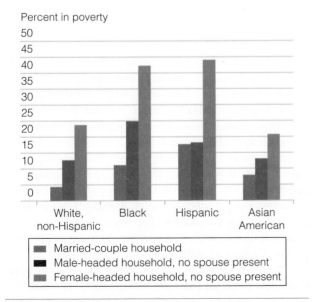

Percent in poverty

FIGURE 8.11 Poverty Status by Family Type and Race

Source: U.S. Census Bureau. 2012. *Historical Income Tables, Table POV-02, People in Families, by Family Structure, Age, and Sex, Iterated by Income-to-Poverty Ratio and Race: 2011*. Washington, DC: U.S. Census Bureau. **www.census.gov**

■ Married-couple household
■ Male-headed household, no spouse present
■ Female-headed household, no spouse present

White, non-Hispanic · Black · Hispanic · Asian American

Frances Twitty/iStockphoto.com

Although most people associate poverty with urban areas, poverty rates outside of metropolitan areas are actually higher than the rates inside metropolitan areas.

MAP 8.1

Mapping America's Diversity: Poverty in the United States

This map shows regional differences in poverty rates (that is, the percentage of poor in different counties). As you can see, poverty is highest in the South, Southwest, and some parts of the upper Midwest. This reflects the higher rates of poverty among Native Americans, Latinos, and African Americans, especially in rural areas. What the map *does not show* is the concentration of poverty in particular urban areas. According to this map, how much poverty is there in your region? Is there poverty that the map does not show?

Source: U.S. Census Bureau. 2010. "Poverty in the United States." **www.census.gov**

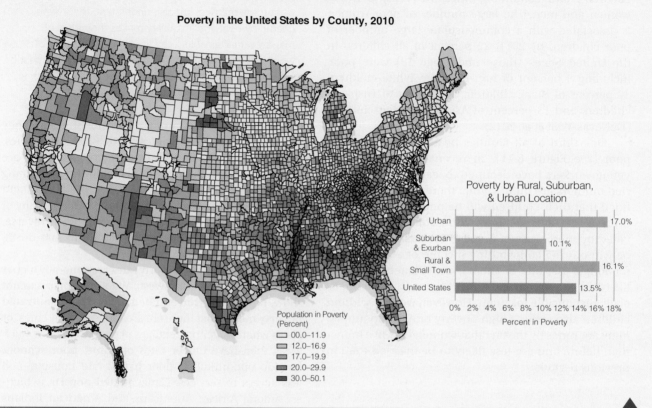

Poverty in the United States by County, 2010

Poverty by Rural, Suburban, & Urban Location

Urban 17.0%
Suburban & Exurban 10.1%
Rural & Small Town 16.1%
United States 13.5%

0% 2% 4% 6% 8% 10% 12% 14% 16% 18%
Percent in Poverty

Population in Poverty (Percent)
□ 00.0–11.9
□ 12.0–16.9
□ 17.0–19.9
□ 20.0–29.9
■ 30.0–50.1

(including Alaska Natives). Among these groups, 10 percent lives in areas where 40 percent or more of the population is poor, compared to 3 percent of poor White Americans and 7 percent of Hispanics (Bishaw 2011).

One marked change in poverty is the growth of poverty in suburban areas. One-third of the nation's poor are now found in suburbs where poverty is growing faster than in major cities (Brookings Institute 2010). Rural poverty also persists in the United States, even though people tend to think of poverty as an urban phenomenon.

Despite the idea that the poor "milk" the system, government supports for the poor are limited. So-called welfare is now largely in the form of food stamps, not cash assistance. Fifteen percent of the public receives food stamps; of those who do, median income is only about $18,000. Half of households receiving food stamps are those with children present; another 18 percent include elderly people, and 20 percent are households where someone is disabled. Considering that the average monthly coupon value is $133, it is hard to understand why federal support is reviled as overly generous and producing sloth and dependence need to squeeze in (U.S> Census Bureau 2012a).

Among the poor are thousands of homeless. Depending on how one defines and measures homelessness, estimates of the number of homeless people vary widely. If you count the number of homeless on any given night, there may be about 444,000 to 842,000 homeless people (depending on the month measured), but measuring those experiencing homelessness over a period of a year, the estimates are about 3.5 million people (National Coalition for the Homeless 2012).

Whatever the actual numbers of homeless people, there has been an increase in homelessness over the past two decades. Families are the fastest-growing segment of the homeless—40 percent; and, children are also 40 percent of the homeless. Moreover, half of the women with children who are homeless have fled from domestic violence (National Coalition Against

Domestic Violence 2001; Zorza 1991). Among homeless people, about 42 percent are African American; another 38 percent are White; 20 percent, Hispanic; 4 percent, Native American; and 2 percent, Asian. A shocking number of the homeless are veterans (about 11 percent of the homeless), including those returning from Iraq and Afghanistan (National Coalition for the Homeless 2012).

There are many reasons for homelessness. The great majority of the homeless are on the streets because of a lack of affordable housing and an increase in poverty, leaving many people with no choice but to live on the street. Add to that problems of inadequate health care, domestic violence, and addiction, and you begin to understand the factors that create homelessness. Some of the homeless are mentally ill (about 16 percent of single, homeless adults); the movement to relocate patients requiring mental health care out of institutional settings has left many without mental health resources that might help them (National Coalition for the Homeless 2012).

Causes of Poverty

Most agree that poverty is a serious social problem. There is far less agreement on what to do about it. Public debate about poverty hinges on disagreements about its underlying causes. Two points of view prevail: Some blame the poor for their own condition, and some look to social structural causes to explain poverty. The first view, popular with the public and many policymakers, is that poverty is caused by the cultural habits of the poor. According to this point of view, behaviors such as crime, family breakdown, lack of ambition, and educational failure generate and sustain poverty, a syndrome to be treated by forcing the poor to fend for themselves. The second view is more sociological, one that understands poverty as rooted in the structure of society, not in the morals and behaviors of individuals.

Blaming the Victim: The Culture of Poverty. Blaming the poor for being poor stems from the myth that success requires only individual motivation and ability. Many in the United States adhere to this view and hence have a harsh opinion of the poor. This attitude is also reflected in U.S. public policy concerning poverty, which is rather ungenerous compared with other industrialized nations. Those who blame the poor for their own plight typically argue that poverty is the result of early childbearing, drug and alcohol abuse, refusal to enter the labor market, and crime. Such thinking puts the blame for poverty on individual choices, not on societal problems. In other words, it blames the victim, not the society, for social problems (Ryan 1971).

The **culture of poverty** argument attributes the major causes of poverty to the absence of work values and the irresponsibility of the poor. In this light, poverty is seen as a dependent way of life that is transferred, like other cultural values, from generation to generation. Policymakers have adapted the culture of poverty argument to argue that the actual causes of poverty are found in the breakdown of major institutions, including the family, schools, and churches.

Is the culture of poverty argument true? To answer this question, we might ask: Is poverty transmitted across generations? Researchers have found only mixed support for this assumption. Many of those who are poor remain poor for only one or two years; only a small percentage of the poor are chronically poor. More often, poverty results from a household crisis, such as divorce, illness, unemployment, or parental death. People tend to cycle in and out of poverty. The public stereotype that poverty is passed through generations is thus not well supported by the facts.

A second question is: Do the poor want to work? The persistent public stereotype that they do not is central to the culture of poverty thesis. This attitude presumes that poverty is the fault of the poor, that poverty would go away if they would only change their values and adopt the American work ethics. What is the evidence for these claims?

Most of the able-bodied poor *do* work, even if only part-time. Moreover, as we saw previously, the number of workers who constitute the *working poor* has increased. You can see why this is true when you calculate the income of someone working full time for minimum wage. Someone working forty hours per week, fifty-two weeks per year, at minimum wage will have an income far below the poverty line. This is the major reason that many have organized a *living wage campaign*, intended to raise the federal minimum wage to provide workers with a decent standard of living.

Current policies that force those on welfare to work also tend to overlook how difficult it is for poor people to retain the jobs they get. Prior to welfare reform in the mid-1990s, poor women who went off welfare to take jobs often found they soon had to return to welfare because the wages they earned were not enough to support their families. Leaving welfare often means losing health benefits, yet incurring increased living expenses. The jobs that poor people find often do not lift them out of poverty. In sum, attributing poverty to the values of the poor is both unproven and a poor basis for public policy.

Structural Causes of Poverty. From a sociological point of view, the underlying causes of poverty lie in the economic and social transformations taking place in the United States. Careful scholars do not attribute poverty to a single cause. There are many causes. Two of the most important are the *restructuring of the economy*, which has resulted in diminished earning power and increased unemployment, and *the status of women in the family and the labor market*, which has contributed to women being overrepresented among the poor. Add to these underlying conditions the federal policies in recent years that have *diminished social support for the poor* in the form of welfare, public housing, and job training. Given these reductions in federal support, it is little wonder that poverty is so widespread.

The restructuring of the economy has caused the disappearance of manufacturing jobs, traditionally an avenue of job security and social mobility for many workers, especially African American and Latino workers (Wilson 1996). The working class has been especially vulnerable to these changes. Economic decline in those sectors of the economy where men have historically received good pay and good benefits means that fewer men are the sole support for their families. Most families now need two incomes to achieve a middle-class way of life. The new jobs that are being created fall primarily in occupations that offer low wages and few benefits; they also tend to be filled by women, especially women of color, leaving women poor and men out of work (McCall 2001; Browne 1999). Such jobs offer little chance to get out of poverty. New jobs are also typically located in neighborhoods far away from the poor, creating a mismatch between the employment opportunities and the residential base of the poor.

Declining wage rates caused by transformations taking place within the economy fall particularly hard on young people, women, and African Americans and Latinos, who are the groups most likely to be among the working poor. The high rate of poverty among women is also strongly related to women's status in the family and the labor market. Divorce is one cause of poverty, because without a male wage in the household, women are more likely to be poor. Women's child-care responsibilities make working outside the home on marginal incomes difficult. Many women with children cannot manage to work outside the home, because it leaves them with no one to watch their children. More women now depend on their own earnings to support themselves, their children, and other dependents. Whereas unemployment has always been considered a major cause of poverty among men, low wages play a major role for women.

The persistence of poverty also increases tensions between different classes and racial groups. William Julius Wilson, one of the most noted analysts of poverty

and racial inequality, has written, "The ultimate basis for current racial tension is the deleterious effect of basic structural changes in the modern American economy on Black and White lower-income groups, changes that include uneven economic growth, increasing technology and automation, industry relocation, and labor market segmentation" (1978: 154). Wilson's comments demonstrate the power of sociological thinking by convincingly placing the causes of both poverty and racism in their societal context, instead of the individualistic thinking that tends to blame the poor for their plight.

Welfare and Social Policy

Current welfare policy is covered by the 1996 Personal Responsibility and Work Reconciliation Act (PRWRA). This federal policy eliminated the long-standing welfare program titled Aid to Families with Dependent Children (AFDC), which was created in 1935 as part of the Social Security Act. Implemented during the Great Depression, AFDC was meant to assist poor mothers and their children. It acknowledged that some people are victimized by economic circumstances beyond their control and deserve assistance. For much of its lifetime, this law supported mostly White mothers and their children; not until the 1960s did welfare come to be identified with Black families.

The new welfare policy gives block grants to states to administer their own welfare programs through the program called **Temporary Assistance for Needy Families (TANF)**. TANF stipulates a lifetime limit of five years for people to receive aid and requires all welfare recipients to find work within two years—a policy known as *workfare*. Those who have not found work within two years of receiving welfare can be required to perform community service jobs for free.

In addition, welfare policy denies payments to unmarried teen parents under eighteen years of age unless they stay in school and live with an adult. It also requires unmarried mothers to identify the fathers of their children or risk losing their benefits (Edin and Kefalas 2005; Hays 2003). These broad guidelines are established at the federal level, but individual states can be more restrictive, as many have been. At the heart of welfare reform is the idea that public assistance creates dependence, discouraging people from seeking jobs. The very title of the new law, emphasizing personal responsibility and work, suggests that poverty is the fault of the poor. Low-income women, for example, are stereotyped as just wanting to have babies to increase the size of their welfare checks—even though research finds no support for this idea (Edin and Kefalas 2005).

Is welfare reform working? Many claim that welfare reform is working because, since passage of the new law, the welfare rolls have shrunk. Since 1996, the year that welfare reform was passed, the number receiving welfare support has declined from twelve million to four million (U.S. Census Bureau 2012a). But having fewer

Is It True?*

*The answers can be found on page 198.

	True	False
1. Income growth has been greatest for those in the middle class in recent years.		
2. The average American household has most of its wealth in the stock market.		
3. Social mobility is greater in the United States than in any other Western nation.		
4. The majority of welfare recipients are African American.		
5. Poor teen mothers do not have the same values about marriage as middle-class people.		
6. Old people are the most likely to be poor.		
7. Poverty in U.S. suburbs is increasing.		

people on welfare does not mean poverty is reduced; in fact, as we have seen, poverty has actually increased since passage of welfare reform. Having fewer people on the rolls can simply mean that people are without a safety net.

Many studies also find that low-wage work does not lift former welfare recipients out of poverty (Hays 2003). Critics of the current policy also argue that forcing welfare recipients to work provides a cheap labor force for employers and potentially takes jobs from those already employed. In the first few years of welfare reform, the nation was also in the midst of an economic boom; jobs were thus more plentiful. But in an economic downturn, those who are on aid or in marginal jobs are vulnerable to economic distress, particularly given the time limits now placed on receiving public assistance (Albelda and Withorn 2002).

Research done to assess the impact of a changed welfare policy is relatively recent. Politicians brag that welfare rolls have shrunk, but reduction in the welfare rolls is a poor measure of the true impact of welfare reform because this would be true simply because people are denied benefits. And because welfare has been decentralized to the state level, studies of the impact of current law must be done on a state-by-state basis. Such studies are showing that those who have gone into workfare programs most often earn wages that keep them below the poverty line. Although some states report that family income has increased following welfare reform, the increases are slight. More people have been evicted because of falling behind on rent. Families also report an increase in other material hardships, such as phones and utilities being cut off. Marriage rates among former recipients have not changed, although more now live with nonmarital partners, most likely as a way of sharing expenses. The number of children living in families without either parent has also increased, probably because parents had to relocate to find work. In some states, the numbers of people neither working

nor receiving aid also increased (Acker et al. 2002; Bernstein 2002).

The public debate about welfare rages on, often in the absence of informed knowledge from sociological research and almost always without input from the subjects of the debate, the welfare recipients themselves. Although stigmatized as lazy and not wanting to work, those who have received welfare actually believe that it has negative consequences for them, but they say they have no other viable means of support. They typically have needed welfare when they could not find work or had small children and were without child care. Most were forced to leave their last job because of layoffs or firings or because the work was only temporary. Few left their jobs voluntarily.

Welfare recipients also say that the welfare system makes it hard to become self-supporting because the wages one earns while on welfare are deducted from an already minimal subsistence. Furthermore, there is not enough affordable day care for mothers to leave home and get jobs. The biggest problem they face in their minds is lack of money. Contrary to the popular image of the conniving "welfare queen," welfare recipients want to be self-sufficient and provide for their families, but they face circumstances that make this very difficult to do. Indeed, studies of young, poor mothers find that they place a high value on marriage, but they do not think they or their boyfriends have the means to achieve the marriage ideals they cherish (Edin and Kefalas 2005; Hays 2003).

Another popular myth about welfare is that people use their welfare checks to buy things they do not need. But research finds that when former welfare recipients find work, their expenses actually go up. Although they may have increased income, their expenses (in the form of child care, clothing, transportation, lunch money, and so forth) increase, leaving them even less disposable income. Moreover, studies find that low-income mothers who buy "treats" for their children (brand-name shoes, a movie, candy,

Is It True? (Answers)

1. FALSE. Income growth has been highest in the top 5 percent of income groups (DeNavas-Walt et al. 2012).

2. FALSE. Eighty percent of all stock is owned by a small percentage of people. For most people, home ownership is the most common financial asset (Oliver and Shapiro 2006).

3. FALSE. The United States has lower rates of social mobility than Canada, Sweden, and Norway, and ranks near the middle in comparison to other Western nations (Beller and Hout 2006).

4. FALSE. Although African Americans are more likely to receive TANF benefits, given their proportion in the population, almost one-third of recipients (32 percent) are White; 22 percent are African American, 15 percent, Hispanic; 4 percent, Native American; 3 percent, Asian, and the remainder indicating mixed or no racial status (U.S. Census Bureau 2012a).

5. FALSE. Research finds that poor teen mothers value marriage and want to be married, but associate marriage with economic security, which they do not think they can achieve (Edin and Kefalas 2005).

6. FALSE. Although those over sixty-five years of age used to be the most likely to be poor, poverty among the elderly has declined; the most likely to be poor are children (DeNavas-Walt et al. 2012).

7. TRUE. Although most of the poor live inside metropolitan areas, poverty in the suburbs has been increasing (DeNavas-Walt et al. 2012).

---

and so forth) do so because they want to be good mothers (Edin and Lein 1997).

Other beneficiaries of government programs have not experienced the same kind of stigma. Social Security supports virtually all retired people, yet they are not stereotyped as dependent on federal aid, unable to maintain stable family relationships, or insufficiently self-motivated. Spending on welfare programs is also a pittance compared with the spending on other federal programs. Sociologists conclude that the so-called welfare trap is not a matter of learned dependency, but a pattern of behavior forced on the poor by the requirements of sheer economic survival (Edin and Kefalas 2005; Hays 2003).

chapter summary

What different kinds of stratification systems exist?
Social stratification is a relatively fixed hierarchical arrangement in society by which groups have different access to resources, power, and perceived social worth. All societies have systems of stratification, although they vary in composition and complexity. *Estate systems* are those in which a single elite class holds the power and property; in *caste systems*, placement in the stratification is by birth; and in *class systems*, placement is determined by achievement.

How do sociologists define class?
Class is the social structural position that groups hold relative to the economic, social, political, and cultural resources of society. It is highly significant in determining one's *life chances*.

How is the class system structured in the United States?
Social class can be seen as a hierarchy, like a ladder, where income, occupation, and education are indicators of class. *Status attainment* is the process by which people end up in a given position in this hierarchy. *Prestige* is the value others assign to people and groups within this hierarchy. Classes are also organized around common interests and exist in conflict with one another.

Is there social mobility in the United States?
Social mobility is the movement between class positions. Education gives some boost to social mobility, but social mobility is more limited than people believe; most people end up in a class position very close to their class of origin. *Class consciousness* is both the perception that a class structure exists and the feeling of shared identification with others in one's class. The United States has not been a particularly class-conscious society because of the belief in upward mobility.

What analyses of social stratification do sociological theorists provide?
Karl Marx saw class as primarily stemming from economic forces; Max Weber had a multidimensional view of stratification, involving economic, social, and political dimensions. Functionalists argue that social inequality motivates people to fill the different positions in society that are needed for the survival of the whole, claiming that the positions most important for society require the greatest degree of talent or training and are

thus most rewarded. Conflict theorists see social stratification as based on class conflict and blocked opportunity, pointing out that those at the bottom of the stratification system are least rewarded because they are subordinated by dominant groups.

How do sociologists explain why there is poverty in the United States?

Culture of poverty is the idea that poverty is the result of the cultural habits of the poor that are transmitted from generation to generation, but sociologists see poverty as caused by social structural conditions, including unemployment, gender inequality in the workplace, and the absence of support for child care for working parents.

What current policies address the problem of poverty?

Current welfare policy, adopted in 1996, provides support through individual states, but recipients are required to work after two years of support and have a lifetime limit of five years' support.

Key Terms

caste system 174
class consciousness 188
class system 174
concentrated poverty 193
conspicuous
        consumption 175
culture of poverty 195
economic restructuring 176
educational attainment 182

estate system 174
false consciousness 188
feminization of poverty 193
ideology 189
income 177
life chances 174
median income 181
meritocracy 187
net worth 177

occupational prestige 182
poverty line 192
prestige 182
social class 174
social differentiation 172
social mobility 187
social stratification 172
socioeconomic status
        (SES) 181

status 172
status attainment 181
Temporary Assistance
        for Needy Families
        (TANF) 196
urban underclass 184
wealth 177

9 Global Stratification

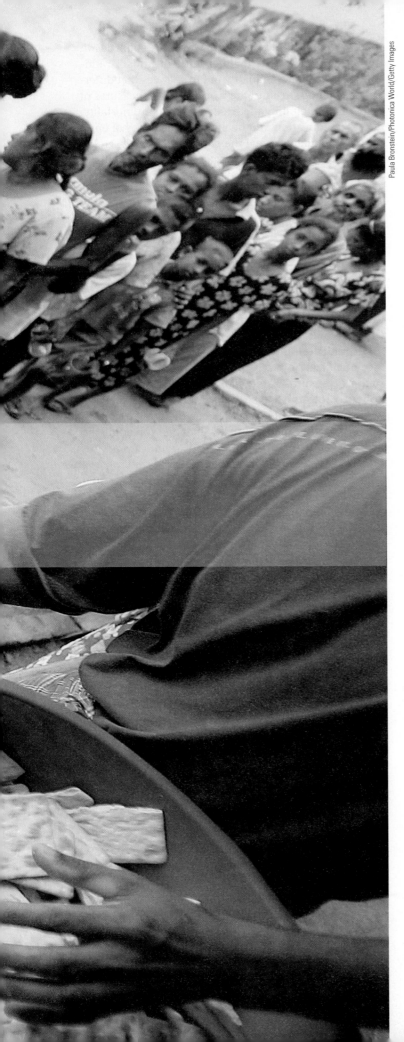

Global Stratification

Theories of Global Stratification

Consequences of Global Stratification

World Poverty

Globalization and Social Change

Chapter Summary

"it takes a village to raise a child"

the saying goes. But it also seems to take a world to make a shirt—or so it seems from looking at the global dimensions of the production and distribution of goods. Try this simple experiment: Look at the labels on your clothing. (If you do this in class, try to do so without embarrassing yourself and others!) What do you see? "Made in Indonesia," "Made in Vietnam," "Made in Malawi," all indicating the linkage of the United States to systems of production around the world. The popular brand Nike, as just one example, has not a single factory in the United States, although its founder and chief executive officer is one of the wealthiest people in the United States. Nike products are manufactured mostly in Southeast Asia.

Taking your experiment further, ask yourself: Who made your clothing? A young person trying to lift his or her family out of poverty? Might it have been a child? The International Labour Organization (ILO) distinguishes child labor (those under age 17) from "employed children" (such as a teenager holding a part-time job or babysitting). **Child labor** specifically refers to "work that deprives children of their childhood, their potential, and their dignity and that is also harmful to mental and physical development" (International Labour Organization 2012). The ILO estimates that about 215 million children around the world are trapped in child labor, almost half of whom are involved in dangerous work and many who are separated from their families and possibly held in slavery (International Labour Organization 2012). This does not mean that a child necessarily made your clothing. In fact, most child labor occurs in agricultural work, although a significant component (25 percent) is in service and manufacturing work.

Data on child labor indicate that our global systems of work are deeply connected to inequality.

Especially in the poorest countries, trying to survive forces people into forms of work—or lack of work—that produce some of the world's greatest injustices—both for children and for adult women and men. As we will see in this chapter, nations are interlocked in a system of global inequality, in which the status of the people in one country is intricately linked to the status of the people in others.

Recall from Chapter 1 that C. Wright Mills identified the task of sociology as seeing the social forces that exist beyond the individual. This is particularly important when studying global inequality. A person in the United States (or western Europe or Japan) who thinks he or she is expressing individualism by wearing the latest style is actually part of a global system of inequality. The adornments available to that person result from a whole network of forces that produce affluence in some nations and poverty in others. Dominant in the system of global stratification are the United States and other wealthy nations. Those at the top of the global stratification system have enormous power over the fate of other nations. Although world conflict stems from many sources, including religious differences, cultural conflicts, and struggles over political philosophy, the inequality between rich and poor nations causes much hatred and resentment. One cannot help but wonder what would happen if the differences between the wealth of some nations and the poverty of others were smaller. In this chapter, we examine the dynamics and effects of global stratification.

learning objectives

- Define global stratification and describe its components
- Compare and contrast different explanations of global stratification
- Describe the various consequences of global stratification
- Explain the causes and consequences of global poverty
- Summarize the impact of globalization for social change

GLOBAL STRATIFICATION

In the world today, there are not only rich and poor people but also rich and poor countries. Some countries are well off, some countries are doing so-so, and a growing number of countries are poor and getting poorer. There is, in other words, a system of **global stratification** in which the units we are considering are countries, much like a system of stratification within countries in which the units are individuals or families.

Just as we can talk about the upper-class or lower-class individuals within a country, we can also talk of the equivalent upper-class or lower-class countries in this world system. One manifestation of global stratification is the great inequality in life chances that differentiates nations around the world. Simple measures of well-being, including life expectancy, infant mortality, access to education and health, and measures of environmental quality reveal the consequences of a global system of inequality. And the gap between the rich and poor is sometimes greater in nations where the average person is least well off. No longer can these nations be understood without considering the global system of stratification of which they are a part.

The effects of the global economy on inequality have become increasingly evident, as witnessed by public concerns about jobs being sent overseas. A coalition of unions, environmentalists, and other groups has also emerged to protest global trade policies that they believe threaten jobs and workers' rights in the United States, as well as contribute to environmental degradation. Such policies also encourage further McDonaldization (see Chapter 6). Thus popular stores such as Gap and Niketown often have been targets of political protests because they symbolize the expansion of global capitalism. Protestors see the growth of such stores as eroding local cultural values and spreading the values of unfettered consumerism around the globe. Protests over world trade policies also have emerged in a student-based movement against companies that manufacture apparel with college logos.

The relative affluence of the United States means that U.S. consumers have access to goods produced around the world. A simple thing, such as a child's toy, can represent this global system. For many young girls in the United States, Barbie is the ideal of fashion and romance. Young girls may have not just one Barbie, but several, each with a specific role and costume. Cheaply bought in the United States, but produced overseas, Barbie is manufactured by those probably not much older than the young girls who play with her and who would need all of their monthly pay to buy just one of the dolls that many U.S. girls collect by the dozens (Press 1996: 12).

The manufacturing of toys and clothing is an example of the global stratification that links the United States and other parts of the world. **Global outsourcing** is the process by which jobs are located overseas even while supporting U.S.-based businesses. Many of the jobs that have been outsourced in this way are semi-skilled jobs, such as data entry, medical transcription, and so forth. But increasingly, outsourced jobs are also found in high-tech industries, software design, market research, and product research activities. Although it is difficult to measure the extent of global outsourcing, it has become a common phenomenon—something you experience when, for example, you engage in a telephone or Internet transaction, such

Global outsourcing is evident when you use your phone to get help on any of a variety of transactions. Where is the person with whom you are speaking? Here, people are working at an outsourcing call center in Mumbai, India.

David Pearson/Alamy

given the lower wages in nations where jobs flow. But this can be at the expense of jobs for workers in the United States (Rajan and Srivastava 2007). Either way, the practice of global outsourcing increasingly links the economies and social systems of nations around the world.

Rich and Poor

One dimension of stratification between countries is wealth. Enormous differences exist between the wealth of the countries at the top of the global stratification system and the wealth of the countries at the bottom. As you can see in Figures 9.1 and 9.2, a very small proportion of the world's population receives a vast share of all income—a visual reminder of the inequality that characterizes our world.

There are different ways to measure the wealth of nations, but the most common is to use the per capita **gross national income (GNI)**. The GNI measures the total output of goods and services produced by residents of a country each year plus the income from nonresident sources, divided by the size of the population. This does not truly reflect what individuals or families receive in wages or pay; it is simply each person's annual share of their country's income, should, in theory, the proceeds be shared equally. But you can use this measure to get a picture of global stratification (see Map 9.1).

Per capita GNI is reliable only in countries that are based on a cash economy. It does not measure informal exchanges or bartering in which resources are exchanged without money changing hands. These

as getting help for your computer or arranging a trip. India, China, and Russia have been major players in the economy of global outsourcing, but other nations, such as Ireland, South Africa, Poland, and Hungary, among others, are increasingly playing an important role. The consequences can be very positive for the economies of the host nations, and the practice lowers personnel costs for U.S.-based companies,

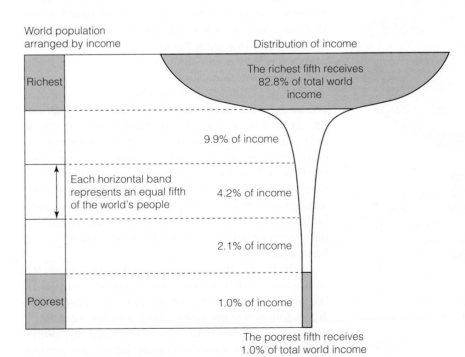

World population arranged by income

Distribution of income

Richest

The richest fifth receives 82.8% of total world income

9.9% of income

Each horizontal band represents an equal fifth of the world's people

4.2% of income

2.1% of income

Poorest

1.0% of income

The poorest fifth receives 1.0% of total world income

FIGURE 9.1 World Income Distribution

Data from: Ortiz, Isabel, and Matthew Cummins. 2011. "Global Inequality: Beyond the Bottom Billion." *Social and Economic Policy Working Paper*. UNICEF, April. **www.unicef.org**

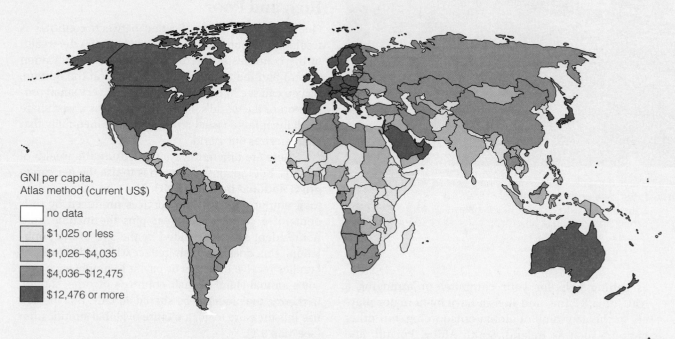

MAP 9.1

Viewing Society in Global Perspective: Rich and Poor

Most nations are linked in a world system that produces wealth for some and poverty for others. The GNI (gross national income), depicted here on a per capita basis for most nations in the world, is an indicator of the wealth and poverty of nations.

Source: The World Bank. 2011. Reprinted by permission. **www.worldbank.org**

GNI per capita, Atlas method (current US$)

- no data
- $1,025 or less
- $1,026–$4,035
- $4,036–$12,475
- $12,476 or more

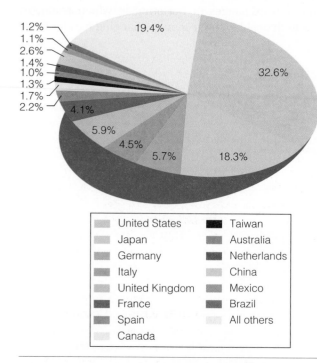

FIGURE 9.2 Who Owns the World's Wealth?

Data from: Davies, James B., Susanna Sandstrom, Anthony Shorrocks, and Edward N. Wolff. 2008. "The World Distribution of Household Wealth." UNU-WIDER, World Institute for Development Economics Research. Helsinki, Finland.

- United States
- Japan
- Germany
- Italy
- United Kingdom
- France
- Spain
- Canada
- Taiwan
- Australia
- Netherlands
- China
- Mexico
- Brazil
- All others

noncash transactions are not included in the GNI calculation, but they are more common in developing countries. As a result, measures of wealth based on the GNI, or other statistics that count cash transactions, are less reliable among the poorer countries and may underestimate the wealth of the countries at the lower end of the economic scale.

The per capita GNI of the United States, which is one of the wealthier nations in the world (though not the wealthiest on a per capita basis), was $48,450 in 2011. The per capita GNI in Burundi, one of the poorest countries in the world, was $1250. But, even compared to other well-to-do, industrialized nations, the United States per capita GNI shows us to be one of the most affluent nations in the world. GNI per capita in Japan, for example, is $45,180; $43,950 in Germany, $37,780 in the United Kingdom, and, in China, only $4930 (World Bank 2012).

Which are the wealthiest nations? Figure 9.3 shows the ten richest and the ten poorest countries in the world (measured by the annual per capita GNI in 2011). Monaco is the richest nation in the world on a per capita basis. Of course, Monaco has a tiny population compared with the United States. The poorest country in the world is the Congo, but note how many of the poorest nations are in sub-Saharan Africa, one

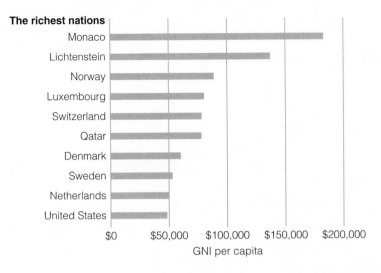

The richest nations

Monaco
Lichtenstein
Norway
Luxembourg
Switzerland
Qatar
Denmark
Sweden
Netherlands
United States

$0 $50,000 $100,000 $150,000 $200,000

GNI per capita

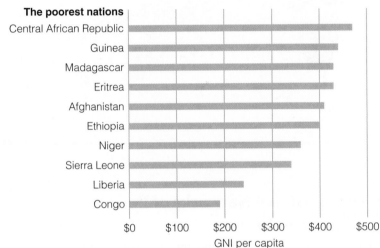

The poorest nations

Central African Republic
Guinea
Madagascar
Eritrea
Afghanistan
Ethiopia
Niger
Sierra Leone
Liberia
Congo

$0 $100 $200 $300 $400 $500

GNI per capita

FIGURE 9.3 The Rich and the Poor: A World View*

*Measured by GNI per capita, in U.S. dollars, for 2011. Data from: The World Bank. 2010. **www.worldbank.org**

of the poorest regions of the world. We will return to this fact in the discussion of world poverty later in this chapter.

The poorest nations are largely rural, have high fertility rates, large populations, and still depend heavily on subsistence agriculture. In very poor countries, the life of an average citizen is meager. Often poor nations are rich with natural resources but are exploited for such resources by more powerful nations. Still, they rank at the bottom of the global stratification system.

Because the poorest nations suffer from extreme poverty, there is terrible human suffering in these places. This also produces instability and the potential for violence. We have witnessed this in events such as mass rape and other forms of violence, including genocide, in some of the world's poorest nations, including the Democratic Republic of the Congo, Burundi, and Rwanda, all part of sub-Saharan Africa where poverty rates are now the highest. We will look more closely at the nature and causes of world poverty later in this chapter.

The wealthiest countries, you will see, are largely industrialized nations or those that are oil-rich. These countries represent the equivalent of the upper class. But simply being one of the wealthiest nations in the world does not mean that all of the nation's population is well off. Some, especially the Scandinavian countries, have low degrees of inequality within. Others, including the United States, have great inequality within, as we have seen in the previous chapter on class inequality. Moreover, inequality between nations has to be seen in relative terms. For example, a very wealthy person in India may have the income of someone in the bottom 5 percent of income earners in the United States, but within India, this can afford the person an expensive mansion and a highly lavish lifestyle relative to other Indian people (Milanovic 2010).

Inequality within nations is measured by something called the Gini coefficient. The **Gini coefficient** is a measure of income distribution within a given population or, in this case, nation. The figure ranges from zero to one, with zero representing a population where there is perfect equality and one, including decimals, indicating a population where just one person has all the money—in other words, the greatest inequality. South Africa has the highest Gini coefficients in the

MAP 9.2

Viewing Society in Global Perspective: The Gini Coefficient

Source: CIA World Factbook.

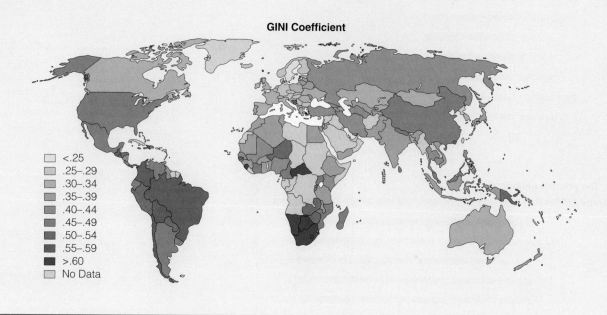

GINI Coefficient

- <.25
- .25–.29
- .30–.34
- .35–.39
- .40–.44
- .45–.49
- .50–.54
- .55–.59
- >.60
- No Data

world; the Scandinavian countries, among the lowest. But the United States ranks very high in the degree of internal inequality among other industrialized nations, as you can see from Figure 9.4. Map 9.2 also gives you a visual image, based on the Gini coefficient, of where inequality within nations appears throughout the world.

Global Networks of Power and Influence

Global stratification involves nations in a large and integrated network of both economic and political relationships. **Power**—meaning the ability of a country to exercise control over other countries or groups of

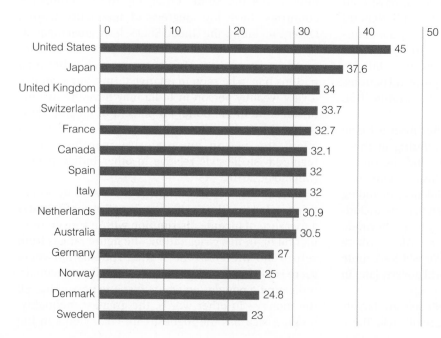

	0	10	20	30	40	50
United States					45	
Japan				37.6		
United Kingdom				34		
Switzerland				33.7		
France				32.7		
Canada				32.1		
Spain				32		
Italy				32		
Netherlands				30.9		
Australia				30.5		
Germany			27			
Norway			25			
Denmark			24.8			
Sweden			23			

FIGURE 9.4 Gini Coefficient among Selected Nations

Note: The Gini coefficient for the whole world is 39.

Data from: Central Intelligence Agency. 2012. *The World Fact Book.* **www.cia.gov**

Servants of Globalization: Who Does the Domestic Work?

Research Question: International migration is becoming an increasingly common phenomenon. Women are one of the largest groups to experience migration, often leaving poor nations to become domestic workers in wealthier nations. What are these women's experiences in the context of global stratification? This is what Rhacel Salazar Parreñas wanted to know.

Research Method: Parreñas studied two communities of Filipina women, one in Los Angeles and one in Rome, Italy, conducting her research through extensive interviewing with Filipina domestic workers in these two locations. She supplemented the interviews with participant observation in church settings, after-work social gatherings, and in employers' homes. The interviews were conducted in English and Tagalog—sometimes a mixture of both.

Research Results: Parreñas found that Filipina domestics experienced many status inconsistencies. They were upwardly mobile in terms of their home country but were excluded from the middle-class Filipino communities in the host nation. Thus they experienced feelings of social exclusion in addition to being separated from their own families.

Conclusions and Implications: The women Parreñas studied are part of a new social for *transnational families*—that is, families whose members live across the borders of nations. These Filipinas provide the labor for more affluent households while their own lives are disrupted by these new global forces. As global economic restructuring evolves, it may be that more and more families will take this form as they adapt to changing economic and social conditions.

Questions to Consider

1. Are there domestic workers in your community who provide child care and other household work for middle- and upper-class households? What is the race, ethnicity, nationality, and gender of these workers? What does this tell you about the division of labor in domestic work and its relationship to global stratification?

2. Why do you think domestic labor is so underpaid and undervalued? Are there social changes that might result in a reevaluation of the value of this work?

Sources: Parreñas, Rhacel Salazar. 2001. *Servants of Globalization: Women, Migration and Domestic Work.* Stanford, CA: Stanford University Press.

---

countries—is a significant dimension of global stratification. Countries can exercise several kinds of power over other countries, including military, economic, and political power. The **core countries** have the most power in the world economic system. These countries control and profit the most from the world system, and thus they are the "core" of the world system. These include the powerful nations of Europe, the United States, Australia, and, increasingly, East Asia.

Surrounding the core countries, both structurally and geographically, are the **semiperipheral countries** that are semi-industrialized and, to some degree,

The gap between the rich and poor worldwide can be staggering. At the same time that many struggle for mere survival, others enjoy the pleasantries of a gentrified lifestyle.

represent a kind of middle class (such as Spain, Turkey, and Mexico). They play a middleman role, extracting profits from the poor countries and passing those profits on to the core countries. At the bottom of the world stratification system, in this model, are the **peripheral countries**. These are the poor, largely agricultural countries of the world. Even though they are poor, they often have important natural resources that are exploited by the core countries. This exploitation, in turn, keeps them from developing and perpetuates their poverty. Often these nations are politically unstable, and, though they exercise little world power, political instability can cause a crisis for core nations that depend on their resources. Military intervention by the United States or European nations is often the result.

See FOR YOURSELF

The Global Economy of Clothing

Look at the labels in your clothes and note where your clothing was made. Where are the products bearing your college logos manufactured and sold? Who profits from the distribution of these goods? What does this tell you about the relationship of *core*, *semiperipheral*, and *peripheral* countries within world systems theory? What further information would reveal the connections between the country where you live and the countries where your clothing is made and distributed? •

Race and Global Inequality

Along with class inequality, there is a racial component to world inequality, which can be seen in several ways. In the richest nations, the population is largely White; in the poorest countries of the world, mostly in Africa, the populations are people of color. Exploitation of the human and natural resources of regions populated by people of color has characterized the history of Western capitalism, with people of color being dominated by Western imperialism and colonialism. The inequities that have resulted are enormous. Patterns of malnutrition and hunger show these inequities.

How did this racial inequality come about? On the surface, global capitalism is not explicitly racist, as were earlier forms of industrial capitalism. Yet, in fact, it is the rapid expansion of the global capital system that has led to the increase in racial inequality between nations. In the new capitalist system, a new **international division of labor** has emerged that is not tied to place but can employ cheap labor anywhere. Cheap labor is usually found in non-Western countries. The exploitation of cheap labor has created a poor and dependent workforce where mostly

people of color work. The profits accrue to the wealthy owners, who are mostly White, resulting in a racially divided world. Some have argued further that multinational corporations' exploitation of the poor peripheral nations has forced an exodus of unskilled workers from the impoverished nations to the rich nations. The flood of third-world refugees into the industrialized nations is thereby increasing racial tensions, fostering violence, and destroying worker solidarity (Sirvananadan 1995).

South Africa, the United States, and Brazil each developed different sets of racial categories. Although all three countries have many people of mixed descent, race is defined differently in each place. In South Africa, the particular history of Dutch and English colonialism led to strongly drawn racial categories that defined people in four separate categories: "White," "coloured" (including indigenous Khoi and San people, as well as people of mixed descent), "Black," and "Indian." Except for Black South Africans, who had no political representation under apartheid, there were three separate parliaments—one for each of the other groups. In the United States, given its history of slavery, the "one drop" rule was used, which defined anyone with any African heritage as Black, thus ruling out any category of mixed race.

Brazil is yet a different case. The Brazilian elite declared Brazil a racial democracy at the early stages of national development. Racial differences were thought not to matter. Yet, instead of creating an egalitarian society free of racism, Afro-Brazilians were still of lower social status and Euro-Brazilians remain at the highest social status, suggesting that color itself stratifies people—a sociological phenomenon sometimes referred to as "colorism" (Telles 2004; Fredrickson 2003; Marx 1997).

THEORIES OF GLOBAL STRATIFICATION

How did world inequality occur? Sociological explanations of world stratification generally fall into three camps: modernization theory, dependency theory, and world systems theory, each explained here (see Table 9.1).

Modernization Theory

Modernization theory views the economic development of countries as stemming from technological change. According to this theory, a country becomes more "modernized" by increased technological development, and this technological development is also dependent on other countries. Modernization theory was initially developed in the 1960s to explain why some countries had achieved economic development and why some had not (Rostow 1978).

table 9.1 Theories of Global Stratification

	Modernization Theory	Dependency Theory	World Systems Theory
Economic Development	Arises from relinquishing traditional cultural values and embracing new technologies and market-driven attitudes and values	Exploits the least powerful nations to the benefit of wealthier nations that then control the political and economic systems of the exploited countries	Has resulted in a single economic system stemming from the development of a world market that links core, semiperipheral, and peripheral nations
Poverty	Results from adherence to traditional values and customs that prevent societies from competing in a modern global economy	Results from the dependence of low-income countries on wealthy nations	Is the result of core nations extracting labor and natural resources from peripheral nations
Social Change	Involves increasing complexity, differentiation, and efficiency	Is the result of neocolonialism and the expansion of international capitalism	Leads to an international division of labor that increasingly puts profit in the hands of a few while exploiting those in the poorest and least powerful nations

© Cengage Learning

Modernization theory sees economic development as a process by which traditional societies become more complex and differentiated. For economic development to occur, modernization theory predicts, countries must change their traditional attitudes, values, and institutions. Economic achievement is thought to derive from attitudes and values that emphasize hard work, saving, efficiency, and enterprise. These values are said by the theory to be found in modern (developed) countries but are lacking in traditional societies. Modernization theory suggests that nations remain underdeveloped when traditional customs and culture discourage individual achievement and kin relations dominate.

As an outgrowth of functionalist theory, modernization theory derives some of its thinking from the work of Max Weber. In *The Protestant Ethic and the Spirit of Capitalism* (1958/1904), Weber saw the economic development that occurred in Europe during the Industrial Revolution as a result of the values and attitudes of Protestantism. The Industrial Revolution took place in England and northern Europe, Weber argued, because the people of this area were hardworking Protestants who valued achievement and believed that God helped those who helped themselves.

Modernization theory is similar to the argument of the culture of poverty, which sees people as poor because they have poor work habits, engage in poor time management, are not willing to defer gratification, and do not save or take advantage of educational opportunities (see Chapter 8). Countries are poor, in other words, because they have poor attitudes and poor institutions.

Modernization theory can partially explain why some countries have become successful. Japan and China are examples of countries that have made huge strides in economic development, in part because of a national work ethic. But the work ethic alone does not explain Japan's success. In sum, modernization theory may partially explain the value context in which some countries become successful and others do not, but it is not a substitute for explanations that also look at the economic and political context of national development. It also rests on an arrogant perspective that the United States and other more economically developed nations have superior values compared to other nations. Critics point out that this perspective blames countries for being poor when other causes of their status in the world may be outside their control. Whether a country develops or remains poor may be the result of other countries exploiting the less powerful. Modernization theory does not sufficiently take into account the interplay and relationships between countries that can affect a country's economic or social condition.

Developing countries, modernization theory says, are better off if they let the natural forces of competition guide world development. Free markets, according to this perspective, will result in the best economic order. But, as critics argue, markets do not develop independently of government's influence. Governments can spur or hinder economic development, especially as they work with private companies to enact export strategies, restrict imports, or place embargoes on the products of nonfavored nations.

Dependency Theory

Although market-oriented theories may explain why some countries are successful, they do not explain why some countries remain in poverty or why some countries have not developed. It is necessary to look at issues outside the individual countries and to examine the connections between them. Drawing on the fact that many of the poorest nations are former colonies of European powers, another theory of world stratification focuses on the processes and results of European colonization and imperialism. This theory, called **dependency theory**, focuses on explaining the persistence of poverty in the world. It holds that the poverty of the low-income countries is a direct result of their political and economic dependence on the wealthy countries. Specifically, dependency theory argues that the poverty of many countries is a result of exploitation by powerful countries. This theory is derived from the work of Karl Marx, who foresaw that a capitalist world economy would create an exploited class of dependent countries, just as capitalism within countries had created an exploited class of workers.

Dependency theory begins by examining the historical development of this system of inequality. As the European countries began to industrialize in the 1600s, they needed raw materials for their factories, and they needed places to sell their products. To accomplish this, the European nations colonized much of the world, including most of Africa, Asia, and the Americas. **Colonialism** is a system by which Western nations became wealthy by taking raw materials from colonized societies and reaping profits from products finished in the homeland. Colonialism worked best for the industrial countries when the colonies were kept undeveloped to avoid competition with the home country. For example, India was a British colony from 1757 to 1947. During that time, Britain bought cheap cotton from India, made it into cloth in British mills, and then sold the cloth back to India, making large profits. Although India was able to make cotton into cloth at a much cheaper cost than Britain, and very fine cloth at that, Britain nonetheless did not allow India to develop its cotton industry. As long as India was dependent on Britain, Britain became wealthy and India remained poor.

Under colonialism, dependency was created by the direct political and military control of the poor countries by powerful developed countries. Most colonial powers were European countries, but other countries, particularly Japan and China, had colonies as well. Colonization came to an end soon after the Second World War, largely because of protests by colonized people and the resulting movement for independence. As a result, according to dependency theory, the powerful countries turned to other ways to control the poor countries and keep them dependent. The powerful countries still intervene directly in the affairs of the dependent nations by sending troops or, more often, by imposing economic or political restrictions and sanctions. But other methods, largely economic, have been developed to control the dependent poor countries, such as price controls, tariffs, and, especially, the control of credit. Indeed, the level of debt that some nations accrue is a major source of global inequality.

The rich industrialized nations, according to dependency theory, are able to set prices for raw materials produced by the poor countries at very low levels so that the poor countries are unable to accumulate enough profit to industrialize. As a result, the poor, dependent countries must borrow from the rich countries. However, debt creates only more dependence. Many poor countries are so deeply indebted to the major industrial countries that they must follow the economic edicts of the rich countries that loaned them the money, thus increasing their dependency. This form of international control has sometimes been called **neocolonialism**, a form of control of the poor countries by the rich countries but without direct political or military involvement.

Multinational corporations are companies that draw a large share of their profits from overseas investments and that conduct business across national borders. They play a role in keeping the dependent nations poor, dependency theory suggests. Although their executives and stockholders are from the industrialized countries, multinational corporations recognize no national boundaries and pursue business where they can best make a profit. Multinationals buy resources where they can get them cheapest, manufacture their products where production and labor costs are lowest, and sell their products where they can make the largest profits.

Many critics fault companies for perpetuating global inequality by taking advantage of cheap overseas labor to make large profits for U.S. stockholders. Companies are, in fact, doing what they should be doing in a market system: trying to make a profit. Nonetheless, dependency theory views the practices of multinationals as responsible for maintaining poverty in the poor parts of the world.

One criticism of dependency theory is that many poor countries (for example, Ethiopia) were never colonies. Some former colonies have also done well. Two of the greatest postwar success stories of economic development are Singapore and Hong Kong. Both of these countries were British colonies—Hong Kong until 1997—and clearly dependent on Britain, yet they have had successful economic development precisely because of their dependence on Britain.

Other former colonies are also improving economically, such as India.

World Systems Theory

Modernization theory examines the factors internal to an individual country, and dependency theory looks to the relationship between countries or groups of countries. Another approach to global stratification is called **world systems theory**. Like the dependency theory, this theory begins with the premise that no nation in the world can be considered in isolation. Each country, no matter how remote, is tied in many ways to the other countries in the world. However, unlike dependency theory, world systems theory argues that there is a world economic system that must be understood as a single unit, not in terms of individual countries or groups of countries. This theoretical approach derives to some degree from the work of the dependency theorists and is most closely associated with the work of Immanuel Wallerstein in *The Modern World System* (1974) and *The Modern World System II* (1980). According to this theory, the level of economic development is explained by understanding each country's place and role in the world economic system.

This world system has been developing since the sixteenth century. The countries of the world are tied together in many ways, but of primary importance are the economic connections in the world markets of goods, capital, and labor. All countries sell their products and services on the world market and buy products and services from other countries. However, this is not a market of equal partners. Because of historical and strategic imbalances in this economic system, some countries are able to use their advantage to create and maintain wealth, whereas other countries that are at a disadvantage remain poor. This process has led to a global system of stratification in which the units are not people but countries.

World systems theory sees the world divided into three groups of interrelated nations: core or first-world countries, semiperipheral or second-world countries, and peripheral or third-world countries. This world economic system has resulted in a modern world in which some countries have obtained great wealth and other countries have remained poor. The core countries control and limit the economic development in the peripheral countries so as to keep the peripheral countries from developing and competing with them on the world market; thus the core countries can continue to purchase raw materials at a low price.

Although world systems theory was originally developed to explain the historical evolution of the world system, modern scholars now focus on the international division of labor and its consequences. This approach is an attempt to overcome some of the shortcomings in world systems theory by focusing on the specific mechanism by which differential profits are attached to the production of goods and services in the world market. A tennis shoe made by Nike is designed in the United States; uses synthetic rubber made from petroleum from Saudi Arabia; is sewn in Indonesia; is transported on a ship registered in Singapore, which is run by a Korean management firm using Filipino sailors; and is finally marketed in Japan and the United States. At each of these stages, profits are taken, but at very different rates.

thinking SOCIOLOGICALLY

What are the major industries in your community? In what parts of the world do they do business, including where their product is produced? How does the *international division of labor* affect jobs in your region? ●

World systems theorists call this global production process a **commodity chain**, the network of production and labor processes by which a product becomes a finished commodity. By following a commodity through its production cycle and seeing where the profits go at each link of the chain, one can identify which country is getting rich and which country is being exploited. As an example, the Gap hoodie that you buy in the United States for about $30 was likely produced from cotton grown in Uzbekistan where workers are paid 2 cents a pound, cut and sewn by workers in Russia who are paid between 39 and 69 dollars a month, and then distributed and sold in the United States (Gordon and Designs 2001).

World systems theory also helps explain the growing phenomenon of international migration. An *international division of labor* means that the need for cheap labor in some of the industrial and developing nations draws workers from poorer parts of the globe. International migration is also the result of refugees seeking asylum from war-torn parts of the world or from countries where political oppression, often against particular ethnic groups, forces some to leave. The development of a world economy, however, is resulting in large changes in the composition of populations around the globe. **World cities**, that is, cities that are closely linked through the system of international commerce, have emerged. Within these cities, families and their surrounding communities often form *transnational communities*, communities that may be geographically distant but socially and politically close. Linked through various communication and transportation networks, transnational communities share information, resources, and strategies for coping with the problems of international migration.

Global stratification often means that consumption in the more affluent nations is dependent on cheap labor in other less affluent nations.

International migration, sometimes legal, sometimes not, has radically changed the racial and ethnic composition of populations not only in the United States but also in many European and Asian nations. Over 200 million people now live outside the country of their birth, some of whom moved because of war and persecution, but many of whom move as work moves around the globe (Eitzen 2009). Many such migrants work in the lowest segments of the labor force. The work of these low-wage laborers is critical to the world economy, but they are often treated with hostility and suspicion, discriminated against, and stereotyped as undeserving and threatening. The United States receives the most international migrants of any nation, but they are common in western Europe, Saudi Arabia, Iran, and other parts of the world (Martin 2001). In many nations, the presence of migrants can lead to political tensions over immigration, even though international faces in world cities are now a major feature of the urban landscape.

It is useful to see the world as an interconnected set of economic ties between countries, and to understand that these ties often result in the exploitation of poor countries. This process of globalization means that countries that were once at the center of this world system—England, for example—no longer occupy such a lofty position. Peripheral countries can also improve their standard of living with investment by core countries, although the benefits do not accrue equally to groups within such nations. Low-wage factories may benefit managers, but not the working class. Even core countries can be hurt by the world system, such as when jobs move overseas. Who

benefits from this world system is differentiated—in all countries—by one's placement, not just in the world class system but also in the class system internal to each country within this global system. World systems theory has provided a powerful tool for understanding global inequality.

CONSEQUENCES OF GLOBAL STRATIFICATION

It is clear that some nations are wealthy and powerful and some are poor and powerless. What are the consequences of this world stratification system? Basic indicators of national well-being can include such things as infant mortality, literacy levels, access to safe water, and the status of women. There are, as we will see, considerable differences in the quality of life in different places in the world.

Population

One of the biggest differences in rich and poor nations is population. The poorest countries have the highest birthrates and the highest death rates. The total *fertility rate*—how many live births a woman will have over her lifetime at current fertility rates—shows that women in the poorest countries have on average almost five children. Because of this high fertility rate, the populations of poor countries are growing faster than the populations of wealthy countries; these countries therefore also have a high proportion of young children.

In contrast, the richest countries have a total population of approximately one billion people—only 15 percent of the world's population. The populations of the richest countries are not growing nearly as fast as the populations of the poorest countries. In the richest countries, women have about two children over their lifetime, and the populations of these countries are growing by only 1.2 percent. Many of the richest countries, including most of the countries of Europe, are actually experiencing population declines. With a low fertility rate, the rich countries have proportionately fewer children, but they also have proportionately more elderly, which can also be a burden on societal resources. Different from the poorest nations, the richest ones are largely urban.

Rapid population growth as a result of high fertility rates can make a large difference in the quality of life of the country. Countries with high birthrates are faced with the challenge of having too many children and not enough adults to provide for the younger generation. Public services, such as schools and hospitals, are strained in high-birthrate countries, especially because these countries are poor to begin with. However, very low birthrates, as many rich countries are now experiencing, can also lead to problems. In countries with low birthrates, there often are not enough young people to meet labor force needs, and workers must be imported from other countries.

Although the data clearly show that poor countries have large populations and high birthrates and rich countries have smaller populations and low birthrates, does this mean that the large population results in the low level of wealth of the country or that high fertility rates keep countries poor?

Scholars are divided on the relationship between the rate of population growth and economic development (Cassen 1994; Demeny 1991). Some theorize that rapid population growth and high birthrates lead to economic stagnation and that too many people keep a country from developing, thus miring the country in poverty (Ehrlich 1968). However, other researchers point out that some countries with very large populations have become developed (Coale 1986). After all, the United States has the third largest population in the world at 309 million people, yet it is one of the richest and most developed nations in the world. China and India, the two nations in the world with the largest populations, are also showing significant economic development. Scholars now believe that even though large population and high birthrates can impede economic development in some situations, in general fertility levels are affected by levels of industrialization, not the other way around. That is, as countries develop, their fertility levels decrease and their population growth levels off (Hirschman 1994; Watkins 1987).

Health and Environment

Significant differences are also evident in the basic health standards of countries, depending on where they are in the global stratification system. The high-income countries have lower childhood death rates, higher life expectancies, and fewer children born underweight. People born today in wealthy countries can expect to live about seventy-seven years, and women outlive men by several years. Except for some isolated or poor areas of the rich countries, almost all people have access to clean water and acceptable sewer systems.

In the poorest countries, the situation is completely different. Many children die within the first five years of life, people live considerably shorter lives, and fewer people have access to clean water and adequate sanitation. In the low-income countries, the problems of sanitation, clean water, childhood death rates, and life expectancies are all closely related. In many of the poor countries, drinking water is contaminated from poor or nonexistent sewage treatment. This contaminated water is then used to drink, to clean eating utensils, and to make baby formula. For adults,

There can be innovative solutions to reduce world poverty, such as this solar panel delivering energy in southern Mozambique.

waterborne illnesses such as cholera and dysentery sometimes cause severe sickness but seldom result in death. However, children under age 5, and especially those under the age of 1, are highly susceptible to the illnesses carried in contaminated water. A common cause of childhood death in countries with low incomes is dehydration brought on by the diarrhea contracted by drinking contaminated water.

Degradation of the environment is a problem that affects all nations, which are linked in one vast environmental system. But global stratification also means that some nations suffer at the hands of others. Overdevelopment is resulting in deforestation, and in the poorest nations, high population and the dependency on agriculture contribute to the depletion of natural resources. In the most industrial nations where the most energy is used, the overproduction of "greenhouse gas"—emission of carbon dioxide from the burning of fossil fuels—is resulting in various threats to our environment.

Although high-income countries have only 15 percent of the world population, together they use more than half of the world's energy. The United States alone uses 20 percent of the world's energy, though it holds only 4 percent of the world's population (see Figure 9.5). Safe water is also crucial; more than 700 million people in 43 different countries are experiencing what the World Bank calls "water stress"—that is, inadequate access to water. The dwindling of water supplies will only be exacerbated by population growth and economic development. The World Bank has, in fact, warned that we are facing a "global water crisis" (World Bank 2010). Clearly, global stratification has some irreversible environmental effects that are felt around the globe.

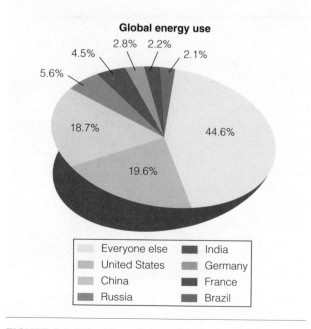

FIGURE 9.5 Who Uses the World's Energy?

Data source: U.S. Energy Information Administration. 2012. **www.eia.gov**

Education and Illiteracy

In the high-income nations of the world, education is almost universal, and the vast majority of people have attended school, at least at some level. Literacy and school enrollment are now taken for granted in the high-income nations, although people in these wealthy nations who do not have a good education stand little chance of success. In the middle- and lower-income nations, the picture is quite different. Elementary school enrollment, virtually universal in wealthy nations, is less common in the middle-income nations and even less common in the poorest nations.

How do people survive who are not literate or educated? In much of the world, education takes place outside formal schooling. Just because many people in the poorer countries never go to school does not mean that they are ignorant or that they are uneducated. Most of the education in the world takes place in family settings, in religious congregations, or in other settings where elders teach the next generation the skills and knowledge they need to survive. This type of informal education often includes basic literacy and math skills that people in these poorer countries need for their daily lives.

The disadvantage of this informal and traditional education is that although it prepares people for their traditional lives, it often does not give them the skills and knowledge needed to operate in the modern world. In an increasingly technological world, this can perpetuate the underdeveloped status of some nations.

Gender Inequality

The position of a country in the world stratification system also affects gender relations within different countries. Poverty is usually felt more by women than by men. Although gender equality has not been achieved in the industrialized countries, compared with women in other parts of the world, women in the wealthier countries are much better off.

The United Nations (UN) is one of the organizations that carefully monitors the status of women globally. The UN has developed an index to assess the progress of women in nations around the world. Called the **gender inequality index**, the measure is a composite of three key components of women's lives: reproductive health, empowerment, and labor market status. Each of these three major components is then measured by particular facts about women's status, such as maternal mortality, educational attainment, and labor force participation (see Figure 9.6). Given how this index is computed, nations with the lowest gender inequality index have the greatest equality between women and men (see Table 9.2). Based on this index, the United Nations has concluded that, around the world, reproductive health—or lack thereof—is

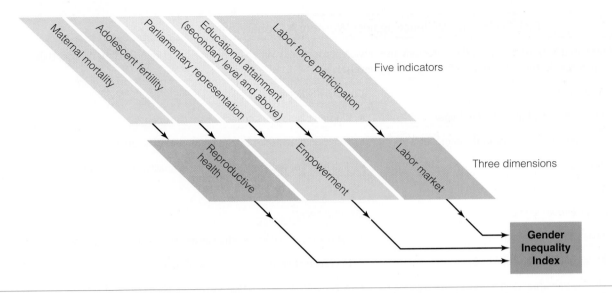

FIGURE 9.6 The Gender Inequality Index

Source: United Nations Development Program. 2010. "Components of the Gender Inequality Index." http://hdr.undp.org. Reprinted with permission.

the greatest contributor to gender inequality (United Nations 2010a).

Reports indicate mixed news with regard to women's status around the world. On the one hand, women's poverty has declined in some of the nations where it has been extreme, particularly in India, China, and some parts of Latin America. But in sub-Saharan Africa, women's poverty has increased. And though women's share of representation in governments has increased, they still hold only 20 percent of parliamentary seats worldwide. Also, although women have achieved near equality in levels of primary education, there are large gaps in the status of women and men in secondary and higher education—a fact that has huge implications for the work that women do when the global economy increasingly demands educational skills (Inter-Parliamentary Union 2012).

Perhaps most distressing is the global extent of violence against women. Violence takes many forms, including violence within the family, rape, sexual harassment, sex trafficking and prostitution, and state-based violence, among other things. The United Nations has concluded that "violence against women persists in every country in the world as a pervasive violation of human rights and a major impediment to achieving gender equality" (United Nations 2006a: 9). Several factors put women at risk of violence, ranging from individual-level risk factors (such as a history of abuse as a child and substance abuse) to societal-level factors, such as gender roles that entrench male dominance and societal norms that tolerate violence as a means of conflict resolution (see Table 9.3). Clearly, the inequalities that mark global stratification have particularly deleterious effects for the world's women.

War and Terrorism

The consequences of global stratification are also found in the international conflicts that bring war and an increased risk of terrorism. Although global inequality is certainly not the only cause of such problems, it contributes to the instability of world peace and the threat

table 9.2 Gender Inequality Index in Selected Countries (2011)

	Gender Inequality Index*	Rank (most equal being first out of 187 countries)
Sweden	.049	1
Netherlands	.052	2
Denmark	.060	3
Switzerland	.067	4
Korea	.099	18
Israel	.145	22
China	.209	38
United States	**.299**	**47**
Libya	.314	51
Haiti	.599	123
India	.617	129
Saudi Arabia	.646	135

*Lower numbers mean more equality

Source: United Nations. 2011. *The Gender Inequality Index*, Table 4. www.undp.org

table 9.3 Risk Factors for Violence against Women: A Global Analysis

The United Nations has studied the frequent use of violence against women in the world and identified the factors that put women at risk. These factors are found at various levels.

Individual Level:	Community Level:
• Frequent use of alcohol and drugs	• Women's isolation and lack of social support
• Membership in marginalized communities	• Community attitudes that tolerate and legitimate male violence
• History of abuse as a child	• High levels of social and economic inequality, including poverty
• Witnessing marital violence in the home	

Family/Relationship Level:	Societal Level:
• Male control of wealth	• Gender roles that entrench male dominance and women's subordination
• Male control of decision making	• Tolerance of violence as a means of conflict resolution
• History of marital violence	• Inadequate laws and policies to prevent and punish violence
• Significant disparities in economic, educational, or employment status	• Limited awareness and sensitivity on the part of officials and social service providers

Source: United Nations. 2006a. *In-Depth Study on All Forms of Violence against Women.* New York: United Nations.

of terrorism. Global stratification generates inequities in the distribution of power between nations. Moreover, globalization has created a world-based capitalist class with unprecedented wealth and power. This is a class that now crosses national borders; thus, some have defined it as a "transnational capitalist class" (Langman and Morris 2002). Coupled with the enormous poverty that exists, the visibility of this class and its association with Western values leads to resentment and conflict. Furthermore, attempts by wealthier nations to control access to the world's natural resources, such as oil, generate much political conflict. Thus the same power and affluence that makes the United States a leader throughout the world makes it a target by those who resent its dominance.

In the Middle East, for example, oil production has created prosperity for some and exposed people in these nations to the values of Western culture. When people from different nations, such as those in the Middle East, study at U.S. universities and travel on business or vacations, they are exposed to Western values and patterns of consumption. As one commentator has noted, "Even those who have remained at home have not escaped exposure to Western culture. In most of the countries of the modern Middle East, Western cultural influences are pervasive. They see Western television programs, they watch Western movies, they listen to Western music, frequently wear Western clothes, and visit Western websites. Even Western foods are locally available. McDonald's are now found in many of the major cities" (Bailey 2003: 341). Moreover, the sexual liberalism of Western

nations and the relative equality of women also add to the volatile mix of nations clashing (Norris and Inglehart 2002).

As a result, some traditional leaders, including religious clerics, define Western culture as a source of degeneracy. Countries such as the United States, where consumerism is rampant, then become the target of those who see this as a threat to their traditional way of life (Ehrlich and Liu 2002). In this sense, global stratification and the dominance of Western culture are inseparable (Bailey 2003). Clashing religious values and the growth of extremist views certainly are a major factor in producing terrorism, but the global dominance of some nations over others is also a factor. At the very least, it explains why those who attack embassies, fly jets into the World Trade Center, and commit other atrocious acts against Americans can still believe they are fighting a noble cause.

Terrorism can be defined as premeditated, politically motivated violence perpetrated against noncombatant targets by people or groups who use their action to try to achieve their political ends (White 2002). Terrorism can be executed through violence or threats of violence and can be executed through various means—suicide bombs, biochemical terror, cyberterror, or other means. Because terrorists operate outside the bounds of normative behavior, it is very difficult to prevent. Although rigid safeguards can be put in place, such safeguards also threaten the freedoms that are characteristics of open, democratic societies. The fact that terrorism is so difficult to stop contributes to the fear that it is intended to generate.

MAP 9.3

Viewing Society in Global Perspective: World Poverty

Source: Central Intelligence Agency. 2009. *World Factbook*

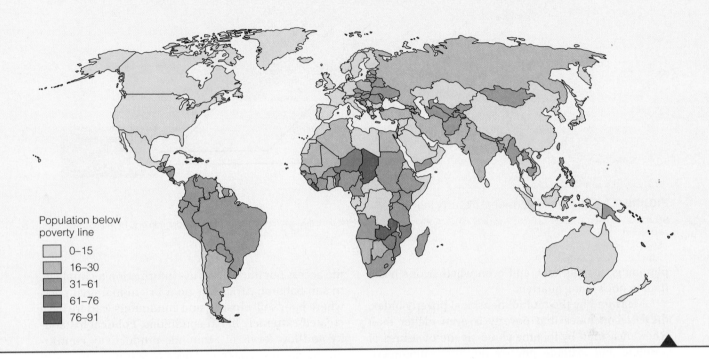

Population below
poverty line

	0–15
	16–30
	31–61
	61–76
	76–91

Inequality is also connected to the context in which terrorism emerges. A study of al Qaeda terrorists finds that the leaders tend to come from middle-class backgrounds, though they often use those who are young, poorly educated, and economically disadvantaged to carry out suicide missions. Families of suicide bombers often receive large cash payments; at the same time, they can feel they have served a sacred cause (Stern 2003). Improving the lives of those who feel collectively humiliated could provide some protection against terrorism.

WORLD POVERTY

One fact of global inequality is the growing presence and persistence of poverty in many parts of the world. There is poverty in the United States, but very few people in the United States live in the extreme levels of deprivation found in some of the poor countries of the world, as seen in Map 9.3. We have seen in Chapter 8 how the poverty line in the United States is calculated. The definition of poverty in the United States identifies **relative poverty**, that is, a measure of poverty relative to the rest of society. Households living in poverty in the United States are poor compared with other Americans, but this would be an inaccurate measure

in a worldwide context because of such huge differences in the standard of living.

The World Bank and United Nations measure world poverty in two ways: **Absolute poverty** is the situation in which people live on less than $1 per day. **Extreme poverty** is defined as living on less than the equivalent of $1.25 per day. Any way you measure it, it is difficult for most Americans to imagine this standard of living. But 1.4 billion people are living in extreme poverty (World Bank 2010).

However, money does not tell the whole story because many people in poor countries do not always deal in cash. In many countries, people survive by raising crops for personal consumption and by bartering or trading services for food or shelter. These activities do not show up in calculations of poverty levels that use amounts of money as the measure. As a result, the United Nations also defines what it calls the multidimensional poverty index (see Figure 9.7).

The **multidimensional poverty index** measures the degree of deprivation in three basic dimensions of human life: health, education, and standard of living. Like the gender inequality index, different components are used to create this measure of poverty, including such measures as nutrition, child mortality, educational attainment, and the availability of water, electricity,

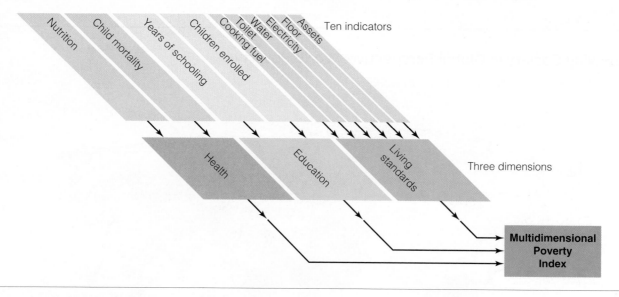

FIGURE 9.7 The Multidimensional Poverty Index

Source: United Nations Development Program. 2010. "Components of the Multidimensional Poverty Index." **http://hdr.undp.org**. Reprinted with permission.

plumbing, cooking fuel, and even whether one has a floor in one's living quarters.

Measured by the multidimensional poverty index, the UN concludes that poverty is even higher than when measured by income alone, as the measures of absolute and extreme poverty do. The multidimensional poverty index also points more directly to the interventions, such as the availability of health clinics and running water, which can significantly improve the lives of millions.

Who Are the World's Poor?

As we have seen, about one-fifth of the world's population lives in poverty, forming what the United Nations calls a *global underclass*. Although world poverty has been decreasing, it still afflicts a huge proportion of the world's people. In a world with a population approaching seven billion, about one billion live in extreme poverty. The decline in world poverty is largely accounted for by the economic growth in East Asia, which historically has been one of the poorest areas of the world. Now East Asia leads the world in poverty reduction. China alone has seen a decline of 400 million people moving out of poverty since the 1980s. On the other hand, in the same time period, poverty in sub-Saharan Africa has increased and is projected to continue rising (World Bank 2012).

The character of poverty differs around the globe. In Asia, the pressures of large population growth leave many without sustainable employment. And, as manufacturing has become less labor intensive with more mechanized production, the need for labor in certain industries has declined. Even though new technologies provide new job opportunities, they also create new forms of illiteracy because many people have neither

the access nor the skills to use information technology. In sub-Saharan Africa, the poor live in marginal areas where poor soil, erosion, and continuous warfare have created extremely harsh conditions. Political instability and low levels of economic productivity contribute to the high rates of poverty. Solutions to world poverty in these different regions require sustainable economic development, as well as an understanding of the diverse regional factors that contribute to high levels of poverty.

Women and Children in Poverty

There is no country in the world in which women are treated as well as men. As with poverty in the United States, women bear a larger share of the burden of world poverty. Some have called this *double deprivation*—in many of the poor countries, women suffer because of their gender and because they disproportionately carry the burden of poverty. For instance, in situations of extreme poverty, women have the burden of taking on much of the manual labor because the men in many cases have left to find work or food. The United Nations concludes that strengthening women's economic security through better work is essential for reducing world poverty.

Because of their poverty, women tend to suffer greater health risks than men. Although women outlive men in most countries, the difference in life expectancy is *less* in the countries in poverty. This is explained by several factors. For one, fertility rates are higher in poor countries. Giving birth is a time of high risk for women, and women in poor countries with poor nutrition, poor maternal care, and the lack of trained birth attendants are at higher risk of dying during and after the birth.

Human Trafficking

The U.S. State Department estimates that there are 12.3 million people worldwide who are enslaved in human trafficking. This includes sexual servitude, forced labor, forced child labor, and other forms of coercive treatment. Human trafficking is a modern form of slavery in which people are used for commercial gain through the use of force, coercion, or fraud (U.S. State Department 2012).

There are many ways to think about human trafficking—including as a moral wrong, as a criminal act by corrupt individuals, and as a human rights issue. As a sociological issue, human trafficking is a complex social structure that is integrally connected to international trade, the social structure of tourism, and the racial, class, and gender inequality that crosses national borders.

Sexual trafficking is a particular form of human trafficking in which women and, often, young girls are bought and sold in an international system of prostitution. Sociologists argue that the male-dominated character of state institutions plays a part in the tolerance of sexual trafficking. Sexual trafficking and sexual tourism are part of a culture in which women's bodies are treated as a commodity. Racial and ethnic inequality also play a part as women of color are sexually exploited based on the racial/gender stereotypes that define them as exotic but also available for the pleasure of men.

As antitrafficking movement has developed that involves a coalition of feminists, various voluntary organizations, the United Nations, some politicians, and others who have organized to stop this practice (Limoncelli 2010).

Aurora Photos/Alamy Limited

Sex trafficking, particularly of young girls, is a common part of the system of global stratification.

High fertility rates are also related to the degree of women's empowerment in society—an often neglected aspect of the discussion between fertility and poverty. Societies where women's voices do not count for much tend to have high fertility rates as well as other social and economic hardships for women, including lack of education, job opportunities, and information about birth control. Empowering women through providing them with employment, education, property, and voting rights can have a strong impact on reducing the fertility rate (Sen 2000).

Women also suffer in some poor countries because of traditions and cultural norms. Most (though not all) of the poor countries are patriarchal, meaning that men control the household. As a result, in some situations of poverty, the women eat after the men, and boys are fed before girls. In conditions of extreme poverty, baby boys may also be fed before baby girls because boys have higher status than girls. As a result, female infants have a lower rate of survival than male infants.

Dan Vincent/Alamy

When children are poor, they may turn to child labor to help support their families. Such is the case with this child laborer working in a municipal dump in Phnom Penh, Cambodia.

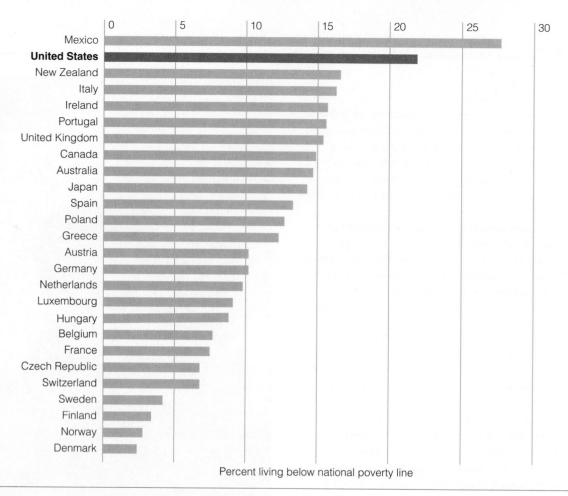

FIGURE 9.8 Child Poverty in the Wealthier Nations*

*Based on 1995 data

Source: UNICEF. 2000a. Child Poverty in Rich Countries 2005. Florence, Italy: United Nations Children's Fund. **www.unicef.org**

A distressing number of children in the world are also poor, including in the most industrialized and affluent nations. Although, as you can see in Figure 9.8, poverty among children in the United States exceeds that of other industrialized nations. Children in poverty do not have the luxury of an education. Schools are usually few or nonexistent in poor areas of the world, and families are so poor that they cannot afford to send their children to school. Children from a very early age are required to help the family survive by working or performing domestic tasks such as fetching water. In extreme situations, children at a young age work as beggars, young boys and girls are sold to work in sweatshops, and young girls are sold into prostitution by their families. This may seem unusually cruel and harsh by Western standards, but it is difficult to imagine the horror of starvation and the desperation that many families in the world must feel that would force them to take such measures to survive. In poor countries, families feel they must have more children for their survival, yet having more children perpetuates the poverty.

Estimates are that there are 215 million children under age 17 in the labor force throughout the world. Most of the children are in Asia, though some are also in sub-Saharan Africa (International Labour Organization 2012). Many of these children work long hours in difficult conditions and enjoy few freedoms, making products (soccer balls, clothing, and toys, for example) for those who are much better off.

Another problem in the very poor areas of the world is homeless children (Mickelson 2000). In many situations, families are so poor that they can no longer care for their children, and the children must go without education or be out on their own, even at young ages. Many of these homeless children end up in the streets of the major cities of Asia and Latin America. In Latin America, it is estimated that there are thirteen million street children, some as young as six years old. Alone, they survive through a combination of begging, prostitution, drugs, and stealing. They sleep in alleys or in makeshift shelters. Their lives are harsh, brutal, and short.

UNDERSTANDING diversity

War, Childhood, and Poverty

"In 2003, surgeons were forced to amputate both of Ali Ismaeel Abbas's arms after an errant U.S. bomb slammed into his Baghdad home during the opening phase of the Iraq war. Pictures of the twelve-year-old, who lost his parents in the attack, soon appeared on TV screens and in newspapers around the world. Since then, Abbas, who was treated in Kuwait, has come to represent a grim reality: All too often the victims of war are innocent children" (McClelland 2003: 20).

In the past ten years alone, UNICEF estimates that over two million children have died in war, with even more injured, disabled, orphaned, or forced into refugee camps (Machel 1996). One estimate is that of all the victims of war, 90 percent are civilian—half of those, children (McClelland 2003).

In the aftermath of war, children are also highly vulnerable to outbreaks of disease. In Iraq, following the war in 2003, many children died of diseases such as anemia and diar-rhea—diseases that can be prevented. Children in Iraq were already living under extreme hardship under the regime of Saddam Hussein. Economic sanctions against Iraq during his regime also produced high infant mortality because of food shortages.

The United Nations has passed resolutions prohibiting the use of children under age 18 in combat. It has linked the threats to children from violence with high rates of poverty around the world. Although reducing poverty would not eliminate the threat of war, it would go a long way toward improving children's lives in war-torn regions.

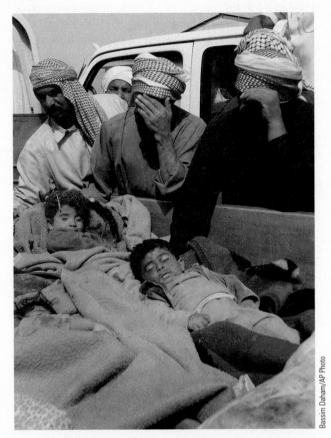

Bassim Daham/AP Photo

War, though it may seem remote to some, affects millions both in the United States and in war-torn countries. Many of those most affected are children. Here, relatives mourn the death of children killed during a U.S. raid in Tikrit, Iraq in 2006.

---

debunking SOCIETY'S MYTHS

MYTH: There are too many people in the world, and there is simply not enough food to go around.
SOCIOLOGICAL PERSPECTIVE: Growing more food will not end hunger. If systems of distributing the world's food were more just, hunger could be reduced. ●

Poverty and Hunger

Poverty is also directly linked to malnutrition and hunger because people in poverty cannot find or afford food. It is estimated that about 925 million people (13.6 percent of the world population) are malnourished, a number that has increased in recent years. Experts attribute the increase to the worldwide economic crisis, the rise in food prices, and neglect by national leaders in paying attention to the agricultural needs of the poor (World Hunger Organization 2012).

Hunger results when there is not enough to feed those in a designated area (such as a region or country). It may be that there is an inadequate supply of food or that households simply cannot afford to purchase enough food to feed themselves. Hunger stifles the mental and physical development of children and leads to disease and death. Although the food supply is plentiful in the world, and is actually increasing faster than the population, the rate of malnutrition remains dangerously high.

Why are people hungry? Is there not enough food to feed all the people in the world? In fact, plenty of food is grown in the world. The world's production of wheat, rice, corn, and other grains is sufficient to adequately feed all the people in the world. Much grain grown in the United States is stored and not used. The problem is that the surplus food does not get to the truly needy. The people who are starving lack what they need for obtaining adequate food, such as arable land or a job that would pay a living wage. In the past, people in most cases grew food crops and were able to feed themselves, but much of the best land today has been taken over by agribusinesses that grow cash crops, such as tobacco or cotton, and subsistence farmers have been forced onto marginal lands on the flanks of the desert where conditions are difficult and crops often do not grow.

Clearly, poverty is a cause of malnutrition, but there are other causes as well. Violence and war within a nation can displace people, leading to large numbers of refugees crowding places where food may not be available to all. Disasters, such as the earthquake that shook Haiti in 2010, can leave people without food and water—a situation complicated when a nation, such as Haiti, is already poor. Even climate change can threaten to create hunger, especially if farming practices cannot adjust to drought, floods, and extreme changes in weather patterns (World Hunger Education Service 2011).

Causes of World Poverty

What causes world poverty, and why are so many people so desperately poor and starving? More to the point, why is poverty decreasing in some areas but increasing in others? We do know what does *not* cause poverty. Poverty is not necessarily caused by too-rapid population growth, although high fertility rates and poverty are related. In fact, many of the world's most populous countries—India and China, for instance—have large segments of their population that are poor, but even with very large populations, these countries have begun to reduce poverty levels. Poverty is not caused by people being lazy or disinterested in working. People in extreme poverty work tremendously hard just to survive, and they would work hard at a job if they had one. It is not that they are lazy; it is that there are no jobs for them.

Poverty is a result of a mix of causes. For one, the areas where poverty is increasing have a history of unstable governments or, in some cases, virtually no effective government to coordinate national development or plans that might alleviate extreme poverty and starvation. World relief agencies are reluctant to work in or send food to countries where the national governments cannot guarantee the safety of relief workers or the delivery of food and aid to where it should go. Food convoys may be hijacked or roads blocked by bandits or warlords.

In many countries with high proportions of poverty, the economies have collapsed and the governments have borrowed heavily to remain afloat. As a condition of these international loans, lenders, including the World Bank and the International Monetary Fund, have demanded harsh economic restructuring to increase capital markets and industrial efficiency. These economic reforms may make good sense for some and may lead these countries out of economic ruin over time, but in the short run, these imposed reforms have placed the poor in a precarious position because the reforms also called for drastically reduced government spending on human services.

Poverty is also caused by changes in the world economic system. Although poverty has been a long-term problem and has many causes, increases in poverty and starvation in Africa and Latin America can be attributed in part to the changes in world markets that favored Asia economically but put sub-Saharan Africa and Latin America at a disadvantage. As the price of products declined with more industrialization in places such as India, China, Indonesia, South Korea, Malaysia, and Thailand, commodity-producing nations in Africa and Latin America suffered. In Latin America, the poor have flooded to the cities, hoping to find work, whereas they did the opposite in Africa, fleeing to the countryside hoping to be able to grow subsistence crops. Governments often had to borrow to provide help to their citizens. Many governments collapsed or found themselves in such great debt that they were unable to help their own people. This has created massive amounts of poverty and starvation.

An often unrecognized cause of poverty is war. War disrupts the infrastructure of a society—its roads, utility systems, water, sanitation, even schools. For countries already struggling economically, this can be devastating. Food production may be disrupted and commerce can be threatened as it may be difficult, even impossible, to move goods in and out of a country. And the loss of life and major injury can mean that there are fewer productive citizens who can work, thus threatening family and community well-being (Pathways to Peace 2009). Moreover, the billions of dollars spent each year on military struggle rob societies of the resources that could be used to address humanitarian needs. Add to this the fact that wars are more likely to occur in nations that are already poor, and you see the impact that war has on world poverty.

In sum, poverty has many causes. It is a major global problem that affects the billions who are living in poverty, but also affects all people in one way or another. In some areas, poverty rates are declining as some countries begin to improve their economic situation; however, in other areas of the world, poverty is increasing, and countries are sinking into financial, political, and social chaos.

GLOBALIZATION AND SOCIAL CHANGE

Globalization is, in some ways, not a new thing. Nations have long been engaged through a global system of trade, travel, and tourism. But what is new about

Population overcrowding, such as on this street in Shanghai, China, strains various natural resources.

David South/Alamy Limited

globalization is the extent to which it permeates daily life for people all over the world and the pace with which globalization is developing. New technologies now allow for extraordinarily fast transactions across tremendous distances, both linking people together in new ways and transferring goods, cultural symbols, and communication systems in ways that were unimaginable not that long ago (Eitzen and Baca Zinn 2012).

Globalization is thus ushering in social changes—some good, some not—that will continue to evolve in the years ahead. As we have seen, globalization has meant that many countries in the world are becoming better off, but many countries remain persistently poor, some very poor. Is the world getting better or worse? What will happen in the future?

There is some good news. In some areas of the world, particularly East Asia, but also in Latin America, many countries have shown rapid growth and are emerging as stronger nations. These countries are sometimes called the **newly industrializing countries (NICs)**, and they include South Korea, Malaysia, Thailand, Taiwan, and Singapore. In these countries, the governments have invested in social and economic development, often with outside help from other nations and corporations. Because some of the NICs have large populations, their success demonstrates that economic development can occur in heavily populated countries. China, for example, has embarked on an aggressive policy of industrial growth, and India is also improving economically.

Yet for all the success stories that globalization has generated, many nations are not making it. These include nations on all continents. In many cases, governments have collapsed or are corrupt, the economy is bankrupt, the standard of living is poor, and people are starving. In many areas of the world, ethnic hatred has led to mass genocide and forced millions from their homes, creating huge numbers of refugees. In Darfur in the western region of Sudan, over 400,000 civilians have been murdered and millions have lost their homes, creating an international outcry demanding that Western governments intervene to stop the violence and provide massive humanitarian aid.

Globalization has also brought the expansion of the system of capitalism, including to nations once hostile to capitalist economics, such as China. This has opened new markets, increased global trade, but also expanded the reach of multinational corporations. The development of such world financial markets may bring prosperity and wealth to many nations and individuals, and it can allow some formerly poor countries to share in the world's wealth. But economic prosperity does not usually filter down to the people at the lower levels of society, and it can force nations into huge amounts of debt, thus allowing poverty and hunger to continue—or even worsen. Thus, while market economies create opportunities for some to become wealthy, both individuals and nations, many nations and individuals do not benefit from this global transformation.

Globalization is a strong force that will continue to shape the future of most nations. Some see globalization simply as the expansion of Western markets and culture into all parts of the world. Western civilization brings positive new values (including democracy and more equality for women), but it can also bring values that may not be seen as positive changes—such as increased consumerism or a change in the nation's sexual mores. Globalization certainly brings new products to remote parts of the world (movies, clothing styles, and other commercial goods), but some see this as a form of imperialism—that is, the domination of Western nations. Resistance to Western globalization and imperialism produces some of the international problems now dominating U.S. and world history, as evidenced in the hostility felt by militant fundamentalist Islamic groups toward the United States.

Globalization has created great progress in the world—including trade, migration, the spread of diverse cultures, the dissemination and sharing of new knowledge, greater freedom for women, travel, and so forth.

Moreover, globalization has not simply extended the values and knowledge of Western culture. Many of the things we now take for granted in our culture originated in non-Western cultures. For example, the decimal system—fundamental to modern math and science—originated in India between the second and sixth centuries and was soon further developed by Arab mathematicians. Western societies certainly get credit for the development of science and technology, but the credit is not theirs alone (Sen 2002).

It is no doubt true that globalization is contributing to the inequality between nations and to the exploitation of some nations and groups by others. Perhaps the solution is not in resisting globalization, but in working so that the benefits of the global economy, global technology, and knowledge reach parts of the world in less exploitative ways. As long as great disparities in standards of living, human rights and basic freedoms, environmental quality, and so forth persist, world conflict is likely to be the result.

chapter summary

What is global stratification?
Global stratification is a system of unequal distribution of resources and opportunities between countries. A particular country's position is determined by its relationship to other countries in the world. The countries in the global stratification system can be categorized according to their per capita *gross national income* or wealth. The world's countries can also be categorized as *first-, second-,* or *third-world* countries, which describe their political affiliation and their level of development. The global stratification system can also be described according to the economic power countries have.

How do systems of power affect different countries in the world?
The countries of the world can be divided into three levels based on their power in the world economic system. The *core countries* are the countries that control and profit the most from the world system. *Semiperipheral countries* are semi-industrialized and play a middleman role, extracting profits from the poor countries and passing those profits on to the core countries. At the bottom of the world stratification system are the *peripheral countries*, which are poor and largely agricultural, but with important resources that the core countries exploit. Most of these nations are populated by people of color, perpetuating racism as part of the world system.

What are the theories of global stratification?
Modernization theory interprets the economic development of a country in terms of the internal attitudes and values. Modernization theory ignores that the development of a country may be due to its economic relationships with other more powerful countries. *Dependency theory* draws on the fact that many of the poorest nations are former colonies of European colonial powers that keep colonies poor and do not allow their industries to develop, thus creating dependency. *World systems theory* argues that no nation can be seen in isolation and that there

is a world economic system that must be understood as a single unit.

What are some of the consequences of global stratification?
The poorest countries have more than half the world's population and have high birthrates, high mortality rates, poor health and sanitation, low rates of literacy and school attendance, and are largely rural. The richest countries have low birthrates, low mortality rates, better health and sanitation, high literacy rates, high school attendance, and largely urban populations. Although women in the wealthy countries are not completely equal to men, they suffer less inequality than do women in poor countries.

How do we measure and understand world poverty?
Relative poverty means being poor in comparison to others. *Absolute poverty* describes the situation where people do not have enough to survive, measured as having the equivalent of $1 per day. *Extreme poverty* is defined as the situation in which people live on less than $1.25 a day. The United Nations has developed a *multidimensional poverty index*—a measure that accounts for health, education, and standard of living. Poverty particularly affects women and children. Children in the very poor countries are forced to work at very early ages and do not have the opportunity for schooling. Street children are a growing problem in many cities of the world. Starvation is also a consequence of the global stratification system.

What is the future of global stratification?
The future of global stratification is varied and depends on the country's position within the world economic system. Some countries, particularly those in East Asia—commonly referred to as *newly industrializing countries*—have shown rapid growth and emerged as developed countries. Many nations, though, are not making it. Governments collapse, countries suffer economic bankruptcy, the standard of living plummets, and people starve.

10 Race and Ethnicity

Key Terms

Race and Ethnicity

Racial Stereotypes

Prejudice, Discrimination, and Racism

Theories of Prejudice and Racism

Diverse Groups, Diverse Histories

Attaining Racial and Ethnic Equality:
The Challenge

Chapter Summary

you might expect a society based on the values of freedom and equality, such as ours, not to be deeply afflicted by racial–ethnic conflict, but think of the following situations:

- In 2009, James von Brunn, an eighty-eight-year-old self-proclaimed White supremacist, gunned down and killed a security guard at the U.S. Holocaust Memorial Museum in Washington, DC. Federal authorities knew James von Brunn was affiliated with various hate groups. The shooter left an anti-Semitic letter in his car parked outside the museum, charging that "Obama was created by Jews." The guard who was shot and killed, Stephen T. Johns, was a thirty-nine-year-old African American guard who worked at the museum. Von Brunn, shot and wounded by other security guards, was charged with murder.

- When Hurricane Katrina struck New Orleans and the Gulf Coast in 2005, hundreds of thousands of people were displaced, and billions of dollars of property destroyed. Many African Americans were disproportionately killed or left homeless. Millions of Americans were shocked by the images of poor people desperate to survive but left without federal or state governmental help for a long time. Both federal and state governments were painfully slow in sending help. Many are still to this day awaiting help in rebuilding.

- Race is still used to exclude jurors in court trials. Recently, Black, Latino, and Vietnamese prospective jurors have been systematically excluded from some California juries in murder and rape trials relative to nonexcluded Whites who are of the same status, education, age, and area of residence. Such instances represent clear occurrences of what is called institutional racism.

- A sorority at a major east coast university was investigated after the sorority sisters posted a photo of their group dressed in sombreros, ponchos, and fake mustaches, also carrying signs that said, "Will mow lawn for weed and beer." Such denigrating and offensive" racial theme parties" periodically appear in the news, angering allies and members of racial-ethnic minority groups.
- In July of 2007, a 38 year-old Indian American man, vice president of a major bank, was attacked on a Lake Tahoe beach as his attackers called him a "terrorist," "relative of Osama bin Laden," and "Indian garbage." The attack broke his eye socket and will give the victim dizzy spells for the rest of his life.

These recent incidents are not pretty. They all have one thing in common—racial-ethnic prejudice and the hatred it can produce. Along with gender and social class, race and ethnicity have fundamental importance in human social interaction and are an utterly integral part of the social institutions in the United States. Race and ethnic prejudice are part of the uglier side of American society.

Of course, racial and ethnic groups do not always interact as enemies. And interracial tension is not always obvious. It can be as subtle as a White person who simply does not initiate interactions with African Americans and Latinos, or an elderly White man who almost imperceptibly leans backward at a cocktail party as a Japanese American man approaches him.

In everyday human interaction, as African American philosopher Cornel West has cogently argued, race matters and still matters a lot (West 1993). What is race, and what is ethnicity? Why does society treat racial and ethnic groups differently, and why is there social inequality—stratification—between these groups? How are these divisions and inequalities able to persist so stubbornly, and how extensive are they? These inequalities are so strong and persistent in American society that sociologists are not convinced by the uninformed recent speculation in the popular press that the second-term election of President Barack Obama, the first African American U.S. president, will lead to the virtual disappearance of prejudice and racism and the achievement of a "postracial" society. In the words of sociologist Lawrence Bobo, such postracial "dreams" must confront American realities (Bobo 2012; cf. Spencer 2012). Such questions and issues fascinate sociologists who do research on racial and ethnic relations and stratification in our society.

learning objectives

- Define race as a social construction
- Define and give examples of stereotype interchangeability
- Define and distinguish prejudice, discrimination, aversive racism, and institutional racism
- Show how conflict theory broadens our understanding of prejudice and discrimination
- Discuss at least one common thread present in the histories of Blacks, Mexican Americans, and Japanese Americans in the United States
- Contrast the civil rights strategy with radical social change strategies toward the attainment of racial–ethnic equality or freedom in the United States

RACE AND ETHNICITY

Within sociology, the terms *ethnicity, race, minority*, and *dominant group* have very specific meanings, different from the meanings these terms have in common usage. These concepts are important in developing a sociological perspective on race and ethnicity.

Ethnicity

An **ethnic group** is a social category of people who share a common culture, for example, a common language or dialect, a common nationality, a common religion, and common norms, practices, customs, and history. Ethnic groups have a consciousness of their common cultural bond. Italian Americans, Japanese Americans, Arab Americans, Polish Americans, Greek Americans, Mexican Americans, and Irish Americans are all examples of ethnic groups in the United States. Ethnic groups are also found in other societies, such as the Pashtuns in Afghanistan or the Shiites and Sunnis in Iraq, whose ethnicity is based on religious differences.

An ethnic group does not exist only because of the common national or cultural origins of a group, however. Ethnic groups develop because of their unique historical and social experiences. These experiences become the basis for the group's *ethnic identity*, meaning the definition the group has of itself as sharing a common cultural bond. Prior to immigration to the United States, Italians, for example, did not necessarily think of themselves as a distinct group with common interests and experiences. Originating from different villages, cities, and regions of Italy, Italian immigrants identified themselves by their family background and community of origin. However, the process of immigration and the experiences Italian Americans faced as a group in the United States, including discrimination, created a new identity for the group, who subsequently began to define themselves as "Italians" (Waters and Levitt 2002; Waters 1990; Alba 1990).

This is a parade on St. Patrick's Day, an Irish holiday. It demonstrates ethnic pride and solidarity.

is "socially constructed" (Higginbotham and Andersen 2012). Although the meaning of race begins with perceived biological/genetic differences between groups (such as differences in physical characteristics like skin color, lip form, and hair texture), on closer examination, the assumption that racial differences are purely biological breaks down. In fact, biologists have pointed out that there is little correspondence between races as defined genetically and the actual naming of the races (Taylor 2012; Morning 2011; Ledger 2009; Lewontin 1996).

debunking SOCIETY'S MYTHS

MYTH: Racial differences are fixed, biological categories.
SOCIOLOGICAL PERSPECTIVE: Race is a social concept, one in which certain physical or cultural characteristics take on social meanings that become the basis for racism and discrimination. The definition of race varies across cultures within a society, across different societies, and at different times in the history of a given society. ●

The social and cultural basis of ethnicity allows ethnic groups to develop more or less intense ethnic identification at different points in time. Ethnic identification may grow stronger when groups face prejudice or hostility from other groups. Perceived or real threats and perceived competition from other groups may unite an ethnic group around common political and economic interests, which as you may recall was a hypothesis early sociological theorist Emile Durkheim (see Chapter 1) advanced. Ethnic unity can develop voluntarily, or it may be involuntarily imposed when ethnic groups are excluded by more powerful groups from certain residential areas, occupations, or social clubs. Or it can develop both voluntarily and also from exclusion and discrimination. These exclusionary practices strengthen ethnic identity.

Defining Race

Like ethnicity, race is primarily, though not exclusively, a socially constructed category. A **race** is a group treated as distinct in society based on certain characteristics, some of which are biological, that have been assigned or attributed social importance. Because of presumed biologically or culturally inferior characteristics (as defined by powerful groups in society), a race is often singled out for differential and unfair treatment. It is not the biological characteristics per se that define racial groups but *how groups have been treated and labeled historically and socially.*

Society assigns people to racial categories, such as Black, White, and so on, not because of science, logic, or fact, but because of opinion and social experience. In other words, how groups are defined racially is a *social* process. This is what is meant when one says that race

The social categories used to divide groups into races are not fixed, and they vary from society to society and at different times in the history of a given society (Morning 2011, 2008; Washington 2006). Within the United States, laws defining who is Black have historically varied from state to state. North Carolina and Tennessee law historically defined a person as Black if he or she had even one great-grandparent who was Black (thus being one-eighth Black—called "octoroon" in the 1890 census; see Table 10.1). In other southern states, having any Black ancestry at all defined one as a Black person—the so-called one-drop rule, that is, one drop of Black blood (Broyard 2007; Malcomson 2000). This one-drop rule still applies to a great extent today in the United States, although its use as a criterion for defining one's race has eroded somewhat.

This is even more complex when we consider the meaning of race in other countries. In Brazil, a light-skinned Black person could well be considered White, especially if the person is of high socioeconomic status; this demonstrates that one's race in Brazil is in part actually *defined* by one's social class. Thus, in parts of Brazil, it is often said that "money lightens" (*o dinheiro embranquece*). In this sense, a category such as social class can become *racialized*. In fact, people in Brazil are considered Black only if they are clearly of African descent and have little or no discernible White ancestry at all. A large percentage of U.S. Blacks would not be considered Black in Brazil (Telles et al. 2011; Telles 2004). Although Brazil is often touted as being a utopia of race "mixing" and racial social equality, nonetheless, as sociologist Edward Telles notes, light-skinned Brazilians continue to be privileged and continue to hold a disproportionate share of the wealth and power. Brazilians of darker skin color have significantly lower earnings, occupational status, and lower access to education (Telles et al. 2011; Villareal 2010; Telles 2004, 1994).

Racialization is a process whereby some social category, such as a social class or nationality, takes on what *society perceives* to be racial characteristics (Harrison 2000; Malcomson 2000; Omi and Winant 1994). The experiences of Jewish people provide a good example of what it means to say that race is a socially constructed category. Jews are more accurately called an *ethnic group* because of common religious and cultural heritage, but in Nazi Germany, Hitler defined Jews as a "race." An ethnic group had thus become *racialized*. Jews were presumed to be biologically inferior to the group Hitler labeled the Aryans—white-skinned, blonde, tall, blue-eyed people. On the basis of this definition—which was supported through Nazi law, taught in Nazi schools, and enforced by the Nazi military—Jewish people were grossly mistreated. They were segregated, persecuted, and systematically murdered in what has come to be called the Holocaust during the Second World War.

Mixed-race people defy the biological categories that are typically used to define race. Is someone who is the child of an Asian mother and an African American father Asian or Black? Reflecting this issue, the U.S. Census's current practice is for a person to have the option of checking several racial categories rather than just one, thus defining one's self as "biracial" or "multiracial" (U.S. Census Bureau 2012; Spencer 2011; Waters 1990), although considerable controversy has arisen over this procedure (Spencer 2011; Harrison 2000). As Table 10.1 shows, the decennial U.S. census (taken every ten years) has dramatically changed its racial and ethnic classifications since 1890, reflecting the fact that society's thinking about racial and ethnic categorization has not remained constant through time (Spencer 2012; Saulny 2011; Washington 2011; Rodriguez 2006; Lee 1993).

Recently, on some college campuses, an organization has been formed called the Multiracial and Biracial Student Association (MBSA). Their goal is to encourage students to define themselves multiracially rather than by only one racial label and to follow through by using multiple racial categories on official forms, as the U.S. census now allows. Very few people did so in the 2010 census; a goal of MBSA has been to encourage multiracial responses on the current census forms. Thus

table 10.1 Comparison of U.S. Census Classifications, 1890–2010

Census Date	White	African American	Native American	Asian American	Other Categories
1890	White	Black, Mulatto, Quadroon, Octoroon	Indian	Chinese Japanese	
1900	White	Black	Indian	Chinese Japanese	
1910	White	Black Mulatto	Indian	Chinese Japanese	Other[b]
1990	White	Black or Negro	Indian (American) Eskimo Aleut	Chinese Japanese Filipino Korean Asian Indian Vietnamese	Hawaiian Guamanian Samoan Asian or Pacific Islander Other
2000 and 2010[a, b]	White	Black or African American	American Indian Alaskan Native	Chinese Japanese Filipino Korean Asian Indian Vietnamese	Native Hawaiian Other Pacific Islander Other

[a]In 2000, for the first time ever, and again in 2010, individuals could select more than one racial category. In 2010, only 5 percent actually did so.
[b]Hispanics were included under "Other" in 1910 and 1920. In 1930 and subsequent years, the category "Mexican" has been listed in addition to the category "Other."

Sources: Lee, Sharon. 1993. "Racial Classification in the U.S. Census: 1890–1990." *Ethnic and Racial Studies* 16(1): 75–94; U.S. Census Bureau. 2003. "Racial and Ethnic Classification Used in Census 2000 and Beyond." Taylor & Francis Ltd.; Rodriguez, Clara E. 2009. "Changing Race." Pp. 22–25 in *Race and Ethnicity in Society: The Changing Landscape*. Belmont, CA: Wadsworth; Silver, Alexandra. 2010. "Brief History of the U.S. Census." *Time* (February 8): 16; Washington, Scott. 2011. "Who Isn't Black? The History of the One-Drop Rule." PhD dissertation, Department of Sociology, Princeton University, Princeton, NJ.

a large variety of multiracial responses are possible: for example, Black-Asian-Greek, Black-White-German, Ghanian-Scottish-Norwegian, American Indian-Irish-Swedish, and many others (Saulny 2011).

Opposition to the multiple categorization of races has arisen upon both scholarly as well as political grounds. Some (Spencer 2011) have argued that advocating simultaneous multiple categorization of races tends to downplay the rich cultural traditions, in the case of Blacks in the United States, including but not limited to language ("Ebonics"), music (jazz, blues, rock, hip-hop, and so on), dance, a vast literature, and many, many others. Some people have argued that multiracial classification will ultimately lead to a "postracial" society and thus a solution of sorts to race problems in the United States. But wiping out single-race categorization will not, and has not, led to less discrimination against minorities of color.

As recent sociological scholars and data analysts have clearly shown, the "postracial dream" conflicts with the hard realities of housing discrimination, higher foreclosure rates during the recession, racial discrimination in education and in standardized testing, differential access to medical care, a lower life expectancy,

and many other forms of discrimination (Bobo 2012; Rugh and Massey 2010). These forms of racial discrimination actually *increased* even within the last decade! Finally, at least one sociological analyst has noted that those who push for multi- or biracial categorization and who envision a "postracial" society are only engaging in a superficial "multirace chic." Spencer notes further that "Those who proclaim that multiracial identity will destroy racial distinctions are living a lie" (Spencer 2012: 70). Political opposition to multiple categorization of the races of a single person also arises because doing so would heavily complicate the matter of counting the number and percentage of a given racial–ethnic group in a census area (census tract). Federal and state governments routinely use such tabulations to determine the health, educational, occupational, and housing needs of a particular racial–ethnic group. Hence, it can be argued, minority groups can and will be shortchanged when it comes to federal and state allocation of funds for these areas of need.

The Significance of Defining Race. The biological characteristics that have been used to define different racial groups vary considerably both within and between groups. Many Asians, for example, are actually lighter skinned than many Europeans and White Americans and, regardless of their skin color, have been defined in racial terms as yellow. Some light-skinned African Americans are also lighter in skin color than some White Americans. Developing racial categories overlooks the fact that human groups defined as races are—biologically speaking—much more alike than they are different.

The biological differences that are presumed to define different racial groups are somewhat arbitrary. Why, for example, do we differentiate people based on skin color and not some other characteristic such as height or hair color? You might ask yourself how a society based on the presumed racial inferiority of red-haired people would compare to other racial inequalities in the United States. The likelihood is that if a powerful group defined another group as inferior because of some biological characteristics, and they used their power to create social institutions that treated this group unfairly, a system of racial inequality would result. In fact, very few biological differences exist between racial groups. Most of the variability in almost all biological characteristics, even blood type and various bodily chemicals, is *within* and not between racial groups (Morning 2011; Rodriguez 2006; Malcomson 2000; Lewontin 1996).

Different groups use different criteria to define racial groups. To American Indians, being classified as an American Indian depends upon proving one's ancestry, but this proof varies considerably from tribe to tribe. Among some American Indians, one must be able to demonstrate at least 75 percent American Indian ancestry to be recognized as such; for other American Indians, demonstrating 50 percent American Indian ancestry is sufficient.

This is Barack Obama, the first African American ever to be elected U.S. president, and for two terms. His father is Black African (Kenyan) and his mother is White American. Why is his race African American?

Scott Olson/Getty Images News/Getty Images

What Exactly Is "Race" Anyway?

Most people still think race is a strict physical, biological category of humans. This is not correct. The notion of race is more social construction than biology. Race is, in part, perceived physical attributes such as skin color, hair texture, lip form, eye form, and so on, but it in greater part is defined by social and cultural attributes. In fact, any biological category of "race" is not a socially identifiable category at all without the social and cultural attributes that society assigns to the various "race" labels. Hence, the notion of race is strongly rooted in society and has taken on its meaning only as people were treated differently throughout time.

Up until the 1950s in the United States, the races were defined as strict physical/biological categories, as follows: Negro (Black), Caucasian (White), Asian (Yellow), American Indian (Red), and finally "Australoid" (Brown). Virtually all the colors of the spectrum! There are still people to this day who define "race" in terms of this archaic color spectrum.

It also matters *who* defines racial group membership. The government makes tribes prove themselves as tribes through a complex set of federal regulations (called the "federal acknowledgment process"); very few are actually given this official status, and the criteria for tribal membership as well as definition as "Indian" or "Native American" have varied considerably throughout American history. Thus, as with African Americans, it has been the state or federal government, and not so much the racial or ethnic group *itself*, that has defined who is a member of the group and who is not!

Official recognition by the government matters. For example, only those groups officially defined as Indian tribes qualify for health, housing, and educational assistance from the Bureau of Indian Affairs (the BIA) or are allowed to manage the natural resources on Indian lands and maintain their own system of governance (Locklear 1999; Brown 1993; Snipp 1989).

This definition of race emphasizes that in addition to physical and also cultural differences, race is created and maintained by the most powerful group (or groups) in society. This definition of race also incorporates presumed group differences in the context of social and historical experience. As a result, who is defined as a race is as much a political question as a biological or cultural one (Brodkin 2006). For example, although they probably did not think of themselves as a race, Irish Americans in the early twentieth century were defined by more powerful White groups as a "race" that was inferior to White people. This was an example of the *racialization* of an ethnic or nationality group. At that time, Irish people were not considered by many even to be White (Ignatiev 1995)! In fact, a century ago, the Irish were called "Negroes/Coloreds/ Blacks/Niggers turned inside out," and Negroes (Black people) were called "smoked Irish" (Malcomson 2000).

The social construction of race has been elaborated in an insightful perspective in sociology known as racial formation theory (Brodkin 2006; Omi and Winant 1994). **Racial formation** is the process by which a group comes to be defined as a race. This definition is supported through official social institutions such as the law and the schools. This concept emphasizes the importance of social institutions in producing and maintaining the meaning of race; it also connects the process of racial formation to the exploitation of so-called racial groups. A good example comes from African American history. During slavery, an African American was defined as being three-fifths of a person (equivalently, as "divested two-fifths the man") for the purposes of deciding how slaves would be counted for state representation in the new federal government and how they would be defined to be taxed as property. The process of defining slaves in this way served the purposes of White Americans, not slaves themselves, and it linked the definition of slaves as a race to the political and economic needs of the most powerful group in society (Higginbotham 1978).

It may surprise readers to know that "Whiteness" is itself a social construction. This only underscores the importance of social constructionism in the definition of race in addition to biological criteria. Historian Nell Painter (2010) contributed to the development of the new field of "Whiteness studies" in demonstrating the racial formation and social construction aspect of defining who is "White." A key point in Painter's argument is her elaboration of how the writer Ralph Waldo Emerson argued that Anglo-Saxons were the "true Whites" and thus superior to other White groups (Irish, Germans, Polish, Italians, and on and on) in culture and intelligence. He even cited IQ test results to back up his argument. (See Chapter 14.)

The process of racial formation also explains how groups such as Asian Americans, American Indians, and Latinos have been defined as races, despite the different experiences and nationalities of the groups composing these three categories. Race, like ethnicity, lumps groups together that may have very different historical and cultural backgrounds, but once they are so labeled, the groups are perceived as a single entity. This reflects a more general principle in the social sciences called **out-group homogeneity effect**, where all members of any out-group are perceived to be similar or even identical to each other, and differences among them are perceived to be minor or nonexistent. This has recently been the case in the United States with Middle Easterners: Egyptians, Lebanese, Syrians, Saudi Arabians, Iranians, Iraqis, Jordanians, Afghans, and many others are classified as one group and called Middle Easterners, or simply "Arabs."

Minority and Dominant Groups

Minorities are racial or ethnic groups, but not all racial or ethnic groups are always considered minorities. Irish Americans, for instance, are not now thought of as minorities, although they certainly were in the early part of the twentieth century. A **minority group** is any distinct group in society that shares common group characteristics and is forced to occupy low status in society because of prejudice and discrimination. The group that assigns a racial or ethnic group to subordinate status in society is called the **dominant group**.

A group may be classified as a minority on the basis of ethnicity, race, sexual preference, age, or class status, for example. A minority group is not necessarily a numerical minority but is a group that holds low status in relation to other groups in society, regardless of the size of the group. In South Africa, Blacks outnumber Whites ten to one, but until Nelson Mandela's election as president and the dramatic change of the country's government in 1994, Blacks were a viciously oppressed and politically excluded social minority under the infamous *apartheid* (pronounced "aparthate" or "apart-hite") system of government. In general, a racial or ethnic minority group has the following characteristics:

1. The minority group possesses characteristics (such as race, ethnicity, sexual preference, age, or religion, and even gender) that are popularly regarded as different from those of the dominant group.
2. The minority group suffers prejudice and discrimination by the dominant group.
3. Membership in the group is frequently ascribed rather than achieved, although either form of status can be the basis for being identified as a minority.
4. Members of a minority group feel a strong sense of group solidarity. There is a "consciousness of kind" or "we" feeling. This bond grows from common cultural heritage and the shared experience of being a recipient of prejudice and discrimination.

---

debunking SOCIETY'S MYTHS

MYTH: Minority groups are those with the least numerical representation in society.
SOCIOLOGICAL PERSPECTIVE: A minority group is any group, regardless of size, that is singled out in society for unfair treatment and that generally occupies a lower status in the society. ●

RACIAL STEREOTYPES

Racial and ethnic inequality is peculiarly resistant to change. Racial and ethnic inequality in society produces racial stereotypes, and these stereotypes become the lens through which members of different groups perceive one another. Over time, these stereotypes become more rigid and unchangeable.

Stereotypes and Salience

In everyday social interaction, people tend to *categorize* other people. Fortunately or unfortunately, we all do this. The most common bases for such categorizations are race, gender, and age. A person *immediately* identifies a stranger as Black, Asian, Hispanic, White, and so on; as a man or woman; and as a child, teenager, adult, or elderly person. Quick and ready categorizations help people process the huge amounts of information they receive about people with whom they come into contact. People quickly assign others to a few categories, saving themselves the task of evaluating and remembering every little discernible detail about a person. People are taught from childhood to treat each person as a unique individual, but research over the years clearly shows that they do not. Instead, people routinely categorize others in some way or another. We process information about others quickly, assigning certain characteristics to them with little actual knowledge of them.

A **stereotype** is an oversimplified set of beliefs about members of a social group or social stratum. It is based on the tendency of humans to categorize a person based on a narrow range of perceived characteristics. Stereotypes are presumed to describe the "typical" member of some social group. They are usually, but not always, incorrect.

Stereotypes based on race or ethnicity are called *racial–ethnic stereotypes.* Here are some common examples of racial–ethnic stereotypes: Asian Americans have been stereotyped as overly ambitious, and academically successful; African Americans often bear the stereotype of being loud and lazy; Hispanics are stereotyped as lazy, oversexed, and for Hispanic men, macho; Jews have been perceived as materialistic. Such stereotypes, presumed to describe the "typical" member of a group, are factually inaccurate for the vast majority of members of a group. *No group in U.S. history has escaped the process of categorization and stereotyping, not even White groups.* For example, Italians have been stereotyped as overly emotional and prone to crime, the Irish as heavy drinkers and political, and so on for virtually any group in U.S. history.

The categorization of people into groups and the subsequent application of stereotypes is based on the **salience principle**, which states that we categorize people on the basis of what appears initially prominent and obvious—that is, salient—about them. Skin color is a salient characteristic; it is one of the first things that we notice about someone. Because skin color is so obvious, it becomes a basis for stereotyping. Gender and age are also salient characteristics of an individual and thus serve as notable bases for group stereotyping.

---

thinking SOCIOLOGICALLY

Observe several people on the street. What are the first things you notice about them (that is, what is salient)? Make a short list of these things. Do these lead you to *stereotype* these people? On what are your stereotypes based? ●

The choice of salient characteristics is culturally determined. In the United States, skin color, hair texture, nose form and size, and lip form and size have become salient characteristics, and these characteristics determine whether we perceive someone as "intelligent" or "stupid," as "attractive" or "unattractive," or even "trustworthy" or "untrustworthy." This was shown in a clever experiment by R. Hunt, who had people rate photographs of other people that varied systematically by skin color, hair texture, and lip form for trustworthiness (Hunt 2005). We use these features to categorize people in our minds on the basis of race. In other cultures, religion may be far more salient than skin color. In the Middle East, whether one is Muslim or Christian is far more important than skin color. Religion in the Middle East is a salient characteristic and takes considerable priority over skin color.

The Interplay among Race, Gender, and Class Stereotypes

Alongside racial and ethnic stereotypes, gender and social class and age are among the most prominent features by which people are categorized. In our society, there is a complex interplay among racial or ethnic, gender, and class stereotypes.

Among *gender stereotypes,* those based on a person's gender, the stereotypes about women are more likely to be negative than those about men. The "typical" woman has been traditionally stereotyped as subservient, overly emotional and talkative, inept at math and science, and so on. Many of these are *cultural stereotypes.* They are conveyed and supported by the cultural media—music, TV, magazines, art, and literature—and also by one's family. Men, too, are painted in crude strokes, although usually not as negatively as women. Men in the media are stereotyped as macho, insensitive, and pigheaded and are portrayed in situation comedies as inept. Stereotypes of men vary, however, depending on their race and class.

Social class stereotypes are based on assumptions about social class status. Upper-class people are stereotyped (by middle-and working-class people) as snooty, aloof, condescending, and phony. Some of the stereotypes held about the middle class (by both the upper class and the working class) are that they are overly ambitious, striving, and obsessed with "keeping up with the Joneses." Finally, stereotypes about working-class people abound: They are perceived by the upper and middle classes as dirty, lazy, unmotivated, violent, and so on. These stereotypes are then used (by the upper and middle classes) as presumed *explanations of why* those perceived are "lower" in their social class.

The principle of **stereotype interchangeability** holds that stereotypes, especially negative ones, are often interchangeable from one social class to another, from one racial or ethnic group to another, from a racial or ethnic group to a social class, or from a social class to a gender. Stereotype interchangeability is sometimes revealed through humor. Ethnic jokes often interchange different groups as the butt of the humor, stereotyping them as dumb and inept. Take the stereotype of African Americans as inherently lazy. This stereotype has also been applied in recent history to Hispanic, Polish, Irish, Italian, and other groups (illustrating interchangeability from one racial–ethnic group to another). It has even been applied to those people perceived as lower social class (showing interchangeability from a racial–ethnic group to a social class). In fact, "laziness" is often used to explain *why* someone is working class or poor.

Middle-class people are more likely to attribute the status of a working-class person to something *internal,* such as "inherent" laziness or lack of willpower (Bryson and Davis 2010; Morlan 2005; Worchel et al. 2000; Krasnodemski 1996). Working-class people are more likely to attribute their status to discrimination or poor opportunities—that is, to an *external* societal factor (Morlan 2005; Kluegel and Bobo 1993; Bobo and Kluegel 1991).

The same kinds of stereotypes have historically been applied to women. Many of the stereotypes applied to women in literature and the media—they are childlike, overly emotional, unreasonable, bad at mathematics, and so on—have also been applied to African Americans, working-class people, the poor, and earlier in the twentieth century, Chinese Americans. This shows stereotype interchangeability among gender, racial groups, and social classes. A common theme is apparent: Whatever group occupies lower social status in society at a given time (whether racial or ethnic minorities, women, or the working class), that group is negatively stereotyped; and, often the same negative stereotypes are used between and among these groups. The stereotype is then used as an "explanation" for the observed behavior of a stereotyped group's member to justify his or her lower status in society. This in turn subjects the stereotyped group to prejudice, discrimination, and racism.

PREJUDICE, DISCRIMINATION, AND RACISM

Many people use the terms *prejudice*, *discrimination*, and *racism* loosely, as if they were all the same thing. They are not. Typically, in common parlance, people also think of these terms as they apply to individuals, as if the major problems of race were the result of individual people's bad will or biased ideas, thus ignoring the social structural and institutional aspects of race in the United States. Sociologists use more refined concepts to understand race and ethnic relations, distinguishing carefully between prejudice, discrimination, and racism.

Prejudice

Prejudice is the evaluation of a social group and the individuals within it, based on conceptions about the social group that are held despite facts that disprove them; the beliefs involve both prejudgment and misjudgment (Jones 1997; Allport 1954). Prejudices are usually defined as negative predispositions or as evaluations that are rarely positive. Thinking ill of people only because they are members of group X is prejudice.

A negative prejudice against someone not in one's own group is often accompanied by a positive prejudice in favor of someone who *is* in one's own group. Thus the prejudiced person will have negative attitudes about a member of an *out-group* (any group other than one's own) and positive attitudes about someone simply because he or she is in one's *in-group* (any group a person considers to be one's own).

Most people disavow racial or ethnic prejudice, yet the vast majority of us carry around some prejudices, whether about racial–ethnic groups, women and men, old and young, upper class and lower class, or straight and gay. Virtually no one is free of prejudice—of both harboring it as well as being the recipient of it. Decades of research have shown definitively that people who are more prejudiced are also more likely to stereotype and categorize others by race or ethnicity or by gender than those who are less prejudiced (Taylor et al. 2013; Jones 1997; Adorno et al. 1950).

Prejudice based on race or ethnicity is called *racial* or *ethnic prejudice*. If you are a Latino and dislike an Anglo only because he or she is White, then this constitutes prejudice: It is a negative judgment or prejudgment based on race and ethnicity and very little else. If the Latino individual attempts to justify these feelings by arguing that "all Whites have the same bad character," then the Latino is using a stereotype as justification for the prejudice. Note that any group can hold prejudice against another group.

Prejudice is also revealed in the phenomenon of **ethnocentrism**, which was examined in Chapter 2 on culture. Ethnocentrism is the belief that one's group is superior to all other groups. The ethnocentric person feels that his or her own group is moral, just, and right, and that an out-group—and thus any member of that out-group—is immoral, unjust, wrong, distrustful, or criminal. The ethnocentric individual uses his or her own in-group as the standard against which all other groups are compared.

Prejudice and Socialization. Where does racial–ethnic prejudice come from? How do moderately or highly prejudiced people end up that way? People are not born with stereotypes and prejudices. Research shows that these attitudes are learned and internalized through the socialization process, including both *primary socialization* (family, peers, teachers) as well as *secondary socialization* (such as the media). Children imitate the attitudes of their parents, peers, and teachers. If a parent complains about "Japs taking away jobs from Americans," then the child grows up thinking negatively about the Japanese, including Japanese Americans. Attitudes about race are formed early in childhood, at about age 3 or 4 (Feagin 2000; Van Ausdale and Feagin 1996; Allport 1954). There is a very close correlation between the racial and ethnic attitudes of parents and those of their children. The more ethnically or racially prejudiced the parent, the more ethnically or racially prejudiced the child will be. This is even true for individuals who insist that they can think for themselves and who think they are not influenced by their parents' prejudice (Taylor et al. 2013).

As we saw in Chapter 2, major vehicles for the communication of racial–ethnic attitudes to both young and old are the media, especially television, magazines, newspapers, and books. For many decades, African Americans, Hispanics, Native Americans, and Asians were rarely represented in the media and then only in negatively stereotyped roles. The Chinese were shown as bucktoothed buffoons who ran shirt laundries in movies, magazines, and television in the 1950s. Japanese Americans were depicted as sneaky and untrustworthy. Hispanics were shown as either ruthless banditos or playful, happy-go-lucky people who took long siestas. American Indians were presented as either villains or subservient characters like the Lone Ranger's faithful sidekick, Tonto. Finally, there is the drearily familiar portrayal of the Black person as subservient, lazy, shuffling, clowning, and bug-eyed, a stereotypical image that persisted from the nineteenth century all the way through the 1950s and early 1960s.

Discrimination

Discrimination is overt negative and unequal treatment of members of some social group or stratum solely because of their membership in that group or stratum. Prejudice is an attitude; discrimination is overt behavior. *Racial-ethnic discrimination* is unequal treatment of a person on the basis of race or ethnicity.

Discrimination in housing has been a particular burden on minorities. Many studies have been able to document discrimination, showing that when two people identical in nearly all respects (age, education, gender, social class, and other characteristics) present themselves as potential tenants for the same housing, if one is White and the other is a minority, the minority person will often be refused housing by a White landlord and the otherwise identical White applicant will not. A minority landlord who refuses housing to a White person while granting it to a minority person of similar social characteristics is also discriminating, but reverse discrimination of this sort is far less frequent and far less of a problem in society (Rugh and Massey 2010; Massey 2005; Feagin 2000; Feagin and Vera 1995; Feagin and Feagin 1993). Finally, it should be noted that during the foreclosure crisis in the recession of the mid-2000s, neighborhoods with higher percentages of Hispanics, Blacks, and even Asians had a faster home foreclosure rate, all else (such as home value) equal, than did predominantly White neighborhoods (Rugh and Massey 2010). In the words of Oliver and Shapiro, Blacks were the "last in, first out" (Oliver and Shapiro 2008: 325). Is this the "postracial" society that some journalists have argued for?

The discrimination affecting the nation's minorities takes a number of forms—for example, income discrimination and discrimination in housing as just noted. Discrimination in employment and promotion and discrimination in education (see Chapters 14 and 15) are two other forms of discrimination.

Although the median income of Black and Hispanic families has increased since 1950, the size of the income gap between these two groups and Whites has remained much the same over time, as can be seen clearly from Figure 10.1. Furthermore, per capita income since 1967 has grown at a faster rate for Whites.

Yet even these median income figures tell only part of the story. In addition to annual income, the net worth of White families has consistently grown faster than that of Black families (Oliver and Shapiro 2006, 2001, 1995). Net worth may well be a better indicator of economic inequality than annual income. Poverty among Blacks decreased since the 1950s, but is now close to the same level as in 1970. The current poverty rate (the percent below the poverty level; see Figure 10.2) is highest for African Americans and Hispanics compared with Whites or Asians. In all these racial groups, children have the highest rate of poverty.

Discrimination in housing is illegal under U.S. law. Nonetheless, for many years, right on down through 2009, banks and mortgage companies often withheld mortgage loans from minorities based on "redlining," an illegal practice in which an entire minority neighborhood is designated as "ineligible for loans." Racial segregation may also be fostered by *gerrymandering*, the calculated redrawing of election districts, school districts, and similar political boundaries to maintain racial segregation. As a result, **residential segregation**, **the spatial separation of racial and ethnic groups into different residential areas**, called "American apartheid" by Massey and Denton (1993; see also Massey 2005; cf. Orfield and Lee 2012 and Rugh and Massey 2010), continues to be a reality in this country (see the box "Doing Sociological Research: American Apartheid").

A somewhat different and unique phenomenon of institutional discrimination is the **digital racial divide**, as alluded to in Chapter 2. The percentage of Black households in the United States with Internet access is only 58 percent; White households, 78 percent; Hispanic, 59 percent; and Asian households, 83 percent

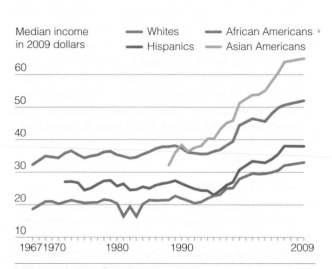

FIGURE 10.1 The Income Gap

Source: DeNavas-Walt, Carmen, Bernadette D. Proctor, and Jessica C. Smith. 2012. *Income, Poverty, and Health Insurance Coverage in the United States: 2011.* Washington, DC: U.S. Department of Commerce. **www.census.gov**

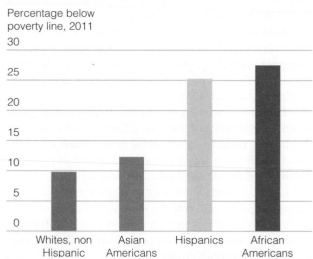

FIGURE 10.2 Poverty among Racial Groups

Source: DeNavas-Walt, Carmen, Bernadette D. Proctor, and Jessica C. Smith. 2012. *Income, Poverty, and Health Insurance Coverage in the United States: April 2010.* Washington, DC: U.S. Department of Commerce. **www.census.gov**

American Apartheid

The term *apartheid* was used to describe the society of South Africa prior to the election of Nelson Mandela in 1994. It refers to the rigid separation of the Black and White races. Sociological researchers Massey and Denton argue that the United States is now under a system of apartheid and that it is based on a very rigid residential segregation in the country.

Research Question: What is the current state of residential segregation? Massey and Denton note that the terms *segregation* and *residential segregation* practically disappeared from the American vocabulary in the late 1970s and early 1980s. These terms were spoken little by public officials, journalists, and even civil rights officials. This was because the ills of race relations in the United States were at the time attributed, though erroneously, to other causes such as a "culture of poverty" among minorities, or inadequate family structure among Blacks, or too much welfare for minority groups. The Fair Housing Act was passed in 1968, and the problem of segregation

and discrimination in housing was declared solved. Yet nothing could be farther from the truth.

Research Methods and Results: Researchers Massey and Denton amassed a large amount of survey data demonstrating that residential segregation not only has persisted in American society but also that it has actually increased since the 1960s. Most Americans vaguely realize that urban America is still residentially segregated, but few appreciate the depth of Black and Hispanic segregation or the degree to which it is maintained by ongoing institutional arrangements and contemporary individual actions. Urban society is thus hypersegregated, or characterized by an extreme form of residential and educational segregation.

Conclusions and Implications: Massey and Denton find that most people think of racial segregation as a faded notion from the past, one that is decreasing over time. Today, theoretical concepts such as the culture of poverty, institutional racism, and welfare are widely

debated, yet rarely is residential segregation considered to be a major contributing cause of urban poverty and the underclass. Massey and Denton argue that their purpose is to redirect the focus of public debate back to race and racial segregation.

Questions to Consider

1. Think about the degree of residential segregation in the neighborhood in which you grew up. How racially and/or ethnically segregated was it? If it was segregated at all, what racial–ethnic groups were living there?

2. Do you think the problem of racial–ethnic residential segregation in the United States is largely "solved"? How so or why not?

3. What are the consequences of residential segregation—for example, for education? For employment?

Sources: Massey, Douglas S., and Nancy A. Denton. 1993. *American Apartheid: Segregation and the Making of the Underclass.* Cambridge, MA: Harvard University Press; Massey, Douglas S. 2005. *Strangers in a Strange Land: Humans in an Urbanizing World.* New York: Norton; Rugh, Jacob S., and Douglas S. Massey. 2010. "Racial Segregation and the American Foreclosure Crisis." *American Sociological Review* 75 (5): 629–651.

(U.S. Census Bureau 2012a). This is a whopping difference. Thus American Blacks (and Hispanics) are thrown a major disadvantage in occupational and particularly educational realms given this divide. It is not possible to exaggerate the importance of the computer and access to the Internet to achievement at all levels of education. This is especially true for college-bound students, who use Internet access for practice SAT and ACT tests and access to college and university officials, admissions officers, and coaches, to say nothing of access to library facilities and sources.

Racism

Racism includes both attitudes and behaviors. A negative attitude taken toward someone simply because he or she belongs to a racial or ethnic group is a prejudice, as has already been discussed. An attitude or prejudice is what you think and feel; a behavior is what you do. **Racism** is the *perception and treatment* of a racial or ethnic group, or member of that group, as intellectually,

socially, and culturally inferior to one's own group. It is more than an attitude; it is institutionalized in society. Racism involves negative attitudes that are sometimes linked with negative behavior.

There are different forms of racism. Researchers (for example, Bobo 2006, 1999; Hunter 2002) have often called obvious, overt racism, such as physical assaults, from beatings to lynchings, **old-fashioned racism**, or *traditional racism* (or *Jim Crow racism*). This form of racism has declined somewhat in our society since the 1950s, though it certainly has not disappeared (Parmelee 2001; Schumann et al. 1997). Instances of racially motivated lynchings still occur in the United States.

Racism can also be subtle, covert, and nonobvious; this is known as **aversive racism** (Kristof 2008; Jones 1997). Consistently avoiding interaction with someone of another race or ethnicity is an example of aversive racism. This form of racism is quite common and has remained at roughly the same level for more than forty years, with perhaps a slight increase (Dovidio in Kristof 2008; Bobo 2006; Dovidio and Gaertner 2005;

Gaertner and Dovidio 2005; Katz et al. 1986). Even when overt forms of racism dissipate, aversive racism tends to persist, because it is less visible than overt racism; people can believe racism has diminished when it has not (Gaertner and Dovidio 2005).

After the Second World War and during the 1950s, a shift to **laissez-faire racism** occurred. This type of racism—also called *symbolic racism* by some (Taylor et al. 2013; Bobo 2006; Bobo and Smith 1998)—involves several elements:

1. The subtle but persistent negative stereotyping of minorities, particularly Black Americans, especially in the media
2. A tendency to blame Blacks themselves for the gap between Blacks and Whites in socioeconomic standing, occupational achievement, and educational achievement
3. Clear resistance to meaningful policy efforts (such as affirmative action, discussed later) designed to ameliorate racially oppressive social conditions and practices in the United States

The last element is rooted in perceptions of threat to maintaining the status quo (Tarman and Sears 2005; Bobo 1999).

A close relative of laissez-faire racism is **color-blind racism**—so named because the individual affected by this type of racism prefers to ignore legitimate racial-ethnic, cultural, and other differences and insists that the race problems in the United States will go away if only race is ignored all together. Accompanying this belief is the opinion that race differences in the United States are merely an illusion and that race is not real. Simply refusing to perceive any differences at all between racial groups (thus being color-blind) is in itself a form of racism (Gallagher 2013; Bonilla-Silva and Baiocchi 2001). This will come as a surprise to many. These types of racism do not necessarily involve explicit or purposeful intent on the part of nonminority individuals to harm minority people.

Color-blindness thus hides what is called **White privilege** behind a mask: It allows Whites to define themselves as politically and racially tolerant as they proclaim adherence to a belief system that does not see or judge individuals by "the color of their skin." They think of skin color as irrelevant. It is not "irrelevant." This view tends to ignore the structured racial dominance in society—White privilege—with its falsely assumed meritocracy and belief that racial barriers have been dismantled. Much of White America now sees a level playing field, yet a majority of Blacks continue to perceive it as still quite uneven. People of any racial background can now wear low-slung hip-hop clothing, listen to gangsta rap music, and root for their favorite majority-Black athletic team. This gives the false impression that racial barriers have fallen, but in fact White status (White privilege) remains (Gallagher 2013; Kristof 2008).

Institutional racism as a form of racism is the negative treatment and oppression of one racial or ethnic group *by society's existing institutions* based on the presumed inferiority of the oppressed group. It is a form of racism that exists at the level of social structure and is in Durkheim's sense external to the individual—thus institutional. Key to understanding institutional racism is seeing that dominant groups have the economic and political power to subjugate the minority group, *even if they do not have the explicit intent* of being prejudiced or discriminating against others. Power, or lack thereof, accrues to groups because of their position in social institutions, not just because of individual attitudes or behavior. The power that resides in society's institutions can be seen in such patterns as persistent economic inequality between racial groups, which is reflected in high unemployment among minorities, lower wages, lower net worth, and different patterns of job placement (Oliver and Shapiro, 2006; Bobo 1999; Bonilla-Silva 1997). Even recently, Hispanics and Blacks tend to be disproportionately excluded from juries in certain locations, such as Los Angeles (Stevenson 2010).

Racial profiling is an example of institutional racism in the criminal justice system. African American and Hispanic people are arrested—and serve longer sentences—considerably more often than are Whites and Asians. In fact, an African American or Hispanic wrongdoer is more likely to be arrested than a White person who commits the exact same crime, even when the White person shares the same age, socioeconomic environment, and prior arrest record as the Black or Hispanic (Doermer and Demuth 2010).

Institutional racism is also seen in educational institutions, such as when schools assign Blacks and Latinos to the lower cognitive ability tracks than Whites with the same test scores, as will be seen in Chapter 14. Also, a fifty-year analysis of data reveals that racial, gender, and social class biases have also existed as part of the SAT itself since the early 1960s (Taylor forthcoming). In these instances, racism is a characteristic of

Minorities are more likely to be arrested than Whites for the same offense. Does this reflect institutional racism rather than any individual prejudice of the arresting police officer?

Mikael Karlsson/Alamy Limited

the institutions and not necessarily of the individuals within the institution. *This is why institutional racism can exist even without prejudice being the cause.*

Consider this: Even if every White person in the country lost all of his or her personal prejudices, and even if he or she stopped engaging in individual acts of discrimination, institutional racism would still persist for some time. Over the years, it has become so much a part of U.S. institutions (hence, the term *institutional racism*) that discrimination can occur even when no single person is deliberately causing it. Existing at the level of social structure instead of at the level of individual attitude or behavior, it is external to the individual personality and is thus a social fact of the sort sociological theorist Emile Durkheim observed (Chapter 1).

---

debunking SOCIETY'S MYTHS

MYTH: The primary cause of racial inequality in the United States is the persistence of prejudice.

SOCIOLOGICAL PERSPECTIVE: Prejudice is one dimension of racial problems in the United States, but institutional racism can flourish even while prejudice is on the decline. Prejudice is an attribute of the individual, whereas institutional racism is an attribute of social structure. ●

THEORIES OF PREJUDICE AND RACISM

Why do prejudice, discrimination, and racism exist? Two categories of theories have been advanced. The first category consists of psychological theories about prejudice. The second category consists of sociological theories of racism, including institutional racism as well as prejudice and discrimination.

Psychological Theories of Prejudice

Two traditional psychological theories of prejudice are the scapegoat theory and the theory of the authoritarian personality. **Scapegoat theory** argues that, historically, members of the dominant group in the United States have harbored various frustrations in their desire to achieve social and economic success (Feagin and Feagin 1993). As a result of this frustration, they vent their anger in the form of aggression. This aggression is directed toward some substitute that takes the place of the original perception of the frustration. Members of minority groups become these substitutes, that is, the scapegoats. The psychological theory that aggression often follows frustration (originally from the frustration–aggression hypothesis of Dollard et al. 1939) is central to the scapegoat principle. For example, a White person who perceived that he or she was denied a job because "too many" Mexican immigrants were being permitted to enter the country would be using Mexican Americans as a scapegoat. He or she may feel negatively (thus prejudiced) toward a specific Mexican American person, even if that person did not have the job in question and had nothing at all to do with the White person not getting the job.

The second theory, an older one, argues that individuals who possess an authoritarian personality are more likely to be prejudiced against minorities than are nonauthoritarian individuals. The **authoritarian personality** (a term coined by Adorno et al. 1950) is characterized by a tendency to rigidly categorize other people, as well as inclinations to submit to authority, strictly conform, be very intolerant of ambiguity, and be inclined toward superstition. The authoritarian person is more likely to stereotype or categorize another and thus readily places members of minority groups into convenient and oversimplified categories or stereotypes. There is research that links strong authoritarianism with high racial–ethnic prejudice, and also high religious orthodoxy and extreme varieties of political conservatism (Bobo and Kluegel 1991; Altemeyer 1988).

Sociological Theories of Prejudice and Racism

Current sociological theory focuses more on explaining the existence of racism, particularly institutional racism as well as other forms of racism, although speculation about the existence of prejudice is also a component. The three sociological theoretical perspectives considered throughout this text have bearing on the study of racism, discrimination, and prejudice: functionalist theory, symbolic interaction theory, and conflict theory.

Functionalist Theory. In a now outdated theory, functionalists argue that for race and ethnic relations to be functional and thus contribute to the harmonious conduct and stability of society, racial and ethnic minorities must assimilate into that society. **Assimilation** is a process by which a minority becomes socially, economically, and culturally absorbed within the dominant society. The assimilation perspective assumes that to become fully fledged members of society, minority groups must adopt as much of the dominant society's culture as possible, particularly its language, mannerisms, and goals for success, and thus give up much of its own culture. Assimilationism stands in contrast to racial cultural **pluralism**—the maintenance and persistence of one's culture, language, mannerisms, practices, art, and so on.

Symbolic Interaction Theory. Symbolic interaction theory addresses two issues: first, the role of social interaction in reducing racial and ethnic hostility, and second, how race and ethnicity are socially constructed.

Symbolic interactionism asks: What happens when two people of different racial or ethnic origins come into contact with each other, and how can such interracial or interethnic contact reduce hostility and conflict? **Contact theory**, which originated with the psychologist Gordon Allport (Allport 1954; Cook 1988), argues that interaction and contact between two groups will reduce prejudice within both groups—but only if three conditions are met:

1. The contact must be between individuals of equal status; the parties must interact on equal ground. A Hispanic cleaning woman and a wealthier White woman who employs her may interact, but their interaction will not reduce prejudice. Instead, their interaction is more likely to perpetuate stereotypes and prejudices on the part of both.
2. The contact between equals must be sustained; short-term contact will not decrease prejudice.
3. Social norms favoring equality must be agreed upon by the participants. Having African Americans and White skinheads interact on a TV talk show, such as the *Jerry Springer Show*, will probably not decrease prejudice; this interaction might well increase it.

Conflict Theory The basic premise of conflict theory is that class-based conflict is an inherent and fundamental part of social interaction. To the extent that racial and ethnic conflict is tied to class conflict, conflict theorists argue that class inequality must be reduced to lessen racial and ethnic conflict in society.

The current "class versus race" controversy in sociology (reviewed in more detail later in this chapter and also in Chapter 8) concerns the question of whether class (namely, economic differences between races) or race is more important in explaining inequality and its consequences or whether they are of equal importance. Those focusing primarily upon class conflict, such as sociologist William Julius Wilson (1996, 1987, 1978), have argued that class and changes in the economic structure are sometimes more important than race in shaping the life chances of different groups. Wilson argues that being disadvantaged in the United States is more a matter of class, although he sees this clearly linked to race. Sociologists focusing primarily on the role of race (Bonilla-Silva and Baiocchi 2001; Feagin 2000; Bonilla-Silva 1997; Willie 1979) argue the opposite: They say that race has been and continues to be more important than class—though class is still important—in explaining and accounting for inequality and conflict in society, and that directly addressing the question of race forthrightly is the only way to solve the country's race problems (see Table 10.2). Wilson has consistently argued, however, that group race differences are clearly causally related to class differences, and that, in addition, race has an effect independent of class.

The "class versus race" controversy is presently giving way to a recent variety of the conflict perspective, called the **intersection perspective**. This perspective argues that both class and race have separate (main) effects as well as combined, or "intersecting," effects of racism, classism (elitism), and also sexism in the oppression of people. Intersection theory posits that any person is socially located in a position that simultaneously involves race, class, and gender, and thus looking at only one of them to explain their status (even with another held constant) is incomplete. This perspective notes that not only are the effects of gender and race intertwined, but also both are intertwined with the effects of class. Class, along with race and gender, are integral components of social structure, according to the intersection perspective (Andersen and Collins 2013; Collins 1990, 1998).

DIVERSE GROUPS, DIVERSE HISTORIES

The different racial and ethnic groups in the United States have arrived at their current social condition through histories that are similar in some ways, yet quite different

table 10.2 Comparing Sociological Theories of Race and Ethnicity

	Functionalism	Conflict Theory	Symbolic Interaction
The Racial Order	Has social stability when diverse racial and ethnic groups are assimilated into society	Is intricately intertwined with class stratification of racial and ethnic groups	Is based on social construction that assigns groups of people to diverse racial and ethnic categories
Minority Groups	Are assimilated into dominant culture as they adopt cultural practices and beliefs of the dominant group	Have life chances that result from the opportunities formed by the intersection of class, race, and gender	Form identity as the result of sociohistorical change
Social Change	Is a slow and gradual process as groups adapt to the social system	Is the result of organized social movements and other forms of resistance to oppression	Is dependent on the different forms of social interaction that characterize intergroup relations

© Cengage Learning

in other respects. Their histories are related because of a common experience of White supremacy, economic exploitation, and political disenfranchisement.

A historical perspective on each group follows, which will aid in understanding how prejudice, discrimination, and racism have operated throughout the history of U.S. society.

Native Americans: The First of This Land

The exact size of the indigenous population in North America at the time of the Europeans' arrival with Columbus in 1492 has been estimated at anywhere from one million to ten million people. Native Americans were here tens of thousands of years before they were "discovered" by Europeans. Discovery quickly turned to conquest, and in the course of the next three centuries, the Europeans systematically drove the Native Americans from their lands, destroying their ways of life and crushing various tribal cultures. Native Americans were subjected to an onslaught of European diseases. Lacking immunity to these diseases, Native Americans suffered a population decline, considered by some to have been the steepest and most drastic of any people in the history of the world. Native American traditions have survived in many isolated places, but what is left is only an echo of the original 500 nations of North America (Snipp 2007, 1989; Nagel 1996; Thornton 1987).

At the time of European contact in the 1640s, there was great linguistic, religious, governmental, and economic heterogeneity among Native American tribes.

The opening of the National Museum of the American Indian (part of the Smithsonian Institution in Washington, DC) was cause for celebration among diverse groups of Native Americans, as well as others.

Jason Reed/Reuters/Landov

Most historical accounts have underestimated the degree of cultural and social variety, however. Between the arrival of Columbus in the Caribbean in 1492 and the establishment of the first thirteen colonies in North America in the early 1600s, the ravages of disease and the encroachment of Europeans caused a considerable degree of social disorganization. Sketchy accounts of American Indian cultures by early colonists, fur traders, missionaries, and explorers underestimated the great social heterogeneity among the various American Indian tribal groups and the devastating effects of the European arrival on Indian society.

By 1800, the number of Native Americans had been reduced to a mere 600,000, and wars of extermination against the Indians were being conducted in earnest. Fifty years later, the population had fallen by another half. Indians were killed defending their land, or they died of hunger and disease when taking refuge in inhospitable country. In 1834, 4000 Cherokee died on a forced march from their homeland in Georgia to reservations in Arkansas and Oklahoma, a trip memorialized as the Trail of Tears. The Sioux (a federation of American Indian tribes) were forced off their lands by the discovery of gold and the new push of European immigration. Their reservation was established in 1889 (quite recently in history), and they were designated as wards, subjecting them to capricious and humiliating governmental policies. The following year, the U.S. Army mistook Sioux ceremonial dances for war dances and moved in to arrest the leaders. A standoff exploded into violence, during which federal troops killed 200 Sioux men, women, and children at the infamous Wounded Knee massacre.

Today, about 55 percent of all Native Americans live on or near a reservation, which is land set aside by the U.S. government for their exclusive use. The other 45 percent live in or near urban areas (U.S. Census Bureau 2012a; Snipp 1989). The reservation system has served the Indians poorly. Many Native Americans now live in conditions of abject poverty, deprivation, and alcoholism, and they suffer massive unemployment (currently more than 50 percent among males—extraordinarily high). They are at the lowest rung of the socioeconomic ladder, with the highest poverty rate. The first here in this land are now last in status, a painful irony of U.S. history.

African Americans

The development of slavery in the Americas is related to the development of world markets for sugar and tobacco. Slaves were imported from Africa to provide the labor for sugar and tobacco production and to enhance the profits of the slaveholders. It is estimated that somewhere between twenty and one hundred million Africans were transported under appalling conditions to the Americas—38 percent went to Brazil; 50 percent to the Caribbean; 6 percent to Dutch, Danish, and Swedish colonies; and only 6 percent to

MAP 10.1

Mapping America's Diversity: Foreign Born Population

This map shows the total number of foreign-born residents per state. Foreign-born includes legal residents (immigrants), temporary migrants (such as students), refugees, and illegal immigrants—to the extent they can be known. Some states have a high number of foreign-born (for example, California, Florida, New York), and other states have fewer (for example, Wyoming, South Dakota, and Vermont).

Where does your region fall in the number of foreign-born to the total population?

Data: U.S. Census Bureau. 2010. *2005–2009 American Community Survey*. ACS Maps. **www.census.gov**

Percent of Total Population Foreign Born

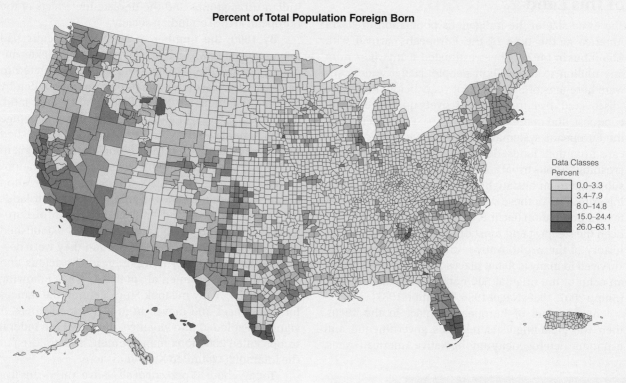

Data Classes
Percent
0.0–3.3
3.4–7.9
8.0–14.8
15.0–24.4
26.0–63.1

the United States (Blassingame 1973; Genovese 1972; Jordan 1969).

Slavery evolved as a form of stratification called a **caste system**, one in which one caste, the slaveholders, profited from the labor of another caste, the slaves. Central to the operation of slavery was the principle that human beings could be chattel (or property). As an economic institution, slavery was based on the belief that Whites are superior to other races, coupled with a belief in a patriarchal social order. The social distinctions maintained between Whites and Blacks were caste-like, with rigid categorization and prohibitions, rather than merely class-like, which suggests more pliant social demarcations. Vestiges of this caste system remain in the United States to this day.

The slave system also involved the domination of men over women—another aspect of the caste system. In this combination of patriarchy and White supremacy, White males presided over their "property" of White women as well as their "property" of Black men and women. This in turn led to gender stratification among the slaves themselves, which reflected the White slaveholder's assumptions about the relative roles of men and women. Black women performed domestic labor for their masters and their own families. White men further exerted their authority in demanding sexual relations with Black women (White 1985; Davis 1981; Raboteau 1978; Blassingame 1973). The predominant attitude of Whites toward Blacks was paternalistic. Whites saw slaves as childlike and incapable of caring for themselves. The stereotypes of African Americans as "childlike" are directly traceable to the system of slavery.

There exists a widespread belief that slaves passively accepted slavery. Scholarship shows this to be false. Instead, the slaves struggled to preserve both their culture and their sense of humanity and to resist, often by open conflict, the dehumanizing effects of a system that defined human beings as mere property (Myers 1998; Blassingame 1973). Slaves revolted against the conditions of enslavement in a variety of ways, from passive means such as work slowdowns and feigned illness to more aggressive means such as destruction of property, escapes, and outright rebellion.

After slavery was presumably ended by the Civil War (1861–1865) and the Emancipation Proclamation

(1863), Black Americans continued to be exploited for their labor. In the South, the system of sharecropping emerged, an exploitative system in which Black families tilled the fields for White landowners in exchange for a share of the crop. With the onset of the First World War and the intensified industrialization of society came the Great Migration of Blacks from the South to the urban North. This massive movement, lasting from the late 1800s through the 1920s, significantly affected the status of Blacks in society because there was now a greater potential for collective action (Marks 1989).

The effect of the phenomenon of migration in general upon the degree of oppression of American Blacks is so strong that one recent researcher (Berlin 2010) identifies four important massive migrations involving Blacks:

1. The horrific trans-Atlantic passage that brought slavery to North America in the seventeenth and eighteenth centuries. With this "middle passage," where Blacks were packed in dreary holds of the slave ships like sardines in a can, came shackles, branding irons, smells of urine and feces, disease, hopelessness, and suicide. Typically, little more than half of the Black Africans aboard the slave ships survived and then only to have their families torn asunder and be sold into a social structure that brought the lifelong oppression of slavery in the United States.

2. The forced movement of a million slaves from the East Coast to the inland South's cotton kingdom in the early 1800s

3. As just noted, the Great Black Migration of six million Blacks from the South to the urban North

4. The current influx of migrants from Africa, South America, and the Caribbean, a movement so large that it accounted for a quarter of Black America's population growth in the last decade of the twentieth century

An interesting footnote to the story of Black internal migration in the United States is the "call to home" phase identified by demographic researchers Hunt and colleagues (2012). This research shows that between 1970 and 2000, Black migration from North to South increased and exceeded White migration from North to South. The data further show that Black migrants favored Southern states where Blacks were the relatively larger share of the state's population.

In the early part of the century, the formation of Black ghettos had a dual effect. It victimized Black Americans with grim urban conditions and encouraged the development of Black resources, including volunteer organizations, settlement houses, social movements, political action groups, and artistic and cultural achievements. During the 1920s, Harlem in New York City became an important intellectual and artistic oasis for Black America—as did organizations and settlement houses at the time in other northern cities, such as the famed Karamu House in Cleveland, Ohio.

The Harlem Renaissance gave the nation great literary figures, such as Langston Hughes, Jessie Fauset, Alain Locke, Arna Bontemps, Zora Neale Hurston, Countee Cullen, Wallace Thurman, and Nella Larsen (Marks and Edkins 1999; Gates et al. 1997; Rampersad 1988, 1986; Bontemps 1972). At the same time, many of America's greatest musicians, entertainers, and artists came to the fore, such as musicians Duke Ellington, Count Basie, Benny Carter, Billie Holiday, Cab Calloway, Louis Armstrong, and painters Hale Woodruff and Elmer Brown. The end of the 1920s and the stock market crash of 1929 brought everyone down a peg or two, Whites as well as Blacks, although in the words of Harlem Renaissance writer Langston Hughes, Black Americans at the time "had but a few pegs to fall" (Hughes 1967).

Latinos

Latinos include Chicanos and Chicanas, Mexican Americans, Puerto Ricans, Cubans, and other recent Latin American immigrants to the United States. They also include Latin Americans who have lived for generations in the United States; many are not immigrants but very early settlers from Spain and Portugal in the 1400s. The population of Latinos has grown considerably over the past few decades, with the largest increase among Mexican Americans. The terms *Hispanic* and *Latino* or *Latina* mask the great structural and cultural diversity among the various Hispanic groups. Diverse Latino groups have been forced by institutional procedures to cause the public to perceive them "as one," for example, by the media, by political leaders, and by the U.S. census system of categorization (Mora 2009).

The use of such inclusive terms also tends to ignore important differences in their respective entries into U.S. society: Chicanos/as through military conquest of the Mexican–American War (1846–1848); Puerto Ricans through war with Spain in the Spanish–American War (1898); and Cubans as political refugees fleeing since 1959 from the Communist dictatorship of Fidel Castro, which the U.S. government vigorously opposed (Telles et al. 2011; Glenn 2002; Bean and Tienda 1987).

Mexican Americans. Before the Anglo (White) conquest, Mexican colonists had formed settlements and missions throughout the West and Southwest. In 1834, the U.S. government ordered the dismantling of these missions, bringing them under tight governmental control and creating a period known as the golden age of the ranchos. Land then became concentrated into the hands of a few wealthy Mexican ranchers, who had been given large land grants by the Mexican government. This economy created a class system within the Chicano community, consisting of the elite ranchers, mission farmers, and government administrators at the top; *mestizos*, who were small farmers and ranchers, as the middle class; a third class of skilled workers; and a bottom class of manual laborers, who were mostly Indians (Maldonado 1997; Mirandé 1985).

With the Mexican–American War of 1846–1848, Chicanos lost claims to huge land areas that ultimately became Texas, New Mexico, and parts of Colorado, Arizona, Nevada, Utah, and California. White cattle ranchers and sheep ranchers enclosed giant tracts of land, thus cutting off many small ranchers, both Mexican and Anglo. Thus began a process of wholesale economic and social exclusion of Mexicans and Mexican Americans from U.S. society, much of which continues to this day, generation after generation (Telles and Ortiz 2008).

It was at this time that Mexicans, as well as early U.S. settlers of Mexican descent, became defined as an inferior race that did not deserve social, educational, or political equality. This is an example of the *racial formation process,* as noted earlier in this chapter (Omi and Winant 1994). Anglos believed that Mexicans were lazy, corrupt, and cowardly, which launched stereotypes that would further oppress Mexicans; these stereotypes were used to justify the lower status of Mexicans and Anglo control of the land that Mexicans were presumed to be incapable of managing (Telles and Ortiz 2008; Moore 1976). As has been noted several times in this chapter, stereotyping has been used in this society as a way of falsely explaining and justifying the lower social status of society's minorities.

During the twentieth century, advances in agricultural technology changed the organization of labor in the Southwest and West. Irrigation allowed year-round production of crops and a new need for cheap labor to work in the fields. Migrant workers from Mexico were exploited as a cheap source of labor. Migrant work was characterized by low earnings, poor housing conditions, poor health, and extensive use of child labor. The wide use of Mexican migrant workers as field workers, domestic servants, and other kinds of poorly paid work continues, particularly in the Southwest (Amott and Matthaei 1996), but now throughout the United States. Barriers to equal educational opportunity for Mexican Americans, such as culturally and linguistically biased standardized tests, continue the process of exclusion of Mexican Americans (Taylor forthcoming; Telles et al. 2011; Telles and Ortiz 2008).

Puerto Ricans. The island of Puerto Rico was ceded to the United States by Spain in 1898. In 1917, the Jones Act extended U.S. citizenship to Puerto Ricans, although it was not until 1948 that Puerto Ricans were allowed to elect their own governor. In 1952, the United States established the Commonwealth of Puerto Rico, with its own constitution. Following the Second World War, the first elected governor launched a program known as Operation Bootstrap, which was designed to attract large U.S. corporations to the island of Puerto Rico by using tax breaks and other concessions. This program contributed to rapid overall growth in the Puerto Rican

Activities such as this Puerto Rican Day Parade in New York City reflect pride in one's group culture and result in greater cohesiveness of the group.

Michel Friang/Alamy Limited

economy, although unemployment remained high and wages remained low. Seeking opportunity, unemployed farmworkers began migrating to the United States. These migrants were interested in seasonal work, and thus a pattern of temporary migration characterized the Puerto Ricans' entrance into the United States (Amott and Matthaei 1996; Rodriguez 1989).

Unemployment in Puerto Rico became so severe that the U.S. government even went so far as to attempt a reduction in the population by some form of population control. Pharmaceutical companies experimented with Puerto Rican women in developing contraceptive pills, and the U.S. government actually encouraged the sterilization of Puerto Rican women. One source notes that more than 37 percent of the women of reproductive age in Puerto Rico had been sterilized by 1974 (Roberts 1997). More than one-third of these women have since indicated that they regret sterilization because they were not made aware at the time that the procedure was irreversible.

Cubans. Cuban migration to the United States is recent in comparison with the many other Hispanic groups. The largest migration has occurred since the revolution led by Fidel Castro in 1959; between then and 1980, more than 800,000 Cubans—one-tenth of the entire island population—migrated to the United States. The U.S. government defined this as a political exodus, facilitating the early entrance and acceptance of these migrants. Many of the first migrants had been middle- and upper-class professionals and landowners under the prior dictatorship of Fulgencio Batista, but they had lost their land during the Castro revolution. In exile in the United States, some worked to overthrow Castro, often with the support of the federal government. Yet many other Cuban immigrants were of modest means and, like other immigrant groups, came seeking freedom from political and social persecution and escape from poverty.

The most recent wave of Cuban immigration came in 1980, when the Cuban government, still under Castro, opened the port of Mariel to anyone who wanted to leave Cuba. In the five months following this action, 125,000 Cubans came to the United States—more than the combined total for the preceding eight years. The arrival of people from Mariel has produced debate and tension, particularly in Florida, a major center of Cuban migration. The Cuban government had previously labeled the people fleeing from Mariel as "undesirable"; some had been incarcerated in Cuba before leaving. They were actually not much different from previous refugees such as the "golden exiles," who were professional and high-status refugees (Portes and Rumbaut 1996). But because the refugees escaping from Mariel had been labeled (stereotyped) as undesirables, and because they were forced to live in primitive camps for long periods after their arrival, they have been unable to achieve much social and economic mobility in the United States—thus ironically reinforcing the initial perception that they were "lazy" and "undesirable." In contrast, the earlier Cuban migrants, who were on average more educated and much more settled, have enjoyed a fair degree of success (Portes and Rumbaut 2001, 1996; Amott and Matthei 1996; Pedraza 1996).

Asian Americans

Like Hispanic Americans, Asian Americans are from many different countries and diverse cultural backgrounds; they cannot be classified as the single cultural rubric of Asians. Asian Americans include migrants from China, Japan, the Philippines, Korea, and Vietnam, as well as more recent immigrants from Cambodia and Laos.

Chinese. Attracted by the U.S. demand for labor, Chinese Americans began migrating to the United States during the mid-nineteenth century. In the early stages of this migration, the Chinese were tolerated because they provided cheap labor. They were initially seen as good, quiet citizens, but racial stereotypes turned hostile when the Chinese came to be seen as competing with White California gold miners for jobs. Thousands of Chinese laborers worked for the Central Pacific Railroad from 1865 to 1868. They were relegated to the most difficult and dangerous work, worked longer hours than the White laborers, and for a long time were paid considerably less than the White workers.

The Chinese were virtually expelled from railroad work near the turn of the twentieth century (in 1890–1900) and settled in rural areas throughout the western states. As a consequence, anti-Chinese sentiment and prejudice ran high in the West. This ethnic antagonism was largely the result of competition between the White and Chinese laborers for scarce jobs. In 1882, the federal government passed the Chinese Exclusion Act, which banned further immigration of unskilled Chinese laborers. Like African Americans, the Chinese and Chinese Americans were legally excluded from intermarriage with Whites (Takaki 1989). The passage of this openly racist act, which was preceded by extensive violence toward the Chinese, drove the Chinese populations from the rural areas into the urban areas of the West. During this period, several Chinatowns were established by those who had been forcibly uprooted and who found strength and comfort within enclaves of Chinese people and culture (Nee 1973).

Japanese. Japanese immigration to the United States took place mainly between 1890 and 1924, after which passage of the Japanese Immigration Act forbade further immigration. Most of these first-generation immigrants, called *issei*, were employed in agriculture or in small Japanese businesses. Many issei were from farming families and wished to acquire their own land, but in 1913, the Alien Land Law of California stipulated that Japanese aliens could lease land for only three years and that lands already owned or leased by them could not be bequeathed to heirs. The second generation of Japanese Americans, or *nisei*, were born in the United States of Japanese-born parents. They became better educated than their parents, lost their Japanese accents, and in general became more "Americanized," that is, culturally assimilated. The third generation, called *sansei*, became even better educated and assimilated, yet still met with prejudice and discrimination, particularly where Japanese Americans were present in the highest concentrations, as on the West Coast from Washington to southern California (Takaki 1989; Glenn 1986).

The Japanese suffered the complete indignity of having their loyalty questioned when the federal government, thinking they would side with Japan after the Japanese attack on Pearl Harbor in December 1941, herded them into concentration camps. By executive order of President Franklin D. Roosevelt, much of the West Coast Japanese American population (many of them loyal second- and third-generation Americans) had their assets frozen and their real estate confiscated by the government. A media campaign immediately followed, labeling Japanese Americans "traitors" and "enemy aliens." Virtually all Japanese Americans in the

During World War II, Japanese Americans, who were full American citizens, were forced into concentration camps. A noon food ("mess") line at one of these camps, Manzanar, is shown here.

United States had been removed from their homes by August 1942, and some were forced to stay in relocation camps until as late as 1946. Relocation destroyed numerous Japanese families and ruined them financially (Takaki 1989; Glenn 1986; Kitano 1976).

In 1986, the U.S. Supreme Court allowed Japanese Americans the right to file suit for monetary reparations. In 1987, legislation was passed, awarding $20,000 to each person who had been relocated and offering an official apology from the U.S. government. One is motivated to contemplate how far this paltry sum and late apology could go in righting what many have argued was the "greatest mistake" the United States has ever made as a government.

Filipinos. The Philippine Islands in the Pacific Ocean fell under U.S. rule in 1899 as a result of the Spanish–American War, and for a while Filipinos could enter the United States freely. By 1934, the islands became a commonwealth of the United States, and immigration quotas were imposed on Filipinos. More than 200,000 Filipinos immigrated to the United States between 1966 and 1980, settling in major urban centers on the West and East Coasts. More than two-thirds of those arriving were professional workers; their high average levels of education and skill have eased their assimilation. By 1985, more than one million Filipinos were in the United States. Within the next thirty years, demographers project that this population will become one of the largest groups of Asian Americans in the United States (Winnick 1990).

Koreans. Many Koreans entered the United States in the late 1960s, after amendments to the immigration laws in 1965 raised the limit on immigration from the Eastern Hemisphere. The largest concentration of Koreans is in Los Angeles. As much as half of the adult Korean American population is college educated, an exceptionally high proportion. Many of the immigrants were successful professionals in Korea; upon arrival in the United States, though, they have been forced to take on menial jobs, thus experiencing downward social mobility and status inconsistency. This is especially true of those Koreans who migrated to the East Coast. However, nearly one in eight Koreans in the United States today owns a business; many own small greengrocer businesses. Many of these stores are located in predominantly African American communities and have become one among several sources of ongoing conflict between some African Americans and Koreans. This has fanned negative feeling and prejudice on both sides—among Koreans against African Americans and among African Americans against Koreans (Chen 1991).

Vietnamese. Among the more recent groups of Asians to enter the United States have been the South Vietnamese, who began arriving following the fall of South Vietnam to the Communist North Vietnamese at the end of the Vietnam War in 1975. These immigrants, many of them refugees who fled for their lives, numbered about 650,000 in the United States in 1975. About one-third of the refugees settled in California. Many faced prejudice and hostility, resulting in part from the same perception that has dogged many immigrant groups before them—that they were in competition for scarce jobs. A second wave of Vietnamese immigrants arrived after China attacked Vietnam in 1978. As many as 725,000 arrived in the United States, only to face discrimination in a variety of locations. Tensions became especially heated when the Vietnamese became a substantial competitive presence in the fishing and shrimping industries in the Gulf of Mexico on the Texas shore. Since that time, however, many communities have welcomed them, and many Vietnamese heads of households have become employed full time (Winnick 1990; Kim 1993).

Middle Easterners

Since the mid-1970s, immigrants from the Middle East have been arriving in the United States. They have come from countries such as Syria, Lebanon, Egypt, and Iran, and more recently, especially Iraq. Contrary to popular belief, the immigrants speak no single language and follow no singular religion and thus are ethnically diverse. Some are Catholic, some are Coptic Christian, and many are Muslim. Many are from working-class backgrounds, but many were professionals—teachers, engineers, scientists, and other such positions—in their homelands. About 65 percent of those residing in this country were born outside the United States; about half are college-educated (Kohut 2007). Like immigrant populations before them, Middle Easterners have formed their own ethnic enclaves in the cities and suburbs of this country as they pursue the often elusive American dream (Abrahamson 2006).

Since the terrorist attacks on the World Trade Center and the Pentagon on September 11, 2001, many male Middle Easterners of several nationalities have become unjustly suspect in this country and are subjected to severe harassment; racially motivated physical attacks; and as already noted, out-and-out racial profiling, if only because they had dark skin and—as with some—wore a turban of some sort on their heads. Most of these individuals, of course, probably have no discernible connection at all with the terrorists. A survey shows that most Muslims in this country believe that the September 11, 2001, terrorist attacks were indeed the cause of the increased racial harassment and violence against them (Kohut 2007). Finally, the U.S. wars with Iraq and Afghanistan have not helped in easing tensions between White Americans and Middle Easterners generally because, as already noted, a majority of White Americans tend to lump Iraqis and Afghanis together with other Middle Easterners.

White Ethnic Groups

The story of White ethnic groups in the United States begins during the colonial period. White Anglo-Saxon Protestants (WASPs), who were originally immigrants from England and to some extent Scotland and Wales, settled in the New World (what is now North America). They were the first ethnic group to come into contact on a large scale with those people already here—namely, Native American Indians. WASPs came to dominate the newly emerging society earlier than any other White ethnic group.

In the late 1700s, the WASPs regarded the later immigrants from Germany and France as foreigners with odd languages, accents, and customs, and derogatory labels (*krauts, frogs*) were applied to them. Tension between the "old stock" and the "foreigners" continued through the Civil War era until around 1860, when the national origins of U.S. immigrants began to change (Handlin 1951). Of all racial and ethnic groups in the United States during that time and since, only WASPs (White Anglo-Saxon Protestants) do not think of themselves as a nationality. The WASPs came to think of themselves as the "original" Americans despite the prior presence of Native American Indians, who the WASPs in turn described and stereotyped as savages. As immigrants from northern, western, eastern, and southern Europe began to arrive, particularly during the mid- to late-nineteenth century, WASPs began to direct prejudice and discrimination against many of these newer groups. Long discriminated against by the male WASP establishment, women began to assert social and political power, challenging the power of male WASPs in the United States. Much of that WASP dominance remains, however, as is evident in their popular use of the terms *race* and *ethnicity* to describe virtually everyone but themselves (Andersen and Collins 2013).

There were two waves of migration of White ethnic groups in the mid- and late-nineteenth century. The first stretched from about 1850 through 1880, and included northern and western Europeans: English, Irish, Germans, French, and Scandinavians. The second wave of immigration occurred from 1890 to 1914, and included eastern and southern European populations: Italians, Greeks, Poles, Russians, and other eastern Europeans, in addition to more Irish. The immigration of Jews to the United States extended for well over a century, but the majority of Jewish immigrants came to the United States during the period from 1880 to 1920.

The Irish arrived in large numbers in the mid-nineteenth century and after as a consequence of food shortages and massive starvation in Ireland. During the latter half of the nineteenth century and in the early twentieth century, the Irish in the United States were abused, attacked, and viciously stereotyped. It is instructive to remember that the Irish, particularly on the East Coast and especially in Boston, underwent a period of ethnic oppression of extraordinary magnitude. A frequently seen sign posted in Boston saloons during that time proclaimed "No dogs or Irish allowed." The sign was not intended as a joke. German immigrants were similarly stereotyped, as were the French and the Scandinavians. It is easy to forget that virtually all immigrant groups have gone through times of oppression and prejudice, although these periods were considerably longer for some groups than for others. As a rule, where the population density of an ethnic group in a town, city, or region was greatest, so too was the amount of prejudice, negative stereotyping, and discrimination to which that group was subjected.

Jewish immigrants were questioned, sometimes brutally, at Ellis Island, the point of entry to the United States for many early European immigrants.

More than 40 percent of the world's Jewish population lives in the United States, making it the largest community of Jews in the world. Most of the Jews in the United States arrived between 1880 and the First World War, originating from the Eastern European countries of Russia, Poland, Lithuania, Hungary, and Romania. Jews from Germany arrived in two phases; the first wave came just prior to the arrival of those from Eastern Europe, and the second came as a result of Hitler's ascension to power in Germany during the late 1930s. Because many German Jews were professionals who also spoke English, they assimilated more rapidly than those from the Eastern European countries. Jews from both parts of Europe underwent lengthy periods of anti-Jewish prejudice, **anti-Semitism** (defined as the hatred of Jewish people), and discrimination, particularly on Manhattan's Lower East Side. Significant anti-Semitism still exists in the United States (Ferber 1999; Essed 1991; Simpson and Yinger 1985).

In 1924, the National Origins Quota Act was passed, one of the most discriminatory legal actions ever taken by the United States in the area of immigration. By this act, the first real establishment of *ethnic quotas* in the United States, immigrants were permitted to enter the country only in proportion to their numbers already existing in the United States. Thus ethnic groups who were already here in relatively high proportions (English, Germans, French, Scandinavians, and others, mostly western and northern Europeans) were allowed to immigrate in greater numbers than were those from southern and eastern Europe, such as Italians, Poles, Greeks, and other eastern Europeans. Hence, the act discriminated against southern and eastern Europeans in favor of western and northern Europeans. It has been noted that the European groups who were discriminated against by the National Origins Quota Act tended to be those with darker skins on average, even though they were White and European.

Immigrants during this period were subject to literacy tests and even IQ tests given in English (Kamin 1974). The act barred anyone who was classified as a convict, lunatic, "idiot," or "imbecile" from immigration. On New York City's Ellis Island, non-English-speaking immigrants, many of them Jews, were given the 1916 version of the Stanford-Binet IQ test in English. Obviously, non-English-speaking people taking this test were unlikely to score high. On the basis of this grossly biased test, governmental psychologist H. H. Goddard classified fully 83 percent of Jews, 80 percent of Hungarians, and 79 percent of Italians as "feebleminded." It did not dawn on Goddard or the U.S. government that the IQ test, in English, probably did not measure something called intelligence, as intended, but instead simply measured the immigrant's mastery of the English language (Taylor forthcoming; 1980; Gould 1999; Kamin 1974).

ATTAINING RACIAL AND ETHNIC EQUALITY: THE CHALLENGE

Race and ethnic relations in the United States have posed a major challenge for the nation, one that is becoming even more complex as the racial–ethnic population becomes more diverse. Even as the nation elected its first African American president in 2008, President Barack Obama, racial inequalities persisted almost unchanged. Intergroup contact has been both negative and positive, obvious and subtle, tragic and helpful. How can the nation respond to its new diversity as well as to the issues faced by racial and ethnic minorities that have been present since the nation's founding? This question engages significant sociological thought and attention to the nation's record of social change with regard to race and ethnic groups.

The White Immigrants Made It: Why Can't They?

Many Americans believe that with enough hard work and loyalty to the dominant White culture of the country, any minority can make it and thus "assimilate" into American society. Quite a few older members of racial and ethnic minority groups still adhere to this belief. It is the often heard argument that African Americans, Hispanics, and Native Americans need only to pull themselves up "by their own bootstraps" to become a success.

thinking SOCIOLOGICALLY

Write down your own racial–ethnic background and list one thing that people from this background have positively contributed to U.S. society or culture. Also list one experience (current or historical) in which people from your group have been victimized by society. Discuss how these two things illustrate the fact that racial–ethnic groups have both been victimized and have made positive contributions to this society. Share your comments with others: What does this reveal to you about the connections between different groups of people and their experiences as racial–ethnic groups in the United States? ●

This *assimilation perspective* dominated sociological thinking a generation ago and is still prominent in U.S. thought (Telles and Ortiz 2008; Alba and Nee 2003; Portes and Rumbaut 2001; Rumbaut 1996b; Glazer 1970). The assimilationist believes that to overcome adversity and oppression, the minority person need only imitate the dominant White culture as much as possible. In this sense, minorities must assimilate "into" White culture and White society. The general assumption is that with each new generation, assimilation becomes more and more likely. But one of the questions asked in this perspective is to what extent groups can maintain some of their distinct cultural values and still be incorporated into the society to which they have moved. One could argue, for example, that the Irish have been able to assimilate quite fully into American culture while still maintaining an ethnic identity—one that is particularly salient around St. Patrick's Day!

Many Asian American groups have followed this pattern and have thus been called by some the "model minority," but this label ignores the fact that Asians are still subject to considerable prejudice, discrimination, racism, and poverty (Ngai 2012; Woo 1998; Lee 1996; Takaki 1989).

There are problems with the assimilation model. First, it fails to consider the time that it takes certain groups to assimilate. Those from rural backgrounds (Native Americans, Hispanics, African Americans, White Appalachians, and some White ethnic immigrants) typically take much longer to assimilate than those from urban backgrounds.

Second, the histories of Black and White arrivals are very different, with lasting consequences. Whites came voluntarily; Blacks arrived in chains. Whites sought relatives in the New World; Blacks were sold and separated from close relatives. For these and other reasons, the experiences of African Americans and Whites as newcomers can hardly be compared, and their assimilation is unlikely to follow the same course.

Third, although White ethnic groups did indeed face prejudice and discrimination when they arrived in America, many entered at a time when the economy was growing rapidly and their labor was in high demand. Thus they were able to attain education and job skills. In contrast, by the time Blacks arrived during the Great Migration to northern industrial areas from the rural South, Whites had already established firm control over labor and used this control to exclude Blacks from better-paying jobs and higher education.

Fourth, assimilation is more difficult for people of color because skin color is an especially salient characteristic, ascribed and relatively unchangeable. White ethnic group members can change their names, which many did (for example, from Levine to Lane; from Bellitto to Bell; many other examples exist), but people of color cannot easily change their skin color.

The assimilation model raises the question of whether it is possible for a society to maintain cultural pluralism, which is defined as different groups in society maintaining their distinctive cultures, while also coexisting peacefully with the dominant group. Some groups have explicitly practiced cultural pluralism: The Amish people of Lancaster County in Pennsylvania and of north-central Ohio—who travel by horse and buggy; use no electricity; and run their own schools, banks, and stores—constitute a good example of a relatively complete degree of cultural pluralism. A somewhat lesser degree of cultural pluralism, but still present, is maintained by "Little Italy" neighborhoods in some U.S. cities and also by certain Black Muslim groups in the United States.

Segregation and the Urban Underclass

Segregation is the spatial and social separation of racial and ethnic groups. Minorities, who are often believed by the dominant group to be inferior, are compelled to live separately under inferior conditions and are given lower-class educations, jobs, and protections under the law. Although desegregation has been mandated by law (thus eliminating *de jure segregation,* or legal segregation), *de facto segregation*—segregation in fact—still exists, particularly in housing and education.

Segregation has contributed to the creation of an **urban underclass**, a grouping of people, largely minorities and the poor, who live at the absolute bottom of the socioeconomic ladder in urban areas (Wilson 2009, 1987; Massey and Denton 1993). Indeed, the level of housing segregation is so high for some groups, especially poor African Americans and Latinos, that it has been termed **hypersegregation**, referring to a pattern of extreme segregation (Rugh and Massey 2010; Massey and Denton 1993; Massey 2005). Currently, the rate of segregation of Blacks and Hispanics in U.S. cities is actually increasing, thus allowing for less and less interaction between White and Black children and White and Hispanic children (Schmitt 2001; Massey and Denton 1993; see the box "Doing Sociological Research: American Apartheid"). In education, the extraordinary realization is that schools are also becoming more segregated, a phenomenon called *resegregation,* because American schools are now more segregated than they were even in the 1980s (Frankenberg and Lee 2002).

Neighborhoods such as this one in Manhattan, New York, are indicative of residential segregation.

Jeff Greenberg/PhotoEdit

As illustrated in this photograph, neighborhoods often exemplify both residential segregation as well as community.

Frances Roberts/Alamy Limited

In a seminal study, W. J. Wilson (1987) attributes the causes of the urban underclass to economic and social structural deficits in society. He rejects the "culture of poverty" explanation, an earlier view that attributes the condition of minorities to their own presumed "cultural deficiencies," a view attributed to writer Oscar Lewis (1966, 1960) and to Moynihan's (1965) so-called Moynihan Report, in which the Black family was seen as deficient in both structure and culture. The problems of the inner city, such as joblessness, crime, teen pregnancy, welfare dependency, and acquired immunodeficiency syndrome (AIDS) are seen to arise from social class inequalities, that is, inequalities in the *structure of society,* and these inequalities have dire behavioral consequences at the individual level, in the form of drug abuse, violence, and lack of education (Wilson 2009, 1996, 1987; Sampson 1987). But despite these disadvantages, many minority individuals nonetheless manage to achieve upward occupational and economic mobility (Newman 1999). Wilson argues that the civil rights agendas need to be expanded and that the major problem of the underclass, joblessness, needs to be addressed by fundamental changes in the economic institution.

This does not mean that race is unimportant, rather that the influence of class is increasing, even though race still continues to remain extremely important, and may be increasing in importance (Bobo 2012). In numerous studies, scholars find that race, in and of itself, still influences such things as income, wealth holdings, occupational prestige, place of residence, educational attainment, and numerous other measures of socioeconomic well-being (Bobo 2012; Oliver and Shapiro 2006; Brown et al. 2005; Pattillo-McCoy 1999).

What is important is understanding the intersecting effects of race and class acting together (Andersen and Collins 2013). Racial–ethnic groups live in what has

been called a *matrix of domination* (Collins 1990). That is, no single factor alone determines one's location in society. Rather, race—together with class, gender, age, even sexual orientation—place one in a system of social advantage and disadvantage. Understanding the *interrelationship* among these social factors is critical to understanding any one of them, including race and ethnicity.

The Civil Rights Movement

A major force behind social change in race relations has been the civil rights movement. The civil rights movement is probably the single most important source for change in race relations in the twentieth century.

The civil rights movement was initially based on the passive resistance philosophy of Martin Luther King Jr., learned from the philosophy of *satyagraha* ("soul firmness and force") of the East Indian Mahatma (meaning "leader") Mohandas Gandhi. This philosophy encouraged resistance to segregation through nonviolent techniques, such as sit-ins, marches, and appealing to human conscience in calls for brotherhood, justice, and equality. Although African Americans had worked for racial justice and civil rights long before this historic movement, the civil rights movement has brought greater civil rights under the law to many groups: women, disabled people, the aged, and gays and lesbians (Andersen 2011).

The major civil rights movement in the United States intensified shortly after the 1954 *Brown v. Board of Education* decision, the famous Supreme Court case that ruled that "separate but equal" in education was unconstitutional. In 1955, African American seamstress and NAACP secretary Rosa Parks made news in Montgomery, Alabama. By prior arrangement with the NAACP, Parks bravely refused to relinquish her seat in the "White-only" section on a segregated bus when asked to do so by the White bus driver. At the time, the majority of Montgomery's bus riders were African American, and the action of Rosa Parks initiated the now-famous Montgomery bus boycott, led by the young Martin Luther King Jr. The boycott, which took place in many cities beyond Montgomery, was successful in desegregating the buses. It got more African American bus drivers hired and catapulted Martin Luther King Jr. to the forefront of the civil rights movement.

Impetus was given to the civil rights movement and the boycott by the unspeakably brutal murder in 1954 of Emmett Till, a Black teenager from Chicago, who was killed in Mississippi merely for whistling at a White woman in a store. After he did so, a group of White men rousted Till from his bed at the home of a relative and beat him until he was dead and unrecognizable as a human being. They then tied a heavy cotton gin fan around his neck, and dumped him into the nearest river. Later, his mother in Chicago allowed a picture of his horribly misshapen head and body in his casket to be published (in *Jet* magazine) so that the public could contemplate the horror vested upon her son. No one was ever prosecuted for the Till murder.

Race and Hurricane Katrina

A brutally devastating hurricane, given the name Katrina, hit New Orleans, Louisiana, and other locations along the country's southern Gulf Coast, such as Biloxi, Mississippi, early in the fall of 2005. Katrina's winds, reaching 150 miles per hour at times, tore apart hundreds of houses, apartment buildings, hospitals, oil wells and derricks, and other structures. Massive flooding devastated New Orleans, with water reaching as high as 20 feet in some locations, stranding people, their pets, and farm animals. Because of the slowness of the federal government's response (it took the president of the United States one full week after the hurricane even to acknowledge the devastation as a national disaster and to visit New Orleans) and because of the extent of the flooding, over 1000 people died from drowning, from direct hits by flying debris, or from lack of medical attention.

The nation was stunned by images of human bodies floating down water-filled streets, the water itself slicked with oil and filled with sewage and other contaminants. In rest homes for the elderly, patients died as a result of lack of electricity and oxygen supplies needed for assisted breathing. In one case, over twenty patients were simply left alone, unattended by administrators and staff

Chris Graythen/Getty Images News/ Getty Images

for a full week. Each and every patient died from neglect and lack of medical care. Many people died simply waiting to be evacuated from their communities or from temporary transfer locations—another effect of the slowness of the state and federal governments to act. The government's Federal Emergency Management Administration (FEMA) failed miserably in its intended role of organizing and coordinating responses to the devastation and getting people to safe locations quickly. FEMA's response was so anemic that its director was forced to resign within weeks.

The most negatively affected areas of New Orleans were those neighborhoods in the lowest-lying areas of the city, areas up to twenty feet or more below sea level. These neighborhoods were primarily poor and Black or Hispanic; many were hypersegregated, that is, almost entirely African American. Clearly, these neighborhoods had the highest potential for flood damage and

the aftermath of contamination and disease. Many have argued that if the White and wealthy had been so concentrated in such neighborhoods, the response of the federal government and the president would probably have been much more rapid and perhaps more effective. To date, a large number of people and residences are *still* without governmental help in rebuilding and recovering.

An additional manifestation of race and class bias was seen as a result of the evacuation of thousands to the New Orleans Superdome and the New Orleans Convention Center. Within days, the Superdome and Convention Center had become cauldrons of misery—no water, no food, unbearable heat, grossly inadequate facilities—including few and clogged toilets and a partially collapsed roof—and an utter lack of medical care for the injured, pregnant women, infants, and the elderly. The individuals evacuated to the Superdome were mainly poor, and they were neglected and ignored for more than a week by the federal government, FEMA, and even the Louisiana state government.

In national polls taken after Hurricane Katrina, there was a racial divide: Three-quarters of African Americans thought that racism affected the response to the poor, but only one-third of White Americans thought so.

The civil rights movement produced many episodes of both tragedy and heroism. In a landmark 1957 decision, President Dwight D. Eisenhower called out the national guard—after initial delay—to assist the entrance of nine Black students into Little Rock Central High School in Little Rock, Arkansas. Sit-ins followed throughout the South in which White and Black students perched at lunch counters until the Black students were served. A number of these people, both Black and White, were beaten bloody for merely attempting this nonviolent protest.

Organized bus trips from North to South to promote civil rights, "freedom rides," forged on despite the murders of freedom riders Viola Liuzzo, a White Detroit housewife; Andrew Goodman and Michael Schwerner, two White students; and James Chaney, a Black student. Their murders have been documented and memorialized in the movie, *Mississippi Burning*, released in 1988. The murders of civil rights workers—especially when they were White—galvanized public support for change.

A Radical Response: Black Power

While the civil rights movement developed throughout the late 1950s and 1960s, a more radical philosophy of change also developed, as more militant leaders grew increasingly disenchanted with the limits of the civil rights agenda (which was perceived as moving too slowly). The militant Black Power movement, taking its name from the book *Black Power* (published in 1967 by political activist Stokely Carmichael, later Kwame Touré, and Columbia University political science professor Charles V. Hamilton), had a more radical critique of race relations in the United States and saw inequality as stemming not just from moral failures but from the institutional power that Whites had over Black Americans (Carmichael and Hamilton 1967).

Before breaking with the Black Muslims (the Black Nation of Islam in the United States) and his religious mentor, Elijah Muhammad, and prior to his assassination in 1965, Malcolm X advocated a form of pluralism, demanding separate business establishments, banks, churches, and schools for Black Americans. He echoed an earlier effort of the 1920s led by Marcus Garvey's back-to-Africa movement, the Universal Negro Improvement Association (UNIA).

The Black Power movement of the late 1960s rejected assimilationism and instead demanded pluralism in the form of self-determination and self-regulation of Black communities. Militant groups such as the Black Panther Party advocated fighting oppression with armed revolution. The U.S. government acted quickly, imprisoning members of the Black Panther Party and members of similar militant revolutionary groups, in some cases killing them outright (Brown 1992).

The Black Power movement also influenced the development of other groups who were affected by the analysis of institutional racism that the Black Power movement developed, as well as by the assertion of strong group identity that this movement encouraged. Groups such as La Raza Unida ("The Race United"), a Chicano organization, encouraged "brown power," promoting solidarity and the use of Chicano power to achieve racial justice.

Likewise, the American Indian Movement (AIM) used some of the same strategies and tactics that the Black Power movement had encouraged, as have Puerto Rican, Asian American, and other racial protest groups. Elements of Black Power strategy were also borrowed by the developing women's movement, and Black feminism was developed upon the realization that women, including women of color, shared in the oppressed status fostered by institutions that promoted racism (Collins 1998, 1990). Overall, the Black Power movement dramatically altered the nature of political struggle, and race and ethnic relations in the United States. It, and the other movements it inspired, changed the nation's consciousness about race and forced even academic scholars to develop a deeper understanding of how fundamental racism is to U.S. social institutions (Branch 2006; Morris 1984).

The Contemporary Challenge: Race-Specific versus Race-Blind Policies. A continuing question from the dialogue between a civil rights strategy and more radical strategies for change is the debate between race-specific versus color-blind programs for change. *Color-blind policies* are those advocating that all groups be treated alike, with no barriers to opportunity posed by race, gender, or other group differences. Equal opportunity is the key concept in color-blind policies.

Race-specific policies are those that recognize the unique status of racial groups because of the long

---

The "Tax" on Being a Minority in America: Give Yourself a True–False Test (An Illustration of White Privilege)

On the following test, give yourself one point for each statement that is true for you personally. When you are done, total up your points. The higher your score (the more points you have), the less "minority tax" you are paying in your own life:

1. My parents and grandparents were able to purchase a house in any neighborhood they could afford.
2. I can take a job in an organization with an affirmative action policy without people thinking I got my job because of my race.
3. My parents own their own home.
4. I can look at the mainstream media and see people who look like me represented in a wide variety of roles.
5. I can choose from many different student organizations on campus that reflect my interests.
6. I can go shopping most of the time pretty well assured that I will not be followed or harassed when I am in the store.
7. If my car breaks down on a deserted stretch of road, I can trust that the law enforcement officer who shows up will be helpful.
8. I have a wide choice of grooming products that I can buy in places convenient to campus and/or near where I live as a student.
9. I never think twice about calling the police when trouble occurs.
10. The schools I have attended teach about my race and heritage and present it in positive ways.
11. I can be pretty sure that if I go into a business or other organization (such as a university or college) to speak with the "person in charge," I will be facing a person of my race.

Your total points: _____

Your racial identity: _____

Your gender: _____

How would you describe your social class? _____

Now gather results from some of your classmates and see if their total points vary according to their own race, gender, and or social class.

Source: Adapted from the *Discussion Guide for Race: The Power of an Illusion*. **www.pbs.org**

---

history of discrimination and the continuing influence of institutional racism. Those advocating such policies argue that color-blind strategies will not work because Whites and other racial–ethnic groups do not start from the same position.

Affirmative action is an example of a race-specific policy for reducing job and educational inequality. Affirmative action means two things. First, it means recruiting minorities from a wide base to ensure consideration of groups that have been traditionally overlooked, while not using rigid quotas based on race or ethnicity. Second, affirmative action can mean taking race into account as one factor among others that can be used in such things as hiring decisions or college admissions. Despite public misunderstandings about affirmative action, establishing specific quotas for minority representation via affirmative action has been ruled unconstitutional.

The practice of affirmative action has, at least until now, been upheld as constitutional in the law—that is, at least as long as the use of race is not the sole factor being considered and quotas are not used. This has been upheld in two U.S. Supreme Court cases: *Regents of the University of California v. Bakke* (1978) and *Grutter v. Bollinger* (2003). Various legal challenges have been made to current constitutional law on affirmative action, but to date the principles articulated in the *Bakke* and *Grutter* cases are the law of the land.

All told, the challenge of race in the United States remains one that continues to affect the distribution of social, economic, and political resources. No single strategy is likely to solve this complex social issue. How will the nation address persistent racial inequality? Debates about racial inequality and how to fix it rage on—in dinner conversations, courtrooms, college classrooms, and other places.

chapter summary

How are race and ethnicity defined?
In virtually every walk of life, race matters. A *race* is a social construction based loosely on physical criteria, whereas an *ethnic group* is a culturally distinct group. A group is *minority* not on the basis of their numbers in a society but on the basis of which group occupies lower average social status.

What are stereotypes, and how are they important?
Stereotyping and *stereotype interchangeability* reinforce racial and ethnic prejudices and thus cause them to persist in the maintenance of inequality in society. Racial and gender stereotypes have similar dynamics in society, and both racial and gender stereotypes receive ongoing support in the media. Stereotypes serve to justify and make legitimate the oppression of groups based on race, ethnicity, class, and gender. Stereotypes such as "lazy" support attributions made to minorities and to working-class people, and attempt to cast blame on the minority in question, thus removing blame from the social structure.

What are the differences between prejudice, discrimination, and racism?
Prejudice is an attitude usually involving negative prejudgment on the basis of race or ethnicity. *Discrimination* is overt, actual behavior involving unequal treatment. *Racism* involves both attitude and behavior. Racism can take on several forms, such as traditional or *old-fashioned racism*, *aversive* or subtle racism, *laissez-faire racism*, *color-blind racism* (which masks *White privilege*), and institutional racism. *Institutional racism* is unequal treatment, carrying with it notions of cultural inferiority of a minority, which has become firmly ingrained into the economic, political, and educational institutions of society. Racial profiling is an example of institutional racism.

Do all minority groups have different histories, or are they similar?
Historical experiences show that different groups have unique histories, although they are bound together by some similarities in the prejudice and discrimination they have experienced.

What are the challenges in attaining racial and ethnic equality?
Not all immigrant groups and minority groups *assimilate* at the same rate, and some groups (U.S. Black Muslims, the Amish) maintain cultural *pluralism*. An *urban underclass* remains entrenched in the United States, and cities remain *hypersegregated* on the basis of race and ethnicity.

What are some of the approaches to attaining racial and ethnic equality?
Approaches include Reverend Martin Luther King's non-violent civil rights strategy, radical social change, and movements such as the Black Power movement, La Raza Unida, and the American Indian Movement (AIM), all of which directly addressed *institutional racism*. *Affirmative action* policies, which are race-specific, rather than race-blind programs, continue to be changed and modified through Supreme Court cases.

Key Terms

affirmative action 252
anti-Semitism 248
assimilation 239
authoritarian
   personality 239
aversive racism 237
caste system 242
color-blind racism 238
contact theory 240
digital racial divide 236

discrimination 235
dominant group 233
ethnic group 228
ethnocentrism 235
hypersegregation 249
institutional racism 238
intersection
   perspective 240
laissez-faire racism 238
minority group 233

old-fashioned racism 237
out-group homogeneity
   effect 232
pluralism 239
prejudice 235
race 229
racial formation 232
racial profiling 238
racialization 230
racism 237

residential segregation 236
salience principle 233
scapegoat theory 239
segregation 249
stereotype 233
stereotype
   interchangeability 234
urban underclass 249
White privilege 238

11 Gender

HOANG DINH NAM/AFP/Getty Images

imagine suddenly becoming a member of the other sex. What would you have to change? First, you would probably change your appearance—clothing, hairstyle, and any adornments you wear. You would also have to change some of your interpersonal behavior. Contrary to popular belief, men talk more than women, are louder, are more likely to interrupt, and are less likely to recognize others in conversation. Women are more likely to laugh, express hesitance, and be polite. Gender differences also appear in nonverbal communication. Women use less personal space, touch less in impersonal settings (but are touched more), and smile more, even when they are not necessarily happy (Mast and Hall 2004; LaFrance 2002; Lombardo et al. 2001; Robinson and Smith-Lovin 2001). Researchers even find that men and women write email in a different style, women writing less opinionated email than men and using it to maintain rapport and intimacy (Colley and Todd 2002; Sussman and Tyson 2000). Finally, you might have to change many of your attitudes because men and women differ significantly on many, if not most, social and political issues (see Figure 11.1). If you are a woman and became a man, perhaps the change would be worth it. You would probably see your income increase (especially if you became a White man). You would have more power in virtually every social setting. You would be far more likely to head a major corporation, run your own business, or be elected to a political office—again, assuming that you are White. Would it be worth it? As a man, you would be far more likely to die a violent death and would probably not live as long as a woman (Kung et al. 2008).

If you are a man who became a woman, your income would most likely drop significantly. More than fifty years after passage of the Equal Pay Act in 1963, men still earn 23 percent more than women, even comparing

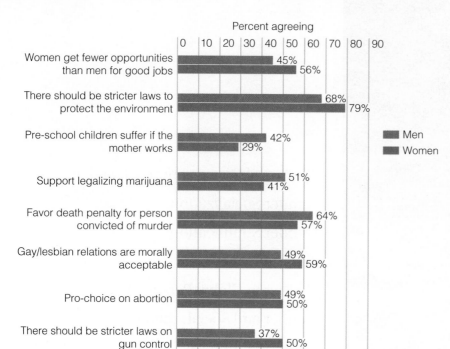

Percent agreeing

Women get fewer opportunities than men for good jobs	Men: 45%	Women: 56%
There should be stricter laws to protect the environment	Men: 68%	Women: 79%
Pre-school children suffer if the mother works	Men: 42%	Women: 29%
Support legalizing marijuana	Men: 51%	Women: 41%
Favor death penalty for person convicted of murder	Men: 64%	Women: 57%
Gay/lesbian relations are morally acceptable	Men: 49%	Women: 59%
Pro-choice on abortion	Men: 49%	Women: 50%
There should be stricter laws on gun control	Men: 37%	Women: 50%

FIGURE 11.1 Hot-Button Issues: The Gender Gap in Attitudes

When people use the term *gender gap*, they are often referring to pay differences between women and men. But, as the data here show, there is also a significant gender gap for some of the important issues of the day.

Data: Mendes, Elizabeth. 2010. (October 28). "New High of 46% of American Support Legalizing Marijuana." *The Gallup Poll*, **www.gallup.com**; Newport, Frank. 2011 (October 13). "In U.S., Support for Death Penalty Falls to 39-Year Low." *The Gallup Poll*, **www.gallup.com**; Saad, Lydia. 2012 (May 14). "U.S. Acceptance of Gay/Lesbian Relations is the New Normal." *The Gallup Poll*, **www.gallup.com**; Saad, Lydia. 2011 (May 23). "Americans Still Split along 'Pro-Choice,' 'Pro-Life' Lines." *The Gallup Poll*, **www.gallup.com**; Jones, Jeffrey M. 2011 (October 26). "Record-Low 26% in U.S. Favor Handgun Ban." Pew Research Center. 2012 (June 4). *American Values Survey*. **www.pewresearch.org**

those working year-round and full time (DeNavas-Walt et al. 2012). You would probably become resentful of a number of things because poll data indicate that women are more resentful than men about things such as the amount of money available for them to live on, the amount of help they get from their mates around the house, how men share child care, and how they look. Women also report being more fearful on the streets than men. However, women are more satisfied than men with their role as parents and with their friendships outside of marriage.

For both women and men, there are benefits, costs, and consequences stemming from the social definitions associated with gender. As you imagined this experiment, you may have had difficulty trying to picture the essential change in your biological identity: But is this the most significant part of being a man or woman? Nature determines whether you are male or female but it is society that gives significance to this distinction. Sociologists see gender as a social fact, because who we become as men and women is largely shaped by cultural and social expectations.

learning objectives

- Understand gender as a social construction
- Explain the process of gender socialization
- Identify different components of gender stratification
- Compare and contrast different theories of gender stratification
- Relate gender inequality in the United States to that in other nations
- Evaluate the different components of change with regard to gender

THE SOCIAL CONSTRUCTION OF GENDER

From the moment of birth, gender expectations influence how boys and girls are treated. Now that it is possible to identify the sex of a child in the womb, gender expectations may begin even before birth. Parents and grandparents might select pink clothes and dolls for baby girls, sports clothing and brighter colors for boys. Even if they try to do otherwise, it will be difficult because baby products are so typed by gender. Much research shows how parents and others continue to treat children in stereotypical ways throughout their childhood. Girls may be expected to cuddle and be sweet, whereas boys are handled more roughly and given greater independence.

see FOR YOURSELF

Changing Your Gender

Try an experiment based on the example of changing gender that opens this chapter.

1. First, make a list of everything you think you would have to do to change your behavior if you were a member of a different gender. Separate the things in your list according to whether they are related to such factors as appearance, attitude, or behavior.

2. Second, for a period of twenty-four hours, try your best to change any of these things that you are willing to do. Keep a log that records how others react to you during this period and how the change makes you feel.

3. When your experiment is over, write a report on what your brief experiment tells you about how gender identities are supported (or not) through social interaction. ●

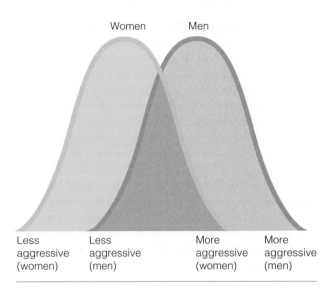

FIGURE 11.2 Gender Differences: Aggression Even when men and women as a whole tend to differ on a given trait, within-gender differences can be just as great as across-gender differences. Some men, for example, are less aggressive than some women.

© Cengage Learning

Defining Sex and Gender

Sociologists use the terms *sex* and *gender* to distinguish biological sex identity from learned gender roles. **Sex** refers to biological identity, being male or female. For sociologists, the more significant concept is **gender**—the socially learned expectations, identities, and behaviors associated with members of each sex. This distinction emphasizes that behavior associated with gender is culturally learned. Gender is a "system of social practices" (Ridgeway 2011: 9) that creates categories of people—men and women—who are defined in relationship to each other on unequal terms.

The definitions that surround these categories stem from culture—apparent especially when we look at other cultures. Across different cultures, the gender expectations associated with men and women vary considerably. In Western industrialized societies, people tend to think of men and women (and masculinity and femininity) in dichotomous terms, even defined as "opposite sexes." The views from other cultures challenge this assumption. Historically, the *berdaches* (pronounced ber-dash) in Navajo society were anatomically normal men who were defined as a third gender considered to be between male and female. Berdaches married other men who were not considered berdaches and were defined as ordinary men. Moreover, neither the berdaches nor the men they married were considered homosexuals, as they would be considered in many of today's Western cultures (Nanda 1998; Lorber 1994).

There can also be substantial differences in the construction of gender across social classes or within subcultures in a given culture. Within the United States, as we will see, there is considerable variation in the experiences of gender among different racial and ethnic groups (Andersen and Collins 2013; Baca Zinn et al. 2010). In addition, even within a given culture, differences among people of a given gender can be greater than differences across gender (see Figure 11.2). Looking at gender sociologically quickly reveals the social and cultural dimensions of something often popularly defined as biologically fixed.

Sex Differences: Nature or Nurture?

Despite the known power of social expectations, many still believe that differences between men and women are biologically determined. Biology is, however, only one component in the differences between men and women. The important question in sociology is not whether biology or culture is more important in forming

men and women, but how culture produces a person's gender identity.

Biological determinism refers to explanations that attribute complex social phenomena to physical characteristics. The argument that men are more aggressive because of hormonal differences (in particular, the presence of testosterone) is a biologically determinist argument. Although people popularly believe that testosterone causes aggressive behavior in men, studies find only a modest correlation between aggressive behavior and testosterone levels. Furthermore, changes in testosterone levels do not predict changes in men's aggression (such as by "chemical castration," the administration of drugs that eliminate the production or circulation of testosterone). What's more, there are minimal differences in the levels of sex hormones between girls and boys during early childhood, yet researchers find considerable differences in the aggression exhibited by boys and girls as children (Fausto-Sterling 2000, 1992).

A person's sex identity is established at the moment of conception when the father's sperm provides either an X or a Y chromosome to the egg at fertilization. The mother contributes an X chromosome to the embryo. The combination of two X chromosomes makes a female, while the combination of an X and a Y makes a male. Under normal conditions, chemical events directed by genes on the sex-linked chromosomes lead to the formation of male or female genitalia.

Being **intersexed** (also known as *hermaphroditism*) is a condition caused by irregularities in the process of chromosome formation or fetal differentiation that produces people with mixed biological sex characteristics. An intersexed infant may be born with ovaries or testes, but with ambiguous or mixed genitals. Or, an intersexed person may be a chromosomal male but have an incomplete penis and no urinary canal.

Case studies of intersexed people reveal the extraordinary influence of social factors in shaping a person's identity (Preves 2003). Parents of intersexed children are usually advised to have their child's genitals surgically assigned to either male or female and also to give the child a new name, a different hairstyle, and new clothes—all intended to provide the child with the social signals judged appropriate to a single gender identity. One physician who has worked on such cases gives the directive to parents that they "need to go home and do their job as child rearers with it very clear whether it's a boy or a girl" (Kessler 1990: 9).

Transgender people are those who live as a gender different from that to which they were assigned at birth (Schilt 2011; Schilt and Westbrook 2009). Although transgender individuals experience pressure to fit within the usual expectations, they challenge the "either/or" way of thinking about gender that characterizes dominant gender norms. Likewise, those who undergo sex changes as adults report enormous pressure from others to be one sex or the other. Managing such identities can be stressful, largely because of the expectations that others have about what are appropriate categories of gender identity. Anyone who crosses these or, in any way, appears to be different from dominant expectations is frequently subject to exclusion and ridicule, showing just how strong gender expectations are (Stryker and Whittle 2006; Gagné and Tewksbury 1998).

From a sociological perspective, biology alone does not determine gender identity. People must adjust to the expectations of others and the social understanding of what it means to be a man or a woman. A person may remain genetically one sex, while socially being the other—or perhaps something in between. In other words, there is not a fixed relationship between biological and social outcomes. If you only see men and women as biologically "natural" states, you miss some of the fascinating ways that gender is formed in society.

Physical differences between the sexes do, of course, exist. In addition to differences in anatomy, boys at birth tend to be slightly longer and weigh more than girls. As adults, men tend to have a lower resting heart rate, higher blood pressure, and higher muscle mass and muscle density. These physical differences contribute to the tendency for men to be physically stronger than women, but this can be altered, depending on level of physical activity. The public now routinely sees displays of women's athleticism and expects great performances from both men and women in world-class events such as the Olympics. Women can achieve a high degree of muscle mass and muscle density through bodybuilding and can win over men in activities that require high levels of endurance, such as the four women who have won the Iditarod—the Alaskan

Women's ability to achieve in athletics, once thought to be limited by their biological makeup, has been radically transformed by new opportunities in sports. In 2011, the University of Connecticut's women's basketball team broke a historic record in *all* sports by winning 89 consecutive games!

dog sled race considered to be one of the most grueling competitions in the world. In other words, until men and women really compete equally in activities from which women have historically been excluded, we may not know the real extent of physical differences between women and men.

Arguments based on biological determinism assume that differences between women and men are "natural" and, presumably, resistant to change. Like biological explanations of race differences, biological explanations of inequality between women and men tend to flourish during periods of rapid social change. They protect the status quo (existing social arrangements) by making it appear that the status of women or people of other races is "natural" and therefore should remain as it is. If social differences between women and men were biologically determined, there would be no variation in gender relations across cultures, but extensive differences are well documented. Moreover, even within the same culture, there can be vast *within-gender* differences. That is, the variation on a given trait, such as aggression or competitiveness, can be as great within a given gender group as the difference across genders. Thus some women are more aggressive than some men, and some men are less competitive than some women (see Figure 11.2).

In sum, we would not exist without our biological makeup, but we would not be who we are without society and culture. As sociologist Cecilia Ridgeway

Cultural Gatekeepers and the Construction of Femininity

Many have noted the distorted images of women that appear in the media. The common argument is that media images present an unrealistic image of women, which shapes women's self-concepts and limits their sense of possibilities for their appearance, their relationships, their careers, and so forth. Femininity is defined in the media by *cultural gatekeepers*—those who make decisions about what images to project. Cultural gatekeepers also have to respond to audience criticism. How they respond is an important part of the institutional process by which media images are sustained.

One sociologist, Melissa Milkie, wanted to explore how images of femininity are constructed in the media, particularly when producers encounter criticism from their audience. As readers of magazines, girls have protested many of the narrow and limiting images in the media, particularly those portraying girls' bodies.

Milkie interviewed ten top editors of leading girls' magazines to find out how they, as cultural gatekeepers, responded to the criticism from girls that images of girls in teen magazines do not reflect what "real girls" are like.

Milkie found that even the top editors think there are institutional limitations on what they can do to respond to girls' criticism. The editors who were very sensitive to the criticisms they received either said there was not much they could do about it or they dismissed the girls' complaints as misguided. They would claim the image was beyond their control, either because of the artistic process, advertisers' needs, or the culture itself. Thus, despite their positions of power, editors believed they could not fully control the images that appear. They pointed to institutional constraints that, in effect, thwarted efforts for change. Some editors simply dismissed the criticisms as girls' misreading the intent or meaning of an image.

Either way, Milkie's research shows how the organizational complexity of media institutions limits how much change is possible in how images of femininity are constructed. Market forces, advertisers, the values of producers, and the values of the public all intertwine in shaping the decisions of cultural gatekeepers. Milkie also shows, however, that people are not passive about what they see in the media, suggesting that how people respond to those images is an important part of the effect of such images in society.

Source: Milkie, Melissa A. 2002. "Contested Images of Femininity: An Analysis of Cultural Gatekeepers' Struggles with the 'Real Girl' Critique." *Gender & Society* 16 (December): 839–859.

---

puts it, gender is a "substantial, socially elaborated edifice constructed on a modest biological foundation" (2011: 9). It is culture that defines certain behaviors as appropriate (or not) for women and for men; these are learned throughout life, beginning with the earliest practices by which people are raised. Understanding the process of *gender socialization* is then a key part of understanding the formation of gender as a social and cultural phenomenon.

GENDER SOCIALIZATION

As we saw in Chapter 4, socialization is the process by which social expectations are taught and learned. Through **gender socialization**, men and women learn the expectations and identities associated with gender in society. The rules of gender extend to all aspects of society and daily life. Gender socialization affects the self-concepts of women and men, their social and political attitudes, their perceptions about other people, and their feelings about relationships with others. Although not everyone is perfectly socialized to conform to gender expectations, socialization is a powerful force directing the behavior of men and women in gender-typical ways.

Even people who set out to challenge traditional expectations often find themselves yielding to the powerful influence of socialization. Women who consciously reject traditional women's roles may still find themselves inclined to act as hostess or secretary in a group setting. Similarly, men may decide to accept equal responsibility for housework, yet they fail to notice when the refrigerator is empty or the child needs a bath—household needs they have been trained to let someone else notice (DeVault 1991). These expectations are so pervasive that it is also difficult to change them on an individual basis. If you doubt this, try buying clothing or toys for a young child without purchasing something that is gender-typed, or talk to parents who have tried to raise their children without conforming to gender stereotypes and see what they report about the influence of such things as children's peers and the media.

The Formation of Gender Identity

One result of gender socialization is the formation of **gender identity**, which is one's definition of oneself as a woman or man. Gender identity is basic to our self-concept and shapes our expectations for ourselves, our abilities and interests, and how we interact with others. Gender identity shapes not only how we think about ourselves and others but also influences numerous behaviors, including such things as the likelihood of drug and alcohol abuse, violent behavior, depression, or even how aggressive you are in driving (Andersen 2011).

One area in which gender identity has an especially strong effect is in how people feel about their appearance. Studies find strong effects of gender identity on body image. Concern with body image begins mostly during adolescence. Thus studies of young children (that is, preschool age) find no gender differences in how boys and girls feel about their bodies (Hendy et al. 2001), but clear differences emerge by early adolescence. At this age, girls report comparing their bodies to others of their sex more often than boys do. By early adolescence, girls report lower *self-esteem* (that is, how well one thinks of oneself) than boys; they also report more negativity about their body image than do boys. This type of thinking among girls is related to lower self-esteem (Jones 2001; Polce-Lynch et al. 2001). Among college students, women also are more dissatisfied with their appearance than are men (Hoyt and Kogan 2001). These studies indicate that idealized images of women's bodies in the media, as well as peer pressures, have a huge impact on young girls' and women's gender identity and feelings about their appearance.

Sociologist Debra Gimlin argues that bodies are "the surface on which prevailing rules of a culture are written" (Gimlin 2002: 6). You see this especially with regard to gender. Men and women alike practice elaborate rituals to achieve particular gender ideals, ideals that are established by the dominant culture.

Changes in gender roles have involved more men in parenting.

David Frazier/The Image Works

thinking SOCIOLOGICALLY

What *gender identities* are reflected in the different products men and women use as part of their daily grooming? How are products for men and for women packaged? Marketed? What color are they? What are their names? How do these artifacts of everyday life reflect *norms* about gender? ●

Sources of Gender Socialization

As with other forms of socialization, there are different agents of gender socialization: family, peers, children's play, schooling, religious training, mass media, and popular culture, to name a few. Gender socialization is reinforced whenever gender-linked behaviors receive approval or disapproval from these multiple influences.

Parents are one of the most important sources of gender socialization. Parents may discourage children from playing with toys that are identified with the other sex, especially when boys play with toys meant for girls.

Research finds that parents are more tolerant of girls not conforming to gender roles than they are for boys (Kane 2006). Fathers, especially, discourage sons from violating gender norms (Martin 2005). And, although fathers are now more involved than in the past in children's care, they are less likely to provide basic care and more likely to be involved with discipline (LaFlamme et al. 2002).

Expectations about gender are changing, although researchers suggest that the cultural expectations about gender may have changed more than people's actual behavior. Thus mothers and fathers now report that fathers should be equally involved in child rearing, but the reality is different. Mothers still spend more time in child-related activities and have more responsibility for children. Furthermore, the gap that mothers perceive between fathers' ideal and actual involvement in child rearing is a significant source of mothers' stress (Milkie et al. 2002).

Gender socialization patterns also vary within different racial–ethnic families. Latinas, as an example, have generally been thought to be more traditional in their gender roles, although this varies by generation and by the experiences of family members in the labor force. Within families, young women and men learn to formulate identities that stem from their gender, racial, and ethnic expectations.

Peers strongly influence gender socialization—sometimes more so than one's immediate family. Peer relationships shape children's patterns of social interaction. The play in which young people engage also shapes analytical skills and their values and attitudes. Studies find that boys and girls often organize their play in ways that reinforce not only gender but also race and age norms (Moore 2001). Peer relationships often reinforce the gender norms of the culture—norms that are typically even more strictly applied to boys than to girls. Thus boys who engage in behavior that is associated with girls are likely to be ridiculed by friends—more so than are girls who play or act like boys (Sandnabba and Ahlberg 1999). In this way, homophobic attitudes, routinely expressed among peers, reinforce dominant attitudes about what it means to "be a man" (Pascoe 2007). Although girls may be called "tomboys," boys who are called "sissies" are more harshly judged. Note, though, that tomboy behavior among girls beyond a certain age may result in the girl being labeled a "dyke."

Children's play is another source of gender socialization. You might think about what you played as a child and how this influenced your gender roles. Typically, though not in every case, boys are encouraged to play outside more; girls, inside. Boys' toys are more machine-like and frequently promote the development of militaristic values; they tend to encourage aggression, violence, and the stereotyping of enemies—values rarely associated with girls' toys. Children's books in schools also communicate gender expectations. Even with publishers' guidelines that discourage stereotyping, textbooks still depict men as aggressive, argumentative, and competitive. Men and boys are also more likely to be featured in children's books, although interestingly, systematic analysis of children's books shows that fathers are not very present

and, when they do appear, they are most often shown as ineffectual (Anderson and Hamilton 2005).

---

see FOR YOURSELF

Visit a local toy store and try to purchase a toy for a young child that is not gender-typed. What could you buy? What could you not buy? What does your experiment teach you about gender role socialization? If you take a child with you, note what toys he or she wants. What does this tell you about the effectiveness of *gender socialization?* ●

Schools are particularly strong influences on gender socialization because of the amount of time children spend in them. Teachers often have different expectations for boys and girls. Studies find, for example, that teachers hold gender stereotypes that women are not as capable in math as men (Riegle-Crumb and Humphries 2012). And we know from other research that when such stereotypes are present, student performance, such as on standardized tests, is negatively affected by the presence of what is called *stereotype threat* (Steele 2010). Earlier studies have also shown that when teachers respond more to boys in school, even if negatively, they heighten boys' sense of importance (American Association of University Women 1998; Sadker and Sadker 1994). Gender inequality is pervasive in schools and at all levels. Even in college, course-taking patterns, selection of majors, and teachers' interaction with students are shaped by gender (Mullen 2010).

Religion is an often overlooked but significant source of gender socialization. The major Judeo-Christian religions in the United States place strong emphasis on gender differences, with explicit affirmation of the authority of men over women. In Orthodox Judaism,

Even with social changes in gender roles, boys and girls tend to engage in play activities deemed appropriate for their gender.

men offer a prayer to thank God for not having created them as a woman or a slave. The patriarchal language of most Western religions and, in some faiths, the exclusion of women from positions of religious leadership signifies the lesser status of women in religious institutions. Any religion, interpreted in a fundamentalist way, can be oppressive to women. Indeed, the most devout believers of any faith tend to hold the most traditional views of women's and men's roles. But the influence of religion on gender attitudes cannot be considered separately from other factors. For many, religious faith inspires a belief in egalitarian roles for women and men; both Christian and Islamic women have organized to resist fundamentalist and sexist practices (Gerami and Lehnerer 2001).

The *media* in its various forms (television, film, magazines, music, and so on) communicate strong—some would even say cartoonish—gender stereotypes. Despite some changes in recent years, television and films continue to depict highly stereotyped roles for women and men. Men on television heavily outnumber women, and women are underrepresented in the leading roles in film (Eschholz et al. 2002). Men are not only more visible but also seen as more formidable, stereotyped in strong, independent roles. Women are more likely now to be portrayed as employed outside of the home and in professional jobs, but it is still more usual to see women depicted as sex objects. In fact, the sexualization of women is so extensive in the media that the American Psychological Association has concluded that there is "massive exposure to portrayals that sexualize women and girls and teach girls that women are sexual objects" (American Psychological Association 2007: 5).

Popular culture has increasingly sexualized even the youngest of girls.

Henry Leutwyler/Contour/Getty Images

see FOR YOURSELF

Gender and Popular Celebrations

Try to buy a friend a birthday card that does not stereotype women or men. Alternatively, try to buy a Father's Day or Mother's Day card and see if you can find one without gender stereotypes. How do the gender images in the cards you see overlap with stereotypes about aging, family, and images of beauty? After doing this experiment, ask yourself how products promoted in the media affect ideas about gender. •

Social scientists debate the extent to which people actually believe what they see on television, but research with children shows that they identify with the television characters they see. Both boys and girls rate the aggressive toys that they see on television commercials as highly desirable. They also judge them as more appropriate for boys' play, suggesting that something as seemingly innocent as a toy commercial reinforces attitudes about gender and violence (Klinger et al. 2001). Even with adults, researchers find that there is a link between viewing sexist images and having attitudes that support sexual aggression, antifeminism, and more

traditional views of women (American Psychological Association 2007). It is easy to see how extensive such images are by just watching the media with a critical eye. Women in advertisements are routinely shown in poses that would shock people if the characters were male. Consider how often women are displayed in ads dropping their pants, skirts, or bathrobe, or are shown squirming on beds. How often are men shown in such poses? Men are sometimes displayed as sex objects in advertising, but not nearly as often as women. The demeanor of women in advertising—on the ground, in the background, or looking dreamily into space—makes them appear subordinate and available to men.

The Price of Conformity

A high degree of conformity to stereotypical gender expectations takes its toll on both men and women. One of the major ways to see this is in the very high rate of violence against women—both in the United States and worldwide. Too frequently, men's power in society is manifested in physical and emotional violence. Violence takes many forms, including rape, sexual abuse, intimate partner violence, stalking, genital mutilation, and honor killings. Around the world, the United Nations is working in various ways to reduce violence against women, including some initiatives to help men examine cultural assumptions about masculinity that promote

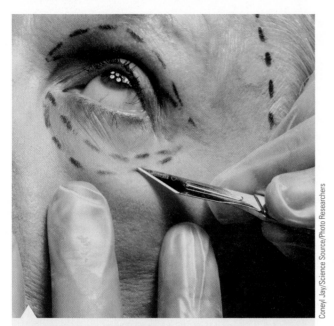

Cosmetic surgery is a rapidly growing, and highly profitable, industry. Some do it to try to eliminate signs of aging; others for aesthetic purposes. The pervasive influence of Western images means that some women in other nations pay for procedures to make them look more "Western."

Men also pay the price of conformity if they too thoroughly internalize gender expectations that say they must be independent, self-reliant, and unemotional.

table 11.1 Facts about Eating Disorders

One in five women struggle with an eating disorder or "disordered eating," referring to eating behaviors that are not classified as anorexia or bulimia but that are considered unhealthy.

Ninety percent of those with eating disorders are women between the ages of 12 and 25.

Of those with anorexia, 10 to 15 percent are men; men are less likely to seek treatment for an eating disorder than are women.

Eating disorders have the highest mortality rate of any mental illness.

Over half (58 percent) of women on college campuses report feeling pressure to lose weight; nearly half (44 percent) of them were of normal weight.

The average female fashion model is 5 foot, 11 inches tall and weighs 110 pounds—almost 50 pounds less than what is judged to be the ideal, healthy weight for someone this height.

The diet industry is worth an estimated $50 billion per year

Sources: Sullivan, Patrick F. 1995. "Mortality in Anorexia Nervosa." *American Journal of Psychiatry*, 152 (July): 1073–1074; Substance Abuse and Mental Health Services Administration (SAMHSA), the Center for Mental Health Services (CMHS), U.S. Department of Health and Human Services; Malinauskas, Brenda et al. 2006. "Dieting Practices, Weight Perceptions, and Body Composition: A Comparison of Normal Weight, Overweight, and Obese College Females." *Nutrition Journal* 5 (March 31): 5–11; Smolak, L. and M. Levine, eds., *The Developmental Psychopathology of Eating Disorders: Implications for Research, Prevention, and Treatment*. Hillsdale, NJ: Lawrence Erlbaum Associates Inc.; Spitzer, Brenda L., Katherine A. Henderson, and Marilyn T. Zivian. 1999. "A Comparison of Population and Media Body Sizes for American and Canadian Women." *Sex Roles* 700 (7/8): 54–565.

violence—a topic we examine further in the next two chapters. For now, it is important to understand that violence against women stems from the attitudes of power and control that gender expectations produce and that can lead *some* men to engage in violent behavior—toward women, as well as toward other men.

Violence by men is only one of the harms to women stemming from dominant gender norms. Adhering to gender expectations of thinness for women and strength for men is related to a host of negative health behaviors, including eating disorders, smoking, and for men, steroid abuse. The dominant culture promotes a narrow image of beauty for women—one that leads many women, especially young women, to be disturbed about their body image. Striving to be thin, millions of women engage in constant dieting, fearing they are fat even when they are well within or below healthy weight standards. Many develop eating disorders by purging themselves of food or cycling through various fad diets—behaviors that can have serious health consequences. Many young women develop a distorted image of themselves, thinking they are overweight when they may actually be dangerously thin. And, despite the known risks of smoking, increasing numbers of young women smoke not only because they think it "looks cool" but also because they think it will keep them thin. Eating disorders can be related to a woman having a history of sexual abuse, but they also come from the promotion of thinness as an ideal beauty standard for women—a standard that can put girls' and women's health in jeopardy (Hesse-Biber 2007).

Eating Disorders: Gender, Race, and the Body

Research Question: "A culture of thinness," "the tyranny of slenderness," "the beauty myth": These are terms used to describe the obsession with weight and body image that permeates the dominant culture, especially for girls and women. Just glance at the covers of popular magazines for women and girls and you will very likely find article after article promoting new diet gimmicks, each bundled with a promise that you will lose pounds in a few days if you only have the proper discipline or use the right products. Moreover, the models on the covers of such magazines are likely to be thin, often dangerously so because being too thin causes serious health problems. Do these body ideals affect all women equally?

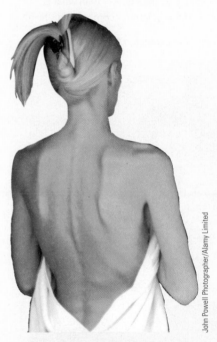

Too much conformity to gender roles can be harmful to your health. Such is the case of anorexic women who starve themselves attempting to meet cultural standards of thinness.

John Powell Photographer/Alamy Limited

Research Method: Meg Lovejoy wanted to know if the drive for thinness is unique to White women and how gendered images of the body might differ for African American and White women in the United States. Her research is based on reviewing the existing research literature on eating disorders, which has generally concluded that, compared with White women, Black women are less likely to develop eating disorders.

Research Results: Black women are less likely than White women to engage in excessive dieting and are less fearful of fat, although they are more likely to be obese and experience compulsive overeating. White women, on the other hand, tend to be very dissatisfied with their body size and overall appearance, with an increasing number engaging in obsessive dieting. Black and White women also tend to distort their own weight in opposite directions: White women are more likely to overestimate their own weight (that is, saying they are fat when they are not); Black women are more likely to underestimate their weight (saying they are average when they are overweight by medical standards). Why?

Conclusions and Implications: Lovejoy concludes that you cannot understand eating disorders without knowing the different stigmas attached to Black and White women in society. She suggests that Black women develop alternative standards for valuing their appearance as a way of resisting mainstream, Eurocentric standards. Black women who do so are then less susceptible to the controlling and damaging influence of the institutions that promote the ideal of thinness as feminine beauty. On the other hand, the vulnerability that Black women experience in society can foster mental health problems that are manifested in overeating. Eating disorders for Black women can also stem from the traumas that result from racism, especially when combined with sexism and other forms of oppression.

Lovejoy and others who have examined this issue conclude that eating disorders must be understood in the context of social structures—gender, race, class, and ethnicity—that affect all women, although in different ways. The cultural meanings associated with bodies differ for different groups in society but are deeply linked to our concepts of ourselves and the basic behaviors—like eating—that we otherwise think of as "natural."

Questions to Consider

1. Pay attention to the music and visual images in popular culture and ask yourself what cultural messages are being sent to different race and gender groups. What messages are being conveyed about appropriate appearance? How do they affect people's body image and their self-esteem?

2. Lovejoy examines eating disorders in the context of gender, race, class, and ethnicity. What cultural meanings are broadcast with regard to age?

3. Is there a "culture of thinness" among your peers? If so, what impact do you think it has on people's self-concept? If not, are there other cultural meanings associated with weight among people in your social groups?

Source: Lovejoy, Meg. 2001. "Disturbances in the Social Body: Differences in Body Image and Eating Problems among African American and White Women." *Gender & Society* 15 (April): 239–261.

Although many men are more likely now than in the past to express intimate feelings, gender socialization discourages intimacy among them, affecting the quality of men's friendships. Although conformity to traditional gender roles denies women access to power, influence, achievement, and independence in the public world, it denies men the more nurturing and other-oriented worlds that women have customarily inhabited. Learning traditional gender roles can also produce physical daring and risk-taking that can result in early death or injury from accidents. The strong undercurrent of violence in today's culture of masculinity can in many ways be attributed to the learned gender roles that put men and women at risk.

Race, Gender, and Identity

Gender identity does not emerge apart from other social factors. Thus, race—in combination with gender (as well as social class)—means that men and women from different racial groups may have different expectations regarding gender roles. African Americans, for example, have more egalitarian gender role beliefs than do Whites, although women in both groups are more egalitarian in their beliefs than men (Vespa 2009). Most interpret this as the result of Black women having more typically been in the labor force than White women, and some suggest that, as a result, the differences in Black and White attitudes toward gender roles is diminishing as White women's labor force participation has increased (Carter et al. 2009). Generally, Latinos and Latinas are more conservative in their gender role beliefs, although, as with African Americans, this difference is declining over time (Kane 2000; Harris and Firestone 1998). And despite the idea that Native Americans are more egalitarian in their outlooks than other groups, they do not differ in gender ideologies from other groups (Harris et al. 2000).

In addition to attitudes, race shapes people's identities. African American women, for example, are socialized to become self-sufficient, aspire to an education, desire an occupation, and regard work as an expected part of a woman's role (Collins 1990). Men's gender identity is also affected by race. Latino men, for example, bear the stereotype of *machismo* and Black men, "the player"—both stereotypes about exaggerated masculinity and sexuality. Latinos and African American men adopt different strategies for defining their identities in the context of such stereotypes. Latinos associate machismo not with sexism but with honor, dignity, and respect (Baca Zinn 1995; Mirandé 1979). And Black men may adopt what has come to be called a "cool pose" to assert a positive presentation of self where they, not others, control their identity. Other Black men may assert an identity of "respectability"—that is, distancing themselves from the negative stereotypes of Black men (Wilkins 2012). The point is that identities have to be negotiated within a context where gender, race, and class stereotypes frame the context wherein women *and* men construct their identities.

Gender Socialization and Homophobia

Homophobia is the fear and hatred of homosexuals. Homophobia plays an important role in gender socialization because it encourages stricter conformity to traditional expectations, especially for men and young boys. Slurs directed against individuals who are gay encourage boys to act more masculine as a way of affirming for their peers that they are not gay (Pascoe 2007). As a consequence, homophobia also discourages so-called feminine traits in men, such as caring, nurturing, empathy, emotion, and gentleness. Men who endorse the most traditional male roles also tend to be the most homophobic (Alden 2001; Burgess 2001; Basow and Johnson 2000). In this way, homophobia is one of the means by which socialization into expected gender roles takes place. The consequence is not only conformity to gender roles, but also a learned hostility toward gays and lesbians.

Homophobia is a learned attitude, as are other forms of negative social judgments about particular groups. Homophobia is also deeply embedded in people's definitions of themselves as men and women. Boys are often raised to be manly by repressing so-called feminine characteristics in themselves. Being called a "fag" or a "sissy" is one of the peer sanctions that socializes a child to conform to particular gender roles. Similarly, pressures on adolescent girls to abandon tomboy behavior are a mechanism by which girls are taught to adopt the behaviors and characteristics associated with womanhood. Being labeled a lesbian may cause those with a strong attraction to women to repress this emotion and direct love only toward men. We can see, therefore, how homophobic ridicule, though it may be in the context of play and joking, has serious consequences for both heterosexual and homosexual men and women. Homophobia socializes most people into expected gender roles, and it produces numerous myths about gays and lesbians—examined in more detail in the following chapter.

The Institutional Basis of Gender

The process of gender socialization tells us a lot about how gender identities are formed, but gender is not just a matter of identity: *Gender is embedded in social institutions.* This means that institutions are patterned by gender, resulting in different experiences and opportunities for men and women. Sociologists analyze gender not just as interpersonal expectations but also as characteristic of institutions. This is what is meant by the term *gendered institution.* This concept means that entire institutions are patterned by gender.

Gendered institutions are the total pattern of gender relations that structure social institutions, including the stereotypical expectations, interpersonal relationships, and the different placement of men and women that are found in institutions. Schools, for example, are not just places where children learn gender roles but are gendered institutions because they are founded on specific gender patterns. Seeing institutions as gendered reveals that gender is not just an attribute of individuals but is also "present in the processes, practices, images and ideologies, and distributions of power in the various sectors of social life" (Acker 1992: 567).

As an example of the concept of gendered institution, think of what it is like to work as a woman in a work organization dominated by men. Women in this situation report that men's importance in the organization is communicated in subtle ways, whereas women are made to feel like outsiders. Important career connections may be made in the context of men's informal interactions with each other—both inside and outside the workplace. Women may be treated as tokens or may think that company policies are ineffective in helping them cope with the particular demands in their lives.

But the point of thinking of institutions as gendered is to think about the gendered characteristics of the institution itself. Work institutions have been structured on the old assumption that men work and women do not. Thus, institutions tend to demand loyalty to work, not family. Even when men and women try to integrate work and family in their lives, gendered institutional practices make this difficult. Gendered institutions thus affect men as well as women, especially if they try to establish more balance between their work and personal lives. To say that work institutions are gendered institutions means that, taken together, there is a cumulative and systematic effect of gender throughout the institution.

Gender is not just a learned role; it is also part of social structure, just as class and race are structural dimensions of society. Notice that people do not think about the class system or racial inequality in terms of "class roles" or "race roles." It is obvious that race relations and class relations are far more than matters of interpersonal interaction. Race, class, and gender inequalities are experienced within interpersonal relationships, but they extend beyond relationships. Just as it would seem strange to think that race relations in the United States are controlled by race–role socialization, it is also wrong to think that gender relations are the result of gender socialization alone. Like race and class, gender is a system of privilege and inequality in which women are systematically disadvantaged relative to men. There are institutionalized power relations between women and men, and men and women have unequal access to social and economic resources.

GENDER STRATIFICATION

Gender stratification refers to the hierarchical distribution of social and economic resources according to gender. Most societies have some form of gender stratification, although the specific form varies from country to country. Comparative research finds that women are more nearly equal in societies characterized by the following traits (Chafetz 1984):

- Women's work is central to the economy.
- Women have access to education.

- Ideological or religious support for gender inequality is not strong.
- Men make direct contributions to household responsibilities, such as housework and child care.
- Work is not highly segregated by sex.
- Women have access to formal power and authority in public decision making.

In Sweden, where there is a relatively high degree of gender equality, the participation of both men and women in the labor force and the household (including child care and housework) is promoted by government policies. Women also have a strong role in the political system, although women still earn less than men in Sweden. In many countries, women and girls have less access to education than men and boys, but that gap is closing. Still, in most countries, the illiteracy rate among women is much higher than among men (United Nations 2010a).

As the preceding list suggests, gender stratification is multidimensional. In some societies, women may be free in some areas of life but not in others. In Japan, for example, women tend to be well educated and participate in the labor force in large numbers. Within the family, however, Japanese women have fairly rigid gender roles. Yet the rate of violence against women in Japan (in the form of rape, prostitution, and pornography) is quite low in relation to other nations, even though women are widely employed as "sex workers" in hostess clubs, bars, and sex joints (Allison 1994). Patterns of gender inequality are most reflected in the wage differentials between women and men around the world, as Figure 11.3 shows.

Gender stratification can be extreme. The public witnessed this in 2012, when Malala Yousafzai, a fourteen-year-old Pakistani girl, was shot in the face by Taliban extremists simply because she advocated for girls' rights to an education. The Taliban, an extremist militia group, strips women and girls of basic human rights. When the Taliban took control of Afghanistan in 1996, women were banished from the labor force, expelled from schools and universities, and prohibited from leaving their homes unless accompanied by a close male relative. The windows of houses where women lived were painted black to keep women literally invisible to the public. This extreme segregation and exclusion of women from public life has been labeled **gender apartheid**. Gender apartheid is also evident in other nations, even if not as extreme as it was under Taliban rule: In Saudi Arabia, women are not allowed to drive; in Kuwait, women were not allowed to vote until 2006.

Sexism and Patriarchy

Gender stratification is supported by beliefs that treat gender inequality as "natural." Sexism defines women as different from and inferior to men; it can be overt,

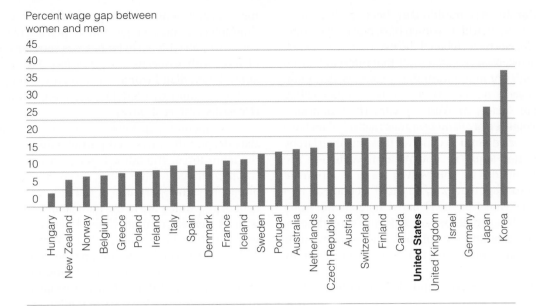

Percent wage gap between women and men

FIGURE 11.3 The Gender Wage Gap in OECD Countries, 2011 This bar graph shows the gap in men's and women's wages in the most economically developed countries. You can see that other than for Japan and Korea, the United States is among the nations with the greatest gap. What does this suggest to you in terms of social policy?

Note: OECD is the Organisation for Economic Co-operation and Development.

Data: From OECD Employment Outlook. **www.oced.org**

but can also be subtle. Like racism, sexism makes gender roles seem natural when they are actually rooted in entrenched systems of power and privilege. In this sense, sexism is both a belief and is anchored in social institutions. For example, the idea that men should be paid more than women because they are the primary breadwinners is a sexist idea, but that idea is also embedded in the wage structure. Because of this institutional dimension to sexism, people no longer have to be individually sexist for there still to be consequences of sexism.

Like racism, sexism generates social myths that have no basis in fact but support the continuing advantage of dominant groups over subordinates. A case in point is the belief that women of color are being hired more often and promoted more rapidly than others. This belief misrepresents the facts. Women rarely take jobs away from men because most women of color work in gender- and race-segregated jobs. The truth is that women, especially women of color, are burdened by obstacles to job mobility that are not present for men, especially White men (Andersen 2011; Padavic and Reskin 2002; Browne 1999). The myth that women of color get all the jobs makes White men seem to be the victims of race and gender privilege. Although there may be occasional cases where a woman of color (or a man of color, for that matter) gets a job that a White man also applied for, gender and race privilege usually favor White men.

Sexism also works to devalue the work that women do—both in dollar terms and in more subjective perceptions. To give a historical example, in jobs that were

once held by men but became dominated by women, wages declined as women became more numerous. When this happens, the prestige of the occupation also tends to fall (Andersen 2011). You can also see this tendency in something labeled the *mommy tax*, referring to the loss of income women experience if they reenter the labor market after staying home to raise children. According to the author who coined this term, a college-educated woman with one child will lose about a million dollars in lifetime earnings as the result of the "mommy tax" (Crittenden 2002). Although the mommy tax afflicts all employed mothers, it has been shown to be especially severe among those who can afford it the least—low-income women (Budig and Hodges 2010).

Sexism emerges in societies structured by **patriarchy**, referring to a society or group in which men have power over women. It can be present in the private sector, such as in families in which husbands have authority over their wives. But patriarchy also marks public institutions when men hold all or most of the powerful positions. Forms of patriarchy vary from society to society, but it is common throughout the world. In some societies, it is rigidly upheld in both the public and private spheres, and women may be formally excluded from voting, holding public office, or working outside the home. In societies like the contemporary United States, patriarchy may be somewhat diminished in the private sphere (at least in some households), but the public sphere continues to be based on patriarchal relations.

Matriarchy has traditionally been defined as a society or group in which women have power over men. Anthropologists have debated the extent to which such societies exist. New research finds that matriarchies do exist, though not in the form the customary definition implies. Based on her study of the Minangkabau matriarchal society in West Sumatra (in Indonesia), anthropologist Peggy Sanday argues that scholars have used a Western definition of power to define matriarchy that does not apply in non-Western societies. The Minangkabau define themselves as a matriarchal society, meaning that women hold economic and social power. However, the Minangkabau are not ruled by women. The people believe that rule should be by consensus, including that of men and women. Thus matriarchy exists but not as a mirror image of patriarchy (Sanday 2002).

In sum, gender stratification is an institutionalized system that rests on specific belief systems supporting the inequality of men and women. Although one could theoretically have a society stratified by gender where women hold power over men, gender stratification has not evolved in this way, as we will see in the next section.

Women's Worth: Still Unequal

Gender stratification is especially obvious in the persistent earnings gap between women and men (see Figure 11.4). Although the gap has closed somewhat since the 1960s, when women earned 59 percent of what

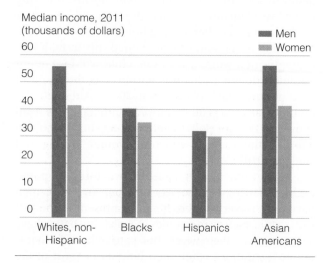

FIGURE 11.4 Median Income by Race and Gender, Full-Time Workers You will often hear that there is a 23 percent gap between women and men wage earners, but these data show that the wage gap varies depending on racial–ethnic background. What facts stand out to you from the data depicted here?

Data: U.S. Census Bureau. 2012. *Detailed Income Tabulations from the Current Population Survey: Selected Characteristics of People 15 Years and Over by Total Money Income in 2009, Work Experience in 2009, Race, Hispanic Origin, and Sex, Table PINC-01.* Washington, DC: U.S. Department of Commerce. **www.census.gov**

men earned, women who work year-round and full time still earn, on average, only 77 percent of what men earn. Women with bachelor degrees earn the equivalent of men with associate's degrees (see Figure 11.5). In 2011, the median income for women working full time and year-round was $37,118; for men, it was $48,202 (DeNavas-Walt et al. 2012).

The income gap between women and men persists despite the increased participation of women in the labor force. The **labor force participation rate** is the percentage of those in a given category who are employed either part time or full time. Fifty-three percent of all women are in the paid labor force compared with 64 percent of men. Since 1960, married women with children have nearly tripled their participation in the labor force. Seventy-one percent of mothers are now in the labor force, including more than half of mothers with infants. Current projections indicate that women's labor force participation will continue to rise, and men's will decline slightly (Bureau of Labor Statistics 2012b).

This pattern of women being in the labor market has long been true for women of color but now also characterizes the experience of White women; the labor force participation rates of White women and women of color have, in fact, converged. More women in all racial groups are also now the sole supporters of their families.

Why do women continue to earn less than men, even when laws prohibiting gender discrimination have been in place for more than fifty years? The Equal Pay Act of 1963 was the first federal law to require that men and women receive equal pay for equal work, an idea that is supported by the majority of Americans. But wage discrimination is rarely overt. Most employers do not even explicitly set out to pay women less than men. Despite good intentions and legislation, however, differences in men's and women's earnings persist. Research reveals four strong explanations for this: human capital theory, dual labor market theory, gender segregation, and overt discrimination.

Human Capital Theory. Human capital theory explains gender differences in wages as resulting from the individual characteristics that workers bring to jobs. Human capital theory assumes that the economic system is fair and competitive and that wage discrepancies reflect differences in the resources (or *human capital*) that individuals bring to their jobs. Factors such as age, prior experience, number of hours worked, marital status, and education are human capital variables. Human capital theory asserts that these characteristics will influence people's worth in the labor market. For example, higher job turnover rates or work records interrupted by child rearing and family responsibilities could negatively influence the earning power of women. Research also finds a significant earnings penalty for women because of motherhood.

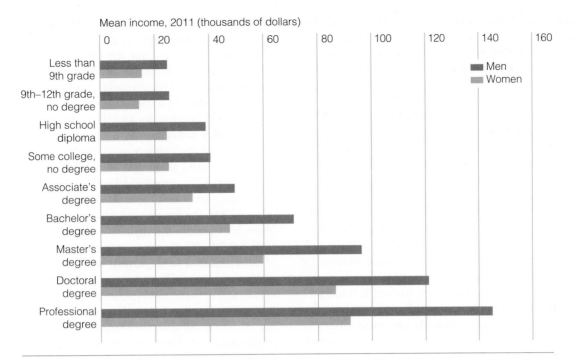

Mean income, 2011 (thousands of dollars)

■ Men
■ Women

- Less than 9th grade
- 9th–12th grade, no degree
- High school diploma
- Some college, no degree
- Associate's degree
- Bachelor's degree
- Master's degree
- Doctoral degree
- Professional degree

FIGURE 11.5 Education, Gender, and Income, 2011 Although the "economic return" on education is high for both men and women at every level of educational attainment, men's income, on average, exceeds that of women. According to this national data, how much education would the average woman have to get to exceed the income of men with a high school diploma? To exceed men with a college degree?

Data: U.S. Census Bureau. 2012. *Detailed Income Tabulations from the Current Population Survey: Selected Characteristics of People 15 Years and Over by Total Money Income in 2009, Work Experience in 2009, Race, Hispanic Origin, and Sex, Table PINC-01.* Washington, DC: U.S. Department of Commerce. **www.census.gov**

Much evidence supports the human capital explanation for the differences between men's and women's earnings because education, age, and experience do influence earnings. But, when you compare men and women who have the same level of education, previous experience, and number of hours worked per

Data show that occupations where women of color predominate also tend to have the lowest wages.

week, women still earn less than men (see Figure 11.5). Although human capital theory explains some of the difference between men's and women's earnings, it does not explain it all. Sociologists have looked to other factors to complete the explanation of wage inequality (Browne 1999).

The Dual Labor Market. A second explanation of discrepancies in men's and women's earnings is **dual labor market theory**, which contends that women and men earn different amounts because they tend to work in different segments of the labor market. The dual labor market reflects the devaluation of women's work because women are most concentrated in low-wage jobs. Although it is hard to untangle cause and effect in the relationship between the devaluation of women's work and low wages in certain jobs, once such an earnings structure is established, it is difficult to change. As a result, although equal pay for equal work may hold in principle, it applies to relatively few people because most men and women are not engaged in equal work.

According to dual labor market theory, the labor market is organized in two different sectors: the *primary market* and the *secondary market*. In the primary labor market, jobs are relatively stable, wages are good, opportunities for advancement exist, fringe benefits are likely, and workers are afforded due process. Working for a major corporation in a management job is an example of this. Jobs in the primary labor market are usually in large organizations where there is greater stability, steady profits, benefits for workers, better wages, and a rational system of management. In contrast, the secondary labor market is characterized by high job turnover, low wages, short or nonexistent promotion ladders, few benefits, poor working conditions, arbitrary work rules, and capricious supervision. Many of the jobs students take—such as waiting tables, bartending, or cooking and serving fast food—fall into the secondary labor market. However, for students, these jobs are usually short term.

Within the primary labor market, there are two tiers. The first consists of high-status professional and managerial jobs with potential for upward mobility, room for creativity and initiative, and more autonomy. The second tier comprises working-class jobs, including clerical work, skilled, and semiskilled blue-collar work. Women and minorities in the primary labor market tend to be in the second tier. Although these jobs may be more secure than jobs in the secondary labor market, they are more vulnerable and do not have as much mobility, pay, prestige, or autonomy as jobs in the first tier of the primary labor market.

There is, in addition, an informal sector of the market where there is even greater wage inequality, no benefits, and little, if any, oversight of employment practices. Individuals may hire such workers as private service workers or under-the-table workers who perform a service for a fee (painting, babysitting, car repairs, and any number of services). Although there are no formal data on the informal sector because much of it tends to be in an underground economy, it is likely that women and minorities form a large segment of this market activity. White men in this sector are also disadvantaged because of the instability and lack of protection in this work.

Gender Segregation. Dual labor market theory explains wage inequality as a function of the structure of the labor market, not the individual characteristics of workers as suggested by human capital theory. Because of the dual labor market, men and women tend to work in different occupations and, when working in the same occupation, in different jobs. This is referred to as **gender segregation**, a pattern in which different groups of workers are separated into occupational categories based on gender. There is a direct association between the number of women in given occupational categories and the wages paid in those jobs. In other words, the greater the proportion of women in a given occupation, the lower the pay (Bureau of Labor Statistics 2012b). Gender segregation is a specific form of **occupational segregation**; segregation in the labor market can also be based on factors such as race, class, age, or any combination thereof.

Despite several decades of legislation prohibiting discrimination against women in the workplace, most women and men still work in gender-segregated occupations. That is, the majority of women work in occupations where most of the other workers are women, and the majority of men work mostly with men. Women also tend to be concentrated in a smaller range of occupations than men. To this day, more than half of all employed women work as clerical workers and sales clerks or in service occupations such as food service workers, maids, health service workers, hairdressers, and child-care workers. Men are dispersed over a much broader array of occupations. Women make up 82 percent of elementary and middle school teachers, 96 percent of secretaries, 90 percent of bookkeepers, and 95 percent of child-care workers—stark evidence of

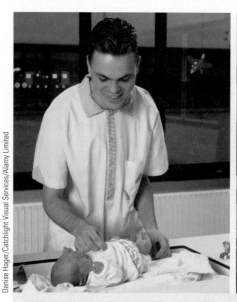
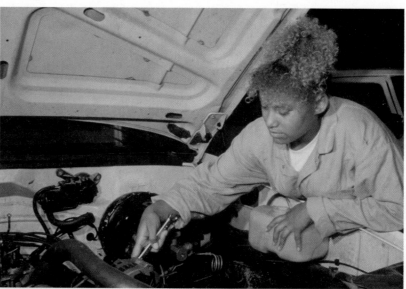

Because gender segregation is so pervasive in the workplace, people may still be surprised when they see women and men in nontraditional occupations.

the persistence of gender segregation in the labor force (Bureau of Labor Statistics 2012b).

Gender segregation also occurs within occupations. Women usually work in different jobs from men, but when they work within the same occupation, they are segregated into particular fields or job types. For example, in sales work, women tend to do noncommissioned sales or to sell products that are of less value than those men sell. Or, among waiters and waitresses, women often work in lower-priced restaurants where they are likely to be tipped less than men (Hall 1993).

Overt Discrimination. A fourth explanation of the gender wage gap is discrimination. **Discrimination** refers to practices that single out some groups for different and unequal treatment. Despite the progress of recent years, overt discrimination continues to afflict women in the workplace. It is argued that men (especially White men), by virtue of being the dominant group in society, have an incentive to preserve their advantages in the labor market. They do so by establishing rules that distribute rewards unequally. Women pose a threat to traditional White male privileges, and men may organize to preserve their own power and advantage (Reskin 1988). Historically, White men used labor unions to exclude women and racial minorities from well-paying, unionized jobs, usually in the blue-collar trades. A more contemporary example is seen in the efforts of some groups to dilute legislation that has been developed to assist women and racial–ethnic minorities. These efforts can be seen as an attempt to preserve group power.

Another example of overt discrimination is the harassment that women experience at work, including *sexual harassment* and other means of intimidation. Sociologists see such behaviors as ways for men to protect their advantages in the labor force. No wonder that women who enter traditionally male-dominated professions suffer the most sexual harassment; the reverse seldom occurs for men employed in jobs historically filled by women. Although men can be victims of sexual harassment, this is rare. Sexual harassment is a mechanism for preserving men's advantage in the labor force—a device that also buttresses the belief that women are sexual objects for the pleasure of men.

Each of these explanations—human capital theory, dual labor market theory, gender segregation, and overt discrimination—contributes to an understanding of the continuing differences in pay between women and men. Wage inequality between men and women is clearly the result of multiple factors that together operate to place women at a systematic disadvantage in the workplace.

The Devaluation of Women's Work

Across the labor market, women tend to be concentrated in those jobs that are the most devalued, causing some to wonder if the fact that women hold these jobs leads to devaluation of the jobs. Why, for example, is pediatrics considered a less prestigious specialty than cardiology? Why are preschool and kindergarten teachers (98 percent of whom are women) paid less than airplane mechanics (98 percent of whom are men)? The association of preschool teaching with children and its identification as "women's work" lowers its prestige and economic value. Indeed, if measured by the wages attached to an occupation, child care is one of the least prestigious jobs in the nation—paying on average only $383 per week in 2011, which would come out to an income below the federal poverty line if you worked every week of the year. Male highway maintenance workers, roofers, and construction workers all make more (Bureau of Labor Statistics 2012b).

Only a small proportion of women work in occupations traditionally thought to be men's jobs (such as the skilled trades). The representation of women in skilled blue-collar jobs has increased, but it is still a very small fraction (typically less than 3 percent) of those in skilled trades such as plumbers, electricians, and carpenters (Bureau of Labor Statistics 2012b). Likewise, very few men work in occupations historically considered to be women's work, such as nursing, elementary school teaching, and clerical work. Interestingly, men who work in occupations customarily thought of as women's work tend to be more upwardly mobile within these jobs than are women who enter fields traditionally reserved for men (Budig 2002; Williams 1992). Gender segregation in the labor market is so prevalent that most jobs can easily be categorized as men's work or women's work. Occupational segregation reinforces the belief that there are significant differences between the sexes. Think of the characteristics of a soldier. Do you imagine someone who is compassionate, gentle, and demure? Similarly, imagine a secretary. Is this someone who is aggressive, independent, and stalwart? The association of each characteristic with a particular gender makes the occupation itself a gendered occupation.

For all women, perceptions of gender-appropriate behavior influence the likelihood of success within institutions. Even something as simple as wearing makeup has been linked to women's success in professional jobs (Dellinger and Williams 1997). When men or women cross the boundaries established by occupational segregation, they are often considered to be gender deviants. They may be stereotyped as gay and have their "true gender identity" questioned. Men who are nurses may be stereotyped as effeminate or gay; women marines may be stereotyped as "butch." Social practices like these serve to reassert traditional gender identities, perhaps softening the challenge to traditionally male-dominated institutions that women's entry challenges (Williams 1995).

As a result, many men and women in nontraditional occupations feel pressured to assert gender-appropriate behavior. Men in jobs historically defined as women's work may feel compelled to emphasize

table 11.2 Is It True?*

	True	False
1. Men are more aggressive than women.		
2. Parents have the most influence on children's gender identities.		
3. Most women hold feminist values.		
4. Being a "stay-at-home" mom is the most satisfying lifestyle for women.		
5. In all racial–ethnic groups, women earn less on average than men.		
6. The wage gap between women and men has closed since the 1970s, largely as the result of women being more likely to enter the labor force.		
7. In terms of wages, middle-class women have most benefited from antidiscrimination policies.		

*The answers can be found on page 274.

© Cengage Learning

their masculinity, or if they are gay, they may feel even more pressure to keep their sexual orientation secret. Such social disguises can make them seem unfriendly and distant, characteristics that can have a negative effect on performance evaluations. Heterosexual women in male-dominated jobs may also feel obliged to squash suspicions that they are lesbians or are excessively mannish. And studies have found that lesbian women are more likely to be open about their sexual identity at work when they work predominantly with women and have women as bosses (Schneider 1984).

Balancing Work and Family

As the participation of women in the labor force has increased, so have the demands of keeping up with work and home life. Research finds that young women and men now want a good balance between work and family life, but they also find that institutions are resistant to accommodating these ideals (Gerson 2010). Men are also more involved in housework and child care than has been true in the past, although the bulk of this work still falls to women—a phenomenon that has been labeled "the second shift" (Hochschild and Machung 1989).

The social speedup that comes from increased hours of employment for both men and women (but especially women), coupled with the demands of maintaining a household, are a source of considerable stress (Jacobs and Gerson 2004). Women continue to provide most of the labor that keeps households running—cleaning, cooking, running errands, driving children around, and managing household affairs. Although more men are engaged in housework and child care, a huge gender gap remains in the amount of such work women and men do. Women are also much more likely to be providing care, not just for children, but also for their older parents. The strains these demands produce have made the home seem more and more like work for many; a large number of women and men report that their days at both work

and home are harried and that they find work to be the place where they find emotional gratification and social support. In this contest between home and work, simply finding time can be an enormous challenge (Hochschild 1997). It is not surprising then that women report stress as one of their greatest concerns (Newport 2000).

THEORIES OF GENDER

Why is there gender inequality? The answer to this question is important, not only because it makes us think about the experiences of women and men, but also because it guides attempts to address the persistence of gender injustice. The major theoretical frameworks in sociology provide some answers, but feminist scholars have also found that traditional perspectives in the discipline are inadequate to address the new issues that have emerged from feminist research.

The Frameworks of Sociology

The major frameworks of sociological theory—functionalism, conflict theory, and symbolic interaction—provide some answers to the question of why gender inequality exists, although, as we will see in the next section, feminist scholars have developed new theories to address women's lives directly.

Functionalist theory traditionally purported that men fill instrumental roles in society whereas women fill expressive roles and presumes that this arrangement works to the benefit of society (see Chapter 1). Functionalism, however, has been criticized for interpreting gender as a fixed role in society and for presuming that such sexist arrangements are functional for society. Although few contemporary functionalist theorists would make such arguments now, functionalism does emphasize people's socialization into prescribed roles as the major impetus behind gender inequality. Thus conditions such as wage inequality, a functionalist might argue, are the result of choices women make that may result in their inequality but that nonetheless involve functional adaptation to the competing demands of family and work roles.

Conflict theorists, in contrast, see women as disadvantaged by power inequities between women and men that are built into the social structure. This includes economic inequity, as well as women's disadvantages in political and social systems. Conflict theorists, for example, see wage inequality as produced from men's historic power to devalue women's work and to benefit as a group from the services that women's labor provides. At the same time, conflict theorists have been much more attuned to the interactions of race, class, and gender inequality because they see all forms of inequality as stemming from the differential access to resources that dominant groups in society have.

Conflict theory also interprets women's inequality as stemming from the system of capitalism. Influenced by the work of Karl Marx, some feminist scholars argue that women are oppressed because they have historically constituted a cheap supply of labor. Women provide a reserve supply of labor, pulled into the labor market when there is a need for underpaid workers and also left doing much of the work that is never paid—that is, housework. Conflict theorists understand women's inequality as largely stemming from economic exploitation, coupled with the power that men hold in virtually all social institutions.

Symbolic interaction theory is less attentive to the economic basis of gender inequality, focusing instead on the immediate realm of social interaction as a site for the ongoing construction of gender as a social relationship. An approach known as **doing gender** (derived from symbolic interaction and its sister perspective, ethnomethodology; see Chapters 2 and 5) interprets gender as something accomplished through the ongoing social interactions people have with one another (West and Fenstermaker 1995; West and Zimmerman 1987). Seen from this framework, people produce gender through the interaction they have with one another and through the interpretations they have of certain actions and appearances. In other words, gender is not something that is an attribute of different people, as functionalists suggest; rather, it is constantly made up and reproduced through social interaction. When you act like a man or act like a woman, you are confirming gender and reproducing the existing social order (Peralta 2002). From this point of view, gender relations would change if large numbers of people behaved differently. This is one reason the theory has been criticized by those with a more macrosociological point of view; they say it ignores the power differences and economic differences that exist based on gender, race, and class. In other words, it does not explain the structural basis of women's oppression (Collins et al. 1995).

All of these sociological theories are useful in explaining different elements of gender in society. Sociological theory has been enriched as well by the growth of feminist theory, a broad label for theories about gender that have grown from the feminist movement.

Feminist Theory

The women's movement, both in the United States and in other nations, has not only introduced widespread changes in women's and men's lives, but it has also transformed how people understand women's and men's lives. Simply put, **feminism** refers to beliefs and action that seek a more just society for women. Feminism is not a single way of thinking, as there are different ways that feminists understand the position of women in society. Feminist thinking has resulted in new forms of social theory, namely what is known as feminist theory. **Feminist theory** has emerged from the women's movement and refers to analyses that seek to understand the position of women in society for the explicit purpose of improving their position in it. But feminist theory also includes analyses of men and the social structure of masculinity, as we will see here.

Various frameworks fall under the rubric of feminist theory, all of them with different, sometimes overlapping, ways of understanding gender in society and the social status of women and men. Each also suggests a particular avenue for social change for the purpose of creating a more just society.

Liberal feminism emerged from a long tradition that began among British liberals in the nineteenth century. Liberal feminism emphasizes individual rights and equal opportunity as the basis for social justice and reform. From this perspective, inequality for women originates in past and present practices that pose barriers to women's advancement, such as laws that historically excluded women from certain areas of work. From a liberal feminist framework, *discrimination* (that is, the unequal treatment of women) is the major source of women's inequality. Removing discriminatory laws and outlawing halting discriminatory practices are the primary ways to improve the status of women. Calls for equal rights for women are the hallmark of a liberal feminist perspective.

Liberal feminism has broad appeal because it is consistent with American values of equality before the law. But it may not be enough to remove some of the structural barriers to women's equality. More radical feminists think that liberal feminism is limited by assuming that social institutions are basically fair if women and men are just treated the same. Radical feminists argue that there cannot be justice for women as long as men hold power in social institutions. Men, for example, control the laws that govern women's reproductive lives, as one example. In this sense, *patriarchy* (that is, the power of men) is the major source of women's oppression and, furthermore, patriarchy is found at the institutional level where men have control, and at the individual level where men's power is manifested

Is It True ? (Answers)

1. FALSE. Generalizations such as this ignore variation occurring within gender categories; moreover, *aggression* is a broad term that can have multiple meanings.

2. FALSE. There are numerous sources of gender socialization; even parents who try to raise their children not to conform too strictly to gender norms will find that peers, the media, schools, and other socialization agents all push people into the expected behaviors associated with gender.

3. TRUE. Although many women do not use the label *feminist* to define themselves, surveys show that the majority of women agree with basic feminist principles. Self-identification as a feminist is most likely among well-educated, urban women (McCabe 2005).

4. FALSE. Surveys show that "stay-at-home" moms experience far more depression, worry, anxiety, and anger than employed women (Mendes et al. 2012); half of women say they would prefer to have a job outside the home (Saad 2012a).

5. TRUE. However, the gap in median income is not as wide within some groups as it is in others. White women, for example, earn 66 percent of what White men earn, but Hispanic women earn 80 percent of what Hispanic men earn (because both have very low earnings on average). Black women earn 82 percent of Black men's earnings, and Asian women, 73 percent of men's earnings. And White and Asian American women, on average, earn more than Black and Hispanic men (U.S. Census Bureau 2012).

6. FALSE. The most significant reason for the decline in the wage gap between women and men is the decline in men's wages; a smaller portion of this closing gap is attributed to changes in women's wages (Mishel et al. 2008).

7. FALSE. Although all women do benefit from equal employment legislation, wage data indicate that the group whose wages have increased the most since the 1970s are women in the top 20 percent of earners. Middle- and working-class women have seen far lower gains, and poor women's wages have been relatively flat over this period of time (Mishel et al. 2008).

in high rates of violence against women. Indeed, many radical feminists see violence against women—in the form of rape, sexual harassment, domestic violence, and sexual abuse—as mechanisms that men use to assert their power in society. Radical feminists think that change cannot come about through the existing system because men control and dominate that system.

Most recently, **multiracial feminism** has developed new avenues of theory for guiding the study of race, class, and gender (Andersen and Collins 2013; Baca Zinn and

table 11.3 Theorizing Gender: Sociological and Feminist Perspectives

	Functionalism	Conflict Theory	Symbolic Interaction Theory	Feminist Theory
Gender Identity	Gender roles are learned through socialization.	Conflict theory focuses on social structures, not individual identities.	Identity constructed through ongoing social interaction and "doing gender."	Gender identity is manifested differently also, depending on other social factors, such as race, class, and age, among others.
Status of Women	Stems from the social roles of women and men in the family.	Stems from their position as a cheap (or free) source of labor and their relative lack of power compared to men.	Stems from the enactment of gender in social interaction.	Liberal feminism advocates equal rights for women; more radical feminists see a need to challenge men's power.
Status of Men	Men hold instrumental roles; women, expressive roles.	Men hold economic advantages in the labor market and hold power in social institutions.	Masculinity is a learned identity that is created and sustained through social interaction.	Men hold power in social institutions and use that power to maintain their advantage.
Social Change	Emerges when social institutions become dysfunctional.	Comes from transformation of economic institutions and change in power structures that advantage of men.	Comes when men and women disrupt existing gender displays.	Will come only when inequalities of race, class, *and* gender are transformed.

Dill 1996; Collins 1990). Multiracial feminism evolves from studies pointing out that earlier forms of feminist thinking excluded women of color from analysis, which made it impossible for feminists to deliver theories that informed people about the experiences of all women. Multiracial feminism examines the interactive influence of gender, race, and class, showing how they together shape the experiences of all women and men. From this perspective, gender is not a singular or uniform experience, but rather intersects with race and class in shaping the experience of women and men. Gender is thus manifested differently, depending on the particular location of a given person or group in a system shaped by gender, race, and class, along with other social identities, such as sexual orientation, ability/disability status, age, nationality, and so forth. Also known as *intersectional theory,* analyses that are situated in multiracial feminist thinking have opened up sociological theory to new ways of thinking that include the multiplicity of experiences that people have in a society as diverse as the United States.

GENDER IN GLOBAL PERSPECTIVE

Increasingly, the economic condition of women and men in the United States is also linked to the fortunes of people in other parts of the world. The growth of a global economy and the availability of a cheaper industrial labor force outside the United States mean that U.S. workers have become part of an international division of labor. U.S.-based multinational corporations looking around the world for less expensive labor often turn to the developing nations and find that the cheapest laborers are women or children. The global division of labor is thus acquiring a gendered component, with women workers, usually from the poorest countries, providing a cheap supply of labor for manufacturing products that are distributed in the richer industrial nations.

Worldwide, women work as much as or more than men. It is difficult to find a single place in the world where the workplace is not segregated by gender. On a worldwide scale, women also do most of the work associated with home, children, and the elderly. Although women's paid labor has been increasing, their unpaid labor in virtually every part of the world exceeds that of men. The United Nations estimates that the value of women's unpaid work (both in the home and in the community) amounts to at least $11 trillion (**www.un.org**).

Despite these general trends, women's situations differ significantly from nation to nation. China is unusual in that couples share household responsibilities far more than those in most other nations. In China, both women and men work long hours in paid employment, and women are encouraged to stay in the labor force when they have children. There are also extensive child-care facilities and a fifty-six-day paid maternity leave in China. Many work organizations have extended this paid leave to six months, although women can lose seniority rights when they are on maternity leave (something that is illegal in the United States).

In contrast, Japan has marked inequality in the domestic sphere. Women are far more likely to leave the labor force after marrying or following childbirth, and Japanese women's identities are more defined by their roles at home, although this is changing for many. Compared with China, Japanese women more closely resemble the pattern that exists in Britain and, to some extent, in the United States, although they are less involved in paid employment than in either of these countries. Ironically, when comparing China, Japan, and Britain, researchers have found that Chinese women are the most discontented with what they perceive as gender injustice, whereas Japanese and British women express greater satisfaction with more limited employment. This may seem surprising, given the greater gender equality of Chinese women with Chinese men. Sociologists explain it as the result of the gap Chinese women see between official ideologies of gender equality and their observations of continuing inequalities in promotions and other benefits of work (Xuewen et al. 1992).

Work is not the only measure by which the status of women throughout the world is inferior to that of men. Women are vastly underrepresented in national parliaments (or other forms of government) everywhere; women's representation in national parliaments is above 25 percent in only nineteen countries of the entire world. Worldwide, women hold only 18 percent of all parliamentary seats. Only twenty-eight nations have ever had a woman as head of state (**www.ipu.org**).

The United Nations has also concluded that violence against women and girls is a global epidemic and one of the most pervasive violations of human rights (United Nations 2012a). Violence against women takes many forms, including rape, domestic violence, infanticide, incest, genital mutilation, and murder (including so-called honor killings, where a woman may be killed to uphold the honor of the family if she has been raped or accused of adultery). Although violence is pervasive, some specific groups of women are more vulnerable than others—namely, minority groups, refugees, women with disabilities, elderly women, poor and migrant women, and women living in countries where there is armed conflict. Statistics on the extent of violence against women are hard to report with accuracy, both because of the secrecy that surrounds many forms of violence and because of differences in how different nations might report violence. Nonetheless, the United Nations estimates that between 20 and 50 percent of women worldwide have experienced violence from an intimate partner or family member.

As we saw in Chapter 9, many factors put women at risk of violence, including cultural norms, women's economic and social dependence on men, and political practices that either provide inadequate legal protection

The End of Men?

In 2012, journalist Hanna Rosin published a book, *The End of Men: And the Rise of Women*, a book that was widely reviewed and earned Rosin appearances on numerous TV talk shows, news reports, and other media outlets. Her argument was fairly straightforward: that women are replacing men as the heads of households; that women's gender roles make them more adaptable to changes in the economy while men's roles make them more rigid and conforming to past ideals; and, that women are now dominating men in many workplaces and in colleges. For Rosin, "the end of men" signals a time when women are surpassing men in many areas of life,

actually leaving men behind who cannot adapt to the new demands of a postindustrial, service-based economy.

Within sociology and beyond, Rosin's book caused quite a stir. Rosin's book points to some major social changes that are affecting both women and men: Women are now a majority in the nation's colleges and universities; women's wages have been rising while men's are falling; women can be found among the nation's highest income earners; and women more often find themselves as the major breadwinner in many families. But these facts are also undercut by realities that Rosin's book glosses over, if acknowledged at all, namely, that women

are still segregated in lower-wage occupations; women major in fields that are not those that produce top earners; women have higher rates of poverty than men; and women remain unequal in every institution in society. Moreover, Rosin's caricature of men as rigid and resistant to change and women as nimble and flexible rests on highly overgeneralized gender stereotypes.

You might ask yourself why Rosin's book struck such a nerve for the public. What changes do you see in your environment that might lead people to think that women have it made and men are falling behind? What specific information would you need to assess the claims that Rosin is making, including information about diverse groups of women and men?

---

or provide explicit support for women's subordination (as in the example of the Taliban given earlier).

GENDER AND SOCIAL CHANGE

Few lives have not been touched by the transformations that have occurred in the wake of the feminist movement (see also Chapter 16). The women's movement has opened work opportunities, generated laws that protect women's rights, spawned organizations that lobby for public policies on behalf of women, and changed public attitudes. Many young women and men now take for granted freedoms struggled for by earlier generations. These include access to birth control, equal opportunity legislation, laws protecting against sexual harassment, increased athletic opportunities for women, more presence of women in political life, and greater access to child care, to name a few changes. These impressive changes occurred in a relatively short period of time.

Indeed, many believe that the gender revolution is over and that there is no further need for feminist change. Some say that the nation is *postfeminist*, a term that means different things to different people. For some, it simply means that the women's movement is over because feminism has outlived the need. For others, it means that second-wave feminism (that of the 1970s and 1980s) does not meet the needs of new generations of women. What has the feminist movement accomplished and what remains to be done?

In some regards, women have reached pinnacles of power and influence unprecedented in U.S. history. Women are highly visible as CEOs of major corporations (see Map 11.1), as Supreme Court justices and presidential cabinet members, as extraordinary and highly paid athletes—all of which would have been highly unusual not that many years ago. Women have also risen to positions of political influence, perhaps signaling a new era for women in the political realm. Women have been especially evident in the new conservative movement, such as in the Tea Party and other organizations.

No doubt there has been substantial progress for women, but the tensions between progressive and conservative politics on women's issues indicate that women's rights are hardly a settled issue. Despite the greater visibility of women, as we have seen, most women still struggle with low wages, managing both work and family, or perhaps struggling alone to support a family. On the conservative side, people feel that the value of women as traditional homemakers is being eroded, perhaps even threatened by women's independence because women in that position need men's economic support. On the progressive side, feminists perceive constant threats to women's reproductive rights and think that the nation has not come far enough in protecting and supporting women's rights. There is currently a complex mix of progressive and conservative politics surrounding gender in the contemporary world. Yet, the changes the feminist movement has inspired have completely transformed many dimensions of women's and men's lives, especially apparent in changed public attitudes.

MAP 11.1

Viewing Society in Global Perspective: Women in Senior Management

Source: Grant Thornton International Business Report. 2013.

Percentage of Women in Senior Management

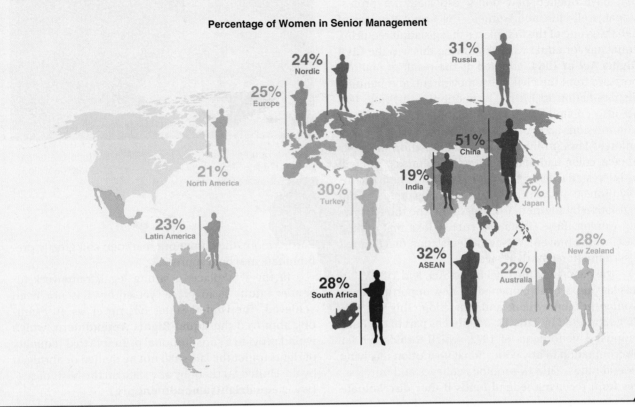

Contemporary Attitudes

Surveys of public opinion about women's and men's lives are good indicators of the changes we have witnessed as the result of the feminist movement. Now, only a small minority of people disapproves of women being employed while they have young children, and both women and men say it is not fair for men to be the sole decision maker in the household. Young women and men have different ideals for their future lives than was true for earlier generations. Kathleen Gerson's research finds that most young adults (those she calls "children of the gender revolution") want a lifelong partner and shared responsibilities for work and family. But Gerson also finds a strong gender divide in how men and women imagine what they would do as a backup plan if their ideals were not realized. Men, much more than women, think that if balancing work and family does not work out in their future, they would fall back on traditional arrangements with wives staying home and husbands working. Women disagree, understanding that they have to be self-reliant in the event that their ideals are not met (Gerson 2010).

Gerson also found that, although both men and women want to share work and family life, institutions have not adjusted to this reality. In other words, resistant institutions have not adjusted to the attitudinal changes and desires for new lifestyles that most women and men embrace.

Attitudinal changes are not, however, complete. Many people want more flexible gender arrangements, but traditional gender norms also remain. Beliefs that there are basic differences between women and men and support for traditional gender norms in many aspects of personal life still prevail (England 2010). Thus there is unevenness in how much gender relations have changed. As a consequence, change has been uneven—revolutionary in some regards and static in others (Greenhouse 2000).

---

debunking SOCIETY'S MYTHS

MYTH: The men most likely to support equality for women are White, middle-class men with a good education.

SOCIOLOGICAL PERSPECTIVE: Although it is true that younger men tend to be more egalitarian than older men, African American men are the most likely to support women's equal rights and the right of women to work outside the home. On most measures of feminist beliefs, African American men tend to be more liberal than White men. ●

Legislative Change

Attitudes are only one dimension of social change. Some of the most important changes have come from laws that now protect against discrimination—laws that have opened new doors, especially for professional, well-educated women. The **Equal Pay Act of 1963** was one of the first pieces of legislation requiring equal pay for equal work. Following this was the **Civil Rights Act of 1964**, adopted as the result of political pressure from the civil rights movement, and banning discrimination in hiring, promotion, and firing; this act also created the Equal Employment Opportunity Commission, an arm of the federal government that enforces laws prohibiting discrimination on the basis of race, color, national origin, religion, or sex. You will be interested to learn that adding "sex" to this law actually protects women by accident. It was added to the legislation by southern Congressmen who thought giving women these rights was such a joke that adding "sex" would prevent passage of legislation that banned discrimination based on race.

The passage of the Civil Rights Act, and Title VII in this law in particular, opened up new opportunities to women in employment and education. This was further supported by **Title IX**, adopted as part of the Educational Amendments of 1972, which forbids gender discrimination in any educational institution receiving federal funds. Title IX prohibits colleges and universities from receiving federal funds if they discriminate against women in any program, including athletics. Adoption of this bill has radically altered the opportunities available to women students and has laid the foundation for many of the coeducational programs that are now an ordinary part of college life. This law has been particularly effective in opening up athletics to women.

Passage of antidiscrimination policies does not, however, guarantee their universal implementation. Has equality been achieved? In college sports, men still outnumber women athletes by more than two to one, and there is still more scholarship support for male athletes than for women. Title IX allows institutions to spend more money on male athletes if they outnumber women athletes, but it also stipulates that the number of male and female athletes should be closely proportional to their representation in the student body. Studies of student athletes show that although there has been improvement in support for women's athletics since the implementation of Title IX, there is still a long way to go toward equity in women's sports (Sigelman and Wahlbeck 1999; Lederman 1992). Title IX is being challenged by some who argue that it has reduced opportunities for men in sports. Proponents of maintaining strong enforcement of Title IX counter this, however, by noting that budget reductions in higher education, not Title IX per se, are responsible for any reduction in athletic opportunities for men.

Young women are the group most likely to support feminist goals. Ending violence against women is a strong theme in contemporary feminism.

Furthermore, they point out that men still greatly predominate in school sports.

In the workplace, a strong legal framework for gender equity is in place, yet equity has not been achieved. The United States has never, as an example, approved the **Equal Rights Amendment**, which would provide a constitutional principle that "equality of rights under the law shall not be denied or abridged by the United States or by any state on the basis of sex" **(www.equalrightsamendment.org)**.

---

debunking SOCIETY'S MYTHS

MYTH: Black women are taking a lot of jobs away from White men.

SOCIOLOGICAL PERSPECTIVE: Sociological research finds no evidence of this claim. Quite the contrary, women of color work in gender- and race-segregated jobs and only rarely in occupations where they compete with White men in the labor market (Padavic and Reskin 2002; Browne 1999). •

One solution to the problem of gender inequality is to have more women in positions of public power. Is increasing the representation of women in existing situations enough? Without reforming the sexism in the institutions, change will be limited and may generate benefits only for groups who are already privileged. Feminists advocate restructuring social institutions to meet the needs of all groups, not just those who already have enough power and privilege to make social institutions work for them. The successes of the women's movement demonstrate that change is possible, but change comes only when people are vigilant about their needs.

chapter summary

How do sociologists distinguish sex and gender?
Sociologists use the term *sex* to refer to biological identity and *gender* to refer to the socially learned expectations associated with members of each sex. *Biological determinism* refers to explanations that attribute complex social phenomena entirely to physical or natural characteristics.

How is gender identity learned?
Gender socialization is the process by which gender expectations are learned. One result of socialization is the formation of *gender identity*. Overly conforming to gender roles has a number of negative consequences for both women and men, including eating disorders, violence, and poor self-concepts. *Homophobia* plays a role in gender socialization because it encourages strict conformity to gender expectations.

What is a gendered institution?
Gendered institutions are those where the entire institution is patterned by gender. Sociologists analyze gender both as a learned attribute and as an institutional structure.

What is gender stratification?
Gender stratification refers to the hierarchical distribution of social and economic resources according to gender. Most societies have some form of gender stratification, although they differ in the degree and kind. Gender stratification in the United States is obvious in the differences between men's and women's wages.

How do sociologists explain the continuing earnings gap between men and women?
There are multiple ways to explain the pay gap. *Human capital theory* explains wage differences as the result of individual differences between workers. *Dual labor market theory* refers to the tendency for the labor market to be organized in two sectors: the primary and secondary markets. *Gender segregation* persists and results in differential pay and value attached to men's and women's work. *Overt discrimination* against women is another way that men protect their privilege in the labor market.

Are men Increasing their efforts in housework and child care?
Many men are now more engaged in housework and child care than was true in the past, although women still provide the vast majority of this labor. Balancing work and family has resulted in social speedup, making time a scarce resource for many women and men.

What is feminist theory?
Different theoretical perspectives help explain the status of women in society. *Functionalist theory* emphasizes how gender roles that differentiate women and men work to the benefit of society. *Conflict theory* interprets gender inequality as stemming from women's status as a supply of cheap labor and men's greater power in social institutions. *Symbolic interaction* sees women and men as "doing gender" in everyday interaction. *Feminist theory*, originating in the women's movement, refers to analyses that seek to understand the position of women in society for the explicit purpose of improving their position in it. Within this broad rubric are both liberal and more radical perspectives, including multiracial feminism, which emphasizes the linkage between gender, race, and class inequality.

When seen in global perspective, what can be observed about gender?
The economic condition of women and men in the United States is increasingly linked to the fortunes of people in other parts of the world. Women provide much of the cheap labor for products made around the world. Worldwide, women work as much or more than men, though they own little of the world's property and are underrepresented in positions of world leadership.

What are the major social changes that have affected women and men in recent years?
Public attitudes about gender relations have changed dramatically in recent years. Women and men are now more egalitarian in their attitudes, although women still perceive high degrees of discrimination in the labor force. A legal framework is in place to protect against discrimination, but legal reform is not enough to create gender equity.

Key Terms

biological determinism 257
Civil Rights Act (1964) 278
discrimination 271
doing gender 273
dual labor market
   theory 269
Equal Pay Act (1963) 278
Equal Rights
   Amendment 278

feminism 273
feminist theory 273
gender 257
gender apartheid 266
gender identity 259
gender segregation 270
gender socialization 259
gender stratification 266

gendered institution 265
homophobia 265
human capital theory 268
intersexed 257
labor force participation
   rate 268
liberal feminism 273
matriarchy 268

multiracial feminism 274
occupational
   segregation 270
patriarchy 267
sex 257
Title IX 278
transgender 258

12 Sexuality

Jonathan Nourok/Stone/Getty Images

a visitor from another planet might conclude that people in the United States are obsessed with sex. Young people watch videos where women gyrate in sexual movements. A stroll through a shopping mall reveals expensive shops selling delicate, skimpy women's lingerie. Popular magazines are filled with images of women in seductive poses trying to sell every product imaginable; even bumper stickers brag about sexual accomplishments. People dream about sex, form relationships based on sex, fight about sex, and spend money to have sex. On the one hand, the United States appears to be a very sexually open society; however, sexual oppression still exists. Gay men, lesbians, and bisexuals are viewed with prejudice and are discriminated against—that is, treated like minority groups, as defined in Chapter 10.

Sexuality, usually thought to be a most private matter, has taken on a public life by being at the center of some of our most heated public controversies. Should marriage be confined to same-sex couples? Should young people be educated about birth control or only encouraged to abstain from sex? Should government require employers to provide insurance coverage for birth control? Who decides whether a woman can choose to have an abortion and under what circumstances?

Sexuality and the issues it spins off are clearly subjects that polarize the public on a range of social issues. On the one hand, sexuality is seen as a private matter but it is also very much on the public agenda. For sociologists, studying sexuality reveals how deeply it is entrenched in social norms, values, and social inequalities. Human sexuality, like other forms of social behavior, is shaped by society and culture.

- Understand the social basis of human sexuality
- Identify current attitudes and behaviors involving sexuality
- Comprehend how sexuality is linked to other forms of inequality
- Compare and contrast theoretical perspectives on sexuality
- Be able to define homophobia and heterosexism and their influence on lesbian and gay experience
- Comprehend sociological underpinnings of current issues regarding sexuality
- Assess how sexuality is being affected by social change

SEX AND CULTURE

Sexual behavior would seem to be utterly natural. Pleasure and sometimes the desire to reproduce are reasons people have sex, but sexual relationships and identities develop within a social context. That social context establishes what sexual relationships mean and how we define our sexual identities, as well as what social supports are given (or denied) to people based on their sexual identity. *Sexuality is socially defined and patterned.*

Sex: Is It Natural?

From a sociological point of view, little in human behavior is purely natural, as we have learned in previous chapters. Behavior that appears to be natural is the behavior accepted by cultural customs and sanctioned by social institutions. People engage in sex not just because it feels good, but also because it is an important part of our social identity. Sexuality creates intimacy between people. But void of a cultural context and the social meanings attributed to sexual behavior, people might not attribute the emotional commitments, psychological interpretations, spiritual meanings, and social significance to sexuality that it has in different human cultures.

Is there a biological basis to sexual identity? This question is debated in both popular and scientific literature. Gay and lesbian people often say their sexual orientation is natural, something they just are, not just something they choose. Two concepts are important in this discussion: *sexual orientation* and *sexual identity*.

Sexual orientation refers to the attraction that people feel for people of the same or different sex. The term *sexual orientation* implies something deeply rooted in a person. **Sexual identity** is the definition of oneself that is formed around one's sexual relationships. Sexual identity is learned in the context of our social relationships and the social structures in which we live. Although sometimes used interchangeably, sexual orientation and sexual identity are not the same thing, nor is one's sexual identity simply based on one's sexual practices. For example, a man may have sex with other men—perhaps even on a regular basis, but not have a sexual identity as being gay. Or a person may have a sexual identity as heterosexual even in the absence of actual sexual relationships. Sexual identity as gay, lesbian, heterosexual, or bisexual emerges in

Sexual relationships, although highly personal, are also shaped by society and culture.

a social context, as we will see in the following discussion on the social construction of sexual identity. As an example, recall from the previous chapter that *transgender* people are those who construct a gender identity different from their biological identity. But this does not predict a particular sexual orientation (Schilt 2011). A transgender person may be straight, lesbian, gay, or bisexual. The point here is to see that sexual identity is not necessarily simply based on biological or "natural" states, even though there is debate about whether there is a biological basis to sexual orientation.

Gay, lesbian, transgender, and bisexual people often say that they do not choose their sexual orientation and that it is something they just "are," as if it were a biological imperative. Part of the debate about this—heated at times—comes from rejecting the idea that being gay is a choice, as if people could change their sexual orientation at will. There are political reasons for rejecting the idea of sexual orientation as a choice because, if it is something inherent in people, then perhaps others will be more accepting of gay, lesbian, and bisexual people.

Perhaps there is some biological basis to sexual orientation, but the evidence is not yet there. Even if a biological influence exists, social experiences are far more significant in shaping sexual identity, even though they are rarely reported in the media with as much acclaim as alleged biological bases to human sexuality (Brookey 2001; Lorber 1994; Connell 1992). Whatever the origins of sexual orientation, there is no doubt that social influences are a very significant part of all people's sexual identity.

Sometimes there is a public claim that scientists have discovered a so-called gay gene, presumably directing the sexual orientation of gays and lesbians. Interestingly, there are never claims about a so-called heterosexual gene because the implicit assumption seems to be that heterosexuality is the natural state, and gay or lesbian behavior is somewhat a mutant form. There is no such scientific evidence.

The evidence usually cited to claim a genetic basis for homosexuality was based on a problematic study of gay brothers (identical twins) who were found to have similar DNA markers on one of their X chromosomes. On closer examination, this research was refuted when it was found that the shared DNA markers found in pairs of gay brothers were no more likely than would be expected by chance (Wickelgren 1999). Moreover, the original study did not control for the environment in which the brothers were raised. They grew up in the same family, so obviously they were raised in the same environment. A true scientific test of the hypothesis that homosexuality is genetically based would require a stricter standard of evidence. One could study identical twins raised together who were the offspring of gay parents to see if, controlling for the environments in which they were raised, both turned out to be gay. To date, there are no such studies of biological relatives raised apart (Hamer et al. 1993).

Even if there is some yet undiscovered basis for sexual orientation, there is extensive evidence of the social influences that shape people's sexual identities. Social and cultural environments play a huge part in creating sexual identities. What interests sociologists is how sexual identity is constructed through social relationships and in the context of social institutions.

The Social Basis of Sexuality

We can see the social and cultural basis of sexuality in numerous ways:

1. Human sexual attitudes and behavior vary in different cultural contexts. If sex were purely natural behavior, sexual behavior would also be uniform among all societies, but it is not. Sexual behaviors considered normal in one society might be seen as peculiar in another. Think about this: In some cultures, women do not believe that orgasm exists, even though it does biologically. In the eighteenth century, European and American writers advised men that masturbation robbed them of their physical powers and that instead they should apply their minds to the study of business. These cultural dictates encouraged men to conserve semen on the presumption that its release would lessen men's intelligence or cause insanity (Schwartz and Rutter 1998; Freedman and D'Emilio 1988).

---

debunking SOCIETY'S MYTHS

MYTH: Over time in a society, sexual attitudes become more permissive.

SOCIOLOGICAL PERSPECTIVE: All values and attitudes develop in specific social contexts; change is not always in a more permissive direction (Freedman and D'Emilio 1988). ●

---

2. Sexual attitudes and behavior change over time. Fluctuations in sexual attitudes are easy to document. For example, in 1968, 68 percent of the American public thought premarital sex was morally wrong, compared to 38 percent who think so now (Gallup 2003; Saad 2010). Teens, as well, have changed their attitudes about sex. In 1977, one-third of teens though it was morally wrong to have sex outside of marriage; now 42 percent think premarital sex is wrong, suggesting that teen tolerance of casual sex has declined in recent years (Lyons 2004, 2002).

   Of course, attitudes do not necessarily predict behavior. Although 42 percent of teens think premarital sex is wrong, by age 19, 70 percent of teens (both men and women) have had sex (Guttmacher Institute 2012). And even though Americans express less tolerance for marital infidelity in public opinion polls, they seem endlessly fascinated by the affairs and sexual escapades of celebrities and high-profile people. At the same time, attitudes toward being unfaithful to one's spouse have become less tolerant. And, people's reports of actually being unfaithful to their spouse have wavered over time—with about 25 percent of men and 15 percent of women now admitting that they have had marital affairs (Carr 2010).

3. Sexual identity is learned. Like other forms of social identity, sexual identity is acquired through socialization and ongoing relationships. Information about sexuality is transmitted culturally and becomes the basis for what we know of ourselves and others. Where did you first learn about sex? For that matter, what did you learn? For some, parents are the source of information about sex and sexual behavior. For many, peers have the strongest influence on sexual attitudes. Long before young people become sexually active, they learn **sexual scripts** that teach us what is appropriate sexual behavior for each gender (Schwartz and Rutter 1998). Children learn sexual scripts by playing roles—playing doctor as a way of exploring their bodies, or hugging and kissing in a way that can mimic heterosexual relationships. Role-playing teaches children social norms about sexuality. The roles learned in youth profoundly influence our sexual attitudes and behavior throughout life.

4. Social institutions channel and direct human sexuality. Social institutions, such as religion, education, or the family, define some forms of sexual expression as more legitimate than others. Debates about same-sex marriage illustrate this concept. Without being able to marry, gay and lesbian couples lose some of the privileges that married couples receive, such as employee benefits and the option to file joint tax returns. Social institutions, such as the media, also influence sexuality through the production and distribution of images that define cultural meanings attributed to sexuality.

5. Sex is influenced by economic forces in society. Sex sells. In the U.S. capitalist economy, sex appeal is used to hawk everything from cars and personal care products to stocks and bonds. In this sense, sex has become a commodity—something bought and sold in the marketplace of society. Sex also can be big business. By one estimate, Americans spend over $10 billion per year in the sex industry, including strip bars, peep shows, phone sex, sex acts, sex magazines, and pornography rentals—and this is likely to be an underestimate (Barton 2006). Moreover, the business of sex means some people may actually be "bought and sold." Think of women who may feel that selling sexual services is their best option for earning a living wage. Sex workers are among some of the most exploited and misunderstood workers. And, as we will see later in this chapter, sex trafficking on a global basis is also more extensive than commonly thought.

6. Public policies regulate sexual and reproductive behaviors. In many ways, the government and other social policies intervene in people's sexual and reproductive decision making. Prohibiting federal spending on abortion, for example, eliminates reproductive choices for women who are dependent on state or federal aid. Government decisions about which reproductive technologies to endorse influence the choices of birth control technology available to men and to women. Government funding, or lack thereof, for sex education can influence how people understand sexual behavior. These facts challenge the idea that sexuality is a private matter

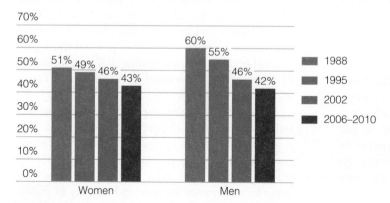

FIGURE 12.1 Sex among Teenagers: A Change over Time The chart shows the change in the percentage of teen men and women (aged 15–19) who have had sex. What do you observe in these data? What sociological factors might you consider to explain what you observe?

Source: Vital and Health Statistics. 2011. *Teenagers in the United States: Sexual Activity, Contraceptive Use, and Childbearing*, 2006–2010. National Survey of Family Growth. Washington, DC: National Center for Health Statistics, U.S. Department of Health and Human Services.

and shows how social institutions can direct sexual behavior. To summarize, human sexual behavior occurs within a cultural and social context.

In sum, culture defines certain sexual behaviors as appropriate or inappropriate. Like other forms of social identity, sexual identity is learned.

CONTEMPORARY SEXUAL ATTITUDES AND BEHAVIOR

Sexual attitudes and behaviors in the United States are a mix of ideas and practices, both of which vary depending on different social factors that shape people's experiences. While on the one hand, many support diverse sexual life-styles and practices, the growth of a conservative movement has also shaped the sexual values and behaviors of others. These differences have also been at the heart of highly contested issues that have shaped U.S. politics in recent years, such as debates about same-sex marriage, sex education in the schools, and policies about reproductive and contraceptive health. This makes sociological research on sexuality all the more fascinating as public attitudes and behaviors shift with changes in the society itself.

Changing Sexual Values

Public opinion is now a mix of both liberal and conservative values about sexuality. Almost two-thirds of Americans (59 percent) think that sex before marriage is morally acceptable, but among Americans who attend church weekly, only 28 percent think this is morally acceptable. But, if you put this question before an international audience, you might be surprised to find that Americans are more conservative on this matter than people in western Europe. In Germany, France, and the United Kingdom, a very large majority of people—in Germany (86 percent), France (88 percent), and the United Kingdom (80 percent)—say that they find this morally acceptable (Newport 2012a; Rheault and Mogahed 2008; Gallup 2003).

On gay rights, more Americans than ever before think that "gay/lesbian relations are morally acceptable" (54 percent in 2012 compared to 42 percent in 2004). Eighty-eight percent think "homosexuals should have equal rights in terms of job opportunities" (Saad 2012c; Newport 2003; see also Figure 12.2). Yet, the public is clearly divided on the issue of same-sex marriage, and it is unclear what direction the law will take in the long run. Several states and the District of Columbia have passed laws making same-sex marriage legal. In 2003, the U.S. Supreme Court ruled (*Lawrence v. Texas*) that private sexual relations are a constitutional liberty, a conclusion widely interpreted as a major victory for gay rights. An even greater victory for supporters of same-sex marriage came in 2013 when the U.S. Supreme Court ruled the Defense of Marriage Act (DOMA) unconstitutional. DOMA, passed by Congress in 2006, had defined marriage as solely between one man and one woman. The ruling means that legally married same sex couples same sex marriage is cannot be denied state and federal benefits (such as the right to inheritance, filing joint tax returns, and so forth).

Attitudes about sex vary significantly depending on various social characteristics. For example, men are more likely than women to think that gay/lesbian relations are morally wrong. Sexual attitudes are also shaped by age. Younger people are more likely than older people to think that gay/lesbian relations are

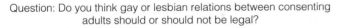

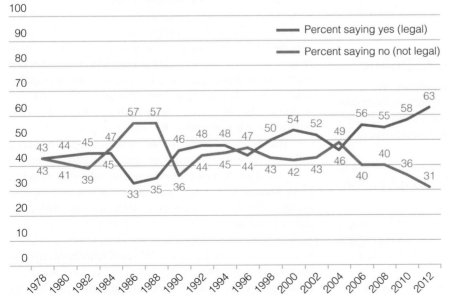

Question: Do you think gay or lesbian relations between consenting adults should or should not be legal?

— Percent saying yes (legal)

— Percent saying no (not legal)

FIGURE 12.2 Support for Gay Rights As you can see, public opinion about whether gay and lesbian relations should be legal has varied significantly over time, but has never been as supported as now. Public opinion will vary, however, depending on how a question about gay/lesbian rights is worded.

Data: Saad, Lydia. 2012 (May 14). "U.S. Acceptance of Gay/Lesbian Relations is the New Normal." *Gallup Poll.* **www.gallup.com**

Note: Question wording was changed from homosexual to gay and lesbian in more recent years.

Gays, lesbians, and their allies have mobilized for social change, fostering pride and celebration as well as a reduction over time in homophobic attitudes.

Marmaduke St. John/Alamy Limited

morally acceptable. These differences likely reflect not only the influence of age, but also historical influences on different generations. Religion also matters. Those who attend church weekly are far less likely to support gay rights compared to those who worship less often (Pelham and Crabtree 2009). Public opinion on matters about sexuality taps underlying value systems, thus generating public conflicts. In general, sexual liberalism is associated with greater education, youth, urban lifestyle, and political liberalism on other social issues.

Sexual Practices of the U.S. Public

Sexual practices are difficult to document. What we know about sexual behavior is typically drawn from surveys. Most of these surveys ask about sexual attitudes, not actual behavior. What people say they do may differ significantly from what they actually do.

As much as sex is in the news, national surveys of sexual practices are rare. Those that have been conducted tell us the following:

- Contrary to public opinion, teens are waiting longer to have sex than was true in the past, though seven in ten teens have had sex by age 19 (Guttmacher Institute 2012).
- Having only one sex partner in one's lifetime is rare (Laumann et al. 1994).

- A significant number of people have extramarital sex; estimates are that 20 percent of all marriages experience at least one instance of infidelity (Guerrero et al. 2010; Parker-Pope 2008).

A small percentage of the public says they personally identify as gay, lesbian, or bisexual, but such identities are notoriously difficult to measure accurately. Black Americans, Hispanics, and Asians are more likely to identify as gay, lesbian, or bisexual, even though homophobic attitudes are more widespread in these populations (Gates and Newport 2012; Durell et al. 2007; Lewis, G. 2003).

SEX AND INEQUALITY: GENDER, RACE, AND CLASS

When you take a sociological view of sexuality, you see how sexuality is linked to other social identities and social systems. Start with gender. Our gender identities link with sexuality in many ways. Indeed, men learn to be men by dissociating themselves from anything that seems "gay" or "sissy" (Pascoe 2007). Boys are raised to be manly by repressing so-called feminine characteristics in themselves. Being called a "fag" or a "sissy" is one peer sanction that socializes a child to conform to particular gender roles. Similarly, verbal attacks on lesbians by using the term *butch* are a mechanism of social control because ridicule can be interpreted as encouraging social conformity to the presumed "normal" gender roles (Pascoe 2007). Homophobia produces many misunderstandings about gay people, such as the misunderstanding that gays have a desire to seduce straight people. There is little evidence that this is true.

Gender also shapes the so-called double standard for men and women. The double standard is the idea that different standards for sexual behavior apply to men and women. The double standard has weakened somewhat over time, particularly in the context of the "hooking up" culture among young people. But it is still very much present, so much so that a national survey of American youth has found that men who report higher numbers of sexual partners are actually more popular than men with fewer partners. The opposite is true for women—that is, women with high numbers of sexual partners are less popular (Kreager and Staff 2009).

The double standard also applies the to "hooking up" culture among young people. *Hooking up* is the term used to describe casual sexual relations ranging from kissing to sexual intercourse, relationships that occur without any particular commitment. Although the popular image of hooking up is that it is totally free and without constraint, there are in fact very gendered norms within this behavior. Women who hook up too frequently or with too many different partners are likely to be judged as "slutty." Men who do the same things are not judged in the same way. They may be seen as "players," but they are not subjected to the same shame or attribution of guilt that is targeted at women.

The double standard rests on a cultural expectation for men to be more sexually active; women, more passive. And the sexual double standard for men and women has many consequences, both for people's identities and reputation, and in people's reactions to sexual violence. Thus, the belief that women who are raped must have brought it on themselves rests on a stereotyped image of women as sexual temptresses who somehow encourage men to rape them. Likewise, there is a stereotype for men—that they are unable to control their sexuality, even though men do not have a stronger sex drive than women. Men are, however, socialized more often to see sex in terms of performance and achievement, whereas women are more likely socialized to associate sex with intimacy and affection.

Sexuality is also integrally tied to race and class inequality in society—a fact that you can see clearly in the sexual stereotypes associated with race and class. Latinas are stereotyped as either "hot" or "virgins"; Latino men are stereotyped as "hot lovers"; African American men are stereotyped as overly virile; Asian American women are stereotyped as compliant and submissive, but passionate. Class relations also produce sexual stereotypes of women and men. Working-class and poor men may be stereotyped as dangerous, whereas working-class women may be disproportionately labeled "sluts."

You could see the influence of social class in sexual stereotypes if you recall the highly publicized case of "Octomom." There was a huge public outcry when a young, unmarried woman on public assistance delivered octuplets, following in vitro fertilization. The young woman, Nadya Suleman, already the mother of

Having celebrities "come out" about being lesbian or gay has empowered others to be able to do so as well.

six children, was completely scorned in the media for having poor judgment as a mother. At the very same time, a reality TV show, *Jon & Kate Plus Eight,* depicted a family with eight children as fun and lovable, not despicable and undeserving as was the case with Nadya Suleman. Nadya Suleman was unmarried, an Iraqi immigrant, and also on public assistance—thus judged more harshly than the heterosexual, middle-class Gosselin family in *Jon & Kate Plus Eight.*

Class, race, and gender hierarchies historically depict people of color and certain women as sexually promiscuous and uncontrollable (Nagel 2003). During slavery, for example, the sexual abuse of African American women was one way that slave owners expressed their ownership of African American people. Access to women slaves' sexuality was seen as a right of the slave owner. Under slavery, racist and sexist images of Black men and women were developed to justify the system of slavery. Black men were stereotyped as lustful beasts whose sexuality had to be controlled by the "superior" Whites. Black women were also depicted as sexual animals who were openly available to White men.

A Black man falsely accused of having had sex with a White woman could be murdered (that is, lynched) without penalty to his killers (Genovese 1972; Jordan 1968). Sexual abuse was also part of the White conquest of American Indians. Historical accounts show that the rape of Indian women by White conquerors was common (Freedman and D'Emilio 1988; Tuan 1984). These patterns also can be seen in the extensive rape of women that often accompanies war and military conquest. Most recently, this has been witnessed in Darfur, where the rape of women is used not just as a horrid act of aggression against women, but also as a way to humiliate and control the Darfur people. Similar violence is occurring in the Democratic Republic of the Congo, where a civil war rages and thousands of women are raped each year as the result of the civil war.

In other contexts, poor women and women of color are the groups most vulnerable to sexual violence and exploitation. Becoming a prostitute, or otherwise working in the sex industry (as a topless dancer, striptease artist, pornographic actress, or other sex-based occupation), is often the last resort for women with limited options to support themselves. Women who sell sex also are condemned for their behavior more so than their male clients—further illustration of how gender stereotypes mix with race and class exploitation. Why, for example, are women, and not their male clients, arrested for prostitution? Although data from the Uniform Crime Reports do not report on the arrest rates of customers, prostitutes themselves claim that only about 10 percent of those arrested for prostitution are customers and that women of color are more likely to be arrested for prostitution than are White women, even though they are a smaller percentage of all prostitutes (Prostitutes Education Network 2009). Although these are not scientific data, they suggest

Sex and Popular Culture

Imagine that some of the classical sociological theorists were reincarnated and today observed sexuality and references to it in everyday life. What observations and comments might they make? *Emile Durkheim* would remind us that marking some behaviors as deviant is how people in society also define what is considered "normal." He might observe young boys calling each other "faggots." Interestingly, were he watching carefully, he would also see that the person being targeted by such comments, including by adults, is probably *not* gay. But the homophobic banter among boys and men is a way of asserting the dominant (and somewhat narrow) norms of what masculinity is presumed to be.

Max Weber would see something else. He would notice that sexuality, particularly heterosexuality, is rampant in popular culture. Clothing styles, popular lingo, and styles of dance are all marked by overt displays of sexuality. He would argue that such cultural displays go hand in hand with an economy that treats sexuality as part of the marketplace. Moreover, the interplay between culture and the economic marketplace leads to social judgments about some sexual styles and identities being more highly valued than others. In other words, Weber would emphasize the multidimensional connection between the economy, culture, and social judgments.

Karl Marx, on the other hand, would be intrigued by the commercial exploitation of sex, or "sexploitation." Sexuality has become a commodity in modern society. It is used to sell things for the benefit of those who own the various industries where fashion, personal care products, and even sex itself, profit some while exploiting others. Marx might even note, were he paying attention to the treatment of women, that sex workers are among the most exploited workers, actually selling their bodies for the benefit of others.

W. E. B. DuBois would add that sexual exploitation is particularly harsh for Black people. The popular practice of college students partying as "pimps and hos" rests on a stereotype of Black men as sexual predators and Black women as promiscuous. These stereotypes have a long history stemming from racism. And Black women are among the most sexually exploited. Their bodies are abused, they are portrayed in sexualized stereotypes in popular culture, and they are those who are most victimized by sexual violence.

Together, these classical theorists provide sociological perspectives on images and practices that are common in everyday life, but are rarely challenged or questioned, unless someone is on the receiving end of narrow, sexist, and racist understandings of sexuality.

---

the role that gender and race play in how laws against prostitution are enforced.

How do sociologists frame these complex connections between sexual identities, sexual stereotypes, and the various forms of social inequality that permeate society? For answers, we turn to sociological theory.

SEXUALITY: SOCIOLOGICAL AND FEMINIST THEORY

How are sexual identities formed? Is marriage only a relationship between a man and a woman? What role does power have in sexual relationships? While these and other questions are debated as moral issues in society, sociologists see these and other questions about sexuality as subjects for sociological study. Sociological theory puts an analytical framework around the study of sexuality, examining its connection to social institutions and current social issues. How do the major sociological theories frame an understanding of sexuality?

Sex: Functional or Conflict-Based?

The three major sociological frameworks—functionalist theory, conflict theory, and symbolic interaction—take divergent paths in interpreting the social basis of human sexuality (see Table 12.1). Added to these is the influence of feminist theory, which has very much transformed sociological understanding of sexuality and its connection to society.

Functionalist theory, with its emphasis on the interrelatedness of different parts of society, tends to depict sexuality in terms of its contribution to the stability of social institutions. Norms that restrict sex to marriage encourage the formation of families. Similarly, beliefs that give legitimacy to heterosexual behavior, but not homosexual behavior, maintain a particular form of social organization in which gender roles are easily differentiated and the nuclear family is defined as the dominant social norm. From this point of view, regulating sexual behavior is functional for society because it prevents the instability and conflict that more liberal sexual attitudes supposedly generate. Functionalists would also explain the call for a return to "family values" as encouraging the uniformity in values necessary for social order.

Conflict theorists see sexuality as part of the power relations and economic inequality in society. *Power* is the ability of one person or group to influence the behavior of another. Power relations in society influence the power that some sexual groups have over others and

table 12.1 Theoretical Perspectives on Sexuality

Interprets	Functionalism	Conflict Theory	Symbolic Interaction	Feminist Theory
Sexual norms	As functional for society because they produce stability in institutions such as the family	As often contested by those who are subordinated by dominant and powerful sexual groups	As emerging and reinforced through social interaction	As established in a system of male domination; producing narrow definitions of women's and men's sexuality
Sexual identity	As learned in the family and other social institutions, with deviant sexual identities contributing to social disorder	As regulated by individuals and institutions that enforce only some forms of sexual behavior as desirable, thus enforcing heterosexism	As socially constructed when people learn the sexual scripts produced in society	As acknowledging that multiple forms of sexual identity are possible with some people crossing the ordinarily assumed boundaries
Sex and social change	As regulating sexual values and norms being important for maintaining traditional and social stability, with too much change resulting in social disorganization	As coming through the activism of people who challenge dominant belief systems and practices	As evolving as people construct new beliefs and practices over time	As sexual values being changed through disrupting taken-for-granted categories of the dominant culture

© Cengage Learning

influence power within sexual relationships ("sexual politics" is discussed later in the chapter). Conflict theorists argue that sexual relations are linked to other forms of stratification, namely, race, class, and gender inequality. According to this perspective, sexual violence (such as rape or sexual harassment) is the result of power imbalances, specifically between women and men.

At the same time, because conflict theorists see economic inequality as a major basis for social conflict, they tie the study of sexuality to economic institutions. They link the international sex trade to poverty, the status of women in society, and the economics of international development and tourism (Altman 2001; Enloe 2001). Still, conflict theorists do not see all sexual relations as oppressive. Sexuality is an expression of great social intimacy. In connecting sexuality and inequality, conflict theorists are developing a structural analysis of sexuality, not condemning sexual intimacy.

Because both functionalism and conflict theory are macrosociological theories (that is, they take a broad view of society, seeing sexuality in terms of the overall social organization of society), they do not tell us much about the social construction of sexual identities. This is where the sociological framework of symbolic interaction is valuable.

Symbolic Interaction and the Social Construction of Sexual Identity

Symbolic interaction theory uses a **social construction perspective** to interpret sexual identity as learned, not inborn. To symbolic interactionists, culture and society shape sexual experiences. Patterns of social approval and social taboos make some forms of sexuality permissible and others not (Lorber 1994; Connell 1992).

The social construction of sexual identity is revealed by **coming out**—the process of defining oneself as gay or lesbian. The process is a series of events and redefinitions in which a person comes to see herself or himself as having a gay identity. In coming out, a person consciously adopts a gay identity either to himself or herself or to others (or both). This is usually not the result of a single experience. If it were, there would be far more self-identified gays and lesbians, because researchers find that a substantial portion of both men and women have some form of homosexual experience at some time in their lives.

The development of sexual identity is not necessarily a linear or unidirectional process, with people moving predictably through a defined sequence of steps or phases. Although they may experience certain milestones in their identity development, some people experience periods of ambivalence about their identity and may switch back and forth between lesbian, heterosexual, and bisexual identity over time (Rust 1995, 1993). Some people may engage in lesbian or gay behavior but not adopt an identity as lesbian or gay. Certainly, many gays and lesbians never adopt a public definition of themselves as gay or lesbian, instead remaining "closeted" for long periods, if not for their entire lifetime.

One's sexual identity may also change. For example, a person who has always thought of himself or

Teens and Sex: Are Young People Becoming More Sexually Conservative?

Research Question: Several national studies have reported a decline in sexual activity among teens. The percentage of sexually active teens has dropped from the early 1990s, rates of teen pregnancy have fallen, teens are having fewer abortions, and the rate of sexually transmitted diseases among teens has declined. Does this herald a growth in sexual conservatism among young people and the success of policies encouraging sexual abstinence?

Research Method: Sociologists Barbara Risman and Pepper Schwartz based their research on a synthesis of all of the national studies on teen sexuality, as well as data from research organizations on the prevalence of teen sexuality.

Research Findings: Most of the change in teen sex activity is attributable to changes in behavior of boys, not girls.

The number of high school boys who are virgins has increased. Girls' behavior has not changed significantly, except among African American girls, whose rates of sexual activity have declined, nearly matching those of White and Hispanic girls. Risman and Schwartz conclude that sexual behavior of boys is then becoming more like girls, the implication being that boys and girls are likely to begin their sexual lives within the context of romantic relationships.

Conclusions and Implications: Although many declare that the changes in teen sexual behavior mean a decline in the sexual revolution, Risman and Schwartz disagree. Certainly fear of AIDS, education about safe sex, and some growth in conservative values have contributed to changes in teen sexual norms. Risman and Schwartz show that numerous factors influence sexual behavior among teens, just as among adults. They suggest that sexuality is a normal part of adolescent social development and conclude that the sexual revolution—along with the revolution in gender norms—is generating more responsible, not more problematic, sexual behavior among young people.

Questions to Consider

1. Are people in your age group generally sexually conservative or sexually liberal? What factors influence young people's attitudes about sexuality?

2. To follow up from question 1, what evidence would you need to find out if young people in your community are more liberal than young people in the past? How would you design a study to investigate this question?

Source: Risman, Barbara, and Pepper Schwartz. 2002. "After the Sexual Revolution: Gender Politics in Teen Dating." *Contexts* 1 (Spring): 16–24.

---

herself as heterosexual may conclude at a later time that he or she is gay, lesbian, possibly bisexual. In more unusual cases, people may undergo a sex change operation, perhaps changing their sexual identity in the process.

Although most people learn stable sexual identities, sexual identity evolves over the course of one's life. Change is, in fact, a normal outcome of the process of identity formation. Changing social contexts (including dominant group attitudes, laws, and systems of social control), relationships with others, political movements, and even changes in the language used to describe different sexual identities all affect people's self-definition.

Feminist Theory: Sex, Power, and Inequality

Feminist theory has had a tremendous impact on the study of sexuality, in large part through acknowledging the diversity in human sexual experience, but just as importantly, in emphasizing the sexual identities and relationships that emerge in the context of power relationships and social institutions that have narrowly defined only certain forms of sexual expression as legitimate and "normal."

Feminist theory has emphasized that sexuality, like other forms of social identity and relationships, exists in the context of social institutions. Social institutions tend to define heterosexuality as the only legitimate form of sexual identity and enforce heterosexuality through social norms and sanctions, including peer pressure, socialization, law and other social policies, and, at the extreme, violence (Rich 1980). In other words, sexuality exists within a context of power relationships that exist within society.

Sexual politics refers to the link between sexuality and power, not just within individual relationships. The feminist movement first linked sexuality to the status of women in society, pointing to the possible exploitation of women within sexual relationships. Sexual politics also refers to the high rates of violence against women and sexual minorities and the privilege and power accorded to those presumed to be heterosexual.

The feminist and gay and lesbian liberation movements have put sexual politics at the center of the public's attention by challenging gender role stereotyping and sexual oppression (D'Emilio 1998). Among other things, this has profoundly changed public knowledge of gay and lesbian sexuality. Gay, lesbian, and feminist scholars have argued, and many now concur, that being lesbian or gay is not the result of psychological deviance or personal maladjustment but is one of several alternatives for intimate social relationships. The political mobilization of many lesbian women and gay men and the willingness of many to make their sexual identity public have also raised public awareness of the civil and personal rights of gays and lesbians. These changes make other changes in intimate relations possible.

Feminist theory also focuses on how sexuality intersects with other systems of power, especially gender, race, and class inequality, and has brought new and critical analyses to the study of sex. One such perspective is *queer theory*.

Queer theory is a perspective that has evolved from recognizing the socially constructed nature of sexual identity and the role of power in defining only some forms of sexuality as "normal"—that is, socially legitimate. Instead of seeing heterosexual or homosexual attraction as fixed in biology, queer theory interprets society as forcing these sexual boundaries, or dichotomies, on people. By challenging the "either/or" thinking that one is either gay or straight, queer theory challenges the idea that only one form of sexuality is normal and all other forms are deviant or wrong. As a result, queer theory has opened up fascinating new studies of gay, straight, bisexual, and transsexual identities and introduced the idea that sexual identity is a continuum of different possibilities for sexual expression and personal identity (Kimmel 2007; Seidman 2003; Williams and Stein 2002).

Queer theory has also linked the study of sexuality to the study of gender, showing how transgressing (or violating) fixed gender categories can reconstruct the possibilities of how all people—men and women, gay, bisexual, transgender, or straight—construct their gender and sexual identity. Transgressing gender categories can show how sex and gender categories are usually constructed in dichotomous categories (that is, opposite or binary types). By violating these constructions, people are liberated from the social constraints that presumably fixed categories of identity create. Thus queer theory emphasizes how performance and play with gender categories can be a political tool for deconstructing fixed sex and gender identities (Rupp and Taylor 2003).

---

thinking SOCIOLOGICALLY

A recent trend among young people on college campuses has been the creation of theme parties called "CEOs and corporate hos." Have such parties occurred on your campus? Why do you think they are so popular? What *sexual scripts* are being played out by such role-playing, and how do these scripts involve gender, race, and class? You might also think about how such parties portray women as sexual objects. ●

A Global Perspective on Sexuality

Cross-cultural studies of sexuality show that sexual norms, like other social norms, develop differently across cultures. Take sexual jealousy. Perhaps you think that seeing your sexual partner becoming sexually involved with another person would naturally evoke jealousy, no matter where it happened. Researchers have found this not to be true. In a study comparing patterns of sexual jealousy in seven different nations (Hungary, Ireland, Mexico, the Netherlands, the United States, Russia, and the former Yugoslavia), researchers found significant cross-national differences in the degree of jealousy when women and men saw their partners kissing, flirting, or being sexually involved with another person (Buunk and Hupka 1987).

Cultures also vary considerably on how they view teen sexuality. Whereas the focus in the United States has been on discouraging sex between teenagers, parents in some other nations see sexuality for teens as part of the normal course of adult development. A study of the Netherlands, for example, has found that two-thirds of Dutch teens (age 15 to 17) are allowed to sleep over with steady girlfriends or boyfriends—with parental approval. Interestingly, the Netherlands also has the lowest rate of teen pregnancy in the world, even with the age at first intercourse there having dropped over the years.

Compare this to the United States where the age at first intercourse has actually been increasing, but where the teen pregnancy rate is one of the highest among industrial nations. What makes the difference? Sociologists point to the cultural values around sexuality. In the United States, teen sexuality is discouraged and seen as risky. In the Netherlands, to the contrary, cultural morals see sexuality as part of developing self-determination, and it is treated with frank discussion, a strong place in public policy for sex education, and an idea that mutual respect is part of healthy sexual relationships (Schalet 2010).

Likewise, tolerance for gay and lesbian relationships varies significantly in different societies around the world. Germany has recently legalized gay and lesbian relationships, allowing them to register same-sex partnerships and have the same inheritance rights as heterosexual couples. The new law does not, however, give them the same tax advantages, nor can same-sex couples adopt children. Cross-cultural studies can make someone more sensitive to the varying cultural norms and expectations that apply to sexuality in different contexts. Different cultures simply view sexuality differently. In Islamic culture, for example, women and men are viewed as equally sexual, although women's sexuality is seen as potentially disruptive and needing regulation (Mernissi 1987).

Sex is also big business, and it is deeply tied to the world economic order. As the world has become more

DOING **sociological research**

Is Hooking Up Bad for Women?

Research Question: The presence of the hookup culture on college campuses is a relatively recent phenomenon and reflects changes in sexual attitudes and behaviors among, especially, young people. Some argue that the hookup culture liberates women from traditional sexual values that constrained women's sexuality. Others argue that this culture is harmful to women, making them sexual objects for men's pleasure (for example, the practice of women making out with other women in public settings, such as bars and campus parties). Is the hookup culture harmful to women or is a sign of their sexual liberation?

Research Method: Several sociologists have examined this question, some using national surveys, others using a more qualitative approach. Paula England and her colleagues, for example, studied sexual activity in a survey of over 14,000 students at 18 different campuses in the United States, exploring students' experiences with hooking up, dating, and relationships. Laura Hamilton and Elizabeth Armstrong used a more qualitative approach, actually residing among students in a so called "party dorm," observing as well as interviewing students for a full year. Leila Rupp and Verta Taylor had their undergraduate students interview other students about the party scene on their campus.

Research Findings: Research finds that both arguments about the effect of hookup cultures on women are true: Some parts of this culture are harmful to women, but women's experiences within the hookup culture also vary and are not uniformly negative.

England, for example, found that the hookup culture is not as wildly rampant as assumed. She found that 72 percent of both men and women participated in at least one hookup, but 40 percent had engaged in three or fewer hookups and only 20 percent of students had engaged in ten or more. England concludes that the popular image of "girls gone wild" in popular culture is simply not true. She also found that hooking up has not replaced committed relationships. But, as Hamilton and Armstrong found, the hookup culture allows women (and men) a chance for sexual exploration. At the same time, the hookup culture does present risks to women—risks of being pushed to drink too much, risks of sexual violence, and loss of self-esteem. But some women say that the hookup culture frees them to pursue education and careers without the emotionally consuming pressures of committed relationships (Hamilton and Armstrong 2009).

Rupp and Taylor have similarly found that the college party scene commonly includes women making out with other women. But, unlike those who argue that this practice sexually objectifies women for the pleasure of men, Rupp and Taylor argue that women who do so are exploring sexuality. But they do so within social boundaries and heterosexual norms. Even though women may engage in same-sex sexual practices, they do not necessarily develop a lesbian identity, although the lines of sexual identity are expanding for women.

Conclusions and Implications: There is not a simple or single answer to the question of whether the hookup culture harms or liberates women. Taken together, these studies reveal a complex portrait of young women's sexuality today. Sexual boundaries are perhaps more fluid than they once were, although they are still marked by sexual double standards and norms of heterosexuality.

Questions to Consider

1. Is there a hookup culture on your campus? If so, are there both positive and negative consequences of the hookup culture for women on your campus? What are they? If there is not such a culture on your campus, why not?

2. How is the hookup culture shaped by such social factors as age, social class, race, or gender? How might it change over time—both as history evolves and as the current generation ages?

Source: Armstrong, Elizabeth A., Laura Hamilton, and Paula England. 2010. "Is Hooking Up Bad for Young Women?" *Contexts* 9 (Summer): 23–27; Hamilton, Laura, and Elizabeth A. Armstrong. 2009. "Gendered Sexuality in Young Adulthood: Double Binds and Flawed Options." *Gender & Society* 23 (October): 589–616; Rupp, Leila J., and Verta Taylor. 2010. "Straight Girls Kissing." *Contexts* 9 (Summer): 28–33.

---

globally connected, an international sex trade has flourished—one that is linked to economic development, world poverty, tourism, and the subordinate status of women in many nations.

Sex trafficking refers to the use of women and girls worldwide as sex workers in an institutional context in which sex itself is a commodity. Sex is marketed in an international marketplace, and women as sex workers are used to promote tourism, cater to business and military men, and support a huge industry of nightclubs, massage parlors, and teahouses (Bales 2010; Sara 2010; Shelley 2010). Through sex trafficking, women—usually very young women—are forced by fraud or coercion into commercial sex acts. Sometimes identified as a form of slavery, sex trafficking can involve a system of debt and bondage, where young women (typically under age 18) are obligated to provide sexual services in exchange for alleged debt for the price of their housing, food, or other living expenses. Sometimes these young women are actually kidnapped; other times, they may simply be duped, initially lured into sex work by promises of marriage, large incomes, or the glamour of travel, but soon trapped in a cycle of debt and/or actual captivity.

The international trafficking of women for sex exploits women—and often children—and puts them at risk for disease and violence.

Related to sex trafficking is **sex tourism**, referring to the practice whereby people travel to particular parts of the world specifically to engage in commercial sexual activity. "Sex capitals" are places where prostitution openly flourishes, such as in Thailand and Amsterdam. Sex is an integral part of the world tourism industry. In Thailand, for example, men as tourists outnumber women by a ratio of three to one. Although certainly not all of them—or even a majority—are going to Thailand solely to explore sex tourism, there are men who go to Thailand, as well as other destinations, explicitly to buy sexual companionship. For example, hostess clubs in Tokyo cater to corporate men. Although these clubs are not houses of prostitution, scholars who have studied them say that they are based on an environment in which women's sexuality is used as the basis for camaraderie among men (Allison 1994). Sex tourism is such a profitable enterprise that the International Labour Organization estimates that somewhere between 2 and 14 percent of the gross domestic product in Thailand, Indonesia, Malaysia, and the Philippines is derived from sex tourism. Much of this lucrative business involves the exploitation of children (U.S. Department of Justice 2009a).

Sex tourism and sexual trafficking are now part of the global economy, contributing to the economic development of many nations and supported by the economic dominance of certain other nations. As with other businesses, the products of the sex industry may be produced in one region and distributed in others. (Think, for example, of the pornographic film industry centered in southern California, but distributed globally.) The sex trade is also associated with world poverty; sociologists have found that the weaker the local economy, the more important the sex trade. The international sex trade is also implicated in problems such as the spread of AIDS worldwide, as well as the exploitation of women who have limited economic opportunities (Altman 2001).

UNDERSTANDING GAY AND LESBIAN EXPERIENCE

Sociological understanding of sexual identity has developed largely through new studies of lesbian and gay experience. Long thought of only in terms of social deviance (see Chapter 7), gays and lesbians have been stereotyped in traditional social science. But the feminist and gay liberation movements have discouraged this approach, arguing that gay and lesbian experience is part of the broad spectrum of human sexuality.

The institutional context for sexuality within the United States, as well as other societies, is one in which homophobia permeates the culture. **Homophobia** is the fear and hatred of lesbians and gays. It is deeply embedded in people's definitions of themselves as men and women; it is manifested in prejudiced attitudes toward gays and lesbians, as well as overt hostility and violence against people suspected of being gay. Homophobia is a learned attitude, as are other forms of negative social judgments about particular groups.

Homophobia produces numerous fears and misunderstandings about gays and lesbians. For example, there is the widespread, though incorrect, belief that children raised in gay and lesbian households will be negatively affected. This is a myth. Sociological research has shown that the ability of parents to form good relationships with their children is far more significant in children's social development than is their parents' sexual orientation (Stacey and Biblarz 2001).

Other myths about gay people are that they are mostly White men with large discretionary incomes who work primarily in artistic areas and personal service jobs (such as hairdressing). This stereotype prevents people from recognizing that gays and lesbians come from all racial–ethnic groups, may be working class or poor, and are employed in a wide range of occupations (Gluckman and Reed 1997). Some lesbians and gays are also elderly, though the stereotype defines gay people as primarily young or middle-aged (Smith 1983). These different misunderstandings reveal just a few of the many unfounded myths about gays and lesbians. Support for these attitudes comes from homophobia, not from actual truth.

thinking SOCIOLOGICALLY

Keep a diary for one week and write down as many examples of *homophobia* and *heterosexism* as you observe in routine social behavior. What do your observations tell you about how heterosexuality is enforced? ●

Heterosexism refers to the institutionalization of heterosexuality as the only socially legitimate sexual orientation. Heterosexism is rooted in the belief that heterosexual behavior is the only natural form

Sexuality and Disability: Understanding "Marginalized" Masculinity

What does it mean to be a "real man"? In a gender-stratified world, masculinity is typically perceived as something one either has or does not. Beliefs about masculinity define "real men" as those who are strong, straight, and sexually powerful. Thus socially constructed beliefs about manhood are deeply tied to assumptions about male sexuality. What happens to men who do not fit this narrowly constructed definition of manhood—men who are marginalized in a social system that only privileges those with particular social characteristics?

Idealized definitions of masculinity shape the experiences of disabled men who may be negatively stereotyped by others as "asexual," "impotent," or "not real men." Discrimination and prejudice against disabled people is pervasive in society. As a result, men with disabilities have to resist the perception that they somehow do not meet the social standards of masculinity. Some may respond by acting "hypermasculine"— that is, working to exaggerate their strength and endurance. Others create different sets of standards for themselves, thus reformulating ideas about masculinity.

Examining the experiences of disabled men with regard to sexuality and gender shows how male privilege is not a universal experience—that is, men's gender privilege is conditioned upon other forms of privilege or lack thereof, such as one's status as abled or disabled. A sociological perspective that recognizes the diversity of social experiences provides a multidimensional lens through which men's sexuality can be understood.

Sources: Coston, Beverly M., and Michael Kimmel. 2012. "Seeing Privilege Where It Isn't: Marginalized Masculinities and the Intersectionality of Privilege." *Journal of Social Issues* 68:97; Gerschick, T. J., and A. S. Miller. 1995. "Coming to Terms: Masculinity and Physical Disability." Pp. 183–204 in *Men's Health and Illness: Gender, Power, and the Body, Research on Men and Masculinities Series*, vol. 8, edited by D. Sabo and D. F. Gordon. Thousand Oaks, CA: Sage Publications.

---

of sexual expression and that homosexuality is a perversion of "normal" sexual identity. Heterosexism is reinforced through institutional mechanisms that project the idea that only heterosexuality is normal. Institutions also provide different benefits to people presumed to be heterosexual. Businesses and communities, for example, rarely recognize the legal rights of people in homosexual relationships, although this is changing. Within an institution, individual beliefs can reflect heterosexist assumptions. Thus a person may be accepting of gay and lesbian people (that is, not be homophobic) but still benefit from heterosexual privileges. At the behavioral level, heterosexist practices can exclude lesbians and gays, such as when coworkers talk about dating the other sex, assuming that everyone is interested in a heterosexual partner. In the absence of institutional supports from the dominant culture, lesbians and gays have invented their own institutional support systems. Gay communities and gay rituals, such as gay pride marches, affirm gay and lesbian identities and provide a support system that is counter to the dominant heterosexual culture. Those who remain "in the closet" deny themselves this support system.

The absence of institutionalized roles for lesbians and gays affects the roles they adopt within relationships. Despite popular stereotypes, gay partners typically do not assume roles as the dominant or submissive sexual partner. They are more likely to adopt roles as equals. Gay couples and lesbian couples are also more likely than heterosexual couples to both be employed, another source of greater equality within the relationship. Researchers have also found that the quality of relationships among gay men is positively correlated with the social support the couple receives from others (Smith and Brown 1997; Metz et al. 1994).

Lesbians and gays are a minority group in our society, denied equal rights and singled out for negative treatment in society. *Minority groups* are not necessarily numerical minorities; they are groups with similar characteristics (or at least perceived similar characteristics) who are treated with prejudice and discrimination (see Chapter 10). As a minority group, gays and lesbians have organized to advocate for their civil rights and to be recognized as socially legitimate citizens. Some organizations and municipalities have enacted civil rights protections on behalf of gays and lesbians, typically prohibiting discrimination in hiring. The Supreme Court ruled in 1996 (*Romer v. Evans*) that gays and lesbians cannot be denied equal protection under the law. The implications of this case for domestic partner benefits, child custody and adoption by gay and lesbian parents, and gay marriage are not yet known, but will likely be determined in the courts in the years ahead.

SEX AND SOCIAL ISSUES

In studying sexuality, sociologists tap into some highly contested social issues of the time. Birth control, reproductive technology, abortion, teen pregnancy, pornography, and sexual violence are all subjects of public concern and are important in the formation of social policy. Debates about these issues hinge in part on attitudes about sexuality and are shaped by race, class, and gender relations. These social issues can generate personal troubles that have their origins in the structure of society—recall the distinction C. Wright Mills made between personal troubles and social issues (see Chapter 1).

Birth Control

The availability of birth control is now less debated than it was in the not-too-distant past, but this important reproductive technology has been strongly related to the status of women in society (Gordon 1977). Reproduction has been controlled by men; to this day, mostly men define the laws and make the scientific decisions about what types of birth control will be available. Women are also most likely seen as being responsible for reproduction because it is a presumed part of their traditional role. At the same time, changes in birth control technology have also made it possible for women to change their roles in society, given that breaking the link between sex and reproduction has freed women from some traditional constraints.

The right to birth control is a recently won freedom. It was not until 1965 that the Supreme Court, in *Griswold v. Connecticut*, defined the use of birth control as a right, not a crime. This ruling originally applied only to married people; unmarried people were not extended the same right until the 1972 Supreme Court decision *Eisenstadt v. Baird*. Today, birth control is routinely available by prescription, but there is heated debate about whether access to birth control should be curtailed for the young—at a time when youths are experimenting with sex at younger ages and risking teenage pregnancy and AIDS. Some argue that increasing access to birth control will only encourage more sexual activity among the young.

Class and race relations also have had a role in shaping birth control policy. In the mid-nineteenth century, increased urbanization and industrialization ended the necessity for large families, especially in the middle class, because fewer laborers were needed to support the family. Early feminist activists such as Emma Goldman and Margaret Sanger also saw birth control as a way of freeing women from unwanted pregnancies and allowing them to work outside the home if they chose. As the birthrate fell among White upper- and middle-class families during this period, these classes feared that immigrants, the poor, and racial minorities would soon outnumber them.

The *eugenics* movement of the early twentieth century grew from the fear of domination by immigrant groups. **Eugenics** sought to apply scientific principles of genetic selection to "improve" the offspring of the human race. It was explicitly racist and class-based, calling for, among other things, the compulsory sterilization of those who eugenicists thought were unfit. Eugenicist arguments appeal to a public that fears the social problems that emerge from race and class inequality: Instead of attributing these problems (such as crime) to the structure of society, eugenicists blame the genetic composition of the least powerful groups in society.

In contemporary society, attitudes about the use of birth control have evolved and have also been very much shaped by greater awareness about the risks of HIV/AIDS infection. Now 62 percent of women aged 15 to 44 use contraceptives, an increase since 1995 (Guttmacher Institute 2012). Among women users, the pill is the most frequent type of birth control used, followed by tubal ligation (sterilization), and then condoms. Similar data are not reported for men—itself a reflection of the belief that contraception is primarily a woman's responsibility.

Among teenagers, the majority (74 percent of women and 82 percent of men) use contraceptives the first time they have sex, and 83 percent of women and 91 percent of men report using contraceptives the last time they had sex—a marked increase from the past (Guttmacher Institute 2012). But fewer than half of teen men who are sexually active use condoms 100 percent of the time. Sexually active teens who do not use contraceptives have a 90 percent chance of becoming pregnant within the first year of initiating sex (Guttmacher Institute 2012).

New Reproductive Technologies

Practices such as surrogate mothering, in vitro fertilization, and new biotechnologies of gene splicing, cloning, and genetic engineering mean that reproduction is no longer inextricably linked to biological parents. A child may be conceived through means other than sexual relations between one man and one woman. One woman may carry the child of another. Offspring may be planned through genetic engineering. A sheep can be cloned (that is, genetically duplicated). So can monkeys. Are humans next? With such developments, those who could not otherwise conceive children (infertile couples, single women, or lesbian couples) are now able to do so, thereby raising new questions. To whom are such new technologies available? Which groups are most likely to sell reproductive services? Which groups most likely to buy? What are the social implications of such changes?

There are no simple answers to such questions, but sociologists would point first to the class, race, and gender dimensions of these issues (Roberts 2012, 1997). Poor women, for example, are far more likely than middle-class or elite women to sell their eggs or offer their bodies as biological incubators. Groups that can afford new, costly methods of reproduction may do so at the expense of women whose economic need places them in the position of selling themselves for financial necessity.

Breakthroughs in reproductive technology raise especially difficult questions for makers of social

New technologies, such as artificial insemination where sperm is injected into an egg, raise new questions about ethics and social policies regarding reproductive rights.

Jochen Tack/Alamy Limited

policy. Developments in the technology of reproduction have ushered in new possibilities and freedoms but also raise questions for social policy. With new reproductive technologies, there is potential for a new eugenics movement. Sophisticated prenatal screenings make it possible to identify fetuses with presumed defects. Might society then try to weed out those perceived as undesirables—the disabled, certain racial groups, certain sexes? Will parents try to produce "designer children"? If boys and girls are differently valued, one sex may be more often aborted, a frequent practice in India and China—two of the most populous nations on Earth. Because of population pressures, state policy in China, for example, encourages families to have only one child. Because girls are less valued than boys, aborting and selling girls is common and has created a U.S. market for the adoption of Chinese baby girls.

There are no traditions to guide us on such questions. Public thinking about sexuality and reproduction will have to evolve. Although the concept of reproductive choice is important to most people, choice is conditioned by the constraints of race, class, and gender inequalities in society. Like other social phenomena, sexuality and reproduction are shaped by their social context.

Abortion

Abortion is one of the most seriously contested political issues. The public is divided on whether they identify as "pro-choice" (41 percent) or "pro-life" (50 percent), but this varies according to various social factors. College graduates and those with postgraduate degrees are more likely to identify as pro-choice. Women, young people, and those of higher incomes are also more likely to identify as pro-choice (Saad 2012b). But identifying with these labels is different than other attitudes toward abortion. Only 20 percent think abortion should be illegal under all circumstances; 26 percent think abortion should be legal in any circumstances; and 51 percent think it should be legal only under certain circumstances (see Figure 12.3). Furthermore, even among those self-identified as "pro-life," 35 percent support abortion in the first trimester, and 37 percent support when the mother's mental health is endangered (Saad 2011). Clearly, there are more complex views involved in thinking about abortion than simple labels can suggest.

The right to abortion was first established in constitutional law by the *Roe v. Wade* decision in 1973. In *Roe v. Wade,* the Supreme Court ruled that at different points during a pregnancy, separate but legitimate rights collide—the right to privacy, the right of the state to protect maternal health, and the right of the state to protect developing life. To resolve this conflict of rights, the Supreme Court ruled that pregnancy occurred in trimesters. In the first, women's right to privacy without interference from the state prevails; in the second, the state's right to protect maternal health takes precedence; in the third, the state's right to protect developing life prevails. In the second trimester, the government cannot

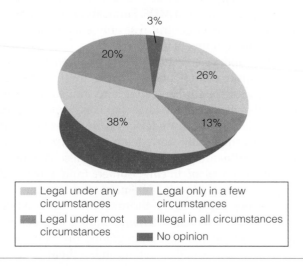

Legend:
- Legal under any circumstances
- Legal under most circumstances
- Legal only in a few circumstances
- Illegal in all circumstances
- No opinion

FIGURE 12.3 Attitudes Toward Abortion Attitudes toward the legality of abortion have been fairly steady over time. As you can see, the majority of Americans support abortion rights at least under some circumstances.

Data: Saad, Lydia. 2011 (August 8). "Plenty of Common Ground Found in Abortion Debate." *The Gallup Poll.* **www.gallup.com**

deny the right to abortion, but it can insist on reasonable standards of medical procedure. In the third, abortion may be performed only to save the life or health of the mother. More recently, the Supreme Court has allowed states to impose restrictions on abortion, but it has not, to date, overturned the legal framework of *Roe v. Wade.*

Data on abortion show that it occurs across social groups, although certain patterns do emerge. The abortion rate has declined since 1980, from a rate of 27.4 per 1000 women to 16.9 in 2008 (among women aged 15 to 44). And, as you can see in Figure 12.4, the number of deaths from illegal abortions plummeted in the years following the *Roe v. Wade* decision. Young women (aged 20 to 29) are the most likely group to get abortions, although the second most likely group is women under 20. Black women are four times more likely to have abortions than White women; poor women are four times more likely to have abortions than other women. You may be surprised to learn that the majority of women having abortions have already had at least one birth (U.S. Census Bureau 2012; Boonstra et al. 2006).

The abortion issue provides a good illustration of how sexuality has entered the political realm. Abortion rights activists and antiabortion activists hold very different views about sexuality and the roles of women. Antiabortion activists tend to believe that giving women control over their fertility breaks up the stable relationships in traditional families. They tend to view sex as something that is sacred, and they are disturbed by changes that make sex less restrictive. This belief has been fueled by the activism of the religious right, where strong passions against abortion have driven issues about sexual behavior directly into the political realm. Abortion rights activists, on the other hand,

see women's control over reproduction as essential for women's independence. They also tend to see sex as an experience that develops intimacy and communication between people who love each other. The abortion debate can thus be interpreted as a struggle over the right to terminate a pregnancy as well as a battle over differing sexual values and a referendum on the nature of men's and women's relationships (Luker 1984).

Pornography and the Sexualization of Culture

Little social consensus has emerged about the acceptability and effects of pornography. Part of this debate is about defining what is obscene. The legal definition of obscenity is one that changes over time and in different political contexts. Public agitation over pornography has divided people into those who think it is solidly protected by the First Amendment, those who want it strictly controlled, those who think it should be banned for moral reasons, and those who think it must be banned because it harms women.

However pornography is legally defined, there is no question that it permeates contemporary culture. Once available only in more "underground" places like X-rated movie houses, pornography is now far more public than in the past. Hotel rooms have a huge array of pornographic films available on the television; pornographic spam appears regularly in people's email inboxes; casual references to pornography are made in popular shows on prime-time TV; images that once would have seemed highly pornographic are now commonly found on widely distributed magazines such as *Maxim, Cosmopolitan,* and others. Highly sexualized

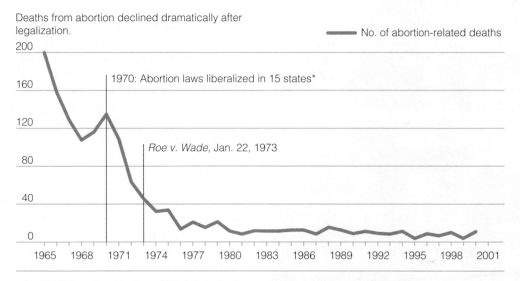

FIGURE 12.4 Deaths from Abortion: Before and After Roe v. Wade

*By the end of 1970, four states had repealed their antiabortion laws, and eleven states had reformed them.

Source: Boonstra, Heather, Rachel Benson Gold, Cory L. Richards, and Lawrence B. Finer. 2006. *Abortion in Women's Lives,* New York: Guttmacher Institute, 2006, p. 13.

expressions and images are so widely seen throughout society that one commentator has said we are experiencing the *pornification of culture* (Levy 2005). Does this indicate that society has become more sexualized?

According to a major report from the American Psychological Association (APA), the answer is yes. The APA defines *sexualization* as including any one of the following conditions:

- People are judged based only on sexual appeal or behavior, to the exclusion of other characteristics.
- People are held to standards that equate physical attractiveness with being sexy.
- People are sexually objectified—meaning made into a "thing" for others' use.
- Sexuality is inappropriately imposed on a person (American Psychological Association 2007).

The APA report then details the specific consequences, especially for young girls, of a culture marked by sexualization. This report shows that overly sexualizing young girls harms them in psychological, physical, social, and academic ways. As examples, young girls may spend more time tending to their appearance than to their academic studies; they may engage in eating disorders to achieve an idealized, but unattainable image of beauty; or they may develop attitudes that put them at risk of sexual exploitation. Although the focus of this report is

Pornographic images have permeated much of the fashion industry as women are often shown in bondage of various sorts.

Ilan Rosen/PhotoStock-Israel/Alamy Limited

on young girls, one cannot help but wonder what effects the "pornification of culture" also has on young boys, as well as adult women and men.

Despite public concerns about pornography, most people believe that pornography should be protected by the constitutional guarantees of free speech and a free press. Yet people also believe that pornography dehumanizes women; women especially think so. Public controversy about pornography is not likely to go away because it taps so many different sexual values among the public.

Teen Pregnancy

Each year about 410,000 teenage girls (under age 19) have babies in the United States. The United States has the highest rate of teen pregnancy among developed nations, even though levels of teen sexual activity around the world are roughly comparable. Teen pregnancy has declined since 1990, a decline caused almost entirely from the increased use of birth control. Contrary to popular stereotypes, the teen birthrate among African American women has declined more than for White women (U.S. Census Bureau 2012). Most teen pregnancies (82 percent) are unplanned, due largely to inconsistent use of birth control (Guttmacher Institute 2012).

Beginning in the early 1980s, the federal government encouraged abstinence policies, putting money behind the belief that encouraging chastity was the best way to reduce teen pregnancy. Under programs that encourage abstinence, young people are encouraged to take "virginity pledges," promising not to have sexual intercourse before marriage. Do such pledges work?

In one of the most comprehensive and carefully controlled studies of abstinence, researchers compared a large sample of teen "virginity pledgers" and "nonpledgers" who were matched on social attitudes such as religiosity and attitudes toward sex and birth control. The study compared the two groups over a five-year period. Results showed that over time there were *no differences* in the number of times those in each group had sex, the age of first sex, or the practice of oral or anal sex. The main difference between the two groups was that pledgers were less likely to use birth control when they had sex. Also, five years after having taken an abstinence pledge, pledgers denied having done so. The researchers concluded that not only are abstinence pledges ineffective (supporting other research findings) but that taking the pledge makes pledgers less likely to protect themselves from pregnancy and disease when having sex (Rosenbaum 2009). Consistent with these findings are other studies that find that abstinence policies account for a very small portion of the decline in teen pregnancy—probably only about 10 percent of the difference (Santelli et al. 2007; Boonstra et al. 2006).

Although the rate of teen pregnancy has declined, so has the rate of marriage for teens who become pregnant. Thus most babies born to teens will be raised by single mothers—a departure from the past when teen mothers

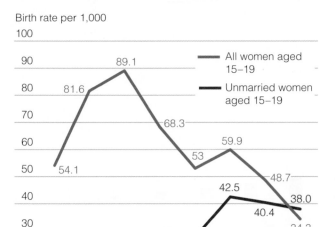

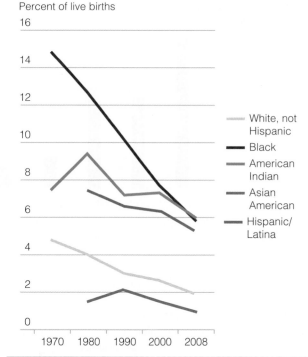

FIGURE 12.5 Teen Births, 1940–2010 In 2010, the rate of births to teen women was lower than at any previous time. What do you think explains this? And how do you explain the trend seen in births for all women aged 15 to 19? Can you explain why births to young, unmarried women have generally increased over time, whereas births among young women overall have declined?

Note: Data for birthrate among unmarried women, aged 15–19, for 2010, is not available; however, most births to women aged 15–19 occur among unmarried women.

Source: National Vital Statistics Report. 2001. *Births to Teenagers in the United States, 1940–2000*, vol. 49. Atlanta, GA: Centers for Disease Control. **www.cdc.gov**; National Vital Statistics Report. 2001. *Births: Preliminary Data for 2010*, vol. 61. Atlanta, GA: Centers for Disease Control. **www.cdc.gov**

FIGURE 12.6 Teenage Childbearing by Race/Ethnicity, 1970–2008 Despite public beliefs to the contrary, teen pregnancy among all racial/ethnic groups has declined substantially. What factors explain this decline?

Source: *Health United States*. 2011. Atlanta, GA: Centers for Disease Control. **www.cdc.gov**

often got married (see Table 12.2). What concerns people about teen parents is that teens are more likely to be poor than other mothers, although sociologists have cautioned that this is because teen mothers are more likely poor *before* getting pregnant (Luker 1996). Teen parents are among the most vulnerable of all social groups.

---

debunking SOCIETY'S MYTHS

MYTH: Providing sex education to teens only encourages them to become sexually active.

SOCIOLOGICAL PERSPECTIVE: Comprehensive sex education actually delays the age of first intercourse; abstinence-only education has not been shown to be effective in delaying intercourse (Risman and Schwartz 2002). ●

Teenage pregnancy correlates strongly with poverty, lower educational attainment, joblessness, and health problems. Teen mothers have a greater incidence of problem pregnancies and are most likely to deliver low-birth-weight babies, a condition associated with

myriad of other health problems. Teen parents face chronic unemployment and are less likely to complete high school than those who delay childbearing. Many continue to live with their parents, although this is more likely among Black teens than among Whites.

Although teen mothers feel less pressure to marry now than in the past, if they raise their children alone, they suffer the economic consequences of raising children in female-headed households—the poorest of all income groups. Teen mothers report that they do not marry because they do not think the fathers are ready for marriage. Sometimes their families also counsel them against marrying precipitously. These young women are often doubtful about men's ability to support them. They want men to be committed to them and their child, but they do not expect their hopes to be fulfilled (Edin and Kefalas 2005). Research shows that low-income single mothers are distrustful of men, especially after an unplanned pregnancy. They think they will have greater control of their household if they remain unmarried. Many teen mothers also express fear of domestic violence as a reason for not marrying (Edin 2000).

Why do so many teens become pregnant given the widespread availability of birth control? Teens typically delay the use of contraceptives until several months after they become sexually active. Teens who do not use contraceptives when they first have sex are twice as likely to get pregnant as those who use contraception.

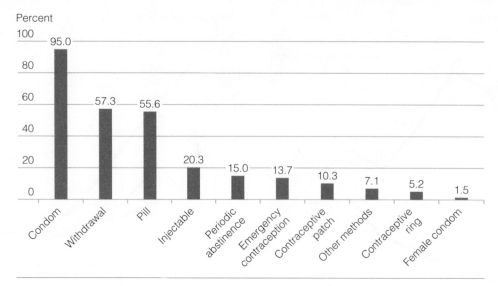

FIGURE 12.7 Contraceptive Use among Teen Women These data include teen (aged 15–19) women's contraceptive use, including only those who have had sexual intercourse. The data indicate contraceptive methods ever used, not necessarily those used regularly. How do social factors influence whether or not a young woman uses contraception?

Source: Center for Health Statistics. 2004. *Teenagers in the United States: Sexual Activity, Contraceptive Use, and Childbearing, 2006–1010, National Survey of Family Growth.* Washington, DC:U.S. Department of Health and Human Services.

In recent years, the percentage of teens using birth control (especially condoms) has increased, although the pill is still the most widely used method (Guttmacher Institute 2012; see Figure 12.7).

Sociologists have argued that the effective use of birth control requires a person to identify himself or herself as sexually active (Luker 1975). Teen sex, however, tends to be episodic. Teens who have sex on a couple of special occasions may not identify themselves as sexually active and may not feel obliged to take responsibility for birth control. Despite many teens initiating sex at an earlier age, social pressure continues to discourage them from defining themselves openly, or even privately, as sexually active.

Teen pregnancy is integrally linked to the gender expectations of men and women in society. Some teen men consciously avoid birth control, thinking it takes away from their manhood. Teen women often romanticize motherhood, thinking that becoming a mother will give them social value they do not otherwise have. For teens in disadvantaged groups, motherhood confers a legitimate social identity on those otherwise devalued by society (Horowitz 1995). Although their hopes about motherhood are not realistic, they indicate how pessimistic the teenagers feel about their lives that are often marked by poverty, a lack of education, and few good job possibilities. For young women to romanticize motherhood is not surprising in a culture where motherhood is defined as a cultural ideal for women, but the ideal can seldom be realized when society gives mothers little institutional or economic support (see Figure 12.8).

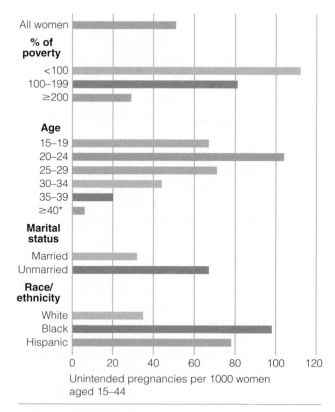

FIGURE 12.8 Teens and Unintended Pregnancy

*Denominator is women 40–44.

Source: Boonstra, H. D., Rachel BensonGold, Cory L.Richards, and Lawrence B.Finer. 2006. *Abortion in Women's Lives.* New York: Guttmacher Institute, 2006, p. 13.

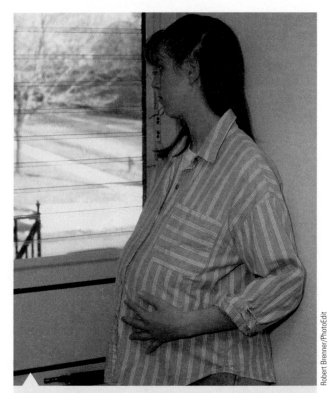

Although teen pregnancy rates have declined, teen mothers are now less likely than in the past to marry.

Robert Brenner/PhotoEdit

Sexual Violence

Before the development of the feminist movement, sexual violence was largely hidden from public view. One great success of the women's movement has been to identify, study, and advocate better social policies to address the problems of rape, sexual harassment, domestic violence, incest, and other forms of sexual coercion. Sexual coercion is not just a matter of sexuality; it is also a form of power relations shaped by the social inequality between women and men. In the forty years or so that these issues have been identified as serious social problems, volumes of research have been published on these different subjects, and numerous organizations and agencies have been established to serve victims of sexual abuse and to advocate reforms in social policy.

Rape and sexual violence were covered in Chapter 7 on deviance and crime, in keeping with the argument that these are forms of deviant and criminal behavior, not expressions of human sexuality. Here we point out that various forms of sexual coercion (rape, domestic violence, and sexual harassment) can best be understood (and therefore changed) by understanding how social institutions shape human behavior and how social interactions are influenced by social factors such as gender, race, age, and class.

Take, for example, the phenomenon now known as *acquaintance rape* (sometimes also called *date rape*). Acquaintance rape is forced and unwanted sexual relations by someone who knows the victim (even if only a brief acquaintance). This kind of rape is common on college campuses, although it is also the most underreported form of rape. Researchers estimate, based on surveys, that 15 to 25 percent of college women experience some form of acquaintance rape (Fisher et al. 2000; see also Chapter 6).

Studies show that although rape is an abuse of power, it is related to people's gender attitudes. Holding stereotypical attitudes about women is strongly related to adversarial sexual beliefs, accepting rape myths, and tolerating violence against women (American Psychological Association 2007).

Violence against women is more likely to occur in some contexts than others, especially in organizations that are set up around a definition of masculinity as competitive, where alcohol abuse occurs, and where women are defined as sexual prey. This is one explanation given for the high incidence of rape in some college fraternities (Armstrong et al. 2006; Stombler and Padavic 1997; Martin and Hummer 1989).

Research on violence against women also finds that Black, Hispanic, and poor White women are more likely to be victimized by various forms of violence, including rape (U.S. Bureau of Justice Statistics 2012). African American and American Indian women and men report the highest incidence of intimate partner violence; Asian Americans and Pacific Islanders have the lowest incidence (Tjaden and Thoennes 2000). Studies also find that Black women are more aware of

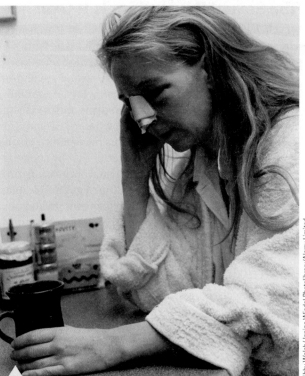

Public awareness of the problem of battered women has increased in recent years, although there is no evidence that battering itself has lessened.

Libby Welsh/Janine Wiedel Photolibrary/Alamy Limited

their vulnerability to rape than are White women and are more likely to organize themselves to resist rape collectively (Stombler and Padavic 1997).

In sum, sociological research on sexual violence shows how strongly sexual coercion is tied to the status of diverse groups of women in society. Rather than explaining sexual coercion as the result of maladjusted men or the behavior of victims, feminists have encouraged a view of sexual coercion that links it to an understanding of dominant beliefs about the sexual dominance of men and the sexual passivity of women. Researchers have shown that those holding the most traditional gender role stereotypes are most tolerant of rapists and least likely to give credibility to victims of rape (Marciniak 1998; Varelas and Foley 1998). Understanding sexual violence requires an understanding of the sociology of sexuality, gender, race, and class in society.

SEX AND SOCIAL CHANGE

As with other forms of social behavior, sexual behavior is not static. Sexual norms, beliefs, and practices emerge as society changes. As we saw in Chapter 11 on gender, some major changes affecting sexual relations come from changes in gender roles. But technological change, as well as the emphasis on consumerism in the United States, also affects sexuality. As you think about sex and social change, you might try to imagine what other social factors influence human sexual behavior.

The Sexual Revolution: Is It Over?

The **sexual revolution** refers to the widespread changes in men's and women's roles and a greater public acceptance of sexuality as a normal part of social development. Many changes associated with the sexual revolution have been changes in women's behaviors. Essentially, the sexual revolution has narrowed the differences in the sexual experiences of men and women. The feminist and the gay and lesbian movements have put the sexual revolution at the center of public attention by challenging gender role stereotyping and sexual oppression, profoundly changing our understanding of gay and lesbian sexuality. The sexual revolution has meant greater sexual freedom, especially for women, but it has not eliminated the influence of gender in sexual relationships.

Technology, Sex, and Cybersex

Technological change has also brought new possibilities for sexual freedom. One significant change is the widespread availability of the birth control pill. Sex is no longer necessarily linked with reproduction; new sexual norms associate sex with intimacy, emotional ties, and physical pleasure (Freedman and D'Emilio 1988). These sexual freedoms are not equally distributed among all groups, however. For women, sex is still more closely tied to reproduction than it is for men because women are still more likely to take the responsibility for birth control.

The popular Bratz dolls are being marketed to young girls, selling an image of women as sexual objects.

Graeme Robertson/Getty Images

Contraceptives are not the only technology influencing sexual values and practices. Now the Internet has introduced new forms of sexual relations as many people seek sexual stimulation from pornographic websites or online sexual chat rooms. Cybersex, as sex via the Internet has come to be known, can transform sex from a personal, face-to-face encounter to a seemingly anonymous relationship with mutual online sex. This introduces new risks; for example, two-thirds of those visiting chat rooms are adults masquerading as children (Cooper and Scherer 1999; Lamb 1998; Wysocki 1998). The Internet has introduced new forms of deviance that are difficult to regulate.

Commercializing Sex

At the same time that there are new sexual freedoms for women, many worry that this will only increase their sexual objectification. Furthermore, sexuality is becoming more and more of a commodity in our highly consumer-based society. Thus girls are being sexualized at younger ages, evidenced by the marketing of thongs to very young girls, the promotion of "sexy" dolls sold to young girls, and the highly sexualized content of media images that young boys and girls consume (Levy 2005). Often, these images are extremely violent and depict women as hypersexual victims of men's aggression, such as the popular video game, *Grand Theft Auto*.

Definitions of sexuality in the culture are heavily influenced by the advertising industry, which narrowly defines what is considered "sexy." Thin women, White women, and rich women are all depicted as more sexually appealing in the mainstream media. Images defining "sexy" also are explicitly heterosexual. The commercialization of sex uses women and, increasingly, men in demeaning ways. Thus, although the sexual revolution has removed sexuality from many of its traditional constraints, the inequalities of race, class, and gender still shape sexual relationships and values.

The combination of sexualization and commodification means that people become "made" into things for others' use. When people are held to narrow definitions of sexual attractiveness or are seen as valuable solely for their sexual appeal, you have social conditions that are ripe for exploitation and damage to people's sense of self-worth and value (American Psychological Association 2007). Thus, even in what seems to be an increasingly "free" sexual society, sexuality is still nested in American culture within a system of power relations—power relations that, despite the sexual revolution, continue to influence how different groups are valued and defined.

chapter summary

In what sense is sexuality, seemingly so personal an experience, a part of social structure?
Sexual relationships develop within a social and cultural context. Sexuality is learned through socialization, is channeled and directed by social institutions, and reflects the race, class, and gender relations in society.

What evidence is there of contemporary sexual attitudes and behavior?
Contemporary sexual attitudes vary considerably by social factors such as age, gender, race, and religion. Sexual behavior has also changed in recent years, with mixed trends in both liberal and conservative sexual values. In general, attitudes on issues of premarital sex and gay and lesbian rights have become more liberal, though this depends on social characteristics such as age, gender, and degree of religiosity, among others.

How is sexuality related to other social inequalities?
Sexuality intertwines with gender, race, and class inequality. This is especially revealed in the sexual stereotypes of different groups, as well as in the double standard applied to men's and women's sexual behaviors, such as in the *hooking up* culture.

What does sociological theory have to say about sexual behavior?
Functionalist theory depicts sexuality in terms of its contribution to the stability of social institutions. *Conflict theorists* see sexuality as part of the power relations and economic inequality in society. *Symbolic interaction* focuses on the social construction of sexual identity.

Feminist theory uncovers the power relationships that frame different sexual identities and behaviors, as well as linking sexuality to other forms of inequality.

How do homophobia and heterosexism influence lesbian and gay experience?
Homophobia is the fear and hatred of gays and lesbians. *Heterosexism* refers to institutional structures that define heterosexuality as the only social legitimate sexual orientation. Both produce relationships of power that define gays and lesbians as a social minority group.

How is sexuality related to contemporary social issues?
Sexuality is related to some of the most difficult social problems—including birth control, abortion, reproductive technologies, teen pregnancy, pornography, and sexual violence. Such social problems can be understood by analyzing the sexual, gender, class, and racial politics of society.

How is sex related to social change?
The *sexual revolution* refers to widespread changes in the roles of men and women and a greater acceptance of sexuality as a normal part of social development. The sexual revolution has been fueled by social movements, such as the feminist movement and the gay and lesbian rights movement. Technological changes, such as the development of the pill, have also created new sexual freedoms. Now, sexuality is influenced by the growth of cyberspace and its impact on personal and sexual interactions. At the same time, sex is treated as a commodity in this society; it is bought and sold and used to sell various products.

Key Terms

coming out 289
eugenics 295
heterosexism 293
homophobia 293

queer theory 291
sex tourism 293
sex trafficking 292
sexual identity 282

sexual orientation 282
sexual politics 290
sexual revolution 302

sexual scripts 284
social construction
perspective 289

13 Families and Religion

suppose you were to ask a large group of people in the United States to describe their families. Many would describe divorced families. Some would describe single-parent families. Some would describe stepfamilies with new siblings and a new parent stemming from remarriage. Others would describe gay or lesbian households, perhaps with children present. Also included would be adoptive families and families with foster children. Others would describe the so-called traditional family with two parents living as husband and wife in the same residence as their biological children. Families have become so diverse that it is no longer possible to speak of "the family" as if it were a single thing.

The traditional *family ideal*—a father employed as the breadwinner and a mother at home raising children—has long been the dominant cultural norm, communicated through a variety of sources, including the media, religion, and the law. Few families now conform to this ideal (see Figure 13.1), and the number of families that ever did is probably fewer than generally imagined (Coontz 1992). Regardless of their form, families now face new challenges, such as managing the demands of family plus work or struggling to meet family needs when work disappears. Many families feel that they are under siege by changes in society that are dramatically

Family diversity is the norm in American society, with no one type of family shaping people's experience.

altering all family experiences. Some of these changes are immediate, such as the loss of work in economically hard times or the strain on families from a family member's illness. But other changes are long-term changes in the social structure of society, such as women's increased work roles and accompanying changes in men's roles; population changes (such as aging and immigration); and even things like the increased reliance on technology, which can alter communication patterns in families.

thinking SOCIOLOGICALLY

What are some of the popular television shows about family life? What do these shows communicate about the *family ideal?* Do they indicate any change in this ideal? ●

Many view the changes taking place in families as positive. Women have new options and greater independence. Fathers are discovering that there can be great pleasure in domestic and child-care responsibilities. Change, however, also brings difficulties: balancing the demands of family and employment, coping with the interpersonal conflicts caused by changing expectations, and striving to make ends meet in families without sufficient financial resources. These changes bring new questions to the sociological study of families.

Family affairs are believed to be private, but as an institution, the family is very much part of the public agenda. Many people believe that "family breakdown" causes society's greatest problems—thus the intense national discussion around so-called family values. Public policies shape family life directly and indirectly, and family life is now being openly negotiated in political arenas, corporate boardrooms, and courtrooms, as well as in the bedrooms, kitchens, and "family" rooms of individual households.

As a social institution, the family is intertwined with other social institutions, such as religion. Religious values and customs tend to be learned first within families, and

religion in the United States influences beliefs about how families should be organized. Many major family events, such as weddings, christenings, and funerals, are observed with religious ceremonies; and, many family

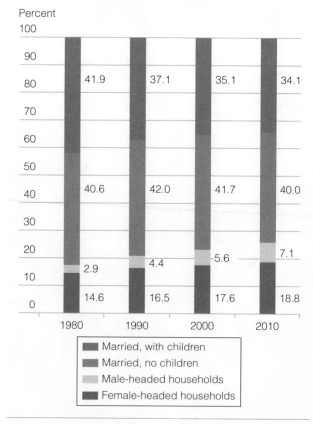

FIGURE 13.1 The Changing Character of U.S. Families, 1980–2010 Over the thirty-year period from 1980 to 2010, the composition of U.S. families has changed considerably. What are the major changes that you can identify from this figure? What sociological factors do you think help explain this change and, similarly, what other changes in society do such changes in the composition of families then create?
Data: U.S. Census Bureau. 2010a. *The 2010 Statistical Abstract.* Washington, DC: U.S. Department of Commerce, p. 54. **www.census.gov**

behaviors, such as reproduction, marriage, divorce, and sexual behavior, are affected by religious values.

This chapter focuses on families because study of the family is so important in society. But we also include a short section on religion—deserving of a chapter of its own—but included here to keep our introduction to sociological thinking relatively brief. The family and religion are, for most people, the first institutions encountered in life. In this chapter, we examine both as social institutions. The section on religion follows the material on the sociology of families.

learning objectives

- Understand the features that define different kinship systems
- Compare and contrast sociological theories of families
- Relate the characteristics that define different types of contemporary family experiences
- Refute some of the myths about contemporary marriage and divorce
- Be able to identify social structures that affect family violence
- Analyze current social policies that affect family experience
- Define how religion is a social institution
- Relate the different measures of religiosity
- Compare and contrast sociological theories of religion
- Identify the sources of religious diversity
- Define the different organizational structures of religion
- Explain how religion is related to social change and vice versa

DEFINING THE FAMILY

The family is a *social institution*, that is, an established social system that emerges, changes, and persists over time. Institutions are "there"; we do not reinvent them every day, although people adapt in ways that make institutions constantly evolve, such as is the case with how families have changed over time. We can define the **family** to refer to a primary group of people—usually related by ancestry, marriage, or adoption—who form a cooperative economic unit to care for offspring and each other and who are committed to maintaining the group over time (adapted from Lamanna and Riedmann 2012: 10).

Families are part of what are more broadly considered to be kinship systems. A **kinship system** is the pattern of relationships that define people's relationships to one another within a family. Kinship systems vary enormously across cultures and over time. In some societies, marriage is seen as a union of individuals; in others, marriage is seen as creating alliances between groups. In some kinship systems, marriages may be arranged, possibly even involving a broker whose job is to conduct the financial transactions and arrange marriage ceremonies

(Croll 1995). In still other kinship systems, maintaining multiple marriage partners may be the norm. Kinship systems can generally be categorized by:

- how many marriage partners are permitted at one time;
- who is permitted to marry whom;
- how descent is determined;
- how property is passed on;
- where the family resides; and,
- how power is distributed.

Polygamy is the practice of men or women having multiple marriage partners. Polygamy usually involves one man having more than one wife, technically referred to as *polygyny*; *polyandry* is the practice of a woman having more than one husband, an extremely rare custom. Within the United States, polygamy is commonly associated with Mormons, even though only a few Mormon fundamentalists (who are estimated to be only about 2 percent of the state population of Utah) now practice polygamy; those who practice polygamy do so without official church sanction (Brooke 1998).

Monogamy is the practice of a sexually exclusive marriage with one spouse at a time. It is the most common form of marriage in the United States and other Western industrialized nations. In the United States, monogamy is a cultural ideal that is prescribed through law and promoted through religious teachings. Lifelong monogamy is not always realized, however, as evidenced by the high rate of divorce and extramarital affairs. Many sociologists characterize modern marriage as *serial monogamy* in which individuals may, over a lifetime, have more than one marriage, but only one spouse at a time (Lamanna and Riedmann 2012).

In addition to defining appropriate marriage partners, kinship systems shape the distribution of property in society by determining descent. **Patrilineal kinship** systems trace descent through the father; **matrilineal kinship** systems, through the mother. **Bilateral kinship** (or bilineal kinship) traces descent through both. You can see the continuing influence of patrilineal kinship in contemporary society by noting how children are typically given the name of the father, not the mother. And even though practices are evolving, it is still quite common for women to take men's names when they marry. **Matrilocal** and **patrilocal** are terms used to describe where a married couple resides. In some societies, married couples are expected to move into the husband's residence—or even the husband's family residence.

Kinship systems also determine whom one can marry. Even without specific laws, society establishes normative expectations about appropriate marriage partners. In general, people in the United States marry people with very similar social characteristics, such as class, race, religion, and educational backgrounds. Interracial marriage, although increasing, is still relatively infrequent (only about 4 percent of married couples; see U.S. Census Bureau 2012a). Even though interracial marriages are not common, a tremendous amount of energy has historically been put into preventing them. Laws have prohibited marriage between various groups, including between Whites and African Americans and between Whites and Chinese, Japanese, Filipinos, Hawaiians, Hindus, and Native Americans. Not until 1967 were laws prohibiting interracial marriage declared unconstitutional by the U.S. Supreme Court (Kennedy 2003; Takaki 1989).

Extended Families

Extended families are the whole network of parents, children, and other relatives who form a family unit. Sometimes extended families, or parts thereof, live together, sharing their labor and economic resources. In some contexts, "kin" may refer to those who are not related by blood or marriage but who are intimately involved in the family support system and are considered part of the family (Stack 1974). *Othermothers* may be a grandmother, sister, aunt, cousin, or a member of the local community, but she is someone who provides extensive child care and receives recognition and support from the community around her (Collins 1990: 119). This term emerged from the experience of African American women, whose historically dual responsibilities in the family and work have meant that they have a history of creating alternative means of providing family care for children. Now this is a common practice for many families, including the Obama "first family": Upon moving to the White House, the Obamas brought Michelle Obama's mother, Marian Robinson, with them to assist with raising the Obama daughters, Malia and Sasha.

The system of *compadrazgo* among Chicanos is another example of an extended kinship system. In this system, the family is enlarged by the inclusion of godparents, to whom the family feels a connection that is the equivalent of kinship. The result is an extended system of connections between "fictive kin" (those who are not related by birth but are considered part of the family) and actual kin that deeply affects family relationships among Chicanos (Baca Zinn and Eitzen 2010).

Nuclear Families

In the **nuclear family**, a married couple resides together with their children. Like extended families, nuclear families develop in response to economic and social conditions. The origin of the nuclear family in Western society is tied to industrialization. Before industrialization, families were the basic economic unit of society. Large household units produced and distributed goods, whether in small communities or large plantation or feudal systems where slaves and peasants provided most of the labor. Production took place primarily in households, and all family members were seen as economically vital. Household and production were united, with no sharp distinction between economic

Diversity can occur within families, as with this mixed-race family.

and domestic life. Women performed and supervised much of the household work, engaged in agricultural labor, and produced clothes and food. The work of women, men, and children was also highly interdependent. Although the tasks each performed might differ, together they were a unit of economic production.

With industrialization, paid labor was performed mostly away from the home in factories and public marketplaces. The transition to wages for labor created an economy based on cash rather than domestic production. Families became dependent on the wages that workers brought home. The shift to wage labor was accompanied by an assumption that men should earn the "family wage" (that is, be the breadwinner). Thus, men who worked as paid laborers were paid more than women, and women became more economically dependent on men. At the same time, a man's status was enhanced by having a wife who could afford to stay at home—a privilege seldom accorded to working-class or poor families. The *family wage system* has persisted and is reflected still in the unequal wages of men and women.

The unique social conditions that racial–ethnic families have experienced also affect the development of family systems. Disruptions posed by the experiences of slavery, migration, and urban poverty affect how families are formed, their ability to stay together, the resources they have, and the problems they face. For example, historically, Chinese American laborers were explicitly forbidden to form families by state laws designed to regulate the flow of labor. Only a small number of merchant families were exempt from the law.

During the westward expansion of the United States, many Mexicans who had settled in the Southwest were displaced. The loss of their land disrupted their families and kinship systems. In the rapidly industrializing

Interracial Dating and Marriage

Picture this: A young couple, stars in their eyes, holding hands, intimacy in their demeanor. Newly in love, the couple imagines a long and happy life together. When you visualize this couple, who do you see? If your imagination reflects the sociological facts, odds are that you did not imagine this to be an interracial couple. Although interracial couples are increasingly common (and have long existed), people are more likely to form relationships with those of their same race—as well as social class, for that matter. What do sociologists know about interracial dating and marriage?

First, patterns of interracial dating are influenced by race, gender, and ethnicity. Among college students, for example, Black men and women are least likely to date (or even hook up) with a person of a different race, although Black men are more likely to do so than Black women. Hispanic and Asian students are more likely to date outside their group than Black students, and White students are least likely to do so of all (McClintock 2010).

These patterns continue in marriage. Interracial marriages are on the rise, although they are still a small percentage of marriages formed. Asian Americans and Hispanics are more likely to marry someone of a different race or ethnicity than are White Americans. African Americans are more likely to do so than Whites, but not as likely as Asians and Hispanics (see Figure 13.2). The most likely interracial marriages are between Black men and White women and between Hispanics and non-Hispanics. National data on Asian American marriage is limited, but studies find that Asian Americans are increasingly likely to marry other Asian Americans, although often someone of a different Asian heritage (Shinagawa and Pang 1996).

People in interracial relationships report negative reactions from their families; the majority of these were not extremely hostile but strong enough to put pressure on the interracial couple (Childs 2005; Dalmage 2000). And although most Blacks and Whites profess to have a color-blind stance toward interracial marriages, when pressed, they raise numerous qualifications and concerns about such pairings (Bonilla-Silva and Hovespan 2000).

Regardless of these attitudes, interracial marriage is on the rise, and attitudes are changing. But the facts about interracial dating and marriage show how something seemingly "uncontrollable," such as love, is indeed shaped by many sociological factors.

Sources: Childs, Erica Chito. 2005. *Navigating Interracial Borders: Black-White Couples and Their Social World*. New Brunswick, NJ: Rutgers University Press; Dalmage, Heather M. 2000. *Tripping on the Color Line: Black-White Multiracial Families in a Racially Divided World*. New Brunswick, NJ: Rutgers University Press; Bonilla-Silva, Eduardo, and Mary Hovespan. 2000. "If Two People Are in Love: Deconstructing Whites' Views on Interracial Marriage with Blacks." Paper presented at the annual meetings of the Southern Sociological Society; Shinagawa, Larry, and Gin Yong Pang. 1996. "Asian American Panethnicity and Intermarriage." *Amerasia Journal* 22 (Spring): 127–152.

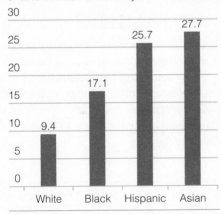

Percent of newlyweds married to someone of different race or ethnicity

- White: 9.4
- Black: 17.1
- Hispanic: 25.7
- Asian: 27.7

FIGURE 13.2 Intermarriage Rates by Race and Ethnicity, 2010 A small percentage of all marriages are interracial/interethnic (4 percent in 2010), but the percentage of newlyweds who are marrying someone of another race or ethnicity is higher—although higher among some groups than others. How would you explain the sociological factors that might affect these group differences in rates of intermarriage?

Source: Wang, Wendy. 2012. "The Rise of Intermarriage: Rates, Characteristics Vary by Race and Gender." Washington, DC: Pew Research Center.

Robin Nelson/Photo Edit

Interracial marriage is increasingly common, although still a small proportion of all marriages.

United States, many Mexican Americans were able to find work in the mines opening in the new territories or by building the railroads spreading from the east toward the Pacific. Employers apparently thought that they had better control over laborers if their families were not there to distract them, so families were typically prohibited from being with their working member. One result was the development of prostitution camps, which followed workers from place to place (Dill 1988). Families continue to be influenced by social structural forces. Some families, particularly those with marginal incomes, find it necessary for the entire family to work to meet the economic needs of the household. Migration to a new land and exposure to new customs also disrupts traditional family values. The ability to form and sustain nuclear families is directly linked to the economic, political, and racial organization of society.

SOCIOLOGICAL THEORY AND FAMILIES

Is the family a source of stability or change in society? Are families organized around harmonious interests, or are they sources of conflict and differential power? How do new family forms emerge, and how do people negotiate the changes that affect families? These questions and others guide sociological theories of the family (see Table 13.1).

Functionalist Theory and Families

Functionalist theorists interpret the family as filling particular societal needs, including socializing the young, regulating sexual activity and procreation, providing physical care for family members, and giving psychological support and emotional security to individuals. According to functionalism, families exist to meet these needs and to ensure a consensus of values in society. In the functionalist framework, the family is conceptualized as a mutually beneficial exchange, wherein women receive protection, economic support, and status in return for emotional and sexual support, household maintenance, and the production of offspring (Glenn 1986). At the same time, men in traditional marriages get the services that women provide—housework, nurturing, food service, and sexual partnership. Functionalists also see families as providing care for children, who are taught the values that society and the family support. In addition, functionalists see the family as regulating reproductive activity, including cultural sanctions about sexuality.

According to functionalist theory, when societies experience disruption and change, institutions such as the family become disorganized, weakening social cohesion. Currently, some analysts interpret the family as "breaking down" under societal strains. Functionalist theory suggests that this breakdown is the result of the disorganizing forces that rapid social change has fostered.

Functionalists also note that, over time, other institutions have begun to take on some functions originally performed solely by the family. For example, as

table 13.1 Theoretical Perspectives on Families

	Functionalism	Conflict Theory	Symbolic Interaction	Feminist Theory
Families	Meet the needs of society to socialize children and reproduce new members	Reinforce and support power relations in society	Emerge as people interact to meet basic needs and develop meaningful relationships	Are gendered institutions that reflect the gender hierarchies in society
	Teach people the norms and values of society	Inculcate values consistent with the needs of dominant institutions	Are where people learn social identities through their interactions with others	Are a primary agent of gender socialization
	Are organized around a harmony of interests	Are sites for conflict and diverse interests of different family members	Are places where people negotiate their roles and relationships with each other	Involve a power imbalance between men and women
	Experience social disorganization ("breakdown") when society undergoes rapid social changes	Change as the economic organization of society changes	Change as people develop new understandings of family life	Evolve in new forms as the society becomes more or less egalitarian

The family is the major institution where socialization of children occurs.

an allowance teaches the child capitalist habits involving money.

Whereas functionalist theory conceptualizes the family as an integrative institution (meaning it has the function of maintaining social stability), conflict theorists depict the family as an institution subject to the same conflicts and tensions that characterize the rest of society. Families are not isolated from the problems facing society as a whole. The struggles brought on by racism, class inequality, sexism, homophobia, and other social conflicts are played out within family life.

Symbolic Interaction Theory and Families

Symbolic interaction emphasizes that meanings people give to their behavior and that of others is the basis of social interaction. Those who study families from this perspective tend to take a more microscopic view of families. A symbolic interactionist might ask how different people define and understand their family experience. Symbolic interactionists also study how people negotiate family relationships, such as deciding who does what housework, how they will arrange child care, and how they will balance the demands of work and family life.

To illustrate, when people get married, they form a new relationship and new identities with specific meanings within society. Some changes may seem very abrupt—a change of name certainly requires adjustment, as does being called a husband or wife. Some changes are more subtle—how one is treated by others and the privileges couples enjoy (such as being a recognized legal unit). Symbolic interactionists see the marriage relationship as socially constructed; that is, it evolves through the definitions that others in society give it as well as through the evolving definition of *self* that married partners make for themselves.

The symbolic interaction perspective understands that roles within families are not fixed, but rather evolve as participants define and redefine their behavior toward each other. Symbolic interaction is especially helpful in understanding changes in the family because it supplies a basis for analyzing new meaning systems and the evolution of new family forms over time.

children now attend school earlier in life and stay in school for longer periods of the day, schools (and other caregivers) have taken on some functions of physical care and socialization originally reserved for the family. Functionalists would say that the diminishment of the family's functions produces further social disorganization because the family no longer carefully integrates its members into society. To functionalists, the family is shaped by the template of society, and such things as the high rate of divorce and the rising numbers of female-headed and single-parent households are the result of social disorganization.

Conflict Theory and Families

Conflict theory interprets the family as a system of power relations that reinforces and reflects the inequalities in society. Conflict theorists also are interested in how families are affected by class, race, and gender inequality. This perspective sees families as the units through which the advantages, as well as the disadvantages, of race, class, and gender are acquired. Conflict theorists view families as essential to maintaining inequality in society because they are the vehicles through which property and social status are acquired (Baca Zinn and Eitzen 2010).

The conflict perspective also emphasizes that families in the United States are shaped by capitalism. The family is vital to capitalism because it produces the workers that capitalism requires. Accordingly, personalities within families are shaped to the needs of a capitalist system; thus families socialize children to become obedient, subordinate to authority, and good consumers. Those who learn these traits become the kinds of workers and consumers that capitalism needs. Families also serve capitalism in other ways; giving a child

Feminist Theory and Families

Feminist theory has contributed new ways of conceptualizing the family by focusing sociological analyses on women's experiences in the family and by making gender a central concept in analyzing the family as a social institution. Feminist theories of the family emerged initially as a criticism of functionalist theory. Feminist scholars argued that functionalist theory assumed that the gender division of labor in the household is functional for society. Feminists have also been critical of functional theory for assuming an inevitable gender division of labor within the family. Feminist critics argue that although functionalists may see the gender division of labor as functional, it is based on stereotypes about men's and women's roles.

Influenced by the assumptions of conflict theory, feminist scholars do not see the family as serving the needs of all members equally. Quite the contrary, feminists have noted that the family is one of the primary institutions producing the gender relations found in society. Feminist theory conceptualizes the family as a system of power relations and social conflict. In this sense, it emerges from conflict theory, but adds that the family is a *gendered institution* (see Chapter 11). Each of the different theoretical perspectives illuminates different features of family experiences.

DIVERSITY AMONG CONTEMPORARY AMERICAN FAMILIES

Today, the family is one of the most rapidly changing of all of society's institutions. Families are systems of social relationships that emerge in response to social conditions that, in turn, shape the future direction of society. There is no static or natural form for the family. Change and variation in families are social facts.

Among other changes, families today are smaller than in the past. There are fewer births, and they are more closely spaced, although these characteristics of families vary by social class, region of residence, race, and other factors. Because of longer life expectancy, childbearing and child rearing now occupy a smaller fraction of parents' adult life. During earlier periods, death (often from childbirth) was more likely to claim the mother than the father of small children; thus men in the past would have been more likely than now to raise children on their own after the death of a spouse. That trend is now reversed; women are now more likely to be widowed with children, and death, once the major cause of early family disruption, has been replaced by divorce (Cherlin 2010; Rossi and Rossi 1990).

Demographic and structural changes have resulted in great diversity in family forms. Compared to thirty years ago, married couples now make up a smaller proportion of households; single-parent households have increased dramatically; and divorced and never-married people make up a larger proportion of the population. Overall, married-couple families make up about half of all households, and single-parent households (typically headed by women), post-childbearing couples, gay and lesbian couples, childless households, and single people are increasingly common (U.S. Census Bureau 2012a). Now people may also spend more

years caring for elderly parents than they did raising their children.

Female-Headed Households

One of the greatest changes in family life is the increase in the number of families headed by women. One-quarter of all children live with one parent, the vast majority of whom (87 percent) live with their mother. Although a large number of those living with one parent do so because of divorce, the largest number live with one parent who has never been married. One-quarter of all households are headed by women, although the number of households headed by single fathers has also increased. The odds of living in a single-parent household are even greater for African American and Latino children (U.S. Census Bureau 2012a).

The two primary causes for the growing number of women heading their own households are the high rate of pregnancy among unmarried teens and the high divorce rate, with death of a spouse also contributing. As discussed in Chapter 12, even though the rate of pregnancy among teenagers has declined, the proportion of teen births occurring outside marriage has increased.

Many people see the increase of female-headed households as representing a breakdown of the family and a weakening of social values. An alternate view, however, is that the rise of female-headed households reflects the growing independence of women, some of whom are making decisions to raise children on their own. Not all female-headed households are women who have never married; many are divorced and widowed women whose circumstances may be quite different from those of a younger, never-married woman.

Some claim that female-headed households are linked to problems such as delinquency, the school dropout rate, children's poor self-image, and other social problems. Sometimes the cause of these troubles is attributed explicitly to the absence of men in the family. Sociologists, however, have not found the absence of men as the sole basis for such problems; rather, it is the presence of economic pressure faced by female-headed households, compared with that of male-headed households, that puts female-headed households under great strain—with the threat of poverty being by far the greatest problem they face. Among households headed by women with children, one-third live below the poverty line, with the rates of poverty highest among Black and Hispanic female-headed households (DeNavas-Walt et al. 2012). It is not the makeup of households headed by women that is a problem but the fact that they are most likely to be poor. This phenomenon was discussed in Chapter 8 as the *feminization of poverty.*

a sociological eye ON THE media

Idealizing Family Life

Cultural norms about motherhood and fatherhood come from many places, but the media is certainly a strong influence on how family ideals—and the ideals for mothers and fathers—are created in society. Media images of the family have certainly changed since the inception of television. In the 1930s, "Hollywood codes"—that is, official rules in Hollywood about what could and could not be seen in movies and, later, on television—meant that families were always shown with two parents, marriage intact, father working, and mom staying at home. Parents never talked about sex; indeed, it appeared as though they never had it because the codes forbade any nudity and required that scenes of passion not excite the audience. As a result, if bedroom scenes between married couples were shown, there were typically only twin beds.

Now, family images on television are more diverse. Programs like *Modern Family*, *The New Normal*, and *Two and a Half Men* show very different images of family life. Still, the media continue to construct an ideal for family life—one that continues to stereotype men and women in family roles. What does research show about these social constructions?

- Most family characters are middle class. Men appearing with children are most likely to be shown outside; they are also more likely to be seen with boys, not girls.
- Fathers are infrequently seen with infants.
- Fathers are shown playing with, reading to, talking with, and eating with children, but not preparing meals, cleaning house, changing diapers, and so forth.
- Women are disproportionately shown in family settings in the media.

What gender stereotypes do such images project? How do they influence people's views of ideal family roles? How realistically do they portray the actual gender division of labor in families? What race and class images confound these results?

Sources: Kaufman, Gayle. 1999. "The Portrayal of Men's Family Roles in Television Commercials." *Sex Roles* 41 (September): 439–458; Coltrane, Scott, and Melinda Messineo. 2000. "The Perpetuation of Subtle Prejudice: Race and Gender Imagery in 1990s Television Advertising." *Sex Roles* 42 (March): 363–389.

Peter 'Hopper' Stone/© ABC/Getty Images

Popular TV shows, such as *Modern Family*, can reflect changing family values, although much of the time they also promote narrow family ideals.

debunking SOCIETY'S MYTHS

MYTH: Absence of fathers is the cause of numerous social problems; if these fathers would just adopt "family values," families would be stronger and children wouldn't get into so much trouble.

SOCIOLOGICAL PERSPECTIVE: The mere presence of men in a family does not in itself prevent family problems. Concerns about father-absent families are typically directed at poor, not middle-class, families. Research on never-married, poor African American fathers finds that they typically want to provide for their children, may spend a lot of time with their children, and want to be good fathers, but find that their life opportunities make it difficult to do so (Coles and Green 2009; Hamer 2001). ●

Although women head the majority of single-parent families, families headed by a single father are increasing, but male-headed households are less likely than female-headed households to experience severe economic problems. Unlike female-headed households where a man is not present to help with housework and children, single fathers commonly get domestic help from women—either girlfriends, daughters, or mothers (Popenoe 2001).

Married-Couple Families

Among married-couple families, the increased participation of women in the paid labor force has added new challenges to family life. Families now are able to sustain a median income level only by having both husband and wife in the paid labor force. As a result, families are experiencing substantial *social speedup*, a term reflecting the common feeling among working parents that there is too much to do and too little time to do it.

Women's labor force participation has created other changes in family life. One is the number of married couples who have *commuter marriages*, when work requires one partner in a dual-career couple to reside in a different city, separated by jobs too distant for a daily commute. The common image of a commuter marriage is one consisting of a prosperous professional couple, each holding important jobs, flashing credit cards, and using airplanes like taxis. However, working-class and poor couples do their share of long-distance commuting: Agricultural workers follow seasonal work; skilled laborers sometimes have to leave their families to find jobs; and many families cross national borders in search of work, often separating one or both parents and children. Although their commute may be less glamorous than that of professional spouses, they are commuting nonetheless. When all types of commuter marriages are included, this form of marriage is more prevalent than is typically imagined.

Stepfamilies

Because of the rise in divorce and remarriage, stepfamilies are now fairly common in the United States. They take numerous forms, including married adults with stepchildren, cohabiting stepparents, and stepparents who do not reside together (Stewart 2001). About 40 percent of marriages involve stepchildren. Stepfamilies may face a difficult period in the transition when two families blend, introducing people to a "new" family. Parents and children may have to learn new roles when they become part of a stepfamily. Children accustomed to being the oldest child in the family, or the youngest, may find that their status in the family group is suddenly transformed. New living arrangements may require children to share rooms, toys, and time with people they perceive as strangers.

In stepfamilies, the parenting roles of mothers and fathers suddenly may expand to include more children, each with his or her needs. Jealousy, competition, and demands for time and attention can make the relationships within stepfamilies tense. The problems are compounded by the absence of norms and institutional support systems for stepfamilies. Without norms to follow, people have to adapt by creating new language to refer to family members and new relationships. Many develop strong relationships within this new kinship system; others find the adjustment extremely difficult, resulting in a high probability of divorce among remarried couples with children (Baca Zinn and Eitzen 2010).

Gay and Lesbian Households

The increased visibility of gay and lesbian families has challenged the traditional understanding of families as only heterosexual. A small but growing number of states have recognized gay marriages as legal. But even without state support, many gay and lesbian couples form long-term, primary relationships that they define as marriage. Like other families, gay and lesbian couples share living arrangements and household expenses, make decisions as partners, and in many cases, raise children (Mezey 2008).

As we saw in Chapter 12, the number of Americans who think same-sex marriage should be legal (53 percent) has increased quite dramatically (see Figure 13.3). Same-sex marriage tends to be more acceptable in the eyes of younger people than older groups, raising the question of whether social support for gay marriages will increase over time or whether young people will shift their values as they age. In the meantime, in most states and municipalities, gays and lesbians who form strong and lasting relationships do so without formal institutional support and, as a result, have had to be innovative in producing new support systems.

Researchers have found that gay and lesbian couples tend to be more flexible and less gender stereotyped in their household roles than heterosexual couples. Lesbian households, in particular, are more egalitarian than are either heterosexual or gay male couples. Money also has less effect on the balance of power in lesbian relationships than is true for heterosexual couples. However, where one partner is the primary breadwinner and the other the primary caregiver for children, the partner staying at home becomes economically vulnerable and less able to negotiate her needs, just as in heterosexual relationships (Sullivan 1996).

The new family forms that lesbian and gay couples are creating mean that they have to actively construct new meanings of such things as motherhood. Researchers find that they do so in ways that are collaborative, including elaborate networks of family and friends (Dalton and Bielby 2000; Dunne 2000). To date, most

Stepfamilies, or blended families, are increasingly common; they can pose challenges for family members who must adapt to new family roles and relationships.

Buccina Studios/Photodisc/Getty Images

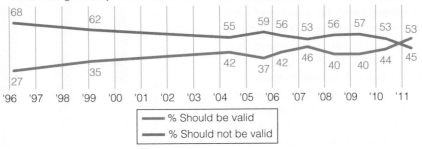

Question Asked: Do you think marriages between same-sex couples should or should not be recognized by the law as valid, with the same rights as traditional marriages?

% Should be valid
% Should not be valid

FIGURE 13.3 Acceptance of Same-Sex Marriage What societal changes do you think have influenced the direction of this line graph representing the opinion of a nationally representative sample of the public?

Source: Newport, Frank. 2011. "For First Time, Majority of Americans Favor Legal Gay Marriage." Princeton, NJ: The Gallup Organization. **www.gallup.com**

gay fathers are those who have children from a previous heterosexual marriage, although the number of gay men adopting children is increasing.

Public debate about gay marriage often centers on the implications for children raised in gay and lesbian families. Research on children in gay and lesbian households finds that, for the most part, there is little difference in outcomes for children raised in gay and lesbian households compared to those raised in heterosexual households. What differences are found are the result of other factors—not just the sexual orientation of parents. The greatest differences are the result of the homophobia that is directed against children in lesbian and gay families, who are very likely to be stigmatized by others. But such children are also less likely to develop stereotypical gender roles and are more open-minded about sexual matters, although they are no more likely to become gay themselves (Stacey and Biblarz 2001). If we lived in a society more tolerant of diversity, the differences that emerge might be viewed as strengths, not deficits.

Single People

Single people, including those never married, widowed, divorced, and separated, today constitute half of the population (over age 15; U.S. Census Bureau 2012a). Some of the increase is a result of the rising number of divorced people, but there has also been an increase in the number of those never married (31 percent of the population over age 15 and one-third of those aged 30 to 34). Men and women are also marrying at a later age—at age 26 on average for women, age 28 for men, compared with age 21 for women and age 23 for men in 1980 (Copen et al. 2012).

Among singles, patterns of establishing intimate relationships have changed significantly. "Hooking up"—a phrase referring to a casual sexual alliance between two people—has supplanted dating as the pattern by which young people get to know each other. Courtship no longer follows preestablished norms. Hooking up is widespread on college campuses and influences the campus culture, although only a minority of students engage in it (about 40 percent of college

women). Hooking up carries multiple meanings. For some, it means kissing; for others, it means sexual-genital play, but not intercourse; for some, it means sexual intercourse. The vagueness of the term contributes to its becoming a shared cultural phenomenon. A majority of college women say that hooking up makes them feel desirable, but also awkward, and they are wary of getting a bad reputation from hooking up too often. The majority of college women still want to meet a spouse while at college (Hamilton and Armstrong 2009; Bogle 2008).

The path to a committed relationship, possibly marriage, involves increasing phases of commitment and sexual exclusivity. Many people find the same sexual and emotional gratification in single life as they would in marriage. Being single is also no longer the stigma it once was, especially for women. And increasing numbers of single people are forming new forms of families. As one example, sociologist Rosanna Hertz has studied women who become mothers by choice outside of marriage. In some cases, these are women who have not found marriage partners or do not want one. Some are lesbian; many are not. But these women still embrace wanting to be mothers, and Hertz has explored how they are forming new family structures, drawing, for example, on kin and friendship networks for familial support (Hertz 2006).

Such changes in family patterns reflect an important sociological conclusion that the form of the family per se does not predict happiness so much as other social factors, including financial resources, the presence of conflict and violence, the extent of personalties beyond the immediate family, and the presence of stressful life problems.

Cohabitation (living together) has become increasingly common. Some of the increase is the result of better census taking, but the increase is also real. Almost two-thirds of married couples now live together before marriage, compared to only 10 percent in 1970. Although some cohabit because they are critical of the existing norms surrounding marriage (Elizabeth 2000), living together has become a common pattern prior to marriage, leading many to ask if living together before marriage stabilizes or destabilizes marriage.

Comparing those who lived together before marriage and those who did not, current research finds that couples report little difference in the quality of marriage. The greatest predictor of marital quality among those who lived together seems to be the presence of children. That is, those couples who had children prior to marriage report lower-equality marriages than those who did not have children prior to marriage. Regardless of cohabitation prior to marriage, however, researchers also find that all couples report lower quality in their marriages over time (Tach and Halpern-Meekin 2009).

In addition to couples living together, a growing number of single people are remaining in their parents' homes for longer periods (see Figure 13.6). Known as the *boomerang generation*, or "accordion families," young people in their twenties are returning home when they would normally be expected to live independently. A much higher proportion of young adults now live with their parents than at any time since the 1940s. Nearly one-third (three in ten) young adults between ages 25 and 34 now live with their parents, compared to only 11 percent in 1980 (Parker 2012). For most, this is for financial reasons, although the particulars differ across social class. Working-class young adults tend to have never left while young people in middle- and upper-middle class families are more likely to "boomerang"—that is, return home to save rent money while perhaps establishing a career or attending school (Klinenberg 2012; Newman 2012).

MARRIAGE AND DIVORCE

Even with the extraordinary diversity of family forms in the United States, the majority of people will still marry at some point in their lives (Figure 13.4). Indeed, the United States has the highest rate of marriage of any Western industrialized nation, as well as a high divorce rate.

Marriage

The picture of marriage as a consensual unit based on intimacy, economic cooperation, and mutual goals is widely shared, although marital relationships also involve a complex set of social dynamics, including cooperation and conflict, different patterns of resource allocation, and a division of labor. Sociologists must be careful not to romanticize marriage to the point that they miss other significant social patterns within marriage.

Gender roles are a significant reality of family life, shaping power dynamics within marriage, as well as the allocation of work, the degree of marital happiness, the likelihood of marital violence, and even the leisure time that each partner has. Although people do not like to think of marriage as a power relationship, gender shapes the power that men and women have within marriage, as it does in other relationships. For one thing, sociologists have long found that the

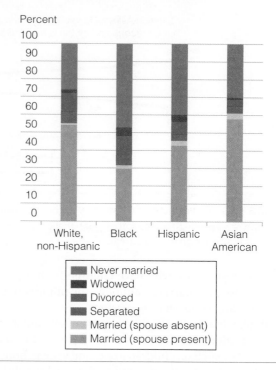

FIGURE 13.4 Marital Status of the U.S. Population by Race What differences in marital status do you observe here, comparing different racial–ethnic groups? What might explain some of the differences you observe?

Data: U.S. Census Bureau. 2012b. *America's Families and Living Arrangements: 2012, Detailed Tables.* Current Population Reports. **www.census.gov**

amount of money a person earns establishes that person's relative power within the marriage, including the ability to influence decisions, the degree of autonomy and independence held by each partner, and the control of expectations about family life. Despite changes in women wanting to work outside the home, men in most marriages (64 percent) are the sole or major earners (Raley et al. 2006). But studies also find that even when wives earn substantially more than their husbands, rare as that is, couples tend to negotiate marital power within the confines of traditional gender expectations (Tichenor 2005).

Within marriage, gender also shapes the division of household labor. Women do far more work in the home and have less leisure time (Sarkasian and Gerstel 2012). Most employed mothers do two jobs—the so-called *second shift* of housework after working all day in a paid job (Hochschild and Machung 1989)—a pattern that exists not just in the United States, but in other nations as well (Figure 13.5). Indeed, sociologists have now even identified the *third shift* of women's work—that is, the greater amount of help that women compared to men give to family and friends, such as assisting those who are sick, preparing holiday celebrations, planning family visits, and so forth (Gerstel 2000). Little wonder that people are feeling that they have less and less time.

Are men more involved in housework than in the past? Yes and no. Men report that they do more

Norway
New Zealand
Japan
Sweden
Germany
Canada
Finland
Australia
Turkey
United Kingdom
France
Korea
United States
Belgium
Spain
Mexico
Poland
Italy

0   10   20   30   40   50   60   70   80

Minutes per day (reported by men)

FIGURE 13.5 Men's and Women's Leisure Worldwide This graph shows the greater number of leisure minutes per day that men report relative to women. What social factors do you think affect men's greater leisure in these different countries?

Source: OECD. 2009. *Society at a Glance—OECD Social Indicators*. www.oecd.org/els/social/indicators/SAG

housework, but they devote only slightly more of their time to housework than in the past. Estimates vary regarding the amount of housework that men do, but studies generally find a large gap between the number of hours women give to housework and child care and the hours men give. Among couples where both partners are employed, only 28 percent equally share the housework. Fathers do more when there is a child in the house under two years of age, but the increase is mostly accounted for by the amount of child care men provide, not the housework they do. The end result is that men have about eleven more hours of leisure per week than women do (Press and Townsley 1998). Interestingly, sociologists have found that the allocation of housework is greatly affected by men's and women's experience in their own families of origin; those from households with a more egalitarian division of labor are likely to carry this into their own relationships (Cunningham 2001).

Despite a widespread belief that young professional couples are the most egalitarian, studies find that there is little difference across social class in the amount of housework that men do. Even women who are earning more than their husbands do more of the housework (Tichenor 2005; Wright et al. 1992). Yet, African American husbands perform a greater share of housework than do White husbands, and Latino households have more diversity in gender roles than stereotypes about machismo would lead us to believe (McLoyd et al. 2000). Within all families, housework and other forms of family

care are negotiated, but typically within the confines of existing gender ideologies (Legerski and Cornwall 2010).

Although marriage can be seen as a romantic and intimate relationship between two people, it can also be seen within a sociological context. Marriage relationships are shaped by a vast array of social factors, not just the commitment of two people to each other. You see this especially when examining marital conflicts. Life events, such as the birth of a child, job loss, retirement, and other family commitments, such as elder care or caring for a child with special needs, all influence the degree of marital conflict and stability (Moen et al. 2001). As conditions in society change, people make adjustments within their relationships, but how well they can cope within a marriage depends on a large array of sociological, not just individual, factors.

Divorce

The United States leads the world not only in the number of people who marry, but also in the number of people who divorce. More than sixteen million people have divorced but not remarried in the population today; more women are in this group than men because women are less likely to remarry following a divorce. Since 1960, the rate of divorce has more than doubled, although it has declined recently since its all-time high in 1980.

You will often hear that one in every two marriages ends in divorce, but this is a misleading statistic. The

DOING sociological research

Men's Caregiving

Research Question: Much research has documented the fact that women do the majority of the housework and child care within families. Why? Many have explained it as the result of gender socialization—women learn early on to be nurturing and responsible for others, while men are less likely to do so. Yet, things are changing, and some men are more involved in the "care work" of family life. What explains whether men will be more engaged in family care work?

Research Method: Sociologists Naomi Gerstel and Sally Gallagher studied a sample of 188 married people. They interviewed ninety-four husbands and ninety-four wives, married to each other; the sample was 86 percent White and 14 percent African American but was too small to examine similarities or differences by race.

Research Results: You might expect that men who had attitudes expressing support for men's family responsibilities would be more involved in family care (defined by Gerstel and Gallagher to include elder care, child care, and various household tasks). But this is not what Gerstel and Gallagher found. Gender attitudes did not influence men's involvement in caregiving. Rather, the characteristics of the men's families were the most influential determinant of their engagement in housework and child care. Men whose wives spent the most time helping kin and men who had daughters were more likely to help kin. Having sons had no influence. But, in addition, men with more sisters tended to spend less time helping with elder parents than men with fewer sisters. Furthermore, men's employment (measured as hours employed, job flexibility, and job stability) did not affect their involvement in care work.

Conclusions and Implications: It is the social structure of the family, not gender beliefs, that shapes men's involvement in family work. As they put it, "It is primarily the women in men's lives who shape the amount and types of care men provide" (Gerstel and Gallagher 2001: 211). This study shows a most important sociological point: Social structure, not just individual attitudes, is the most significant determinant of social behavior.

Questions to Consider

1. Who does the work in your family? Is it related to the social organization of your family, as Gerstel and Gallagher find in other families?
2. Do you think that men's gender identity changes when they become more involved in care work? What hinders and/or facilitates men's engagement in this kind of work?

Source: Gerstel, Naomi, and Sally Gallagher. 2001. "Men's Caregiving: Gender and the Contingent Character of Care." *Gender & Society* 15 (April): 197–217.

---

marriage rate is 6.8 marriages per 1000 people and the divorce rate, 3.4 per 1000 people (U.S. Census Bureau 2012a). At first glance, it appears that there are half as many divorces as marriages. But the marriage rate is the number of marriages formed in a year and does not include the number of continuing marriages; thus divorce is not as widespread as one in every two of all marriages.

Still, the rate of divorce is high and has risen since 1950, though it has been declining since 1980. The likelihood of divorce is also not equally distributed across all social groups. Divorce is more likely for couples who marry young, while in their teens or early twenties. Second marriages are more likely than first marriages to end in divorce. Divorce is somewhat higher among low-income couples, a fact reflecting the strains that financial problems create. Divorce is also somewhat higher among African Americans than among Whites, partially because African Americans make up a disproportionate part of lower-income groups. Hispanics have a lower rate of divorce than either Whites or Blacks, probably the result of religious influence. Recently, the divorce rate among Asian Americans has also risen, interpreted as the possible shedding of cultural taboos (Armas 2003). This explanation seems supported by the fact that Asian Americans born in the United States are more likely to be divorced than Asians who immigrated (McLoyd et al. 2000).

Is It True?*

	True	False
1. Half of all marriages end in divorce.		
2. Children who grow up in gay or lesbian families are likely to become gay.		
3. Single people are a larger proportion of the population than was true in the past.		
4. Children are better off growing up in a home where mothers are not employed.		
5. Women and men find great satisfaction in trying to balance family and work.		

*The answers can be found on the next page.

A number of factors contribute to the current high rate of divorce in the United States. Demographic changes (shifts in the composition of the population) are part of the explanation. The rise in life expectancy, for example, has an effect on the length of marriages. In earlier eras, people died younger, and thus the average length of marriages was shorter. Some marriages that earlier would have ended with the death of a spouse may now be dissolved by divorce. Still, cultural factors also contribute to divorce.

In the United States, individualism is a cultural norm, placing a high value on a person's satisfaction within marriage. The cultural orientation toward individualism may predispose people to terminate a marriage in which they are personally unhappy. In other cultural contexts (including this society years ago), marriage, no matter how difficult, may have been seen as an unbreakable bond, regardless of whether one was unhappy. But even with the American belief in individualism, people still value the ideal of lifelong romantic love and the security of a long-term partner, whether or not they are able to actually achieve this (Hull et al. 2010).

Changes in women's roles also are related to the rate of divorce. Women today are now less financially dependent on husbands than in the past, even though they still earn less. As a result, the economic interdependence that once bound women and men as a marital unit is no longer as strong. Although most married women would be less well off without access to their husband's income, they could probably still support themselves. This can make it possible for people to end marriages that they find unsatisfactory.

For people in unhappy marriages, divorce, though painful and financially risky, can be a positive option (Kurz 1995). The belief that couples should stay together for the sake of the children is now giving way to a belief, supported by research, that a marriage with protracted conflict is more detrimental to children than divorce. Although there are periodic public outcries about the negative effect of divorce on children, many other factors influence their long-term psychological and social adjustment. Few children feel relieved or pleased by divorce; feelings of sadness, fear, loss, and anger are common, along with desires for reconciliation and feelings of conflicting loyalties. But most children adjust reasonably well after a year or so. Moreover, children's adjustment is influenced most by factors that precede the divorce. The single most important factor influencing children's poor adjustment is marital violence and prolonged discord (Arendell 1998; Cherlin et al. 1998; Furstenberg 1998; Amato and Booth 1996; Stewart et al. 1997). The emotional strain on children is significantly reduced if the couple remains amicable. If both parents remain active in the upbringing of the children, the evidence shows that children do not suffer from divorce; especially important is the ability of the mother to be an effective parent after a divorce. Her ability to be effective can be influenced by the resources she has and her ongoing relationship with the father (Buchanan et al. 1996; Simons 1996; Furstenberg and Nord 1985).

In the aftermath of divorce, many fathers become distant from their children. Sociologists have argued that the tradition of defining men in terms of their role as breadwinners minimizes the attachment they feel for their children. If the family is then disrupted, they may feel that their primary responsibility, as financial provider, is lessened, leaving them with a diminished sense of obligation to their children.

FAMILY VIOLENCE

Generally speaking, the family is depicted as a private sphere where members are nurtured and protected, existing away from the influences of the outside world.

Although this is the experience of many, families also can be locales for violence, disruption, and conflict. Family violence, hidden for many years, is a phenomenon that has recently been the subject of much sociological research.

Domestic Violence and Abuse

Estimates of the extent of domestic violence are hard to come by and notoriously unreliable because the majority of cases of domestic violence go unreported. The National Centers for Disease Control estimates that 33 percent of women will be raped, physically assaulted, or stalked by an intimate partner in their lifetime (Black et al. 2011). Men also experience partner violence, although far less frequently. Women who experience violence are also twice as likely as men to be injured. Violence also occurs in gay and lesbian relationships, although silence around the issue may be even more pervasive given the marginalized status of gays and lesbians. Men living with male partners are just as likely to be raped, assaulted, or stalked as are women living with men, but the incidence of violence against women by women partners is about half as likely as heterosexual violence. Researchers conclude that this is because most domestic violence is committed by men. Violence is usually accompanied by emotionally abusive and controlling behavior. Jealous and dominating partners are the most likely perpetrators of domestic violence (Tjaden and Thoennes 2000; West 1998; Renzetti 1992).

One of the most common questions asked about domestic violence is why victims stay with their abuser. First, despite the belief that battered women do not leave their abusers, the majority do leave—at least for a period of time—and they seek ways to prevent further victimization. But some do not leave, and others leave and then return. Why? The answers are complex and stem from sociological, psychological, and economic problems. Victims tend to believe that the batterer will change, but they also find they have few options; they may perceive that leaving will be more dangerous, because violence can escalate when the abuser thinks he (or she) has lost control. Many women are unable to support their children and meet their living expenses without a husband's income. Mandatory arrest laws in cases of domestic violence can exacerbate this problem because they may, despite their intentions, discourage a woman from reporting violence for fear her batterer will lose his job (Miller 1997). Sociological analyses of violence in the family have led to the conclusion that women's relative powerlessness in the family is at the root of high rates of violence against women. Because most violence in the family is directed against women, the imbalance of power between men and women in the family is the source of most domestic violence. Because women are relatively powerless within the society, they may not have the resources to leave their marriage.

Child Abuse

Violence within families also victimizes many children who experience child abuse. Not all forms of child abuse are alike. Some people consider repeated spanking to be abusive; others think of this as legitimate behavior. Child abuse, however, is behavior that puts children at risk and may include physical violence and neglect. As with battering, the exact incidence of child abuse is difficult to know. In 2010, 3.6 million children were reported to child protective services at least once around the nation. Given the difficulty of measuring the actual extent of child abuse, this could be a misleading number. But what is known of those instances reported is that the most frequent reports involve children between birth and one year of age.

Victimization is almost evenly distributed between boys (49 percent of reports) and girls (51 percent of reports). White children are 45 percent of those where there are reports; African American children, 22 percent; and Hispanic children, 21 percent. Given the proportion of African Americans in the total population, you can see that they are overrepresented in these reports of abuse, as are Hispanics. This could be because of higher rates of victimization in these populations, but it could also be that they are simply less likely to be able to hide abuse because of the connection to social service agencies, policing, and other ways that official state agencies are more likely engaged in these communities. The most common forms of child abuse are neglect (78 percent of reports); physical abuse (18 percent); and sexual abuse (9 percent). Whereas men are the most likely perpetrators of domestic and sexual violence, women are just as likely to be the perpetrators of child abuse as are men (Children's Bureau 2011).

Research on child abuse finds a number of factors associated with abuse, including chronic alcohol use by a parent, unemployment, and isolation of the family. Sociologists point to the absence of social supports—in the form of social services, community assistance, and cultural norms about the primacy of motherhood—as related to child abuse, because most abusers are those with weak community ties and little contact with friends and relatives (Baca Zinn and Eitzen 2010).

Incest

Incest is a particular form of child abuse involving sexual relations between people who are closely related. A history of incest has been related to a variety of other problems, such as drug and alcohol abuse, runaways, delinquency, and various psychological problems, including the potential for violent partnerships in adult life. Studies find that fathers and uncles are the most frequent incestuous abusers, and that incest is most likely in families where mothers are debilitated (such as by mental illness or alcoholism). In such families, daughters often take on the mothering role, being taught to comply with men's demands to hold the family together. Scholars have linked women's powerlessness within families to the dynamics surrounding incest (Herman 1981).

Elder Abuse

The National Center on Elder Abuse estimates that between one and two million elders are abused in the United States, but it is difficult to gauge the true extent of the problem. Elder abuse tends to be hidden in the privacy of families, and victims are reluctant to talk about their situations, so estimates are only approximations. What is known is that reports of elder abuse have increased. Whether this reflects an actual increase or more reporting is open to speculation (National Center on Elder Abuse 2012; Teaster 2000).

Why are the elderly abused? One explanation is that caring for the elderly is very stressful for the caregiver—usually a daughter who may be employed in addition to caring for the elderly person. Research finds that abusers are most likely to be middle-aged women and (sadly) the daughter of the victim—the person most likely to be caring for the older person. Sons, however, are most likely to be engaged in direct physical abuse, accounting for almost half of the known physical abusers. Sometimes the physical abuser is a husband, where the abuse is a continuation of abusive behavior in the marriage. The same factors that affect family life in any generation contribute to the problem of elder abuse (Teaster 2000).

CHANGING FAMILIES IN A CHANGING SOCIETY

Like other social institutions, the family is in a constant state of change, particularly as new social conditions arise and as people in families adapt to the changed conditions of their lives. Some changes affect only a given family—the individual changes that come from the birth of a new child, the loss of a partner, divorce, migration, and other life events. These changes are what C. Wright Mills referred to as "troubles" (see Chapter 1). Some may even be happy events; the point is that they are changes that happen at the individual level, as people adjust to the presence of a new child, adjust to a breakup with a long-term partner, or grieve the loss of a spouse.

As Mills would have pointed out, many microsociological events that people experience in families have their origins in the broader macrosociological changes affecting society as a whole. These may be long-term changes (such as the changes in women's roles we have been noting), or they may be particular to a given time in history. For example, what impact do difficult economic times, such as depressions and recessions, have on families?

This is a question that has been examined through sociological research. One such study involved a rural community where the primary industry employing local residents closed. What happens to families under such widespread community strain? Many have shown how economic problems can lead to a broken family. But, interestingly, Jennifer Sherman looked at a community under economic strain and found that families were more likely to stay together when men in the family were flexible in their gender roles, such as taking on additional household work and child care when mothers are carrying the economic load (Sherman 2009).

This study indicates that people have to be adaptive in families, depending on the circumstances they face. Those who are rigid in their outlook and roles may have trouble adjusting to social change. In another study (examined earlier in Chapter 11), sociologist Kathleen Gerson studied young adults and their expectations for family and work life. Although she has found that young adults want relationships where both partners combine work and family, she found that men and women differ significantly in what she calls their "fallback position"—that is, what they would do if their ideal egalitarian arrangements either are not realized or fail. Men, Gerson found, are far more likely than women to say they would "fall back" on a traditional arrangement, that is, men working and women staying home. But women say that their fallback is to be self-reliant. Gerson concludes that both men and women will need to be more flexible in their gender and family roles if they are to weather the changes that will likely impact families in the future (Gerson 2010).

Global Changes in Family Life

Changes in the institutional structure of families are also being affected by the process of globalization. The increasing global basis of the economy means that people often work long distances from other family members—a phenomenon that occurs at all points on the social class spectrum, although the experience of such global mobility varies significantly by social class. A corporate executive may accumulate thousands—even millions—of first-class flight miles, crossing the globe to conduct business. A regional sales manager may spend most nights away from a family, likely staying in modestly priced motels and eating in fast-food franchises along the way. Truckers may sleep in the cabs of their tractor trailers after logging extraordinary numbers of hours of driving in a given week. Laborers may move from one state or country to the next, following the pattern of the harvest, living in camps away from families, and being paid by the amount they pick.

Global patterns of work and migration have created a new family form, the **transnational family**, defined as families where one parent (or both) lives and works in one country while his or her children remain in the country of origin. A good example is found in Hong Kong, where most domestic labor is performed by Filipina women who work on multiple-year contracts managed by the government, typically on a live-in basis. They leave their children in the Philippines, usually cared for by a relative, and send money home; the meager wages they earn in Hong Kong far exceed the average income of workers in the Philippines. This pattern is so common that the average Filipino migrant worker supports five people at home; one in five Filipinos directly depends on migrant workers' earnings (Parreñas 2001; Constable 1997).

One need not go to other nations to see such transnational patterns in family life. In the United States, Caribbean women and African American women have had a long history of having to leave their children with others while they sought employment in different regions of the country. Central American and Mexican women may come to work in the United States while their children stay behind. Mothers may return to see their children whenever they can, or alternatively, children may spend part of the year with their mothers, part with other relatives.

Mothers in transnational families have to develop new concepts of their maternal role, because their situation means giving up the idea that biological mothers should raise their own children. Many have expanded their definition of motherhood to include breadwinning, traditionally defined as the role of fathers. Transnational women also create a new sense of home, one not limited to the traditional understanding of "home" as a single place where mothers, fathers, and their children reside (Hondagneu-Sotelo 2001; Alicea 1997; Das Gupta 1997; Hondagneu-Sotelo and Avila 1997).

FAMILIES AND SOCIAL POLICY

Family social policies are the subject of intense national debate. How people think about families—and whether they consider certain forms of relationships as family—very much shapes public debates about family policies (Powell et al. 2010). Should gay marriages be recognized by the state? What responsibility does society have to help parents balance the demands of work and family? Many issues on the front lines of national social policy engage intense discussions of families. Some claim the family is breaking down. Others celebrate the increased diversity among families. Many blame the family for the social problems our society faces. Drugs, low educational achievement, crime, and violence are often attributed to a crisis in "family values," as if rectifying these attitudes is all it will take to solve our nation's difficulties.

The family is the only social institution that typically takes the blame for all of society's problems. Is it reasonable to expect families to solve social problems? Families are afflicted by most of the structural problems that are generated by racism, poverty, gender inequality, and class inequality. Expecting families to solve the problems that are the basis for their own difficulties is like asking a poor person to save us from the national debt.

Balancing Work and Family

Balancing the multiple demands of work and family is one of the biggest challenges for most families. With more parents employed, it is difficult to take time from one's paid job to care for newborn or newly adopted children, tend to sick children, or care for elderly parents or other family members. As more families include two earners, more people feel pulled in multiple directions, always strategizing to find the time to get everything done. Work institutions are structured on a gendered model of the male breadwinner, where family and work are assumed to be separate, non-intersecting spheres. But now there is significant "spillover" between family and work—work seeping into the home and home also affecting people's work (Moen 2003).

The **Family and Medical Leave Act (FMLA)**, adopted by Congress in 1993, is meant to provide help for these conflicts. It requires employers to grant employees a total of twelve weeks in unpaid leave to care for newborns, adopted children, or family members with a serious health condition. The FMLA is the first law to recognize the need of families to care for children and other dependents. A number of conditions, however, limit the effectiveness of the FMLA, not the least of which is that the leave is unpaid, making it impossible for many employed parents. Many workers in firms where there are family-friendly policies worry that taking advantage of these policies will harm their prospects for career advancement (Blair-Loy and Wharton 2002). Currently, only 15 percent of workers have child-care benefits available to them from employers (Long 2007). Among industrialized nations, the United States provides the least in support for maternity and child-care policies (see Table 13.2).

Child Care

Family leave policies, much as they are needed, also do not address the ongoing needs for child care. Half of all working families have child-care expenses; the other half either have unpaid relatives or friends providing care or they arrange their work schedules to coincide with school hours. On average, child care equals about 9 percent of family earnings—the second largest expense in the household budget (following rent or mortgage). But low-income families pay an even higher

table 13.2 Maternity Leave Benefits: A Comparative Perspective

Country	Length of Maternity Leave	Percentage of Wages Paid in Covered Period	Provider of Coverage
Zimbabwe	90 days	60–75%	Employer
Cuba	18 weeks	100%	Social Security
Iran	90 days	66.7% for 16 weeks	Social Security
China	90 days	100%	Employer
Saudi Arabia	10 weeks	50 or 100%	Employer
Canada	17–18 weeks	55% for 15 weeks	Unemployment insurance
Germany	14 weeks	100%	Social Security to a ceiling; employer pays difference
France	16–26 weeks	100%	Social Security
Italy	5 months	80%	Social Security
Japan	14 weeks	60%	Social Security or health insurance
Russian Federation	140 days	100%	Social Security
Sweden	14 weeks	450 days, 100% paid	Social Security
United Kingdom	14–18 weeks	90% for 6 weeks; flat rate thereafter	Social Security
United States	12 weeks	n/a	n/a

Source: United Nations. 2000. *The World's Women 2000: Trends and Statistics.* New York: United Nations, pp. 140–143.

percent of their earnings on child care—14 percent of earnings (Giannarelli and Barsimantov 2000).

Many parents struggle to find good and affordable child care for their children, some relying on relatives for care; others, on paid providers; and some, a combination of both. In the United States, one-half of three-year-olds and two-thirds of four-year-olds now spend much of their time in child-care centers. But the national approach is one of patching together different programs and primarily relying on private initiatives for care. Compare this with France. Although participation is voluntary, almost all parents in France enroll young children in the *école maternelle* system, where a place is guaranteed to every child aged 3 to 6. These child-care centers are integrated with the school system and are seen as a form of early education. Moreover, in the United States, childcare costs match tuition costs at public universities, but child care in France is seen as a social responsibility and is paid by the government. National norms about whether families are a private or public responsibility clearly shape social policy (Clawson and Gerstel 2002; Folbre 2001).

Care work— work that sustains life, including child care, elder care, housework, and other forms of household labor—is increasingly provided to middle- and upper-class families by women of color and immigrant women (Duffy 2011). Nannies, cleaners, and personal attendants now do much of the domestic work that wives and mothers once provided. These trends raise many new questions for sociologists, such as how work is negotiated outside of the public labor market, how

Percent of U.S. population living in multi-generational households

FIGURE 13.6 Multigenerational Households over Time Living in an extended family household was common before World War II, but, as you can see, the trend was a decline in this form of living arrangement until fairly recently. Sociologists explain the decline as the result of several changes in society, including the growth of nuclear family-centered suburbs, a decline in immigration in the post-World War II period, and a general rise in the affluence of the population. This trend reversed around 1980, as you can see in this line graph. What changes in society do you think have been influencing the rise in multigenerational households? Is there evidence of these trends in your own family? Why?

Source: Pew Research Center. 2010. "The Return of the Multi-Generational Family Household. Data from U.S. Decennial Census Data, 1940–2000." **ww.pewsocialtrends.org**

"mothering" is defined when it is provided by multiple people, how domestic workers care for their own families, and what work conditions exist in the lives of domestic laborers (Hondagneu-Sotelo 2001). Clearly, new social policies are needed to address the needs of diverse families (Moen et al. 2004).

Elder Care

The shrinking size of families means that the proportion of elderly people is growing faster than the number of younger potential caretakers. As life expectancy has increased and people live longer, elder care becomes a greater and greater need. Family members provide almost all long-term care for the elderly—work that is often taken for granted (Meyer 1994; Glazer 1990).

Women, who shoulder much of the work of elder care, can now expect to spend more years as the child of an elderly parent than as the mother of children under eighteen. The effects of the burden of care are apparent in the stress that women report from this role. Women also believe they are better at elder care than their husbands and brothers, but with the rapid increase in the older population that lies ahead, these social norms may have to change. As the U.S. population ages, social policies will likely need to respond to this growing need.

Because families are so diverse, different families need different social supports. Family leave policies that give parents time off to care for their children or sick relatives are helpful but of little use to people who cannot afford to take time off work without pay. Greater employer support for child care can help men and women meet family needs. Some policies will benefit some groups more than others—one reason why policymakers need to be sensitive to the diversity of family experiences. Social policies cannot solve all the problems that families face, but they can go a long way toward creating the conditions under which diverse family units can thrive.

DEFINING RELIGION

Religion has a profound effect on society and human behavior. This is easily observed in daily life outside the family. Church steeples dot the landscape. Invocations to a religious deity occur at the beginning of many public gatherings. The news frequently reports on events generated by religious conflict.

Religious beliefs have led to conflict, but religious beliefs have also been the soul of some of the most liberating social movements, including the civil rights movement and other human rights movements around the world. Some of life's sweetest moments are marked by religious celebration, and some of its most bitter conflicts persist because of unshakable religious conviction. Religion is both an integrative force in society and the basis for many of our most deeply rooted social conflicts.

Paul Burns/Digital Vision/Getty Images

Religious socialization is a powerful source of people's values; most of the time, children take on the religious orientation of their family of origin.

Sociologists study religion as both a belief system and a social institution. The belief systems of religion have a powerful hold on what people think and how they see the world. The patterns and practices of religious institutions are among the most important influences on people's lives. Sociologists are interested in several questions about religion: How are religious belief and practice related to other social factors, such as social class, race, age, gender, and level of education? How are religious institutions organized? How does religion influence social change? In using sociology to understand religion, what is important is not what one believes about religion, but one's ability to examine religion objectively in its social and cultural context.

What is religion? Most people think of it as a category of experience separate from the mundane acts of everyday life, perhaps involving communication with a deity or communion with the supernatural (Johnstone 1992). Sociologists define **religion** as an institutionalized system of symbols, beliefs, values, and practices by which a group of people interprets and responds to what they feel is sacred and that provides answers to questions of ultimate meaning (Johnstone 1992; Glock and Stark 1965). The elements of this definition bear closer examination:

1. **Religion is institutionalized.** Religion is more than just beliefs: It is a pattern of social action organized around the beliefs, practices, and symbols that people develop to answer questions about the meaning of existence. As an institution, religion presents itself as larger than any single individual; it persists over time and has an organizational structure into which members are socialized.
2. **Religion is a feature of groups.** Religion is built around a community of people with similar beliefs.

It is a cohesive force among believers because it is a basis for group identity and gives people a sense of belonging to a community or organization. Religious groups can be formally organized, as in the case of bureaucratic churches, or they may be more informally organized, ranging from prayer groups to cults. Some religious communities are extremely close-knit, as in convents; other communities are more diffuse, such as people who identify themselves as Protestant but attend church only on Easter.

3. **Religions are based on beliefs that are considered sacred.** The **sacred** is that which is set apart from ordinary activity for worship, seen as holy, and protected by special rites and rituals. The sacred is distinguished from the **profane**, which is of the everyday world and specifically not religious (Chalfant et al. 1987; Durkheim 1947/1912). Each religion defines what is to be considered sacred; most religions have sacred objects and sacred symbols. The holy symbols are infused with special religious meaning and inspire awe.

A **totem** is an object or living thing that a religious group regards with special reverence. A statue of Buddha is a totem and so is a crucifix hanging on a wall. Among the Zuni (a Native American group), fetishes are totems; these are small, intricately carved animal objects representing different dimensions of Zuni spirituality. A totem is important not for what it is, but for what it represents. To a Christian taking communion, a piece of bread is defined as the flesh of Jesus; eating the bread unites the communicant mystically with Christ. To a nonbeliever, the bread is simply that—a piece of bread (McGuire 2001). Likewise, Native Americans hold certain ground to be sacred and are deeply offended when the holy ground is disturbed by industrial or commercial developers who see only potential profit.

4. **Religion establishes values and moral proscriptions for behavior.** A proscription is a constraint imposed by external forces. Religion typically establishes proscriptions for the behavior of believers, some of them quite strict. For example, the Catholic Church defines living together as sexual partners outside marriage as a sin. Often religious believers come to see such moral proscriptions as simply "right" and behave accordingly. At other times, individuals may consciously reject moral proscriptions, although they may still feel guilty when they engage in a forbidden practice. Of course, what people believe and what they do can be very contradictory, perhaps best exemplified by the various scandals recently reported involving sexual abuse of young boys by Catholic priests.

5. **Religion establishes norms for behavior.** Religious belief systems establish social norms about how the faithful should behave in certain situations. Worshipers may be expected to cover their heads in a temple, mosque, or cathedral, or to wear certain clothes. Such behavioral expectations may be quite strong. The next time you are at a gathering where a prayer is said before a meal, note how many people bow their heads, even though some of those present may not believe in the deity being invoked.

6. **Religion provides answers to questions of ultimate meaning.** The ordinary beliefs of daily life are **secular** beliefs and may be institutionalized, but they are specifically not religious. Science, for example, generates secular beliefs based on particular ways of thinking—logic and empirical

David Silverman/Getty Images

Mario Tama/Getty Images News/Getty Images

Religious spirituality takes many forms but produces feelings of awe and reverence among believers, as in this Orthodox Christian baptism and this Jain ceremony of soaking in vermillion in recognition of a sacred tradition.

observations are at the root of scientific beliefs. Religious beliefs, in contrast, often have a supernatural element. They emerge from spiritual needs and may provide answers to questions that cannot be probed with the profane tools of science and reason. Think of the difference in how religion and science explain the origins of life. Whereas science explains this as the result of biochemical and physical processes, different religions have other accounts of the origin of life.

see **FOR YOURSELF**

For one week, keep a daily log, noting every time you see an explicit or implicit reference to religion. At the end of the week, review your notes and ask yourself how religion is connected to other social institutions. Based on your observations, how do you interpret the relationship between the *sacred* and the *secular* in this society? ●

THE SIGNIFICANCE OF RELIGION IN THE UNITED STATES

The United States is one of the most religious societies in the world. Two-thirds of Americans think religion can solve all or most of society's problems. Most (80 percent) say they depend on God to make decisions in their daily lives, and a majority (60 percent) think God has set the course of their lives (Shieman 2010). Religion is, for millions of people, the strongest component of their individual and group identity. Much of the world's most celebrated art, architecture, and music has its origins in religion, whether in the classical art of western Europe, the Buddhist temples of the East, or the gospel rhythms of contemporary rock.

Religion is also strongly related to a number of social and political attitudes. Religious identification is a good predictor of how traditional a person's beliefs will be. People who belong to religious organizations that encourage intolerance of any form are most likely to be racially prejudiced. However, there is not a simple relationship between religious belief and prejudice, because religious principles are also often the basis for lessening racial prejudice. Those with deeper religious involvement tend to have more traditional gender attitudes; *homophobia* has also been linked to religious belief, although some religious congregations have actively worked to encourage the participation of gays and lesbians.

The Dominance of Christianity

Despite the U.S. Constitution's principle of the separation of church and state, Christian religious beliefs and practices dominate U.S. culture. Indeed, Christianity is often treated as if it were the national religion. It is commonly said that the United States is based on a Judeo-Christian heritage, meaning that our basic cultural beliefs stem from the traditions of the pre-Christian Old Testament of the Bible (the Judaic tradition) and the Gospels of the New Testament. The dominance of Christianity is visible everywhere. State-sponsored colleges and universities typically close for Christmas break, not Yom Kippur. Christmas is a national holiday, but not Ramadan, the most sacred holiday among Muslims. Despite the dominance of Christianity, however, the pattern of religion in the United States is a mosaic one.

Measuring Religious Faith

Religiosity is the intensity and consistency of practice of a person's (or group's) faith. Sociologists measure religiosity both by asking people about their religious beliefs and by measuring membership in religious organizations and attendance at religious services (see Figure 13.7). The majority of people in the United States identify themselves as Protestant or Catholic, though there is great religious diversity within the United States (see Figure 13.8).

Forms of Religion

Religions can be categorized in different ways according to the specific characteristics of faiths and how religious groups are organized. In different societies and among different religious groups, the form religion takes reflects differing belief systems and reflects and

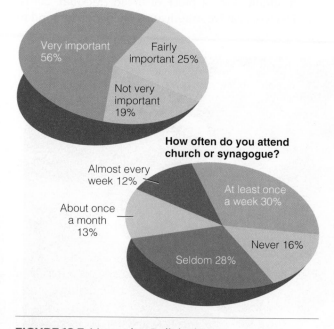

How important is religion?

Very important 56%
Fairly important 25%
Not very important 19%

How often do you attend church or synagogue?

Almost every week 12%
About once a month 13%
At least once a week 30%
Never 16%
Seldom 28%

FIGURE 13.7 Measuring Religiosity

Source: Newport, Frank. 2009. "This Christmas, 78% of the Public Identify as Christian." Gallup Poll. **www.gallup.com**; Newport, Frank. 2008. "Easter Season Finds a Religious, Largely Christian Nation." Gallup Poll. **www.gallup.com**

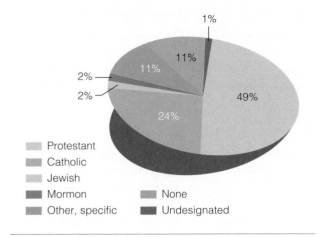

1%

11%

11%

2%

2%

49%

24%

- Protestant
- Catholic
- Jewish
- Mormon
- Other, specific
- None
- Undesignated

FIGURE 13.8 Religious Identification in the United States

Source: Gallup Poll. 2006. "Religion." Princeton, NJ: Gallup Organization. **www.gallup.com**

supports other features of the society. Believing in one god or many, worshiping in small or large groups, and associating religious faith with gender roles all contribute to the social organization of religion and its relationship to the rest of society.

Religiosity varies significantly among different groups in society. Church membership and attendance is higher among women than men and more prevalent among older than younger people. African Americans are more likely than Whites to belong to and attend church. On the whole, church membership and attendance fluctuate over time; membership has decreased slightly since 1940, but attendance has remained largely the same since. Large, national religious organizations, such as the mainline Protestant denominations, have lost many members, whereas smaller, local congregations have increased membership.

In recent years, there has been a decrease in the number of people who think that religion can answer all or almost all of today's problems. Changes in immigration patterns have also affected religious patterns in the United States, with Muslims, Buddhists, and Hindus now accounting for several million believers (Haddad et al. 2003; Niebuhr 1998). One of the greatest changes has been a tremendous increase in the number identifying as evangelical Protestants, but Islam has also been one of the fastest growing religions in the United States in recent years (Gallagher 2003; Dudley and Roozen 2001).

One basic way to categorize religions is by the number of gods or goddesses adherents worship. **Monotheism** is the worship of a single god. Christianity and Judaism are monotheistic in that both Christians and Jews believe in a single god who created the universe. Monotheistic religions typically define god as omnipotent (all-powerful) and omniscient (all-knowing). **Polytheism** is the worship of more than one deity. Hinduism, for example, is extraordinarily complex, with millions of

gods, demons, sages, and heroes—all overlapping and entangled in religious mythology; within Hinduism, the universe is seen as so vast that it is believed to be beyond the grasp of a single individual, even a powerful god (Grimal 1963).

Religions may also be patriarchal or matriarchal. **Patriarchal religions** are those in which the beliefs and practices of the religion are based on male power and authority. Christianity is a patriarchal religion; the ascendancy of men is emphasized by the role of women in the church, the instruction given on relations between the sexes, and even the language of worship itself. **Matriarchal religions** are based on the centrality of female goddesses, who may be seen as the source of food, nurturance, and love, or who may serve as emblems of the power of women (McGuire 2001). In societies based on matriarchal religions, women are more likely to share power with men in the society at large. Likewise, in highly sexist, patriarchal societies, religious beliefs are also likely to be patriarchal.

SOCIOLOGICAL THEORIES OF RELIGION

The sociological study of religion probes how religion is related to the structure of society. Recall that one basic question sociologists ask is, "What holds society together?" Coherence in society comes from both the social institutions that characterize society and the beliefs that hold society together. In both instances, religion plays a key role. From the functionalist perspective of sociological theory, religion is an integrative force in society because it has the power to shape collective beliefs. In a somewhat different vein, the sociologist Max Weber saw religion in terms of how it supported other social institutions. Weber thought that religious belief systems provided a cultural framework that supported the development of specific social institutions in other realms, such as the economy. From yet a third point of view, based on the work of Karl Marx and conflict theory, religion is related to social inequality in society (see Table 13.3).

Emile Durkheim: The Functions of Religion

Emile Durkheim argued that religion is functional for society because it reaffirms the social bonds that people have with each other, creating social cohesion and integration. Durkheim believed that the cohesiveness of society depends on the organization of its belief system. Societies with a unified belief system are highly cohesive; those with a more diffuse or competing belief system are less cohesive.

Religious **rituals** are symbolic activities that express a group's spiritual convictions. Making a pilgrimage to Mecca, for example, is an expression of religious faith and

table 13.3 Theoretical Perspectives on Religion

	Functionalism	Conflict Theory	Symbolic Interaction
Religion and the social order	Is an integrative force in society	Is the basis for intergroup conflict; inequality in society is reflected in religious organizations, which are stratified by factors such as race, class, or gender	Is socially constructed and emerges with social and historical change
Religious beliefs	Provide cohesion in the social order by promoting a sense of collective consciousness	Can provide legitimation for oppressive social conditions	Are socially constructed and subject to interpretations; can also be learned through religious conversion
Religious practices and rituals	Reinforce a sense of social belonging	Define in-groups and out-groups, thereby defining group boundaries	Are symbolic activities that provide definitions of group and individual identity

© Cengage Learning

a reminder of religious belonging. In Durkheim's view, religious rituals are vehicles for the creation, expression, and reinforcement of social cohesion. Groups performing a ritual are expressing their identity as a group. Whether the rituals of a group are highly elaborated or casually informal, they are symbolic behaviors that sustain group awareness of unifying beliefs. Lighting candles, chanting, or receiving a sacrament are behaviors that reunite the faithful and help them identify with the religious group, its goals, and its beliefs (McGuire 2001). Durkheim believed that religion binds individuals to the society in which they live by establishing what he called a **collective consciousness**, the body of beliefs common to a community or society that gives people a sense of belonging. In many societies, religion establishes the collective consciousness and creates in people the feeling that they are part of a common whole.

Durkheim's analysis of religion suggests some of the key ideas in symbolic interaction theory, particularly in the significance he gave to symbols in religious behavior. Symbolic interaction theory sees religion as a socially constructed belief system, one that emerges in different social conditions. From the perspective of symbolic interaction, religion is a meaning system that gives people a sense of identity, defines one's network of social belonging, and confers one's attachment to particular social groups and ways of thinking.

Max Weber: The Protestant Ethic and the Spirit of Capitalism

Theorist Max Weber also saw a fit between the religious principles of society and other institutional needs. In his classic work, *The Protestant Ethic and the Spirit of Capitalism*, Weber argued that the Protestant faith supported the development of capitalism in the Western world. He began by noting a seeming contradiction: How could a religion that supposedly condemns extensive material consumption coexist in a society (such as the United States) with an economic system based on the pursuit of profit and material success?

Weber argued that these ideals were not as contradictory as they seemed. As the Protestant faith developed, it included a belief in predestination—one's salvation is predetermined and a gift from God, not something earned. This state of affairs created doubt and anxiety among believers, who searched for clues in the here and now about whether they were among the chosen—called the "elect." According to Weber, material success was taken to be one clue that a person was among the elect and thus favored by God, which drove early Protestants to relentless work as a means of confirming (and demonstrating) their salvation. As it happens, hard work and self-denial—the key features of the **Protestant ethic**—lead not only to salvation but also to the accumulation of capital. The religious ideas supported by the Protestant ethic therefore fit nicely with the needs of capitalism.

Emile Durkheim theorized that public rituals provide cohesion in society.

Andrew Aitchison/Alamy Limited

According to Weber, these austere religionists stockpiled wealth, had an irresistible motive to earn more (that is, eternal salvation), and were inclined to spend little on themselves, leaving a larger share for investment and driving the growth of capitalism (Weber 1958/1904).

Karl Marx: Religion, Social Conflict, and Oppression

Durkheim and Weber concentrated on how religion contributes to the cohesion of society. Religion can also be the basis for conflict, as we see in the daily headlines of newspapers. In the Middle East, differences between Muslims and Jews have caused decades of political instability. These conflicts are not solely religious, but religion plays an inextricable part. Certainly religious wars, religious terrorism, and religious genocide have contributed to some of the most violent and tragic episodes of world history. The image of religion in history has two incompatible sides: piety and contemplation on the one hand, battle flags on the other. In the United States, domestic conflicts over ethical issues such as abortion, assisted suicide, and school prayer evolve from religious values even though they are played out in the secular world of politics and public opinion. Conflict theory illuminates many of the social and political conflicts that engage religious values.

The link between religion and social inequality is also key to the theories of Karl Marx. Marx saw religion as a tool for class oppression. According to Marx, oppressed people develop religion, with the urging of the upper classes, to soothe their distress (Marx 1972/1843). The promise of a better life hereafter makes the present life more bearable, and the belief that "God's will" steers the present life makes it easier for people to accept their lot. To Marx, religion is a form of *false consciousness* (see Chapter 8) because it prevents people from rising up against oppression. He called religion the "opiate of the people" because it encourages passivity and acceptance.

Marx saw religion as supporting the status quo and being inherently conservative (that is, resisting change and preserving the existing social order). To Marx, religion promotes stratification because it supports a hierarchy of people on Earth and the subordination of humankind to divine authority. Christianity, for example, supported the system of slavery. When European explorers first encountered African people, they regarded them as godless savages, and they justified the slave trade by arguing that slaves were being converted to the Christian way of life. Principles of Christianity thus legitimated the system of slavery in the eyes of the slave owners and allowed them to see themselves as good people, despite their enslavement of other human beings.

At the same time, religion can be the basis for liberating social change. In the civil rights movement in the United States and in Latin American liberation movements, the words and actions of religious organizations have been central in mobilizing people for change. This does not undermine Marx's main point, however, because there remains ample evidence of the role of religion in generating social conflict and resisting social change.

Symbolic Interaction: Becoming Religious

Recall that symbolic interaction theory states that people act toward things on the basis of the meaning things have for them and that those meanings emerge through social interaction. This can explain much about human behavior that is based in religious ideas. Thus, seen from outside the faith, religious practices (kneeling in church, wearing a yamaka, making a pilgrimage to Mecca, or chanting) may seem peculiar or different, but within the faith, these and other religious practices carry meaning—meaning that is deeply important to religious believers. But not only does symbolic interaction explain particular religious behaviors, it can also explain how other behaviors may be based in the meanings that religion holds for people. For example, we have earlier said that even something as reprehensible as suicide bombings can be understood if you understand how religious zealots interpret the meaning of religious texts.

Symbolic interaction theory can also help you understand how people become religious, a process sociologists call *religious socialization*. Religious socialization may be a slow and gradual process, such as in how children learn religious values over time. Religious socialization can also be more dramatic, as when a person joins a cult or some other extreme religious group. And people may reinterpret religious beliefs when they question their religious faith or even switch to a new religion, such as when a Christian converts to Judaism. The emphasis on meaning that is typical of symbolic interaction helps explain how the same religion can be interpreted differently by different groups or in different times. How, for example, could Christian beliefs be used by some to make slavery seem legitimate, while at the same time it provided the belief system that helped others survive slavery as well as to fight against it? And now, religion is very much a factor in shaping people's political behavior—how they think about social issues, as well as how they vote.

Symbolic interaction thus sees religious belief—and its meaning to different people—as essential for understanding many forms of social behavior. Symbolic interaction also helps explain how different religious beliefs and practices emerge in social and historical contexts—contexts that shape what religion means to people.

DIVERSITY AND RELIGIOUS BELIEF

The world is marked by diverse religious beliefs. Christianity has the largest membership, followed by Islam. But Hindus, Jews, Confucianists, Buddhists, and observers

MAP 13.1

Mapping America's Diversity: Religious Diversity in the United States

The Simpson's Diversity Index is a measure developed by the Association of Statisticians of American Religious Bodies. It calculates the likelihood of two individuals within a given county belonging to different religious groups. Thus the lower the index, noted in the darkest shade, the less religious diversity in that county; conversely, the higher the index noted in the lightest shade, the more religious diversity exists in that location. According to this map, where is there the most and least religious diversity?

Do you think that affects such things as political values and other social issues in these different regions of the country?

Source: Association of Statisticians of American Religious Bodies. 2012. **www.asarb.org**

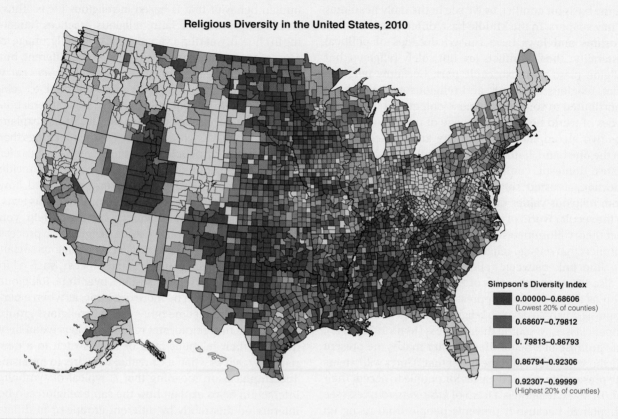

Religious Diversity in the United States, 2010

Simpson's Diversity Index

- 0.00000–0.68606 (Lowest 20% of counties)
- 0.68607–0.79812
- 0. 79813–0.86793
- 0.86794–0.92306
- 0.92307–0.99999 (Highest 20% of counties)

---

of folk religions also comprise the world's religions (see Map 13.2). In the United States, religious identification varies with a number of social factors, including age, income level, education, and political affiliation. Younger people are more likely than older people to express no religious preference. Those in higher income brackets are more likely to identify as Catholic or Jewish than those in lower income brackets, who are more likely to identify as Protestant, although these trends vary among Protestants by denomination. Fundamentalist Protestants, for example, are most likely to come from lower-income groups.

The Influence of Race and Ethnicity

Race is one of the most significant indicators of religious orientation. African Americans are much more likely than Whites, Hispanics, or Asian Americans to say that religion is very important in their lives. And, although most African Americans identify as Protestant (75 percent), a small, but growing number are Catholic (5 percent; Pew Research Center 2009).

Many urban African Americans have also become committed Black Muslims, which involves strict regulation of dietary habits and prohibition of many activities, such as alcohol use, drug use, gambling, use of cosmetics, and hair straightening. The emphasis among Black Muslims on self-reliance and traditional African identity has earned it a fervent following, although the actual number of Black Muslims in the United States is relatively small. For many African Americans, religion has been a defense against the damage caused by racism. Churches have served as communal centers, political units, and sources of social and community support, making churches among the most important institutions within the African American community

(Gilkes 2000). Religion also has been a strong force in Latino communities, with the largest number identifying as Catholic. However, there are a growing number of Latino Protestants, both in mainstream Protestant denominations and in fundamentalist groups.

Asian Americans have a great variety of religious orientations, in part because the category "Asian American" is constructed from so many different Asian cultures. Hinduism and Buddhism are common among Asians, but so is Christianity. As with all groups whose family histories include immigration, religious belief and practice among Asian Americans frequently changes between generations. The youngest generation may not worship as their parents and grandparents did, although some aspects of the inherited faith may be retained. Within families, the discontinuity with a religious past brought on by cultural assimilation can be a source of tension between grandparents, parents, and children. Within the United States, Asian Americans often mix Christian and traditionally Buddhist, Confucian, or Hindu beliefs, resulting in new religious practices (Pew Forum on Religion & Public Life 2012).

Muslims are a growing segment of U.S. society; two-thirds of Muslims are former immigrants, but a substantial portion (35 percent) are native born, either African American or others who have converted to this faith or were raised in a Muslim household. Despite stereotypes about Muslim conservatism, Muslim Americans are actually more liberal than the general public on many issues—for example, they are more likely to vote Democratic. At the same time, however, Muslim Americans are generally less tolerant of homosexuality than the public at large (Pew Research Center 2007). And the majority say that being Muslim in the United States has become more difficult since 9/11. Interestingly, studies find that younger Muslims (those under thirty) tend to be more observant than older Muslims— perhaps explained by the heightened identity that has emerged since 9/11.

Among Latinos, religious identity tends to be strong. About two-thirds of Latinos are Catholic, half of whom also identify as charismatic. Among Latinos of all faiths, people tend to worship in places that are primarily comprised of other Hispanics. This makes religion for Latinos a strong ethnic, as well as religious, identity (Pew Forum on Religion & Public Life 2007).

RELIGIOUS ORGANIZATIONS

Sociologists have organized their understanding of the various religious organizations into three types: churches, sects, and cults. These are *ideal types* in the sense that Max Weber used the term. That is, the ideal types convey the essential characteristics of some social entity or phenomenon, even though they do not explain every feature of each entity included in the generic category.

Churches are formal organizations that tend to see themselves, and are seen by society, as the primary and legitimate religious institutions; the term can be broadly applied to formal religious organizations, including temples and mosques. Such religious organizations tend to be integrated into the secular world to a degree that sects and cults are not. They are sometimes closely tied to the state. Many churches are organized as complex bureaucracies with a division of labor and different roles for groups within, including a formally trained clergy and professional staff. However, some churches may be smaller, less formal, with devoted, but less formally trained, clergy. A new phenomenon for churches is the development of *megachurches*— those with memberships numbering into the thousands. These are increasingly common. Not only do megachurches have huge attendance but they also may broadcast on huge screens, possibly even televising church services.

Sects are groups that have broken off from an established church. They emerge when a faction within an established religion questions the legitimacy or purity of the group from which they are separating. Many sects form as offshoots of existing religious organizations. Sects tend to place less emphasis on

Religious diversity in the United States has become a common feature of everyday life, particularly in some urban areas.

Ricki Rosen/Corbis

organization (as in churches) and more emphasis on the purity of members' faith. The Shakers, for example, were formed by departing from the Society of Friends (the Quakers). They retain some Quaker practices, such as simplicity of dress and a belief in pacifism, but have departed from Quaker religious philosophy. The Shakers believe that the second coming of Christ is imminent, but that Christ will appear in the form of a woman (Kephart 1993). Sects tend to admit only truly committed members, refusing to compromise their beliefs. Some sects hold emotionally charged worship services, although others, like the Amish, are more stoic. The Shakers, for example, have such emotional services that they shake, shout, and quiver while "talking with the Lord," earning them their name. The only bodily contact permitted among the Shakers is during the unrestrained religious rituals; they are celibate (do not have sexual relations) and gain new members only through adoption of children or recruitment of newcomers (Kephart 1993).

Cults, which are like sects in their intensity, are religious groups devoted to a specific cause or charismatic leader. Many cults arise within established religions and sometimes continue to peaceably reside within the parent religion simply as a fellowship of people with a particular, often mystical, dogma. As they are developing, it is common for tension to exist between cults and the society around them. Cults tend to exist outside the mainstream of society, arising when believers think that society is not satisfying their spiritual needs and attracting those who feel a longing for meaningful attachments. Internally, cults seldom develop an elaborate organizational structure but are instead close-knit communities held together by personal attachment and loyalty to the cult leader.

Cults form around leaders with great **charisma**, a quality attributed to individuals believed by their followers to have special powers (Johnstone 1992). Typically, followers are convinced that the charismatic leader has received a unique revelation or possesses supernatural gifts. Although there are exceptions, cult leaders are usually men, probably because men are more likely to be seen as having the characteristics associated with charismatic leadership.

---

debunking SOCIETY'S MYTHS

MYTH: People who join extreme religious cults are maladjusted and have typically been quickly brainwashed by cult leaders.

SOCIOLOGICAL PERSPECTIVE: Conversion to a religious cult usually involves a gradual process of resocialization wherein the convert voluntarily develops new associations with others and develops a new worldview based on these new relationships. ●

RELIGION AND SOCIAL CHANGE

What is the role of religion in social change? Durkheim saw religion as promoting social cohesion; Weber saw it as culturally linked to other social institutions; Marx assessed religion in terms of its contribution to social oppression. Is religion a source of oppression, or is it a source of personal and collective liberation from worldly problems? There is no simple answer to this question. Religion has had a persistent conservative influence on society, but it has also been an important part of movements for social justice and human emancipation.

The role of religious organizations in social change has of late become a question of public policy. Should faith-based organizations receive government support for work they do in helping people, or does this violate the constitutional separation of church and state? These constitutional issues will ultimately be settled by law, but sociological research sheds light on the implications of such organizations. Though liberals fear that faith-based organizations will infuse religion into government too much and conservatives hope that faith-based initiatives will support their political agenda, research shows that faith-based initiatives enhance the participation of traditionally disadvantaged groups in the democratic process (Kniss 2003; Wood 2002).

The public debate about faith-based initiatives is occurring at a time when evangelical groups have increased membership and influence and have affiliated with conservative political causes, dramatically increasing the influence of religion on politics. At the same time, there has been a decrease in the importance of religion to many people. As a social institution, religion is in transition. Religion, like other aspects of society, is also becoming more commercialized. A large self-help industry has developed in religious publication, and religious music is increasingly successful as a form of enterprise. All sorts of religious products are bought and sold in what sociologists now call a "spiritual marketplace" (Roof 1999; Wuthnow 1998). Clearly, religion influences social change, but it is also influenced by the same changes that affect other social institutions.

At the same time, religion continues to have an important role in liberation movements around the world. Throughout the world, liberation theologians have used the prestige and organizational resources of the Catholic Church to develop a consciousness of oppression among poor peasants and working-class people. Likewise, in the United States, churches have had a prominent role in the civil rights movement (Marx 1967/1867; Morris 1984). Churches supplied the infrastructure of the developing Black protest

The Rise of Religious Fundamentalism

Headlines in the daily news make it obvious that religious fundamentalism is an increasing force in today's world. Why is this happening now? Max Weber thought that the process of modernization would lead to a more secular society but, in fact, religious fundamentalism has been on the rise in recent years—both in the United States and abroad. How do you explain this? And is it contrary to sociological logic?

Fundamentalist religious movements seem to be surging across the globe. You see the influence of religious fundamentalism in the terrorist assaults on the United States, but in a different vein, religious fundamentalism has also been influencing domestic politics. What is meant by fundamentalism in a sociological sense?

Religious fundamentalists are those who are "true believers." They have very literal interpretations of religious texts and tend to be highly certain of their religious worldview. Religious fundamentalists tend to see the world in simplistic either/or terms—dividing people into either good or evil, godly or demonic. Such divisive imagery reduces the complexity of human life into simplistic categories—categories that can fuel hate and conflict (Anthony et al. 2002). When such religious fanaticism is intertwined with the power of a state government, religiously inspired leaders can use the power of the military and government propaganda to wield extraordinary power.

Religious fundamentalists tend to hold their belief as absolute—that is, having little doubt about the truth of their beliefs. Often, religious fundamentalists have a messianic and millennial view, believing that a moment will come when all truth is revealed—or a messiah appears. Religious fundamentalism is often associated with intolerance of other beliefs, and it can, though does not always, become associated with violence. Religious extremism is now associated with terrorism in the Middle East, but it has fueled other horrendous acts in various parts of the world, including mass executions and genocide, enslavement, and other heinous crimes against humanity.

All religions, taken to an extreme, are dangerous social forces because they can drive adherents to think they are doing sacred work even when they are engaging in violent, murderous behavior.

It is easy to see the acts of religious extremists as the work of misled individuals, but those who study religious extremism know that it has social origins. Religious extremism is learned, usually within a narrowly circumscribed social world, such as the *madrassas*—religious camps in Pakistan (and other areas) where young boys are taught a strict interpretation of Islam. For young boys uprooted from families by war, detached from other social contacts, and with no other education, it is easy to be socialized into a narrow worldview that gives them a cause to fight for (Rashid 2000).

Fundamentalist religious organizations tend to have charismatic leadership and to make sharp boundaries between believers and nonbelievers. Churches, temples, or mosques have an authoritarian and hierarchical organization, one that is typically patriarchal—that is based on the power of men. Religions in this vein typically enforce strict behavioral rules, such as dress, comportment, perhaps hairstyle, or eating and drinking habits. And, as we have seen with religious fundamentalists in the United States in recent years, they can become political activists—morally fervent in trying to achieve their particular worldview. In most cases, religious fundamentalists are also very conservative on matters involving the family and appropriate gender roles—stemming largely from their literal interpretation of religious texts (Emerson and Hartmann 2006).

Religious fundamentalism in the United States surged starting in the 1970s, very likely as a counterreaction to the marginalization of religion that had emerged during the social movements of the 1960s. But it has been fueled as well by a counterreaction to increasing secularization and the search for a firmly anchored identity and community in an increasingly modernizing world where forces of commerce tend to shape social relations. Seen in this light, perhaps Weber was partially right—that the tendency of modernization is toward secularization, but secularization then can produce a counterreaction in the form of the rise of religious fundamentalism.

Also, where there is a lack of modernism—and where people perceive that their traditional way of life is being overtaken by Western influences—religious extremism can come from trying to defend a traditional way of life (Stern 2003; Pain 2002; Khashan and Kreidie 2001).

Finally, religious extremist movements tend to be highly patriarchal—that is, based on the power of men and the subordination of women. This is true not only in the extremist factions of contemporary Islamic movements but also in extremist segments of the Christian right in the United States (Antrobus 2002; Ferber 1998). When religious extremism links with militaristic and patriarchal values, it becomes extremely dangerous.

movements of the 1950s and 1960s, and the moral authority of the church was used to reinforce the appeal to Christian values as the basis for racial justice. Now they continue to be important places for the mobilization of Black politics and provide an important source of community support—often when other institutions have abandoned the Black community (Zuckerman 2002).

MAP 13.2

Viewing Society in Global Perspective: World Religions

Source: www.wadsworth.com/religion_d/special_features/popups
/maps/matthews_world/images/w001.jpg.

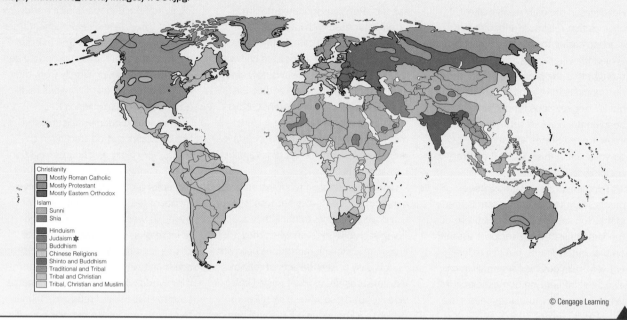

Christianity
- Mostly Roman Catholic
- Mostly Protestant
- Mostly Eastern Orthodox

Islam
- Sunni
- Shia

- Hinduism
- Judaism ✡
- Buddhism
- Chinese Religions
- Shinto and Buddhism
- Traditional and Tribal
- Tribal and Christian
- Tribal, Christian and Muslim

© Cengage Learning

The role of women is also changing in most religious organizations. Women have long been denied the right to full participation in many faiths. Some religions still refuse to ordain women as clergy, but the public generally supports the ordination of women. Women now make up a large portion of divinity students. Whereas traditional religious images of women have provided the basis for the subordination of women, those stereotypes are eroding. In sum, religion is a force of both social change and social stability.

chapter summary

How are different kinship systems defined?
All societies are organized around a *kinship system*, varying in how many marriage partners are allowed, who can marry whom, how descent is determined, family residence, and power relations within the family. *Extended family* systems develop when there is a need for extensive economic and social cooperation. The *nuclear family* is the result of the rise of Western industrialization that separated production from the home.

What does sociological theory contribute to our understanding of families?
Functionalism emphasizes that families have the function of integrating members to support society's needs. *Conflict theorists* see the family as a power relationship, related to other systems of inequality. *Symbolic interaction* takes a more microscopic look at families, emphasizing how different family members experience and define their family experience. *Feminist theory* emphasizes the family as a gendered institution and is critical of perspectives that take women's place in the family for granted.

What changes characterize the diversity in contemporary families?
One of the greatest changes in families has been the increase in female-headed households, which are most likely to live in poverty. The increase in women's labor force participation has also affected families, resulting in dual roles for women. Stepfamilies face unique problems stemming from the blending of two households. Gay and lesbian households are also more common and challenge traditional heterosexual definitions of the family. Single people make up an increasing portion of the population, due in part to the later age when people marry.

Is marriage declining?
The United States has both the highest marriage rate and the highest divorce rate of any industrialized nation. The high divorce rate is explained as the result of a cultural orientation toward individualism and personal gratification, as well as structural changes that make women less dependent on men within the family.

Why is family violence such a problem?
Family violence takes several forms, including partner violence, child abuse, incest, and elder abuse. Power relationships within families, as well as gender differences in the division of labor, help explain domestic violence.

What major changes are affecting contemporary families?
Changes at the global level are producing new forms of families—*transnational families*—where at least one parent lives and works in a nation different from the children. Social policies designed to assist families should recognize the diversity of family forms and the interdependence of the family with other social conditions and social institutions.

What are the elements of a religion?
Sociologists are interested in religion because of the strong influence it has in society. *Religion* is an institutionalized system of symbols, beliefs, values, and practices by which a group of people interprets and responds to what they feel is sacred and that provides answers to questions of ultimate meaning.

How do sociologists measure the significance of religion for people, and what forms does religion take?
The United States is a deeply religious society. Christianity dominates the national culture, even though the U.S. Constitution specifies a separation between church and state. *Religiosity* is the measure of the intensity and practice of religious commitment.

How do the different sociological theories analyze religion?
Durkheim understood religions and religious rituals as creating social cohesion. Weber saw a fit between the ideology of the *Protestant ethic* and the needs of a capitalistic economy. Religion is also related to social conflict. Marx saw religion as supporting societal oppression and encouraging people to accept their lot in life. *Symbolic interaction theory* focuses on the process by which people become religious. Religious conversion involves a dramatic transformation of religious identity and involves several phases through which individuals learn to identify with a new group and lose other existing social ties.

What diversity exists in religious faith and practice?
The United States is a diverse religious society. Protestants, Catholics, Jews, and, increasingly, Muslims, make up the major religious faiths in the United States. Religious extremism can emerge in any religion and is generated by certain societal characteristics.

How is a religion organized?
Churches are formal religious organizations. They are distinct from *sects*, which are religious groups that have withdrawn from an established religion. *Cults* are groups that have also rejected a dominant religious faith, but they tend to exist outside the mainstream of society.

How has religion been affected by social change?
In recent years, there has been an enormous growth in conservative religious groups. Religion is a conservative influence in society, but religion also has an important part in movements for human liberation, including the civil rights movement and the move to ordain women in the church.

Key Terms

bilateral kinship **307**
charisma **332**
church **331**
collective consciousness **328**
cult **332**
extended families **308**
family **307**
Family and Medical Leave Act (FMLA) **322**

kinship system **307**
matrilineal kinship **307**
matriarchal religion **327**
matrilocal **307**
monogamy **307**
monotheism **327**
nuclear family **308**
patriarchal religion **327**

patrilineal kinship **307**
patrilocal **307**
polygamy **307**
polytheism **327**
profane **325**
Protestant ethic **328**
religion **324**

religiosity **326**
ritual **327**
sacred **325**
sect **331**
secular **325**
totem **325**
transnational family **321**

14 Education and Health Care

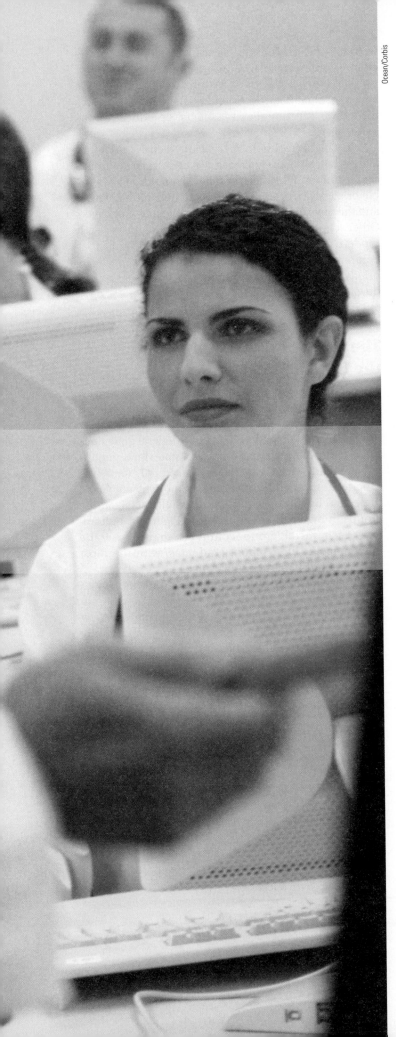

the United States is thought to have the best education and health care systems in the world. Compared to underdeveloped countries, Americans are expected to have brighter futures because American schools produce better-educated students and because Americans have access to the best medical coverage for overall better health. What about comparisons within the United States? Do all Americans get a quality education? If not, who does and who does not? And why do some benefit from amazingly sophisticated health care procedures while others suffer from poor-quality care?

These questions ask you to examine the character of education and health care as social institutions—institutions that, like other social institutions, have a social structure. The social structure of educational institutions means that, for some, schooling provides a path to a good job; for others, schooling provides minimal knowledge with little opportunity. The social structure of health care institutions does not provide equal access to equally good care. Indeed, as sociologist Jill Quadagno has pointed out, health care in other nations "is done on the basis of clinical need, not ability to pay," as is the case in the United States (Quadagno 2005: 2).

Both education and health care have thus, of late, been major topics of public policy and debate. Teacher unions are under fire. The Affordable Care Act went through a heated confirmation process in Congress. The questions still remain: How do we improve schools? How much are students learning? Should there be universal health care insurance? Is the United States falling behind other nations in its educational and health care systems?

In this chapter, we examine these two critical social institutions, shedding light on these questions. We begin with education, providing an analysis of some of the key insights generated by sociological research and theory.

learning objectives

- Understand the role of school within society
- Identify the various theoretical perspectives on education
- Argue the reasons why schools and education matter in society
- Outline the inequalities within education by race and class
- Summarize the principal ideas regarding educational reform
- Understand health care issues in the United States
- Outline the inequalities within health care in the United States
- Identify the various theoretical perspectives on health care
- Summarize the principal ideas regarding health care reform

SCHOOLING AND SOCIETY

Education in any society is about the transmission of the society's knowledge. In some societies, such as the United States, education is highly formalized—indeed, even regulated by government (at least for public institutions). In other societies, education may be less formal, perhaps provided solely through the transmission of knowledge by elders or family members (for example, home schooling is the norm in some protected religious communities). But in U.S. society, education involves teaching formal knowledge, such as reading, writing, and arithmetic, as well as cultural knowledge, such as morals, values, and ethics. Education prepares the young for entry into society and is thus a form of socialization. Sociologists refer to the more formal, institutionalized aspects of education as **schooling**.

In a highly technological society such as the United States, education is increasingly necessary for future opportunities. Why then do some of our schools resemble prisons where students are searched upon entering and the physical environment is dilapidated and bleak? Other schools look like beautiful campuses, places with modern facilities and sophisticated scientific and other equipment. To put these inequities in perspective, let us briefly look at how education in the United States has developed over time.

The Rise of Education in the United States

During the nineteenth century, education was considered a luxury, available only to White, male children of the upper classes and not required for most jobs (Cookson and Persell 1985). In 1900, federal guidelines made education compulsory, yet state laws requiring attendance were generally enforced only for White Americans through eighth grade. In more recent years, there has been consistent gain for all racial–ethnic groups in high school completion. Figure 14.1 shows how each racial group continues to increase the percentage of high school graduates, with the most gains among Hispanic Americans.

There are three kinds of education in the United States: public education, private education, and home-schooling. Among public schools, there are now *charter schools*—those that receive public funds, but that are not subject to the same rules and regulations as other

Inequality in education is very apparent in the physical facilities of wealthy and poor schools. This inequality is further reflected in educational opportunities within schools.

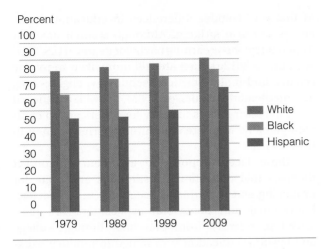

Figure 14.1 Percentage of 18- to 24-Year-Olds Who Completed High School by Race: 30-Year Trends

Source: Chapman, C., Laird, J., Ifill, J., and Kewal Ramani, A. 2011. *Trends in High School Dropout and Completion Rates in U.S., 1972–2009*. Washington, DC: National Center for Education Statistics, U.S. Department of Education .

public schools. Charter schools are attended by choice, sometimes by a lottery system. Many have a specialized curriculum. Although charter schools are still accountable to local and state school authorities, many think they offer an alternative to failing public schools. The number of students enrolled in charter schools has dramatically increased, from just under a half million in the 2000–2001 school year to over 1.6 million in the 2009–2010 school year (National Center for Education Statistics 2012).

Private schools are another type of education in the United States. Currently, private schools make up one-quarter of all schools and are attended by about 9 percent of all students (National Assessment of Educational Progress 2012). The majority of private schools (about 68 percent) are religious (or parochial) schools, where students get religious instruction as part of the educational curriculum (National Center for Education Statistics 2009).

In *homeschooling*, children are educated at home, most often by a mother. Homeschooling now educates about 3 percent of school-aged children, a small, but increasing, number. Over three-quarters of home-schooled students are White; the majority come from low- to median-income levels (under $50,000 per year). The most common reasons parents give for home-schooling their child are criticisms of the environment and academic instruction in the public schools and the desire for a religious education (National Center for Education Statistics 2003). Some studies have shown that homeschooled children score higher than publicly schooled students on standardized subject tests, but any differences between homeschooled and publicly schooled children in academic achievement disappear over time (Jones and Gloeckner 2004).

Education in Global Perspective

Debates about public versus private schooling often center on how well the public schools are doing in educating the nation's young people. For years, the United States has been heralded as having the best school system in the world. Once at the top of the list of national scores in math and science, the United States now ranks thirty-fifth out of forty nations in student math achievement scores and twenty-ninth out of forty on science tests (Darling-Hammond 2010).

What explains this decline in the nation's standing? Most of it has to do with the enormous inequality that characterizes U.S. schools and the failure of American education to address the needs of poor, African American, and Latino students whose scores thus bring down the national averages. Unlike other nations, the United States spends more on students from higher socioeconomic backgrounds, leaving those in poor and racially segregated schools disadvantaged on educational achievement (Darling-Hammond 2010).

These facts mean that inequality in education (examined further in a later section of this chapter) is strongly linked to the nation's standing in the global community. In other nations, students spend more time in school, raising questions about whether the school year should be extended for U.S. students. An extended school year is one possible solution. Other recommendations have been made, including funding for early education, increased teacher training, and better resources for all students across race and class lines. The 2003 No Child Left Behind Act (NCLB) was one attempt to increase student achievement by holding teachers and school officials accountable for test scores. Recent debate over the effectiveness of this initiative has left more questions than answers. The debates about U.S. education in a global context have no easy answers, but can be well informed by the research of sociology.

THE SOCIOLOGY OF EDUCATION: THEORETICAL PERSPECTIVES

As for other social institutions, sociological theory provides perspectives that illuminate public concerns about education. Questions about the purposes of education, how education is organized, and who education serves are addressed in different ways by the major theoretical perspectives in sociology.

Functionalist Theory

Functionalist theory in sociology argues that education accomplishes the following consequences, or "functions," for a society. First is *socialization*. As we have already seen, socialization takes place in the family, but the family is not the sole location of socialization. Schools also have a socializing influence through passing on "book knowledge" in the form of information and skills. But schools also pass on cultural heritage and history, including values, beliefs, habits, and norms—in short, culture. Some of this is explicit in the schools, such as learning a language or the music and art of one's culture. But it can also be implicit—such as guiding students through norms around punctuality, discipline, and manners.

Occupational training is another function of education, especially in an industrialized society such as the United States. In less complex societies, jobs and training may be passed from parent to child. A significant number of occupations and professions today are still passed on from parent to offspring, particularly among the upper classes (such as a father passing on a law practice to his son) or among certain highly skilled occupations (plumbers, ironworkers, and electricians). Most jobs today, however, require at least a high school education, and increasingly, a college or postgraduate degree. In today's highly technological society, higher levels of education are increasingly necessary to secure a job with a livable wage. The high cost of education coupled with less availability of jobs adds another dimension to education's functional role of training students for a specific occupation.

Social control, or the regulation of deviant behavior, is also a function of education, although a less obvious one. Such indirect, subtle consequences emerging from the activities of institutions are called **latent functions** of the institution. Increased urbanization and immigration beginning in the late nineteenth century were accompanied by rises in crime, overcrowding, homelessness, and other urban ills. One perceived benefit of compulsory education (that is, one latent function) was that it kept young people off the streets and out of trouble. There is a *hidden curriculum* in schools—a latent function of education; that is, schools not only "function" to give skills and training, but they also teach students norms, identities, and other forms of social learning that are not part of the formal curriculum.

Conflict Theory

In contrast to functionalist theory, which emphasizes how education unifies and stabilizes society, conflict theory emphasizes the power and inequality that are part of education as a social institution. Inequality in education occurs along numerous lines, with class, race, and gender among the most influential. Thus, the higher one's social class, the more likely one will have higher educational attainment. Racial differences in education have also produced what is called the *achievement gap*—with volumes of research and heated public debate about the causes and consequences

of this gap. Gender differences in educational outcomes are also striking. Although women are now 52 percent of those earning bachelor degrees (U.S. Census Bureau 2012a), men are still more likely to pursue science and math. Science, technology, engineering, and mathematic majors, or STEM, are widely believed to be necessary fields of study for Americans to compete globally. Men are still outnumbering women in these fields.

These facts support the argument of conflict theorists that educational institutions are a site for producing and reproducing inequality in our society. Educational and behavioral expectations are reinforced in schools, from preschool through college and beyond. Education is designed to produce workers for the continued growth of a capitalist economy. Those people in society given the opportunity for educational advancement are the same ones awarded opportunities within the economic structure of the United States.

Photo Alto/Frederic Cirou / Jupiter images

Contrary to the impression given in this photo, girls are frequently underrepresented in scientific and technical classes in school, particularly in upper grades and college.

DOING sociological research

Homeroom Security

Research Question: Random searches, zero-tolerance policies, metal detectors, surveillance cameras, security guards—the presence of these things in our public schools makes it seem as though the school system is a police state. What effect does such a strong security system have on today's students? Is this excessive discipline necessary and effective? These are the questions that drove sociologist Aaron Kupchik to study the climate of punishment in today's schools.

Research Method: Kupchik and his research assistants spent two years observing classrooms, hallways, and disciplinary meetings in four high schools located in two different states. In each state, two high schools were largely White and middle class; two, mostly poor with students predominantly from racial–ethnic minority backgrounds. In addition to participant observation, Kupchik interviewed 100 people, including students, parents, teachers, security staff, police, and administrators, and analyzed data from questionnaires that were given to all juniors in each of the four schools. This was a very comprehensive research design.

Research Results: Kupchik found that an atmosphere of punishment, not of learning, predominates in each of these schools and guides the interaction between students, faculty, and staff. Although poor, minority students are most likely subjected to such punishment, this climate also characterizes the treatment of White, middle-class schools. Students in each school perceived that the rules were unfair. The primary finding coming from this research is that the practices of punishment and discipline far outweigh the threat of actual wrongdoing. No doubt, Kupchik argues, high schools are sites of bullying, crime, and victimization, but the level of security in schools now is disproportionate to the actual risk of crime and wrongdoing.

Conclusion and Implications: Ironically, Kupchik concludes that increased security and surveillance in schools actually increases student wrongdoing. Why? Because students will follow rules if they think they are fair, but will thwart them if they perceive the rules as unfair. Moreover, the fixation on rules and punishment overlooks the real problems students face—the context in which student wrongdoing actually emerges. With the priority given to punishment in schools, students' actual needs are not being met, and taxpayer dollars may not be used to greatest effect.

Don Tremain/Alamy Limited

Questions to Consider

1. What kind of security was in place in the high school you attended? Was it effective? Did it decrease or increase misbehavior?

2. Why has there been such an emphasis on what Kupchik calls "homeroom security" in recent years? Do enhanced security practices in the schools address student needs?

Source: Kupchik, Aaron. 2010. *Homeroom Security: School Discipline in an Age of Fear.* New York: New York University Press.

Schools are also systems of power—not just on the level of teacher–student relationships, but also as a social system. School boards, principals, parents, teachers, and unions all vie for power and control in a system that ties them together—not just in stability and cooperation as functionalist theory presumes, but also in conflict and through power dynamics, the insight of conflict theory.

Symbolic Interaction Theory

Symbolic interaction theory focuses on how people interpret social interaction—in other words, some of the subjective dimensions of education and schooling. This is well illustrated in what has come to be known as the **teacher expectancy effect**—the effect of teacher expectations on a student's actual performance. When students and teachers interact, certain expectations arise on the part of both. The teacher may expect or anticipate certain behaviors, good or bad, from students. Through the operation of the teacher expectancy effect, these expectations can actually create the very behavior in question. Thus fulfilled, the behavior is actually caused by the expectation rather than the other way around. For example, if a White teacher expects Latino boys to perform below average on a math test relative to White students, over time, the teacher may act in ways that encourage the Latino boys to get below-average math test scores. The point is that what the teacher expects students to do effects what they will do. Teachers' expectations can dramatically influence how much a student learns *independent* of the student's actual ability.

Insights into the teacher expectancy effect come from symbolic interaction theory. In a classic study, Rosenthal and Jacobson (1968) told teachers of several grades in an elementary school that certain children in their class were academic "spurters," who would increase their performance that year. The rest of the students were called "nonspurters." The researchers selected the "spurters" list completely at random, unbeknownst to the teachers. The distinction had no relation to an ability test the children took early in the school year, although the teachers were told (falsely) that it did. At the end of the school year, although all students improved somewhat on the **achievement test**, those labeled "spurters" made greater gains than those designated "nonspurters." Although more recent sociological research has attempted to replicate this study, true replication is nearly impossible. Studies continue to show evidence that teacher expectations influence outcomes, but this is mediated by many factors beyond labeling of students (Jussim and Harber 2005).

How are expectations converted into performance? The powerful mechanism of the **self-fulfilling prophecy**, in which merely applying a label has the effect of justifying it, affects performance (Taylor et al. 2013). In other words, if a student is defined (labeled) as a certain type, the student often becomes that type. You can see how such a process might also be deeply affected by race, class, and gender stereotypes.

A very good example of this is the concept of **stereotype threat**. This refers to the fact that perceived negative stereotypes about one's group can actually affect one's academic performance. This has been demonstrated by recent research on why minorities and women are leaving the STEM fields of study (Beasley and Fischer 2012). The understanding that stereotypes exist about women and minorities regarding educational abilities in math and science leads to anxiety for the students. Women and minorities fear judgment about their performance in science, engineering, and math courses, and are more likely to leave those majors as a result. One of the brilliant insights of symbolic interaction theory is that the meaning attributed to a behavior can be a powerful predictor of what a person becomes. Each of the three core sociological theories offers an important perspective on education (see Table 14.1).

DOES SCHOOLING MATTER?

How much does schooling really matter? Does more schooling actually lead to a better job, more annual income, and enhanced opportunities? Parents and students who invest money and time into higher education want to see the payoff at the end. But how much schooling matters, and for whom, is not just about individual success or failure. How education is organized as a social institution and how education is related to systems of inequality in society are the larger sociological questions.

The self-fulfilling prophecy occurs when students believe they cannot perform well academically because their teachers have low expectations of them. Often these low expectations are based on stereotypes about race, class, and gender, and not based on actual student performance. Unfortunately, when students accept these lower expectations and believe them to be true, their performance confirms it.

RonTech2000/iStockphoto.com

Education Linked to Future Success

One way that sociologists measure a person's social class or socioeconomic status (SES) is to determine the person's amount of schooling, income, and type of occupation (see Chapter 8 on class stratification). Sociologists call these the *indicators* of SES. In the general population, there is a strong relationship between formal education and occupation. Although the relationship is not perfect, it is true that the higher a person's occupational status, the more formal education he or she is likely to have received. Thus, on average, doctors, lawyers, professors, and nuclear physicists spend many more years in school than garbage collectors and shoe shiners. This relationship is strong enough that you can often, although not always, guess a person's level of educational attainment just by knowing his or her occupation. There are indeed instances of laborers, such as taxi drivers, who

table 14.1 Sociological Theories of Education

	Functionalism	Conflict Theory	Symbolic Interaction
Education in society	Fulfills certain societal needs for socialization and training; "sorts" people in society according to their abilities	Reflects other inequities in society, including race, class, and gender inequality, and perpetuates such inequalities by tracking practices, for example	Emerges depending on the character of social interaction between groups in schools
Schools	Inculcate values needed by the society	Are hierarchical institutions reflecting conflict and power relations in society	Are sites where social interaction between groups (such as teachers and students) influences chances for individual and group success
Social change	Means that schools take on functions that other institutions, such as the family, originally fulfilled	Threatens to put some groups at continuing disadvantage in the quality of education	Can be positive as people develop new perceptions of formerly stereotyped groups

have PhDs, but they are relatively rare. Also rare is the reverse: the self-educated, self-made individual who completed only high school and is now the CEO of a major corporation.

Schools are stratifying institutions; that is, they sort people into different categories, based on social factors such as one's social class, race, gender, even perceived social worth. Schools themselves are stratified institutions, but they also build on the stratification that exists in the society at large. This is the case not just in the United States, but in other societies as well.

In England, for example, schools are explicitly stratified. Students in England take a general examination at the end of the compulsory high school education; the results determine whether the student leaves school or goes on to college study. In China, the educational system is about subject-specific knowledge. The "gaokao" system is the college entrance exam program for all Chinese students. How students perform on the exam is nearly the only determinant for how and where students are placed. And in Japan, an examination given at age 12 determines even more rigidly a child's subsequent educational opportunities. Students who wish to continue their education at a college or university must score high enough to gain admission to prep schools. Exams in these countries create an explicit stratifying mechanism for sorting students into an educational system that tracks them into successful professions—or not.

In the United States, educational stratification is not so explicit and, until recently, was not based solely on test scores—an issue discussed later in this chapter. But education is a stratified system, and it produces social mobility at the same time that it reproduces inequalities otherwise found in society.

Look, for example, at the connection between income, education, and social factors such as gender. Although, the higher one's education, the higher one's income on the whole, it is nonetheless true that the average income for women is less than the average income for men at *each* education level. Men with professional degrees (law, medicine, and so forth) earn a median annual income of around $101,000 on average; women with that same education earn only about $66,000—less than two-thirds of what a man earns. Of course, some of this is because women tend to get professional degrees in different fields than men. Still, the connection between education, income, and gender holds across all levels of education. Thus a man with no graduate education but only a bachelor's degree earns more than a woman with a master's degree. And men with some college, but no bachelor's degree, earn almost the same as women with a bachelor's degree (U.S. Census Bureau 2012b; see Figure 14.2).

Education and Social Mobility

Education has traditionally been viewed in the United States as the way out of poverty and low social standing—that is, as the main route to upward *social mobility*. The assumption has been that a person can overcome modest beginnings by staying in school.

There is some truth to this. Those with more education do have higher earnings, better jobs, and more perceived social worth. But much sociological research has demonstrated that the effect of education on a person's eventual job and income greatly depends

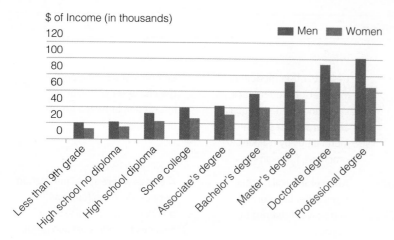

FIGURE 14.2 Education, Income, and Gender

Source: U.S. Census Bureau. 2012b. "Annual Social and Economic Supplement." Washington, DC: U.S. Census Bureau, Current Population Reports. **www.census.gov**

on the *social class* that the person was born into (Bowles et al. 2005).

Class and race also work together to "protect" the upper classes from downward social mobility. Education is used by the upper classes to avoid downward mobility by such means as sending their children to elite private secondary schools. Among middle-class Whites, education considerably improves the chances of getting middle-class jobs, yet access to upper-class positions is limited. Among those of the working class, chances of getting a good education are not impossible, but they involve a lot of social support, financial aid, and, sometimes, just plain luck. For the chronically unemployed—the underclass—chances of getting a good education are minimal. In sum, education is strongly affected by social class origins. Occupation and income are heavily influenced by social class and by education. These interrelationships are summarized in Figure 14.3, which shows that social class origin affects occupation and income both directly and indirectly by way of education.

Testing and Accountability

Social class not only influences the level of education one is likely to receive, but—and perhaps this will be a surprise—it also affects one's test scores on

standardized exams. On average, students from lower-income families have lower scores on exams such as the Scholastic Assessment Test (SAT) and the American College Testing (ACT) program. As shown in Figure 14.4, there is a smooth and dramatic increase in average (mean) SAT score as family income increases, for both SAT verbal and math scores. In this sense, one's SAT score is a proxy, or substitute, measure of one's social class: Within a certain range, you can guess someone's likely SAT score from knowing only the income and social class of his or her parents! As you can see from Figure 14.4, each additional $10,000 in family income is worth about 10 to 15 points on either the SAT verbal or the SAT math tests.

One possible reason for this is access to test preparation courses. These courses typically cost money and may be inaccessible to some students. Devine-Eller (2012) finds that as household income goes up, the likelihood of participating in test preparation courses also goes up for ninth through eleventh graders. This research also shows that students are much more likely to participate in a test preparation course if they are active in school activities, have parents who are highly educated themselves, and have parents that are actively involved in school (Devine-Eller 2012). The idea of **cultural capital** in this context suggests that certain types of parents will have access to knowledge and information about preparing their student for college entrance exams. Beyond simply the ability to pay for test preparation courses, parents with knowledge and experience regarding college admissions are able to provide better opportunity for their children.

Less help prepping for standardized tests may diminish a student's chance of getting into the best colleges or universities. The intersection of race and class also contributes to the inequality of educational attainment. Statistics about SAT scores indicate that White students typically score higher in critical reading, mathematics, and writing than Blacks and Latinos (see Table 14.2). These patterns indicate that college entrance exams continue to stratify student access to educational success.

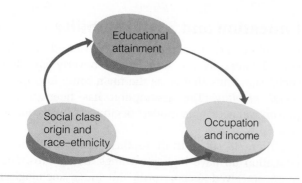

FIGURE 14.3 Relationship of Social Class, Race–Ethnicity, Education, Occupation, and Income

© Cengage Learning

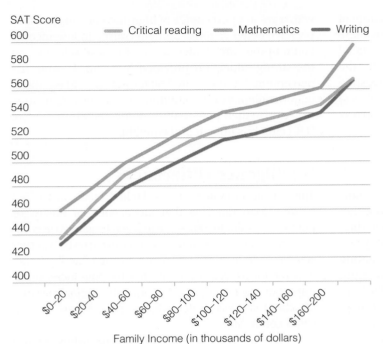

SAT Score

— Critical reading — Mathematics — Writing

Family Income (in thousands of dollars)

FIGURE 14.4 SAT Scores and Family Income

Source: The College Board. Copyright © 2010. *National Report on College-Bound Seniors*. Reproduced with permission. **www.collegeboard.com**

The educational system in the United States appears to allow for some social mobility as the result of education, but clearly not as much as people believe. A good education is essential for a good job, but the odds of getting such an education are considerably shaped by one's social class of origin, one's race—and, to a lesser extent—one's gender. The work of Jonathan Kozol (2006) highlights how so many of the poorest children in the United States lack the very basic necessities for a good education. Support and scholarship programs intended to aid those with greater disadvantages in education help, but without such intervention, the forces of social stratification are reproduced in the educational system.

EDUCATION AND INEQUALITY

Education has reduced many inequalities in society. More high school diplomas are awarded to all race and class groups, and more minorities and women attend and graduate from two- and four-year colleges. Nonetheless, many inequalities still exist in U.S. education. These can be shown in numerous ways. Studies of inequality and poverty, specifically in U.S. urban centers, highlight how schools are organized, often along race and class lines, leaving many of our most marginalized students without quality education (Kozol 2006). The educational debate is really one of nature versus nurture. One side argues that environmental conditions and

table 14.2 Average SAT Scores by Ethnicity and Gender

Students who described themselves as:	Critical Reading		Mathematics		Writing	
	Men	Women	Men	Women	Men	Women
American Indian or Alaska Native	487	484	508	479	459	474
Asian, Asian American, or Pacific Islander	520	519	605	577	520	532
Black or African American	426	432	436	422	408	428
Mexican or Mexican American	459	451	486	451	444	451
Puerto Rican	456	452	468	438	437	448
Other Hispanic, Latino, or Latin American	460	449	484	446	444	449
White	530	526	555	519	508	523
Other	494	494	534	498	484	498

Source: The College Board. Copyright © 2010. *National Report on College-Bound Seniors*. Reproduced with permission. **www.collegeboard.com**

social structural arrangements influence educational abilities. The other side of the argument suggests there are genetic differences in intelligence and educational ability. Sociological research generally falls on the side of environment and social structure, suggesting that the environmental factors, such as class and race segregation, are as much responsible for intelligence as genetic inheritance.

Segregation and Resegregation

In 1954, in a landmark decision, the U.S. Supreme Court ruled in the case of **Brown v. Board of Education** that separate but equal in all public facilities, including schools, was unconstitutional. Although it took years before school districts actually began implementing this decision, and only then with substantial pressure from the federal government, some measure of school desegregation followed the *Brown* decision. Although school desegregation was not as great as was hoped, progress was made. By the 1980s, many school districts, particularly in the American South, had made significant gains in integrating schools by race.

Now, however, that historic change is reversing, and the nation is retreating to highly segregated schools. Researchers have found that American schools are now more racially segregated by race and by class than was the case in the 1980s, in every part of the country for both African Americans and Latinos. Communities have reversed the desegregation orders of the 1970s to create neighborhood schools. Although parents prefer their child not have to be bussed far from home to attend school, the result is that students attend a very homogenous school where all students come from their neighborhood. This means wealthy neighborhoods have wealthy schools and poor neighborhoods have poor schools.

School segregation is problematic on many counts, one of which is the isolation of groups from one another and the resulting loss of friendship, interracial understanding, and comingling. But segregated schools that are heavily minority or poor are also generally of very poor quality—as the *Brown* decision noted. In other words, segregation breeds inequality, and even a cursory look at segregated schools that are predominantly minority and/or poor will reveal this. Unqualified teachers, ill-equipped science labs, a weak curriculum, and a prison-like atmosphere prevail in such schools, thus robbing students who may be perfectly smart from achieving the kind of education that will lead to a good job (Kozol 2006). Little wonder then that young African American men are more likely to end up in prison than to graduate from college (Carson and Sabol 2012; National Center for Education Statistics 2012).

The current criticism of the American educational system centers on these disparate conditions between wealthy, predominantly White, suburban schools and poor, predominantly minority, urban schools. Communities with a greater percentage of high-income families simply have better schools. Students raised in low-income communities are denied access to these schools and the strong education provided by them. Despite the perception that the United States is a fair and equitable nation, millions of U.S. children are lacking opportunity to live up to their potential. The educational structure of U.S. schools is not fair and equitable.

Intelligence Differences

The education system in the United States has relied heavily upon the idea that intelligence, or ability, or potential is a single trait—one that can be gauged according to the numerical results of standardized tests. Whether standardized tests are a strong measure of ability—and, in some forms, achievement—has become increasingly important as testing has become the major measure of school success under educational reforms like No Child Left Behind.

There are three major criticisms regarding the use of standardized tests. First, the tests tend to measure only limited ranges of abilities (such as quantitative aptitude or verbal aptitude) while ignoring other cognitive endowments such as creativity, musical ability, spatial perception, or even political skill and athletic ability (Zwick 2004; Freedle 2003).

Second, the tests possess at least some degree of cultural and gender bias—and also a strong social class bias. As a result, they may perpetuate rather than reduce inequality between different cultural, racial, gender, and class groups. Many studies show that although standardized ability tests are somewhat capable of predicting future school performance for White men, *most* studies show less accurate forecasts for the success of minorities, especially Hispanics, African Americans, and American Indians; they also predict school performance less accurately for women than for men (Taylor forthcoming). In other words, the extent to which the tests accurately predict later college grades is compromised for minorities, women, and people of working-class origins.

Third, SATs actually do not predict school performance very well for all groups. For example, SAT scores are only modestly accurate predictors of college grades even for White students (Zwick 2004). This fact is not well known. Grade point average in high school (and school class rank as well) is also only a modestly accurate predictor of success in college. High school grades are about as accurate as the SATs in predicting college grades—maybe even a little better (Alon and Tienda 2007).

In general, average scores for tests such as the SAT differ across different groups: Whites score higher on average than minorities, and the higher a person's social class, the higher his or her test score is likely to be. This is where the intelligence debate begins. The segregation of schools discussed previously indicates clear reasons for poor test scores among some students. Lack of resources in schools, inadequate teacher training, and unsafe conditions are environmental factors that likely contribute to below-average academic performance.

Still, occasional claims are made that differences in test scores are somehow genetically inherited. A notorious example was the publication of a book, *The Bell Curve* (1994). The book caused a major stir, one that is still ongoing among educators, lawmakers, teachers, public officials, policymakers, and the general public.

Authors of *The Bell Curve*, Herrnstein and Murray, argued that the distribution of intelligence in the general population closely approximates a bell-shaped curve (called the *normal distribution*). But they primarily argued that there is one basic, fundamental kind of intelligence, not several independent kinds of intelligences, and that it is genetically inherited. Using great masses of data, they claimed that fundamental intelligence is about 70 percent genetically inherited and only 30 percent determined by environment. Therefore, they argue, intelligence is determined primarily by one's genes rather than by one's social and educational environment.

Herrnstein and Murray used studies of identical twins who were separated early in life as the basis for their argument But critics point out that some of the identical twins in the studies cited by Herrnstein and Murray (and used by other researchers) were actually not very separated at all. Some were separated for longer periods during their lives and had fewer similarities in their social and educational environments. Once this is taken into account, in fact, the more separated the identical twins were, the less similar they were in intelligence. This shows the effect of their differing social environments (such as attending different schools or living in very different neighborhoods) more than the effect of their identical gene. Indeed, experts conclude that genetic inheritance actually accounts only for about 30 percent, not 70 percent, of intelligence (Taylor forthcoming).

Although there is *some* small genetic basis to intelligence, as long as society is marked by the inequalities that we can sociologically observe, then group differences in ability must be seen within that context. Understanding the drastic race and class differences in schools across this country highlights access to educational opportunity as opposed to intellectual ability.

---

debunking SOCIETY'S MYTHS

MYTH: Intelligence is mostly determined by genetic inheritance.

SOCIOLOGICAL PERSPECTIVE: Intelligence is a complex concept not easily measured by one thing and is likely shaped as much by environmental factors as by genetic endowment (Taylor forthcoming). ●

School Tracking and Individualized Education Plans. **Tracking** (also called *ability grouping*) is the separating of students within schools according to some measure of ability (Oakes 2005). Tracking has taken place for more than seventy years. As early as first grade, children are likely divided into high-track, middle-track, and lower-track groups, or some variation thereof. Perhaps you were assigned to one of these tracks in elementary, junior high, or high school. In high school, the high-track students take college preparatory courses in math and science and read Shakespeare. Middle-track students take courses in business administration and typing. Lower-track students take vocational courses in auto mechanics, masonry, or dental hygiene. Although this kind of tracking is now on the decline in the United States, versions of it still exist.

The original idea behind tracking is that students would get a better education and be better prepared for life after high school if they are grouped early according to ability. Theoretically, students in all tracks learn faster because the curriculum is tailored to their ability level, and the teacher can concentrate on smaller, more homogenous groups.

Advocates of *detracking* give the opposite argument. Detracking is based on the belief that combining students of varying cognitive abilities benefits the students more than tracking, especially by the time students get to junior high and high school. Students of high and low ability can thus learn from each other; the high-ability students are not seen to be "held back" by students with less ability, but are enriched by their presence.

How does tracking work when placing students with learning disabilities or special educational requirements? For years, American students with disabilities were isolated from mainstream students and given an entirely separate curriculum. The Individuals with Disabilities Education Act, however, outlines the federal guidelines for providing quality education for students with disabilities.

Since the act was amended in 1997, new trends focus on the need for **individualized education programs** (IEPs), which outline specific types of learning that target specific needs. Not long ago, students with special education needs were taken out of the main classroom; more recently, education research highlights the value of mixed-ability classrooms, including both IEP and mainstream students.

Which approach is better? Most researchers and educators who have studied tracking agree that not all students should be mixed together in the same classes. The differences between students can be too great and their needs too dissimilar. Some degree of tracking has always had advocates based on its presumed benefits for all students. This presumption is under attack. One of the most consistent research findings on tracking is that students in the higher tracks receive positive effects, but that the lower-track students suffer negative effects. To begin with, students in the lower tracks learn less because they are, quite simply, taught less. They are asked to read less and do less homework. High-track students are taught more; furthermore, they are consistently rewarded by teachers and administrators for their academic abilities (Oakes 2005). At the elementary level, mixed-ability classrooms are more common. As students progress through middle and high schools, IEPs are developed to address needs for each student and for each subject in school. Structurally, this can be very challenging for teachers and school administrators. Individualized lesson plans could benefit all students, not just those with learning disabilities. Teachers, however, must find ways to teach the same material to a classroom full of students, all of whom learn differently.

thinking SOCIOLOGICALLY

Were you in a *tracked* elementary school? What were the tracks? Did you get the impression that teachers devoted different amounts of actual time to students in different tracks? Did teachers "look down" on those in the lower tracks? What about the students—did they treat some tracks as "better" or "worse" than others (were they perceived as differing in prestige)? Based on your recollections, what does this tell you about tracking and social class? ●

Who gets assigned to which tracks? Research shows that track assignment is not solely based on the performance in cognitive ability tests. Social class and race are involved. Students with the same test scores often get assigned to different tracks because of differences in their social class and race. Few administrators or teachers consciously and deliberately assign students to tracks based on these criteria, but it occurs nevertheless. Researchers have consistently found that when following two students with identical scores on cognitive ability tests, the student of higher social class is more likely than the student of lower social class status to get assigned to the higher track.

This inequality is at the root of the American education debate. A core American value states that all people are created equal. Through a fair and equitable educational system, all students would have equal access to opportunity and success. Sociological research examines the social institution of education to better understand the consequences for students and how to better improve those consequences.

EDUCATIONAL REFORM

There are clearly major challenges facing the educational system in the United States. Calls for reform are many and are coming from parents, communities, teachers and administrators, as well as presidents and politicians. The major reform in public education to date has been the *No Child Left Behind* (NCLB) Act of 2001.

The goal of NCLB was, in part, an attempt to narrow the achievement and test score gap between White students and students of color in U.S. public schools. Of course, this act does not address fundamental problems of public education, including racial segregation and wealthy White students attending private educational institutions. But the NCLB Act has meant restructuring

Schools in the United States are rapidly resegregating by race.

education with an emphasis on "accountability"—that is, measured assessments of where and how schools are succeeding or failing. Much of the emphasis in NCLB has then been on *high-stakes testing*. Students cannot graduate or move on to the next grade without reaching certain levels of proficiency, as measured on standardized exams. NCLB also calls for teachers to be replaced based on their students' test scores, and schools that were seen as underachieving were threatened with closure.

Despite assertions by politicians that this law was having a positive and "dramatic" effect, the results have unfortunately shown that wide gaps persist in verbal and math test scores. And the gap has actually *widened* during the period when the NCLB law was in effect. Some of the gap in test scores, as we have seen, can be attributed to mismeasurement and cultural bias in the tests, but the problems in the education system run deep and cannot be measured by test scores alone.

Under the Obama administration, a new reform is being initiated. Under the "Race to the Top" plan, states compete for large sums of money (in the millions of dollars) to assist them in school reform. The program targets four specific areas:

1. Adopting standards and assessments that will prepare students to succeed in college and the workplace and to compete in a global economy
2. Developing good measures of student success that can be used to inform teachers and administrators about improving instruction
3. Recruiting, rewarding, and retaining the best teachers and principals
4. Improving the lowest-achieving schools

The Race to the Top educational reform is yet to be evaluated fully. Efforts to improve the educational system in the United States speak to our need, both globally and domestically, to create learners that will succeed in moving the United States forward.

Educational reform begins with a clear understanding of education as an institution, with structural elements that create and reinforce inequality. Continued research and governmental commitment will help create a more balanced, fair, and successful model for educating Americans.

HEALTH CARE IN THE UNITED STATES

Like education, health care in the United States is an institution. We examine health care here both for the type of care available in the United States and for the differing access to health care among different groups of people. Despite global recognition for having the best medical treatments available, questions remain as to whether or not the quality of health care in the United States has declined over the last two decades. Why are

costs for medical insurance so high, and why do so many people not have health insurance? Why are some Americans at greater risk for illness than others? These questions are at the core of current political struggles about health care, but they are also informed by sociological research and theory.

Generally speaking, the citizens of the United States are quite healthy in relation to the rest of the world. And the nation's health care is some of the best in the world. But, as we will see, there are very great discrepancies among people within the United States in terms of how healthy they are and their access to health care. Although health is a physiological phenomenon, it also has social dimensions. The field of medical sociology studies these social dimensions of health and illness, the social organization of health care institutions, and the inequality of access to quality health care.

Health and Illness

Illness and how to treat disease have advanced greatly over the course of American history. Scientific breakthroughs in the natural sciences brought us to a remarkable time in Western medicine when Americans have access to diagnosis, treatment, and cures for so many diseases once believed to be fatal. Underdeveloped countries are far behind American medical schools and hospitals in availability of treatments, diagnostic tests, and social support for the sick. Medical doctors are still among the most revered professionals, with much higher salaries than average income earners. Overall, the modern American system of health care and medicine is a model of success.

There are, however, criticisms of the U.S. health care system. Much like our education institutions, medical institutions are social structures that create different experiences for different groups of people; the system is not perfect. The criticism largely revolves around unequal access to good health care. The debate over affordable health care and equality of care has dominated the recent political landscape. Inequality in health care is addressed later in this chapter.

Another general critique of American medicine is about the overall model of how we treat disease. Because of all those technological advances in science, the assumption is that the most up-to-date treatments are *better* for the patient and have *better* success. In many cases, this is true. We would much rather put antibiotic cream on an open wound, than "bleed" it with an unsterile cut. Non-Western techniques, however, are not entirely without merit. There are examples of people traveling to India to practice yoga to cure nerve disorders. Right here in the United States, acupuncturists are successfully treating people for everything from chronic back pain to breach pregnancies. One complaint of our health care system is that these alternative medical practices are not always endorsed by physicians and are rarely paid for by insurance. Recent estimates are that Americans paid nearly $34 billion in one year on complementary and alternative medicine (National Institutes of Health 2007).

The lack of emphasis on prevention within American health institutions is another failing of the system, according to some critics. Medicine in this country is mostly a disease model in which patients are first diagnosed and then seek treatment for the illness. Despite evidence that prevention of many illnesses is possible, the system is structurally set up for treatment rather than prevention. Most health insurance does not reimburse, for example, for a health club or gym membership for someone at high risk for diabetes. The cost of managing diabetes far outweighs the cost of supporting an active lifestyle to prevent diabetes; yet, health care institutions are designed to respond to disease rather than to avoid it.

If health care institutions turned attention to prevention, several costly and difficult health problems could be significantly reduced in the United States. **Obesity** is a major health concern in the United States, and a contributing factor for heart disease, stroke, diabetes, and some cancers. Recently, the Centers for Disease Control and Prevention classified obesity as an epidemic, with one in three adults and one in six children classified as obese (Centers for Disease Control 2011). Obesity occurs when more calories are consistently ingested than are burned through physical activity. This, however, is oversimplifying the problem and puts too much emphasis on individual choices. The epidemic of obesity, costing the United States nearly $150 billion annually, is a social problem well beyond the scope of an individual trouble (Centers for Disease Control 2011). Individuals do not simply lack self-control when eating. Instead, environmental factors have created a society focused on food, where what we eat, when we eat, and how much we eat are contributing to Americans' obesity.

One environmental factor is the unavailability of fresh, affordable, and healthy foods for many. Many people in low-income inner-city neighborhoods, for example, have to go miles before reaching a grocery store with fresh, affordable produce. Convenience stores, vending machines, and fast-food restaurants dominate the city landscape, certainly in poorer areas. These places provide inexpensive, calorie-rich foods that fill people up. Unfortunately, these foods are also rich in fat and lack key nutrients needed for healthy bone development and childhood growth. The racial and socio-economic characteristics of these neighborhoods mean that some racial groups are more likely than others to be obese, especially among children (see Figure 14.5). A steady diet of high-fat, high-sugar, and highly processed foods increases the likelihood of obesity.

Another contributing factor to the obesity epidemic in the United States is our change in lifestyle. We are less active than in previous cultures, many of us driving to work rather than walking. Communities

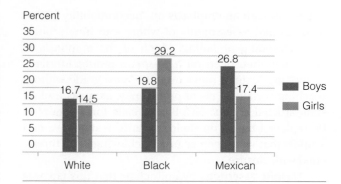

FIGURE 14.5 Percent of Children Aged 12–19 Who Are Obese, 2007–2008

Source: Centers for Disease Control and Prevention. 2008. National Center for Health Statistics.

are built in suburbs, not within walking distance of workplaces. Even among those who live in urban areas near their workplace, transportation options eliminate the need for walking. Americans also work more hours than ever before, finding less time for physical exercise. Lack of active movement coupled with an abundance of fattening food choices makes obesity a national health concern.

Scientific knowledge, based in nutrition, chemistry, biology, and physiology, provides the know-how to battle obesity. Health professionals are aware of the benefits of healthy foods and physical activities, yet the emphasis remains on treating diseases that are likely consequences of obesity. Given the social and cultural dimensions of overeating, the struggle to reduce the number of obese Americans is more challenging. New governmental guidelines for nutrition, state and federal initiatives for physical activity, and media emphasis on weight loss are all part of a good start for

Super-sized food is contributing to the problem of obesity.

When Should Treatment Stop?: Issues for End-of-Life Care

Have you ever heard the term *sandwich generation*? This refers to people who are at a point in their lives where they are caring for children while also caring for aging parents. For many families, this is a very difficult time. Financial, emotional, and mental distress often accompany the confusion of how best to care for an aging or dying family member. Sociological examination of the current health care system reveals little input from the medical community in preparing for end of life. Because the disease model currently in place in the United States emphasizes treatment, few doctors will guide patients through the end of life. Insurance companies often do not cover the cost of *palliative care* (not to cure or fix, but simply to keep comfortable) or *hospice care* (to keep comfortable through the dying process). Instead, most Americans die in the hospital with tubes and monitors attached. Yet, research reveals that most people prefer to die at home with family around them (Ko et al. 2013). Further research provides that the caring of elderly typically falls to women (Jordan and Cory 2010), and that there are clear cultural differences among different American families that should be considered when reforming end-of-life care (Ko et al. 2013; Cravey and Mitra 2011). Sociologists examine the cultural expectations and structural inequality in caring for dying family members. At what point are doctors and hospitals no longer needed?

reversing the obesity trend. Health care institutions can also be part of the solution by including healthy eating and exercise as part of an overall prevention-focused medical plan.

The Social Organization of Health Care

Health care now is a vast institution, including not only hospitals and doctors but also many auxiliary sectors, such as nursing homes, rehabilitation centers, drop-in clinics, and various "alternative" health care services, such as homeopathy, wellness centers, even exercise and nutrition centers. But the colossal factors in the organization of health care institutions are the for-profit insurance and pharmaceutical companies. Health is big business, and the connection between for-profit companies, the government, and health care lies at the heart of current debates about health care.

The United States is one of the few industrialized nations that does not provide universal health care to its citizens. The 2012 passing of the **Affordable Care Act** aims to address the problem of too many uninsured Americans. Health care in the United States is a labyrinth of health care deliverers, for-profit insurance companies and government programs that provide health care for the aged and for the poor. Political debate over entitlement programs like **Medicare** (which provides health insurance to older Americans) and **Medicaid** (which provides health insurance to poor Americans) threatened the passing of the Affordable Care Act. The new policy is only recently in place and will likely be evaluated over the next several years.

The key components of the Affordable Care Act advocate for better protection for consumers of health insurance; more affordable insurance for all Americans; better access to care, including some preventative medicine; and a stronger Medicare program that reduces cost while increasing coverage for older Americans. Primarily, the Act is the government's approach to addressing issues of unequal access and an attempt to get more Americans health insurance to cover the cost of care.

The critics of the new Affordable Care Act claim that insurance costs will rise for companies that employ workers, leading them to cut back on jobs and force many companies to fail. Another analysis argues that, under the new policy, physicians will be unable to collect payment for much of their work, leading to fewer quality doctors in practice. Despite the passing of the new law, debate continues in the political arena and among health care professionals over how to best provide care for people in the United States.

The American health care system has been compared to those of other Western countries and revealed some clear differences. For example, in European countries like France and Germany, health care is much more unified in approach, allowing patients to experience more cohesive care from diagnosis to treatment to cure (Reid 2010). In the United States, a specialist, a doctor who concentrates on one specific area of medical care, is desirable for almost any illness. Primary care physicians are not expected to treat disease, but rather to simply manage good health and then refer patients to a specialist when needed. This contributes to the confusion, the high cost, and the ineffectiveness of American health care. Patients often complain that diagnostic test results are not shared between doctors or are not done at all. Multiple doctors may be involved in diagnosis and treatment, and they are not in agreement or are not communicating effectively with one another.

The confusion and frustration in managing care is challenging in the best of cases. For fully insured Americans with high education and good incomes, navigating through the health care system is often complicated and difficult. For the millions of Americans that are not insured, have less education, and are financially vulnerable, an illness can be devastating in more ways than one. Medical options are not equally available to all Americans. Health care institutions recreate the structural inequality of society.

---

see FOR YOURSELF

Mapping Food

Identify two neighborhoods in your community that differ by their social class and/or racial composition. Draw a map of each neighborhood and then take a drive through each with your map in hand. Mark every place where you see some kind of food outlet, and mark whether it is a major grocery chain, a convenience store, fast-food outlet, or other provider of meals. You might also note what kind of transportation is needed to get to each location. When you have finished, what patterns do you see about the availability of healthy food in each neighborhood? If you lived in either, how far would you have to go to purchase fresh, good-quality food? Can you get there without a car? What does your experiment suggest about class and race disparities in health outcomes? ●

HEALTH AND INEQUALITY

Prominent problem areas in the U.S. health care system include the following:

- **Unequal distribution of health care by race–ethnicity, social class, or gender.** Health care is more readily available and more readily delivered to White or middle-class individuals in urban and suburban areas than to minorities. The lack of health care delivery to Native American populations is particularly serious. Likewise, men and women receive unequal treatment for certain types of medical conditions, with women more likely than men to receive truncated treatment.
- **Unequal distribution of health care by region.** Each year, many in the United States die because they live too far away from a doctor, hospital, or emergency room. Doctors and hospitals are concentrated in cities and suburbs; they are much less likely to be situated in isolated rural areas. Rural people in Appalachia and some parts of the South and Midwest may have to travel 100 miles or more to get to a doctor or emergency room.
- **Inadequate health education of inner-city and rural parents.** Many inner-city and rural parents do not understand the importance of immunizing

their children against smallpox, tuberculosis, and other illnesses, and they are often suspicious of immunization programs. This hesitancy is reinforced by the depersonalized and inadequate health care that residents of low-income communities often encounter when care is available at all.

---

debunking SOCIETY'S MYTHS

MYTH: The health care system works with the best interests of clients in mind.
SOCIOLOGICAL PERSPECTIVE: The health care system is structured along the same lines as other social institutions, thus reflecting similar patterns of inequality in society. ●

Race and Health Care

Racial disparities in health mean that African Americans are more likely than Whites to fall victim to various diseases, including cancer, heart disease, stroke, and diabetes. Death of the mother during childbirth is almost three times as likely among African American women as among White women (Centers for Disease Control 2011), and although the occurrence of breast cancer is lower among African American women than White women, the *mortality rate* (death rate) for breast cancer in African American women is considerably higher than it is for White women (Centers for Disease Control 2011).

Hispanics, like African Americans, Native Americans, and other minorities, are also significantly less healthy than Whites (Centers for Disease Control 2011). Hispanics contract tuberculosis at a rate seven times that of Whites. Other indicators of health, such as infant mortality, reveal a picture for Hispanics similar to that of African Americans and Native Americans.

Although differences in culture, diet, and lifestyle account for some of the racial disparities in health care, it is well established in study after study that African Americans and Latinos simply do not receive medical attention as early as Whites. When they do get treatment, the stage of their illness is often more advanced and the treatment they receive is not of the same quality. African Americans and Hispanics, especially when they are poor, are less likely than Whites to have a regular source of medical care (see Figure 14.6). When they do, it is likely to be a public health facility or an outpatient clinic. Because of language barriers as well as other cultural differences, Hispanics are less likely than other minority groups to use available health services, such as hospitals, doctors' offices, and clinics (National Center for Health Statistics 2012).

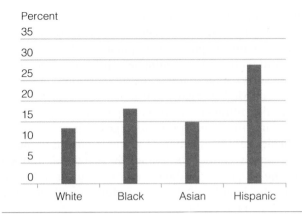

Percent

35	
30	
25	
20	
15	
10	
5	
0	

White   Black   Asian   Hispanic

Figure 14.6 People without Health Insurance Coverage, 2010–2011 This chart shows the percentage of people in different race groups who did not have health insurance. Why do you think lack of coverage would be particularly high among Hispanics?

Source: U.S. Census Bureau. 2012b. *Current Population Survey 2011 and 2012 Annual Social and Economic Supplement.* **www.census.gov**

Social Class and Health Care

In the United States, social class has a pronounced effect on health and the availability of health services. The lower the social class status of a person or family, the less access available to adequate health care (National Center for Health Statistics 2012). Consequently, the lower one's social class, the less long one will live. People with higher incomes who are asked to rate their own health tend to rate themselves higher than people with lower incomes. Almost 50 percent of those in the highest income bracket rate their health as excellent, whereas only 25 percent in the lowest income bracket do so. In fact, the effects of social class are nowhere more evident than in the distribution of health and disease, showing up dramatically in the rates of infant mortality, stillbirths, tuberculosis, heart disease, cancer, arthritis, diabetes, and a variety of other illnesses. The reasons lie partly in personal habits that are themselves partly dependent on one's social class. For example, those with lower socioeconomic status smoke more often, and smoking is the major cause of lung cancer and a significant contributor to cardiovascular disease.

Social circumstances also have an effect on health. Poor living conditions, elevated levels of pollution in low-income neighborhoods, and lack of access to health care facilities all contribute to the high rate of disease among low-income people. Another contributing factor is the stress caused by financial troubles. Research has consistently shown correlations between psychological stress and physical illness (Taylor et al. 2013). The poor are more subject to psychological stress than the middle and upper classes, and it shows up in their comparatively high level of illness.

Medicaid is the government program that provides medical care in the form of health insurance for the poor, welfare recipients, and the disabled. The program is funded through tax revenues. The costs covered per individual vary from state to state because the state must provide funds to the individual in addition to the funds that are provided by the federal government. Medicaid, together with Medicare, are as close as the United States has come to the ideal of universal health insurance.

Gender and Health Care

Although women live longer on average than men, national health statistics show that hypertension is more common among men than women until age 55, when the pattern reverses. This may reflect differences in the social environment men and women experience, with women finding their situation to be more stressful as they advance toward old age. Under age 35, men are more likely to be overweight than women; after that, women are more likely to be overweight. Women have a higher likelihood of contracting chronic disease than men, although men are more likely to be disabled by disease (National Center for Health Statistics 2012).

Health and Disability

The *disability rights movement*, a movement that has defined disabled people as a social group with rights similar to other minority groups in society, has transformed how people think about disability, challenging many preconceived ideas. For example, within a social context, there is a tendency for people to see someone with a disability solely in terms of that social status—what sociologists call a stigma. A **stigma** is a social identity that develops when a person is socially devalued by others because of some identifiable characteristic. When someone is stigmatized, that identity tends to override all other identities, and the person is treated accordingly.

Understanding the social dynamics associated with disabilities has resulted from the efforts of the disability rights movement. The movement has called attention to the social realities of disabilities, even questioning the very language used to identify people with disabilities—for example, using the term *physically challenged* rather than the more negative connotation of *disabled*.

One of the most significant achievements of the disability rights movement is the Americans with Disabilities Act (ADA), passed by Congress in 1990. This law prohibits discrimination against disabled people. The ADA legislates that disabled people may not be denied access to public facilities; thus the presence of such things as ramps, wheelchair access on buses and stairways, handicapped parking spaces, and chirping sounds in crosswalk lights for blind pedestrians, all

The disability rights movement has opened up new opportunities to those who face the challenge of disability.

social changes that are now so prevalent that you might even take them for granted. They have resulted, however, from the social mobilization of those who saw a need for social change.

The Americans with Disabilities Act also requires employers and schools to provide "reasonable accommodations" such that those with disabilities are not denied access to employment and education. For many students with various learning disabilities, this has meant making accommodations for taking tests with extended time or in settings where the test taker is not subject to as much distraction as in a crowded classroom. Individualized education programs (IEPs) were discussed in the education section of this chapter. The increased awareness of disability rights has transformed society in ways that have opened up new opportunities for those who, years ago, would have found themselves with less access to education and jobs and, therefore, more isolated in society.

Age and Health Care

As people age, their health care needs are no doubt likely to increase. And, until recently, many of the nation's elderly were also likely to be low income. Although class status varies among the nation's elderly, all older people at this point are beneficiaries of the national *Medicare* program. Medicare was begun in 1965, under the administration of President Lyndon Johnson. It provides medical insurance, including hospital care, prescription drug plans, and other forms of medical care for all individuals age 65 or older. The new Affordable Care Act aims to strengthen Medicare benefits.

Medicare is partially funded through payroll taxes whereby both employees and employers pay a small percentage of employee wages to cover some of the cost of this large (and costly) federal program. But, with so many people in the population now living longer, and with the now-aging baby boomer population being such a large share of the total population, many wonder if Medicare can be sustained in the near future. With the number of workers paying payroll taxes shrinking, the elderly population growing, and the cost of health care rising, there is a looming fear that Medicare simply cannot be financially sustained. Though not the sole basis for the nation's challenges in health care, the health needs of the older population are clearly a major challenge.

THEORETICAL PERSPECTIVES OF HEALTH CARE

The sociology of medicine is anchored in the same major theoretical perspectives that we have studied throughout this book: functionalist theory, conflict theory, and symbolic interaction theory (see Table 14.3).

Functionalist Theory

Functionalism argues that any institution, group, or organization can be interpreted by looking at its positive and negative functions in society. Positive functions contribute to the harmony and stability of society. The positive functions of the health care system are the prevention and treatment of disease. Ideally, this would mean the delivery of health care to the entire population without regard to race, ethnicity, social class, gender, age, or any other characteristic. At the same time, the health care system is notable for a number of negative functions, those that contribute to disharmony and instability of society.

Functionalism also emphasizes the systematic way that various social institutions are related to each other, together forming the relatively stable character of society. You can see this with regard to how the health care system is entangled with government through such things as federal regulation of new drugs and procedures. The government is also deeply involved in health care through scientific institutions such as the National Institutes of Health, a huge government agency that funds new research on various matters of health and health care policy. As a social institution, health care is also one of the nation's largest employers and thus is integrally tied to systems of work and the economy.

Conflict Theory

Conflict theory stresses the importance of social structural inequality in society. From the conflict perspective, the inequality inherent in our society is

table 14.3 Theoretical Perspectives on the Sociology of Health

	Functionalism	Conflict Theory	Symbolic Interaction
Central point	The health care system has certain functions, both positive and negative.	Health care reflects the inequalities in society.	Illness is partly socially constructed.
Fundamental problem uncovered	The health care system produces some negative functions.	Excessive bureaucratization of the health care system and privatization lead to excess cost.	Patients and health professionals serve specific roles. What is determined as illness is cultural context specific.
Policy implications	Policy should decrease negative functions of health care system for minority groups, the poor, and women.	Policy should improve access to health care for minority racial–ethnic groups, the poor, and women.	Determining something as disease will make insurance reimbursement more likely.

© Cengage Learning

responsible for the unequal access to medical care. Minorities, the lower classes, and the elderly, particularly elderly women, have less access to the health care system in the United States than Whites, the middle and upper classes, and the middle-aged. Restricted access is further exacerbated by the high costs of medical care.

Excessive bureaucratization is another affliction of the health care system that adds to the alienation of patients. The U.S. health care system is burdened by endless forms for both physician and patient, including paperwork to enter individuals into the system, authorize procedures, dispense medicines, monitor progress, and process payments. Long waits for medical attention are normal, even in the emergency room. Prolonged waits have reached alarming proportions in the emergency rooms of many urban hospitals in the United States and can only deepen the alienation of patients.

Symbolic Interaction Theory

Symbolic interactionists hold that illness is partly (although obviously not totally) socially constructed (Armstrong 2003). The definitions of illness and wellness are culturally relative—the social context of a condition partly determines whether or not it is sickness. Consider the example of alcoholism and other addictions. During the era of Prohibition, people who drank were considered deviants and lacking moral fortitude. Now, however, alcoholism is a diagnosable disease, listed in the *Diagnostic Statistical Manual* as an illness. The medicalization of alcoholism refers to how Americans culturally and socially label abuse of alcohol as a disease that requires treatment. This has profound consequences for how people with alcoholism are treated. People who are *ill* receive more sympathy and more care than those who are labeled *deviant*.

Symbolic interaction also highlights the roles played within the health care institution. There is a hierarchy that puts medical doctors at the top and medical assistants, nursing staff, and orderlies at the bottom. Patients take on the role of a child, with little agency in how treatment is administered. The diagnosis, the treatment plan, and the prognosis are managed with little input from the patient. Insurance companies and pharmaceutical companies play an entirely

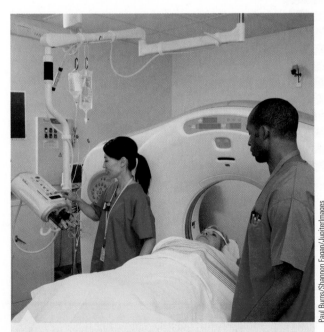

New medical technologies add to the quality of health care, but also to the cost.

Paul Burns/Shannon Fagan/JupiterImages

different role, one that oversees the availability of medical care by determining what procedures or treatments will be financially covered.

The symbolic interaction approach to studying health care institutions focuses on the roles of the patient and medical professionals and on the cultural context within which disease is labeled and treated. Table 14.3 outlines the theoretical perspectives of health care and illness.

HEALTH CARE REFORM

Currently, the cost of medical care in the United States is approximately 18 percent of our gross domestic product, making health care the nation's third leading industry. The United States tops the list of all countries in per person expenditures for health care (see Figure 14.7). Other countries spend considerably less money and deliver a level of health care at least as good. For example, Sweden and the United Kingdom spend roughly half as much per capita as

the United States, and Turkey spends a bit more than one-third as much.

The Cost of Health Care

One of the challenges of health care is sheer cost. Most health care is provided by a fee-for-service principle in which the patient is responsible for paying the fees the health care provider charges. Patients with health insurance are able to pass on health expenses, either in full or partially, to the insurance company. But the cost for health care services is high, in some cases, astronomically expensive. Hospital care can cost thousands, even millions, of dollars for any extended stay. Sophisticated procedures require expensive machinery and technicians, and the nation needs to invest in medical research that allows practitioners to stay abreast of new technologies and new treatments for a wide array of medical conditions.

But most sectors of the health care system (hospitals, pharmaceutical companies, even physician's

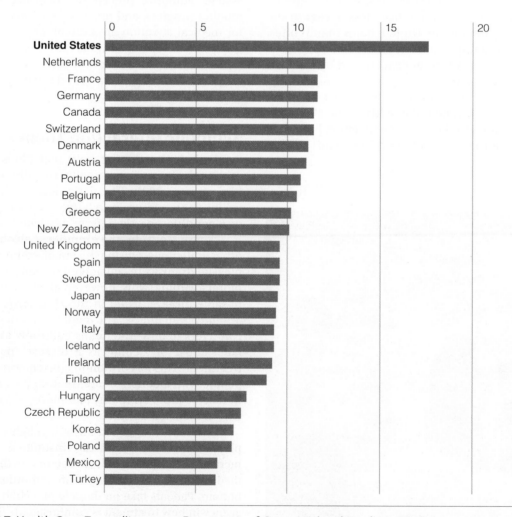

FIGURE 14.7 Health Care Expenditures as a Percentage of Gross National Product, 2010

Source: Organisation for Economic Co-operation and Development (OECD). 2012. http://www.oecd.org/unitedstates/BriefingNoteUSA2012.pdf

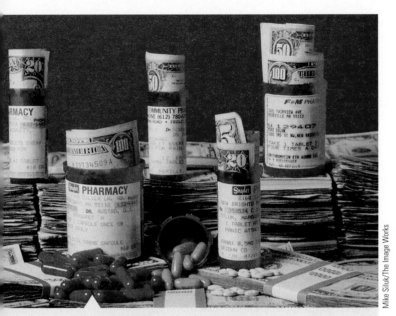

The high cost of prescription drugs is indicative of the problems generated by a profit-based health care system.

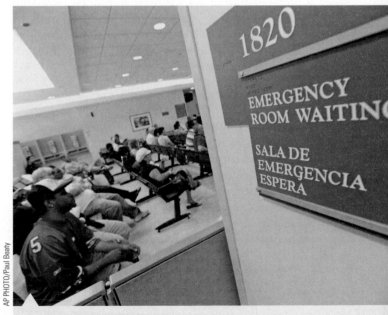

Mike Siluk/The Image Works

Health Care for All?

Why does the United States not provide health care to its citizens in line with other Western nations? Sociologists offer several explanations for the nation's reluctance to develop a national health care plan similar to that of other Western nations. First, there is an antigovernment attitude among many in the United States that fuels resistance to a national health care system. Second, analysts argue that, unlike in other Western nations, there is a relatively weak labor movement in the United States, resulting in more limited state-based benefits for workers. Third, racial politics have also shaped the nation's health care system; federal social welfare programs are associated in many people's minds with racial groups, and this, too, fuels the politics of health care reform. And, finally, the health care system in this country is fundamentally structured on private, for-profit interests (Quadagno 2005).

Taken together, these factors mean that millions of people in the United States have no health insurance and are thus left vulnerable should they become sick. In 2011, 48.6 million people in the United States had no health insurance (DeNavas-Walt et al. 2012). For many, the main source for medical care is a hospital emergency room, often called the "doctor's office of the poor." This is a very expensive way to deliver routine health care—and there is rarely any follow-up care or comprehensive and preventative treatment.

office practices) are structured as for-profit businesses. Physicians, for example, may have to raise their rates to cover the high cost of malpractice insurance where annual insurance premiums (costs) have skyrocketed. The cost of these insurance premiums is passed along to consumers (patients) and has contributed to the rise in the overall cost of health care.

Adding to the high cost of health care is the role of big pharmaceutical companies. Spending for prescription drugs in the United States has increased from $40 billion in 1990 to a whopping $250 billion in 2009! And there is little sign that this spending will do anything but go further up. Prescription drugs are one of the fastest-growing components of health care costs. The rise in spending on drugs is partially attributed to increased use, but other factors include the actual cost of the drugs, the availability of new drugs for various maladies, and, without question, the cost of advertising directly to the public. The money spent on advertising directly to consumers has doubled since 1999 (Kaiser Family Foundation 2010). You can see this yourself as hardly an hour goes by on television without an advertisement for some kind of prescription drug.

The health care crisis in the United States is largely a question of cost, but it also entails a debate over the nation's responsibility for the health of its citizens. Who should pay for the soaring costs of health care? Who receives the benefits of such sophisticated medicine? Should there be universal health care for all? These questions are at the heart of the current national debate about health care reform.

AP PHOTO/Paul Beaty

For many without health insurance, emergency rooms are the primary source of health care.

The Affordable Care Act was finally passed into law in June 2012 after the Supreme Court ruled it did not violate the Constitution. The debate over whether or not "Obamacare" was a form of socialism continued through the entire 2012 presidential election.

Jose Gil/Shutterstock.com

see FOR YOURSELF

Youth and Health Insurance

Young people (aged 18 to 34) are the age group least likely to be covered by health insurance. Identify a group of young people you know and ask them if they are covered by health insurance. If they are insured, where does their insurance come from? Who pays? If they are not insured, ask them why not and whether they think this is important. Do they support a national health insurance program?

Having conducted your interviews, ask yourself how social factors such as the age, race, ethnicity, gender, and educational/occupational status of those you interviewed might have affected what people say about their insurance. Do you think any or all of these social characteristics are related to the likelihood that these people are covered by health insurance and whether these characteristics are related to their attitudes about coverage? What are the implications of your results for public support for new health care policies? ●

chapter summary

What is the importance of the education institution?
Education is the social institution that is concerned with the formal transmission of society's knowledge. It is therefore part of the socialization process. Although the U.S. education system has long produced students at the top of the world's educational achievements, the United States is falling behind other nations on standardized test scores.

How does sociological theory inform our understanding of education?
Functionalism interprets education as having various purposes for society, such as socialization, occupational training, and social control. *Conflict theory* emphasizes the power relationships within educational institutions, as well as how education serves the powerful interests in society. *Symbolic interaction theory* focuses on the subjective meanings that people hold. These meanings influence educational outcomes.

How does education link to success?
The number of years of formal education for individuals has important, but in many ways modest, effects on their ultimate occupation and income. Social class origin affects the extent of educational attainment (the higher the social class origins, the more education is

ultimately attained), as well as occupation and income (higher social class origin likely means a more prestigious occupation and more income).

Does the educational system perpetuate or reduce inequality?
Although the education system in the United States has traditionally been a major means for reducing racial, gender, and class inequalities among people, the education institution has perpetuated these inequalities. Segregation of schools and communities keep minority and poor children in schools that lack resources for success.

What current reforms are guiding education?
The No Child Left Behind Act emphasized accountability in the schools, largely through testing. Current educational reform is occurring under a program called Race to the Top, in which reform focuses on achieving educational standards, assessing school progress, and developing strong measures of student and teacher success.

How does the United States compare to other nations in the area of health care?
The United States is one of the only industrialized nations that does not provide universal health care for

its citizens, even though the care available is some of the best in the world. The health care system is organized according to social patterns, including that disease itself is influenced by social facts, such as race, gender, and social class.

How does sociological theory inform our understanding of health and health care?
Functionalism interprets the health care system in terms of the systematic way that health care institutions are related to each other. Conflict theory addresses the inequalities that occur within the health care system.

Symbolic interaction analyses the interpretations that can affect people's health care, such as the tendency to place patients in a sick role and label some ailments as disease and others not.

What is the health care crisis in the United States?
High costs and questions about universal health care have created a policy crisis today in the U.S. health care system. The passage of the Affordable Care Act has attempted to address some of the problems on universal health care, but the bill remains controversial.

Key Terms

achievement test 342
Affordable Care Act 351
Brown v. Board of Education 346
cultural capital 344
individualized education programs 347

latent functions 340
Medicaid 351
Medicare 351
obesity 350

schooling 338
self-fulfilling prophecy 342
stereotype threat 342

stigma 353
teacher expectancy effect 341
tracking 347

15 Economy and Politics

AP Photo/Bebeto Matthews

because you are reading this book, chances are that you are in school, probably seeking a college degree that will help you find a good job. Your hopes and dreams probably rest on getting a decent education, discovering what you plan to do in a job, and gaining the skills you need to succeed. You might be wondering how your education will prepare you for work and how you will match your interests to the current job market.

On one level, these are individual issues that involve your interests and talents, your motivation, and how well prepared you are to study and work. But behind these individual matters lie other social structures—structures that shape the options you face, the resources you have to pursue your dreams, and whether good jobs are available to you. The background structures that frame your individual life are the result of social institutions, and how those institutions are structured by social stratification.

Finding work situates you within economic institutions—that is, the economy, like other social institutions, has a particular social structure that includes an organized system of social roles, norms, and values. Social institutions extend beyond us, but they shape day-to-day life. Typically, people do not think about such structures and may not even be aware of the institutional structures that shape their lives. Other times, social institutions become very apparent to people, such as has happened during the economic problems of recent years.

Challenges in the contemporary economy closely link the economy to another major social institution: the political system. Rising national debt, heated debates over social entitlement programs, who pays taxes and how much, conflicts over the role of government in job creation, and other issues closely link the social structure of the economic and political institutions. Much of the time, these connections may be taken for granted, although at times of economic peril, people can become more aware of such economic and political forces.

In recent years, for example, thousands have lost their jobs because of economic conditions, many of whom suffer from long-term unemployment. Some have lost financial resources that they have worked a lifetime to save. Young people entering the labor market for the first time, even as college graduates, now find themselves facing possible joblessness—through no fault of their own—merely because they exited school at a time when broader social forces were shaping their life opportunities.

Although understanding the institutional forces at work at any given time does not reduce the suffering that people may be experiencing, such an understanding can help you think about social changes that can transform people's experience in society. How is the economy organized and how is it connected to political institutions? Such questions guide the content of this chapter.

learning objectives

- Compare different types of economic systems
- Identify the components of change in the contemporary global economy
- Identify the different organizational components of the workplace
- Explain the conditions affecting diverse groups in the workplace
- Compare and contrast sociological theories of work
- Compare different types of political systems
- Define different forms of power and authority and the organization of bureaucracies
- Analyze patterns of political participation
- Comprehend the organization of the military as a social institution

ECONOMY AND SOCIETY

All societies are organized around an economic base. The **economy** of a society is the system by which goods and services are produced, distributed, and consumed. We first look at the economic significance of the historic transformation from agriculturally-based societies to industrial and now postindustrial societies.

The Industrial Revolution

In Chapter 5, we discussed the evolution of different types of societies. Recall that one of the most significant of these changes was, first, the development of agricultural societies and, later, the far-ranging impact of the Industrial Revolution. Now, the Industrial Revolution is giving way to the growth of postindustrial societies—a development in the economic system with far-reaching consequences for how society is organized.

North Wind Picture Archives/Alamy Limited

The Industrial Revolution transformed labor, moving it from the household to the factory or other sites where work was mechanized and oriented to mass production.

The Industrial Revolution is usually pinpointed as beginning in mid-eighteenth-century Europe, soon thereafter spreading throughout other parts of the world. The Industrial Revolution led to numerous social changes because Western economies became organized around the mass production of goods. The Industrial Revolution led to the creation of factories, which separated work and family by relocating the place where most people were employed.

We still live in a society that is largely industrial, but that is quickly giving way to a new kind of social organization: the postindustrial society. Whereas industrial societies are primarily organized around the production of goods, **postindustrial societies** are organized around the provision of information and services. Thus, in the United States, we have moved from being a manufacturing-based economy to an economy centered on the provision of services. *Service* is a broad term meant to encompass a wide range of economic activities now common in the labor market. It includes banking and finance, retail sales, hotel and restaurant work, and health care; it also includes parts of the information technology industry—not electronics assembly, but areas such as software design and the exchange of information (through the Internet, publishing, video production, and the like).

Comparing Economic Systems

The three major economic systems found in the world today are *capitalism, socialism,* and *communism.* These are not totally distinct, that is, many societies have a mix of these economic systems. **Capitalism** is an economic system based on the principles of market competition, private property, and the pursuit of profit. Within capitalist societies, stockholders own corporations—or a share of the corporation's wealth. Under capitalism, owners keep a surplus of what is generated by the economy; this is their *profit,* which may be in the form of money, financial assets, and other commodities.

Socialism is an economic institution characterized by state (government) ownership and management of the basic industries; that is, the means of production are the property of the state, not of individuals. Modern socialism emerged from the writings of Karl Marx, who predicted that capitalism would give way to egalitarian, state-dominated socialism, followed by a transition to stateless, classless communism.

Many European nations, for example, have strong elements of socialism that mix with the global forces of capitalism. Sweden supports an extensive array of state-run social services, such as health care, education, and social welfare programs, but Swedish industry is capitalist. Other world nations are more strongly socialist, although they are not immune from the penetrating influence of capitalism. The People's Republic of China was formerly a strongly socialist society that is currently undergoing great transformation to capitalist principles, including state encouragement of a market-based economy, the introduction of privately owned industries, and increased engagement in the international capitalist economy.

Communism is sometimes described as socialism in its purest form. In pure communism, industry is not the private property of owners. Instead, the state is the sole owner of the systems of production. Communist philosophy argues that capitalism is fundamentally unjust because powerful owners take more from laborers (and society) than they give, and use their power to maintain the inequalities between the worker and owner classes. Communist theorists in the nineteenth century declared that capitalism would inevitably be overthrown as workers worldwide united against owners and the system that exploited them. Class divisions were supposed to be erased at that time, along with private property and all forms of inequality. History has not borne out these predictions.

THE CHANGING GLOBAL ECONOMY

One of the most significant developments of modern times is the creation of a global economy, affecting work in the United States and worldwide. The concept of the **global economy** acknowledges that all dimensions of the economy now cross national borders, including investment, production, management, markets, labor, information, and technology (Altman 2001; Carnoy et al. 1993). Economic events in one nation now can have major reverberations throughout the world. When the economies of any major nation are unstable, the effects are felt worldwide.

Multinational corporations—those that draw a large share of their revenues from foreign investments and conduct business across national borders—have become increasingly powerful, spreading their influence around the globe. The global economy links the lives of millions of Americans to the experiences of other people throughout the world. You can see the internationalization of the economy in everyday life: Status symbols such as high-priced sneakers are manufactured for just a few cents in China. The Barbie dolls that young girls accumulate are inexpensive by U.S. standards, yet it would require one month's wages for the Indonesian or Chinese worker who makes the doll to buy it for her child.

In the global economy, the most developed countries control research and management, and assembly-line work is performed in nations with less privileged positions in the global economy. A single product, such as an automobile, may be assembled from parts made all over the world—the engine assembled in Mexico, tires manufactured in Malaysia, and electronic parts constructed in China. The relocation of manufacturing to wherever labor is cheap has led to the emergence of the **global assembly line**, a new international division of labor in which research and development is conducted in the United States, Japan, Germany, and other major world powers, and the assembly of goods is done primarily in underdeveloped and poor nations—mostly by women and children.

Related to the global assembly line is the phenomenon known as outsourcing. **Outsourcing** is the transfer of a specialized task from one organization to another that occurs for cost saving; often, the work is transferred to a different nation, as you have likely witnessed when calling someone for help with your computer. The person who answers may well be working in India or somewhere else and is part of an economy that is deeply entangled with that in the United States.

Within the United States, the development of a global economy has also created anxieties about foreign workers, particularly among the working class. Because it is easier to blame foreign workers for unemployment in the United States than it is to understand the complex processes that have produced this phenomenon, U.S. workers have been prone to **xenophobia**, the fear and hatred of foreigners. Campaigns to "buy American" reflect this trend, although the concept of buying American is becoming increasingly antiquated in a global economy.

When buying a product from a U.S. company, it is likely that the parts, if not the product itself, were built overseas. In a global economy, distinctions between U.S. and foreign businesses blur. Moreover, the label "Made in U.S.A." does not necessarily mean that the product was made by well-paid workers in the United States. In the garment industry, sweatshop workers, many of whom are recent immigrants and primarily women, are likely to have stitched the clothing that bears such a label. Moreover, these workers are likely to be working under exploitative conditions.

The development of a global economy is part of the broad process of **economic restructuring**, which refers to the contemporary transformations in the basic structure of work that are permanently altering the workplace. This process includes the changing composition of the workplace, deindustrialization, and use of enhanced technology. Some changes are *demographic*—that is, resulting from changes in the population. The labor force is becoming more diverse, with women and people of color becoming the majority of those employed. Other changes are driven by *technological developments*. For example, the economy is based less on its earlier manufacturing base and more on service industries.

thinking SOCIOLOGICALLY

Identify a job you once held (or currently hold) and make a list of all the ways that workers in this segment of the labor market are being affected by the various dimensions of *economic restructuring*: demographic changes, globalization of the economy, and technological change. What does your list tell you about how social structure shapes people's individual work experiences? ●

A More Diverse Workplace

A more diverse workplace is becoming a common result of economic restructuring. Today's workforce is both older, includes more women, and is more racially and ethnically diverse than ever before. People in the older age group (age 55 and older) have lower labor force participation rates than middle-age people, yet they comprise, at least for the time being, a somewhat larger percentage of workers than was true in the past—due in large part to the size of the baby boomer population.

As the U.S. population becomes more diverse, diversity in the labor market will continue. In fact, the White, non-Hispanic share of the labor market has fallen to 68 percent of the labor force in 2010, compared to 76 percent as recently as 1990. White, non-Hispanics in the labor market are expected to fall to 62 percent by the year 2020. Meanwhile, Asian and Hispanic workers—and all women—are expected to become an increased proportion of the labor market (Toossi 2012).

The workforce is also growing more slowly than in the past, even though the U.S. population is growing (Toossi 2012). These basic facts about population change in the workforce will shape the experience of generations to come.

thinking SOCIOLOGICALLY

Think about the labor market in the region where you live. What racial and ethnic groups have historically worked in various segments of this labor market, and how would you now describe the racial and ethnic *division of labor*? ●

These changes in the social organization of work and the economy are creating a more diverse labor force, but much of the growth in the economy is projected to be in service industries, where, for the better jobs, education and training are required. People without these skills will not be well positioned for success. Manufacturing industries, where racial minorities have in the past maintained a foothold on employment, are now in decline. New technologies and corporate layoffs have reduced the number of entry-level corporate jobs that recent college graduates have always used as a starting point for career mobility. Many college graduates are employed in jobs that do not require a college degree. College graduates, however, do still have higher earnings than those with less education.

Deindustrialization

Deindustrialization refers to the transition from a predominantly goods-producing economy to one based on the provision of services. This does not mean that goods are no longer produced, but that fewer workers in the United States are required to produce goods because machines can do the work people once did and because many goods-producing jobs have moved overseas.

Deindustrialization is most easily observed by looking at the decline in the number of jobs in the manufacturing sector of the U.S. economy since the Second World War. The manufacturing sector includes workers who actually produce goods. At the end of the war in 1945, the majority of workers (51 percent) in the United States were employed in manufacturing-based jobs. Now manufacturing accounts for only about 10 percent of the total labor force (U.S. Department of Labor 2012).

The *service sector* employs the other 90 percent, including two segments: the actual delivery of services (such as food preparation, cleaning, or child care) and the transmission and processing of information (such as banking and finance, computer operation, clerical work, and even education workers, such as teachers). Parts of the service sector are higher-wage and prestigious jobs, such as physicians, lawyers, financial professionals, and so forth, but huge parts of the service sector—and those with the largest occupational growth—are low-wage, semiskilled, and unskilled jobs. This lower end of the service sector employs many women, people of color, and immigrants.

The human cost of deindustrialization can be severe. Deindustrialization has led to **job displacement**, the permanent loss of certain job types that occurs when employment patterns shift. When a manufacturing plant shuts down, many people may lose their jobs at the same time, and whole communities can be affected. Among the areas hardest hit by deindustrialization are communities that were heavily dependent on a single industry, such as steel towns or automobile-manufacturing cities such as Detroit and Cleveland. But this also hits rural communities hard because those communities often have only one major employer, such as a textile plant.

Job displacement hits people in both rural and inner cities hard because emerging new industries tend to be located in suburban, not urban or rural, areas—a phenomenon called **spatial mismatch**. It is then no surprise that poverty is highest in central cities and rural America. This process has also been especially hard on young people, especially African American and Latino youth. Unless young people have the educational and technical skills for employment in a new economy, or if they live in areas hard hit by job displacement, they have little opportunity for getting a good start in the now global economy. You see this in the extremely high unemployment rates for Black and Hispanic teens (see Figure 15.3 on page 371; Wilson 2009, 1996).

Technological Change

Coupled with deindustrialization, rapidly changing and developing technologies are bringing major changes in work, including how it is organized, who does it, and how much it pays. One of the most influential technological developments of the twentieth century has been the invention of the semiconductor. Computer technology has made possible workplace transactions that would have seemed like science fiction just a few years ago. Electronic information can be transferred around the world in less than a second. Employees can provide work for corporations located on another continent; thus, a woman in Southeast Asia or the Caribbean can type a book manuscript for a publishing house in New York. Some argue that the computer chip has as much significance for social change as the earlier inventions of the wheel and the steam engine.

Increasing reliance on the rapid transmission of electronic data has produced *electronic sweatshops,* a term referring to the back offices found in many industries, such as airlines, insurance firms, and mail-order houses, where workers at computer terminals process thousands of transactions in a day. Workers' performance is likely monitored by a computer, conjuring up images of "Big Brother" invisibly watching. Computers can measure how fast cashiers ring up groceries and how fast ticket agents book reservations. Records derived from computer monitoring then become the basis for job performance evaluation.

Technological innovation in the workplace is a mixed blessing. **Automation**—the process by which human labor is replaced by machines—eliminates many repetitive and tiresome tasks, and it makes rapid communication and access to information possible. But our increasing dependence on technology may make workers subservient to machines. Robots can do the spot welding on automobiles; some are even used for human surgery. Sophisticated robots are capable of highly complex tasks, enabling them to assemble finished products or flip burgers in fast-food restaurants. Will robots replace human workers? Robots are expensive to buy, but employers can see other savings because "the robot hamburger-flipper would need no lunch or bathroom breaks, would not take sick days, and most certainly would neither strike nor quit" (Rosengarten 2000: 4).

Deskilling is a process whereby the level of skill required for performing certain jobs declines over time. Deskilling may result when a job is automated or when a more complex job is divided into a sequence of easily

Automation means that machines such as these auto assembly-line robots can now supply the labor that human workers originally provided.

DOING **sociological research**

Precarious Work: The Shifting Conditions of Work in Society

The American labor force has historically provided most people (though not all) with fairly steady work. Once in the labor market, a person could count on a relatively stable job over the course of a lifetime, often in the same company. Now that is more rare than common, resulting in a new phenomenon, labeled by sociologist Arne Kalleberg as "precarious work." Precarious work is defined as work that is uncertain, unpredictable, and risky from the point of view of a worker (Kalleberg 2009).

Research Question: What are the social conditions that have made work more precarious and made workers less secure in their employment?

Research Methods: Kalleberg has studied these questions using a macrolevel approach, drawing from secondary data sources, such as information from the U.S. Department of Labor, U.S. Department of Education, the General Social Survey, the Economic Policy Institute, and his own analyses of other published research studies.

Research Results: Kalleberg finds that the growth of precarious work produces increased stress for workers, thus also affecting families and personal relationships. He documents several reasons for this particular transformation in employment:

1. The expansion and institutionalization of nonstandard employment relations, such as temporary work and contract labor;
2. A general decline in job stability, meaning that people do not remain with the same employer over time;
3. An increasing tendency for employers to hire workers from outside of the work organization, rather than developing skills and talents from within;
4. Growth in involuntary job loss, especially among prime-age white men in white-collar occupations;
5. Growth in long-term unemployment;
6. A shift of risk from employers to employees, particularly in the decline of employee benefits;
7. Decline of unions and worker protections as employers have sought greater flexibility in management practices and the protection of profit margins.

Conclusions and Implications: Work is central to people's identity. As these various and multiple social, economic, and political forces have aligned, work has become more precarious and thus eroded people's sense of security. Kalleberg argues that new forms of work arrangements will be needed to ensure that employees, not just employers, have a commitment to the economy and society.

Questions to Consider

1. Is the work you plan to do precarious? Why or why not?
2. How do you see the social factors that Kalleberg identifies as influencing the work of people in your social networks?

Sources: Kalleberg, Arne L. 2009. "Precarious Work, Insecure Workers: Employment Relations in Transition." *American Sociological Review* 4 (February): 1–22; Kalleberg, Arne L. 2012. "The Social Contract in an Era of Precarious Work." *Pathways* (Fall): 3–6.

---

performed units. With deskilling, workers are paid less and have less control over their tasks. Jobs may become routine and boring. Deskilling contributes to polarization of the labor force. The best jobs require increasing levels of skill and technological knowledge, whereas people at the bottom of the occupational hierarchy may become stuck in dead-end positions and become alienated from their work (Apple 1991).

Along with deskilling has come an increasing reliance on temporary or **contingent workers** who do not hold regular jobs, but whose employment is dependent on demand. This includes contract workers, temporary workers, on-call workers (those called only when needed), the self-employed, and day laborers. Women are more likely than men to be employed in these jobs, and women are concentrated in the least desirable jobs—those with the lowest pay and least likelihood of providing benefits. Considering race, Whites are more likely to be independent contractors or self-employed, whereas Blacks and Hispanics are more likely to be found in temporary and part-time work (Cook 2000; Kalleberg et al. 2000; Hudson 1999).

Immigration

One of the most significant changes in the U.S. labor force has been the increased presence of immigrant labor. There are approximately 38 million foreign-born people in the United States—13 percent are U.S. residents. There are also about 10.5 million undocumented immigrants, so-called illegal immigrants. Immigrants constitute more than one-third of the labor force in fields such as building cleaning and maintenance; agriculture; meat, poultry, and fish production; and construction. Eighty-two percent of immigrant households have at least one worker present, more than is true for native-born households (where the figure is 73 percent).

Immigrants have far lower wages than native-born citizens, and they also experience higher rates of poverty. Although immigrants have been concentrated in particular states (California, Florida, Arizona, and Texas, among others), the recent trend has been that immigrants have moved into other geographic regions and smaller towns, thus transforming the character of hundreds of American communities (Camaroto 2007).

Immigration has not only changed the composition of the work force, but it has also stimulated intense political debate. Should the nation restrict immigration? Should a guest worker program be created? What rights do immigrants have? These issues engage different interest groups—including immigrants themselves and the organizations that support them, business leaders, nonimmigrant workers, and local communities where immigrants are employed. These interests often diverge, creating conflict and certainly creating questions for policymakers about the best way to handle immigration.

Popular wisdom holds that the bulk of new immigrants, particularly Hispanic immigrants, are illegal, poor, and desperate, but the facts show otherwise. In fact, the proportion of professionals and technicians among legal immigrants exceeds the proportion of professionals in the labor force as a whole. This conclusion, though, is based on formal immigration data that exclude undocumented immigrants, most of whom are working class. Still, those who migrate are usually not the most downtrodden in their home country; seldom are the poorest able to migrate. Even undocumented immigrants tend to have higher levels of education and occupational skills than the typical workers in their homeland. Immigrants include both the most educated and the least educated segments of the population.

---

debunking SOCIETY'S MYTHS

MYTH: Immigrants take jobs away from American citizens.
SOCIOLOGICAL PERSPECTIVE: Immigrants and native workers do not tend to compete for the same jobs; studies also find that the average U.S. workers' wages actually increase because of immigration (Peri and Sparber 2009; Cortes 2008). ●

Some European nations have created *guest worker programs* to provide the labor that immigrants provide. Guest worker programs allow immigrants to work in a nation for a limited period of time without fear of deportation, but they must return to their nation of origin at the end of their time limit. Critics argue that such programs threaten jobs for U.S. workers and allow employers to exploit this unequal class of workers. Supporters say guest worker programs allow for better regulation of immigration and avoid some of the problems of illegal immigration. It is unclear what direction the United States will take in immigration policy, but what is clear is that the economy is highly dependent on immigrant labor.

SOCIAL ORGANIZATION OF THE WORKPLACE

Most people think of work as an activity for which a person is paid, but work also includes the labor that people do without pay. Unpaid jobs such as housework, child care, and volunteer activities make up much of the work done in the world. Sociologists define **work** as productive human activity that creates something of value, either goods or services. Given this definition of work, housework, though unpaid, is defined as work, even though it is not included in the official measures of productivity used to indicate national work output.

Arlie Hochschild (1983) has introduced the concept of emotional labor to address some forms of work that are common in a service-based economy. **Emotional labor** is work specifically intended to produce a desired state of mind in a client and often involves putting on a false front before clients. Many jobs require some handling of other people's feelings, and emotional labor is performed where inducing or suppressing a feeling in the client is one of the primary work tasks. Airline flight attendants perform emotional labor—their job is to please the passenger and, as Hochschild suggests, to make passengers feel as though they are guests in someone's living room.

The Division of Labor

The **division of labor** is the systematic *interrelatedness* of different tasks that develop in complex societies. When different groups engage in different economic activities, a division of labor is said to exist. In a relatively simple division of labor, one group may be responsible for planting and harvesting crops, whereas another group is responsible for hunting game. As the economic system becomes more complex, the division of labor becomes more elaborate.

In the United States, the division of labor is affected by gender, race, class, and age—the major axes of stratification. The *class division of labor* can be observed by looking at the work done by people with different educational backgrounds, because education is a fairly reliable indicator of class. People with more education tend to work in higher-paid, higher-prestige occupations. Class also leads to perceived distinctions in the value of manual labor versus mental labor. Those presumed to be doing mental labor (management and professional positions) tend to be paid more and have more job prestige than those presumed to be doing manual labor. Class thus produces stereotypes about the working class; manual labor is presumed to be the inverse of mental labor, meaning it is presumed to require no thinking. By extension, workers who do manual labor may be incorrectly assumed not to be very smart, regardless of their intelligence.

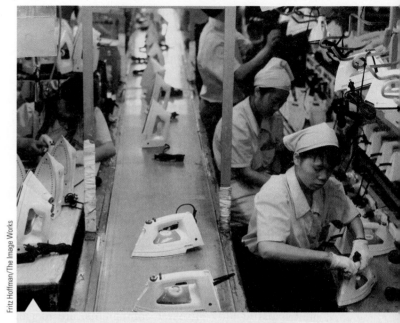

Race and gender segregation in the labor market mean that women of color are concentrated in occupations where most other workers are also women of color.

The *gender division of labor* refers to the different work that women and men do in society. In societies with a strong gender division of labor, the belief that some activities are women's work (for example, secretarial work) and other activities men's work (for example, construction) contributes greatly to the propagation of inequality between women and men, especially because cultural expectations usually place more value (both social and economic) on men's work. This helps explain why librarians and social workers are typically paid less than electricians despite the likelihood that women have higher education.

Also structuring the division of labor is race and ethnicity—the *racial–ethnic division of labor*. Even with civil rights and equal employment opportunity laws in place, race and ethnicity structure how people are distributed in jobs. Although racial–ethnic groups are distributed over a wide array of jobs, including as professional workers, there remains a concentration of people of color and immigrant labor in the lower-wage, low-status positions in the division of labor. This work is as necessary for society's survival as high-end jobs, but it is devalued and, often, underappreciated labor.

The **glass ceiling** is the term used to describe the limits to advancement that women, as well as racial–ethnic people and minorities, experience at work. Many barriers to the advancement of women and minorities have been removed, yet invisible barriers still persist. The existence of the glass ceiling is well documented. Although there has been an increase in the number of managers who are women and minorities, most top management jobs are still held by White men, who are far more likely than other groups to control a budget, participate in hiring and promotion, and have subordinates who report to them. Women and racial minorities remain clustered at the bottom of managerial hierarchies (see Figure 15.1).

The Occupational System and the Labor Market

Jobs are organized into an *occupational system*—the array of jobs that together constitute the labor market. Within the occupational system, people are distributed in patterns that reflect the race, class, and gender organization of society. Jobs vary in their economic rewards, their perceived value and prestige, and the opportunities they hold for advancement.

The Dual Labor Market. The labor market can be seen as comprising two major segments, the *primary labor market* and the *secondary labor market*, a phenomenon known as the **dual labor market** (see also Chapter 11). The primary labor market offers jobs with relatively high wages, benefits, stability, good working conditions, opportunity for promotion, job protection, and due process for workers (meaning workers are treated according to established rules and procedures that are allegedly fairly administered). Blue-collar and service workers in the primary labor market are often unionized, which leads to better wages and job benefits. High-level corporate jobs and unionized occupations fall into this segment of the labor market.

The secondary labor market is characterized by low wages, few benefits, high turnover, poor working conditions, little opportunity for advancement, no job protection, no retirement plan, and perhaps arbitrary

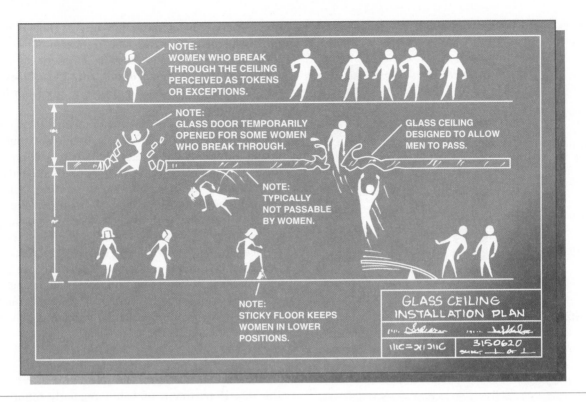

FIGURE 15.1 The Glass Ceiling Whimsically depicted as an architectural drawing by Norman Andersen, the *glass ceiling* refers to the structural obstacles still inhibiting upward mobility for women workers. Most employed women remain clustered in low-status, low-wage jobs that hold little chance for mobility. Those who make it to the top often report being blocked and frustrated by patterns of exclusion and gender stereotyping.

© Cengage Learning

treatment of workers. Many service jobs, such as waiting tables, nonunionized assembly work, and domestic work, are in the secondary labor market. Women and minority workers are the most likely groups to be employed in the secondary labor market. This particular structure in the labor market can explain much about race, gender, and class inequalities in work.

Occupational Distribution. Occupational distribution describes the pattern by which workers are located in the labor force. Workers are dispersed throughout the occupational system in patterns that vary greatly by race, class, and gender, revealing a certain **occupational segregation** on the basis of such characteristics. Women are most likely to work in technical, sales, and administrative support, primarily because of their heavy concentration in clerical work. This is now true for both White women and women of color. White men are most likely found in managerial and professional jobs, whereas African American and Hispanic men are most likely employed as operators and laborers—among the least well paid and least prestigious in the occupational system (Bureau of Labor Statistics 2011a).

Changes in occupational segregation are noticeable over time. For example, in 1960, 38 percent of all Black women were employed as private domestic

workers; by the 1990s, this had declined to less than 2 percent. Over time, there has also been some increase in the number of women employed in working-class jobs traditionally held only by men. Today, women are 12 percent of precision production, craft, and repair workers, compared with 1 percent in 1970 (U.S. Department of Labor 2012). This is a significant increase in the number of women in these jobs, although women are still a small proportion of all such workers.

Occupational Prestige and Earnings. *Occupational prestige* is the perceived social value of an occupation in the eyes of the general public (see Chapter 8). Sociologists have found a strong correlation between occupational prestige and the race and gender of people employed in given jobs. African American and Latino men are disproportionately found in jobs that have the lowest occupational prestige scores; White and Asian American men hold the jobs with the highest occupational prestige, followed by White women, Asian American women, African American women, and Latinas (Xu and Leffler 1992; Stearns and Coleman 1990). The gender composition of jobs and their occupational prestige is also linked: Jobs that employ mostly women are lower in prestige than those that employ more men. Indeed, jobs often lose their prestige as many women enter a given profession.

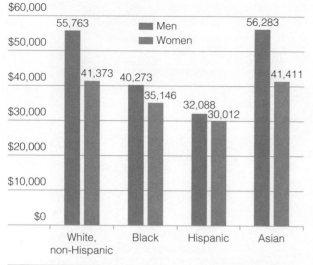

Median income for full-time workers

$60,000

$50,000

$40,000

$30,000

$20,000

$10,000

$0

■ Men
■ Women

55,763   41,373   40,273   35,146   32,088   30,012   56,283   41,411

White, non-Hispanic   Black   Hispanic   Asian

FIGURE 15.2 The Income Gap These median income figures are for full-time workers only. Does anything surprise you about what you see in the graph? What features of the labor market might explain some of these race and gender patterns?

Data: U.S. Census Bureau. 2012. Table PINC-01. *Selected Characteristics of People 15 Years and Over, by Total Money Income in 2011, Work Experience in 2011, Race, Hispanic Origin, and Sex.* Washington, DC: U.S. Department of Commerce. **www.census.gov**

Likewise, the prestige of jobs increases as more men enter the field.

Beyond what people think about jobs, what really matters is what the jobs pay. Sociologists have extensively documented that earnings from work are highly dependent on race, gender, and class, as shown in Figure 15.2. White men earn the most, with a gap between men's and women's earnings among all groups. African American women and Hispanic men and women earn the least. Occupations in which White men are the numeric majority tend to pay more than occupations in which women and minorities are a majority of the workers.

DIVERSE GROUPS/DIVERSE WORK EXPERIENCES

Data on characteristics of the U.S. labor force typically are drawn from official statistics reported by the U.S. Department of Labor. The labor force now includes approximately 139 million people (U.S. Department of Labor 2012). However, who works, where, and how varies considerably for different groups in the population.

One of the most dramatic changes in the labor force since the Second World War has been the increase in the number of women employed. Since 1948, the employment of women has increased from 35 to 60 percent of all women; women now constitute almost half (47 percent) of all workers. Other changes in the labor force include that racial minorities are the fastest

growing segment of the labor force. Although White, non-Hispanic workers are still 70 percent of the workforce, Hispanics, Asians, and African Americans are the fastest-growing segments of the labor force. Hispanics and Asians have the fastest growth in the labor force, largely because of immigration.

These trends, however, do not mean—as popularly believed—that minorities and women are routinely taking jobs from White men. Jobs where White men have predominated are precisely those that have been declining because of economic restructuring, such as in the manufacturing sector. It is jobs in areas of the labor market that are race- and gender-segregated that are increasing, such as fast-food work and other low-wage service jobs. The conflicts that exist about work—such as the belief immigrant and foreign workers are taking U.S. jobs or that women and minorities are taking away White men's jobs—stem from social structural transformations in the economy, not the individual behaviors of people affected by these changes.

Unemployment and Joblessness

The U.S. Department of Labor regularly reports the **unemployment rate**, defined as the percentage of those not working but officially defined as looking for work. At the end of 2012, the unemployment rate was 7.8—that is, 7.8 percent of the labor force. Full employment is considered to be around 4 percent, but unemployment has been high as the result of the economic recession that hit the nation in 2008. To date, the United States is a long way from recovering its prerecession unemployment rate.

The official unemployment rate does not include all people who are jobless. It includes only those who meet the official definition of unemployment—those who do not have a job and who have looked for work in the period being reported. Thus the official measure excludes people who earned money at any job during the time prior to the data being collected; it excludes people who have given up looking for full-time work (so-called *discouraged workers*); those who have settled for part-time work; people who are ill or disabled; those who cannot afford the child care, transportation, or other necessities for getting to work; and, those who work only a few hours a week even though their economic position may differ little from that of someone with no work at all. Migrant workers and other transient populations are also undercounted in the official statistics. Workers on strike are also counted as employed, even if they receive no income while they are on strike. Because so many people are excluded from the official definition of unemployment, the official unemployment rate seriously underestimates actual joblessness.

Moreover, the people most likely to be left out of the unemployment rate are those for whom unemployment runs the highest—the youngest and oldest workers, women, and racial minority groups. These groups are also those most likely to have left jobs that do not qualify them for unemployment insurance, because to be eligible for unemployment you have to have worked a certain period of time and earned enough wages to qualify for a claim.

Official unemployment rates also ignore **under-employment**—the condition of being employed at a skill level below what would be expected given a person's training, experience, or education. This condition can also include working fewer hours than desired. A laid-off autoworker flipping hamburgers at a fast-food restaurant is underemployed and so is a person with a law degree who drives a taxi.

Even given the problems in measuring unemployment, the highest rates of unemployment are among Black men, followed by Hispanic men (see Figure 15.3). When the government reports the national unemployment rate, it is a safe bet that unemployment among African Americans will be at least twice the national rate—a pattern that has persisted over time. Although not regularly reported by the Department of Labor, unemployment among Native Americans is also staggeringly high—the greatest barrier being the simple absence of jobs. For the one-quarter of Native Americans living on reservations there are few jobs available; those that do exist are often held by Whites (Snipp 1996). In urban areas, Native Americans face many of the same problems as urban African Americans and Hispanics. Were the unemployment rate that is found among Native Americans, African Americans, and Hispanics (with the exception of Cuban Americans) the national unemployment rate, it would be called a major economic depression.

People often attribute unemployment to the individual failings of workers, claiming that unemployed people do not try hard enough to find jobs or prefer a welfare check to hard work and a paycheck. This leads some to attribute unemployment to the "laziness" of unemployed individuals rather than to actual, factual structural conditions in society (Morlan 2005). This viewpoint reflects the common myth that *anyone* who works hard enough and puts forward sufficient effort can succeed.

As the economic recession has clearly shown, economic and structural conditions beyond an individual's control usually result in unemployment and joblessness. During the recession, hundreds of thousands of workers either lost jobs or found themselves in highly precarious working situations. In times like a major recession, it is easier to see how structural changes in the economy affect people at the individual level. (Recall the distinction between *troubles* and *issues* that C. Wright Mills made and that was examined in Chapter 1 of this book.)

Sociologists examine how changes in the social organization of the economy trickle into the day-to-day reality of people's lives. Although people do not ordinarily interpret their situations as shaped by processes such as deindustrialization, corporate downsizing, or spatial mismatch, the fact is that these structural conditions have enormous influence on how we live. Hence, unemployment is rarely caused by personality "laziness."

Sexual Harassment

Workplaces, like other social institutions, are shaped by power relationships among workers. One consequence for women workers is the possibility of sexual harassment. **Sexual harassment** is legally defined as unwanted physical or verbal sexual behavior that occurs in the context of a relationship of unequal power and that is experienced as a threat to the victim's job or educational activities (Saguy 2003). Sexual harassment is of two forms, according to the law. *Quid pro quo sexual harassment* forces sexual compliance in exchange for an employment or educational benefit.

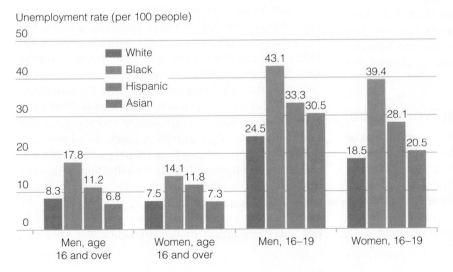

Unemployment rate (per 100 people)

Legend: White, Black, Hispanic, Asian

Men, age 16 and over: 8.3, 17.8, 11.2, 6.8
Women, age 16 and over: 7.5, 14.1, 11.8, 7.3
Men, 16–19: 24.5, 43.1, 33.3, 30.5
Women, 16–19: 18.5, 39.4, 28.1, 20.5

FIGURE 15.3 Unemployment by Race, Age, and Gender, 2011
Unemployment varies significantly among different groups. What patterns do you see depicted in this graph? What does this suggest about the need for new social policies?

Data: Bureau of Labor Statistics. 2012. *Employment and Income.* Washington, DC: U.S. Department of Labor. **www.bls.gov**

A professor who suggests to a student that going out on a date or having sex would improve the student's grade is engaging in quid pro quo sexual harassment. The other form of sexual harassment recognized by law is the creation of a *hostile working environment,* in which unwanted sexual behaviors are a continuing condition of work. This kind of sexual harassment may not involve outright sexual demands, but includes unwanted behaviors such as touching, teasing, sexual joking, and other kinds of sexual behavior and comment.

Sexual harassment was first made illegal by Title VII of the 1964 Civil Rights Act, which identifies sexual harassment as a form of sex discrimination. In 1986, the U.S. Supreme Court upheld the principle that sexual harassment violates federal laws against sex discrimination (*Meritor Savings Bank* v. *Vinson*). The law defines sexual harassment as discriminatory because it places workers at a disadvantage on the basis of their sex. Same-sex harassment also falls under the law, as does harassment directed by women against men. The law also makes employers liable for financial damages if they do not have policies appropriate for handling complaints or have not educated employees about their paths of redress.

Fundamentally, sexual harassment is an abuse of power when perpetrators use their position to exploit subordinates. The true extent of sexual harassment is difficult to estimate because it tends to be underreported. Surveys indicate though that as many as one-half of all employed women experience some form of sexual harassment at some time in their working lives. Men are sometimes the victims of sexual harassment, although far less frequently than women—about 3 percent of all cases. There is some evidence that women of color are more likely to be harassed than White women (Gruber 1982). Same-gender harassment also occurs; when men are harassed by other men, these men react more severely than do men who have been harassed by women (DuBois et al. 1998). Most studies find that typically neither women nor men are aware of the proper channels for reporting sexual harassment. Women are less likely than men to report sexual harassment, primarily because they believe that nothing will be done to stop the behavior (Andersen 2011; Cortina et al. 1998; VanRoosmalen and McDaniel 1998).

Gays and Lesbians in the Workplace

The increased willingness of lesbians and gay men to be open about their sexual identity has resulted in more attention paid to their experience in the workplace. Surveys find that a large majority of the U.S. population (89 percent) endorse the general concept that lesbians and gay men should have equal rights in job opportunities, but when asked about specific occupations, a significant proportion hold prejudices that gays should not be elementary school teachers (43 percent), high school teachers (36 percent), or clergy (47 percent) (Gallup Poll 2008).

Thus, although there is increasing acceptance of lesbians and gays generally, persistent homophobia can affect the self-esteem, productivity, and general well-being of gay and lesbian workers. Gays and lesbians often fear they will suffer adverse career consequences if heterosexual coworkers know they are gay. Shielding themselves from antagonism or rejection may make them appear distant and isolate them from social networks. These behaviors can actually have a negative effect on their performance reviews because fellow workers may find them unfriendly and withdrawn. Research finds, however, that the relationships of gay employees with their coworkers are less stressful when the employees are open about their identity (Schneider 1984). The work organization is also improved for lesbian, transgender, and gay employees when employee assistance programs encourage open communication, policies are sensitive to lesbian and gay needs, and discriminatory practices can be identified and stopped (Sussal 1994).

Disability and Work

Not too many years ago, people with disabilities were not thought of as a social group; rather, disability was thought of as an individual frailty or perhaps a stigma. Sociologist **Irving Zola** (1935–1994) was one of the first to suggest that people with disabilities face issues similar to those of minority groups. Instead of using a medical model that treats disability like a disease and sees individuals as impaired, conceptualizing people with disabilities as a minority group enabled people to think about the social, economic, and political environment that this population faces. Instead of seeing people with disabilities as pitiful victims, this approach emphasizes the group rights of the disabled, illuminating things such as access to employment and education (Zola 1993, 1989).

Now those with disabilities have the same legal protections afforded to other minority groups. Key to these rights is the Americans with Disabilities Act (ADA), adopted by Congress in 1990. Building on the Civil Rights Act of 1964 and earlier rehabilitation law, in particular the Rehabilitation Act of 1973, the Americans with Disabilities Act protects people with disabilities from discrimination in employment and stipulates that employers and other providers (such as schools and public transportation systems) must provide "reasonable accommodation." People with disabilities must be qualified for the jobs or activities for which they seek access, meaning that they must be able to perform the essential requirements of the job or program without accommodation to the disability. For students,

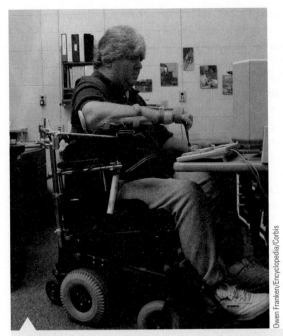

The Americans with Disabilities Act provides legal protection for workers with disabilities, giving them rights to reasonable accommodations by employers and access to education and jobs.

disabled, including in their job application, wages and benefits, advancement, and employer-sponsored social activities. The law applies to state and local governments, as well as employers. The ADA also legislates that public buses, trains, and light rail systems must be accessible to riders with disabilities; airlines are excluded from this requirement. The law also requires businesses and public accommodations to be accessible and requires telephone companies to provide services that allow people with speech or hearing impairments to communicate by telephone.

SOCIOLOGICAL THEORIES OF ECONOMY AND WORK

Sociological research continues to find that there are large differences in the experiences that different people have at work. Why? We have to turn to theory to answer such a question. The major theoretical perspectives identified in this book also provide the frameworks for the social structure of work. Each view-point—functionalism theory, conflict theory, and symbolic interaction—offers a unique analysis of work and the economic institution of which it is a part (see Table 15.1).

reasonable accommodation includes provision of adaptive technology, exam assistants, and accessible buildings.

The law prohibits employers with fifteen or more employees from discriminating against job applicants who are disabled or current employees who become

Functionalism

Functionalism interprets work and the economy as a functional necessity for society. Functionalists argue that society "sorts" people into occupations, with the more able sorted into prestigious occupations that pay more because they are more valuable—more

table 15.1 Theoretical Perspectives on Work

	Functionalism	Conflict Theory	Symbolic Interaction
Defines work as:	Functional for society because work teaches people the values of society and integrates people within the social order; more "talented" people rank higher	Generating class conflict because of the unequal rewards associated with different jobs	Organizing social bonds between people who interact within work settings
Views work organizations as:	Functionally integrated with other social institutions	Producing alienation, especially among those who perform repetitive tasks	Interactive systems within which people form relationships and create beliefs that define their relationships to others
Interprets changing work systems as:	An adaptation to social change	Based in tensions arising from power differences between different class, race, and gender groups	The result of the changing meanings of work resulting from changed social conditions
Explains wage inequality as:	Motivating people to work harder	Reflecting the devaluation of different classes of workers	Producing different perceptions of the value of different occupations

© Cengage Learning

"functional"—for society. Functionalist theory also explains that when society changes too rapidly, work institutions generate social disorganization—perhaps creating **alienation**, a feeling of powerlessness and separation from society (Chapter 6).

According to functionalist theories, workers are paid according to their value, which is derived from the characteristics they bring to the job: education, experience, training, and motivation to work. As we saw in Chapter 8, functionalist theorists see inequality as what motivates people to work. From this point of view, the high wages and other rewards associated with some jobs are the incentive for people to spend long years in training and garnering experience; otherwise, the jobs would go unfilled. To functionalists, then, differential wages are a source of motivation and a means to ensure that the most talented workers fill jobs essential to society and that different wages reflect the differently valued characteristics (education, years' experience, training, and so forth) that workers bring to a job.

Conflict Theory

Conflict theorists strongly disagree with the functionalist point of view, arguing that many talented people are thwarted by the systems of inequality they encounter in society. Thus, far from ensuring that the most talented will fill the most important jobs, conflict theorists see that some of the most essential jobs are, in fact, the most devalued and underrewarded. From a conflict perspective, wage inequality is one way that systems of race, class, and gender inequality are maintained. Factors such as the dual labor market, overt

race and gender discrimination, and persistent and unequal judgments about what work is appropriate for what groups shape the inequalities that are found in work.

Conflict theorists view the transformations taking place in the workplace as the result of inherent tensions in the social systems, tensions that arise from the power differences between groups vying for social and economic resources. Class conflict is then a major element of the social structure of work. Conflict theorists see the class division of labor as the source of unequal rewards for workers. Conflict theorists analyze the fact that some forms of work are more highly valued than others, both in how the work is perceived by society and how it is rewarded. As noted long ago by social-economic theorist **Thorstein Veblen** (1994/1899), mental labor has always been more highly valued than manual labor.

Symbolic Interaction Theory

Symbolic interaction theory brings a different perspective to the sociology of work. *Symbolic interaction theorists* would be interested in what work means to people and how social interactions in the workplace form social bonds. Some symbolic interaction studies examine how new workers learn their roles and how workers' identities are shaped by the social interactions in the workplace. Symbolic interaction theorists also note the creative ways that people deal with routinized jobs, perhaps performing exaggerated displays of routine tasks to humanize otherwise boring work (Leidner 1993).

POWER, POLITICS, AND THE STATE

The reach of government—and the politics associated with it—is extensive, indicating a sociological lesson you have learned throughout this book: Social structures that extend beyond our immediate day-to-day life influence the things we do and how we behave every day. Sometimes this becomes apparent to people; other times, it is less tangible, but the influence of social structure and social forces is everywhere, as you have seen in the preceding discussion of work and the economy.

Sociologists use the term **state** to refer to the organized system of power and authority in society. The state is an abstract concept that includes the institutions that represent official power in society, including the government, the legal system (law, courts, and the prison system), the police, and the military. Theoretically, the state exists to regulate social order, ranging from individual behavior and interpersonal conflicts to international affairs. Less powerful groups in the society may see the state more as an oppressive force than as a protector of individual rights. However,

The OccupyAmerica movement was a vivid reminder to the nation of the impact of the economic recession on the lives of millions.

Daryl Lang/Shutterstock.com

they may still turn to the state to rectify injustice, for example, by advocating for civil rights or seeking state-based rights and protections for people with disabilities.

The state has a central role in determining the rights and privileges of various groups. The state determines who is a citizen and who is not. A case in point is the ongoing battle in Arizona and other state legislatures over whether United States-born children of foreign-born and undocumented Hispanic parents are full-fledged citizens. Furthermore, the state may be called upon to resolve conflicts between management and labor (such as in airline strikes), and the state may pass legislation determining the benefits of different groups or make decisions that extend rights to various groups, such as the right for same-sex marriage.

Numerous institutions make up the state, including the government, the legal system, the police, and the military. The *government* creates laws and procedures that regulate and guide a society. The *military* is the branch of government responsible for defending the nation against domestic and foreign conflicts. The *court system* is designed to punish wrongdoers and adjudicate disputes. Court decisions also determine the guiding principles or laws of human interaction. *Law* is a fundamental type of formal social control that outlines what is permissible and what is forbidden. The *police* are responsible for enforcing law in the community and for maintaining public order. The *prison system* is the institution responsible for punishing those who have broken the law. Under the U.S. Constitution, these state institutions treat people equally, although sociologists have documented how often this is not the case (see Chapter 7).

Sociological analyses of the state focus on several different questions. One important issue is the relationship between the state and inequality in society. State policies can have very different impacts on different groups, as we will see later in this chapter. Another issue explored by sociological theory is the connection between the state and other social institutions—the state and religion or the state and the family.

The State and Social Order

Throughout this book, we have seen that a variety of social processes contribute to order in society, but none so explicitly and unambiguously as the official system of power and authority in society. In making laws, the state decrees which actions are or are not legitimate. Punishments for illegitimate actions are enforced, and systems for administering punishment are maintained. The state also influences public opinion through its power to regulate the media.

Some states, such as the United States, are **democracies**—that is, there is a representative government with elections by the population and, typically, a multiparty political system. Democracy can be compared to states that are **authoritarian**—that is, where power is concentrated in the hands of a very few individuals who rule through centralized power and control. An authoritarian state can become **totalitarian**, an extreme form of authoritarianism where the state has total control over all aspects of public and, to the extent possible, private life. Such was the case under Nazi Germany where the Nazi Party and Hitler, in particular, had absolute power. Saddam Hussein's regime in Iraq was a more contemporary example. Under such repressive states as authoritarian and totalitarian states, if there are elections, they are often corrupt. Such states are also likely to circulate **propaganda**, disseminated with the intention to justify the state's power. *Censorship* is another means by which the state can direct public opinion and try to enforce a singular way of thinking—that of the dominant group.

The state's role in maintaining public order is also apparent in how the state manages dissent. Protest movements perceived by those in power to challenge state authority or disrupt society may be repressed through state action, such as by surveillance, imprisonment, or police or military force. Even in democratic societies, social control can be exercised in multiple ways, whether through such means as the power to intercept email or shut down social media—or, in a more minor vein in the United States, increasing surveillance through, for example, the increasingly common cameras at traffic intersections.

Global Interdependence and the State

The global character of modern society means that political systems, along with economic systems, are now elaborately entangled—a phenomenon that can be observed daily in the newspaper. In late 2010, when massive youth-led pro-democracy demonstrations broke out in Tunisia (on December 17, 2010), demonstrations in Egypt quickly followed, resulting in the ouster of Egypt's then-dictator, Hosni Mubarak. Within days, similar demonstrations broke out in the neighboring countries of Algeria, Jordan, Yemen, Sudan, Libya, and now a strong resistance movement in Syria.

This world interdependence is also occurring at a time when there is increasing nationalism in some nations. **Nationalism** is the strong identity associated with an extreme sense of allegiance to one's culture or nation, often to the exclusion of interdependent relations with others. Nationalism can become a political movement, such as when groups subordinated by external nations use their original national culture as the basis for resisting oppression. It can also be a political movement when a group identifying itself as "a nation" (regardless of its official status as such) tries to become a dominant force in the world. Before its near-destruction by U.S. military forces, the Taliban in Afghanistan was an example of this.

POWER, AUTHORITY, AND BUREAUCRACY

The concepts of power and authority are central to sociological analyses of the state. **Power** is the ability of one person or group to exercise influence and control over others. The exercise of power can be seen in relationships ranging from the interaction of two people (husband and wife, police officer and suspect) to a nation (or social movement within a nation) threatening or dominating other nations. Sociologists are most interested in how power is structured in society: who has it, how it is used, and how it is built into institutions such as the state. In the United States, a society that is heavily stratified by race, class, and gender, power is structured into basic social institutions in ways that reflect these inequalities. Moreover, institutionalized power in society influences the social dynamics within individual and group relationships.

The exercise of power may be persuasive or coercive. For example, a strong political leader may persuade the nation to support a military invasion or a social policy through popular appeal. Or power may be exerted by sheer force. Generally speaking, groups with the greatest material resources have the greatest power, but not always. A group may by sheer size be able to exercise power, through, for example, organized social protests. But smaller groups may also be able to exercise power, such as in armed uprisings.

Power can be legitimate—accepted by the members of society as right and just—or it can be illegitimate. **Authority** is power perceived by others as legitimate and formal. Authority emerges not from the exercise of power, but from the belief of constituents that the power is legitimate. In the United States, the source of the president's domestic power is not just his status as commander of the armed forces but also the belief by most people that his power is legitimate. The law is also perceived by most as a legitimate system of authority. In contrast, *coercive power* is achieved through force, often against the will of the people being so forced. A dictatorship often relies on its ability to exercise coercive power through its control of the military or state police, at least until both the military and/or the police, sometimes in conjunction with a popular social movement or widespread demonstration, overthrow the dictator. This is precisely what happened early in 2011 with Egypt and the other countries that followed suit.

Types of Authority

Max Weber (1864–1920), the German classical sociologist, postulated that three types of authority exist in society: traditional, charismatic, and rational–legal (Weber 1978/1921). **Traditional authority** stems from long-established patterns that give certain people or groups legitimate power in society. A monarchy is an example of a traditional system of authority. Within a monarchy, kings and queens rule, not necessarily because of their appeal or because they have won elections, but because of long-standing traditions within the society.

Charismatic authority is derived from the personal appeal of a leader. Charismatic leaders are often believed to have special gifts, even magical powers, and their presumed personal attributes inspire devotion and obedience. Charismatic leaders often emerge from religious movements, but they come from other realms also. President Obama is for many a charismatic person and leader, admired not only for being the first African American president of the United States but also for his brilliance and his ability to inspire so many people, especially young people.

Rational–legal authority stems from rules and regulations, typically written down as laws, procedures, or codes of conduct. This is the most common form of authority in the contemporary United States. People obey not because national leaders are charismatic or because of social traditions, but because there is a legal system of authority established by formalized rules and regulations.

---

thinking SOCIOLOGICALLY

Observe the national evening news for one week, noting the people featured who have some kind of *authority*. List each of them and note their area of influence. What form of authority would you say each represents: *traditional*, *charismatic*, or *rational–legal*? How is the kind of authority that a person has reflected in his or her position in society (that is, race, class, gender, occupation, education, and so on)? ●

The Growth of Bureaucracies

According to Weber, rational–legal authority leads inevitably to the formation of bureaucracies. As we noted in Chapter 6, a **bureaucracy** is a formal organization characterized by an authority hierarchy, a clear division of labor, explicit rules, and impersonality. Bureaucratic power comes from the accepted legitimacy of the rules, not personal ties to individuals. The rules may change, but they do so through formal, bureaucratic procedures. People who work within bureaucracies are selected, trained, and promoted based on how well they apply the rules. Those who establish the rules are unlikely to be the same people who administer them. Bureaucracies are hierarchical, and the bureaucratic leadership may be quite remote. Power in bureaucracies is dispersed downward through the system to those who actually carry out the bureaucratic functions. It is an odd feature of bureaucracy that those with the least power to influence how the rules are formulated—those at the bottom of the hierarchy—are often the most adamant about strict adherence to the rules; their job evaluation may rest on their enforcement.

Within bureaucracies, personal temperament and individual discretion are not supposed to influence the application of rules. But bureaucracy has another face, as we saw in Chapter 6. Rank-and-file bureaucratic workers frequently exercise discretion in applying rules and procedures, "working the system," perhaps by personalizing the interaction or dodging bureaucratic stipulations. But most of the time, dealing with an elaborate bureaucracy—even an electronic one like voice mail—can be very frustrating.

THEORIES OF POWER

Does the state act in the interests of its different constituencies or does it merely reflect the needs of the most powerful? In other words, how is power exercised in society? This question has spawned much sociological study and debate and has resulted in several theoretical models of state power. Sociologists have developed four theoretical models to answer this question: the *pluralist model,* the *power elite model,* the *autonomous state model,* and *feminist theories of the state.* Each begins with a different set of assumptions and arrives at different conclusions (see Table 15.2).

The Pluralist Model

The **pluralist model** interprets power in society as derived from the representation of diverse interests of different groups in society. This model assumes that in democratic societies, the system of government works to balance the different interests of groups in society. An **interest group** can be any constituency in society organized to promote its own agenda, including large, nationally based groups such as the American Association of Retired Persons (AARP) and the National Rifle Association (NRA). Some interest groups are organized around professional and business interests, such as the American Medical Association (AMA) and the Tobacco Institute. Others concentrate on one political or social goal, such as NORML, working to reform marijuana laws. According to the pluralist model, interest groups achieve power and influence through their organized mobilization of concerned people and groups.

The pluralist model has its origins in functionalist theory. The pluralist model sees power as broadly diffused across the public with people who want to effect a change or express their points of view needing only to mobilize to do so. The pluralist model suggests that members of diverse groups can participate equally in a representative and democratic government. As seen from this model, various special interest groups compete for government attention and action. The pluralist model sees special interest groups as an

table 15.2 Theories of Power in Society

	Pluralism	Power Elite	Autonomous State	Feminist Theory
Interprets the state as:	Representing diverse and multiple groups in society	Representing the interests of a small, but economically dominant, class	Taking on a life of its own, perpetuating its own form and interests	Masculine in its organization and values (that is, based on rational principles and a patriarchal structure)
Interprets political power as:	Derived from the activities of interest groups and as broadly diffused throughout the public	Held by the ruling class	Residing in the organizational structure of state institutions	Emerging from the dominance of men over women
Interprets social conflict as:	The competition between diverse groups that mobilize to promote their interests	Stemming from the domination of elites over less powerful groups	Developing between states, as each vies to uphold its own interests	Resulting from the power men have over women
Interprets social order as:	The result of the equilibrium created by multiple groups balancing their interests	Coming from the interlocking directorates created by the linkages among those few people who control institutions	The result of administrative systems that work to maintain the status quo	Resulting from the patriarchal control that men have over social institutions

© Cengage Learning

The pluralist model of the state sees diverse interest groups as mobilizing to influence policy and gain political power.

committees has now grown to almost 4000. PACs have enormous influence on the political process. They are now so powerful that many people are critical of the influence they have on elections. In the 2012 presidential election, counting all PAC contributions, PACs spent over $546 million to influence voters, most of it to oppose candidates (www.latimes.com).

The Power Elite Model

The **power elite model** originated in the work of **Karl Marx** (1818–1883) and developed from the framework of conflict theory. According to Marx, the dominant or ruling class controls all the major institutions in society; the state itself is simply an instrument by which the ruling class exercises its power. The Marxist view of the state emphasizes the power of the upper class over the lower classes, the small group of elites over the rest of the population. The state, according to Marx, is not a representative, rational institution, but an expression of the will of the ruling class (Marx 1972/1843).

Marx's theory was elaborated much later by C. Wright Mills (1956), who popularized the term *power elite*. Mills attacked the pluralist model, arguing that the true power structure consists of people well positioned in three areas: the economy, the government, and the military. These three institutions are considered the bastions of the power elite, although some have argued that Mills overemphasized the role of the military (Domhoff 2002). While sharing common beliefs and goals, the power elite shape political agendas and outcomes in the society along the narrow lines of their particular collective interests.

The power elite model posits a strong link between government and business, a view supported by the strong hand government takes in directing the economy and by the role of military spending as a principal component of U.S. economic affairs. The power elite model also emphasizes how power overlaps between influential groups.

Interlocking directorates are organizational linkages created when the same people sit on the board of directors for numerous corporations. People in elite circles may serve on the boards of several major companies, universities, and foundations at the same time. People drawn from the same elite group receive most of the major government appointments; thus, the same relatively small group of people tends to the interests of all these organizations and the interests of the government. These interests naturally overlap and reinforce each other.

Members of the upper class do not need to occupy high office themselves to exert their will, as long as they are in a position to influence people who are in power (Domhoff 2002). The majority of the power elite are White men, which means that the interests and outlooks of White men dominate the national agenda.

integral part of the political system, even though they are not an official part of government. In the pluralist view, special interest groups make government more responsive to the needs and interests of different people, an especially important function in a highly diverse society.

The pluralist model helps explain the importance of **political action committees (PACs)**, groups of people who organize to support candidates they feel will represent their views. In 1974, Congress passed legislation enabling employees of companies, members of unions, professional groups, and trade associations to support political candidates with money they raise collectively. The number of political action

The Autonomous State Model

A third view of power developed by sociologists, the **autonomous state model**, interprets the state as its own major constituent. From this perspective, the state develops interests of its own, which it seeks to promote independently of other interests and the public that it allegedly serves. The state does not reflect the needs of the dominant groups, as Marx and power elite theorists would contend. It is an administrative organization with its own needs, such as maintenance of its complex bureaucracies and protection of its special privileges (Rueschmeyer and Skocpol 1996; Skocpol 1992; Evans et al. 1985).

The huge government apparatus now in place in the United States is a good illustration of autonomous state theory. The government provides a huge array of social support programs, including Social Security, unemployment benefits, agricultural subsidies, public assistance, and other economic interventions intended to protect citizens from the vagaries of a capitalist market system (Collins 1988). The purpose of these programs is to serve people in need. Autonomous state theory argues that the government has grown into a massive, elaborate bureaucracy, run by bureaucrats more absorbed in their own interests than in meeting the needs of the people. As a consequence, government can become paralyzed in conflicts between revenue-seeking state bureaucrats and those who must fund them. This can lead to revolt against the state, as in the tax revolts appearing sporadically throughout the country (Lo 1990; Collins 1988).

Feminist Theories of the State

Feminist theorists diverge from the preceding theoretical models by seeing men as having the most power in society. The pluralists see power as widely dispersed through the class system, power elite theorists see political power directly linked to upper-class interests, and autonomous state theorists see the state as relatively independent of class interests.

Some feminist theorists argue that all state institutions reflect men's interests; they see the state as fundamentally patriarchal, its organization embodying the fixed principle that men are more powerful than women. Feminist theories of the state conclude that despite the presence of a few powerful women, the state is devoted primarily to men's interests, and moreover, the actions of the state will tend to support gender inequality (Haney 1996; Blankenship 1993). One historical example would be laws denying women the right to own property once they married. Such laws protected men's interests at the expense of women.

Evidence that "the state is male" (MacKinnon 2006, 1983) is easy to observe by looking at powerful political circles. Despite the recent inclusion of more women in powerful circles and the presence of some notable women as major national figures, most of the powerful are men. The U.S. Senate is 80 percent men; groups that exercise state power, such as the police and military, are predominantly men. Moreover, these institutions are structured by values and systems that can be described as culturally masculine—that is, based on hierarchical relationships, aggression, and force. Feminist theory begins with the premise that an understanding of power cannot be sound without a strong analysis of gender (Haney 1996).

GOVERNMENT: POWER AND POLITICS IN A DIVERSE SOCIETY

The terms *government* and *state* are often used interchangeably. More precisely, the government is one of several institutions that make up the state. The **government** includes those institutions that represent the population, making rules that govern the society. The government of the United States is a *democracy*; therefore, it is based on the principle of representing all people through the right to vote.

The actual makeup of the government, however, is far from representative of society. Not all people participate equally in the workings of government, neither as elected officials nor as voters, nor do their interests receive equal attention. Women, the poor and working class, and racial–ethnic minorities are less likely to be represented by government than are White middle- and upper-class men. Sociological research on political power has concentrated on inequality in government affairs and demonstrated large, persistent differences in the political participation and representation of various groups in society.

Diverse Patterns of Political Participation

One would hope that all people in a democratic society would be equally eager to exercise their right to vote and be heard. That is far from the case. Among democratic nations, the United States has one of the *lowest* voter turnouts (see Figure 15.4). In the 2012 presidential election, the percentage of eligible voters who went to the polls was only 57 percent of the population, less than the all-time high of 62 percent in the 2008 presidential election. A turnout of 50 percent or less is more typical of U.S. national elections. Voter turnout in congressional and local elections is even lower.

Generally, older, better-educated, and financially better-off people are the most likely to vote. One of the biggest changes in voting is the change stemming from diversity within the U.S. population. Historically, racial–ethnic minority groups have had lower voter turnout

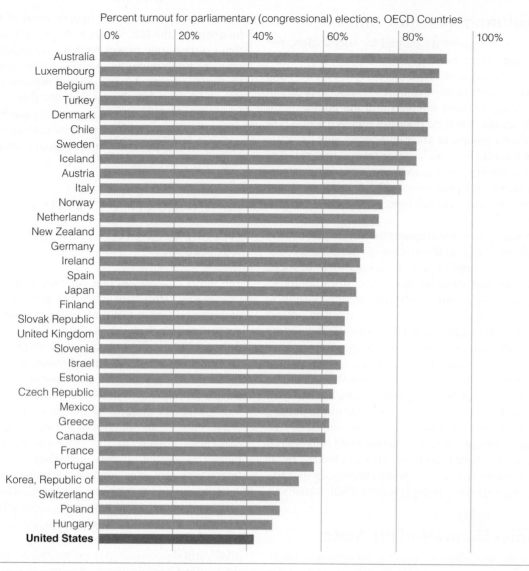

Percent turnout for parliamentary (congressional) elections, OECD Countries

Australia
Luxembourg
Belgium
Turkey
Denmark
Chile
Sweden
Iceland
Austria
Italy
Norway
Netherlands
New Zealand
Germany
Ireland
Spain
Japan
Finland
Slovak Republic
United Kingdom
Slovenia
Israel
Estonia
Czech Republic
Mexico
Greece
Canada
France
Portugal
Korea, Republic of
Switzerland
Poland
Hungary
United States

FIGURE 15.4 International Voter Turnout, 2009–2011 As you can see, the United States has lower voter turnout in national elections that other industrialized nations.

Source: Institute for Democracy and Electoral Assistance. 2012. "Voter Turnout." Stockholm: Sweden. **www.idea.it/vt**

than White Americans. But in the 2012 presidential election, more African Americans, Hispanics, and Asian Americans voted than ever before. In fact, African American turnout was 13 percent of all votes cast in the 2012 presidential election—more than their representation in the total U.S. population (see Figure 15.5; Taylor, P. 2012). The White share of eligible voters has been falling for some time, but is now becoming so low that candidates have to appeal to more diverse groups if they expect to be elected.

Not only do social factors influence the likelihood of voting, but they also influence how people vote. African Americans, Asian Americans, and Latinos, with the exception of Cuban Americans, tend to be markedly Democratic. Fully 93 percent of Black Americans (98 percent of Black women) voted for Obama in 2012, as did 73 percent of Asian Americans and 71 percent of Hispanics (www.nytimes.com).

Gender, income, education, and religion also affect voter behavior. The **gender gap** refers to the differences between men and women in political attitudes and behavior. Women tend to have more liberal views than men on a variety of social and political issues and are more likely to vote Democratic. And women are now more likely to vote than men. Although women tend to be more liberal than men, the recent political rise of the conservative Tea Party movement has witnessed still further participation of women in politics.

Political Power: Who's in Charge?

A democratic government is supposed to be representative of the people in the nation. Is this the case in the United States? Hardly. Women and racial-ethnic minorities are vastly underrepresented in our government. Moreover, most members are from

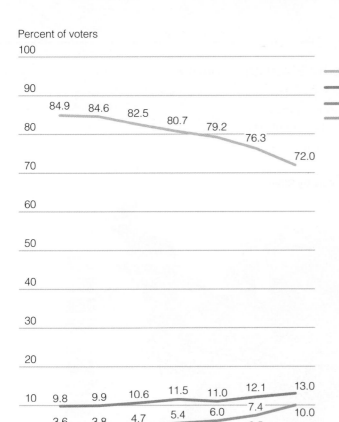

Percent of voters

	1988	1992	1996	2000	2004	2008	2012
White	84.9	84.6	82.5	80.7	79.2	76.3	72.0
Black	9.8	9.9	10.6	11.5	11.0	12.1	13.0
Hispanic	3.6	3.8	4.7	5.4	6.0	7.4	10.0
Asian		1.2	1.6	1.8	2.3	2.5	3.0

FIGURE 15.5 The Changing Electorate? Demographic Composition of Voters by Race and Ethnicity, 1988 to 2012

Source: Lopez, Mark 2009. "Dissecting the 2008 Electorate: Most Diverse in History." Pew Research Center. **www.pewresearch.com**; Edison Research.

Note: In 2012, 2 percent of votes identified as "other;" thus, numbers do not add up to 100 percent. Whites include only non-Hispanic Whites. Blacks include only non-Hispanic Blacks. Asians include only non-Hispanic Asians. Native Americans and mixed-race groups are not shown. Asian share not available prior to 1992.

upper-middle-class or upper-class back-grounds; the vast majority have law, politics, and business as their prior occupation; very few were blue-collar workers before coming to Congress.

Simply getting into politics requires a substantial investment of money. The total cost of the 2012 elections was a record $5.8 billion dollars! The two presidential candidates (Barack Obama and Mitt Romney) together spent $1.3 billion. Even running for the House of Representatives is likely to cost you about a million dollars, twice what it was just a few years ago (Center for Responsive Politics 2012; Bimbaum 2004).

Candidates depend on contributions from individuals and groups to finance their election campaigns, with wealthy individuals among the largest campaign contributors, especially to presidential elections (Center for Responsive Politics 2012). The largest contributors to political campaigns are typically PACs. Much of the money given by individuals and PACs goes to incumbents, who have an overwhelming edge in elections and already sit on the committees where public policy is hammered out. This picture of elites and business interests funneling money to candidates, who return to the same donors for more money when the next campaign rolls around, has shaken the faith of many Americans in the political system.

There is little belief that the political process is a democratic and populist mechanism by which the "little people" can select political leaders to represent them. In fact, national surveys in 2012 showed that only 13 percent of U.S. citizens have a great deal of confidence in Congress, compared to 37 percent with a great deal of confidence in the Supreme Court and in the presidency (Jones 2012).

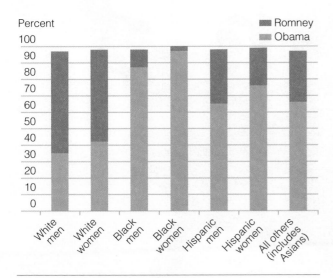

FIGURE 15.6 Who Voted How? Voters by Gender and Race, 2012 Presidential Election

Source: Edison Research, CNN Exit Polls.

MAP 15.1 AND 15.2

Mapping America's Diversity: Electoral Vote by State and County

The electoral vote is usually only reported state by state (blue for Democrats, red for Republicans), as you can see in the top map. But, as you can see in bottom map, if you shade the outcome by proportion of the vote at a county-by-county level, you get a somewhat different picture of the U.S. electorate.

Source: Top map © Cengage Learning; Lower map from Professor Mark Newman, University of Michigan

2012 Electoral Vote by State

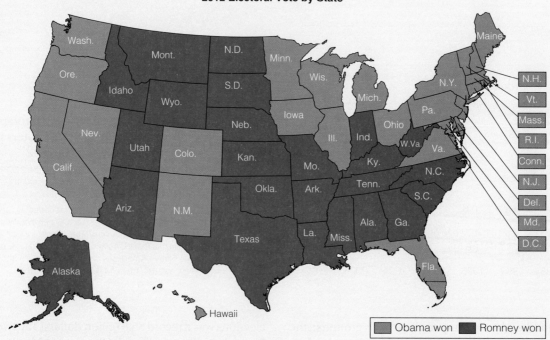

Obama won | Romney won

2012 Electoral Vote by County

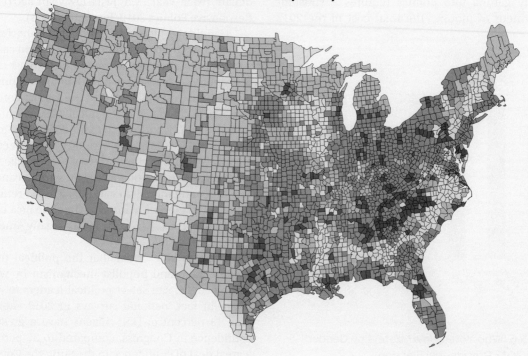

The Tea Party and the American Dream

Contemporary American politics have in recent years been shaped by the rise of a new movement: the Tea Party. Who is in it? What do they want? What has produced this new political phenomenon? In the short time since the emergence of the Tea Party, sociologists have studied it—its members, its funders, and its social origins. Sociologists thus provide perspective on this new phenomenon beyond what you will see in the mass media.

The Tea Party membership is largely comprised of middle-class and upper-middle-class people who are overwhelmingly White, and also includes large numbers of women. Indeed, many of the movement's heroes are women—deeply conservative women such as Sarah Palin, Michelle Bachmann, and others, many of whom embrace the language of feminism—at least in terms of the empowerment of women—even while they oppose many of the goals of the feminist movement.

Sociologists have found that Tea Partiers are frustrated by economic and political systems that they see as having abandoned them. They think that the government is distant from the needs of people like them. Much opposed to government spending, Tea Partiers see themselves as the taxpayers whose money is being drained off to support the lives of others. Although most support government programs like Social Security and Medicare, they are strongly opposed to other so-called "entitlement" programs, especially food stamps, housing subsidies, and Pell grants. They tend to believe in a nostalgic past where people could achieve the American dream through hard work and self-sufficiency and where they perceive limited government. They believe there is too much government waste and are critical of an expanded "welfare state." In sum, they see the American dream as broken and they want it restored, but not through government action.

Sociological research on Tea Party members has found that they tend to be more authoritarian than the general public—that is, they believe in strict discipline and resent the free rein that they believe characterizes contemporary youth. Curiously, although being more authoritarian, they are also strongly libertarian—meaning (consistent with their antigovernment stance) they oppose state-based restrictions on individual behavior.

One of the interesting things about the Tea Party is that this is an activist movement by groups who have been more complacent politically in the past. What has mobilized them? Certainly, a perceived sense of disenfranchisement is a strong motivation for political action. But researchers have found that, although this is a grassroots movement, it is heavily funded by extremely wealthy individuals who support the right-wing, antigovernment agenda of the Tea Party. The mobilization of the Tea Party has also been fueled by the attention and support given to the movement by the conservative media, including FOX-TV and various right-wing blogs and other media outlets. Tea Party members tend not to be critical of the very top; instead, they focus their ire on those perceived as benefiting from government largesse. In this sense, regardless of the individual reasons why someone might become active in the Tea Party, the movement is providing support for pro-business, antigovernment elites—a long-standing position of right-wing movements and organizations.

Sources: Drier, Peter. 2012. "The Battle for the Republican Soul: Who Is Drinking the Tea Party?" *Contemporary Sociology* 41 (November): 756–762; Fetner, Tina. 2012. "The Tea Party: Manufactured Dissent or Complex Social Movement?" *Contemporary Sociology* 41 (November): 762–766; McVeigh, Ricky. 2012. "Making Sense of the Tea Party." *Contemporary Sociology* 41 (November): 766–769; Skocpol, Theda, and Vanessa Williamson. 2012. *The Tea Party and the Remaking of Republican Conservatism*. New York: Oxford University Press; Stein, Arlene, and Marcy Westerling. 2012. "The Politics of Broken Dreams." *Contexts* 11 (Summer): 8–10; Braunstein, Ruth. 2011. "Who Are 'We the People'"? *Contexts* 10 (May): 72–73; DiMaggio, Anthony. 2011. *The Rise of the Tea Party: Political Discontent and Corporate Media in the Age of Obama*. New York: Monthly Review Press.

The Tea Party has galvanized a new conservative movement in U.S. politics.

Diversity in the Power Elite

As society has become more diverse, has it made a difference in the makeup of the power elite? Various groups—women, racial–ethnic groups, lesbians, and gays—have vied for more representation in the halls of power, but have their efforts succeeded? If they make it to power, does this change the corporations, military, or government—the major institutions composing the power elite?

Sociologists Richard L. Zweigenhaft and G. William Domhoff examined these questions by analyzing the composition of boards of directors and chief executive officers (CEOs) of the largest banks and corporations in the United States, as well as analyzing Congress, presidential cabinets, and the generals and admirals who form the military elite. In addition, they examined the political party preferences and the political positions of people found among the power elite. Do women and minorities bring new values into power, thereby changing society as

they move into powerful positions, or do their values match those of the traditional power elite or become absorbed by a system more powerful than they are? Zweigenhaft and Domhoff's study looks as well at whether those who do make it into the power elite are within the innermost circles or whether they are marginalized.

They find that women, Jews, gays, lesbians, Black Americans, and Hispanics have become more numerous within the power elite, but only to a small degree. The power elite is still overwhelmingly White, wealthy, Christian, and male. Women and other minorities who make it into the power elite also tend to come from already privileged backgrounds, as measured by their social class and education. Among African Americans and Latinos, skin color continues to make a difference, with darker-skinned Blacks and Hispanics less likely to achieve prominence compared with lighter-skinned

people. Furthermore, Zweigenhaft and Domhoff find that the perspectives and values of women and minorities who rise to the top do not differ substantially from their White male counterparts. Some of this is explained by the common class origins of those in the power elite. The researchers also attribute the managing of one's identity to avoid challenging the system as a sorting factor that perpetuates the dominant worldview and practices of the most powerful.

The authors of this study conclude that "the irony of diversity" is that greater diversity may have strengthened the position of the power elite because its members appear to be more legitimate through their inclusion of those previously left out. But, by including only those who share the perspectives and values of those already in power, little is actually changed.

Sources: Zweigenhaft, Richard L, and G. William Domhoff. 2006. *Diversity in the Power Elite: Have Women and Minorities Reached the Top?* New Haven: Yale University Press; Domhoff, G. William. 2002. *Who Rules America?* New York: McGraw-Hill.

Women and Minorities in Government

Although there have been some gains in the number of women and minorities in government, they are still underrepresented—both at the federal and state levels. As a result of the 2012 elections, a record 20 women are now in the U.S. Senate (out of 100) and 81 in the House of Representatives (out of 435). In the Congress, there are now 43 African Americans (one in the Senate); 31 Latinos (three in the Senate); 12 Asian Americans; and seven openly gay or bisexual members, including the first openly lesbian member of the Senate.

There has been an increase in religious diversity in the Congress, but the vast majority of senators and representatives are Protestant (56 percent) and Catholic (31 percent), although the Congress includes people of different faiths, including Jewish (6 percent) and small numbers of various other religious backgrounds, such as Muslims, Hindus, Buddhist, and one self-identified atheist, among others (Pew Research Center 2012). Although the 113th Congress is the most diverse in U.S. history, there is a long way to go before Congress truly represents the diversity in the population.

Researchers offer several explanations for why women and racial–ethnic minorities continue to be underrepresented in government. Certainly, prejudice plays a role. It was not long ago, in the 1960 Kennedy-Nixon election, that Kennedy became the first Catholic president elected. As recently as the 2000 presidential election, Joseph Lieberman was the first Jewish candidate to appear on a major national ticket. When Mitt Romney ran in 2012, he was the first Mormon candidate to appear on a national ballot; 18 percent of the public said they would not vote for a Mormon as president, even if qualified (Newport 2012b). And we have witnessed much prejudice directed toward President Obama on the false assumption that he is "Muslim" and false accusations of not being a U.S. citizen, although he was born in Hawaii.

Gender and racial prejudice run just as deep in the public mind. Although the percentage of Americans who now say they would vote for a woman for president has climbed to 92 percent (53 percent in 1969), a substantial number (42 percent) also say they think a man would make a better president than a woman (Simmons 2001).

Individual prejudice alone, however, cannot account for the lack of representation. Societal causes are a major

MAP 15.3

Viewing Society in Global Perspective: Women Heads of State

Does it surprise you that the United States fares poorly, relative to much of the rest of the world, when it comes to women's political leadership at the national level? What do you think explains this difference around the world?

Source: Council of Women World Leaders; Center for Asia-Pacific Women in Politics, guide2womenleaders.com. Martin K.I. Christensen.

Countries That Have Ever Had a Female Head of State or Government

Head of state

Head of government

Head of state and government

factor in the successful elections of women and people who are better represented in local political office. Women and minority candidates receive a great deal of political support from local groups, but at the national level, they do not fare as well. The power of incumbents, most of whom are White men, is also a disadvantage to any new office seeker.

THE MILITARY AS A SOCIAL INSTITUTION

Social institutions are stable systems of norms and values that fulfill certain functions in society. The military is a social institution whose function is to defend the nation against external (and sometimes internal) threats. A strong military is often considered an essential tool for maintaining peace. The military arm of the state is among the most powerful and influential social institutions in almost all societies. In the United States, the military is the largest single employer. Approximately 3.6 million men and women serve in the U.S. military, one million on active duty and the rest in the reserves. This does not include the many hundreds of thousands who are employed in industries that support the military, nor does it include civilians who work for the Department of Defense and other military-affiliated agencies (U.S. Census Bureau 2012a).

The military is one of the most hierarchical social institutions, and its hierarchy is extremely formalized. People who join the military are explicitly labeled

with rank, and if promoted, they pass through a series of additional well-defined levels (ranks), each with clearly demarcated sets of rights and responsibilities. An explicit line exists between officers (lieutenants and higher ranks) and enlisted personnel, and officers have many privileges that others do not. Higher ranks are also entitled to absolute obedience from the ranks below them, with elaborate rituals created to remind both dominants and subordinates of their status.

As in other social institutions, military enlistees are carefully socialized to learn the norms of the culture they have joined. Military socialization places a high premium on conformity and eliminates individuality. All new recruits have their heads shaved, are issued identical uniforms, are allowed to retain very few of their personal possessions, and they are endlessly harassed by the infamous "D.I." (drill instructor). They must quickly learn new, strictly enforced codes of behavior.

Most of the military is a part of the institution of government, but there has also been *privatization of the military*—meaning that an increasing number of military functions have been paid on a contract basis to private, for-profit employers. Under this development, the military becomes like a business, with people and corporations reaping profits on activities that once would have been not for profit. The privatization of the military can include companies that provide specific services (such as security), as well as engineering and building contracts. Critics of this trend warn that it will sacrifice safety and national security for the sake of corporate and individual profit and could lure the brightest people away from traditional military service if they see economic gains from private military service (Singer 2007).

Race and the Military

The greatest change in the military as a social institution is the representation of racial minority groups and women within the armed forces. Picture a U.S. soldier. Whom do you see? At one time, you would have almost certainly pictured a young White male, possibly wearing army green camouflage and carrying a weapon. Today, the image of the military is much more diverse. Drawing on the cultural images you have stored in your mind, you are just as likely to picture a young African American man in a military dress uniform with a stiffly starched shirt and a neat and trim appearance or perhaps a woman wearing a flight helmet in the cockpit of a fighter plane.

African Americans have served in the military for almost as long as the U.S. armed forces have been in existence. Except for the Marines, which desegregated in 1942, the armed forces were officially segregated until 1948, when President Harry Truman signed an executive order banning discrimination in the armed services. Although much segregation continued after this order, the desegregation of the armed forces is often credited with promoting more positive interracial relationships and increased awareness among Black Americans of their right to equal opportunities than has been the case in society at large. Until that time, the widespread opinion among Whites was that to allow Black and White soldiers to serve side by side would destroy soldiers' morale.

Currently, 17 percent of active military personnel are African American and 11 percent are Hispanic (who fall into various other racial–ethnic categories); almost 4 percent are Asian Americans; 1.7 percent are American Indian or Alaskan Native; and 2 percent identified as multiracial. A recent trend has been the high rate of enlistment among African American women. Enlistees have many reasons to join the military, but the desire for education and job training is certainly among the strongest motivators, along with wanting to serve one's country (U.S. Department of Defense 2010).

Within the military today, there is a policy of equal pay for equal rank. African Americans and Latinos, however, are overrepresented in lower-ranking support positions. Often, they are excluded from the higher-status, technologically based positions—those most likely to bring advancement and higher earnings both in the military and beyond. Most minorities remain in positions with little supervisory responsibility, such as service and supply jobs. Although the number of racial minorities in officers' positions has been increasing, they are still underrepresented and are less likely to get there via the route of military academies, as is the case

for White officers (Segal and Segal 2004). Still, for both Whites and racial minorities, serving in the military leads to higher earnings relative to one's nonmilitary peers.

Women in the Military

The military academies did not open their enrollment to women until 1976. Since then, the armed services have profoundly changed their admission policies, and in 1996, the Supreme Court ruled (in *United States v. Virginia*) that women cannot be excluded from state-supported military academies such as the Citadel and the Virginia Military Institute (VMI). This was a landmark decision that opened new opportunities for women who want the rigorous physical and academic training that military academies provide (Kimmel 2000).

The involvement of women in the military has reached an all-time high in recent years. Women are 14 percent of enlisted military personnel. Now, almost 204,000 women are on active duty in the Untied States. The Air Force has the highest proportion of women (22 percent), followed by the Navy (19 percent), the Army (15 percent), and the Marines (7 percent; U.S. Department of Defense 2010). Women now comprise one-third of entering students in the Coast Guard academy.

The former exclusion of women from military service was rationalized by the popular conviction that women should not serve in combat. Despite this attitude, women have been fighting in active combat to defend the nation and were officially made eligible for combat role sin 2013.

The presence of women in the military has transformed the armed forces, but it also has raised new issues for military personnel. Fully half of military personnel are married, and there has been an increase in the number of dual-military couples (7 percent of active-duty members are in dual-military marriages). Family separations, frequent moves (on average every three years), risk of injury or death, and living in a foreign country are only some of the challenges that military personnel face in trying to manage their lives (U.S. Department of Defense 2010; Segal and Segal 2004).

For women in the military, the highly gendered organization of which they are a part is also a challenge. Indeed, recent reports have documented an alarmingly high rate of sexual assault and sexual harassment against women in the military—by other military personnel. A recent Pentagon report found that one-third of women in the military (and 6 percent of men) experienced sexual harassment, including unwanted crude and offensive behavior, unwanted sexual attention, and sexual coercion; the same report found that 5 percent of women in the military experienced some form of unwanted sexual contact, such as rape, unwanted sodomy, or indecent assault (U.S. Department of Defense 2012). Periodic scandals involving rape, sexual harassment, and other forms of

The men and women who serve in the armed forces, such as this young woman returning from Iraq, are often separated from families and loved ones for long periods of time.

intimidation against women in the military (including the military academies) reveal that, although certainly not all military men engage in these behaviors, institutions organized around such masculine characteristics as aggression, domination, and hierarchy put women at risk.

Gays and Lesbians in the Military

Gays and lesbians have long served in military duty, despite the policies that have attempted to exclude them. The military has admitted that there always have been gays and lesbians in all branches of the U.S. armed forces, but homophobia is a pervasive part of military culture (Becker 2000; Myers 2000).

President Clinton in the early 1990s introduced a "don't ask, don't tell" military policy, now repealed, by which recruiting officers could not ask about sexual preference. The Obama administration in the fall of 2010 ended the "don't ask, don't tell" policy. However, it remains unclear whether gays and lesbians will be permitted to live openly as gay while also pursuing careers in the armed services. Supporters of the ban on gays in the military often use arguments similar to the arguments used before 1948 to defend the racial segregation of fighting units. As in 1938, they claim that the morale of soldiers will drop if forced to serve alongside gay men and women, national security will be threatened,

and known homosexuals serving in the military will upset the status quo and destroy the fighting spirit of military units.

Military Veterans

Now, almost two million veterans have returned from the wars in Iraq and Afghanistan. Add to that the veterans of the Gulf War, Vietnam War, Korean War, and the living veterans of World War II, and it totals 22 million veterans living in the United States (U.S. Census Bureau 2012a). For all veterans, the return home—though joyful—also has risks, risks that result from social, as well as physical, needs for recovery and adaptation.

The changed nature of combat in the two Iraq and Afghanistan wars has meant that returning veterans have more complex forms of physical and emotional injury. Exposure to repeated blasts of IEDs (improvised explosive devices) has resulted in more traumatic brain injuries. Having an all-volunteer army has also produced more frequent redeployment, resulting in longer-term exposure to war trauma, as well as greater exposure to blasts and other forms of violence. And changes in military and medical technology have also increased survival rates from injuries that would have killed military personnel in the past.

All of these factors have meant increased risks for returning veterans, including not only difficult recoveries from physical injuries, but also high rates of mental health disorders, a high risk of suicide, depression, and/or drug and alcohol addiction. In addition to these social problems, veterans face various adaptation challenges as they transition back into the civilian workforce—that is, if they find work. Veterans and their families and partners also have to adapt to new family roles, perhaps even including a readjustment as parents because their children will have matured in their absence. Adding to this complexity is the fact that there may be a significant readjustment to a new physical or mental disability. Managing the health problems that may have developed during deployment produces new forms of stress on preexisting relationships (Institute of Medicine of the National Academies 2010).

When veterans return home, often the social supports they need are not strong. One consequence has been an increase in the number of homeless veterans— a figure that has doubled since 2010 (Zoroya 2012). And African American veterans, for whom the military has been a path for social mobility, may face the additional fact that social institutions fail them again in the form of unemployment and persistent racism (Fleury-Steiner 2011).

The situation for U.S. veterans shows how critical social institutions are in the lives of these men and women. But for all members of society, the support—or lack thereof—provided by social institutions is a critical backdrop to the character of everyday life.

chapter summary

How are societies economically organized?

Societies are organized around an economic base. The *economy* is the system on which the production, distribution, and consumption of goods and services are based. *Capitalism* is an economic system based on the pursuit of profit, market competition, and private property. *Socialism* is characterized by state ownership of industry; *communism* is the purest form of socialism.

How has the global economy changed?

As capitalism has spread throughout the world, *multinational corporations* conduct business across national borders. A number of countries have undergone *deindustrialization,* or changeover from a goods-producing economy to a services-producing one. This has caused many heavy-industry jobs in U.S. cities to vanish, thus increasing the unemployment rate in those cities. Changes in information technology, plus increased *automation,* have resulted in the further elimination of jobs in both the United States and abroad.

What is the social organization of work?

Sociologists define *work* as human activity that produces something of value. Some work is judged to be more valuable than other work. *Emotional labor* is work that is intended to produce a desired state of mind in a client. The *division of labor* is the differentiation of work roles in a social system. In the United States, there is a class, gender, and racial division of labor. The labor market in the United States is described as a *dual labor market.* Jobs in the primary sector of the labor market carry better wages and working conditions, whereas those in the secondary labor market pay less and have fewer job benefits. Women and minorities are disproportionately employed in the secondary labor market. Patterns of occupational distribution also show tremendous segregation by race and gender in the labor market. Race and gender also affect the occupational prestige, as well as the earnings, of given jobs.

How is diversity reflected in the workplace?

The workplace is becoming more diverse with greater numbers of racial–ethnic groups, women, and an older workforce. Official *unemployment rates* underestimate the actual extent of joblessness. Women and minorities often encounter the *glass ceiling*—a term used to describe the limited mobility of women and minority workers in male-dominated organizations. In addition, women more often than men face *sexual harassment* at work—defined as the unequal imposition of sexual requirements in the context of a power relationship. Homophobia in the workplace also negatively affects the working experience of gays and lesbians. New protections are in place for disabled workers through the *Americans with Disabilities Act* (ADA).

What is the state?

The *state* is the organized system of power and authority in society. It comprises different institutions, including the government, the military, the police, the law and the courts, and the prison system. The state is supposed to protect its citizens and preserve society, but it often protects the status quo, sometimes to the disadvantage of less powerful groups in the society. States can also be organized as *democracies,* as *authoritarian,* or as *totalitarian.*

How do sociologists define power and authority?

Power is the ability of a person or group to influence another. *Authority* is power perceived to be legitimate and formal. There are three kinds of authority: *traditional authority,* based on long-established patterns; *charismatic authority,* based on an individual's personal appeal or charm; and *rational–legal authority,* based on the authority of rules and regulations (such as law).

What theories explain how power operates in the state?

Sociologists have developed four theories of power. The *pluralist model* sees power as operating through the influence of diverse interest groups in society. The *power elite model* sees power as based on the interconnections between the state, industry, and the military. *Autonomous state theory* sees the state as an entity in itself that operates to protect its own interests. *Feminist theorists* argue that the state is patriarchal, representing primarily men's interests.

How well does the government represent the diversity of the U.S. population?

An ideal democratic government would reflect and equally represent all members of society. The makeup of the U.S. government does not reflect the diversity of the general population. African Americans, Latinos, Native Americans, Asians, and women are underrepresented within the government. Political participation also varies by a number of social factors, including income, education, race, gender, and age. African Americans and Latinos, however, are overrepresented in the military, in part because of the opportunity the military purports to offer groups otherwise disadvantaged in education and the labor market; however, both are underrepresented at the levels of high-level commissioned officers. There is an increased presence of women in the military; however, prejudice and discrimination continue against lesbians and gays in the military.

Key Terms

alienation 374
authoritarian 375
authority 376
automation 365
autonomous state
    model 379
bureaucracy 376
capitalism 363
charismatic authority 376
communism 363
contingent worker 366
deindustrialization 364
democracy 375
division of labor 367

dual labor market 368
economic
    restructuring 364
economy 362
emotional labor 367
gender gap 380
glass ceiling 368
global assembly
    line 363
global economy 363
government 379
interest group 377
interlocking
    directorate 378

job displacement 365
multinational
    corporations 363
nationalism 375
occupational
    segregation 369
outsourcing 363
pluralist model 377
political action
    committee (PAC) 378
postindustrial society 362
power 376
power elite model 378
propaganda 375

rational–legal
    authority 376
sexual harassment 371
socialism 363
spatial mismatch 365
state 374
totalitarian 375
traditional
    authority 376
underemployment 371
unemployment
    rate 370
work 367
xenophobia 363

16 Environment, Population, and Social Change

ANDREW MILLS/Star Ledger/Corbis

THE ENVIRONMENT

increasingly, many people worry about whether we can preserve the earth's resources as we know them. Scientists and others are warning us that the polar ice cap is melting, causing ocean levels to rise. Climate patterns are changing and, although the specific effects are being debated, people worry that more severe storms, extreme heat, or perhaps bone-chilling cold will become more common. In some parts of the world, population growth outpaces the ability to feed people. In other places, including in the United States, water can no longer be assumed to be available or safe to drink. And air pollution is so bad in some parts of the world that people routinely wear masks. Is our current lifestyle sustainable?

Sustainability, in fact, has become an organizing cry—a cry for new social policies that will protect the Earth's environment and the people who live within it. Thus new movements have developed—movements to eat local food, to support the creation of urban community gardens, or to recycle used products by transforming them into something else. These and other developments signal the public's concern with the environment and the related phenomena of population, pollution, and social change.

You might think that studying such things as sustainable energy and environmental pollution is solely the work of scientists and engineers. No doubt, these are critical scientific problems, but as both scientists and engineers will tell you, social issues are just as important in understanding how we can preserve the Earth's resources. What lifestyles consume the highest amounts of energy? What social and cultural changes are needed to protect our environment and not deplete the Earth's natural resources? How is social inequality related to the degradation of the environment? And are there just too many people for the world to sustain human society as we know it?

These and other questions drive the substance of this chapter—a chapter that looks at the sociological issues that come from studying population and the environment. Although *environmental sociology* has been a long-standing field, it has particular urgency as people have now become more attuned to the potential crises that our planet faces because of climate change, environmental pollution, and population pressure.

learning objectives

- Identify the social dimensions of environmental change
- Explain how inequality affects environmental quality for different groups
- Understand the basic processes of population change
- List the changes that affect population diversity in the United States
- Explain theories of population growth
- Describe the different components and sources of social change
- Compare and contrast sociological theories of social change
- Analyze the social implications of globalization and modernization

A CLIMATE IN CRISIS? ENVIRONMENTAL SOCIOLOGY

Human beings, animals, and plants all depend on one another, as well as on the physical environment, for their survival. **Environmental sociology** is the scientific study of the interdependencies that exist between humans and our physical environment. A *human ecosystem* is any system of interdependent parts that involves human beings in interaction with one another and the physical environment. A city is a human ecosystem; so is a rural farmland community. In fact, the entire world is a human ecosystem.

The examination of ecosystems has demonstrated two things: The supply of many natural resources is finite, and if one element of an ecosystem is disturbed, the entire system is affected. For much of the history of humankind, the natural resources of the Earth were so abundant compared with the amounts humans used that they may as well have been infinite. No more. Some resources, such as certain fossil fuels, are simply non-renewable and may be gone soon. Other resources, such as timber or seafood, are renewable only if we do not plunder the sources of supply so

recklessly that they disappear. Some natural resources are so abundant that they still seem infinite, such as the planet's stock of air and water. But, at this stage of our societal development, we are learning that without more vigilance, we can even destroy the near-infinite resources (Gore 2006). Such an awareness puts social behavior squarely at the center of thinking about and solving our environmental problems in society.

Society at Risk? Air, Water, and Energy

You might think of environmental issues as primarily the work of physical and natural scientists, but even they have concluded that the social sciences are critical to solving our environmental problems. A scientist or engineer might invent a new way to heat our homes and power our cars, but without understanding the social dimensions of issues like energy, pollution, water usage, and other environmental behaviors, we cannot make progress in maintaining and improving our environmental sustainability. The challenges we face in protecting the environment are many. Gaseous wastes are gnawing away at the ozone layer, and our buried chemical wastes are trickling into the water table creating underground pools of poison. Pollution has damaged the Earth's surface water so badly that worldwide underground water reserves are being mined faster than nature can replenish them (Barlow and Clarke 2002).

The most threatening forms of pollution are the poisoning of the planet's air and water. Air pollution is not only ugly and uncomfortable, it is also deadly. The skies of all major cities around the world are stained with pollution hazes, and in cities that rest within

China is the world's leading producer of carbon-based emissions.

MAP 16.1

Viewing Society in Global Perspective: Global Warming Predictions

Scientists can use long-term data to predict air temperature increases in the future. Given what you have learned about *climate change* and

what you see in this map, what social changes do you think could occur as part of this environmental change?

Source: U.S. National Oceanic and Atmospheric Administration (NOAA). **www.nasa.gov/vision/earth/everydaylife/climate_class.html**

Surface Air Temperature Increase 1960 to 2060

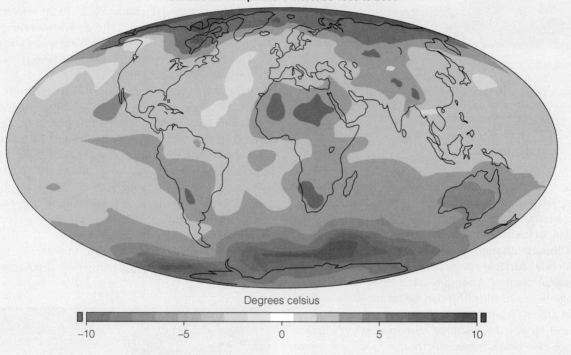

Degrees celsius

-10　　　　　-5　　　　　0　　　　　5　　　　　10

geological basins, such as Mexico City and Los Angeles, the concentrations of pollutants can rise so high that pollution-sensitive individuals cannot leave their homes or must wear masks when they go outdoors.

The invention of the automobile has transformed society. Today, many cannot imagine getting around without a car. We have designed many of our cities and, especially, our suburbs, in ways that require people to drive. A huge portion of the pollutants released into the air comes from the exhaust pipes of motor vehicles. The major component of this exhaust is carbon monoxide, a highly toxic substance. Also found in exhaust fumes are nitrogen oxides, the substances that give smog its brownish-yellow color. The action of sunlight causes these oxides to combine with hydrocarbons, also emitted from exhausts, forming a host of health-threatening substances. But we have become so dependent on automobiles for transportation that even with the now greater awareness of the consequences of driving all those gas-guzzling cars, it is, in many cases, difficult to design transportation systems that rely less on cars.

Other issues arise from the pollution generated by cars and our many commercial products. Chemicals called chlorofluorocarbons (CFCs) are used as a coolant in refrigerators, in the manufacture of plastics, and as an aerosol can propellant. CFCs released into the air find their way to the ozone layer in the upper atmosphere, where they eliminate the highly reactive ozone. The ozone layer is a shield that blocks dangerous ultraviolet light, and as this shield is destroyed, more ultraviolet light gets through, causing an increase in sunburn, skin cancer, and other illnesses. As the sun's energy pours onto the Earth, some is reflected from the Earth's surface to the Earth's atmosphere—the so-called *greenhouse effect*. Of the reflected energy, a portion is captured by carbon dioxide in the Earth's atmosphere, while the rest radiates into space. If the amount of solar energy trapped by carbon dioxide rises, the temperature increases—resulting in *global warming*.

Rather small changes in the average temperature of the Earth can have dramatic consequences. A few degrees of difference can cause greater melting in the

ENVIRONMENT, POPULATION, AND SOCIAL CHANGE ＜ **393**

Miners used to use canaries, which have a fragile respiratory system, to signal gas leaks in the mines. If the canary died, it was a signal of danger to the miners. Polar bears may be the new "miner's canary" in that the melting of their habitat is a warning of the rise in temperature of polar water—an indication of global warming.

Arctic regions, which raises the level of the sea, affecting water, land, and weather systems worldwide. Today, we see images of polar bears drowning because of the breakup of ice floes, that is, the melting of the polar cap.

Climate change is the systematic increase in worldwide surface temperatures and the resulting ecological change. Although some deny that climate change is the result of human behavior, there is little doubt among scientists that climate change is happening and largely the result of human activity. Climate change also poses numerous threats to society as we know it. With climate change will come more extreme weather patterns. People in coastal areas will be prone to rising coastal waters and storm surges, all too vividly seen during Superstorm Sandy in the fall of 2012. Some will have too much water; others, too little. Scientists see climate change as a serious problem (National Research Council 2012). Does the public think so?

Outside the United States, including in less-developed nations, the public shows more concern about climate change than is true within the United States (Brechin 2003). Despite overwhelming scientific evidence of climate change, some in the United States deny its existence, thus thwarting policy changes that could address its causes and consequences, such as programs that would make us less reliant on fossil fuels. But denial about climate change is only part of the problem. Even when people have information about the potential effects of climate change, they often ignore taking action. Why? The "social organization of denial," according to sociologist Kari Norgaard, comes from people holding unpleasant emotions—such as the fear of flooding or devastation—at a distance (Norgaard 2006). Although some may deny climate change for purely political reasons or for lack of information, for many, the sheer unpleasantness of facing such catastrophic change is more than people can willingly admit or face.

What, for example, would we do without water? Most Americans have come to think of water as abundant, free, and safe. But the safety and availability of water is now threatened. Thousands of rural water wells have been abandoned due to contamination. Households served by municipal water systems are also endangered; fully 20 percent of the country's public water systems do not meet the minimum toxicity standards set by the government. Although many have assumed that the nation's water is plentiful, it is highly questionable whether that can still be assumed (Fishman 2011). In the western and southwestern United States, the groundwater supply is being depleted at a rapid pace, making water one of the causes of political conflict between different states in the region (Espeland 1998).

Threats to our water supply also spill into other issues. As people have become concerned about water quality, drinking water from plastic bottles has increased. But where do the bottles go? Estimates are that about 40 million plastic water bottles go into the nation's trash *every day*—only about 12 percent of which are recycled (Lianos 2005). From production to disposal, water bottles reveal that technical know-how merges with human behavior, creating a complex system of environmental challenges.

The nation's water is also threatened by the chemical pollutants that industries discharge into

This woman was told by a gas company in Powder River Basin, Wyoming, that the drilling for gas near her house "would never cause you to lose your water." Shortly after drilling began near her home, her well water turned into a muddy methane slurry, which she unhappily holds in a glass in this photo.

Bottled water (not counting other plastic containers) produces 1.5 million tons of waste every year, only a small percentage of which is recycled.

and outrage can force the government to crack down on major polluters.

We are racing through our nonrenewable natural resources and destroying much that could be renewable. Addressing this problem also requires looking at some of the inequalities that are revealed when we examine such things as energy usage. On a global scale, the use of natural resources is not evenly shared around the world. The United States, which is a little under 5 percent of the world's population, consumes 20 percent of the world's energy and emits about 20 percent of the carbon dioxide emissions from fossil fuels. But China now exceeds the United States in the release of carbon-based emissions, sending 7,706 million metric tons of carbon dioxide emissions into the atmosphere; the United States ranks second in the world, spewing 5,424 million metric tons of carbon dioxide emissions into the atmosphere per year (U.S. Energy Information Administration 2011; see Figure 16.1). How much is a metric ton? The average car now weighs about two tons, so 5 million metric tons would be the weight of about two and a half million cars sent into the atmosphere, if that were even possible!

rivers, lakes, and the oceans, including solid wastes, sewage, nondegradable by-products, synthetic materials, toxic chemicals, and radioactive substances. Add the polluting effects of sewage systems of towns and large cities, detergents, oil spills, pesticide runoff, and runoff from mines, and the enormity of the problem is clear.

Federal and state statutes now prohibit industry from polluting the nation's water, but the pollution continues. Why? The answer is economic, political, and sociological. Industries that contribute to a vigorous economy have traditionally met with little interference from the government. Public awareness

see FOR YOURSELF

The Wasteful Society

For just one day (a full twenty-four-hour period), make a list of everything that you use up or discard. Include everything that you throw away, including garbage, waste from cooking and eating, gasoline in your car, and so forth. At the end of the day, list the things you discarded. Indicate whether there were any alternatives to discarding these things. How might one reduce the amount of waste produced in society generally? ●

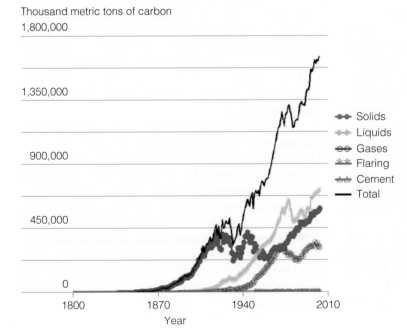

Thousand metric tons of carbon

FIGURE 16.1 U.S. Carbon Dioxide (CO_2) Emissions, 1800–2010 You can vividly see in this graph the increased emission from fossil fuels and other substances. What historical events and changes are reflected in the ups and downs that you see here? What changes would be needed in society to produce a decrease in the future?

Source: Carbon Dioxide Information Analysis Center. 2007. http://cdiac.ornl.gov/

Disasters: At the Interface of Social and Physical Life

Even while the normal practices of everyday life threaten the Earth, periodic hazards also come from natural disasters. Floods, earthquakes, tsunamis, hurricanes, tornadoes—these are just a few of the disasters that disrupt communities, families, and public health. Many disasters are forces of nature; some are predictable, some are not. Other disasters, such as chemical spills, explosions, and huge forest fires, are more directly attributable to human actions. But, either way, disasters are not solely the result of physical or natural factors.

According to sociologists who study them, disasters juxtapose physical events, such as floods, hurricanes, earthquakes, and the like with vulnerable populations (Tierney 2007). And, although people think of disasters as "nature's wrath," the impact of disasters is often the result of social behavior too. Hurricane Katrina is a good example. Although the hurricane itself emerged from nature, neglect of an inadequate levy system made communities in low-lying areas more vulnerable than others. Katrina affected many, but it had a disproportionate impact on poor and African American communities, in areas where the inadequacy of flood control projects exposed largely African American communities to the risk of serious flooding.

Time and time again, human behavior is implicated in the impact of natural disasters. Overdevelopment can destroy natural environments, such as barrier islands, that mitigate the effects of a natural disaster. During the 1930s, the devastation wrecked by the Dust Bowl resulted from agricultural practices (the overproduction of wheat) that had stripped the prairies in the southern Plains of natural grasses. Although drought brought on the devastating dust storms, had humans not destroyed the grasslands, it is doubtful that the consequences would have been so dire (Egan 2006).

Social systems are also disrupted in the aftermath of disasters. The 2011 nuclear power plant leakage in Japan that occurred following the major earthquake and *tsunami* (the resulting massive ocean wave) showed how vulnerable social systems can be to natural disasters. In the United States, the massive Gulf Oil Spill in 2010 spewed thousands of gallons of crude oil into the Gulf. The spill was so large and disastrous that large numbers of shrimpers and fishermen were forced out of business. The spill rapidly polluted major marshes surrounding the Gulf, killing off much flora, fauna, and wildlife, including birds of several species and all varieties of fish, shrimp, and mussels.

Who is most vulnerable during disasters is also shaped by social factors (Tierney et al. 2001). The poor and the elderly are often the most vulnerable. During the infamous Chicago heat wave of 1993, temperatures soared above 105 degrees and over 700 people

Researchers find that, in the aftermath of disaster or other unexpected events, people tend to come together to provide assistance to each other, as happened in Watertown, Massachusetts, when the FBI and police were on a massive hunt for the terrorist who killed three people and maimed many others by exploding two bombs near the finish line of the annual Boston marathon.

died. Research by sociologist Eric Klinenberg (2002) has found that social factors influenced these very high mortality rates. The isolation of elderly people, retrenchment of social services, and little institutional support in poor neighborhoods meant that those most likely to die from this disaster were the poor, the elderly, African Americans (because of their concentration in poor neighborhoods), and women (because they are more of the older population).

Government responses to disasters also show the consequences of human behavior for understanding the social dimensions of otherwise natural disasters. The slow work of the federal bureaucracy, as well as partisan politics, both play a role in social responses to disasters. Social stereotypes also figure in how victims of disasters are portrayed. Following Hurricane Katrina in New Orleans, African Americans were depicted in media coverage as wild looters and thieves—an image not seen so much when the predominantly White, working and middle-class communities in New York and New Jersey were so profoundly disrupted following Superstorm Sandy.

Now, given climate change, people even wonder if disasters will be more frequent. The warming of the Earth's oceans could produce more frequent and more damaging hurricanes. Heat waves could become more frequent, overpowering power supplies as people try to cool their homes. Drought could intensify, fueling political struggles over who controls water supplies in the driest areas of our nation. Some populations will be more vulnerable than others, showing once again how factors such as race, social class, age, and gender shape the impacts and portrayals of so-called natural disasters.

Black Women in the Environmental Justice Movement

Black women are at the forefront of social movements to challenge *environmental racism*. If they live in racially segregated neighborhoods, particularly low-income neighborhoods, toxic wastes are common and pose a threat to the health of these communities. Black women have become increasingly aware of these threats to their health and that of their families, and organizing environmental justice organizations has been one response.

A team of sociologists in Atlanta, Georgia (Antoinette Gomez, Fatemeh Shafiei, and Glenn Johnson) have studied Black women's activism in

environmental justice and examined what motivates these women to become movement activists. Their research finds that most of the women who are activists in this movement are mothers who are particularly concerned about the health and welfare of their families.

The women cite such things as having no "green" space available where their children can play, and they are concerned about the toxic wastes that permeate play spaces. They also attribute the health problems in their communities (asthma, skin disorders, and respiratory and sinus problems) to the chemicals in their neighborhoods. They also say that their families and their

religious beliefs and church organizations have encouraged their engagement with environmental justice, and they relate their concerns about environmental degradation to other social problems in their neighborhoods—drugs, violence, and crime.

These may not be the women people typically think of as political leaders, but their work in organizing around environmental justice shows that leadership comes in many forms—and sometimes in quite ordinary places.

Source: Gomez, Antoinette M., Fatemeh Shafiei, and Glenn Johnson. 2011. "Black Women's Involvement in the Environmental Justice Movement: An Analysis of Three Communities in Atlanta, Georgia." *Race, Gender, and Class* 18: 189–214.

ENVIRONMENTAL INEQUALITY AND ENVIRONMENTAL JUSTICE

Many argue that of all environmental problems facing the United States today, the most urgent is the dumping of hazardous wastes, if only for the sheer noxiousness of the materials being dumped. Since 1970, the production of toxic wastes increased ninefold (Weeks 2011). Of course, any degradation in the Earth's well-being affects everyone, but who is most vulnerable to pollutants and toxic waste dumping reveals patterns of social inequality.

Environmental racism is the pattern whereby toxic wastes and other pollutants are disproportionately found in minority and poor neighborhoods, a pattern with clear health consequences (Brulle and Pellow 2006). Research has determined that it is virtually impossible that dumps are being placed so often in communities of minority and lower socioeconomic status by chance alone (Bullard and Wright 2009). Is class or race to blame? Likely both, although researchers conclude that patterns of toxic waste dumping are not solely explainable by social class (Mohai and Saha 2007). Wealthier communities are better able to resist dumping in their neighborhoods, and housing discrimination and other race-related disparities are strongly linked to toxic waste being more present in minority areas, as illustrated in Figure 16.2.

Studies find that Native American, Hispanic, and particularly African American populations reside disproportionately closer to toxic sources than do Whites. Such patterns are *not* explainable by social class differences alone. That is, when communities of the same socioeconomic characteristics but different racial-ethnic compositions are compared, Native Americans, Hispanics, and African Americans of a given socioeconomic level live closer to toxic dumps than do Whites of the *same* socioeconomic level (Bullard and Wright 2010; Mohai and Saha 2007; Holmes 2000).

Take a look around your own neighborhood. Are there industrial waste sites nearby? Where are toxic products being disposed? For that matter, is there recycling available and to whom? You are likely to find patterns of waste disposal that are significantly linked to the class and racial composition of your neighborhood.

Within minority and poor neighborhoods, many groups have mobilized to protest and stop dumping in their communities. The *environmental justice movement* is the broad term used to refer to the social action that communities have taken to ensure that toxic waste dumping and other forms of pollution do not fall disproportionately on groups because of their race, class, or gender (Pellow 2004).

Environmental justice encompasses a wide array of programs for change. Developing more organic methods of growing food—indeed, encouraging community gardens where people can grow their own food—and other "green" programs are important

FIGURE 16.2 *Environmental racism* refers to the pattern whereby people living in predominantly minority communities are more likely exposed to toxic dumping and other forms of pollution. Nuclear waste and testing in the American Southwest, for example, have been located in areas predominantly inhabited by Native Americans. In other areas, African Americans and Latinos are exposed to the effects of industrial waste.

Visual concept by Norman Andersen

trends to promote social change for a more sustainable society. But social change is hard to accomplish, in part because it takes more than individual effort. As we will see, social change requires individual action, but it is also collective, that is, a fundamentally *social* process.

ARE THERE TOO MANY PEOPLE? POPULATION STUDIES

Studies of the environment raise fundamental questions about how human societies relate to the physical and natural world. Population growth and density are responsible for some of the challenges we face with our environment: urban overcrowding and sprawl, traffic jams, pollution, and the threat of diminishing or tarnished Earth resources. Many wonder if we can sustain the current way of living, given the size of the national and world populations. Are there simply too many people for our planet to support?

There are seven billion people living in this world. What do we know about how population is shaped? When a baby is born, what are his or her odds of survival beyond the first year? How many others will be born the following year? Will the population of people born in a given year influence the future of society simply because of the size of this age group?

These questions can be studied through the sociology of population. The scientific study of population is called **demography**. Demography includes studying the size, distribution, and composition of human populations as well as studying population changes over time, both those of the past and those predicted for the future.

Basic population facts drive many of the experiences and attitudes of some people. Young people may feel insecure about their future; decisions about having or not having children may loom; young people will likely have to care for older people; and, hotly debated topics like immigration are likely to continue to shape national politics. And the decisions people make—both personal and national—will ripple forward for years to come. Will there be a need for more senior centers? Will minority children get a good education? If there is a decline in the number of middle-aged people, who will take care of the old? Will environmental resources hold up in such a way as to maintain current lifestyles?

Counting People: Demographic Processes

Demography draws on huge bodies of data generated by a variety of sources. One major source is the U.S. Census Bureau. A **census** is a head count of the entire population of a country, usually done at regular intervals. The U.S. census is conducted every ten years, as required by the U.S. Constitution; the latest was conducted in 2010. The census attempts to enumerate every individual and to obtain information such as gender, race, ethnicity, age, education, occupation, and other social factors. The census is updated annually through a much smaller sample of the population that can then be used to track changes more frequently, although in less detail, than the decennial census (conducted every ten years).

The current population of the United States is more than 313 million—a milestone when the three-hundredth-million mark was passed in 2011. By mid-century, the U.S. population is not only predicted to be larger (439 million) but also older and more diverse. White Americans are expected to decline in population; Hispanics and Asians are expected to nearly triple and African Americans to double in size by 2050. And, as soon as 2030, one in five Americans will be over 65. The older population is expected to double in size by mid-century (U.S. Census Bureau 2012a).

Even with this detail, however, it is known that the census undercounts a small percentage of the country's population. It is simply impossible to have every single person complete the census form that is distributed. Who would be most likely undercounted? You can probably guess. Those most likely to have been undercounted by the census are the homeless, immigrants, minorities living in poor neighborhoods, and others of low social status. In general, the lower your overall social status (such as by income, occupation, race–ethnicity, gender, immigrant status, or other measures), the less likely you are to be counted in the U.S. census. Even given this, note how many more of those in the population are from minority groups.

The constitutional requirement for a census was included to ensure fair apportionment of representatives in the federal government. Undercounting specific groups of people leaves them underrepresented in government. The estimated undercount for the entire U.S. population overall is only about 2 percent, yet the undercount for African Americans nationally has been estimated to be as high as 20 percent, and for Hispanics as high as 25 percent (NAACP Legal Defense and Education Fund 2010; Harrison 2000).

Although counting people may seem tedious and dry to you, it can actually be a fiercely debated topic. Now (and beginning in the 2000 census), people are allowed to select multiracial (or "mixed race") as a response regarding their racial and ethnic identity on the census questionnaire. If you are, for example, Hispanic and Black or Black and White, how should the census "count" you in a racial or ethnic category—a category that will later be used to determine such facts as you have seen in this book, such as income distribution by race or the ethnic makeup of neighborhoods. The use of the multiracial response option gives individuals

an opportunity to define themselves as mixed race (see Figure 16.3). One argument against this option is that it subtracts from the number of people who would have otherwise indicated only one category, thus further undercounting African Americans, Hispanics, and Native Americans. Currently, only 3 percent of people responding to the census indicate a multiracial response, but this has increased since 2000, when it was first measured (at 2.4 percent) and is expected to increase substantially again in 2020 (Humes et al. 2011; Morning 2008).

The world population is currently seven billion people, and it is expected to grow to nine billion by 2050. Half of that growth is expected to be in only eight countries, presented here in declining order of their share of growth: India, Pakistan, Nigeria, Democratic Republic of the Congo, Bangladesh, Uganda, United States of America, Ethiopia, and China. Note that most of the growth will be in less-developed areas of the world, as you can also see in Figure 16.4. Most of these countries are places with high rates of poverty, but even within the United States, the bulk of projected population growth will be among Hispanics and Asian Americans (United Nations 2012b; Shrestha and Heisler 2011). Barring some major catastrophe, such as a health epidemic, this means that these nations will have a higher **population density**, defined as the number of people per unit of area, usually per square mile.

The total number of people in a society at any given moment is determined by only three variables: births, deaths, and migrations. These three variables show different patterns for different racial and ethnic groups, different social strata, and both genders. Births add to the total population, and deaths subtract from it. Migration into a society from outside, called **immigration**, adds to the population, whereas **emigration**, the departure of people from a society (also called *out-migration*), subtracts from the population.

Birthrate. The **crude birthrate** (or **birthrate**) of a population is the number of babies born each year for every 1000 members of the population or, alternatively, the number of births divided by the total population, multiplied by 1000. It is labeled crude because it does not take into account age or sex differences:

$$\text{Crude birth rate (CBR)} = \frac{\text{number of births}}{\text{total population}} \times 1000$$

The birthrate for the entire world population is now approximately 28 births per 1000 people in any given year. Nations vary considerably in their birthrates, with the highest being Niger, with 51.1 births per 1000 people, and the lowest, Japan, with only 7.4 births per 1000 people (U.S. Census Bureau 2012a). The birthrate for the United States is approximately 14 births per 1000 people now, lower than at any other time in U.S. history. By way of comparison, the birth rate in 1910 was 30.1, but has declined rather steadily since then, with the

→ **NOTE: Please answer BOTH Question 5 about Hispanic origin and Question 6 about race. For this census, Hispanic origins are not races.**

5. Is this person of Hispanic, Latino, or Spanish origin?

☐ **No,** not of Hispanic, Latino, or Spanish origin

☐ Yes, Mexican, Mexican Am., Chicano

☐ Yes, Puerto Rican

☐ Yes, Cuban

☐ Yes, another Hispanic, Latino, or Spanish origin – *Print origin, for example, Argentinean, Colombian, Dominican, Nicaraguan, Salvadoran, Spaniard, and so on.* ↗

6. What is this person's race? *Mark* ☒ *one or more boxes.*

☐ White

☐ Black, African Am., or Negro

☐ American Indian or Alaska Native – *Print name of enrolled or principal tribe.* ↗

☐ Asian Indian   ☐ Japanese   ☐ Native Hawaiian

☐ Chinese   ☐ Korean   ☐ Guamanian or Chamorro

☐ Filipino   ☐ Vietnamese   ☐ Samoan

☐ Other Asian – *Print race, for example, Hmong, Laotian, Thai, Pakistani, Cambodian, and so on.* ↗   ☐ Other Pacific Islander – *Print race, for example, Fijian, Tongan, and so on.* ↗

☐ Some other race – *Print race.* ↗

FIGURE 16.3 The Census Counts Race This is the form that the U.S. Census Bureau uses in its decennial census to tally the racial–ethnic composition of the U.S. population. How would you answer? Does this adequately measure your racial identity? If so, why? If not, how might you revise it, and would that be correct for others?

Source: U.S. Census Bureau. 2010. *Overview of Race and Hispanic Origin: 2010.* Washington, DC: U.S. Department of Commerce. **www.census.gov**

exception of the years just after World War II when the population now called baby boomers was born (National Center for Health Statistics 2012).

The effects of birthrates are somewhat cumulative. For example, minorities tend to be overrepresented at the lower end of the socioeconomic scale, compounding the likelihood of a high birthrate. Similarly, religious and cultural differences affect the birthrate.

Catholics, for example, have a higher birthrate than non-Catholics of the same socioeconomic status. Hispanic Americans have a high likelihood of being Catholic, another factor that contributes to the higher birthrate among Hispanic Americans. Projections that the United States will have a significantly greater proportion of minorities are based on births, deaths, and migration rates.

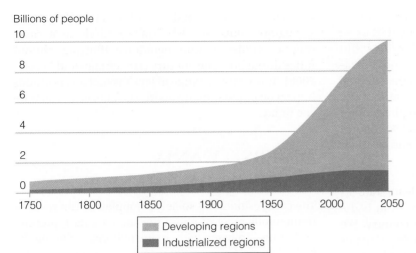

FIGURE 16.4 World Population Growth, 1750–2050 As you can see from this graph, population in the most developed parts of the world is expected to remain somewhat flat or even decline, while population in the less-developed areas will increase dramatically. What implications does this have for feeding the world and protecting people's health?

Data: United Nations Population Division. 2011. *World Population Prospects*, 2010 revision. **www.prb.org**

Death Rate. The **crude death rate** (or **death rate**) of a population is the number of deaths each year per 1000 people, or the number of deaths divided by the total population, times 1000:

$$\text{Crude death rate (CDR)} = \frac{\text{number of deaths}}{\text{total population}} \times 1000$$

The death rate can be an important measure of the overall standard of living for a population. In general, the higher the standard of living enjoyed by a country, or a group within the country, the lower the death rate. The death rate of a population also reflects the quality of medicine and health care. Poor medical care, which goes along with a low standard of living, will correlate with a high death rate. The death rate can be an important indicator of a population's overall standard of living. In general, the higher the standard of living, the lower the death rate.

In nations with a poor standard of living, infant mortality is typically high. The **infant mortality rate** is measured by the number of deaths per year of infants less than one year old for every 1000 live births. In the United States, the overall infant mortality rate is generally low (6.1 in 2010), although not compared to other industrialized nations, as you can see in Figure 16.5.

Infant mortality rates, a measure of the chances of the very survival of members of the population, are important to compare across racial–ethnic groups and across social class strata. The relatively higher

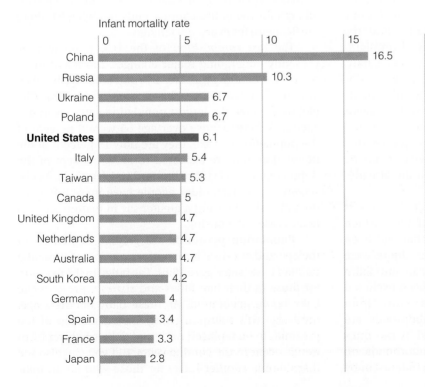

FIGURE 16.5 Infant Mortality Rates in Industrialized Nations You may be surprised to see that the United States ranks rather high in its rate of *infant mortality* relative to other industrialized nations. What might explain this?

Source: U.S. Census Bureau. 2012a. *The 2012 Statistical Abstract*. Washington, DC: U.S. Department of Commerce. **www.census.gov**

rate of infant mortality in the United States stems in large part from the poverty and inequality that exist, especially among racial and ethnic minorities, but also the White poor. Infant mortality is a good indicator of the overall quality of life, as well as the survival chances for members of that racial or class group. There are also many other causes of higher infant mortality, such as presence of toxic wastes, malnutrition of the mother, inadequate food, and outright starvation.

Migration. Joining the birthrate and death rate as factors in determining the size of a population is the migration of people into and out of the country. We see the impact of migration in current policy debates about immigration. Who should be allowed into a country? Should those who have immigrated illegally be given amnesty and allowed to stay? Should children who came to the United States at a very young age, but now have never known another country, be allowed to attend college by paying in-state tuition? These questions stem from population changes that are rather dramatically shaping the nation's future.

Migration affects society in many ways. Immigration has ebbed and flowed over the years, but some waves of immigration have, at certain times, had a huge impact on society. Of course, the United States has always been a land of immigrants. Only American Indians and Mexicans, settled in what is now the American Southwest, are indigenous people to this land. Of course, one could hardly call African Americans who came here as slaves "immigrants," as their entry was forced. Still, our nation has become a diverse mix of peoples, given the different origins of our population.

In 1924, National Origins Quota Act encouraged immigration from northern and western Europe (England, France, Germany, Switzerland, and the Scandinavian countries), but discouraged immigration from eastern and southern Europe (Greece, Italy, Poland, Turkey, and eastern European Jews generally, among others; see Chapter 10). Despite this openly discriminatory law, millions of eastern Europeans successfully made the journey to Ellis Island and then the U.S. mainland, only to face prejudice, discrimination, and the accusation that they were taking jobs that would have otherwise gone to the already-present White majority.

Currently, immigration to the United States is affected most by the Hart-Cellar Act of 1965, which abolished the national origins quotas that had been mandated in 1920. This meant that the doors were open for immigrants from Asia, Africa, and Latin and Central America—places that had been excluded from the prior policies that favored those from northern and western Europe. Neighborhoods are now invigorated and culturally enriched by mosques or Buddhist temples; by whole neighborhoods of, Vietnamese Catholics, Koreans, or Asian Indians; or by

war refugees from Somalia and Bosnia. And, whereas immigrants once settled almost entirely in a small number of cities, now immigration is affecting communities throughout the country (Hirschman and Massey 2008). This simple change in law has had an enormous impact on population diversity, the effects of which we see today.

DIVERSITY AND POPULATION CHANGE

The composition of a society's population can reveal a tremendous amount about the society's past, present, and future. To begin with, many nations, including the United States, have a striking imbalance in the number of men and women, with many fewer men than would be expected. The **sex ratio** is the number of males per 100 females, or the number of males divided by the number of females, times 100.

$$\text{Sex ratio} = \frac{\text{number of males}}{\text{number of females}} \times 100$$

A sex ratio above 100 indicates there are more males than females in the population; below 100 indicates there are more females than males. A ratio of exactly 100 indicates the number of males equals the number of females.

In almost all societies, there are more boys born than girls, but because males have a higher infant mortality rate and a higher death rate after infancy, there are usually more females in the overall population. In the United States, approximately 105 males are born for every 100 females, thus giving a sex ratio for live births of 105. After factoring in male mortality, the sex ratio for all ages for the entire country ends up being 94; there are 94 males for every 100 females.

The *age composition* of the U.S. population is presently undergoing major changes. More and more people are entering the sixty-five and older age bracket. This trend is known as the *graying of America*. The elderly are now the largest population category in our society. Whereas those over age 65 were 8 percent of the population in 1950, they are now 13 percent of the population and are expected to be 20 percent of the population by 2050 (Shrestha and Heisler 2011). As our society gets grayer, older people have more influence on national policy and a greater say in matters such as health care and housing.

Population pyramids are graphic depictions of the age and sex distribution of a given population at a point in time (see Figure 16.6). The bulge in the pyramid for those in their late fifties and sixties represents the baby boom generation. As these baby boomers age, the bulge will continue rising toward the top of the pyramid, to be replaced underneath by whatever birth trends occur in the coming years. But you can also see that there is another bulge for those who are in their

The End of the White Majority?

As the U.S. population is becoming more diverse, many are saying that White people will no longer be the majority. Even after the 2012 presidential election, when the votes of racial–ethnic minorities helped to reelect President Obama (along with 39 percent of the White vote), many pundits pontificated about whether White people were losing their historic hold on national power. Conservative commentator Bill O'Reilly even declared, "The white establishment is the new minority" (Fox News, November 6, 2012). What's true here?

First, even with the population projections indicating that Whites will become a smaller share of the U.S. population, Whites will still constitute a numerical majority (74 percent anticipated by 2050).

Second, and perhaps more importantly, from a sociological perspective, the terms *majority* and *minority* do not refer to numbers alone, as you learned in Chapter 10. Sociologically speaking, *majority* refers to a group that holds political, social, and economic power over others—and that group can, indeed, be a small percentage of the population (as we saw in apartheid South Africa when Whites, who had total rule over the whole population, were a mere 10 percent of the population). It remains true, even with the increased presence of people of color in social, political, cultural, and economic institutions, Whites are still the dominant group—in terms of power, privilege, and prestige.

Third, White is not a monolithic category. White people are diverse by many social–demographic characteristics—including age, gender, social class, region of residence—and all of these social facts affect the degree to which White people actually hold any power at all! So the next time you hear someone saying that White people are the new minority, you should have the sociological tools to challenge that assertion.

---

twenties and thirties, baby boomlets—that is, children of baby boomers. You might ask yourself how these generational structures are likely to affect society and its institutions over time as these "bulges" in population move upward.

The bulges you are seeing in Figure 16.6 are *birth cohorts*. A **cohort** consists of all the people born within a given period. A cohort can include all people born within the same year, decade, or other time period. Over time, cohorts either stay the same size or get smaller owing to deaths, but can never grow larger. If we have knowledge of the death rates for this population, we can predict quite accurately the size of the cohort as it passes through the stages of life from infancy to old age. This enables us to predict things such as how many people will enter the first grade in a given period,

how many are likely to enroll in college, and how many will arrive at retirement decades down the road. Administrators of social entities such as schools and pension funds can make preparations on the basis of cohort predictions.

thinking SOCIOLOGICALLY

In what age *cohort* would you place yourself? Are there particular ways that you think being in this cohort influences your behaviors and values? ●

The United States has long had a diverse population. Even as early as the first U.S. census in 1790, African Americans were 20 percent of the U.S. population, higher than now (at 13 percent). Most were slaves,

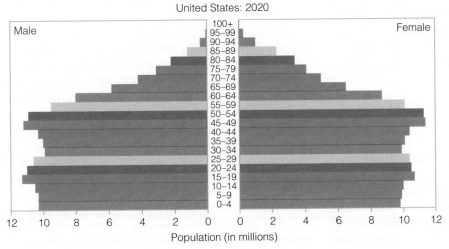

United States: 2020

Male / Female

Population (in millions)

FIGURE 16.6 The Age–Sex Pyramids A graphic depiction, such as this population pyramid, can capture social realities—such as generational differences—that might drive important issues for public policy. Here, you can see two primary "bulges" in the U.S. population. How do you think these two populations might have a different stake in national debates about issues such as Medicare, taxes, Social Security, and the like?

Source: U.S. Census Bureau. 2012. "Annual Estimates of the Resident Population by Sex and Five-Year Age Group for the United States." www.census.gov

brought by forced labor, although there was a class of free people (8 percent of the Black population in 1790). Diverse immigrant groups have long characterized the mosaic of the U.S. population, but at no other time has our nation seen as much diversity as now.

Racial and ethnic diversity is being driven by two major changes, one being immigration. But racial–ethnic groups also vary in basic population matters such as fertility and mortality. Thus, were you to make separate population pyramids for White, Black, Asian, American Indian, and Hispanic people, you would see that these populations are significantly younger and thus more likely to be a larger proportion of the population into the future. Just as examples, the median age of the White, non-Hispanic population is forty-one years of age; for African Americans, it is thirty-one years of age, for Native Americans thirty years, and for Hispanics, twenty-seven years. For Asians, median age is thirty-five (U.S. Census Bureau 2012a).

These facts result in what is now being called a population that is "majority minority," realizing that, in a sociological sense, minority refers not just to numbers in a population but also to the unequal treatment of diverse groups. Thus, by the year 2050, it is expected that whites will be 74 percent of the population (compared to 81 percent in 2000)—still a numerical majority, but less so than in the recent past. African Americans are expected to remain somewhat steady at 13 percent of the total U.S. population. Hispanics and Asians will grow the most, from 12 to 30 percent from 2000 to 2050. Asian Americans are predicted to grow from 4 percent to almost 8 percent. Those labeled as "other" in the U.S. census, including American Indians, Alaskans, Aleuts, and those identifying as mixed race will also double their current representation from 8 to 14 percent (U.S. Census Bureau 2012a).

These population data will mean significant changes in not just the composition of the U.S. population, but also in the social issues that the nation faces. Will government be more representative? Will intergroup tensions increase or decrease as the nation grapples with a more diverse population? What strains will a younger racial–ethnic population put on school systems? And what must people learn to be able to work, live, and learn among such population diversity? These and other questions will stem from the diversity in population growth that we are already witnessing.

THEORIES OF POPULATION GROWTH

Among the major problems facing modern-day civilization is the specter of uncontrolled population growth. As noted earlier, the world population increases by approximately 270,000 people every day. Some view overpopulation as an epochal catastrophe about to roll over us like a tidal wave. Others dispute whether the problem exists at all, explaining that there is no scientific consensus on the *carrying capacity* of the planet, the number of people the planet can support on a sustained basis, and that technological advances that have dependably met our needs in the past can be counted on to do so in the future as the number of mouths to feed continues to grow.

Is the world overpopulated? Is the United States? What can we expect from the future? If the less optimistic scenarios turn out to be accurate, these could be the most important questions facing humankind.

Malthusian Theory

Over 250 ago, **Thomas R. Malthus** (1798/1926), a Scotch clergyman, portended disastrous population growth, arguing that a population tends to grow faster than the subsistence needed to sustain it. Malthus noted that populations tend to grow not by *arithmetic increase*, adding the same number of new individuals each year, but by *exponential increase*, in which the number of individuals added each year grows, with the larger population generating an even larger number of births with each passing year. Exponential increase, in contrast to arithmetic increase, causes a population to grow ever faster. Malthus predicted widespread catastrophe and famine. **Malthusian theory** is the idea that a population tends to grow faster than the subsistence needed to sustain it.

Malthus reasoned that the only checks on population growth were famine, disease, and war. In Malthus's time, disease could reach apocalyptic scales. The outbreak of bubonic plague in Europe from 1334 to 1354 eliminated one-third of the population; a smallpox epidemic in 1707 wiped out three-fourths of the populations of Mexico and the West Indies. Wars took a toll on European men, with deaths in battle causing semipermanent gaps in the population pyramids of European populations. Along with "positive" checks on population growth, Malthus acknowledged what he called *preventive checks*, such as sexual abstinence, but he knew that sexual abstinence was unlikely to be the behavior change that halted uncontrolled population growth.

Malthusian theory actually predicted rather well the population of many early and agrarian societies. He failed, however, to foresee three revolutionary developments that derailed his predictions of growth and catastrophe. Technological advances have permitted the production of more food, resulting in subsistence levels higher than Malthus would have predicted. Medical science has fought off diseases that Malthus expected to periodically wipe out entire nations. And the development and widespread use of contraceptives in many countries have kept the birthrate at a level lower than Malthus would have thought possible.

The technological victories of this century have not completely erased the specter of Malthus. Viral epidemics warn us that disease can still wipe out huge populations. Heartrending pictures of swollen, starving babies remind us that famine and starvation continue to destroy human populations in some parts of the world just as they have for thousands of years. Overall, Malthus's theory has served as a warning that subsistence and natural resources are limited. The Malthusian doomsday has not yet occurred, but some believe that Malthus's warning was not in error, just premature.

The "Population Bomb"

More recently, a modern Malthus has appeared in the form of Paul Ehrlich, along with his wife Anne Ehrlich. Ehrlich's book, *The Population Bomb* (Ehrlich 1968), was the first in a series of writings in which he argues that many of the dire earlier predictions of Malthus were not far from wrong. The growth in world population, according to Ehrlich, is a time bomb ready to go off in the near future, with dismal consequences. Ehrlich asserts that the sheer mathematics of population growth worldwide are sufficient to demonstrate that world population cannot possibly continue to expand at its present rates.

The Ehrlichs point out that worldwide population growth has outgrown food production and that massive starvation must inevitably follow. They were among the earliest modern thinkers to argue that the quality of the environment, especially the availability of clean air and water, was a critical factor in the growth and health of populations. They pointed to mass starvation in parts of Africa, hunger and poverty in the United States among Black and Hispanic populations, increased homelessness in cities, rampant extinctions of plant and animal species, and the irrecoverable destruction of environments such as the rain forests, as evidence of the massive threats if population growth and environmental degradation are not checked.

Demographic Transition Theory

Demographic transition theory proposes that countries pass through a consistent sequence of population patterns linked to the degree of development in the society and end with a situation in which the birthrates and death rates are both relatively low (Davis 1945). Overall, the population level is predicted to eventually stabilize, with little subsequent increase or decrease over the long term.

There are three main stages to population change, according to demographic transition theory (see Figure 16.7). Stage 1 is characterized by a high birthrate and high death rate. The United States during its colonial period was in this stage. Women were bearing children at a younger age, and it was not uncommon for a woman to have twelve or thirteen children—a very high birthrate. However, infant mortality was also high, as was the overall death rate (both for infants and their

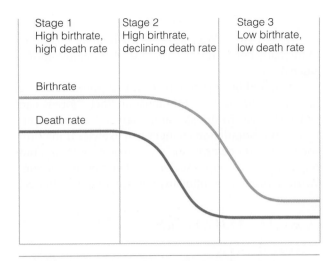

FIGURE 16.7 Demographic Transition Theory

Sources: Kingsley Davis. 1945. "The World Demographic Transition." *Annals of the American Academy of Political and Social Sciences* 237: 1–11; Coale, Ansley. 1986. "Population Trends and Economic Development." Pp. 96–104 in *World Population and the U.S. Population Policy: The Choice Ahead*, edited by J. Menken. New York: W. W. Norton; Weeks, John R. 2011. *Population: An Introduction to Concepts and Issues*, 11th ed. Belmont, CA: Wadsworth Publishing Co.

mothers) owing to primitive medical techniques and unhealthy sanitary conditions.

Stage 2 in the demographic transition is characterized by a high birthrate but a declining death rate. Hence, the overall level of the population increases. The United States entered stage 2 in the second half of the nineteenth century as industrialization took hold in earnest. The norms of the day continued to encourage large families, thereby causing high birthrates, while advances in medicine and public sanitation whittled away at the infant mortality rate and the overall death rate. Life expectancy increased, and the population grew in size.

The characteristics of stage 2 did not apply across all social groups or social classes, however. Minorities at the time (Blacks, and especially Native Americans and Chinese in the Midwest and West) were less likely to benefit from medical advances. The infant mortality and overall mortality rates for Blacks, Native Americans, and Chinese remained high, while their life expectancy remained considerably shorter than life expectancy for the White population. Lack of access to quality medical care was particularly devastating to Native Americans, who had very high death rates for all ages during this period. Demographic transition theory is not completely accurate for different racial–ethnic groups within the same society.

Stage 3 of the demographic transition is characterized by a low birthrate and low death rate. The overall level of the population tends to stabilize in stage 3. Medical advances continue, the general prosperity of the society is reflected in lowered death rates, and cultural changes take place, such as a reduction in family size. The United States entered this stage prior to the Second World War, and with the notable exception of the baby boom, has exhibited stage 3 demographics since then.

It should be apparent by now that population size has an important social dimension. Social forces can cause changes in the size and character of the population, and population changes can likewise transform society. In other words, society shapes population, but population also shapes society—a dynamic interaction. We now turn our attention to processes of social change.

A MULTIDIMENSIONAL PROCESS

We are currently witnessing extraordinary changes in society—both at home and around the world. Consider the following: A gigabyte of information can travel from China to the United States in less time than it takes you to read this paragraph. The reach of electronic communication is extraordinary and has shaped revolutions, such as the Arab Spring in the Middle East—even presidential elections in the United States as candidates increasingly use electronic networks to target specific messages to particular voters. Was all this imaginable fifty years ago? Even thirty years ago? Not really.

Social change is the alteration of social interactions, institutions, stratification systems, and elements of culture over time. Societies are in a constant state of flux. Some changes are rapid, such as the accelerating technological changes like the use of email, Facebook, Twitter, and perhaps new systems that will be invented even before this book is published. Other changes are more gradual, such as the increasing urbanization that characterizes the contemporary world. Sometimes people adapt quickly to change, such as the enthusiastic embrace of electronic communication by young

people, especially. Other times, people resist change or are slow to adapt to new possibilities. The speed of social change varies from society to society and from time to time within the same society.

Microchanges are subtle alterations in the day-to-day interactions between people. A fad "catching on" is an example of a microchange, such as the flip-flops that are now commonly worn. Not that long ago, flip-flops were just inexpensive sandals, worn largely by poor people. They were introduced into the United States after World War II, modeled on the Japanese sandal, the "zori" (Fortini 2005). Now they have become a common fashion statement.

Take the popularity of bungee jumping. Although not as widespread as some previous fads, this highly dangerous recreation is one of a group of "extreme sports," such as "extreme skiing," that have become popular across the country. Bungee jumping and other such "sports" have caused quite a few serious injuries and deaths, but have also provided thrilling footage for soft drink commercials. This may account for why a large number of youths have suddenly developed a taste for putting themselves in bone-smashing danger.

Macrochanges are gradual transformations that occur on a broad scale and affect many aspects of society. In the process of *modernization*, societies absorb the changes that come with new times and shed old ways. One frequently noted trend accompanying modernization is that societies develop greater differentiation between the division of labor and elaboration in systems of social inequality. Large or small, fast or slow, social change generally has in common the following characteristics (Lenksi 2005):

1. *Social change is uneven.* The various parts of a society do not all change at the same rate; some parts lag behind others. This is the principle of **culture lag**, a term coined by sociological theorist William

Social norms about dress and human activity sometimes make social change more evident when there is historical contrast.

F. Ogburn (1922), whereby there is a delay between when social conditions change and when cultural adjustments are made (see also Chapter 2). Often the first change is a development in material culture (such as a technological change in computer hardware), which is followed by a change in non-material culture (meaning the habits and mores of the culture).

2. *The onset and consequences of social change are often unforeseen.* Television pioneers, who envisioned a mode of mass communication more compelling than radio, could not predict television would become such a dominant force in determining the interests and habits of youth and the activities and structure of the family.

3. *Social change often creates conflict.* Change often triggers conflicts along lines of race and ethnicity, social class, gender, and age. The spread of Western culture into other parts of the world, made possible by the ease of communication, has been also said to foster terrorism as fundamentalist factions resist the changes brought.

4. *The direction of social change is not random.* Change has "direction" relative to a society's history. A populace may want to make a good society better, or it may rebel against a status quo regarded as unendurable. Whether change is wanted or resisted, when it occurs, it takes place within a specific social and cultural context.

Social change cannot erase the past. As a society moves toward the future, it carries along its past, its traditions, and its institutions (Lenski 2005). A generally satisfied populace that strives to make a good society better obviously wishes to preserve its past, but even when a society is in revolt against a status quo that is intolerable, the social change that occurs must be understood in the context of the past as much as the future.

Sources of Social Change

The causes of social change are many and varied but fall into several broad areas, including cultural diffusion; technological innovation; the mobilization of people through social movements and collective behavior; and, sometimes, war and revolution. We examine each in the following sections.

Cultural Diffusion. **Cultural diffusion** (as noted in Chapter 2) is the transmission of cultural elements from one society or cultural group to another. Cultural diffusion can occur by trade, migration, mass communications media, and social interaction. Anthropologist Ralph Linton (1936) long ago alerted us to the fact that much of what many people regard as "American" originally came from other lands—cloth (developed in Asia), clocks (invented in Europe), coins (developed in Turkey), and much more.

Expressions and cultural elements found in the English-speaking United States have been harvested from all over the world. Barbecued ribs, originally eaten by Black slaves in the South after the ribs were discarded by White slave owners who preferred meatier parts of the pig, are now a delicacy enjoyed throughout the United States by virtually all ethnic and racial groups. One theorist, Robert Farris Thompson (1993), points out that an exceptionally large range of elements in material and nonmaterial culture that originated in Africa have diffused throughout virtually all groups and subcultures in the United States, including aspects of language, music, dance, art, dress, decorative styles, and even forms of greeting. For example, the expressions *uh-huh* (yes) and *unh-unh* (no) come from West Africa. Cultural diffusion occurs not only from one place to another (such as from West Africa to the United States) but also across time, such as from a community in the past to many diverse ethnic groups in the present.

The immigration of Latino groups into the United States over time has dramatically altered U.S. culture by introducing new food, music, language, slang, and many other cultural elements. By a similar token, popular culture in the United States has diffused into many other countries and cultures: Witness the adoption of American clothing styles, rock, rap, hip-hop, and Big Macs in countries such as Japan, Germany, Russia, and China. The Coca-Cola logo can be found in grocery shops worldwide, from the rain forests of Brazil to the ice floes of Norway.

Technological Innovation and the Cyberspace Revolution. Technological innovations can be strong catalysts of social change. The historical movement from agrarian societies to industrialized societies has been tightly linked to the emergence of technological innovations and inventions (see Chapter 5). Inventions often come about because they answer a need in the society that promises great rewards. The waterwheel promised agrarian societies greater power to raise crops despite dry weather, while also saving large amounts of time and labor. It is possible to trace a timeline from the use of the waterwheel to the use of the large hydro-electric dams that power industrialized societies, and along the way find evidence of how each major advance changed society.

In today's world, the most obvious technological change transforming society is the rise of the computer and the subsequent development of desktop computing since the 1980s. The invention and development of the Internet and the resulting communication is now called *cyberspace*, which includes the use of computers for communication between people and communication between people and computers. YouTube videos can go "viral" and reach thousands, perhaps even millions, in a very short period of time. Unique in its vastness and lack of a required central location, the Internet has

Who Cares and Why? Fair Trade and Organic Food

Research Question: There has been a notable increase in the public's use of farmers' markets, a greater presence of organic food sections even in mainstream grocery stores, and other indications of the public's growing concern with where one's food comes from and how it is grown. Social movements to enhance awareness of food production have influenced some of this behavior, as has a corporate response to the public's interest in local, safe, and sustainable food production. Do people purchase food (and other products) because of their political and ethical values, and what are those values? This question forms the basis for a study of "ecolabels" by Philip Howard and Patricia Allen.

Research Method: Howard and Allen mailed a survey to 1000 randomly selected respondents, asking them to rate five different reasons why they would select food with different "ecolabels." They identified five different labels: humane (meat, dairy, and eggs coming from animals who have not been treated cruelly); living wage (provides wages to workers above poverty level); locally grown; small-scale (supports small farms or businesses); and "made in the USA." They also collected data on

various demographic variables, such as age, income, level of education, gender, and place of residence. They analyzed the results using sophisticated statistical techniques of regression analysis.

Research Results: First, the researchers note that their respondents were more likely to be women, older, white, higher income, and well-educated than the demographic composition of their random sample. This is an important caveat in interpreting the results, because the results are not generalizable to the whole population. One-third of their respondents reported purchasing local foods frequently; many fewer bought organic food regularly.

The three most popular interests in purchasing food were buying local, humane treatment, and providing a living wage for food production workers, but there were differences by demographic group. Buying local was even more important for rural residents. For those who buy organic food, humane reasons topped their preferences. Women were more interested in ecolabeling than men; higher-income people were less likely to care about living wage than were lower-income respondents. Older respondents were more concerned

about the influence of corporations of food production.

Conclusions and Implications: Consumers want the food they buy to reflect their political and ethical judgments. Of course, there are implications of these conclusions for marketing. But, from a sociological perspective, you can also see the influence of demographic variables on the decisions people make about purchasing their food. And, although it was not specifically examined in this study, social movements to "buy local," protect animals, and advocate for food safety have also influenced consumer preferences, meaning that there have been significant changes over time in the food choices that people have.

Questions to Consider

1. Examine your own behavior. What influences what you buy to eat?
2. Do political and ethical values influence your choices?
3. To what degree are your choices influenced by corporations and marketing?
4. Does your social location in particular demographic groups influence your eating habits?

Source: Howard, Philip H., and Patricia Allen. 2010. "Beyond Organic and Fair Trade? An Analysis of Ecolabel Preferences in the United States." *Rural Sociology* 75: 244–269.

---

very rapidly become so much a part of human communication and social reality that it pervades and has transformed literally every social institution—educational, economic, political, familial, and religious.

The path by which technology is introduced into society often reflects the predominant cultural values in that society. Some cultural values may prevent a technological innovation from changing a society. For example, anthropologists have noted that new technologies introduced into an agrarian society very often meet with resistance even though the new technology might greatly benefit the society. The Yanomami, an agrarian society existing deep in the rain forests of South America, live without electricity, automobiles, guns, and other items of material culture associated with industrialized societies. The Yanomami place great positive cultural value on their way of hunting and engaging in war. The recent introduction of steel into Yanomami

culture, however, may have introduced major social changes and changed them into a more warlike society (Tierney 2000).

Social Movements and Collective Behavior. Social change does not develop in the abstract. Change comes from the actions of human beings. A **social movement** is a group that acts with some continuity and organization to promote or resist social change in society. What would the United States be like had the civil rights movement not been inspired by Mahatma Gandhi's liberation movement in India? How would contemporary politics be different had the African National Congress and other movements for the liberation of Black South Africans not dismantled apartheid? How is the world currently affected by the development of a more fundamental Islamic religious movement in the Middle East?

These and countless other examples show the significance of social movements for the many changes affecting our world. Persistence has been the case for the civil rights movement in the United States, a movement that not only has transformed American society but also has inspired similar movements throughout the world. Some social movements are *transnational social movements*, in which an organization crosses national borders, such as the reactionary terrorist group al Qaeda or even the Taliban. Examples of social movements abound: the civil rights movements, the women's movement, the environmental movements, and such contemporary movements as the Tea Party and, in African and Middle Eastern countries, the so-called Arab Spring.

Social movements involve groups, sometimes quite large and well organized, that act with some continuity and organization to promote or resist change in society (Turner and Killian 1993). Social movements tend to persist over time more than collective behavior. Social movements are both spontaneous and structured, although within movements, tensions typically exist between spontaneity and structure. Unlike everyday organizations, movements thrive on spontaneity and often must swiftly develop new strategies and tactics in the quest for change. During the civil rights movement, students improvised the technique of sit-ins, which quickly spread because they succeeded in gaining attention for the activists' concerns.

Social movements may aim to change individual behaviors, such as the New Age movement that focuses

Social movements such as the disability rights movement can raise public awareness and result in new forms of social behavior.

on personal transformation. Other movements aim to change some aspect of society, such as the gay and lesbian movements that have sought to end discrimination and change public attitudes. Some movements use reform strategies, such as by trying to change laws. Others may be more radical, seeking change in the basic institutions of society. Other social movements are reactionary—that is, organized to resist change or to reinstate an earlier social order that participants perceive to be better.

Social change is reflected not only in the methods of playing tennis today, but also in the presence of more minorities in professional tennis.

Social movements do not typically develop out of thin air. For a movement to begin, there must be a preexisting communication network (Freeman 1983). The importance of a preexisting communication network is well illustrated by the beginning of the civil rights movement, which most historians date to December 1, 1955, the day Rosa Parks was arrested in Montgomery, Alabama, for refusing to give up her seat to a White man on a municipal bus. Although she is typically understood as simply having been too tired to give up her seat that day, Rosa Parks had been an active member of the movement against segregation in Montgomery. When Rosa Parks refused to give up her seat according to plan, the movement stood ready to mobilize. News of her arrest spread quickly via networks of friends, kin, church, and school organizations (Morris 1999; Robinson 1987).

Celebrities can also advance the cause of social movements by bringing visibility to the movement, such as Michael J. Fox's leadership in the movement to advance awareness and knowledge about Parkinson's disease. As movements develop, they quickly establish an organizational structure. The shape of the movement's organization may range from formal bureaucratic structures to decentralized, interpersonal, and egalitarian arrangements. Many movements combine both. Examples of what are now large bureaucracies are the National Organization for Women, the National Association for the Advancement of Colored People (NAACP), Amnesty International, Greenpeace, the Jewish Defense League, and the National Rifle Association. As social movements become institutionalized, they are most likely to take on bureaucratic form.

Fads come and go but can also define the culture of an age. Hula hoops, now popular again, first became a fad in the 1950s.

Related to social movements is **collective behavior**, behavior that occurs when the usual conventions that guide social behavior are disrupted for some reason and people establish new, usually sudden, norms in response to an emerging situation (Turner and Killian 1993). Although collective behavior may emerge spontaneously, it can be predicted. Some phenomena defined as collective behavior are whimsical and fun, such as fads, fashions, and certain crowds (flash mobs, for example). Other collective behaviors can be terrifying, as in panics or riots. Whether whimsical or not, collective behavior is innovative, sometimes revolutionary; it is this feature that links collective behavior to social change.

Collective behavior is group, not individual, behavior. A lone gunman, for example, who opens fire in a crowded movie theater is engaged in individual behavior, but the crowd that gathers following this unexpected event is engaged in collective behavior. Collective behavior involves new and emerging relationships that arise in unexpected circumstances. It represents the often novel, but dynamic and changing character of society.

War and Revolution. A **revolution** is the overthrow of state or the total transformation of central state institutions. A revolution thus results in far-reaching social change. Numerous sociologists have studied revolutions and identified the conditions under which revolutions are likely to occur. Revolutions can sometimes break down a state and various disenfranchised groups. An array of groups in a society may be dissatisfied with the status quo and organize to replace established institutions. Dissatisfaction alone is not enough to produce a revolution, however. The opportunity must exist for the group to mobilize en masse. Thus revolutions can result when structured opportunities are created, such as through war or an economic crisis or mobilization through a *social movement*, as we will see later in this chapter.

Social structural conditions that often lead to revolution can include a highly repressive state—so repressed that a strong political culture develops out of resistance to state oppression. A major economic crisis can also produce revolution—as can the development of a new economic system, such as capitalism—that transforms the world economy.

War and severe political conflict result in large and far-reaching changes for both the conquering society, or a region within a society (as in civil war), and for the conquered. The conquerors can impose their will on the conquered and restructure many of their institutions, or the conquerors can exercise only minimal changes.

The U.S. victory over Japan and Germany in the Second World War resulted in societal changes in each country. The war transformed the United States into a mass-production economy and affected family

ClassicStock/Alamy Limited

structure (father's absence increasing and women not previously employed joining the labor force) and education (men of college age went off to war in large numbers). Many in the armed forces who returned from the war were educated under a scholarship plan called the GI Bill.

The war also transformed Germany in countless ways, given the vast physical destruction brought on by U.S. bombs and the worldwide attention brought to anti-Semitism and the Nazi Holocaust. The cultural and structural changes in Japan were extensive, as well. The decimation of the Jewish population in Germany and other nations throughout Europe resulted in the massive migration of Jews to the United States. The Vietnam War also resulted in many social changes, including the migration of Vietnamese to the United States. If this history of war is any indication, we might shortly expect a wave of migration of Iraqis and Afghanis to the United States.

THEORIES OF SOCIAL CHANGE

How do we explain social change? Some believe that change is cyclical, that is, recurring at regular intervals. Cyclical theories build on the idea that societies have a life cycle, like seasonal plants, or at least a life span, like humans. Arnold J. Toynbee, a social historian and a principal theorist of cyclical social change, argues that societies are born, mature, decay, and sometimes die (Toynbee and Caplan 1972). For at least part of his life, Toynbee believed that Western society was fated to self-destruct because he thought energetic social builders would be replaced by entrenched elites who ruled by force. Then, society would wither under these sterile regimes. Some believe that societies become more decrepit, only to be replaced by more youthful societies.

Sociological theories of change are more nuanced than the idea that there are regular life cycles in society. Each of the major sociological perspectives takes a somewhat different approach to theorizing how social change occurs.

Functionalist Theory

Recall from previous chapters that functionalist theory builds on the postulate that all societies, past and present, possess basic elements and institutions that perform certain functions permitting a society to survive and persist. A *function* is a consequence of a social element that contributes to the continuance of a society. For example, the function of an institution such as the family is to provide the society with sufficient population to assure its continuance.

The early theorists Herbert Spencer (1882) and Emile Durkheim (1964/1895) both argued that as societies move through history, they become more complex. Spencer argued that societies move from

"homogeneity to heterogeneity." Durkheim similarly argued that societies move from a state of **mechanical solidarity**, a cohesiveness based on the similarity among its members, to **organic solidarity**, a cohesiveness based on difference; a division of labor that exists among its members joins them together, because each depends on the others to perform specialized tasks (see Chapter 5). Through the creation of specialized roles, structures, and institutions, societies thus move from a condition of relative undifferentiation to higher social differentiation.

According to functional theorists, societies that are structurally simple and homogeneous, such as foraging or pastoral societies, where all members engage in similar tasks, move to societies more structurally complex and heterogeneous, such as agricultural, industrial, and postindustrial societies, where great social differentiation exists in the division of labor among people who perform many specialized tasks. The consequence (or function) of increased differentiation and division of labor is a higher degree of stability and cohesiveness in the society, brought about by mutual dependence (Parsons 1966, 1951a).

Conflict Theory

Karl Marx (1967/1867), the founder of conflict theory, theorized that societies change and social change has direction, the central principle in Spencer's social evolutionary theory, but Marx emphasized the role of economics. He argued that societies could indeed "advance," and that advancement was to be measured by the movement from a class society to a society with no class structure. Marx believed that, along the way, class conflict was inevitable.

The central notion of conflict theory, as we have seen, is that conflict is inherently built into social relations (Dahrendorf 1959). For Marx, social conflict, particularly between the two major social classes—working class versus upper class, proletariat versus bourgeoisie—was not only inherent in social relations but was indeed the driving force behind all social change. Marx believed that the most important causes of social change were the tensions between social groups, especially those defined along social class lines. Different classes have different access to power, with the relatively lower class carrying less power.

Although the groups Marx originally referred to were indeed social classes, subsequent interpretations include conflict between any socially distinct groups that receive unequal privileges and opportunities. However, be aware that the distinction between class and other social variables is necessarily murky. For example, conflict between Whites and minorities is at least partly (but not wholly) class conflict, because minorities are disproportionately represented among the less well-off classes.

Racial and ethnic conflict in the United States involves far more than class differences alone: Many cultural differences exist between Whites and Native Americans, Latinos, Blacks, and Asians. Furthermore, cultural differences exist *within* broadly defined ethnic groups as well. We have pointed out earlier in this book that there are broad differences in norms and heritage among Chinese Americans, Japanese Americans, Vietnamese Americans, and so on—all often grouped rather coarsely as Asian Americans. The same error is often made with "Hispanic" groups—Hondurans, Guatemalans, Mexicans, Puerto Ricans, Panamanians, and others all lumped together.

The central idea of conflict theory is the notion that social groups will have competing interests regardless of how they are defined. Conflict is an inherent part of the social scene in any society.

Symbolic Interaction Theory

Because symbolic interaction focuses on more micro-level behaviors, it does not explain the wide-scale social changes that other theorists analyze. But it still contributes to our understanding of processes of change, especially in how it emphasizes the meaning that people attach to social behavior. Symbolic interaction, in analyzing change, might, for example, look at the quixotic character of fads and fashion and how people adapt and respond to these behaviors. Fads, for example, emerge, tend to spread rapidly, but then typically vanish—what sociologist Joel Best calls "emerging, surging, and purging" (Best 2006). How, for example, did UGG boots become such a symbol of social status? Fads, such as this one, come and go—perhaps passing quickly, even when they are widely popular. Perhaps by the time this book is published, UGGs will be considered passé and some wholly unpredictable other status symbol will produce changes in how people dress.

Symbolic interaction theory thus looks at change in terms of how people define social behaviors and how those definitions emerge. Of course, with fads, much of that definition comes through the work of marketing and the sponsorship of the corporate world. But some changes come from the ordinary behaviors of people as they reformulate their ideas and attitudes. Changes in public attitudes and how people think are signs of social change, as symbolic interaction theory would point out.

GLOBALIZATION AND MODERNIZATION: SHAPING OUR LIVES

Globalization is the increased interconnectedness and interdependence of numerous societies around the world. No longer can the nations of the world be viewed as separate and independent societies. The irresistible current trend has been for societies to develop deep dependencies on each other, with interlocking economies and social customs. In Europe, this trend proceeded as far as developing a common currency, the *euro*, for all nations participating in the newly constructed common economy.

As the world becomes increasingly interconnected, does this mean we are moving toward a single, homogeneous culture? In such a culture, electronic communications, computers, and other developments would erase the geographic distances between cultures and, eventually, the cultural differences. Early in 2011, the African and Middle Eastern youth anti-dictator rebellions in several countries (among them Egypt, Tunisia, Libya, and others; see Chapter 15) was no doubt greatly aided by electronic communication across international borders via email, Facebook, and other such electronic media, understood and used by the younger generations but in little use by their elders in these countries.

As societies become more interconnected, cultural diffusion between them creates common ground, while cultural differences may become more important as the relationships among nations becomes more intimate. The different perspectives on globalization are represented by three main theories that we will review: modernization theory, world systems theory, and dependency theory, which are included in Table 16.1.

As societies grow and change, they become more modern in a general sense. **Modernization** is a process of social and cultural change initiated by industrialization and followed by increased social differentiation and division of labor. Societies can, of course, experience social change without industrialization.

Social movements, such as the Arab Spring represented by the uprising in Cairo, Egypt in 2012, can produce dramatic social change.

Mohamed Elsayyed/Shutterstock.com

table 16.1 Theories of Social Change

	Functionalist Theory	Conflict Theory	Symbolic Interaction Theory
How do societies change?	Societies change from simple to complex and from an undifferentiated to a highly differentiated division of labor.	Conflict is inherent in social relations, and society changes from a class-based to a classless society.	Social change occurs when new meaning systems develop around people's behaviors and attitudes.
What is the primary cause of social change?	Technological innovation and globalization make society more differentiated but still stable.	Economic inequality drives social change.	Changes in people's attitudes and beliefs drive social change.
What is the impact of social change on individuals?	Individuals remain integrated into the whole because society seeks equilibrium.	Individuals are faced with conflict, but the powerless may organize to drive social change.	People have to adapt to new understandings that emerge from social change.

© Cengage Learning

Modernization is a specific type of social change that industrialization tends to bring about. The change toward an industrialized society can have positive consequences, such as improved transportation and a higher gross national product, or negative consequences, such as pollution, elevated stress, and increases in certain job discrimination.

Modernization has three general characteristics (Berger et al. 1974):

1. *Modernization is typified by the decline of small, traditional communities.* The individuals in foraging or agrarian societies live in small-scale settlements with their extended families and neighbors. The primary group is prominent in social interaction. Industrialization causes an overall decline in the importance of primary group interactions and an increase in the importance of secondary groups, such as colleagues at work.
2. *With increasing modernization, a society becomes more bureaucratized.* Interactions come to be shaped by formal organizations. Traditional ties of kinship and neighborhood feeling decrease, and members of the society tend to experience feelings of uncertainty and powerlessness.
3. *There is a decline in the importance of religious institutions.* With the mechanization of daily life, people begin to feel that they have lost control of their own lives: People may respond by building new religious groups and communities (Wuthnow 1994).

From Community to Society

The German sociologist **Ferdinand Tönnies** (1855–1936) formulated a theory of modernization that still applies to today's societies (Tönnies 1963/1887). Tönnies

viewed the process of modernization as a progressive loss of *gemeinschaft* (German for "community"), a state characterized by a sense of common feeling, strong personal ties, and sturdy primary group memberships, along with a sense of personal loyalty to one another. Tönnies argued the Industrial Revolution, which emphasized efficiency and task-oriented behavior, destroyed the sense of community and personal ties associated with an earlier rural life. At the crux of this was a society organized on the basis of self-interest, which caused the condition of *gesellschaft* (German for "society"), a kind of social organization characterized by a high division of labor, less prominence of personal ties, the lack of a sense of community among the members of society, and the absence of a feeling of belonging—maladies often associated with modern urban life.

The United States since the 1940s has become a gesellschaft, with social interaction less intimate and less emotional, although certain primary groups such as the family and the friendship group still permit strong emotional ties. However, Tönnies noted that the role of the family is considerably less prominent in a gesellschaft than in a gemeinschaft. Patriarchy is less prominent, yet more public, and more women are employed outside the home. In the large cities that characterize the gesellschaft, people live among strangers and pass people on the street who are unfamiliar. In a gemeinschaft, people have already seen most of the people they encounter. The level of interpersonal trust is considerably less in a gesellschaft. Social interaction tends to be even more confined within ethnic, racial, and social class groups. To find personal contact and to satisfy the need for intimate interaction, individuals often join small church groups, training groups, or personal awareness groups or movements.

Urbanization

One consequence of the elaboration of society is the growth of cities. Of course, cities have always been important sites for population density, commerce, and a bustling social life. But the growth and development of cities is a relatively recent occurrence in the course of human history. Scholars locate the development of the first city around 3500 B.C. (Flanagan 1995).

The study of the urban, the rural, and the suburban is the task of *urban sociology*, a subfield of sociology that examines the social structure and cultural aspects of the city compared with rural and suburban centers. These comparisons involve what urban sociologist Gideon Sjøberg (1965) calls the *rural–urban continuum*, those structural and cultural differences that exist as a consequence of differing degrees of urbanization. **Urbanization** is the process by which a community acquires the characteristics of city life and the "urban" end of the rural–urban continuum.

Early German sociological theorist Georg Simmel (1902/1950) argued that urban living had profound social psychological effects on individuals. Thus he was among the early theorists who argued that social structure could affect individuals. He argued that urban life has a quick pace and is stimulating, but as a consequence of this intense style of life, individuals become insensitive to people and events that surround them. Urban dwellers tend to avoid the emotional involvement that, according to Simmel, was more likely found in rural communities. Interaction tends to be characterized as economic rather than social, and close, personal interaction is frowned upon and discouraged. Yet, urban dwelling can increase the likelihood of other ills: Early theorist Emile Durkheim noted that the suicide rate per 10,000 people was greater in more urbanized areas than in rural areas (Durkheim 1951/1897).

The sociologist Louis Wirth (1928), focusing on Chicago in the 1920s, also argued that the city was a center of distant, cold interpersonal interaction, and as a result, urban dwellers experienced alienation, loneliness, and powerlessness. One positive consequence of all this, according to Wirth and Simmel, was the liberating effect that arises from the relative absence of close, restrictive ties and interactions. Thus city life offered individuals a certain feeling of freedom.

A contrasting view of urban life is offered by Herbert Gans (1982/1962), who studied people in Boston in the late 1950s and concluded that many city residents develop strong loyalties to others and are characterized by a sense of community. Such subgroupings he referred to as the *urban village*, which is characterized by several "modes of adaptation," among them *cosmopolites*—typically students, artists, writers, and musicians, who together form a tightly knit community and choose urban living to be near the city's cultural facilities. A

second category are the *ethnic villagers*, people who live in ethnically and racially segregated neighborhoods. Such *urban enclaves* tend to develop their own unique identities, such as San Francisco's Chinatown or Miami's Little Havana.

Social Inequality, Powerlessness, and the Individual

Another product of modernization, along with mass society, is pronounced social stratification, according to theorists such as Karl Marx (1967/1867) and Jurgen Habermas (1970). In their view, the personal feelings of powerlessness that accompany modernization are due to social inequalities related to race, ethnicity, class, and gender stratification. Marx argued that inequalities are the inevitable product of the capitalist system. Habermas argued that inequalities are the cause of social conflict.

The social structural conditions that arise from modernization, such as increased social stratification, affect individual people's lives. Building a stable personal identity is difficult in a highly modernized society that presents individuals with complex and conflicting choices about how to live. Many individuals flounder among lifestyles while searching for personal stability and a sense of self. According to Habermas, individuals in highly modernized environments are more likely than their less modernized peers to experiment with new religions, social movements, and lifestyles in search of a fit with their conception of their own "true self." These individual responses to social structural conditions reveal how the social structure can affect personality.

The influential social theorist Herbert Marcuse (1964) has argued that modernized society fails to meet the basic needs of people, among them the need for a fulfilling identity. In this respect, modern society and its attendant technological advances are not stable and rational, as is often argued, but unstable and irrational. The technological advances of modern society do not increase the feeling of having control over one's life, but instead reduce that control and lead to feelings of powerlessness.

This powerlessness leads to the *alienation* of individuals from society—individuals experience feelings of separation from the group or society. This alienation is most likely to affect people traditionally denied access to power, such as racial minorities, women, and the working class. This alienation from the highly modernized, technological society is, in Marcuse's view, one of the most pressing problems of civilization today. Marcuse argues that, despite the popular view that technology is supposed to yield efficient solutions to the world's problems, it may be more accurate to say that technology is a primary cause of many problems in modern society.

chapter summary

What is environmental sociology?

Any society is a human ecosystem with interacting and interdependent forces, consisting of human populations, natural resources, and the state of the environment. *Climate change* is the systematic increase in worldwide surface temperatures and the resulting ecological change. Although some deny that climate change is the result of human behavior, there is little doubt among scientists that climate change is largely the result of human activity.

What is meant by environmental racism?

Environmental racism is the pattern whereby toxic dumps are found more frequently in or very near African American, Hispanic, and Native American communities. Disasters also have a social dimension that can make certain population more vulnerable than others.

What are the basic dimensions of population development?

Demography is the scientific study of population. The total number of people in a society at any given moment is determined by only three variables: births, deaths, and migrations.

How is diversity becoming more important in population change?

Diversity in a population occurs along lines of sex, age, as well as race and ethnicity. Currently, the United States is rapidly becoming more diverse because of immigration, but also fertility and mortality rates that vary among different groups.

How do sociologists explain population growth?

Malthusian theory warns us about the dangers of exponential population growth that only such calamities as famine and war could prevent. Others have also warned that there could be a population bomb, with the size of the population outpacing the capacity to support people. *Demographic transition theory* postulates that societies pass through sequences of population patterns linked to developmental stages affecting the birth and death rates.

What are the different sources of social change?

Social change is that process by which social interaction, the social stratification system, and entire institutions in a society change over time. Sources of social change include *cultural diffusion*, technological innovation, *social movements/collective behavior*, and war and *revolution*.

What sociological theories explain social change?

Functionalist theories explain that societies move or evolve from the structurally simple to the structurally complex. *Conflict theories* predict that social conflict is an inherent part of any social structure and that conflict between social class strata or racial–ethnic groups can bring about social change. *Symbolic interaction theory* studies microlevel processes by which people attribute new meanings to beliefs and behaviors.

What changes are brought about by globalization and modernization?

Globalization is the increased interconnectedness and interdependence of societies around the world. It is being coupled with *modernization*, a process of social and cultural change initiated by industrialization and resulting in increased social differentiation and a strong division of labor. Modernization can produce feelings of powerlessness among individuals as society becomes more stratified and complex.

Key Terms

birthrate (or crude birthrate) 399
census 398
climate change 394
cohort 403
collective behavior 410
cultural diffusion 407
culture lag 406

death rate (or crude death rate) 401
demographic transition theory 405
demography 398
emigration 399
environmental racism 397

environmental sociology 392
globalization 412
immigration 399
infant mortality rate 401
Malthusian theory 404
mechanical solidarity 411
modernization 412

organic solidarity 411
population density 399
population pyramid 402
revolution 410
sex ratio 402
social change 406
social movements 408
urbanization 414

Glossary

absolute poverty the situation in which individuals live on less than $365 a year, or $1.00 a day.

achieved status a status attained by effort.

achievement test test intended to measure what is actually learned rather than potential.

adult socialization the process of learning new roles and expectations in adult life.

affirmative action a method for opening opportunities to women and minorities that specifically redresses past discrimination by taking positive measures to recruit and hire previously disadvantaged groups.

Affordable Care Act the nation's health care reform law that has extended some health care insurance to larger segments of the U.S. population.

age cohort an aggregate group of people born during the same time period.

age discrimination different and unequal treatment of people based solely on their age.

age prejudice a negative attitude about an age group that is generalized to all people in that group.

age stereotype preconceived judgments about what different age groups are like.

age stratification the hierarchical ranking of age groups in society.

ageism the institutionalized practice of age prejudice and discrimination.

alienation the feeling of powerlessness and separation from one's group or society.

altruistic suicide the type of suicide that can occur when there is excessive regulation of individuals by social forces.

anomic suicide the type of suicide occurring when there are disintegrating forces in the society that make individuals feel lost or alone.

anomie the condition existing when social regulations (norms) in a society break down.

anticipatory socialization the process of learning the expectations associated with a role one expects to enter in the future.

anti-Semitism the belief or behavior that defines Jewish people as inferior and that targets them for stereotyping, mistreatment, and acts of hatred.

ascribed status a status determined at birth.

assimilation process by which a minority becomes socially, economically, and culturally absorbed within the dominant society.

attribution error error made in attributing the causes for someone's behavior to their membership in a particular group, such as a racial group.

attribution theory the principle that dispositional attributions are made about others (what the other is "really like") under certain conditions, such as out-group membership.

authoritarian personality a personality characterized by a tendency to rigidly categorize people and to submit to authority, rigidly conform, and be intolerant of ambiguity.

authoritarian state where power is concentrated in the hands of a very few individuals who rule through centralized power and control.

authority power that is perceived by others as legitimate.

automation the process by which human labor is replaced by machines.

autonomous state model a theoretical model of the state that interprets the state as developing interests of its own, independent of other interests.

aversive racism subtle, nonovert, and nonobvious racism.

beliefs shared ideas held collectively by people within a given culture.

bilateral kinship a kinship system where descent is traced through the father and the mother.

biological determinism explanations that attribute complex social phenomena to physical characteristics.

bioterrorism a form of terrorism involving the dispersion of chemical or biological substances intended to cause widespread disease and death.

birthrate the number of babies born each year for every 1000 members of the population.

Brown v. Board of Education the 1954 Supreme Court decision that ruled separate but equal public facilities to be unconstitutional.

bureaucracy a type of formal organization characterized by an authority hierarchy, a clear division of labor, explicit rules, and impersonality.

capitalism an economic system based on the principles of market competition, private property, and the pursuit of profit.

caste system a system of stratification (characterized by low social mobility) in which one's place in the stratification system is determined by birth.

census a count of the entire population of a country.

charisma a quality attributed to individuals believed by their followers to have special powers.

charismatic authority authority derived from the personal appeal of a leader.

child labor work that deprives children of their childhood, their potential, and their dignity, and that is harmful to mental and physical development.

church a formal organization that sees itself and is seen by society as a primary and legitimate religious institution.

Civil Rights Act of 1964 federal law prohibiting discrimination on the basis of race, color, national origin, religion, or sex.

class see social class.

class consciousness the awareness that a class structure exists and the feeling of shared identification with others in one's class with whom one perceives common life chances.

class system the organized pattern of social class in society.

climate change the systematic increase in worldwide surface temperatures and the resulting ecological change.

coalition an alliance formed by two or more individuals or groups against another individual or one or more groups to achieve certain ends.

coercive organization organizations for which membership is involuntary; examples are prisons and mental hospitals.

cohort (birth cohort) see age cohort.

collective behavior (action) behavior that occurs when the usual conventions are suspended and people collectively establish new norms of behavior in response to an emerging situation.

collective consciousness the body of beliefs that are common to a community or society and that give people a sense of belonging.

colonialism system by which Western nations became wealthy by taking raw materials from other societies (the colonized) and reaping profits from products finished in the homeland.

color-blind racism ignoring legitimate racial, ethnic, and cultural differences between groups, thus denying the reality of such differences.

coming out the process of defining oneself as gay or lesbian.

commodity chain the network of production and labor processes by which a product becomes a finished commodity. By following the commodity chain, it is evident which countries gain profits and which ones are being exploited.

communism an economic system where the state is the sole owner of the systems of production.

concentrated poverty refers to geographic areas where large percentages of people are poor.

concept any abstract characteristic or attribute that has the potential to be measured.

conflict theory a theoretical perspective that emphasizes the role of power and coercion in producing social order.

conspicuous consumption the ostentatious display of goods to mark one's social status.

contact theory the theory that prejudice will be reduced through social interaction with those of different race or ethnicity but of equal status.

content analysis the analysis of meanings in cultural artifacts such as books, songs, and other forms of cultural communication.

contingent worker those who do not hold regular jobs, but whose employment is dependent upon demand.

controlled experiment a method of collecting data that can determine whether something actually causes something else.

core countries (core nations) within world systems theory, those nations that are more technologically advanced.

correlation the degree of positive (direct) or negative (inverse) association between two variables.

counterculture subculture created as a reaction against the values of the dominant culture.

covert participant observation the form of participant observation wherein the observed individuals are not told that they are being studied.

crime one form of deviance; specifically, behavior that violates criminal laws.

criminology the study of crime from a scientific perspective.

cross-tabulation a table that shows how the categories of two variables are related.

crude birthrate the number of babies born each year for every 1000 members of the population.

crude death rate the number of deaths each year per 1000 members of the population.

cult a religious group devoted to a specific cause or charismatic leader.

cultural capital (also known as *social capital*) cultural resources that are socially designated as being worthy (such as knowledge of elite culture) and that give advantages to groups possessing such capital.

cultural diffusion the transmission of cultural elements from one society or cultural group to another.

cultural hegemony the pervasive and excessive influence of one culture throughout society.

cultural relativism the idea that something can be understood and judged only in relationship to the cultural context in which it appears.

culture the complex system of meaning and behavior that defines the way of life for a given group or society.

culture lag the delay in cultural adjustments to changing social conditions.

culture of poverty the argument that poverty is a way of life and, like other cultures, is passed on from generation to generation.

culture shock the feeling of disorientation that can come when one encounters a new or rapidly changed cultural situation.

cyberspace interaction interaction occurring when two or more people share a virtual reality experience via electronic communication and interaction with each other.

cyberterrorism the use of the computer to commit one or more terrorist acts.

data the systematic information that sociologists use to investigate research questions.

data analysis the process by which sociologists organize collected data to discover what patterns and uniformities are revealed.

death rate the number of deaths each year per 1000 people.

debriefing a process whereby a researcher explains the true purpose of a research study to a subject (respondent); usually done after completion of the study.

debunking looking behind the facades of everyday life.

deductive reasoning the process of creating a specific research question about a focused point, based on a more general or universal principle.

deindividuation the feeling that one's self has merged with a group.

deindustrialization the transition from a predominantly goods-producing economy to one based on the provision of services.

democracy system of government based on the principle of representing all people through the right to vote.

demographic transition theory argument that countries pass through a consistent sequence of population patterns linked to the degree of development and ending with a low birth and death rate.

demography the scientific study of population.

dependency theory the global theory maintaining that industrialized nations hold less-industrialized nations in a dependent, thus exploitative, relationship that benefits the industrialized nations at the expense of the less-industrialized ones.

dependent variable the variable that is a presumed effect (*see also* independent variable).

deviance behavior that is recognized as violating expected rules and norms.

deviant career continuing to be labeled as deviant even after the initial (primary) deviance may have ceased.

deviant community groups that are organized around particular forms of social deviance.

deviant identity the definition a person has of himself or herself as a deviant.

differential association theory theory that interprets deviance as behavior one learns through interaction with others.

digital divide the persistence of inequality in people's access to electronic information.

digital racial divide pattern whereby Blacks and Hispanics are on average less likely than Whites to use digital modes of communication and information, particularly the Internet.

discrimination overt negative and unequal treatment of the members of some social group or stratum solely because of their membership in that group or stratum.

disengagement theory theory predicting that as people age, they gradually withdraw from participation in society and are simultaneously relieved of responsibilities.

diversity the variety of group experiences that result from the social structure of society.

division of labor the systematic interrelation of different tasks that develops in complex societies.

doing gender a theoretical perspective that interprets gender as something accomplished through the ongoing social interactions people have with one another.

dominant culture the culture of the most powerful group in society.

dominant group the group that assigns a racial or ethnic group to subordinate status in society.

dual labor market the division of the labor market into two segments—the primary and secondary labor markets.

dual labor market theory a theory that contends that the labor market is divided into two segments—the primary and secondary labor markets.

dyad a group consisting of two people.

economic restructuring contemporary transformations in the basic structure of work that are permanently altering the workplace, including the changing composition of the workplace, deindustrialization, the use of enhanced technology, and the development of a global economy.

economy the system on which the production, distribution, and consumption of goods and services is based.

educational attainment the total years of formal education.

egoistic suicide the type of suicide that occurs when people feel totally detached from society.

elite deviance the wrongdoing of powerful individuals and organizations.

emigration (*versus immigration*) migration of people from one society to another (also called out-migration).

emotional labor work that is explicitly intended to produce a desired state of mind in a client.

empirical refers to something that is based on careful and systematic observation.

Enlightenment the period in eighteenth- and nineteenth-century Europe characterized by faith in the ability of human reason to solve society's problems.

environmental racism the dumping of toxic wastes with disproportionate frequency at or very near areas with high concentrations of minorities.

environmental sociology the scientific study of the interdependencies that exist between humans and our physical environment.

Equal Pay Act of 1963 first legislation requiring equal pay for equal work.

Equal Rights Amendment a constitutional principle, never passed, guaranteeing that equality of rights under the law shall not be denied or abridged on the basis of sex.

estate system a system of stratification in which the ownership of property and the exercise of power is monopolized by an elite or noble class that has total control over societal resources.

ethnic group a social category of people who share a common culture, such as a common language or dialect, a common religion, or common norms, practices, and customs.

ethnocentrism the belief that one's in-group is superior to all out-groups.

ethnomethodology a technique for studying human interaction by deliberately disrupting social norms and observing how individuals attempt to restore normalcy.

eugenics a social movement in the early twentieth century that sought to apply scientific principles of genetic selection to "improve" the offspring of the human race.

evaluation research research assessing the effect of policies and programs.

expressive needs needs for intimacy, companionship, and emotional support.

extended families the whole network of parents, children, and other relatives who form a family unit and often reside together.

extreme poverty the situation in which people live on less than $275 a year, or $1.25 a day.

false consciousness the thought resulting from subordinate classes internalizing the view of the dominant class.

family a primary group of people—usually related by ancestry, marriage, or adoption—who form a cooperative economic unit to care for any offspring (and each other) and who are committed to maintaining the group over time.

Family and Medical Leave Act (FMLA) federal law requiring employers of a certain size to grant leave to employees for purposes of family care.

feminism a way of thinking and acting that advocates a more just society for women.

feminist theory analyses of women and men in society intended to improve women's lives.

feminization of poverty the process whereby a growing proportion of the poor are women and children.

folkways the general standards of behavior adhered to by a group.

formal organization a large secondary group organized to accomplish a complex task or set of tasks.

forms of racism types or versions of racism, such as traditional racism, aversive racism, institutional racism, and others.

functionalism a theoretical perspective that interprets each part of society in terms of how it contributes to the stability of the whole society.

game stage the stage in childhood when children become capable of taking a multitude of roles at the same time.

game theory a mathematical theory that regards human interaction as a game, thus characterized by strategies, rewards and punishments, and winners and losers.

gemeinschaft German for *community*, a state characterized by a sense of common feeling among the members of a society, including strong personal ties, sturdy primary group memberships, and a sense of personal loyalty to one another; associated with rural life.

gender socially learned expectations and behaviors associated with members of each sex.

gender apartheid the extreme segregation and exclusion of women from public life.

gender gap gender differences in behavior, such as voting behavior.

gender identity one's definition of self as a woman or man.

gender inequality index measure of three key components of women's lives, including reproductive health, empowerment, and labor market status.

gender segregation the distribution of men and women in different jobs in the labor force.

gender socialization the process by which men and women learn the expectations associated with their sex.

gender stratification the hierarchical distribution of social and economic resources according to gender.

gendered institutions the total pattern of gender relations that structure social institutions, including the stereotypical expectations, interpersonal relationships, and the different placement of men and women that are found in institutions.

generalization applying information obtained on a small sample of units (such as people) to a larger population of the units.

generalized other an abstract composite of social roles and social expectations.

gesellschaft a type of society in which increasing importance is placed on the secondary relationships people have—that is, less intimate and more instrumental relationships.

Gini coefficient measure of income distribution within a given population.

glass ceiling popular concept referring to the limits that women and minorities experience in job mobility.

global assembly line an international division of labor where research and development is conducted in an industrial country (for example, United States, Germany, Japan), and the assembly of goods is done primarily in underdeveloped and poor nations, mostly by women and children.

global culture the diffusion of a single culture throughout the world.

global economy term used to refer to the fact that all dimensions of the economy now cross national borders.

global outsourcing process by which jobs are located overseas even while supporting U.S.-based businesses.

global stratification the systematic inequalities between and among different groups within nations that result from the differences in wealth, power, and prestige of different societies relative to their position in the international economy.

globalization increased economic, political, and social interconnectedness and interdependence among societies in the world.

government those state institutions that represent the population and make rules that govern the society.

gross national income (GNI) the total output of goods and services produced by residents of a country each year plus the income from nonresident sources, divided by the size of the population.

group a collection of individuals who interact and communicate, share goals and norms, and who have a subjective awareness as "we".

group size effect the effect upon the person of groups of varying sizes.

groupthink the tendency for group members to reach a consensus at all costs.

hate crime assaults and other malicious acts (including crimes against property) motivated by various forms of bias, including that based on race, religion, sexual orientation, ethnic and national origin, or disability.

Hawthorne effect the effect of the research process itself on the groups or individuals being studied; hence, the act of studying them often itself changes them.

heterosexism institutional structures that define heterosexuality as the only social legitimate sexual orientation.

homophobia the fear and hatred of gays and lesbians.

human capital theory a theory that explains differences in wages as the result of differences in the individual characteristics of the workers.

hypersegregation a pattern of extreme racial, ethnic, and/or social class residential segregation, such that nearly all individuals in an area are of one such group.

hypothesis a statement about what one expects to find in research.

identity how one defines oneself.

ideology a belief system that tries to explain and justify the status quo.

imitation stage the stage in childhood when children copy the behavior of those around them.

immigration (*versus emigration*) the migration of people into a society from outside it (also called in-migration).

impression management a process by which people attempt to control how others perceive them.

imprinting a process whereby a newly hatched or newborn member of a species attaches itself to the first object "seen" by it, whether or not it is the mother, and whether it is an animal, human, or a physical object.

income the amount of money brought into a household from various sources during a given year (wages, investment income, dividends, etc.).

independent variable a variable that is the presumed cause of a particular result (*see* dependent variable).

index crimes the FBI's tallying of violent crimes of murder, manslaughter, rape, robbery, and aggravated assault, plus property crimes.

indicator something that points to or reflects an abstract concept.

Individualized Education Plan (IEP) a list of programs and services available at a school to help students successfully learn; programs include options for students with both learning disabilities and physical disabilities.

inductive reasoning the process of arriving at general conclusions from specific observations.

infant mortality rate the number of deaths per year of infants under age 1 for every 1000 live births.

informant in covert participant observation research, a single group member who provides "inside" information about the group being studied.

informed consent a formal acknowledgment by research subjects (respondents) that they understand the purpose of the research and agree to be studied.

institutional racism racism involving notions of racial or ethnic inferiority that have become ingrained into society's institutions.

instrumental needs emotionally neutral, task-oriented (goal-oriented) needs.

interest group a constituency in society organized to promote its own agenda.

interlocking directorate organizational linkages created when the same people sit on the boards of directors of a number of different corporations.

internalization a process by which a part of culture becomes incorporated into the personality.

international division of labor system of labor whereby products are produced globally, while profits accrue only to a few.

intersectional perspective analysis that interprets class and race and gender as having separate *as well as* combined effects in shaping people's experiences.

intersexed person a person born with the physical characteristics of both sexes.

issues problems that affect large numbers of people and have their origins in the institutional arrangements and history of a society.

job displacement the permanent loss of certain job types when employment patterns shift, as when a manufacturing plant shuts down.

kinship system the pattern of relationships that defines people's family relationships to one another.

labeling theory a theory that interprets the responses of others as most significant in understanding deviant behavior.

labor force participation rate the percentage of those in a given category who are employed.

laissez-faire racism maintaining the status quo of racial groups by persistent stereotyping and blaming of minorities themselves for achievement and socioeconomic gaps between groups.

language a set of symbols and rules that, when put together in a meaningful way, provides a complex communication system.

latent functions subtle, unintended consequences of an institutional element of which the participants (the people) are usually unaware.

law the written set of guidelines that define what is right and wrong in society.

liberal feminism a feminist theoretical perspective asserting that the origin of women's inequality is in traditions of the past that pose barriers to women's advancement.

life chances the opportunities that people have in common by virtue of belonging to a particular class.

life course the connection between people's personal attributes, the roles they occupy, the life events they experience, and the social and historical context of these events.

life expectancy the average number of years individuals and particular groups can expect to live.

looking-glass self the idea that people's conception of self arises through reflection about their relationship to others.

macroanalysis analysis of the whole of society, how it is organized and how it changes.

Malthusian theory after T. R. Malthus, the principle that a population tends to grow faster than the subsistence (food) level needed to sustain it.

mass media channels of communication that are available to very wide segments of the population.

master status some characteristic of a person that overrides all other features of the person's identity.

material culture the objects created in a given society.

matriarchal religion a religion based on the centrality of a female goddess or goddesses.

matriarchy a society or group in which women have power over men.

matrilineal kinship kinship systems in which family lineage (or ancestry) is traced through the mother.

matrilocal a pattern of family residence in which married couples reside with the family of the wife.

McDonaldization the increasing and ubiquitous presence of the fast-food model in vast numbers of organizations.

mean the sum of a set of values divided by the number of cases from which the values are obtained; an average.

mechanical solidarity unity based on similarity, not difference, of roles.

median the midpoint in a series of values that are arranged in numerical order.

median income the midpoint of all household incomes.

Medicaid a governmental assistance program that provides health care assistance for the poor, including the elderly.

medicalization of deviance explanations of deviant behavior that interpret deviance as the result of individual pathology or sickness.

Medicare a governmental assistance program established in the 1960s to provide health services for older Americans.

meritocracy a system in which one's status is based on merit or accomplishments.

microanalysis analysis of the smallest, most immediately visible parts of social life, such as people interacting.

minority group any distinct group in society that shares common group characteristics and is forced to occupy low status in society because of prejudice and discrimination.

mode the most frequently appearing score among a set of scores.

modernization a process of social and cultural change that is initiated by industrialization and followed by increased social differentiation and division of labor.

modernization theory a view of globalization in which global development is a worldwide process affecting nearly all societies that have been touched by technological change.

monogamy the marriage practice of a sexually exclusive marriage with one spouse at a time.

monotheism the worship of a single god.

mores strict norms that control moral and ethical behavior.

multidimensional poverty index measure of poverty that accounts for health, education, and the standard of living.

multinational corporation corporations that conduct business across national borders.

multiracial feminism form of feminist theory noting the exclusion of women of color from other forms of theory and centering its analysis in the experiences of all women.

nationalism the strong identity associated with an extreme sense of allegiance to one's culture or nation.

neocolonialism a form of control of poor countries by rich countries, but without direct political or military involvement.

net worth the value of one's financial assets minus debt.

newly industrializing countries (NICs) countries that have shown rapid growth and have emerged as developed countries.

nonmaterial culture the norms, laws, customs, ideas, and beliefs of a group of people.

nonverbal communication communication by means other than speech, as by touch, gestures, use of distance, eye movements, and so on.

non-zero sum game a game in which the amount of reward is not equal to the amount of loss; thus reward plus loss is not zero, as with a zero-sum game.

normative organization an organization having a voluntary membership and that pursues goals; examples are the PTA or a political party.

norms the specific cultural expectations for how to act in a given situation.

nuclear family family in which a married couple resides together with their children.

obesity a classification of someone who is severely overweight and at risk for a variety of health problems.

occupational prestige the subjective evaluation people give to jobs as better or worse than others.

occupational segregation a pattern in which different groups of workers are separated into different occupations.

old-fashioned racism overt and obvious expressions of racism, such as physical assaults, lynchings, and other such acts against a minority.

organic solidarity unity based on role differentiation, not similarity.

organic metaphor refers to the similarity early sociologists saw between society and other organic systems.

organizational culture the collective norms and values that shape the behavior of people within an organization.

organizational ritualism a situation in which rules become ends in themselves rather than means to an end.

organized crime crime committed by organized groups, typically involving the illegal provision of goods and services to others.

out-group homogeneity effect the tendency for an in-group member to perceive members of any out-group as similar or identical to each other.

outsourcing transferring a specialized task or job from one organization to a different organization, usually in another country, as a cost-saving device.

overt participant observation the form of participant observation wherein the observed individuals are told that they are being studied.

participant observation a method whereby the sociologist becomes both a participant in the group being studied and a scientific observer of the group.

patriarchal religion religion in which the beliefs and practices of the religion are based on male power and authority.

patriarchy a society or group where men have power over women.

patrilineal kinship a kinship system that traces descent through the father.

patrilocal a pattern of family residence in which married couples reside with the family of the husband.

peers those of similar status.

percentage the number of parts per hundred.

peripheral countries (nations) poor countries, largely agricultural, having little power or influence in the world system.

personal crimes violent or nonviolent crimes directed against people.

personality the cluster of needs, drives, attitudes, predispositions, feelings, and beliefs that characterize a given person.

play stage the stage in childhood when children begin to take on the roles of significant people in their environment.

pluralism pattern whereby groups maintain their distinctive culture and history.

pluralist model a theoretical model of power in society as coming from the representation of diverse interests of different groups in society.

polarization shift a shift of group opinion in terms of risk, either an increase or a decrease in degree of risk, from before discussion to after discussion.

political action committees (PACs) groups of people who organize to support candidates they feel will represent their views.

polygamy a marriage practice in which either men or women can have multiple marriage partners.

polytheism the worship of more than one deity.

Ponzi scheme a criminal method (a type of "pyramid scheme") of using new investors' funds to pay off original investors under the guise that the funds are being legitimately invested in stocks and bonds.

popular culture the beliefs, practices, and objects that are part of everyday traditions.

population a relatively large collection of people (or other unit) that a researcher studies and about which generalizations are made.

population density the number of people per square mile.

population pyramid graphic depictions of the age and sex distribution of a given population at a point in time.

positivism a system of thought that regards scientific observation to be the highest form of knowledge.

postindustrial society a society economically dependent upon the production and distribution of services, information, and knowledge.

poverty line the figure established by the government to indicate the amount of money needed to support the basic needs of a household.

power a person or group's ability to exercise influence and control over others.

power elite model a theoretical model of power positing a strong link between government and business.

preindustrial society one that directly uses, modifies, and/or tills the land as a major means of survival.

prejudice the negative evaluation of a social group, and individuals within that group, based upon conceptions about that social group that are held despite facts that contradict it.

prestige the value with which different groups or people are judged.

primary group a group characterized by intimate, face-to-face interaction and relatively long-lasting relationships.

profane that which is of the everyday, secular world and is specifically not religious.

propaganda information disseminated by a group or organization (such as the state) intended to justify its own power.

property crimes crimes involving theft of or harm to property without bodily harm to the victim(s).

Protestant ethic belief that hard work and self-denial lead to salvation.

proxemic communication meaning conveyed by the amount of space between interacting individuals.

psychoanalytic theory a theory of socialization positing that the unconscious mind shapes human behavior.

qualitative research research that is somewhat less structured than quantitative research but that allows more depth of interpretation and nuance in what people say and do.

quantitative research research that uses numerical analysis.

queer theory a theoretical perspective that recognizes the socially constructed nature of sexual identity.

race a social category, or social construction, that we treat as distinct on the basis of certain characteristics, some biological, that have been assigned social importance in the society.

racial formation process by which groups come to be defined as a "race" through social institutions such as the law and the schools.

racial profiling the use of race alone as a criterion for deciding whether to stop and detain someone on suspicion of having committed a crime.

racialization a process whereby some social category, such as a social class or nationality, is assigned what are perceived to be race characteristics.

racism the perception and treatment of a racial or ethnic group, or member of that group, as intellectually, socially, and culturally inferior to one's own group.

random sample a sample that gives everyone in the population an equal chance of being selected.

rate parts per some number (for example, per 10,000; per 100,000).

rational–legal authority authority stemming from rules and regulations, typically written down as laws, procedures, or codes of conduct.

reference group any group (to which one may or may not belong) used by the individual as a standard for evaluating her or his attitudes, values, and behaviors.

reflection hypothesis the idea that the mass media reflect the values of the general population.

relative poverty a definition of poverty that is set in comparison to a set standard.

reliability the likelihood that a particular measure would produce the same results if the measure were repeated.

religion an institutionalized system of symbols, beliefs, values, and practices by which a group of people interprets and responds to what they feel is sacred and that provides answers to questions of ultimate meaning.

religiosity the intensity and consistency of practice of a person's (or group's) faith.

replication study research that is repeated exactly, but on a different group of people at a different point in time.

research design the overall logic and strategy underlying a research project.

resegregation the process by which once integrated schools become more racially segregated.

residential segregation the spatial separation of racial and ethnic groups in different residential areas.

resocialization the process by which existing social roles are radically altered or replaced.

revolution the overthrow of a state or the total transformation of central state institutions.

risky shift (also polarization shift) the tendency for group members, after discussion and interaction, to engage in riskier behavior than they would while alone.

rite of passage ceremony or ritual that symbolizes the passage of an individual from one role to another.

ritual a symbolic activity that expresses a group's spiritual convictions.

role behavior others expect from a person associated with a particular status.

role conflict two or more roles associated with contradictory expectations.

role modeling imitation of the behavior of an admired other.

role set all roles occupied by a person at a given time.

role strain conflicting expectations within the same role.

sacred that which is set apart from ordinary activity, seen as holy, and protected by special rites and rituals.

salience principle categorizing people on the basis of what initially appears prominent about them.

sample any subset of units from a population that a researcher studies.

Sapir–Whorf hypothesis a theory that language determines other aspects of culture because language provides the categories through which social reality is defined and perceived.

scapegoat theory argument that dominant group aggression is directed toward a minority as a substitute for frustration with some other problem.

schooling socialization that involves formal and institutionalized aspects of education.

scientific method the steps in a research process, including observation, hypothesis testing, analysis of data, and generalization.

secondary group a group that is relatively large in number and not as intimate or long in duration as a primary group.

sect groups that have broken off from an established church.

secular the ordinary beliefs of daily life that are specifically not religious.

segregation the spatial and social separation of racial and ethnic groups.

self our concept of who we are, as formed in relationship to others.

self-concept a person's image and evaluation of important aspects of oneself.

self-fulfilling prophecy the process by which merely applying a label changes behavior and thus tends to justify the label.

semiperipheral countries semi-industrialized countries that represent a kind of middle class within the world system.

serendipity unanticipated, yet informative, results of a research study.

sex used to refer to biological identity as male or female.

sex ratio (gender ratio) the number of males per 100 females.

sex tourism practice whereby people travel to engage in commercial sexual activity.

sex trafficking refers to the practice whereby women, usually very young women, are forced by fraud or coercion into commercial sex acts.

sexual harassment unwanted physical or verbal sexual behavior that occurs in the context of a relationship of unequal power and that is experienced as a threat to the victim's job or educational activities.

sexual identity the definition of oneself that is formed around one's sexual relationships.

sexual orientation the attraction that people feel for people of the same or different sex.

sexual politics the link feminists argue exists between sexuality and power, and between sexuality and race, class, and gender oppression.

sexual revolution the widespread changes in men's and women's roles and a greater public acceptance of sexuality as a normal part of social development.

sexual scripts the ideas taught to us about what is appropriate sexual behavior for a person of our gender.

significant others those with whom we have a close affiliation.

social capital *see* cultural capital.

social change the alteration of social interaction, social institutions, stratification systems, and elements of culture over time.

social class the social structural hierarchical position groups hold relative to the economic, social, political, and cultural resources of society.

social construction perspective a theoretical perspective that explains identity and society as created and learned within a cultural, social, and historical context.

social control the process by which groups and individuals within those groups are brought into conformity with dominant social expectations.

social control agents those who regulate and administer the response to deviance, such as the police or mental health workers.

social control theory theory that explains deviance as the result of the weakening of social bonds.

social Darwinism the idea that society evolves to allow the survival of the fittest.

social differentiation the process by which different statuses in any group, organization, or society develop.

social facts social patterns that are external to individuals.

social institution an established and organized system of social behavior with a recognized purpose.

social interaction behavior between two or more people that is given meaning.

social learning theory a theory of socialization positing that the formation of identity is a learned response to social stimuli.

social media the term used to refer to the vast networks of social interaction that new media have created.

social mobility a person's movement over time from one class to another.

social movement a group that acts with some continuity and organization to promote or resist social change in society.

social network a set of links between individuals or other social units such as groups or organizations.

social organization the order established in social groups.

social sanctions mechanisms of social control that enforce norms.

social stratification a relatively fixed hierarchical arrangement in society by which groups have different access to resources, power, and perceived social worth; a system of structured social inequality.

social structure the patterns of social relationships and social institutions that make up society.

socialism an economic institution characterized by state ownership and management of the basic industries.

socialization the process through which people learn the expectations of society.

socialization agents those who pass on social expectations.

society a system of social interaction that includes both culture and social organization.

socioeconomic status (SES) a measure of class standing, typically indicated by income, occupational prestige, and educational attainment.

sociological imagination the ability to see the societal patterns that influence individual and group life.

sociology the study of human behavior in society.

spatial mismatch the pattern whereby jobs are located in a geographic area located away from groups who need work.

spurious correlation a false correlation between X and Y, produced by their relationship to some third variable (Z) rather than by a true causal relationship to each other.

state the organized system of power and authority in society.

status an established position in a social structure that carries with it a degree of prestige.

status attainment the process by which people end up in a given position in the stratification system.

status inconsistency exists when the different statuses occupied by the individual bring with them significantly different amounts of prestige.

status set the complete set of statuses occupied by a person at a given time.

stereotype an oversimplified set of beliefs about the members of a social group or social stratum that is used to categorize individuals of that group.

stereotype interchangeability the principle that negative stereotypes are often interchangeable from one racial group (or gender or social class) to another.

stereotype threat the effect of a negative stereotype about one's self upon one's own test performance.

stigma an attribute that is socially devalued and discredited.

Stockholm syndrome a process whereby a captured person identifies with the captor as a result of becoming inadvertently dependent upon the captor.

structural strain theory a theory that interprets deviance as originating in the tensions that exist in society between cultural goals and the means people have to achieve those goals.

subculture the culture of groups whose values and norms of behavior are somewhat different from those of the dominant culture.

symbolic interaction theory a theoretical perspective claiming that people act toward things because of the meaning things have for them.

symbols things or behavior to which people give meaning.

taboos those behaviors that bring the most serious sanctions.

tactile communication patterns of touch, influenced by gender, that express emotional support, assert power, or express sexual interest.

taking the role of the other the process of imagining oneself from the point of view of another.

teacher expectancy effect the effect of a teacher's expectations on a student's actual performance, independent of the student's ability.

Temporary Assistance for Needy Families (TANF) federal program by which grants are given to states to fund welfare.

tenure a guarantee of continuing employment in an organization.

terrorism the unlawful use of force or violence against people or property to intimidate or coerce a government or population in furtherance of political or social objectives.

Title IX legislation that prohibits schools that receive federal funds from discriminating based on gender.

total institution an organization cut off from the rest of society in which individuals are subject to strict social control.

totalitarian state an extreme form of authoritarianism where the state has total control over all aspects of public and private life.

totem an object or living thing that a religious group regards with special awe and reverence.

tracking grouping, or stratifying, students in school on the basis of ability test scores.

traditional authority authority stemming from long-established patterns that give certain people or groups legitimate power in society.

transgendered those who deviate from the binary (that is, male or female) system of gender.

transnational family families where one parent (or both) lives and works in one country while the children remain in their country of origin.

triad a group consisting of three people.

triadic segregation the tendency for a triad to separate into a dyad and an isolate.

troubles privately felt problems that come from events or feelings in one individual's life.

underemployment the condition of being employed at a skill level below what would be expected given a person's training, experience, or education.

unemployment rate the percentage of those not working, but officially defined as looking for work.

urban underclass a grouping of people, largely minority and poor, who live at the absolute bottom of the socioeconomic ladder in urban areas.

urbanization the process by which a community acquires the characteristics of city life.

utilitarian organization a profit or nonprofit organization that pays its employees salaries or wages.

validity the degree to which an indicator accurately measures or reflects a concept.

values the abstract standards in a society or group that define ideal principles.

variable something that can have more than one value or score.

verstehen the process of understanding social behavior from the point of view of those engaged in it.

victimless crimes violations of law not listed in the FBI's serious crime index, such as gambling or prostitution.

wealth the monetary value of everything one actually owns.

White privilege the ability for Whites to maintain an elevated status in society that masks racial inequality.

work productive human activity that produces something of value, either goods or services.

world cities cities that are closely linked through the system of international commerce.

world systems theory theory that capitalism is a single world economy and that there is a worldwide system of unequal political and economic relationships that benefit the technologically advanced countries at the expense of the less technologically advanced.

xenophobia the fear and hatred of foreigners.

zero-sum game an interpersonal game in which for all concerned, the total amount of reward (winnings) exactly equals the total amount of punishment (losses).

References

Aberle, David F., Albert K. Cohen, A. Kingsley Davis, Marion J. Levy Jr., and Francis X. Sutton. 1950. "The Functional Prerequisites of a Society." *Ethics* 60 (January): 100–111.

Abrahamson, Mark. 2006. *Urban Enclaves: Identity and Place in the World,* 2nd ed. New York: Worth.

Acker, Joan. 1992. "Gendered Institutions: From Sex Roles to Gendered Institutions." *Contemporary Sociology* 21 (September): 565–569.

Acker, Joan, Sandra Morgen, and Lisa Gonzales. 2002. "Welfare Restructuring, Work & Poverty: Policy from Oregon." Working Paper, Center for the Study of Women in Society, Eugene, OR.

Acs, Gregory, and Seth Zimmerman. 2008. *U.S. Intergenerational Mobility from 1984 to 2004.* Pew Charitable Trust Economic Mobility Project. **www.pewtrusts.org**

Adorno, T. W., Else Frenkel-Brunswik, D. J. Levinson, and R. N. Sanford. 1950. *The Authoritarian Personality.* New York: Harper and Row.

Alba, Richard. 1990. *Ethnic Identity: The Transformation of Ethnicity in the Lives of Americans of European Ancestry.* New Haven, CT: Yale University Press.

Alba, Richard, and Gwen Moore. 1982. "Ethnicity in the American Elite." *American Sociological Review* 47 (June): 373–383.

Alba, Richard D., and Victor Nee. 2003. *Remaking the American Mainstream: Assimilation and Contemporary Immigration.* Cambridge, MA: Harvard University Press.

Albas, Daniel, and Cheryl Albas. 1988. "Aces and Bombers: The Post-Exam Impression Management Strategies of Students." *Symbolic Interaction* 11: 289–302.

Albelda, Randy, and Ann Withorn (eds.). 2002. *Lost Ground: Welfare Reform, Poverty, and Beyond.* Boston, MA: South End Press.

Alden, Helena L. 2001. "Gender Role Ideology and Homophobia." Paper presented at the Annual Meeting of the Southern Sociological Society.

Aldrich, Howard, and Martin Ruef. 2006. *Organizations Evolving.* Thousand Oaks, CA: Sage.

Alexander, Michelle. 2010. *The New Jim Crow: Mass Incarceration in the Age of Colorblindness.* New York: The New Press.

Alicea, Marixsa. 1997. "'What Is Indian About You?' A Gendered, Transnational Approach to Ethnicity." *Gender & Society* 11 (October): 597–626.

Algars, Monica, Pekka Santtila, and N. Kenneth Sandnabba. 2010. "Conflicted Gender Identity, Body Dissatisfaction, and Disordered Eating in Adult Men and Women." *Sex Roles* 63: 118–125.

Alliance for Board Diversity. 2011. *Missing Pieces: Women and Minorities on Fortune 500 Boards.* **www.theabd.org**

Allison, Anne. 1994. *Nightwork: Sexuality, Pleasure, and Corporate Masculinity in a Tokyo Hostess Club.* Chicago, IL: University of Chicago Press.

Allport, Gordon W. 1954. *The Nature of Prejudice.* Reading, MA: Addison-Wesley.

Alon, Sigal, and Marta Tienda. 2007. "Diversity, Opportunity, and the Shifting Meritocracy in Higher Education." *American Sociological Review* 72 (August): 487–511.

Altemeyer, Bob. 1988. *Enemies of Freedom: Understanding Right-Wing Authoritarianism.* San Francisco, CA: Jossey-Bass.

Altman, Dennis. 2001. *Global Sex.* Chicago, IL: University of Chicago Press.

Amato, Paul R., and Alan Booth. 1996. "A Prospective Study of Divorce and Parent-Child Relationships." *Journal of Marriage and the Family* 58(2): 356–365.

American Association of University Women. 1998. *Gender Gaps: Where Schools Still Fail Our Children.* Washington, DC: American Association of University Women.

American Association of University Women. 2010. "Back to School after All These Years." *AAUW Outlook* 104 (Fall).

American Psychological Association. 2007. *Report of the APA Task Force on the Sexualization of Girls.* Washington, DC: American Psychological Association.

Amott, Teresa L., and Julie A. Matthaei. 1996. *Race, Gender, and Work: A Multicultural History of Women in the United States,* 2nd ed. Boston, MA: South End Press.

Andersen, Margaret L. 2011. *Thinking about Women: Sociological Perspectives on Sex and Gender,* 9th ed. Boston, MA: Allyn and Bacon.

Andersen, Margaret L., and Patricia Hill Collins. 2013. *Race, Class and Gender: An Anthology,* 8th ed. Belmont, CA: Wadsworth/Cengage.

Anderson, David A., and Mykol Hamilton. 2005. "Gender Role Stereotyping of Parents in Children's Picture Books: The Invisible Father." *Sex Roles: A Journal of Research* 52(3–4): 145–151.

Anderson, Elijah. 1976. *A Place on the Corner.* Chicago, IL: University of Chicago Press.

Anderson, Elijah. 1990. *Streetwise: Race, Class, and Change in an Urban Community.* Chicago, IL: University of Chicago Press.

Anderson, Elijah. 1999. *Code of the Street: Decency. Violence, and the Moral Life of the Inner City.* New York: W. W. Norton.

Anderson, Elijah. 2011. *The Cosmopolitan Canopy: Race and Civility in Everyday Life.* W. W. Norton.

Anthony, Dick, Thomas Robbins, and Steven Barrie-Anthony. 2002. "Cult and Anticult Totalism: Reciprocal Escalation and Violence." *Terrorism and Political Violence* 14 (Spring): 211–239.

Antrobus, Peggy. 2002. "Feminism as Transformational Politics: Towards Possibilities for Another World." *Development* 45 (June): 46–52.

Apple, Michael W. 1991. "The New Technology: Is It Part of the Solution or Part of the Problem in Education?" *Computers in the Schools* 8 (April–October): 59–81.

Arendell, Terry. 1998. "Divorce American Style." *Contemporary Sociology* 27 (May): 226–228.

Arendt, Hannah. 1963. *Eichmann in Jerusalem: A Report on the Banality of Evil.* New York: Viking Press.

Armas, Genaro C. 2003. "Asian American Divorce Rate Up." *San Francisco Chronicle:* A4.

Armstrong, Elizabeth M. 2003. *Conceiving Risk, Bearing Responsibility: Fetal Alcohol Syndrome and the Diagnosis of Moral Disorder.* Baltimore, MD: Johns Hopkins University Press.

Armstrong, E. A., L. Hamilton, and B. Sweeney. 2006. "Sexual Assault on Campus: A Multilevel, Integrative Approach to Party Rape." *Social Problems* 53: 483–499.

Armstrong, Elizabeth A., Laura Hamilton, and Paula England. 2010. "Is Hooking Up Bad for Young Women?" *Contexts* 9 (Summer): 23–27.

Arnett, J. 2010. "Emerging Adulthood." *New York Times Magazine* (August 18). **www.nytimes.com**

Arnett, J., and J. Tanner, eds. 2010. *Growing into Adulthood: The Lives and Contexts of Emerging Adults.* Washington, DC: American Psychological Association.

Asch, Solomon. 1951. "Effects of Group Pressure upon the Modification and Distortion of Judgments." In *Groups, Leadership, and Men,* edited by H. Guetzkow. Pittsburgh, PA: Carnegie Press.

Asch, Solomon. 1955. "Opinions and Social Pressure." *Scientific American* 19 (July): 31–35.

Association of Statisticians of American Religious Bodies. 2012. "U.S. Religion Census." **www.asarb.org**

Atkins, Celeste. 2011. "Big Black Mammas: The Intersection of Race, Gender and Weight in the U.S." Unpublished manuscript, Pima Community College, AZ.

Babbie, Earl. 2013. *The Practice of Social Research,* 13th ed. Belmont, CA: Wadsworth/Cengage.

Baca Zinn, Maxine. 1995. "Chicano Men and Masculinity." Pp. 33–41 in *Men's Lives,* 3rd ed., edited by Michael S. Kimmel and Michael A. Messner. Boston, MA: Allyn and Bacon.

Baca Zinn, Maxine, and D. Stanley Eitzen. 2010. *Diversity in Families,* 9th ed. Boston: Allyn and Bacon.

Baca Zinn, Maxine, Pierrette Hondagneu-Sotelo, and Michael Messner. 2010. *Gender through the Prism of Difference.* New York: Oxford University Press.

Baca Zinn, Maxine, and Bonnie Thornton Dill. 1996. "Theorizing Difference from Multiracial Feminism." *Feminist Studies* 22 (Summer): 321–331.

Bailey, Garrick. 2003. *Humanity: An Introduction to Cultural Anthropology,* 6th ed. Belmont, CA: Wadsworth.

Bales, Kevin. 2010. *The Slave Next Door: Human Trafficking and Slavery in America Today.* Berkeley, CA: University of California Press.

Bandura, Albert, and R. H. Walters. 1963. *Social Learning and Personality Development.* New York: Holt, Reinhart, and Winston.

Barber, Benjamin R. 1995. *Jihad vs. McWorld: How Globalism and Tribalism Are Reshaping the World.* New York: Random House.

Barlow, Maude, and Tony Clarke. 2002. "Who Owns Water?" *The Nation* (September 2/9): 11–14.

Barton, Bernadette. 2006. *Stripped: Inside the Lives of Exotic Dancers.* New York: New York University Press.

Basow, Susan A., and Kelly Johnson. 2000. "Predictors of Homophobia in Female College Students." *Sex Roles* 42 (March): 391–404.

Baumeister, Roy F., and Brad J. Bushman. 2008. *Social Psychology and Human Nature.* Belmont, CA: Wadsworth.

Bean, Frank D., and Marta Tienda. 1987. *The Hispanic Population of the United States.* New York: Russell Sage Foundation.

Beasley, Maya, and Mary Fischer. 2012. "Why They Leave: The Impact of Stereotype Threat on the Attrition of Women and Minorities from Science, Math, and Engineering Majors." *Social Psychology of Education* 15(4): 427–448.

Becker, Elizabeth. 2000. "Harassment in the Military Is Said to Rise." *The New York Times* (March 10): 14.

Becker, Howard S. 1963. *Outsiders: Studies in the Sociology of Deviance.* New York: Free Press.

Belknap, Joanne. 2001. *The Invisible Woman.* Belmont, CA: Wadsworth.

Belknap, Joanne, Bonnie S. Fisher, and Francis T. Cullen. 1999. "The Development of a Comprehensive Measure of the Sexual Victimization of College Women." *Violence Against Women* 5 (February): 185–214.

Bell, Derrick. 1992. *Faces at the Bottom of the Well: The Permanence of Racism.* New York: HarperCollins.

Bellah, Robert (ed.). 1973. *Emile Durkheim on Morality and Society: Selected Writings.* Chicago, IL: University of Chicago Press.

Beller, Emily, and Michael Hout. 2006. "Intergenerational Social Mobility: The United States in Comparative Perspective." *The Future of Children* 16 (Fall): 19–36.

Ben-Yehuda, Nachman. 1986. "The European Witch Craze of the Fourteenth-Seventeenth Centuries: A Sociologist's Perspective." *American Journal of Sociology* 86: 1–31.

Benedict, Ruth. 1934. *Patterns of Culture.* Boston, MA: Houghton Mifflin.

Berger, Peter L. 1963. *Invitation to Sociology: A Humanistic Perspective.* Garden City, NY: Doubleday Anchor.

Berger, Peter L., Brigitte Berger, and Hansfried Kellner. 1974. *The Homeless Mind: Modernization and Consciousness.* New York: Vintage Books.

Berger, Peter L., and Thomas Luckmann. 1967. *The Social Construction of Reality: A Treatise in the Sociology of Knowledge.* Garden City, NY: Anchor Books.

Berkman, Lisa F., Thomas Glass, Ian Brissette, and Teresa E. Seeman. 2000. "From Social Integration to Health: Durkheim in the New Millennium." *Social Science and Medicine* 51 (September): 843–857.

Berlin, Ira. 2010. *The Making of African America: The Four Great Migrations.* New York: Viking.

Bernstein, Nina. 2002. "Side Effect of Welfare Law: The No-Parent Family." *The New York Times* (July 29): A1.

Berscheid, Ellen, and Harry R. Reis. 1998. "Attraction and Class Relationships." Pp. 193–281 in *The Handbook of Social Psychology,* 4th ed., edited by Daniel T. Gilbert, Susan T. Fiske, and Gardner Lindzey. New York: Oxford University Press.

Best, Joel. 1999. *Random Violence: How We Talk about New Crimes and New Victims.* Berkeley, CA: University of California Press.

Best, Joel. 2001. *Damned Lies and Statistics: Untangling Numbers from the Media, Politicians, and Activists.* Berkeley, CA: University of California Press.

Best, Joel. 2006. *Flavor of the Month: Why Smart People Fall for Fads.* Berkeley, CA: University of California Press.

Best, Joel. 2008. *Social Problems,* 2nd ed. New York: W.W. Norton.

Best, Joel. 2011. *Everyone's a Winner: Life in Our Own Congratulatory Culture.* Berkeley, CA: University of California Press.

Bimbaum, Jeffrey H. 2004. "Cost of Congressional Campaigns Skyrockets." *The Washington Post* (October 3): A8.

Binns, Allison. 2003. *White Gold, Weed, and Blow: The Drug Trades of Afghanistan, Columbia, and Mexico in Comparative Historical Perspective.* Senior thesis, Princeton University, Princeton, NJ.

Bishaw, Alemayehi. 2011. *Areas with Concentrated Poverty, 2006–2010.* Washington, DC: U.S. Census Bureau.

Black, M. C., K. C. Basile, M. J. Breiding, S. G. Smith, M. I. M. L. Walters, M.T. Merrick, J. Chen, and M. R. Stevens. 2011. *The National Intimate Partner and Sexual Violence Survey: 2010 Summary Report.* Atlanta, GA: National Center for Injury Prevention and Control, Centers for Disease Control and Prevention.

Blair-Loy, Mary, and Amy S. Wharton. 2002. "Employees' Use of Work-Family Policies and the Workplace Social Context." *Social Forces* 80 (3): 813–845.

Blake, C. Fred. 1994. "Footbinding in Neo-Confucian China and the Appropriation of Female Labor." *Signs* 19 (Spring): 676–712.

Blankenship, Kim. 1993. "Bringing Gender and Race In: U.S. Employment Discrimination Policy." *Gender & Society* 7 (June): 204–226.

Blassingame, John. 1973. *The Slave Community: Plantation Life in the Antebellum South.* New York: Oxford University Press.

Blau, Peter M., and Otis Dudley Duncan. 1967. *The American Occupational Structure.* New York: Wiley.

Blau, Peter M., and W. Richard Scott. 1974. *On the Nature of Organizations.* New York: Wiley.

Blinde, Elaine M., Diane E. Taub, and Lingling Han. 1994. "Sport as a Site for Women's Group and Social Empowerment: Perspectives from the College Athlete." *Sociology of Sport Journal* 11 (March): 51–59.

Blumer, Herbert, 1969. *Studies in Symbolic Interaction.* Englewood Cliffs, NJ: Prentice Hall.

Bobo, Lawrence D. 1999. "Prejudice as Group Position: Microfoundations of a Sociological Approach to Racism and Race Relations." *Journal of Social Issues* 55 (3): 445–492.

Bobo, Lawrence D. 2006. "The Color Line, the Dilemma, and the Dream: Race Relations in America at the Close of the Twentieth Century." Pp. 87–95 in *Race and Ethnicity in Society: The Changing Landscape,* edited by Elizabeth Higginbotham and Margaret L. Andersen. Belmont, CA: Wadsworth.

Bobo, Lawrence. 2012 (January). "Post Racial Dreams, American Realities." Paper read before the Department of Sociology, Princeton University, Princeton, NJ.

Bobo, Lawrence, and James R. Kluegel. 1991. "Modern American Prejudice: Stereotypes, Social Distance, and Perceptions of Discrimination toward Blacks, Hispanics, and Asians." Paper presented before the American Sociological Association, Cincinnati, OH.

Bobo, L., and R. A. Smith. 1998. "From Jim Crow Racism to Laissez-Faire Racism: An Essay on the Transformation or Racial Attitudes in America." Pp. 182–220 in *Beyond Pluralism: Essays on the Conception of Groups and Group Identities in America,* edited by W. Katkin, N. Landsmand, and A. Tyree. Urbana, IL: University of Illinois Press.

Bocian, Debbie Gruenstein, Wei Li, and Keith S. Ernst. 2010. *Foreclosures by Race and Ethnicity: The Demographics of a Crisis.* Durham, NC: Center for Responsible Lending.

Boeringer, Scott B. 1999. "Associations of Rape Supportive Attitudes with Fraternal and Athletic Participation." *Violence against Women* 5 (January): 81–90.

Boggs, Vernon W. 1992. *Salsiology: Afro-Cuban Music and the Evolution of Salsa in New York City*. Westport, CT: Greenwood Press.

Bogle, Kathleen. 2008. *Hooking Up: Sex, Dating, and Relationships on Campus*. New York: New York University Press.

Bonilla-Silva, Eduardo. 1997. "Rethinking Racism: Toward a Structural Interpretation." *American Sociological Review* 62 (3): 465–480.

Bonilla-Silva, Eduardo, and Gianpaolo Baiocchi. 2001. "Anything but Racism: How Sociologists Limit the Significance of Racism." *Race and Society* 4 (2): 117–131.

Bonilla-Silva, Eduardo, and Mary Hovespan. 2000. "If Two People Are in Love: Deconstructing Whites' Views on Interracial Marriage with Blacks." Paper presented at the annual meeting of the Southern Sociological Society.

Bontemps, Arna (ed.). 1972. *The Harlem Renaissance Remembered*. New York: Dodd, Mead.

Boonstra, Heather, Rachel Benson Gold, Cory L. Richards, and Lawrence B. Finer. 2006. *Abortion in Women's Lives*. New York: Guttmacher Institute.

Bourdieu, Pierre. 1984. *Distinction: A Social Critique of the Judgement of Taste*, translated by Richard Nice. Cambridge, MA: Harvard University Press.

Bowles, Samuel, Herbert Gintis, and Melissa Osborn Groves (eds.) 2005. "Introduction." Pp. 1–22 in *Unequal Chances: Family Background and Economic Success*. Princeton, NJ: Princeton University Press.

Branch, Taylor. 1998. *Pillar of Fire: America in the King Years, 1963–1965*. New York: Simon and Shuster.

Branch, Taylor. 2006. *At Canaan's Edge: America in the King Years, 1965–1968*. New York: Simon and Schuster.

Brand, Jennie E., and Yu Xie. 2010. "Who Benefits Most from College? Evidence for Negative Selection in Heterogeneous Economic Returns to Higher Education." *American Sociological Review* 75(2): 273–302.

Braunstein, Ruth. 2011. "Who Are 'We the People'"? *Contexts* 10 (May): 72–73.

Brechin, Steven R. 2003. "Comparative Public Opinion and Global Climatic Change and the Kyoto Protocol: The U.S. versus the World?" *International Journal of Sociology and Social Policy* 23: 106–134.

Brehm, S. S., R. S. Miller, D. Perlman, and S. M. Campbell. 2002. *Intimate Relationships*, 3rd ed. Boston, MA: McGraw-Hill.

Bricker, Jesse, Arthur B. Kennickell, Kevin B. Moore, and John Sabelhaus. 2012. "Changes in U.S. Family Finances from 2007 to 2010: Evidence from the Survey of Consumer Finances." *Federal Reserve Bulletin* 98 (June). **www .federalreserve.gov**

Brislen, William, and Clayton D. Peoples. 2005. "Using a Hypothetical Distribution of Grades to Introduce Social Stratification." *Teaching Sociology* 33 (January): 74–80.

Britt, Chester L. 1994. "Crime and Unemployment among Youths in the United States, 1958–1990: A Time Series Analysis." *American Journal of Economics and Sociology* 53 (January): 99–109.

Brodkin, Karen. 2006. "How Did Jews Become White Folks?" Pp. 59–66 in Elizabeth Higginbotham and Margaret L. Andersen, eds., *Race and Ethnicity in Society: The Changing Landscape*. Belmont, CA: Wadsworth.

Brooke, James. 1998. "Utah Struggles with a Revival of Polygamy." *The New York Times* (August 23): A12.

Brookey, Robert Alan. 2001. "Bio-Rhetoric, Background Beliefs, and the Biology of Homosexuality." *Argumentation and Advocacy* 37: 171–183.

Brookings Institute. 2010. *The Suburbanization of Poverty: Trends in Metropolitan America, 2000-2008*. Washington, DC: The Brookings Institute. **www.brookings.edu**

Brooks-Gunn, Jeanne, Wen-Jui Han, and Jane Waldfogel. 2002. "Maternal Employment and Child Cognitive Outcomes in the First Three Years of Life: The NICHD Study of Early Child Care." *Child Development* 73 (July-August): 1052–1072.

Brown, Cynthia. 1993. "The Vanished Native Americans." *The Nation* 257 (October 11): 384–389.

Brown, Elaine. 1992. *A Taste of Power: A Black Woman's Story*. Pantheon: New York.

Brown, Michael K., Martin Carnoy, Troy Duster, Elliott Currie, David B. Oppenheimer, Marjorie Schulz, and David Wellman. 2005. *Whitewashing Race: The Myth of a Color-Blind Society*. Berkeley, CA: University of California Press.

Brown, Roger. 1986. *Social Psychology*, 2nd ed. New York: Free Press.

Browne, Irene (ed.). 1999. *Latinas and African American Women at Work: Race, Gender, and Economic Inequality*. New York: Russell Sage Foundation.

Broyard, Bliss. 2007. *One Drop: My Father's Hidden Life*. New York: Little, Brown.

Brulle, Robert J., and David N. Pellow. 2006. "Environmental Justice: Human Health and Environmental Inequalities." *Annual Review of Public Health* 27: 103–123.

Bryant, Susan L., and Lillian M. Range. 1997. "Type and Severity of Child Abuse and College Students' Lifetime Suicidality." *Child Abuse and Neglect* 21 (December): 1169–1176.

Bryson, Bethany P., and Alexander K. Davis. 2010. "Conquering Stereotypes in Research on Race and Gender." *Sociological Forum* 25 (1): 161–166.

Buchanan, Christy M., Eleanor Maccoby, and Sanford M. Dornsbusch. 1996. *Adolescents after Divorce*. Cambridge, MA: Harvard University Press.

Budig, Michelle J. 2002. "Male Advantage and the Gender Composition of Jobs: Who Rides the Glass Escalator?" *Social Problems* 49 (May): 257–277.

Budig, Michelle J., and Melissa J. Hodges. 2010. "Differences in Disadvantage: Variation in the Motherhood Penalty across White Women's Earnings Distribution." *American Sociological Review* 75 (October): 705–728.

Bullard, Robert D., and Beverly Wright. 2009. *Race, Place, and Environmental Justice after Hurricane Katrina*. Boulder, CO: Westview Press.

Bullard, Robert D., and Beverly Wright. 2010. "Race, Place, and Environment." Pp. 381–390 in *Race and Ethnicity in Society: The Changing Landscape*, 3rd ed., edited by Elizabeth Higginbotham and Margaret L. Andersen. Belmont, CA: Wadsworth.

Bullock, Heather E., Karen Fraser, and Wendy R. Williams. 2001. "Media Images of the Poor." *The Journal of Social Issues* 57 (Summer): 229–246.

Bureau of Labor Statistics. 2012a. Employment Projections: Civilian Labor Force Participation Rates by Age, Sex, Race, and Ethnicity. Washington, DC: U.S. Department of Labor. **www .bls.gov**

Bureau of Labor Statistics. 2012b. *Employment and Earnings*. Washington, DC: U.S. Department of Labor. **www .bls.gov**

Burgess, Samuel H. 2001. "Gender Role Ideology and Anti-Gay Opinion in the United States." Paper presented at the Annual Meeting of the Southern Sociological Association, Atlanta, GA.

Burkholder, Richard. 2003. "Iraq and the West: How Wide Is the Morality Gap?" The Gallup Poll, Princeton, NJ. **www .gallup.com**

Burris, Val. 2000. "The Myth of Old Money Liberalism: The Politics of the Forbes 400 Richest Americans." *Social Problems* 47 (Summer): 360–378.

Burt, Cyril. 1966. "The Genetic Determination of Differences in Intelligence: A Study of Monozygotic Twins Reared Together and Apart." *British Journal of Psychology* 57: 137–153.

Bushman, Brad P. 1998. "Primary Effects of Media Violence on the Accessibility of Aggressive Constructs in Memory." *Personality and Social Psychology Bulletin* 24: 537–545.

Butsch, Richard. 1992. "Class and Gender in Four Decades of Television Situation Comedy: Plus ça Change . . ." *Critical Studies in Mass Communication* 9: 387–399.

Buunk, Bram, and Ralph B. Hupka. 1987. "Cross-Cultural Differences in the Elicitation of Sexual Jealousy." *Journal of Sex Research* 23 (February): 12–22.

Camaroto, Steven. 2007. "Immigrants in the United States, 2007: A Profile of America's Foreign-Born Population." Washington, DC: Center for Immigration Studies. **www.cis.org**

Campbell, Anne. 1987. "Self-Definition by Rejection: The Case of Gang Girls." *Social Problems* 34 (December): 451–466.

Cantor, Joanne. 2000. "Media Violence." *Journal of Adolescent Health* 27 (August): 30–34.

Carbon Dioxide Information Analysis Center. 2007. **http://cdiac.ornl.gov**

Carmichael, Stokely, and Charles V. Hamilton. 1967. *Black Power: The Politics of Liberation in America*. New York: Vintage.

Carnoy, Martin, Manuel Castells, Stephen S. Cohen, and Fernando Henrique Cardoso. 1993. *The New Global Economy in the Information Age*. University Park, PA: The Pennsylvania State University Press.

Carr, Deborah. 2010. "Cheating Hearts." *Contexts* 9 (Summer): 58–60.

Carroll, J. B., 1956. *Language, Thought, and Reality: Selected Writings of Benjamin Lee Whorf*. Cambridge, MA: MIT Press.

Carter, J. S., Mamadi Corra, and Shannon K. Carter. 2009. "The Interaction of Race and Gender: Changing Gender-Role Attitudes, 1974–2006." *Social Science Quarterly* 90(1): 196–211.

Carter, Timothy S. 1999. "Ascent of the Corporate Model in Environmental-Organized Crime." *Crime, Law and Social Change* 21: 1–30.

Carson, E. Ann, and William J. Sabol. 2012. *Prisoners in 2011*. Washington, DC: Bureau of Justice Statistics, U.S. Department of Justice.

Cassen, Robert. 1994. "Population and Development: Old Debates, New Conclusions." *U.S.-Third World Policy Perspectives* 19: 282.

Cassidy, J., and P. R. Shaver (eds.). 1999. *Handbook of Attachment: Theory, Research, and Critical Applications*. New York: Guilford.

Cavalier, Elizabeth. 2003. "'I Wear Dresses, I Wear Muscles': Media Images of Women Soccer Players." Paper presented at the annual meeting of the Southern Sociological Society, New Orleans, Louisiana.

Centeno, Miguel A., and Eszter Hargittai. 2003. "Defining a Global Geography." *The American Behavioral Scientist* 44 (10).

Center for Responsive Politics. 2012. *Stats at a Glance*. **www.opensecrets.org**

Centers for Disease Control and Prevention. 2011. "National Center for Chronic Disease Prevention and Health Promotion, Division of Nutrition, Physical Activity, and Obesity." Atlanta, GA: Centers for Disease Control and Prevention.

Centers, Richard. 1949. *The Psychology of Social Classes*. Princeton, NJ: Princeton University Press.

Central Intelligence Agency. 2012. *The World Fact Book*. **www.cia.gov**

Chafetz, Janet. 1984. *Sex and Advantage*. Totowa, NJ: Rowman and Allanheld.

Chalfant, H. Paul, Robert E. Beckley, and C. Eddie Palmer. 1987. *Religion in Contemporary Society,* 2nd ed. Palo Alto, CA: Mayfield.

Chang, Jung. 1991. *Wild Swans: Three Daughters of China*. New York: Simon & Schuster.

Chapman, C., J. Laird, J. Ifill, and KewalRamani, A. 2011. *Trends in High School Dropout and Completion Rates in U.S., 1972–2009*. Washington, DC: U.S. Department of Education, National Center for Education Statistics.

Chen, Elsa, Y. F. 1991. "Conflict between Korean Greengrocers and Black Americans." Unpublished senior thesis, Princeton University.

Cherlin, Andrew J. 2010. "Demographic Trends in the United States: A Review of Research in the 2000s." *Journal of Marriage and Family* 72 (June): 403–419.

Cherlin, Andrew J., P. Lindsay Chase-Lansdale, and Christine McRae. 1998. "Effects of Parental Divorce on Mental Health throughout the Life Course." *American Sociological Review* 63 (April): 219–249.

Children's Bureau. 2011. "Child Maltreatment 2010." Washington, DC: U.S. Department of Health and Human Services.

Childs, Erica Chito. 2005. *Navigating Interracial Borders: Black-White Couples and Their Social World*. New Brunswick, NJ: Rutgers University Press.

Cicourel, Aaron V. 1968. *The Social Organization of Juvenile Justice*. New York: Wiley.

Clark, Kenneth B., and Mamie P. Clark. 1947. "Racial Identification and Preference in Negro Children." pp. 602–611 in *Readings in Social Psychology,* edited by T. M. Newcomb and E. L. Hartley. New York: Holt.

Clawson, Dan, and Naomi Gerstel. 2002. "Caring for Our Young: Child Care in Europe and the United States." *Contexts* 1 (Fall–Winter): 28–35.

Clawson, Rosalee A., and Rakuya Trice. 2000. "Poverty As We Know It: Media Portrayals of the Poor." *The Public Opinion Quarterly* 64 (Spring): 53–64.

Cloud, John. 2012. "Preventing Mass Murder: Can We Identify Dangerous Men before They Kill?" *Time* Magazine (August 6): 33.

Coale, Ansley. 1986. "Population Trends and Economic Development." Pp. 96–104 in *World Population and the U.S. Population Policy: The Choice Ahead,* edited by J. Menken. New York: W. W. Norton.

Cole, David. 1999. "The Color of Justice." *The Nation* (October 11): 12–15.

Coles, Roberta, and Charles Green, eds. 2009. *The Myth of the Missing Black Father*. New York: Columbia University Press.

College Board. 2010. *National Report on College-Bound Seniors*. **www .collegeboard.com**

Colley, Ann, and Zazie Todd. 2002. "Gender-Linked Differences in the Style and Content of E-mails to Friends." *Journal of Language and Social Psychology* 21 (December): 380–393.

Collins, Patricia Hill. 1990. *Black Feminist Theory: Knowledge, Consciousness and the Politics of Empowerment*. Cambridge, MA: Unwin Hyman.

Collins, Patricia Hill. 1998. *Fighting Words: Black Women and the Search for Justice*. Minneapolis, MN: University of Minnesota Press.

Collins, Patricia Hill. 2004. *Black Sexual Politics: African Americans, Gender, and the New Racism*. New York: Routledge.

Collins, Patricia Hill, Lionel A. Maldonado, Dana Y. Takagi, Barrie Thorne, Lynn

Weber, and Howard Winant. 1995. "On West and Fenstermaker's 'Doing Difference.'" *Gender & Society* 9 (August): 491–505.

Collins, Randall. 1988. *Theoretical Sociology*. San Diego, CA: Harcourt Brace Jovanovich.

Collins, Randall. 1994. *Four Sociological Traditions*. New York: Oxford.

Collins, Randall, and Michael Makowsky. 1972. *The Discovery of Society*. New York: Random House.

Coltrane, Scott, and Melinda Messineo. 2000. "The Perpetuation of Subtle Prejudice: Race and Gender Imagery in 1990s Television Advertising." *Sex Roles* 42 (March): 363–389.

Connell, R. W. 1992. "A Very Straight Gay: Masculinity, Homosexual Experience, and the Dynamics of Gender." *American Sociological Review* 57 (December): 735–751.

Conrad, Peter, and Joseph W. Schneider. 1992. *Deviance and Medicalization: From Badness to Sickness,* expanded ed. Philadelphia, PA: Temple University Press.

Constable, Nicole. 1997. *Maid to Order in Hong Kong: Stories of Filipina Workers*. Ithaca, NY: Cornell University Press.

Cook, Christopher D. 2000. "Temps Demand a New Deal." *The Nation* (March 27): 13–19.

Cook, Karen S., and Alexandra Gervasi. 2006. "Homans and Emerson on Power: Out of the Skinner Box." In *George C. Homans: History, Theory, and Method,* edited by Javier Trevino. New York: Paradigm Publishers.

Cook, S. W. 1988. "The 1954 Social Science Statement and School Segregation: A Reply to Gerard." Pp. 237–256 in *Eliminating Racism: Profiles in Controversy,* edited by D. A. Taylor. New York: Plenum.

Cooksey, Elizabeth C., and Ronald R. Rindfuss. 2001. "Patterns of Work and Schooling in Young Adulthood." *Sociological Forum* 16 (December): 731–755.

Cookson, Peter W., Jr., and Caroline Hodges Persell. 1985. *Preparing for Power: America's Elite Boarding Schools*. New York: Basic Books.

Cooley, Charles Horton. 1902. *Human Nature and Social Order*. New York: Scribner's.

Cooley, Charles Horton. 1967 [1909]. *Social Organization*. New York: Schocken Books.

Coontz, Stephanie. 1992. *The Way We Never Were*. New York: Basic Books.

Copen, Casey E., Kimberly Daniels, Jonathan Vespa, and William D. Mosher. 2012. "First Marriages in the United States: Data from the 2006–2010 National Survey of Family Growth." *National Health Statistics Reports,* no. 29, March 22. **www.cdc.gov**

Cortes, Patricia. 2008. "The Effect of Low-Skilled Immigration on U.S. Prices: Evidence from CPI Data." *Journal of Political Economy* 116: 381–422.

Cortina, Lilia M., Suzanna Swan, Louise F. Fitzgerald, and Craig Walo. 1998. "Sexual Harassment and Assault: Chilling the Climate for

Women in Academia." *Psychology of Women Quarterly* 22 (September): 419–441.

Coser, Lewis. 1977. *Masters of Sociological Thought.* New York: Harcourt Brace Jovanovich.

Coston, Beverly M., and Michael Kimmel. 2012. "Seeing Privilege Where It Isn't" Marginalized Masculinities and the Intersectionality of Privilege. *Journal of Social Issues* 68 (March): 97–111.

Craig, Maxine Leeds. 2002. *Ain't I a Beauty Queen? Black Women, Beauty and the Politics of Race.* New York: Oxford University Press.

Crane, Diana (ed.). 1994. *The Sociology of Culture: Emerging Theoretical Perspectives.* Cambridge, England: Blackwell.

Cravey, Tiffany, and Aparna Mmitra. 2011. "Demographics of the Sandwich Generation by Race and Ethnicity in the United States." *Journal of Socio-Economics* 40(3): 306–311.

Crittenden, Ann. 2002. *The Price of Motherhood: Why the Most Important Job in the World Is the Least Valued.* New York: Holt.

Croll, Elisabeth. 1995. *Changing Identities of Chinese Women.* London: Zed Books.

Cunningham, Mick. 2001. "The Influence of Parental Attitudes and Behaviors on Children's Attitudes toward Gender and Household Labor in Early Adulthood." *Journal of Marriage and Family* 63 (February): 111–122.

Currie, Dawn. 1997. "Decoding Femininity: Advertisements and Their Teenage Readers." *Gender & Society* 11 (August): 453–477.

D'Alessio, Stewart J., and Lisa Stolzenberg. 2003. "Race and the Probability of Arrest." *Social Forces* 81 (June): 1381–1397.

D'Emilio, John. 1998. *Sexual Politics, Sexual Communities: The Making of a Homosexual Minority in the United States, 1940–1970,* rev. ed. Chicago, IL: University of Chicago Press.

Demos and Young Invincibles. 2011. *The State of Young America: The Databook.* New York: Demos.

Dahrendorf, Rolf. 1959. *Class and Class Conflict in Industrial Society.* Stanford, CA: Stanford University Press.

Dalmage, Heather M. 2000. *Tripping on the Color Line: Black-White Multiracial Families in a Racially Divided World.* New Brunswick, NJ: Rutgers University Press.

Dalton, Susan E., and Denise D. Bielby. 2000. "'That's Our Kind of Constellation': Lesbian Mothers Negotiate Institutionalized Understandings of Gender within the Family." *Gender & Society* 14 (February): 36–61.

Darley, John M., and R. H. Fazio. 1980. "Expectancy Confirmation Processes Arising in the Social Interaction Sequence." *American Psychologist* 35: 867–881.

Darling-Hammond, Linda. 2010. *The Flat World and Education: How America's Commitment to Equity Will Determine Our Future.* New York: Teachers College Press.

Das Gupta, Monisha. 1997. "'A Chambered Nautilus': The Contradictory Nature of Puerto Rican Women's Role in the Construction of a Transnational Community." *Gender & Society* 11 (October): 627–655.

Davies, James B., Susanna Sandstrom, Anthony Shorrocks, and Edward N. Wolff. 2008. "The World Distribution of Household Wealth." Helsinki, Finland: UNU-WIDER, World Institute for Development Economics Research.

Davis, Angela. 1981. *Women, Race, and Class.* New York: Random House.

Davis, James A., and Tom Smith. 1984. *General Social Survey Cumulative File, 1972–1982.* Ann Arbor, MI: Inter-University Consortium for Political and Social Research.

Davis, Kingsley. 1945. "The World Demographic Transition." *Annals of the American Academy of Political and Social Sciences* 237: 1–11.

Davis, Kingsley, and Wilbert E. Moore. 1945. "Some Principles of Stratification." *American Sociological Review* 10 (April): 242–247.

DeChamplain, Pierre. 2010. "Married to the Mafia." Pp. 78–79 in *Mysteries of History: Secret Societies. U. S. News and World Report,* special ed.

Deegan, Mary Jo. 1988. "W. E. B. DuBois and the Women of Hull-House, 1895–1899." *The American Sociologist* 19 (Winter): 301–311.

Dellinger, Kristen, and Christine L. Williams. 1997. "Makeup at Work: Negotiating Appearance Rules in the Workplace." *Gender & Society* 11 (April): 151–177.

Demeny, Paul. 1991. "Tradeoffs between Human Numbers and Material Standards of Living." Pp. 408–421 in *Resources, Environment, and Population: Present Knowledge, Future Options,* edited by Kingsley Davis and Michail S. Bernstam. New York: Population Council.

DeNavas-Walt, Carmen, Bernadette D. Proctor, and Cheryl Hill Lee. 2010. *Income, Poverty, and Health Insurance Coverage in the United States: April 2010.* Washington, DC: U.S. Census Bureau. **www.census.gov**

DeNavas-Walt, Carmen, Bernadette Proctor, and Jessica C. Smith. 2012. *Income, Poverty, and Health Insurance Coverage in the United States: 2011.* Washington, DC: U.S. Department of Commerce. **www.census.gov**

DeVault, Marjorie. 1991. *Feeding the Family: The Social Organization of Caring as Gendered Work.* Chicago, IL: University of Chicago Press.

Devine-Eller, Audrey. 2012. "Timing Matters: Test Preparation, Race, and Grade Level." *Sociological Forum* 27(2): 458–480.

Dill, Bonnie Thornton. 1988. "Our Mothers' Grief: Racial Ethnic Women and the Maintenance of Families." *Journal of Family History* 13 (October): 415–431.

Dillard, J.L. 1972. *Black English: Its Historical Usage in the United States.* New York: Random House.

Dillon, Sam. 2009. "No Child Law Is Not Closing a Racial Gap." *The New York Times* (April 26): A15–16.

DiMaggio, Anthony. 2011. *The Rise of the Tea Party: Political Discontent and Corporate Media in the Age of Obama.* New York: Monthly Review Press.

DiMaggio, Paul J., and Walter W. Powell. 1991. "Introduction." Pp. 1–38 in *The New Institutionalism in Organizational Analysis,* edited by W. W. Powell and P. J. DiMaggio. Chicago, IL: University of Chicago Press.

Dines, Gail, and Jean M. Humez. 2002. *Gender, Race, and Class in Media,* 2nd ed. Thousand Oaks, CA: Sage.

Dixit, Avinash K., and Susan Sneath. 1997. *Games of Strategy.* New York: W. W. Norton.

Doermer, Jill K., and Stephen Demuth. 2010. "The Independent and Joint Effects of Race/Ethnicity, Gender, and Age on Sentencing Outcomes in U. S. Federal Courts." *Justice Quarterly* 27 (1): 1–27.

Dolan, Jill. 2006. "Blogging on Queer Connections in the Arts and the Five Lesbian Brothers." *GLQ* 12: 491–506.

Dollard, John, Neal E. Miller, Leonard W. Doob, O. H. Mowrer, and Robert R. Sears. 1939. *Frustration and Aggression.* New Haven, CT: Yale University Press.

Domhoff, G. William. 2002. *Who Rules America?* New York: McGraw-Hill.

Dominguez, Silvia, and Celeste Watkins. 2003. "Creating Networks for Survival and Mobility: Social Capital among African-American and Latin-American Low-Income Mothers." *Social Problems* 50 (1): 111–135.

Dovidio, John F., and Samuel L. Gaertner. 2005. "Color Blind or Just Plain Blind? The Pernicious Nature of Contemporary Racism." *The Non-Profit Quarterly* 12 (4): 22–27.

Dovidio, John F. and Samuel L. Gaertner. 2008. "Aversive Racism." In *Encyclopedia of Race, Ethnicity and Society.* Thousand Oaks, CA: Sage.

Dowd, James J., and Laura A. Dowd. 2003. "The Center Holds: From Subcultures to Social Worlds." *Teaching Sociology* 31 (January): 20–37.

Downey, Douglas B., and Benjamin G. Gibbs. 2010. "How Schools Really Matter." *Contexts* 9 (Spring): 50–54.

Drier, Peter. 2012. "The Battle for the Republican Soul: Who Is Drinking the Tea Party?" *Contemporary Sociology* 41 (November): 756–762.

DuBois, Cathy L., Deborah E. Knapp, Robert H. Faley, and Gary A. Kustis. 1998. "An Empirical Examination of Same and Other Gender Sexual Harassment in "The Workplace." *Sex Roles* 9–10 (November): 731–749.

DuBois, W. E. B. 1901. "The Freedmen's Bureau." *Atlantic Monthly* 86: 354–365.

DuBois, W. E. B. 1903. *The Souls of Black Folk: Essays and Sketches.* Chicago: A. C. McClurg and Co.

Dubrofsky, Rachel. 2006. "The Bachelor: Whiteness in the Harem." *Critical Studies in Media Communication* 23: 39–56.

Dudley, Carl S., and David A. Roozen. 2001. *Faith Communities Today: A Report on Religion in the United States Today.* Hartford, CT: Hartford Institute for Religion Research, Hartford Seminary.

Due, Linnea. 1995. *Joining the Tribe: Growing Up Gay & Lesbian in the '90s.* New York: Doubleday.

Duffy, Mignon. 2011. *Making Care Count: A Century of Gender, Race, and Paid Care Work.* New Brunswick, NJ: Rutgers University Press.

Duneier, Mitchell. 1999. *Sidewalk.* New York: Farrar, Strauss and Giroux.

Dunne, Gillian A. 2000. "Opting into Motherhood: Lesbians Blurring the Boundaries and Transforming the Meaning of Parenthood and Kinship." *Gender & Society* 14 (February): 11–35.

Dunne, Mairead, and Louise Gazely. 2008. "Teachers, Social Class, and Underachievement." *British Journal of Sociology of Education* 29: 451–463.

Durell, Megan, Catherine Chiong, and Juan Battle. 2007. "Race, Gender Expectations, and Homophobia: A Quantitative Analysis," *Race, Gender & Class* 14: 299–317.

Durkheim, Emile. 1951 [1897]. *Suicide.* Glencoe, IL: Free Press.

Durkheim, Emile. 1947 [1912]. *Elementary Forms of Religious Life.* Glencoe, IL: Free Press.

Durkheim, Emile. 1950 [1938]. *The Rules of Sociological Method.* Glencoe, IL: Free Press.

Durkheim, Emile. 1964 [1895]. *The Division of Labor in Society.* New York: Free Press.

Edin, Kathryn. 2000. "What Do Low-Income Single Mothers Say about Marriage?" *Social Problems* 47 (February): 112–133.

Edin, Kathyrn, and Maria Kefalas. 2005. *Promises I Can Keep: Why Poor Women Put Motherhood before Marriage.* Berkeley, CA: University of California Press.

Edin, Kathryn, and Laura Lein. 1997. *Making Ends Meet: How Single Mothers Survive Welfare and Low-Wage Work.* New York: Russell Sage Foundation.

Egan. Timothy. 2006. *The Worst Hard Time: The Story of Those Who Survived the Great American Dust Bowl.* Boston: Houghton Mifflin.

Ehrlich, Paul. 1968. *The Population Bomb.* New York: Ballantine Books.

Ehrlich, Paul R., and Jianguo Liu. 2002. "Some Roots of Terrorism." *Population and Environment,* vol. 24, no. 2, Human Sciences Press.

Eitzen, D. Stanley. 2009. "Dimensions of Globalization." Pp. 37–42 in *Globalization: The Transformation of Social World,* edited by D. Stanley Eitzen and Maxine Baca Zinn. Belmont, CA: Cengage.

Eitzen, D. Stanley. 2012. *Fair and Foul: Beyond the Myths and Paradoxes of Sport.* Lanham: MD: Rowman and Littlefield.

Eitzen, D. Stanley, and Maxine Baca Zinn. 2012. *In Conflict and Order,* 13th ed. Upper Saddle River: Pearson.

Elizabeth, Vivienne. 2000. "Cohabitation, Marriage, and the Unruly Consequences of Difference." *Gender & Society* 14 (February): 87–110.

Emerson, Michael O., and David Hartmann. 2006. "The Rise of Religious Fundamentalism." *Annual Review of Sociology* 21: 127–144.

Emerson, Rana A. 2002. "Where My Girls At?: Negotiating Black Womanhood in Music Videos." *Gender & Society* 16 (February): 115–135.

England, Paula. 2010. "The Gender Revolution: Uneven and Stalled." *Gender & Society* 24 (April): 149–166.

Enloe, Cynthia. 2001. *Bananas, Beaches, and Bases: Making Feminist Sense of International Politics,* updated ed. Berkeley, CA: University of California Press.

Epstein, Marina, and L. Monique Ward. 2011. "Exploring Parent-Adolescent Communication About Gender: Results from Adolescent and Emerging Adult Samples." *Sex Roles* 65: 108–118.

Erikson, Eric. 1980. *Identity and the Life Cycle.* New York: W.W. Norton.

Erikson, Kai. 1966. *Wayward Puritans: A Study in the Sociology of Deviance.* New York: Wiley.

Erikson, Kai. 1994. *A New Species of Trouble: The Human Experience of Modern Disasters.* New York: W. W. Norton.

Esbensen-Finn, Aage, Elizabeth Piper Deschenes, and Thomas L. Winfree Jr. 1999. "Differences between Gang Girls and Gang Boys: Results from a Multisite Survey." *Youth and Society* 31 (September): 27–53.

Eschholz, Sarah, Jana Bufkin, and Jenny Long. 2002. "Symbolic Reality Bites: Women and Racial/Ethnic Minorities in Modern Film." *Sociological Spectrum* 22 (July–August): 299–334.

Espeland, Wendy Nelson. 1998. *The Struggle for Water: Politics, Rationality, and Identity in the American Southwest.* Chicago: University of Chicago Press.

Essed, Philomena. 1991. *Understanding Everyday Racism.* Newbury Park, CA: Sage.

Etzioni, Amatai. 1975. *A Comparative Analysis of Complex Organization: On Power, Involvement, and Their Correlates,* rev. ed. New York: Free Press.

Etzioni, Amatai, John Wilson, Bob Edwards, and Michael W. Foley. 2001. "A Symposium on Robert D. Putnam's *Bowling Alone: The Collapse and Revival of American Community*." *Contemporary Sociology* 30 (May): 223–230.

Evans, Peter B., Dietrich Ruesschemeyer, and Theda Skocpol. 1985. *Bringing the State Back In.* Cambridge, MA: Cambridge University Press.

Fausto-Sterling, Anne. 1992. *Myths of Gender: Biological Theories about Women and Men.* New York: Basic Books.

Fausto-Sterling, Anne. 2000. *Sexing the Body: Gender Politics and the Construction of Sexuality.* New York: Basic Books.

Feagin, Joe R. 2000. *Racist America: Roots, Future Realities, and Racial Reparations.* New York: Routledge.

Feagin, Joe R. 2007. *Systemic Racism: A Theory of Oppression.* New York: Routledge.

Feagin, Joe R., and Clairece B. Feagin. 1993. *Racial and Ethnic Relations,* 4th ed. Englewood Cliffs, NJ: Prentice Hall.

Feagin, Joe R., and Hernan Vera 1995. *White Racism.* New York: Routledge.

Federal Bureau of Investigation. 2010. *Uniform Crime Reports.* Washington, DC: U.S. Department of Justice. **www.fbi.gov**

Federal Bureau of Investigation. 2011. *Uniform Crime Reports.* Washington, DC: U.S. Department of Justice. **www.fbi.gov**

Federal Bureau of Investigation. 2012. *Uniform Crime Reports.* Washington, DC: U. S. Department of Justice.

Federal Election Commission. 2008. "Growth in PAC Financial Activity Slows." **www.fec.gov**

Fein, Melvyn L. 1988. "Resocialization: A Neglected Paradigm." *Clinical Sociology*.6: 88–100.

Ferber, Abby. 1998. *White Man Falling: Race, Gender, and White Supremacy.* Lanham, MD: Rowman and Littlefield.

Ferber, Abby. 1999. "What White Supremacists Taught a Jewish Scholar about Identity." *The Chronicle of Higher Education* (May 7): 86–87.

Ferdinand, Peter. 2000. *The Internet, Democracy, and Democratization.* London: Frank Cass Publishers.

Fernald, Anne, and Hiromi Morikawa. 1993. "Common Themes and Cultural Variations in Japanese and American Mothers' Speech to Infants." *Child Development* 64 (June): 637–656.

Festinger, Leon, Stanley Schachter, and Kurt Back. 1950. *Social Pressures in Informal Groups: A Study of Human Factors in Housing.* Stanford, CA: Stanford University Press.

Fetner, Tina. 2012. "The Tea Party: Manufactured Dissent or Complex Social Movement?" *Contemporary Sociology* 41 (November): 762–766.

Fishbein, Allan J., and Patrick Woodall. 2006. "Women are Prime Targets for Subprime Lending: Women Are Disproportionately Represented in High-Cost Mortgage Market." Washington, DC: Consumer Federation of America.

Fisher, Bonnie S., Francis T. Cullen, and Michael G. Turner. 2000. *Sexual Victimization of College Women.* Washington, DC: Bureau of Justice Statistics.

Fishman, Charles. 2011. *The Big Thirst: The Secret Life and Turbulent Future of Water.* New York: Free Press.

Fitzgibbon, Marian, and Melinda Stolley. 2000. "Minority Women: The Untold Story." **www.pbs.org**

Fitzgibbon, Marian, and Melinda Stolley. 2002. "Minority Women: The Untold Story," NOVA Online. **http://www.pbs.org/wgbh/nova/thin/minorities.html**

Flanagan, William G. 1995. *Urban Sociology: Images and Structure.* Boston, MA: Allyn and Bacon.

Fleisher, Mark. 2000. *Dead End Kids: Gang Girls and the Boys They Know.* Madison, WI: University of Wisconsin Press.

Fleury-Steiner, Ben. 2012. *Disposable Heroes: The Betrayal of African American Veterans.* Lanham, MD: Rowman and Littlefield.

Folbre, Nancy. 2001. *The Invisible Heart: Economics and Family Values.* New York: New Press.

Fortini, Amanda. 2005. "The Great Flip-Flop Flap." *Slate*, July 22. **www.slate.com**

Frankenberg, Erica, and Chungmei Lee. 2002. *Race in American Public Schools: Rapidly Resegregating School Districts.* Los Angeles: Civil Rights Project, UCLA.

Frazier, E. Franklin. 1957. *The Black Bourgeoisie.* New York: Collier Books.

Fredrickson, George M. 2003. *Racism: A Short History.* Princeton, NJ: Princeton University Press.

Freedle, Roy O. 2003. "Correcting the SATs Ethnic and Social Class Bias: A Method of Re-Estimating SAT Scores." *Harvard Educational Review* 73 (Spring): 1–43.

Freedman, Estelle B., and John D'Emilio. 1988. *Intimate Matters: A History of Sexuality in America.* New York: Harper & Row.

Freeman, Jo. 1983. "A Model for Analyzing the Strategic Options of Social Movement Organizations." Pp. 193–210 in *Social Movements of the Sixties and Seventies,* edited by Jo Freeman. New York: Longman.

Freud, Sigmund. 1923 [1960]. *The Ego and the Id,* translated by Joan Riviere. New York: W.W. Norton.

Fried, Amy. 1994. "'It's Hard to Change What We Want to Change': Rape Crisis Centers as Organizations." *Gender & Society* 4 (December): 562–583.

Friedman, Thomas L. 1999. *The Lexus and the Olive Tree.* New York: Farrar, Strauss, and Giraux.

Fry, Richard. 2012. *A Record One-in-Five Households Now Owe Student Loan Debt.* Washington, DC: Pew Research Center. **www.pewsocialtrends.org**

Fryberg, Stephanie. 2003. "Really: You Don't Look Like an American Indian: Social Representations and Social Group Identities." PhD diss., Department of Psychology, Stanford University.

Frye, Marilyn. 1983. *The Politics of Reality.* Trumansburg, NY: The Crossing Press.

Fukuda, Mari. 1994. "Nonverbal Communication within Japanese and American Corporations." Unpublished manuscript, Princeton University.

Furstenberg, Frank. 1998. "Relative Risk: What Is the Family Doing to Our Children?" *Contemporary Sociology* 27 (May): 223–225.

Furstenberg, Frank F., Jr., and Christine Winquist Nord. 1985. "Parenting Apart: Patterns of Childrearing after Marital Disruption." *Journal of Marriage and the Family* 47 (November): 898–904.

Gaertner, Samuel L., and John F. Dovidio. 2005. "Understanding and Addressing Contemporary Racism: From Aversive Racism to the Common In-group Identity Model." *Journal of Social Issues* 61 (3): 615–639.

Gagné, Patricia, and Richard Tewksbury. 1998. "Conformity Pressures and Gender Resistance among Transgendered Individuals." *Social Problems* 45 (February): 81–101.

Gallagher, Charles A. 2013. "Color-Blind Privilege." Pp. 91–95 in *Race, Class and Gender: An Anthology,* 7th ed., edited by Margaret L. Andersen and Patricia Hill Collins. Belmont CA: Wadsworth/Cengage.

Gallagher, Sally K. 2003. *Evangelical Identity and Gendered Family Life.* New Brunswick, NJ: Rutgers University Press.

Gallup, George H., Jr. 2003. "Current Views on Premarital, Extramarital Sex." Princeton, NJ: Gallup Organization. **www.gallup.com**

Gallup Organization. 2010. "Tobacco and Smoking." The Gallup Poll. Princeton, NJ: The Gallup Organization. **www.gallup.com**

Gallup Poll. 2006. "Religion." Princeton, NJ: Gallup Organization. **www.gallup.com**

Gallup Poll. 2008. "Homosexual Relations." Princeton, NJ: Gallup Poll. **www.gallup.com**

Gallup Poll. 2012. "Guns." Princeton, NJ: The Gallup Organization. **www.gallup.com**

Gamson, Joshua. 1998. *Freaks Talk Back: Tabloid Talk Shows and Sexual Nonconformity.* Chicago, IL: University of Chicago Press.

Gamson, William, and Andre Modigliani. 1974. *Conceptions of Social Life: A Text-Reader for Social Psychology.* Boston, MA: Little, Brown.

Gans, Herbert. 1982 [1962]. *The Urban Villagers: Group and Class in the Life of Italian Americans.* New York: Free Press.

Garfinkel, Harold. 1967. *Studies in Ethnomethodology.* Englewood Cliffs, NJ: Prentice Hall.

Garrow, David J. 1981. *The FBI and Martin Luther King.* New York: W. W. Norton.

Gastil, John. 1990. "Generic Pronouns and Sexist Language: The Oxymoronic Character of Masculine Generics." *Sex Roles* 23 (December): 629–643.

Gates, Gary J., and Frank Newport. 2012 (October 18). "Special Report: 3.4% of U.S. Adults Identify as LGBT." *Gallup Poll.* **www.gallup.com**

Gates, Henry Louis, Jr., and Nellie Y. McKay, eds. 1997. *The Norton Anthology of African American Literature.* New York: W. W. Norton.

Genovese, Eugene. 1972. *Roll, Jordan, Roll: The World the Slaves Made.* New York: Pantheon.

Gentile, Douglas A., Lindsay C. Mathieson, and Nicki R. Crick. 2011. "Media, Violence Associations with the Form and Function of Aggression among Elementary School Children." *Social Development* 20: 213–232.

Gerami, Shahin, and Melodye Lehnerer. 2001. "Women's Agency and Household Diplomacy: Negotiating Fundamentalism." *Gender & Society* 15 (August): 556–573.

Gersch, Beate. 1999. "Class in Daytime Talk Television." *Peace Review* 11 (June): 275–281.

Gerschick, T. J., and A. S. Miller. 1995. "Coming to Terms: Masculinity and Physical Disability." Pp. 183–204 in *Men's Health and Illness: Gender, Power, and the Body,* Research on Men and Masculinities Series, vol. 8, edited by D. Sabo and D. F. Gordon. Thousand Oaks, CA: Sage Publications.

Gerson, Kathleen. 2010. *The Unfinished Revolution: How a New Generation is Shaping Family, Work, and Gender in America.* New York: Oxford University Press.

Gerstel, Naomi. 2000. "The Third Shift: Gender and Care Work Outside the Home." *Qualitative Sociology* 23: 467–483.

Gerstel, Naomi, and Sally Gallagher. 2001. "Men's Caregiving: Gender and the Contingent Character of Care." *Gender & Society* 15 (April): 197–217.

Gerth, Hans, and C. Wright Mills (eds.). 1946. *From Max Weber: Essays in Sociology.* New York: Oxford University Press.

Giannarelli, Linda, and James Barsimantov. 2000. *Child Care Expenses of America's Families.* Washington. DC: Urban Institute.

Giddings, Paula. 1994. *In Search of Sisterhood: Delta Sigma Theta and the Challenge of the Black Sorority Movement.* New York: William Morrow.

Giddings, Paula. 2008. *A Sword among Lions: Ida B. Wells and the Campaign against Lynching.* New York: Amistad.

Gilbert, D. T., and P. S. Malone. 1995. "The Correspondence Bias." *Psychological Bulletin* 117: 21–38.

Gilbert, Daniel R., Susan T. Fiske, and Gardner Lindzey (eds.). 1998. *The Handbook of Social Psychology,* 4th ed. New York: McGraw-Hill.

Gilens, Martin. 1996. "Race and Poverty in America: Public Misperceptions and the American News Media." *The Public Opinion Quarterly* 60 (Winter): 515–541.

Gilkes, Cheryl Townsend. 2000. *"If It Wasn't for the Women ..." Black Women's Experience and Womanist Culture in Church and Community.* Maryknoll, NY: Orbis Books.

Gimlin, Debra. 1996. "Pamela's Place: Power and Negotiation in the Hair Salon." *Gender & Society* 10 (October): 505–526.

Gimlin, Debra. 2002. *Body Work: Beauty and Self-Image in American Culture.* Berkeley, CA: University of California Press.

Gitlin, Todd. 2002. *Media Unlimited: How the Torrent of Images and Sounds Overwhelms Our Lives.* New York: Metropolitan Books.

Glassner, Barry. 1999. *Culture of Fear: Why Americans Are Afraid of the Wrong Things.* New York: Basic Books.

Glaze, Lauren E., and Laura M. Maruschak. 2008. "Parents in Prison and Their Minor Children." Washington, DC: U.S. Bureau of Justice Statistics. **www.ojp.usdoj.gov**

Glazer, Nathan. 1970. *Beyond the Melting Pot: The Negroes, Puerto Ricans, Jews, Italians, and Irish of New York City.* Cambridge, MA: MIT Press.

Glazer, Nona. 1990. "The Home as Workshop: Women as Amateur Nurses and Medical Care Providers." *Gender & Society* 4: 479–499.

Glenn, Evelyn Nakano. 1986. *Issei, Nisei, War Bride: Three Generations of*

Japanese American Women in Domestic Service. Philadelphia, PA: Temple University Press.

Glenn, Evelyn Nakano. 2002. *Unequal Freedom: How Race and Gender Shaped American Citizenship and Labor*. Cambridge, MA: Harvard University Press.

Glock, Charles, and Rodney Stark. 1965. *Religion and Society in Tension*. Chicago, IL: Rand McNally.

Gluckman, Amy, and Betsy Reed (eds.). 1997. *Homo Economics: Capitalism, Community, and Lesbian and Gay Life*. New York: Routledge.

Goffman, Alice. 2009. "On the Run: Wanted Men in a Philadelphia Ghetto." *American Sociological Review* 74 (June): 339–357.

Goffman, Erving. 1959. *The Presentation of Self in Everyday Life*. Garden City, NY: Doubleday.

Goffman, Erving. 1961. *Asylums: Essays on the Social Situation of Mental Patients and Other Inmates*. Garden City, NY: Anchor.

Goffman, Erving. 1963. *Stigma: Notes on the Management of Spoiled Identity*. Englewood Cliffs, NJ: Prentice Hall.

Goffman, Erving. 1974. *Frame Analysis*. New York: Harper and Row.

Goldberg, Abbie, Deborah Kashy, and JuliAnna Smith. 2012. "Gender-Typed Play Behaviors in Early Childhood: Adopted Children with Lesbian, Gay, and Heterosexual Parents." *Sex Roles* 67: 503–515.

Gomez, Antoinette M., Fatemeh Shafiei, and Glenn Johnson. 2011. "Black Women's Involvement in the Environmental Justice Movement: An Analysis of Three Communities in Atlanta, Georgia." *Race, Gender, and Class* 18: 189–214.

Gordon, Jesse, and Knickerbocker Designs. 2001. "The Sweat behind the Shirt." *The Nation* (March 10): 14ff.

Gordon, Linda. 1977. *Woman's Body/ Woman's Right*. New York: Penguin.

Gore, Albert. 2006. *An Inconvenient Truth: The Planetary Emergency of Global Warming and What We Can Do about It*. Emmaus, PA: Rodale.

Gottfredson, Michael R., and Travis Hirschi. 1990. *A General Theory of Crime*. Stanford, CA: Stanford University Press.

Gottfredson, Michael R., and Travis Hirschi. 1995. "National Crime Control Policies." *Society* 32 (January–February): 30–36.

Gould, Stephen Jay. 1999. "The Human Difference." *The New York Times* (July 2).

Gramsci, Antonio. 1971. *Selections from the Prison Notebooks of Antonio Gramsci*, edited by Quintin Hoare and Geoffrey Nowell. London: Lawrence and Wishart.

Granovetter, Mark. 1973. "The Strength of Weak Ties." *American Journal of Sociology* 78 (May): 1360–1380.

Granovetter, Mark. 1974. *Getting a Job: A Study of Contacts and Careers*. Cambridge, MA: Harvard University Press.

Granovetter, Mark S. 1995. "Afterward 1994: Reconsiderations and a New Agenda." Pp. 139–182 in *Getting a Job,* 2nd ed.,

by Mark S. Chicago, IL: University of Chicago Press.

Grant, Don Sherman, II, and Ramiro Martínez Jr. 1997. "Crime and the Restructuring of the U.S. Economy: A Reconsideration of the Class Linkages." *Social Forces* 75 (March): 769–799.

Greenhouse, Steven. 2000. "Poll of Working Women Finds Them Stressed." *The New York Times,* April 3.

Grimal, Pierre (ed.). 1963. *Larousse World Mythology*. New York: Putnam.

Grindstaff, Laura. 2002. *The Money Shot: Trash Class, and the Making of TV Talk Shows*. Chicago, IL: University of Chicago Press.

Gruber, James E. 1982. "Blue-Collar Blues: The Sexual Harassment of Women Autoworkers." *Work and Occupations* 3 (August): 271–298.

Guerrero, Laura K., Peter A. Andersen, and Walid A. Afifi. 2010. *Close Encounters: Communication in Relationships*, 3rd ed. Thousand Oaks, CA: Sage.

Gurin, Patricia, E. L. Dey, Sylvia Hurtado, and G. Gurin. 2002. "Diversity and Higher Education: Theory and Impact on Educational Outcomes." *Harvard Educational Review* 72: 330–366.

Guttierez y Muhs, Gabriella, Yolanda Flores Niemann, Carmen G. Gonzalez, and Angela P. Harris (eds.). 2012. *Presumed Incompetent: The Intersections of Race and Class for Women in Academia*. Logan, UT: Utah State University Press.

Guttmacher Institute. 2012. *Facts on American Teens' Sexual and Reproductive Health*. New York: Guttmacher Institute. **www .guttmacher.org**

Habermas, Jürgen. 1970. *Toward a Rational Society: Student Protest, Science, and Politics*. Boston, MA: Beacon Press.

Haddad, Yvonne Yazbeck, Jane I. Smith, and John L. Esposito (eds.). 2003. *Religion and Immigration: Christian, Jewish, and Muslim Experiences in the United States*. Walnut Creek, CA: AltaMira Press.

Hall, Edward T. 1966. *The Hidden Dimension*. New York: Doubleday.

Hall, Edward T., and Mildred Hall. 1987. *Hidden Differences: Doing Business with the Japanese*. New York: Anchor Press/ Doubleday.

Hall, Elaine J. 1993. "Waitering/Waitressing: Engendering in the Work of Table Servers." *Gender & Society* (September): 329–346.

Hamer, Dean H., Stella Hu, Victoria L. Magnuson, Nan Hu, and Angela M. L. Pattatucci. 1993. "A Linkage between DNA Markers on the X Chromosome and Male Sexual Identification." *Science* 261 (July): 321–327.

Hamer, Jennifer. 2001. *What It Means to Be Daddy*. New York: Columbia University Press.

Hamilton, Laura, and Elizabeth A. Armstrong. 2009. "Gendered Sexuality in Young Adulthood: Double Binds and Flawed Options." *Gender & Society* 23 (October): 589–616.

Hamilton, Mykol C. 1988. "Using Masculine Generics: Does Generic He Increase

Male Bias in the User's Imagery?" *Sex Roles* 19 (December): 785–799.

Handlin, Oscar. 1951. *The Uprooted*. Boston, MA: Little, Brown.

Haney, C., C. Banks, and P. G. Zimbardo. 1973. "Interpersonal Dynamics in a Simulated Prison." *International Journal of Criminology and Penology* 1: 69–97.

Haney, Lynne. 1996. "Homeboys, Babies, Men in Suits: The State and the Reproduction of Male Dominance." *American Sociological Review* 61 (October): 759–778.

Harden, Blaine. 2012. *Escape from Camp 14: One Man's Remarkable Odyssey from North Korea to Freedom in the West*. New York: Viking.

Hargittai, Eszther. 2008. "The Digital Reproduction of Inequality." Pp. 936–944 in *Social Stratification*, edited by David Grusky. Boulder, CO: Westview Press.

Harp, Dustin, and Mark Tremayne. 2006. "The Gendered Blogosphere: Examining Inequality Using Network and Feminist Theory." *Journalism & Mass Communication Quarterly* 83: 247–264.

Harris, Angel L. 2006. "I Don't Hate School: Revisiting 'Oppositional Culture' Theory of Blacks' Resistance to Schooling." *Social Forces* 85: 797–834.

Harris, Marvin. 1974. *Cows, Pigs, Wars, and Witches: The Riddles of Culture*. New York: Vintage.

Harris, Richard J., and Juanita M. Firestone. 1998. "Changes in Predictors of Gender Role Ideologies among Women: A Multivariate Analysis." *Sex Roles* 38 (3–4): 239–252.

Harris, Richard J., Juanita M. Firestone, and Mary Bollinger. 2000. "Gender Role Attitudes: Native Americans in Comparative Perspective." *Free Inquiry in Creative Sociology* 28 (2): 63–76.

Harrison, Deborah, and Tom Trabasso. 1976. *Black English: A Seminar*. Hillsdale, NJ: Erlbaum Associates.

Harrison, Roderick. 2000. "Inadequacies of Multiple Response Race Data in the Federal Statistical System." Manuscript. Joint Center for Political and Economic Studies and Howard University, Department of Sociology, Washington, DC.

Hastorf, Albert, and Hadley Cantril. 1954. "They Saw a Game: A Case Study." *Journal of Abnormal and Social Psychology* 40 (2): 129–134.

Hauan, Susan M., Nancy S. Landale, and Kevin T. Leicht. 2000. "Poverty and Work Effort among Urban Latino Men." *Work and Occupations* 27 (May): 188–222.

Hays, Sharon. 2003. *Flat Broke with Children: Women in the Age of Welfare Reform*. New York: Oxford University Press.

Hearnshaw, Leslie. 1979. *Cyril Burt: Psychologist*. Ithaca, NY: Cornell University Press.

Heider, Fritz. 1958. *The Psychology of Interpersonal Relations*. New York: John Wiley and Sons, Inc.

Heimer, Karen. 1997. "Socioeconomic Status, Subcultural Definitions and

Violent Delinquency." *Social Forces* 75 (March): 799–833.

Hendy, Helen M., Cheryl Gustitus, and Jamie Leitzel-Schwalm. 2001. "Social Cognitive Predictors of Body Images in Preschool Children." *Sex Roles* 44 (May): 557–569.

Henry, Kathy. 2008. "Warrior Woman—Ida B. Wells Barnett." *Ezinearticles.* **ezinearticles.com**

Henslin, James M. 1993. "Doing the Unthinkable." Pp. 253–262 in *Down to Earth Sociology*, 7th ed., edited by James M. Henslin. New York: Free Press.

Herman, Judith. 1981. *Father-Daughter Incest.* Cambridge, MA: Harvard University Press.

Herring, Cedric. 2009. "Does Diversity Pay?: Race, Gender, and the Business Case for Diversity." *American Sociological Review* 74 (2): 208–224.

Herrnstein, Richard J., and Charles Murray. 1994. *The Bell Curve: Intelligence and Class Structure in American Life.* New York: Free Press.

Hertz, Rosanna. 2006. *Single by Chance, Mothers by Choice: How Women are Choosing Parenthood without Marriage and Creating the New American Family.* New York: Oxford University Press.

Hesse-Biber, Sharlene Hagy. 2007. *The Cult of Thinness,* 2nd ed. New York: Oxford University Press.

Higginbotham, A. Leon, Jr. 1978. *In the Matter of Color: Race and the American Legal Process.* New York: Oxford University Press.

Higginbotham, Elizabeth. 2001. *Too Much to Ask: Black Women in the Era of Integration.* Chapel Hill, NC: University of North Carolina Press.

Higginbotham, Elizabeth, and Margaret L. Andersen, eds. 2012. *Race and Ethnicity in Society: The Changing Landscape,* 3rd edition. Belmont, CA: Wadsworth/Cengage.

Hill, Lori Diane. 2001. "Conceptualizing Educational Attainment Opportunities of Urban Youth: The Effects of School Capacity, Community Context and Social Capital." PhD diss., University of Chicago.

Hirschi, Travis. 1969. *Causes of Delinquency.* Berkeley: University of California Press.

Hirschman, Charles. 1994. "Why Fertility Changes." *Annual Review of Sociology* 20: 203–223.

Hirschman, Charles, and Douglas S. Massey. 2008. "Places and Peoples: The New American Mosaic." Pp. 1–21 in *New Faces in New Places: The Changing Geography of American Immigration,* ed. by Douglass Massey. New York: Russell Sage Foundation.

Hochschild, Arlie Russell. 1983. *The Managed Heart: Commercialization of Human Feelings.* Berkeley, CA: University of California Press.

Hochschild, Arlie Russell. 1997. *The Time Bind: When Work Becomes Home and Home Becomes Work.* New York: Metropolitan Books.

Hochschild, Arlie Russell. 2003. *The Commercialization of Intimate Life: Notes from Home and Work.* Berkeley: University of California Press.

Hochschild, Arlie Russell, with Anne Machung. 1989. *The Second Shift: Working Parents and the Revolution at Home.* New York: Viking.

Hofstadter, Richard. 1944. *Social Darwinism in American Thought.* Philadelphia, PA: University of Pennsylvania Press.

Hollander, Jocelyn A. 2002. "Resisting Vulnerability: The Social Reconstruction of Gender in Interaction." *Social Problems* 49 (November): 474–496.

Hollingshead, August B., and Frederick C. Redlich. 1958. *Social Class and Mental Illness: A Community Study.* New York: Wiley.

Holmes, Schuyler. 2000. "Environmental Racism and Classism in Toxic Waste Dumping in Cleveland, Ohio." Junior thesis. Princeton University.

Hondagneu-Sotelo, Pierrette. 2001. *Doméstica: Immigrant Workers Cleaning and Caring in the Shadows of Affluence.* Berkeley, CA: University of California Press.

Hondagneu-Sotelo, Pierrette, and Ernestine Avila. 1997. "'I'm Here, but I'm There': The Meanings of Latina Transnational Motherhood." *Gender & Society* 11(5): 548–571.

Hornung, Carlton A. 1977. "Social Status, Status Inconsistency and Psychological Stress." *American Sociological Review* 42 (August): 623–638.

Horowitz, Ruth. 1995. *Teen Mothers: Citizens or Dependents?* Chicago, IL: University of Chicago Press.

Horton, Hayward Derrick, Beverlyn Lundy Allen, Cedric Herring, and Melvin E. Thomas. 2000. "Lost in the Storm: The Sociology of the Black Working Class, 1850 to 1990." *American Sociological Review* 65 (February): 128–137.

Howard, Philip H., and Patricia Allen. 2010. "Beyond Organic and Fair Trade? An Analysis of Ecolabel Preferences in the United States." *Rural Sociology* 75: 244–269.

Hoyt, Wendy, and Lori R. Kogan. 2001. "Satisfaction with Body Image and Peer Relationships for Males and Females in a College Environment." *Sex Roles* 45 (August): 199–215. **www.aauw.org/learn/publications/outlook/backissues.cfm**

Hudson, Ken. 1999. "No Shortage of 'Nonstandard' Jobs." *Briefing Paper.* Washington, DC: Economic Policy Institute, December.

Huesmann, L. R., J. Moise, C. D. Podolski, and J. D. Eron. 2003. "Longitudinal Relations between Childhood Exposure to Media Violence and Adult Aggression and Violence, 1977–1992." *Developmental Psychology* 39: 201–221.

Hughes, Langston. 1967. *The Big Sea.* New York: Knopf.

Hull, Kathleen E., Ann Meier, and Timothy Ortyl. 2010. "The Changing Landscape of Love and Marriage." *Contexts* 9 (Spring): 32–37.

Humes, Karen R., Nicholas A. Jones, and Roberto R. Martinez. 2011. *Overview of Race and Hispanic Origin 2010.* Washington, DC: U.S. Census Bureau. **www.census.gov**

Hunt, Matthew O., Larry L. Hunt, and William W. Falk. 2012. "'Call to Home?' Race, Region, and Migration to the U. S. South, 1970–2000." *Sociological Forum* 27 (1): 117–141.

Hunt, Ruth. 2005. "Subjective Rating by Skin Color Gradation, Gender, and Other Characteristics." Junior Project, Princeton University, manuscript.

Hunter, Margaret. 2002. "Rethinking Epistemology, Methodology, and Racism: Or, Is White Sociology Really Dead?" *Race and Society* 5 (2): 119–138.

Ignatiev, Noel. 1995. *How the Irish Became White.* New York: Routledge.

Inglehart, Ronald, and Wayne E. Baker. 2000. "Modernization, Cultural Change, and the Persistence of Traditional Values." *American Sociological Review* 65 (February): 19–51.

Institute of Medicine of the National Academies. 2010. *Returning Home from Iraq and Afghanistan: Preliminary Assessment of Readjustment Needs of Veterans, Service Members, and Their Families.* Committee on the Initial Assessment of Readjustment Needs of Military Personnel, Veterans, and Their Families. Washington, DC: National Academies Press.

Inter-Parliamentary Union. 2012. *Women in National Parliaments 2012.* **www.ipu.org**

International Labour Organization. 2012. *What is Child Labor.* **www.ilo.org**

Irwin, Katherine. 2001. "Legitimating the First Tattoo: Moral Passage through Informal Interaction." *Symbolic Interaction* 24 (March): 49–73.

Jackson, Pamela Braboy. 2000. "Stress and Coping among Black Elites in Organizational Settings." Unpublished manuscript.

Jackson, Pamela B., Peggy A. Thoits, and Howard F. Taylor. "The Effects of Tokenism on America's Black Elite." Paper read before the American Sociological Association, Los Angeles, CA, August 1994.

Jackson, Pamela B., Peggy A. Thoits, and Howard F. Taylor. 1995. "Composition of the Workplace and Psychological Well-Being: The Effects of Tokenism on America's Black Elite." *Social Forces* 74 (December): 543–557.

Jacobs, David, Zhenchao Qian, Jason T. Charmichael, and Stephanie L. Kent. 2007. "Who Survives Death Row: An Individual and Contextual Analysis." *American Sociological Review* 72 (August): 610–632.

Jacobs, Jerry A., and Kathleen Gerson. 2004. *The Time Divide: Work, Family, and Gender Inequality.* New York: Oxford University Press.

Janis, Irving L. 1982. *Groupthink: Psychological Studies of Policy Decisions and Fiascos,* 2nd ed. Boston, MA: Houghton Mifflin.

Jennings, M. K., and R. G. Niemi. 1974. *The Political Character of Adolescence.* Princeton, NJ: Princeton University Press.

Johnston, David Cay. 2000. "Corporations' Taxes are Falling Even as Individual Burden Rises." *The New York Times* (February 20): A1ff.

Johnstone, Ronald I. 1992. *Religion in Society: A Sociology of Religion*, 4th ed. Englewood Cliffs, NJ: Prentice Hall.

Jones, Diane Carlson. 2001. "Social Comparison and Body Image: Attractiveness Comparison to Models and Peers among Adolescent Girls and Boys." *Sex Roles* 45 (November): 645–664.

Jones, E. E., and R. E. Nisbett. 1972. "The Actor and the Observer: Divergent Perceptions of Causes of Behavior." Pp. 79–94 in *Attribution: Perceiving the Causes of Behavior*, edited by E. E. Jones, D. E. Kanouse, H. H. Kelley, R. E. Nisbett, S. Valins, and B. Weiner. Morristown, NJ: General Learning Press.

Jones, James M. 1997. *Prejudice and Racism*, 2nd ed. New York: McGraw-Hill.

Jones, Jeffrey M. 2012. "Confidence in U.S. Public Schools at New Low." *The Gallup Poll*. Princeton, NJ: Gallup Organization.

Jones, Paul, and Gene Gloeckner. 2004. "A Study of Home School Graduates and Traditional School Graduates." *The Journal of College Admission* (Spring): 17–20.

Jordan, Catheleen, and David Cory. 2010. "Boomers, Boomerangs, and Bedpans." *National Social Science Journal* 34(1): 79–84.

Jordon, Winthrop D. 1968. *White Over Black: American Attitudes Toward the Negro 1550–1812.* Chapel Hill, NC: University of North Carolina Press.

Jordan, Winthrop D. 1969. *The White Man's Burden: Historical Origins of Racism in the United States.* New York: Oxford University Press.

Joseph, Jay. 2010. "The Genetics of Political Attitudes and Behavior: Claims and Refutations." *Ethical Human Psychology and Psychiatry* 12 (3): 199–216.

Jucha, Robert. 2002. *Terrorism.* Belmont, CA: Wadsworth.

Jussim, Lee, and Kent D. Harber. 2005. "Teacher Expectations and Self-Fulfilling Prophesies: Knowns and Unknowns, Resolved and Unresolved Controversies." *Personality and Social Psychology Review* 9: 131–155.

Kadushin, Charles. 1974. *The American Intellectual Elite.* Boston, MA: Little, Brown.

Kaiser Family Foundation. 2010. *Prescription Drug Trends.* Menlo Park, CA: Kaiser Family Foundation. **www.kff.org**

Kalleberg, Arne L. 2009. "Precarious Work, Insecure Workers: Employment Relations in Transition." *American Sociological Review* 4 (February): 1–22.

Kalleberg, Arne L. 2012. "The Social Contract in an Era of Precarious Work." *Pathways* (Fall): 3–6.

Kalleberg, Arne L., Barbara F. Reskin, and Ken Hudson. 2000. "Bad Jobs in America: Standard and Nonstandard Employment Relations and Job Quality in the United States." *American Sociological Review* 65 (April): 256–278.

Kalof, Linda. 1999. "The Effects of Gender and Music Video Imagery on Sexual Attitudes." *Journal of Social Psychology* 139 (June): 378–385.

Kamin, Leon J. 1974. *The Science and Politics of IQ.* Potomac, MD: Lawrence Erlbaum.

Kane, Emily. 2000. "Racial and Ethnic Variations in Gender-Related Attitudes." *Annual Review of Sociology* 26: 419–439.

Kane, Emily W. 2006. "'No Way My Boys Are Going to Be Like That!'": Parents' Responses to Children's Gender Nonconformity." *Gender & Society* 20(2):149–176.

Kang, Miliann. 2010. *The Managed Hand: Race, Gender and the Body in Beauty Service Work.* Berkeley, CA: University of California Press.

Kanter, Rosabeth Moss. 1977. *Men and Women of the Corporation.* New York: Basic Books.

Kaplan, Elaine Bell. 1996. *Not Our Kind of Girl: Unraveling the Myths of Black Teenage Motherhood.* Berkeley, CA: University of California Press.

Kaplan, Mark S., Nathalie Huguet, Benston H. McFarland, and Jason T. Newsom. 2007. "Suicide among Male Veterans: A Prospective Population-Based Study." *Journal of Epidemiology and Community* Health 61 (July): 619–624.

Katz, I., J. Wackenhut, and R. G. Hass. 1986. "Racial Ambivalence, Value Duality, and Behavior." Pp. 35–60 in *Prejudice, Discrimination, and Racism,* edited by J. F. Dovidio and S. L. Gaertner. New York: Academic Press.

Kaufman, Gayle. 1999. "The Portrayal of Men's Family Roles in Television Commercials." *Sex Roles* 41 (September): 439–458.

Kendall, Lori. 2002. "'Oh No! I'm a Nerd!': Hegemonic Masculinity on an Online Forum." *Gender & Society* 14 (April): 256–274.

Kennedy, Randall. 2003. *Interracial Intimacies: Sex, Marriage, Identity and Adoption.* New York: Pantheon.

Kephart, W. H. 1993. *Extraordinary Groups: An Examination of Unconventional Life,* rev. ed. New York: St. Martin's Press.

Kerr, N. L. 1992. "Issue Importance and Group Decision Making." Pp. 68–88 in *Group Process and Productivity,* edited by S. Worchel, W. Wood, and J. A. Simpson. Newbury Park, CA: Sage.

Kessler, Suzanne J. 1990. "The Medical Construction of Gender: Case Management of Intersexed Infants." *Signs* 16 (Autumn): 3–26.

Khashan, Hilal, and Lina Kreidie. 2001. "The Social And Economic Correlates of Islamic Religiosity." *World Affairs* 1654 (Fall): 83–96.

Kim, Elaine H. 1993. "Home Is Where the *Han* Is: A Korean American Perspective on the Los Angeles Upheavals." Pp. 215–235 in *Reading Rodney King/Reading Urban Uprising,* edited by Robert Gooding-Williams. New York: Routledge.

Kimmel, Michael. 2000. "Saving the Males: The Sociological Implications of Virginia Military Institute and the Citadel." *Gender & Society* 14 (August): 494–516.

Kimmel, Michael. 2007. *The Sexual Self: The Social Construction of Sexual Scripts.* Nashville, TN: Vanderbilt University Press.

Kimmel, Michael. 2008. *Guyland: The Perilous World Where Boys Become Men.* New York: Harper.

Kimmel, Michael S., and Michael A. Messner. 2004. *Men's Lives,* 6th ed. Boston, MA: Allyn and Bacon.

Kistler, Michelle E., and Moon J. Lee. 2010. "Does Exposure to Sexual Hip-Hop Music Videos Influence the Sexual Attitudes of College Students?" *Mass Communication and Society* 12 (January): 67–86.

Kitano, Harry. 1976. *Japanese Americans: The Evolution of a Subculture,* 2nd ed. New York: Prentice Hall.

Kitsuse, John I., and Aaron V. Cicourel. 1963. "A Note on the Uses of Official Statistics." *Social Problems* 11 (Fall): 131–139.

Kleinfeld, Judith. 1999. "Student Performance: Males versus Females." *Public Interest* 134 (Winter). **www.nationalaffairs.com**

Klinenberg, Eric. 2002. *Heat Wave: A Social Autopsy of Disaster in Chicago.* Chicago: University of Chicago Press.

Klinenberg, Eric. 2012. *Going Solo: The Extraordinary Rise and Surprising Appeal of Living Alone.* New York: The Penguin Press.

Klinger, Lori J., James A. Hamilton, and Peggy J. Cantrell. 2001. "Children's Perceptions of Aggression and Gender-Specific Content in Toy Commercials." *Social Behavior and Personality* 29: 11–20.

Kluegel, J. R., and Lawrence Bobo. 1993. "Dimensions of Whites' Beliefs about the Black-White Socioeconomic Gap." Pp. 127–147 in *Race and Politics in American Society,* edited by P. Sniderman, P. Tetlock, and E. Carmines. Stanford, CA: Stanford University Press.

Kniss, Fred. 2003. "Church and State." *Contexts* 2 (Spring) 62–63.

Knoke, David. 1992. *Political Networks: The Structural Perspective.* New York: Cambridge University Press.

Ko, Eunjeong, Sunhee Cho, Ramona Perez, Younsook Yeo, and Helen Palomino. 2013. "Good and Bad Death: Exploring the Perspectives of Older Mexican Americans." *Journal of Gerontological Social Work* 56(1): 6–25.

Kochen, M. (ed.). 1989. *The Small World.* Norwood, NJ: Ablex Press.

Kochhar, Rakesh, Richard Fry, and Paul Taylor. 2011. "Wealth Gaps Rise to Record Highs between Whites, Blacks, Hispanics." Washington, DC: Pew Research Center. **www.pewsocialtrends.org**

Kocieniewski, David, and Robert Hanley. 2000. "Racial Profiling Was Routine, New Jersey Says." *The New York Times* (November 28): 1.

Kohut, Andrew. 2007. "Muslim Americans: Middle Class and Mostly Mainstream." Washington, DC: Pew Research Center.

Konishi, Hideo, and Debraj Ray. 2003. "Coalition Formation as a Dynamic

Process." *Journal of Economic Theory* 110 (May): 1–41.

Kozol, Jonathan. 2006. *The Shame of the Nation: The Restoration of Apartheid Schooling in America*. New York: Broadway.

Krasnodemski, Memory. 1996. "Justified Suffering: Attribution Theory Applied to Perceptions of the Poor in America." Unpublished senior thesis, Princeton University.

Kreager, David, and Jeremy Staff. 2009. "The Sexual Double Standard and Adolescent Acceptance." *Social Psychology Quarterly* 72 (June): 143–164.

Kristof, Nicholas D. 2008. "Racism without Racists." *The New York Times* (October 4). **www.nytimes.com**

Kroll, Luisa, and Kerry A. Dolan. 2012. "The Forbes 400: The Richest People in America." *Forbes* (September 19).

Kuhn, Harold W., and Sylvia Nasar. 2002. *The Essential John Nash*. Princeton, NJ: Princeton University Press.

Kung, H. C., D. L. Hoyert, J. Q. Xu, and S. L. Murphy. 2008. *Deaths: Final Data for 2005. National Vital Statistics Reports* 56 (10). Hyattsville, MD: National Center for Health Statistics.

Kupchik, Aaron. 2010. *Homeroom Security: School Discipline in an Age of Fear*. New York: New York University Press.

Kurz, Demie. 1995. *For Richer for Poorer: Mothers Confront Divorce*. New York: Routledge.

Lacy, Karen R. 2007. *Blue-Chip Black: Race, Class and Status in the New Black Middle Class*. Berkeley, CA: University of California Press.

LaFlamme, Darquise, Andree Pomerrleau, and Gerard Malcuit. 2002. "A Comparison of Fathers' and Mothers' Involvement in Childcare and Stimulation Behaviors during Free-Play with Their Infants at 9 and 15 Months." *Sex Roles* 11–12 (December): 507–518.

LaFrance, Marianne. 2002. "Smile Boycotts and Other Body Politics." *Feminism & Psychology* 12 (August): 319–323.

Lamanna, Mary Ann, and Agnes Riedman. 2012. *Marriage and Families: Making Choices in a Diverse Society*, 11th ed. Belmont, CA: Wadsworth.

Lamb, Michael. 1998. "Cybersex: Research Notes on the Characteristics of the Visitors to Online Chat Rooms." *Deviant Behavior* 19 (April–June): 121–135.

Lamont, Michèle. 1992. *Money, Morals, and Manners: The Culture of the French and the American Upper-Middle Class*. Chicago, IL: University of Chicago Press.

Langhinrichsen-Rohling, Jennifer. Peter Lewinsohn, Paul Rohde, John Seeley, Candice M. Monson, Kathryn A. Meyer, and Richard Langford. 1998. "Gender Differences in the Suicide-Related Behaviors of Adolescents and Young Adults." *Sex Roles* 39 (December): 839–854.

Langman, Lauren, and Douglas Morris. 2002. "Internetworked Social Movements: The Promises and Prospects for Global Justice." Paper presented at the International

Sociological Association, Brisbane, Australia.

Lareau, Annette. 2003. *Unequal Childhoods: Class, Race, and Family Life*. Berkeley, CA: University of California Press.

Laumann, Edward O., John H. Gagnon, Robert T. Michael, and Stuart Michaels. 1994. *The Social Organization of Sexuality: Sexual Practices in the United States*. Chicago, IL: University of Chicago Press.

Lawson, Helene M., and Kira Leck. 2006. "Dynamics of Internet Dating." *Social Science Computer Review* 24 (Summer): 189–208.

Lederman, Douglas. 1992. "Men Outnumber Women and Get Most of the Money in Big-Time Sports Programs." *The Chronicle of Higher Education* 38 (April): A1ff.

Ledger, Kate. 2009. "Sociology and the Gene." *Contexts* 8 (3): 16–20.

Lee, Hangwoo. 2006. "Privacy, Publicity, and Accountability of Self Presentation in an On-line Discussion Group." *Sociological Inquiry* 76: 1–22.

Lee, Matthew T., and M. David Ermann. 1999. "Pinto 'Madness' as a Flawed Landmark Narrative: An Organizational and Network Analysis." *Social Problems* 46 (February): 30–47.

Lee, Richard M., Harold D. Grotevant, Wendy L. Hellerstedt, and Megan R. Gunnar. 2006. "Cultural Socialization in Families with Internationally Adopted Children." *Journal of Family Psychology* 20: 571–580.

Lee, Sharon M. 1993. "Racial Classification in the U.S. Census: 1890–1990." *Ethnic and Racial Studies* 16 (1): 75–94.

Lee, Stacey J. 1996. *Unraveling the "Model Minority" Stereotype: Listening to Asian American Youth*. New York: Teacher's College Press.

Legerski, Elizabeth Miklya, and Marie Cornwall. 2010. "Working-Class Job Loss, Gender, and the Negotiation of Household Labor." *Gender & Society* 24 (August): 447–474.

Leidner, Robin. 1993. *Fast Food, Fast Talk: Service Work and the Routinization of Everyday Life*. Berkeley, CA: University of California Press.

Lemert, Edwin M. 1972. *Human Deviance, Social Problems, and Social Control*. Englewood Cliffs, NJ: Prentice Hall.

Lengermann, P. M., and Niebrugge-Brantley, J. 1998. *The Woman Founders: Sociology and Social Theory, 1830–1930*. Boston: McGraw-Hill.

Lenski, Gerhard. 2005. *Ecological-Evolutionary Theory: Principles and Applications*. Boulder, CO: Paradigm Publishers.

Lenski, Gerhard E. 1954. "Status Crystallization: A Non-Vertical Dimension of Social Status." *American Sociological Review* 19 (August): 405–413.

Levine, Linda. 2012. *An Analysis of the Distribution of Wealth across Households*. Congressional Research Service. **www.crs.gov**

Levy, Ariel. 2005. *Female Chauvinist Pigs: Women and the Rise of Raunch Culture*. New York: Free Press.

Lewin, Tamar. 2002. "Study Links Working Mothers to Slower Learning." *The New York Times* (July 17): A14.

Lewin, Tamar. 2010. "Baby Einstein Founder Goes to Court." *The New York Times*, (January 13): A15.

Lewis, Amanda. 2003. *Race in the Schoolyard: Negotiating the Color Line in Classrooms and Communities*. New Brunswick, NJ: Rutgers University Press.

Lewis, Gregory B. 2003. "Black-White Differences in Attitudes toward Homosexuality and Gay Rights." *The Public Opinion Quarterly* 67 (April): 59–78.

Lewis, Oscar. 1960. *Five Families: Mexican Case Studies in the Culture of Poverty*. New York: Basic Books.

Lewis, Oscar. 1966. "The Culture of Poverty." *Scientific American* 215 (October): 19–25.

Lewontin, Richard. 1996. *Human Diversity*. New York: W. H. Freeman.

Lianos, Miguel. 2005 (March 3). "Plastic Bottles Pile Up as Mountains of Waste." **www.nbcnews.com**

Lieberson, Stanley. 1980. *A Piece of the Pie: Black and White Immigrants Since 1880*. Berkeley, CA: University of California Press.

Limoncelli, Stephanie. 2010. *The Politics of Trafficking: The First International Movement to Combat the Sexual Exploitation of Women*. Palo Alto, CA: Stanford University Press.

Lin, Nan. 1989. "The Small World Technique as a Theory Construction Tool." Pp. 231–238 in *The Small World*, edited by M. Kochen. Norwood, NJ: Ablex Press.

Linton, Ralph. 1936. *The Study of Man*. New York: Appleton Century Crofts.

Lo, Clarence. 1990. *Small Property versus Big Government: Social Origins of the Property Tax Revolt*. Berkeley, CA: University of California Press.

Locklear, Erin M. 1999. "Where Race and Politics Collide: The Federal Acknowledgment Process and Its Effects on Lumsee and Pequot Indians." Unpublished senior thesis, Princeton University.

Lombardo, William K., Gary A. Cretser, and Scott C. Roesch. 2001. "For Crying Out Loud—The Differences Persist into the '90s." *Sex Roles* 45 (December): 529–547.

Long, George I. 2007. *Employer-Provided "Quality of Life" Benefits for Workers in Private Industry, 2007*. Washington, DC: Bureau of Labor Statistics. **www.bls.gov**

Lopez, Mark. 2009. *Dissecting the 2008 Electorate: Most Diverse in History*. Pew Research Center. **www.pewresearch .com**

Lorber, Judith. 1994. *Paradoxes of Gender*. New Haven, CT: Yale University Press.

Lorenz, Conrad. 1966. *On Aggression*. New York: Harcourt Brace Jovanovich.

Lovejoy, Meg. 2001. "Disturbances in the Social Body: Differences in Body Image and Eating Problems among African American and White Women." *Gender & Society* 15 (April): 239–261.

Lucal, Betsy. 1994. "Class Stratification in Introductory Textbooks: Relational or Distributional Models?" *Teaching Sociology* 22 (April): 139–150.

Luker, Kristin. 1975. *Taking Chances: Abortion and the Decision Not to Contracept.* Berkeley, CA: University of California Press.

Luker, Kristin. 1984. *Abortion and the Politics of Motherhood.* Berkeley, CA: University of California Press.

Luker, Kristin. 1996. *Dubious Conceptions: The Politics of Teenage Pregnancy.* Cambridge, MA: Harvard University Press.

Lyons, Linda. 2002. "Teen Attitudes Contradict Sex-Crazed Stereotype." *The Gallup Poll.* January 29, **www.gallup.com**

Lyons, Linda. 2004. "Teens: Sex Can Wait." Princeton, NJ: The Gallup Poll. **www.gallup.com**

Machel, Graca. 1996. *Impact of Armed Conflict on Children.* New York: UNICEF/United Nations.

MacKinnon, Catherine. 1983. "Feminism, Marxism, Method, and the State: An Agenda for Theory." *Signs* 7 (Spring): 635–658.

MacKinnon, Catherine A. 2006. "Feminism, Marxism, Method, and the State: An Agenda for Theory." Pp. 829–868 in *The Canon of American Legal Thought,* edited by D. Kennedy and W. F. Fisher. Princeton: Princeton University Press.

MacKinnon, Neil J., and Tom Langford. 1994. "The Meaning of Occupational Prestige Scores: A Social Psychological Analysis and Interpretation." *Sociological Quarterly* 35 (May): 215–245.

Mackintosh, N. J. 1995. *Cyril Burt: Fraud or Framed?* Oxford, England: Oxford University Press.

Maimon, David, and Danielle C. Kuhl. 2008. "Social Control and Youth Suicidality: Situating Durkheim's Ideas in a Multilevel Framework." *American Sociological Review* 73 (December): 921–943.

Malcomson, Scott L. 2000. *One Drop of Blood: The American Misadventure of Race.* New York: Farrar, Strauss, and Giroux.

Maldonado, Lionel, A. 1997. "Mexicans in the American System: A Common Destiny." In *Ethnicity in the United States: An Institutional Approach,* edited by William Velez. Bayside, NY: General Hall.

Malinauskas, Brenda et al. 2006. "Dieting Practices, Weight Perceptions, and Body Composition: A Comparison of Normal Weight, Overweight, and Obese College Females." *Nutrition Journal* 5 (March 31): 5–11.

Maltby, Lauren E., M. Elizabeth L. Hall, Tamara L. Anderson, and Keith Edwards. 2010. "Religion and Sexism: The Moderating Role of Participant Gender." *Sex Roles* 62: 615–622.

Malthus, Thomas Robert. 1798 [1926]. *First Essay on Population 1798.* London: Macmillan.

Mandel, Daniel. 2001. "Muslims on the Silver Screen." *Middle East Quarterly* 8 (Spring): 19–30.

Mantsios, Gregory. 2010. "Media Magic: Making Class Invisible." Pp. 386–394 in *Race, Class, and Gender: An Anthology,* edited by Margaret L. Andersen and Patricia Hill Collins. Belmont, CA: Wadsworth.

Marciniak, Liz-Marie. 1998. "Adolescent Attitudes toward Victim Precipitation of Rape." *Violence and Victims* 12 (Fall): 287–300.

Marcuse, Herbert. 1964. *One-Dimensional Man.* Boston, MA: Beacon Press.

Margolin, Leslie. 1992. "Deviance on Record: Techniques for Labeling Child Abusers in Official Documents." *Social Problems* 39 (February): 58–70.

Marks, Carole. 1989. *Farewell, We're Good and Gone: The Great Black Migration.* Bloomington, IN: Indiana University Press.

Marks, Carole, and Deana Edkins. 1999. *The Power of Pride: Stylemakers and Rulebreakers of the Harlem Renaissance.* New York: Crown.

Martin, Karin A. 2005. "William Wants a Doll. Can He Have One? Feminists, Child Care Advisors, and Gender-Neutral Child Rearing." *Gender & Society* 19 (August): 456–479.

Martin, Patricia Yancey, and Robert Hummer. 1989. "Fraternities and Rape on Campus." *Gender & Society* 3 (December): 457–473.

Martin, Susan F. 2001. "Heavy Traffic: International Migration in an Era of Globalization." *Brookings Review* 19 (Fall): 41–44.

Martineau, Harriet. 1837. *Society in America.* London: Saunders and Otley.

Martineau, Harriet. 1838. *How to Observe Morals and Manners.* London: Charles Knight and Co.

Marx, Anthony. 1997. *Making Race and Nation: A Comparison of the United States, South Africa, and Brazil.* New York: Cambridge University Press.

Marx, Karl. 1967 [1867]. *Capital.* F. Engels (ed.). New York: International Publishers.

Marx, Karl. 1972 [1843]. "Contribution to the Critique of Hegel's *Philosophy of Right*." Pp. 11–23 in *The Marx-Engels Reader,* edited by Robert C. Tucker. New York: W. W. Norton.

Massey, Douglas S. 2005. *Strangers in a Strange Land: Humans in an Urbanizing World.* New York: Norton.

Massey, Douglas S., and Nancy A. Denton. 1993. *American Apartheid: Segregation and the Making of the Underclass.* Cambridge, MA: Harvard University Press.

Mast, Marianne Schmid, and Judith A. Hall. 2001. "Gender Differences and Similarities in Dominance Hierarchies in Same-Gender Groups Based on Speaking Time." *Sex Roles* 44 (May): 537–556.

Mast, Marianne S., and Judith A. Hall. 2004. "When is Dominance Related to Smiling? Assigned Dominance, Dominance Preference, Trait Dominance, and Gender as Moderators." *Sex Roles* 50 (March): 387–399.

Mauer, Marc. 1999. *Race to Incarcerate.* New York: The New Press.

Mazumder, Bhashkar. 2008. "Intergenerational Economic Mobility in the US: 1940 to 2000." *Journal of Human Resources* 43 (January): 139–172.

McCabe, Janice. 2005. "What's in a Label? The Relationship between Feminist Self-Identification and 'Feminist' Attitudes among U.S. Women and Men." *Gender & Society* 19(4): 480–505.

McCall, Leslie. 2001. *Complex Inequality: Gender, Class, and Race in the New Economy.* New York: Routledge.

McClelland, Susan. 2003. "A Grim Toll on the Innocent." *Maclean's* (May 12): 20.

McClintock, Elizabeth Aura. 2010. "When Does Race Matter? Race, Sex, and Dating at an Elite University." *Journal of Marriage and the Family* 72 (February): 45–72.

McCollum, Chris. 2002. "Relatedness and Self-Definition: Two Dominant Themes in Middle-Class Americans' Life Stories." *Ethos* 30: 113–139.

McCord, William. 1991. "The Asian Renaissance." *Society* 28 (September-October): 50–61.

McGuire, Meredith. 2001. *Religion: The Social Context,* 5th ed. Belmont, CA: Wadsworth.

McLoyd, Vonnie C., Ana Mari Cauce, David Takeuchi, and Leon Wilson. 2000. "Marital Processes and Parental Socialization in Families of Color: A Decade Review of Research." *Journal of Marriage and the Family* 62 (November): 1070–1093.

McVeigh, Ricky. 2012. "Making Sense of the Tea Party." *Contemporary Sociology* 41 (November): 766–769.

Mead, George Herbert. 1934. *Mind, Self, and Society.* Chicago. IL: University of Chicago Press.

Meier, Barry. 2013. "Maker Hid Data about Design Flaw in Hip Implant." *The New York Times* (January 13): B1 and B6.

Mendes, Elizabeth, Lydia Saad, and Kyley McGeeney. 2012. "Stay at Home Moms Report More Depression, Sadness, Anger." *The Gallup Poll.* Princeton, NJ: Gallup Organization, May 18. **www.gallup.com**

Meredith, Martin. 2003. *Elephant Destiny: Biography of an Endangered Species in Africa.* New York: HarperCollins.

Mernissi, Fataima. 1987. *Beyond the Veil: Male-Female Dynamics in Modern Muslim Society.* Bloomington, IN: Indiana University Press.

Merton, Robert K. 1957. *Social Theory and Social Structure.* New York: Free Press.

Merton, Robert K. 1968. "Social Structure and Anomie." *American Sociological Review* 3: 672–682.

Merton, Robert, and Alice K. Rossi. 1950. "Contributions to the Theory of Reference Group Behavior." Pp. 279–334 in *Continuities in Social Research Studies, Scope and Method of "The American Soldier,"* edited by Robert K. Merton and Paul F. Lazarsfeld. New York: Free Press.

Messerschmidt, James W. 1997. *Crime as Structured Action: Gender, Race, Class and Crime in the Making.* Thousand Oaks, CA: Sage.

Messner, Michael A. 1992. *Power at Play: Sports and the Problem of Masculinity.* Boston, MA: Beacon Press.

Messner, Michael A. 2002. *Taking the Field: Women, Men, and Sports.*

Minneapolis, MN: University of Minnesota Press.

Messner, Michael A. 2009. *It's All for the Kids: Gender, Families, and Youth Sports*. Berkeley: University of California Press.

Messner, Steven F. 2011. *Crime and the American Dream*. Belmont, CA: Wadsworth/Cengage.

Metz, Michael A., B. R. Rosser-Simon, and Nancy Strapko. 1994. "Differences in Conflict-Resolution Styles among Heterosexual, Gay, and Lesbian Couples." *Journal of Sex Research* 31: 293–308.

Meyer, Madonna Harrington. 1994. "Gender, Race, and the Distribution of Social Assistance: Medicaid Use among the Elderly." *Gender & Society* 8 (March): 8–28.

Mezey, Nancy J. 2008. *New Choices, New Families: How Lesbians Decide about Motherhood*. Baltimore, MD: Johns Hopkins University Press.

Mickelson, Roslyn Arlin (ed.). 2000. *Children on the Streets of the Americas: Globalization, Homelessness, and Education in the United States, Brazil, and Cuba*. New York: Routledge.

Milanovic, Brandon. 2010. *The Haves and Have Nots: A Brief and Idiosyncratic History of Global Inequality*. New York: Basic Books.

Milgram, Stanley. 1974. *Obedience to Authority: An Experimental View*. New York: Harper & Row.

Milkie, Melissa A. 1999. "Social Comparisons, Reflected Appraisals, and Mass Media: The Impact of Pervasive Beauty Images on Black and White Girls' Self-Concepts." *Social Psychology Quarterly* 62 (June): 190–210.

Milkie, Melissa A. 2002. "Contested Images of Femininity: An Analysis of Cultural Gatekeepers' Struggles with the 'Real Girl' Critique." *Gender & Society* 16 (December): 839–859.

Milkie, Melissa A., Suzanne M. Bianchi, Marybeth J. Mattingly, and John P. Robinson. 2002. "Gendered Division of Childrearing: Ideals, Realities, and the Relationship to Parental Well-being." *Sex Roles* 47 (July): 21–38.

Milkie, Melissa, and Pia Peltola. 1999. "Playing All the Roles: Gender and the Work–Family Balancing Act." *Journal of Marriage and the Family* 61 (May): 476–490.

Miller, Eleanor. 1986. *Street Women*. Philadelphia: Temple University Press.

Miller, Eleanor. 1991. "Jeffrey Dahmer, Racism, Homophobia, and Feminism." *SWS Network News* 8 (December): 2.

Miller, Susan L. 1997. "The Unintended Consequences of Current Criminal Justice Policy." Talk presented at Research on Women Series, University of Delaware, Newark, DE.

Mills, C. Wright. 1956. *The Power Elite*. New York: Oxford University Press.

Mills, C. Wright. 1959. *The Sociological Imagination*. New York: Oxford University Press.

Milner, Murray, Jr. 2004. *Freaks, Geeks, and Cool Kids: American Teenagers, Schools, and the Culture of Consumption*. New York: Routledge.

Miner, Horace. 1956. "Body Ritual among the Nacirema." *American Anthropologist* 58: 503–507.

Mintz, Beth, and Michael Schwartz. 1985. *The Power Structure of American Business*. Chicago, IL: University of Chicago Press.

Mirandé, Alfredo. 1979. "Machismo: A Reinterpretation of Male Dominance in the Chicano Family." *The Family Coordinator* 28: 447–449.

Mirandé, Alfredo. 1985. *The Chicano Experience*. Notre Dame, IN: Notre Dame University Press.

Mishel, Lawrence, Jared Bernstein, and Heidi Shierholz. 2008. *State of Working America 2008/2009*. Washington, DC: Economic Policy Institute.

Misra, Joy, Stephanie Moller, and Maria Karides. 2003. "Envisioning Dependency: Changing Media Depictions of Welfare in the 20th Century." *Social Problems* 50 (November): 482–504.

Mizruchi, Ephraim H. 1983. *Regulating Society: Marginality and Social Control in Historical Perspective*. New York: Free Press.

Mizruchi, Mark S. 1992. *The Structure of Corporate Political Action: Interfirm Relations and Their Consequences*. Cambridge, MA: Harvard University Press.

Moen, Phyllis. 2003. *It's about Time: Couples and Careers*. Ithaca, NY: Cornell University Press.

Moen, Phyllis, with Donna Dempster-McClain, Joyce Altobelli, Wipas Wimonsate, Lisa Dahl, Patricia Roehling, and Stephen Sweet. 2004. *The New "Middle" Workforce*. The Bronfenbrenner Life Course Center and Cornell Careers Institute. **www .lifecourse.cornell**

Moen, Phyllis, Jungmeen E. Kim, and Heather Hofmeister. 2001. "Couples' Work/Retirement Transitions, Gender, and Marital Quality." *Social Psychology Quarterly* 64 (March): 55–71.

Mohai, Paul, and Robin Saha. 2007. "Racial Inequality in the Distribution of Hazardous Waste: A National-Level Reassessment." *Social Problems* 54 (3): 343–370.

Montada, L., and M. Lerner Jr. (eds.). 1998. *Responses to Victimization and Beliefs in a Just World*. New York: Plenum.

Montgomery, James D. 1992. "Job Search and Network Composition: Implications of the Strength of Work Ties Hypothesis." *American Sociological Review* 57 (October): 586–596.

Moore, Gwen. 1979. "The Structure of a National Elite Network." *American Sociological Review* 44 (October): 673–692.

Moore, Joan. 1976. *Hispanics in the United States*. Englewood Cliffs, NJ: Prentice Hall.

Moore, Joan W., and John M. Hagedorn. 1996. "What Happens to Girls in Gangs?" Pp. 205–218 in *Gangs in America*, edited by C. Ronald Huff. Thousand Oaks, CA: Sage.

Moore, Robert B. 1992. "Racist Stereotyping in the English Language." Pp. 317–328 in *Race, Class, and Gender: An Anthology*, 2nd ed., edited by Margaret L. Andersen

and Patricia Hill Collins. Belmont, CA: Wadsworth.

Moore, Valerie A. 2001. "'Doing' Racialized and Gendered Age to Organize Peer Relations: Observing Kids in Summer Camp." *Gender & Society* 15 (December): 835–858.

Mora, Christina. 2009. *DeMuchos, Uno: The Institutionalization of Latino Panethnicity in the United States, 1960–1990*. PhD diss., Princeton University, Princeton, NJ.

Moreland, Richard L., and Scott R. Beach. 1992. "Exposure Effects in the Classroom: The Development of Affinity among Students." *Journal of Experimental Social Psychology* 28: 255–276.

Morin, Rich, and Seth Motel. 2012. "A Third of Americans Now Say They Are in the Lower Classes." *Pew Social & Demographic Trends*. Washington, DC: Pew Research Center.

Morlan, Patricia A. 2005. "Are We Taught to Blame the Poor? Attributions for Poverty: Sociodemographic Predictors and the Effects of Higher Education." Senior thesis, Princeton University.

Morning, Ann. 2005. "Race." *Contexts* 4 (Fall): 44–46.

Morning, Ann. 2008. "Reconstructing Race in Science and Society: Biology Textbooks, 1952–2002." *American Journal of Sociology* 114 (Suppl.): S106–S137.

Morning, Ann. 2011. *The Nature of Race: How Scientists Think and Teach about Human Differences*. New York: New York University Press.

Morris, Aldon. 1984. *The Origins of the Civil Rights Movement: Black Communities Organizing for Change*. New York: Free Press.

Morris, Aldon D. 1999. "A Retrospective on the Civil Rights Movement: Political and Intellectual Landmarks." *Annual Review of Sociology* 25: 517–539.

Moskos, Peter. 2008. *Cop in the Hood: My Year in Policing Baltimore's Eastern District*. Princeton, NJ: Princeton University Press.

Moynihan, Daniel P. 1965. *The Negro Family: The Case for National Action*. Washington, DC: Office of Policy Planning and Research, U.S. Department of Labor.

Mullen, Ann L. 2010. *Degrees of Inequality: Culture, Class, and Gender in American Higher Education*. Baltimore, MD: Johns Hopkins University Press.

Myers, Steven Lee. 2000. "Survey of Troops Finds Antigay Bias Common in Service." The *New York Times* (March 24): 1.

Myers, Walter D. 1998. *Amistad Affair*. New York: NAL/Dutton.

Myerson, Allen R. 1998. "Rating the Bigshots: Gates vs. Rockefeller." *The New York Times* (May 24): 4.

Myrdal, Gunnar. 1944. *An American Dilemma: The Negro Problem and Modern Democracy*, 2 Vols. New York: Harper and Row.

NAACP Legal Defense and Education Fund. 2010. *Annual Report*. Washington, DC: NAACP Legal Defense Fund.

Nagel, Joane. 1996. *American Indian Ethnic Renewal: Red Power and the Resurgence of Identity and Culture*. New York: Oxford University Press.

Nagel, Joane. 2003. *Race, Ethnicity, and Sexuality: Intimate Intersections, Forbidden Frontiers*. New York: Oxford University Press.

Nakao, Keiko, and Judith Treas. 1994. "Updating Occupational Prestige and Socioeconomic Status Scores: How the New Measures Measure Up." Pp. 1–72 in Peter Marsden (ed.), *Sociological Methodology 1994*. Oxford: Basil Blackwell, Inc.

Nanda, Serena. 1998. *Neither Man Nor Woman: The Hijras of India*. Belmont, CA: Wadsworth.

National Assessment of Educational Progress. 2012. *Private and Other Nonpublic Schools and the Nation's Report Card*. Washington, DC: National Center for Educational Statistics, U.S. Department of Education.

National Center for Education Statistics. 2003. *Homeschooling in the United States*. **www.nces.ed.gov**

National Center for Education Statistics. 2009. *Characteristics of Private Schools in the United States: Results from the 2007–08 Private School Universe*. Washington, DC: U.S. Department of Education.

National Center for Education Statistics. 2012. *The Condition of Education, 2012*. Washington, DC: U.S. Department of Education.

National Center for Health Statistics. 2010a. *Health United States 2009*. Atlanta, GA: Centers for Disease Control and Prevention.

National Center for Health Statistics. 2010b. "Smoking, Alcohol Use, and Illicit Drug Use Reported by Adolescents." National Health Statistics Reports. Hyattsville, MD: U. S. Department of Health and Human Services.

National Center for Health Statistics. 2012. "Health, United States 2011." Hyattsville, MD: U.S. Department of Health and Human Services.

National Center on Elder Abuse. 2012. **www.ncea.aoa.gov**

National Coalition against Domestic Violence. 2001. **www.ncadv.org**

National Coalition for the Homeless. 2012. "Fact Sheet." **www.nationalhomeless.org**

National Institutes of Health. 2007. *National Center for Complementary and Alternative Medicine*. Bethesda, MD: U.S. Department of Health and Human Services.

National Research Council. 2012. *Advancing the Science of Climate Change*. Washington, DC: The National Academies Press. **www.nas-sites.org**

Nee, Victor. 1973. *Longtime Californ': A Documentary Study of an American Chinatown*. New York: Pantheon Books.

Neighbors, L. A., and J. Sobal. 2007. "Prevalence and Magnitude of Body Weight and Shape Dissatisfaction among University Students." *Eating Behaviors* 8: 429–439.

Newman, Katherine. 1999. *No Shame in My Game: The Working Poor in the Inner City*. New York: Russell Sage Foundation/Vintage Books.

Newman, Katherine S. 2012. *The Accordion Family: Boomerang Kids, Anxious Parents, and the Private Toll of Global Competition*. Boston: Beacon Press.

Newman, Katherine S., Cybelle Fox, David Harding, Jal Mehta, and Wendy Roth. 2006. *Rampage: Social Roots of School Shootings*. New York: Basic Books.

Newport, Frank. 2000. "Women's Most Pressing Concerns Today are Money, Family, Health, and Stress." *The Gallup Poll Monthly* (March): 40–41.

Newport, Frank. 2003 (May 15). "Six Out of 10 Americans Say Homosexual Relations Should Be Legal." *The Gallup Poll*. Princeton, NJ: Gallup Organization. **www.gallup.com**

Newport, Frank. 2008. "Easter Season Finds a Religious, Largely Christian Nation." *The Gallup Poll*. Princeton, NJ: Gallup Organization. **www.gallup.com**

Newport, Frank. 2009. "This Christmas, 78% of the Public Identify as Christian." *The Gallup Poll*. Princeton, NJ: Gallup Organization. **www.gallup.com**

Newport, Frank. 2011. "For First Time, Majority of Americans Favor Legal Gay Marriage." Princeton, NJ: The Gallup Organization. **www.gallup.com**

Newport, Frank. 2012a. "Americans, Including Catholics, Think Birth Control is Morally OK." *Gallup Poll* (May 22). **www.gallup.com**

Newport, Frank. 2012b. "Bias against Mormon Presidential Candidate Same as in 1967." *The Gallup Poll*. Princeton, NJ: Gallup Organization.

Newport, Frank. 2012c. "Young Adults Admit Too Much Time on Cell Phones, Internet." *The Gallup Poll*. Princeton, NJ: Gallup Organization.

Ngai, Mae M. 2012. "Impossible Subjects: Illegal Aliens and the Making of Modern America" Pp. 192–196 in *Race and Ethnicity in Society: The Changing Landscape*, 3rd ed., edited by Elizabeth Higginbotham and Margaret L. Andersen. Belmont, CA: Wadsworth/Cengage.

Niebuhr, Gustav. 1998. "Markup of American Religion Is Looking More Like Mosaic, Data Say." *The New York Times* (April 12): 14.

Nolan, Patrick, and Gerhard Lenski. 2008. *Human Societies*. New York: Paradigm Publishers.

Norgaard, Kari M. 2006. "'People Want to Protect Themselves a Little Bit': Emotions, Denial, and Social Movement Nonparticipation." *Sociological Inquiry* 76 (3): 372–396.

Norris, Pippa, and Ronald Inglehart. 2002. "Islamic Culture and Democracy: Testing the 'Clash of Civilizations' Thesis." *Comparative Sociology* 1: 235–263.

Oakes, Jeannie. 2005. *Keeping Track: How Schools Structure Inequality*, 2nd ed. New Haven, CT: Yale University Press.

Oakes, Jeannie, Karen Hunter Quartz, Steve Ryan, and Martin Lipton. 2000. *Becoming Good American Schools. The Struggle for Civic Virtue in Education Reform*. San Francisco: Jossey-Bass.

Ogburn, William F. 1922. *Social Change with Respect to Cultural and Original Nature*. New York: B. W. Huebsch.

O'Leary, Carol. 2002. "The Kurds of Iraq: Recent History, Future Prospects." *Middle East Review of International Affairs* 6 (December 2002): 17–29.

Oliver, Melvin, and Thomas M. Shapiro. 1995. *Black Wealth/White Wealth: A New Perspective on Racial Inequality*. New York: Routledge.

Oliver, Melvin L., and Thomas M. Shapiro. 2001. "Wealth and Racial Stratification." Pp. 222–240 in *America Becoming: Racial Trends and Their Consequences*, edited by Neil Smelser, William Julius Wilson, and Faith Mitchell. Washington, DC: National Academies Press.

Oliver, Melvin L., and Thomas M. Shapiro. 2006. *Black Wealth/White Wealth: A New Perspective on Racial Inequality*, 10th ed. New York: Routledge.

Oliver, Melvin L., and Thomas M. Shapiro. 2008. "Sub-Prime as Black Catastrophe." *The American Prospect*, September 22.

Ollivier, Michele. 2000. "'Too Much Money off Other People's Backs': Status in Late Modern Societies." *Canadian Journal of Sociology* 25 (Fall): 441–470.

Omi, Michael, and Howard Winant. 1994. *Racial Formation in the United States*, 2nd ed. New York: Routledge.

O'Neil, John. 2002. "Parent Smoking and Teenage Sex." *The New York Times*, Sept. 3, p. F7.

Orfield, Gary, and Chungmei Lee. 2012. "Historic Reversals, Accelerating Resegregation, and the Need for New Integration Strategies." Los Angeles: Civil Rights Project, UCLA. **www.civilrightsproject.ucla.edu**

Organization for Economic Co-operation and Development (OECD). 2009. *Society at a Glance—OECD Social Indicators*. **www.oecd.org/els/social/indicators/SAG**

Ortiz, Isabel, and Matthew Cummins. 2011. "Global Inequality: Beyond the Bottom Billion." *Social and Economic Policy Working Paper*. UNICEF, April. **www.unicef.org**

Padavic, Irene, and Barbara Reskin. 2002. *Women and Men at Work*, 2nd ed. Thousand Oaks, CA: Sage.

Padgett, Tim, Anthony Esposito, and Aaron Nelson. 2011. "The Chilean Miners." *Time* (January 3): 105–11.

Page, Charles H. 1946. "Bureaucracy's Other Face." *Social Forces* 25 (October): 89–94.

Page, Scott E. 2007. *The Difference: How the Power of Diversity Creates Better Groups, Firms, Schools, and Societies*. Princeton, NJ: Princeton University Press.

Pager, Devah. 2007. *Marked: Race, Crime, and Finding Work*. Chicago: University of Chicago Press.

Pain, Emil. 2002. "The Social Nature of Extremism and Terrorism." *Social Sciences* 33: 55–68.

Painter, Nell Irvin. 2010. *The History of White People*. New York: W. W. Norton.

Park, Robert E., and Ernest W. Burgess. 1921. *Introduction to the Science of Society*.

Chicago, IL: University of Chicago Press.

Parker, Kim. 2012. *The Boomerang Generation: Feeling OK about Living with Mom and Dad.* Washington, DC: Pew Research Center. **www .pewsocialtrends.org**

Parker-Pope, Tara. 2008. "Love, Sex, and the Changing Landscape of Infidelity." *The New York Times*, October 27.

Parmelee, L. F. 2001. "A Roper Center Data Review: Mending the Fabric: Race Relations in Black and White." *Public Perspective* 12: 22–31.

Parreñas, Rhacel Salazar. 2001. *Servants of Globalization: Women, Migration, and Domestic Work.* Stanford, CA: Stanford University Press.

Parsons, Talcott (ed.). 1947. *Max Weber: The Theory of Social and Economic Organization.* New York: Free Press.

Parsons, Talcott. 1951a. *The Social System.* Glencoe, IL: Free Press.

Parsons, Talcott. 1951b. *Toward a General Theory of Action.* Cambridge, MA: Harvard University Press.

Parsons, Talcott. 1966. *Societies: Evolutionary and Comparative Perspectives.* Englewood Cliffs, NJ: Prentice Hall.

Pascoe, C. J. 2007. *Dude, You're a Fag: Masculinity and Sexuality in High School.* Berkeley, CA: University of California Press.

Pathways to Peace. 2009. "Impact on War on Poverty." **www.icrc.org**

Pattillo-McCoy, Mary. 1999. *Black Picket Fences: Privilege and Peril among the Black Middle Class.* Chicago, IL: University of Chicago Press.

Paulus, P. B., T. S. Larey, and M. T. Dzindolet. 2001. "Creativity in Groups and Teams." Pp. 319–338 in *Groups at Work: Theory and Research*, edited by M.E. Turner. Mahwah, NJ: Erlbaum.

Pedraza, Silvia. 1996. "Cuba's Refugees: Manifold Migrations." Pp. 263–279 in *Origins and Destinies: Immigration, Race, and Ethnicity in America*, edited by Silvia Pedraza and Rubén Rumbaut. Belmont, CA: Wadsworth.

Pelham, Brett, and Steve Crabtree. 2009. "Religiosity and Perceived Intolerance of Gays and Lesbians." Princeton, NJ: Gallup Poll, March 10. **www.gallup.com**

Pellow, David N. 2004. "The Politics of Illegal Dumping: An Environmental Justice Framework." *Qualitative Sociology* 27(4): 511–525.

Peralta, Robert L. 2002. *Getting Trashed in College: Doing Alcohol, Doing Gender, Doing Violence.* Ph.D. dissertation, University of Delaware.

Peri, Giovanni, and Chad Sparber. 2007. "Task Specialization, Immigration, and Wages." *American Economic Journal: Applied Economics* 1 (July): 135–169.

Perlmutter, David. 2008. *Blogwars: The New Political Battleground.* New York: Oxford University Press.

Perrow, Charles. 1986. *Complex Organization: A Critical Essay,* 3rd ed. New York: Random House.

Perrow, Charles. 1994. "The Limit of Safety: The Enhancement of a Theory of Accidents." *Journal of Contingencies and Crisis Management* 22: 212–220.

Perrow, Charles. 2007. *Organizing America: Wealth, Power, and Origins of American Capitalism.* Princeton: Princeton University Press.

Perry-Jenkins, Maureen, Rena L. Repetti, and Anne C. Crouter. 2000. "Work and Family in the 1990s." *Journal of Marriage and the Family* 62 (November): 981–998.

Pescosolido, Bernice A., Elizabeth Grauerholz, and Melissa A. Milkie. 1997. "Culture and Conflict: The Portrayal of Blacks in U.S. Children's Picture Books through the Mid- and Late-Twentieth Century." *American Sociological Review* 62 (June): 443–464.

Petersen, Trond, Ishak Saporta, and Mark-David L. Seidel. 2000. "Offering a Job: Meritocracy and Social Networks." *American Journal of Sociology* 106 (November): 763–816.

Pettigrew, Thomas F. 1992. "The Ultimate Attribution Error: Extending Allport's Cognitive Analysis of Prejudice." Pp. 401–419 in *Readings about the Social Animal*, edited by Elliott Aronson. New York: Freeman.

Pew Forum on Religion & Public Life. 2007. *Changing Faiths: Latinos and the Transformation of American Religion.* Washington, DC: Pew Research Center. **www.pewforum.org**

Pew Forum on Religion & Public Life. 2009. *A Religious Portrait of African-Americans.* Pew Forum on Religion & Public Life. 2009. Washington, DC: Pew Research Center. **www.pewforum.org**

Pew Forum on Religion & Public Life. 2012. *Asian Americans: A Mosaic of Faiths.* Washington, DC: Pew Research Center. **www.pewforum.org**

Pew Internet and American Life Project. 2012. "Demographics of Internet Users." **www.pewinternet.com**

Pew Research Center. 2007. "Muslim Americans: Middle Class and Mostly Mainstream." Washington, DC: Pew Research Center. **www.pewresearch.org**

Pew Research Center. 2009. *Dissecting the 2008 Election: Most Diverse in U.S. History.* Washington, DC: Pew Research Center. **www.pewresearch.org**

Pew Research Center. 2010. "The Return of the Multi-Generational Family Household. Data from U.S. Decennial Census Data, 1940–2000." **www .pewsocialtrends.org**

Pew Research Center. 2011. "No Shift Toward Gun Control After Tucson Shootings." Washington, DC: Pew Research Center. **www.pewresearch.org**

Pew Research Center. 2012. *Faith on the Hill: The Religious Composition of the 113th Congress.* Washington, DC: Pew Research Center. **www.pewresearch .org**

Polce-Lynch, Mary, Barbara J. Myers, Wendy Kliewer, and Christopher Kilmartin. 2001. "Adolescent Self-Esteem and Gender: Exploring Relations to Sexual Harassment, Body Image, Media Influence, and Emotional Expression." *Journal of Youth and Adolescence* 30 (April): 225–244.

Popenoe, David. 2001. "Today's Dads: A New Breed?" *The New York Times* (June 19): A22.

Portes, Alejandro. 2002. "English-Only Triumphs, but the Costs Are High." *Contexts* 1 (February): 10–15.

Portes, Alejandro, and Rubén G. Rumbaut. 1996. *Immigrant America: A Portrait,* 2nd ed. Berkeley, CA: University of California Press.

Portes, Alejandro, and Rubén G. Rumbaut. 2001. *Legacies: The Story of the Immigrant Second Generation.* Berkeley, CA: University of California Press.

Powell, Brian, Catherine Bolzendahl, Claudia Geist, and Lala Carr Steelman. 2010. *Counted Out: Same-Sex Relations and Americans' Definitions of Family.* New York: Russell Sage Foundation.

Press, Andrea. 2002. "The Paradox of Talk." *Contexts* 1 (Fall–Winter): 69–70.

Press, Eyal. 1996. "Barbie's Betrayal." *The Nation* (December 30): 11–16.

Press, Julie E., and Eleanor Townsley. 1998. "Wives and Husbands' Reporting: Gender, Class, and Social Desirability." *Gender & Society* 12 (April): 188–218.

Preves, Sharon E. 2003. *Intersex and Identity: The Contested Self.* New Brunswick, NJ: Rutgers University Press.

Price-Glynn, Kim. 2010. *Strip Club: Gender, Power, and Sex Work.* New York: New York University Press.

Prostitutes Education Network. 2009. **www .bayswan.org**

Prus, Robert, and C. R. D. Sharper. 1991. *Road Hustler.* New York: Kaufman and Greenberg.

Puffer, Phyllis. 2009. "Durkheim Did Not Say 'Normlessness': The Concept of Anomic Suicide for Introductory Sociology Courses." *Southern Rural Sociology* 24: 200–222.

Punch, Maurice. 1996. *Dirty Business: Exploring Corporate Misconduct; Analysis and Cases.* Thousand Oaks, CA: Sage.

Putnam, Robert D. 2000. *Bowling Alone: The Collapse and Revival of American Community.* New York: Simon and Schuster.

Quadagno, Jill. 2005. *One Nation, Uninsured: Why the U.S. Has No National Health Insurance.* New York: Oxford University Press.

Raboteau, Albert J. 1978. *Slave Religion: The "Invisible Institution" in the Antebellum South.* New York: Oxford University Press.

Rajan, Ramkishen S., and Sadhana Srivastava. 2007. *Harvard Asia Pacific Review* 9 (Winter): n.p.

Raley, Sara B., Marybeth J. Mattingly, and Suzanne M. Bianchi. 2006. "How Dual Are Dual-Income Couples? Documenting Change from 1970 to 2001." *Journal of Marriage and Family* 68(1): 11–28.

Rampersad, Arnold. 1986. *The Life of Langston Hughes: Vol. I: 1902–1941. I, Too, Sing America.* New York: Oxford University Press.

Rampersad, Arnold. 1988. *The Life of Langston Hughes: Vol. II: 1941–1967.*

I Dream A World. New York: Oxford University Press.

Rashid, Ahmed. 2000. *Taliban: Militant Islam, Oil, and Fundamentalism in Central Asia.* New Haven, CT: Yale University Press.

Rasmussen Reports. 2009. "62% Say Today's Children Will Not Be Better Off than Their Parents." **www.rasmussenreports.com**

Read, Jen'nan Ghazal. 2003. "The Sources of Gender Role Attitudes among Christian and Muslim Arab-American Women." *Sociology of Religion* 64 (Summer): 207–222.

Read, Piers Paul. 1974. *Alive: The Story of the Andes Survivors.* Philadelphia, PA: Lippincott.

Reich, Robert. 2010. "The Root of Economic Fragility and Political Anger." **www.robertreich.org**

Reid, T. R. 2010. *The Healing of America: A Global Quest for Better, Cheaper, and Fairer Health Care.* New York, NY: Penguin Books.

Reiman, Jeffrey H. 2007. *The Rich Get Richer and the Poor Get Prison,* 8th ed. Boston, MA: Allyn and Bacon.

Reiman, Jeffrey, and Paul Leighton. 2009. *The Rich Get Richer and the Poor Get Prison: A Reader.* Upper Saddle River, NJ: Pearson.

Renzetti, Claire. 1992. *Violent Betrayal: Partner Abuse in Lesbian Relationships.* Newbury Park, CA: Sage.

Reskin, Barbara. 1988. "Bringing the Men Back In: Sex Differentiation and the Devaluation of Women's Work." *Gender & Society* 2 (March): 58–81.

Rheault, Magail, and Dalia Mogahed. 2008. "Moral Issues Divide Westerners from Muslims in the West." *The Gallup Poll.* Princeton, NJ: Gallup Organization. **www.gallup.com**

Rich, Adrienne. 1980. "Compulsory Heterosexuality and Lesbian Existence." *Signs* 5 (Summer): 631–660.

Ridgeway. Cecilia L. 2011. *Framed by Gender: How Gender Inequality Persists in the Modern World.* New York: Oxford University Press.

Riegle-Crumb, Catherine and Melissa Humphries. 2010. "Exploring Bias in Math Teachers' Perceptions of Students' Ability by Gender and Race/Ethnicity." *Gender & Society* 26 (April): 290–322.

Rindfuss, Ronald R., Elizabeth C. Cooksey, and Rebecca L. Sutterlin. 1999. "Young Adult Occupational Achievement: Early Expectations versus Behavioral Reality." *Work and Occupations* 26 (May): 220–263.

Risman, Barbara, and Pepper Schwartz. 2002. "After the Sexual Revolution: Gender Politics in Teen Dating." *Contexts* 1 (Spring): 16–24.

Ritzer, George. 2010. *The McDonaldization of Society,* 6th ed. Thousand Oaks, CA: Sage Publications.

Rivas-Drake, Deborah. 2011. "Ethnic-Racial Socialization and Adjustment among Latino College Students: The Mediating Roles of Ethnic Centrality, Public Regard, and Perceived Barriers to Opportunity." *Journal of Youth and Adolescence* 40: 606–619.

Robbins, Thomas. 1988. *Cults, Convents, and Charisma: The Sociology of New Religious Movements.* Beverly Hills, CA: Sage.

Roberts, Dorothy. 1997. *Killing the Black Body: Race, Reproduction and the Meaning of Liberty.* New York: Vintage Books.

Roberts, Dorothy. 2012. *Fatal Invention: How Science, Politics, and Big Business Re-Create Race in the Twenty-First Century.* New York: New Press.

Robertson, Tatsha, and Garrance Burke. 2001. "Fighting Terror: Concerned Family;" "Shock, Worry for Family of U.S. Man Captured with Taliban." *The Boston Globe* (December 4): A1.

Robinson, Dawn T., and Lynn Smith-Lovin. 2001. "Getting a Laugh: Gender, Status, and Humor in Task Discussions." *Social Forces* 80 (September): 123–158.

Robinson, Jo Ann Gibson. 1987. *The Montgomery Bus Boycott and the Women Who Started It.* Knoxville, TN: The University of Tennessee Press.

Rodriguez, Clara E. 1989. *Puerto Ricans: Born in the U.S.A.* Boston, MA: Unwin Hyman.

Rodriguez, Clara E. 2009. "Changing Race." Pp. 22–25 in *Race and Ethnicity in Society: The Changing Landscape,* 3rd ed., edited by Elizabeth Higginbotham and Margaret L. Andersen. Belmont, CA: Wadsworth.

Roethlisberger, Fritz J., and William J. Dickson. 1939. *Management and the Worker.* Cambridge, MA: Harvard University Press.

Romain, Suzanne. 1999. *Communicating Gender.* Mahwah, NJ: Erlbaum.

Roof, Wade Clark. 1999. *Spiritual Marketplace.* Princeton, NJ: Princeton University Press.

Rosenbaum, Janet Elise. 2009. "Patient Teenagers? A Comparison of the Sexual Behavior of Virginity Pledgers and Matched Nonpledgers." *Pediatrics* 123 (January): 110–120.

Rosenfeld, Michael J., and Reuben J. Thomas. 2012. "Searching for a Mate: The Rise of the Internet as a Social Intermediary." *American Sociological Review* 77 (August): 523–547.

Rosengarten, Danielle. 2000. "Modern Times." *Dollars & Sense* (September): 4.

Rosenhan, David L. 1973. "On Being Sane in Insane Places." *Science* 179 (January 19): 250–258.

Rosenthal, Robert, and Lenore Jacobson. 1968. *Pygmalion in the Classroom: Teacher Expectations and Pupils' Intellectual Development.* New York: Holt, Rinehart and Winston.

Rosin, Hanna. 2012. *The End of Men: And the Rise of Women.* New York: Riverhead Books.

Ross, L. 1977. "The Intuitive Psychologist and His Shortcoming: Distortions in the Attribution Process." Pp. 174–221 in *Advances in Experimental Social Psychology,* vol. 10, edited by L. Berkowitz. New York: Academic Press.

Rossi, Alice S., and Peter H. Rossi. 1990. *Of Human Bonding: Parent-Child Relations across the Life Course.* New York: Aldine de Gruyter.

Rostow, W.W. 1978. *The World Economy: History and Prospect.* Austin, TX: University of Texas Press.

Royster, Deirdre. 2003. *The Invisible Hand: How White Networks Exclude Black Men from Blue-Collar Jobs.* Berkeley, CA: University of California Press.

Ruef, Martin, Howard Aldrich, and N. Carter. 2003. "The Structure of Foundation Teams: Homophily, Strong Ties, and Isolation among U.S. Entrepreneurs." *American Sociological Review* 68: 195–222.

Rueschmeyer, Dietrich, and Theda Skocpol. 1996. *States, Social Knowledge, and the Origins of Modern Social Policies.* Princeton, NJ: Princeton University Press.

Rugh, Jacob S., and Douglas S. Massey. 2010. "Racial Segregation and the American Foreclosure Crisis." *American Sociological Review* 75 (October): 629–651.

Rumbaut, Rubén. 1996b. "Prologue." Pp. xvi–xix in *Origins and Destinies: Immigration, Race, and Ethnicity in America,* edited by Silvia Pedraza and Rubén Rumbaut. Belmont, CA: Wadsworth.

Rupp, Leila J., and Verta Taylor. 2003. *Drag Queens at the 801 Cabaret.* Chicago, IL: University of Chicago Press.

Rupp, Leila J., and Verta Taylor. 2010. "Straight Girls Kissing." *Contexts* 9 (Summer): 28–33.

Rust, Paula. 1995. *Bisexuality and the Challenge to Lesbian Politics.* New York: New York University Press.

Rust, Paula C. 1993. "'Coming Out' in the Age of Social Constructionism: Sexual Identity Formation among Lesbian and Bisexual Women." *Gender & Society* 7 (March): 50–77.

Ryan, Charlotte. 1996. "Battered in the Media: Mainstream News Coverage of Welfare Reform." *Radical America* 26 (August): 29–41.

Ryan, Suzanne, Jennifer Manlove, and Sandra L. Hofferth. 2006. "State-Level Welfare Policies and Nonmarital Subsequent Childbearing." *Population Research and Policy Review* 25 (1): 103–126.

Ryan, William. 1971. *Blaming the Victim.* New York: Pantheon.

Saad, Lydia. 2003. "Pondering 'Women's Issues,' Part II." *The Gallup Poll.* Princeton, NJ: The Gallup Organization. **www.gallup.com**

Saad, Lydia. 2010. "Four Moral Issues Sharply Divide Americans." *The Gallup Poll.* Princeton, NJ: The Gallup Organization. **www.gallup.com**

Saad, Lydia. 2011 (August 8). "Plenty of Common Ground Found in Abortion Debate." *The Gallup Poll.* Princeton, NJ: Gallup Organization. **www.gallup.com**

Saad, Lydia. 2012a. "In U.S., Half of Women Prefer a Job Outside the Home." *The Gallup Poll.* Princeton, NJ: Gallup Organization, September 7. **www.gallup.com**

Saad, Lydia. 2012b. (May 29). "In U.S., Non-religious, PostGrads are Highly 'Pro-Choice." *The Gallup Poll.* Princeton, NJ: Gallup Organization. **www.gallup.com**

Saad, Lydia. 2012c. (May 14). "U.S. Acceptance of Gay/Lesbian Relations is the New Normal." *The Gallup Poll.* Princeton, NJ: Gallup Organization. **www.gallup.com**

Sabol, William J., and Heather Couture. 2008. "Prison Inmates at Mid Year 2007." Washington, DC: U.S. Bureau of Justice Statistics. **www.ojp.usdoj.gov**

Sadker, Myra, and David Sadker. 1994. *Failing at Fairness: How America's Schools Cheat Girls.* New York: Scribner's.

Saguy, Abigail. 2003. *What Is Sexual Harassment? From Capitol Hill to the Sorbonne.* Berkeley, CA: University of California Press.

Sampson, Robert J. 1987. "Urban Black Violence: The Effect of Male Joblessness and Family Disruption." *American Journal of Sociology* 93: 348–382.

Sanchez, Lisa Gonzalez. 1999. "Reclaiming Salsa." *Cultural Studies* 13 (April): 237–250.

Sanchez-Jankowski, Martin. 1991. *Islands in the Street: Gangs and American Urban Society.* Berkeley: University of California Press.

Sanday, Peggy. 2002. *Women at the Center: Life in a Modern Matriarchy.* Ithaca, NY: Cornell University Press.

Sandnabba, N. Kenneth, and Christian Ahlberg. 1999. "Parents' Attitudes and Expectations about Children's Cross-Gender Behavior." *Sex Roles* 40 (February): 249–263.

Santelli, John, et al. 2007. "Explaining Recent Declines in Adolescent Pregnancy in the United States: The Contribution of Abstinence and Increased Contraceptive Use." *American Journal of Public Health* 97(1): 150–156.

Sapir, Edward. 1921. *Language: An Introduction to the Study of Speech.* New York: Harcourt Brace.

Sara, Siddharth. 2010. *Sex Trafficking: Inside the Business of Modern Slavery.* New York: Columbia University Press.

Sarkasian, Natalia, and Naomi Gerstel. 2012. *Nuclear Family Values, Extended Family Lives: The Power of Race, Class, and Gender.* New York: Russell Sage Foundation.

Saulny, Susan. 2011. "Black? White? Asian? More Younger Americans Choose All of the Above." *The New York Times* (January 30): 1 and 18.

Sawhill, Isabel, and Sara McClanahan (eds.). 2006. "Introducing the Issue." *The Future of Children* 16 (Fall): 3–17.

Schacht, Steven P. 1996. "Misogyny on and off the 'Pitch': The Gendered World of Male Rugby Players." *Gender & Society* 10 (October): 550–565.

Schaffer, Kay, and Song Xianlin. 2007. "Unruly Spaces: Gender, Women's Writing and Indigenous Feminism in China." *Journal of Gender Studies* 16 (1): 17–30.

Schalet, Amy. 2010. "Sex, Love, and Autonomy in the Teenage Sleepover." *Contexts* 9 (Summer): 16–21.

Schilt, Kristen. 2011. *Just One of the Guys? Transgender Men and the Persistence of Gender Inequality.* Chicago: University of Chicago Press.

Schilt, Kristen, and Laurel Westbrook. 2009. "Doing Gender, Doing Heteronormativity." *Gender & Society* 23 (August): 440–464.

Schlosser, Eric. 2001. *Fast Food Nation: The Dark Side of the All-American Meal.* New York: Houghton Mifflin.

Schmitt, Eric. 2001. "Segregation Growing among U.S. Children." *The New York Times* (May 6): 28.

Schmitt, Frederika E., and Patricia Yancey Martin. 1999. "Unobtrusive Mobilization by an Institutionalized Rape Crisis Center: 'It Comes From the Victims.'" *Gender & Society* 13: 364–384.

Schneider, Beth. 1984. "Peril and Promise: Lesbians' Workplace Participation." Pp. 211–230 in *Women Identified Women,* edited by Trudy Darty and Sandee Potter. Mountain View, CA: Mayfield.

Schumann, Howard, Charlotte Steeh, Lawrence Bobo, and Maria Krysan. 1997. *Racial Attitudes in America: Trends and Interpretations.* Cambridge, MA: Harvard University Press.

Schwartz, John, and Matthew L. Wald. 2003. "NASA's Failings Go Far beyond Foam Hitting Shuttle, Panel Says." *The New York Times* (June 7): 1 and 12.

Schwartz, Pepper, and Virginia Rutter. 1998. *The Gender of Sexuality.* Thousand Oaks, CA: Sage.

Scott, Ellen K., Andrew S. London, and Nancy A. Myers. 2002. "Dangerous Dependencies: The Intersection of Welfare Reform and Domestic Violence." *Gender & Society* 16 (December): 878–897.

Segal, David R., and Mady Wechsler Segal. 2004. "America's Military Population." *Population Bulletin* 59 (December).

Seidman, Steven. 2003. *The Social Construction of Sexuality.* New York: Oxford University Press.

Sen, Amartya. 2000. "Population and Gender Equity." *The Nation* (July 24–31): 16–18.

Sen, Amartya. 2002. "How to Judge Globalism." *The American Prospect* (Winter, special supplement): A2–A6.

Sharp, Susan F., Toni L. Terling-Watt, Leslie A. Atkins, Jay Trace Gilliam, and Anna Sanders. 2000. "Purging Behavior in a Sample of College Females: A Research Note on General Strain Theory and Female Deviance." *Deviant Behavior* 22: 171–188.

Shelley, Louise. 2010. *Human Trafficking: A Global Perspective.* Cambridge, MA: Cambridge University Press.

Sherman, Jennifer. 2009. "Bend to Avoid Breaking: Job Loss, Gender Norms, and Family Stability in Rural America." *Social Problems* 56 (November): 599–620.

Shibutani, Tomatsu. 1961. *Society and Personality: An Interactionist Approach to Social Psychology.* Englewood Cliffs, NJ: Prentice Hall.

Shieman, Scott. 2010. "Socioeconomic Status and Beliefs about God's Influence in Everyday Life." *Sociology of Religion* 71: 25–51.

Shinagawa, Larry, and Gin Yong Pang. 1996. "Asian American Panethnicity and Intermarriage." *American Journal* 22 (Spring): 127–152.

Shrestha, Laura B., and Elayne J. Heisler. 2011. "The Changing Demographic Profile of the United States." *CRS Report for Congress.* Washington, DC: Congressional Research Service, March 31.

Sigelman, Lee, and Paul J. Wahlbeck. 1999. "Gender Proportionality in Intercollegiate Athletics: The Mathematics of Title IX Compliance." *Social Science Quarterly* 80 (September): 518–538.

Signorielli, Nancy. 2009. "Race and Sex in Prime Time: A Look at Occupations and Occupational Prestige." *Mass Communication and Society* 12: 332–352.

Silver, Alexandra. 2010. "Brief History of the U.S. Census." *Time* (February 8): 16.

Silverthorne, Zebulon A., and Vernon L Quinsey. 2000. "Sexual Partner Age Preferences of Homosexual and Heterosexual Men and Women." *Archives of Sexual Behavior* 29 (No. 1): 67–76.

Simmel, Georg. 1902/1950. "The Number of Members as Determining the Sociological Form of the Group." *The American Journal of Sociology* 8 (July): 1–46.

Simmons, Wendy W. 2001. "Majority of Americans Say More Women in Political Office Would Be Positive for the Country." *The Gallup Poll.* Princeton, NJ: Gallup Organization. **www.gallup.com**

Simon, David R. 2007. *Elite Deviance,* 8th ed. Boston, MA: Allyn and Bacon.

Simons, Ronald L. 1996. "The Effect of Divorce on Adult and Child Adjustment." Pp. 3–20 in *Understanding Differences between Divorced and Intact Families,* edited by Robert L. Simons. Thousand Oaks, CA: Sage.

Singer, Peter W. 2007. *Corporate Warriors: The Rise of the Privatized Military Industry.* Ithaca, NY: Cornell University Press.

Sirvananadan, A. 1995. "La trahison des clercs. (Racism)." *New Statesman and Society* 8: 20–22.

Sjøberg, Gideon. 1965. *The Preindustrial City: Past and Present.* New York: Free Press.

Skocpol, Theda. 1992. *Protecting Soldiers and Mothers: The Origins of Social Policy in the United States.* Cambridge, MA: Belknap Press.

Skocpol, Theda, and Vanessa Williamson. 2012. *The Tea Party and the Remaking of Republican Conservatism.* New York: Oxford University Press.

Smelser, Neil J. 1992. "Culture: Coherent or Incoherent." Pp. 3–28 in *Theory of Culture,* edited by R. Münch and N. J. Smelser. Berkeley, CA: University of California Press.

Smith, Andrew. 2009. "Nigerian Scam E-Mails and the Charms of Capital." *Cultural Studies* 23 (1): 27–47.

Smith, Barbara. 1983. "Homophobia: Why Bring It Up?" *Interracial Books for Children Bulletin* 14: 112–113.

Smith, M. Dwayne, Joel A. Devine, and Joseph F. Sheley. 1992. "Crime and Unemployment: Effects Across Age and Race Categories." *Sociological Perspectives* 35 (Winter): 551–572.

Smith, Richard B., and Robert A. Brown. 1997. "The Impact of Social Support on Gay Male Couples." *Journal of Homosexuality* 33: 39–61.

Smith, Sandra. 2007. *Lone Pursuit: Distrust and Defensive Individualism among the Black Poor.* New York: Russell Sage Foundation.

Smolak, L. and M. Levine, eds., 1996. *The Developmental Psychopathology of Eating Disorders: Implications for Research, Prevention, and Treatment.* Hillsdale, NJ: Lawrence Erlbaum Associates Inc.

Snipp, C. Matthew. 1989. *American Indians: The First of This Land.* New York: Russell Sage Foundation.

Snipp, C. Matthew. 1996. "The First Americans: American Indians." Pp. 390–404 in *Origins and Destinies: Immigration, Race, and Ethnicity in America,* edited by Sylvia Pedraza and Rubén G. Rumbaut. Belmont, CA: Wadsworth.

Snipp, Matthew. 2007. "An Overview of American Indian Populations." Pp. 38–48 in *American Indian Nations: Yesterday, Today, and Tomorrow,* edited by George Horse Capture, Duane Champaign, and Chandler Jackson. Walnut Creek, CA: Altamira Press.

Sotirovic, Mira. 2000. "Effects of Media Use on Audience Framing and Support for Welfare." *Mass Communication & Society* 2–3 (Spring–Summer): 269–296.

Sotirovic, Mira. 2001. "Media Use and Perceptions of Welfare." *Journal of Communication* 51 (December): 750–774.

Spencer, Herbert. 1882. *The Study of Sociology.* London: Routledge.

Spencer, Ranier. 2011. *Reproducing Race: The Paradox of Generation Mix.* Las Vegas, Nevada: Lynne Rienner.

Spencer, Ranier. 2012. "Mixed-Race Chic." Pp. 67–70 in Elizabeth Higginbotham and Margaret L. Andersen, eds., *Race and Ethnicity in Society: The Changing Landscape,* 3rd ed. Belmont, CA: Wadsworth/Cengage.

Spitzer, Brenda L., Katherine A. Henderson, and Marilyn T. Zivian. 1999. "A Comparison of Population and Media Body Sizes for American and Canadian Women." *Sex Roles* 700 (7/8): 545–565.

Spitzer, Steven. 1975. "Toward a Marxian Theory of Deviance." *Social Problems* 22: 638–651.

Squires, Gregory. 2007. "Demobilization of the Individualistic Bias: Housing Market Discrimination as a Contributor to Labor Market and Economic Inequality." *The Annals of the American Academy of Political and Social Science* 609: 200–214.

Stacey, Judith, and Timothy J. Bibliarz. 2001. "(How) Does the Sexual Orientation of Parents Matter?" *American Sociological Review* 66 (April): 159–183.

Stack, Carol. 1974. *All Our Kin: Strategies for Survival in a Black Community.* New York: Harper Colophon Books.

Stearns, Linda Brewster, and Charlotte Wilkinson Coleman. 1990. "Industrial and Labor Market Structures and Black Male Employment in the Manufacturing Sector." *Social Science Quarterly* 71 (June): 285–298.

Steele, Claude. M. 2010. *Whistling Vivaldi and Other Clues to How Stereotypes Affect Us.* New York: W. W. Norton.

Steffensmeier, Darrell, and Stephen Demuth. 2000. "Ethnicity and Sentencing Outcomes in U.S. Federal Courts: Who Is Punished More Harshly?" *American Sociological Review* 65 (October): 705–729.

Steger, Manfred B. 2009. *Globalization: A Very Short Introduction.* New York: Oxford University Press.

Stein, Arlene, and Marcy Westerling. 2012. "The Politics of Broken Dreams." *Contexts* 11 (Summer): 8–10.

Stern, Jessica. 2003. *Terror in the Name of God: Why Religious Militants Kill.* New York: Ecco.

Stern, Kenneth S. 1996. *A Force upon the Plain.* New York: Simon and Schuster.

Sternheimer, Karen. 2007. "Do Video Games Kill?" *Contexts* 6 (Winter): 13–17.

Sternheimer, Karen. 2011. *Celebrity Culture and the American Dream: Stardom and Social Mobility.* New York: Routledge.

Stevens, Scott P. 2011. *Games People Play: Game Theory in Life, Business and Beyond.* On-Line Lecture Series: Chantilly, VA: Great Courses.

Stevenson, Bryan. 2010. "Race Still Used to Exclude Jurors." *Race and Justice News* (December 6).

Stewart, Abigail J., Anne P. Copeland, Nia Lane Chester, Janet E. Malley, and Nicole B. Barenbaum. 1997. *Separating Together: How Divorce Transforms Families.* New York: Guilford Press.

Stewart, Susan D. 2001. "Contemporary American Stepparenthood: Integrating Cohabiting and Nonresident Stepparents." *Population Research and Policy* 20 (August): 345–364.

Stombler, Mindy, and Irene Padavic. 1997. "Sister Acts: Resisting Men's Domination in Black and White Fraternity Little Sister Programs." *Social Problems* 44 (May): 257–275.

Stone, Pamela. 2007. *Opting Out: Why Women Really Quit Careers and Head Home.* Berkeley, CA: University of California Press.

Stoner, J. A. F. 1961. "A Comparison of Individual and Group Decisions Involving Risk." Unpublished M. A. thesis, MIT.

Stryker, Susan, and Stephen Whittle (eds.). 2006. *The Transgender Studies Reader.* New York: Routledge.

Substance Abuse and Mental Health Services Administration (SAMHSA), the Center for Mental Health Services (CMHS), U.S. Department of Health and Human Services.

Sudnow, David N. 1967. *Passing On: The Social Organization of Dying.* Englewood Cliffs, NJ: Prentice Hall.

Sullivan, Maureen. 1996. "Rozzie and Harriet? Gender and Family Patterns of Lesbian Coparents." *Gender & Society* 12 (December): 747–767.

Sullivan, Patrick F. 1995. "Mortality in Anorexia Nervosa." *American Journal of Psychiatry* 152 (July): 1073–1074.

Sullivan, Teresa A., Elizabeth Warren, and Jay Lawrence Westbrook. 2000. *The Fragile Middle Class: Americans in Debt.* New Haven, CT: Yale University Press.

Sumner, William Graham. 1906. *Folkways: A Study of the Sociological Importance of Usages, Manners, Customs, Mores, and Morals.* Boston: Ginn.

Sussal, Carol M. 1994. "Empowering Gays and Lesbians in the Workplace." *Journal of Gay and Lesbian Social Services* 1: 89–103.

Sussman, N. M., and D. H. Tyson. 2000. "Sex and Power: Gender Differences in Computer-Mediated Interactions." *Computers in Human Behavior* 16 (July): 381–394.

Sutherland, Edwin H. 1940. "White Collar Criminality." *American Sociological Review* 5 (February): 1–12.

Sutherland, Edwin H., and Donald R. Cressey. 1978. *Criminology,* 10th ed. New York: Lippincott.

Swidler, Ann. 1986. "Culture in Action: Symbols and Strategies." *American Sociological Review* 51: 273–286.

Switzer, J. Y. 1990. "The Impact of Generic Word Choices: An Empirical Investigation of Age- and Sex-Related Differences." *Sex Roles* 22: 69–82.

Szasz, Thomas S. 1974. *The Myth of Mental Illness.* New York: Harper & Row.

Tach, Laura, and Sarah Halpern-Meekin. 2009. "How Does Premarital Cohabitation Affect Trajectories of Marital Quality?" *Journal of Marriage and Family* 71 (May): 298–317.

Takaki, Ronald. 1989. *Strangers from a Different Shore: A History of Asian Americans.* New York: Penguin.

Tannen, Deborah. 1990. *You Just Don't Understand: Women and Men in Conversation.* New York: William Morrow.

Tarman, C., and D. O. Sears. 2005. "The Conceptualization and Measurement of Symbolic Racism." *Journal of Politics* 67: 731–761.

Tatum, Beverly. 1997. *Why Are All the Black Kids Sitting Together at the Cafeteria?* New York: Basic Books.

Taylor, Howard F. 1980. *The IQ Game: A Methodological Inquiry into the Heredity-Environment Controversy.* New Brunswick, NJ: Rutgers University Press.

Taylor, Howard F. 1992. "The Structure of a National Black Leadership Network: Preliminary Findings." Unpublished manuscript, Princeton University.

Taylor, Howard F. 2002. "Deconstructing the Bell Curve: Racism, Classism and Intelligence in America," Pp. 60–76 in *2001 Race Odyssey: African Americans and Sociology,* edited by Bruce R. Hare. Syracuse, NY: Syracuse University Press.

Taylor, Howard F. 2012. "Defining Race." Pp. 7–13 in *Race and Ethnicity in Society: The Changing Landscape,* 3rd ed., edited by in Elizabeth Higginbotham and Margaret L. Andersen. Belmont, CA: Wadsworth/Cengage.

Taylor, Howard F. Forthcoming. *The SAT Triple Whammy: Race, Gender and Social Class Bias.* Princeton University, Princeton, NJ.

Taylor, Paul. 2012. "The Growing Electoral Clout of Blacks is Driven by Turnout,

Not Demographics." *Pew Social & Demographic Trends*. Washington, DC: Pew Research Center.

Taylor, Shelley E., Letitia Ann Peplau, and David O. Sears. 2013. *Social Psychology*, 13th ed. Upper Saddle River, NJ: Prentice Hall.

Teaster, Pamela B. 2000. "A Response to the Abuse of Vulnerable Adults: The 200 Survey of State Adult Protective Services." Washington, DC: National Center on Elder Abuse. **www .elderabusecenter.org**

Telles, Edward E. 1994. "Residential Segregation and Skin Color in Brazil." *American Sociological Review* 57 (April): 186–197.

Telles, Edward E. 2004. *Race in Another America: The Significance of Skin Color in Brazil*. Princeton, NJ: Princeton University Press.

Telles, Edward E., and Vilma Ortiz. 2008. *Generations of Exclusion: Mexican Americans, Assimilation, and Race*. New York: Russell Sage Foundation Press.

Telles, Edward, Mark Q. Sawyer, and Gaspar Rivera-Salgado, eds. 2011. *Just Neighbors? Research on African American and Latino Relations in the United States*. New York: Russell Sage Foundation.

Tenenbaum, Harriet R. 2009. "'You'd Be Good at That': Gender Patterns in Parent–Child Talk about Courses." *Social Development* 18: 447–463.

Thakkar, Reena R., Peter M. Gutierrez, Carly L. Kuczen, and Thomas R. McCanne. 2000. "History of Physical and/or Sexual Abuse and Current Suicidality in College Women." *Child Abuse and Neglect* 24 (October): 1345–1354.

The Sentencing Project. 2009. Washington, DC. **www.sentencingproject.org**

Thoits, Peggy A. 2009. "Sociological Approaches to Mental Illness." In *A Handbook for the Study of Mental Health*, edited by Teresa L. Scheid and Tony L. Brown. Cambridge: Cambridge University Press.

Thomas, William I. 1931. *The Unadjusted Girl*. Boston, MA: Little, Brown.

Thomas, William I. 1966 [1931]. "The Relation of Research to the Social Process." Pp. 289–305 in *W. I. Thomas on Social Organization and Social Personality*, edited by Morris Janowitz. Chicago, IL: University of Chicago Press.

Thomas, William I., with Dorothy Swaine Thomas. 1928. *The Child in America*. New York: Knopf.

Thompson, Robert Farris. 1983. *Flash of the Spirit: African and Afro-American Art and Philosophy*. New York: Random House.

Thompson, Robert Farris. 1993. *Face of the Gods: Art and Alters of Africa and the African Americas*. New York: Random House.

Thorne, Barrie. 1993. *Gender Play: Girls and Boys in School*. New Brunswick, NJ: Rutgers University Press.

Thornton, Russell. 1987. *American Indian Holocaust and Survival: A Population History*. Norman, OK: University of Oklahoma Press.

Thornton, Russell. 2001. "Trends among American Indians in the United States."

Pp. 135–169 in *America Becoming: Racial Trends and Their Consequences*, vol. I, edited by Neil J. Smelser, William Julius Wilson, and Faith Mitchell. Washington, DC: National Academies Press.

Tichenor, Veronica. 2005. "Maintaining Men's Dominance: Negotiating Identity and Power When She Earns More." *Sex Roles* 53(3–4): 191–205.

Tienda, Marta, and Haya Stier. 1996. "Generating Labor Market Inequality: Employment Opportunities and the Accumulation of Disadvantage." *Social Problems* 43 (May): 147–165.

Tierney, Kathleen J., Michael K. Lindell, and Ronald W. Perry. 2001. *Facing the Unexpected: Disaster Preparedness and Response in the United States*. Washington, DC: National Academies Press.

Tierney, Kathleen J. 2007. "From the Margins to the Mainstream? Disaster Research at the Crossroads." *Annual Review of Sociology* 33: 503–525.

Tierney, Patrick. 2000. *Darkness in El Dorado: How Scientists and Journalists Devastated the Amazon*. New York: W.W. Norton.

Tilly, Louise, and Joan Scott. 1978. *Women, Work, and Family*. New York: Holt, Rinehart, and Winston.

Tjaden, Patricia, and Nancy Thoennes. 2000. *Extent, Nature, and Consequences of Intimate Partner Violence*. Washington, DC: National Institute of Justice and the Centers for Disease Control and Prevention.

Tönnies, Ferdinand. 1963 [1887]. *Community and Society (Gemeinschaft and Gesellschaft)*. New York: Harper & Row.

Toossi, Mitra. 2012. "Labor Force Projections to 2020: A More Slowly Growing Workforce." *Monthly Labor Review* (January): 43–64.

Toynbee, Arnold J., and Jane Caplan. 1972. *A Study of History*. New York: Oxford University Press.

Travers, Jeffrey, and Stanley Milgram. 1969. "An Experimental Study of the Small World Problem." *Sociometry* 32: 425–443.

Treiman, Donald J. 2001. "Occupations, Stratification and Mobility." Pp. 297–313 in *The Blackwell Companion to Sociology*, edited by Judith R. Blau. Malden, MA: Blackwell.

Tuan, Yi-Fu. 1984. *Dominance and Affection: The Making of Pets*. New Haven, CT: Yale University Press.

Tuchman, Gaye. 1979. "Women's Depiction by the Mass Media." *Signs* 4 (Spring): 528–542.

Tumin, Melvin M. 1953. "Some Principles of Stratification." *American Sociological Review* 18 (August): 387–393.

Turner, Jonathan. 1974. *The Structure of Sociological Theory*. Homewood, IL: Dorsey Press.

Turner, Ralph, and Lewis Killian. 1993. *Collective Behavior*, 4th ed. Englewood Cliffs, NJ: Prentice Hall.

Turner, Terence. 1969. "Tchikrin: A Central Brazilian Tribe and Its Symbolic Language of Body Adornment." *Natural History* 78 (October): 50–59.

Uleman, J. S., L. S. Newman, and G. B. Moskowitz. 1996. "People as Flexible Interpreters: Evidence and Issues From Spontaneous Trait Inference." Pp. 211–279 in *Advances in Experimental Social Psychology*, vol. 28, edited by Mark Zanna. Boston, MA: Academic Press.

Ullman, Sarah E., George Karabatsos, and Mary P. Koss. 1999. "Alcohol and Sexual Assault in a National Sample of College Women." *Journal of Interpersonal Violence* 14 (June): 603–625.

UNICEF. 2000. *Child Poverty in Rich Nations. Florence, Italy: United Nations Children's Fund*. New York: United Nations. **www.unicef-icdc.org**

United Nations. 2000. *The World's Women 2000: Trends and Statistics*. New York: United Nations, pp. 140–143.

United Nations. 2006. *In Depth Study on All Forms of Violence against Women*. New York: United Nations.

United Nations. 2010. *The Gender Inequality Index*. New York: United Nations. **www.undp.org**

United Nations. 2012a. *WomenWatch*. New York: United Nations. **www.un .org/womenwatch**

United Nations. 2012b. "World Population to Increase by 2.6 Billion over Next 45 Years, with All Growth Occurring in Less Developed Regions." Press Release, POP918. **www.un.org**

United Nations Development Program. 2010. "Components of the Gender Inequality Index." **www.hdr.undp.org**

United Nations Development Program. 2013. Human Development Report: International Human Development Indicators. **http://hdr.undp.org /en/data/map/**

United Nations Population Division. 2011. *World Population Prospects*, 2010 revision. New York: United Nations. **www.prb.org**

United States National Library of Medicine. 2012. *Tox-Map: Environmental Health e-Maps*. **http://toxmap.nlm.nih.gov /toxmap/main/index.jsp**

U.S. Bureau of Justice Statistics. 2012. *Criminal Victimization in the United States 2012, Statistical Tables*. Washington, DC: U.S. Bureau of Justice Statistics.

U.S. Bureau of Labor Statistics. 2011. "Data Bases, Tables, and Calculations, By Subject." Washington, DC: U.S. Bureau of Justice Statistics. **www .bls.gov**

U.S. Bureau of Labor Statistics. 2012a. *American Time Use Survey Summary*. Washington, DC: U.S. Department of Labor.

U.S. Bureau of Labor Statistics. 2012b. "A Profile of the Working Poor, 2008." Washington, DC: U.S. Department of Labor. **www.bls.gov**

U.S. Census Bureau. 2003. "Racial and Ethnic Classification Used in Census 2000 and Beyond." Washington, DC: U.S. Census Bureau. **www .census.gov**

U.S. Census Bureau. 2012a. *The 2012 Statistical Abstract*. Washington, DC: U.S. Census Bureau. **www.census.gov**

U.S. Census Bureau. 2012b. *Current Population Reports: Annual Social and Economic Supplement*. Washington,

DC: U.S. Census Bureau. **www .census.gov**

U.S. Census Bureau. 2012c. *Historic Income Tables: Households, Table H-5; Net Worth and Asset Ownership of Households: 2010.* Washington, DC: U.S. Census Bureau. **www.census.gov**

U.S. Census Bureau. 2012d. *America's Families and Living Arrangements: 2012, Detailed Tables.* Current Population Reports. Washington, DC: U.S. Census Bureau. **www.census.gov**

U.S. Courts. 2012. *Bankruptcy Statistics.* Washington, DC: U.S. Courts. **www .uscourts.gov**

U.S. Department of Defense. 2010. *Demographics 2010: Profile of the Military Community.* Washington, DC: U.S. Department of Defense.

U.S. Department of Defense. 2012. "Fiscal Year 2011 Annual Report on Sexual Assault in the Military." **www .defense.gov**

U.S. Department of Justice. 2009. "Child Sex Tourism." Washington, DC: U.S. Department of Justice. **www.usdoj.gov**

U.S. Department of Labor. 2012. *Employment and Earnings.* Washington, DC: U.S. Department of Labor. **www .bls.gov**

U.S. Department of State. 2011. *Annual Report on Intercountry Adoption.* Washington, DC: U.S. Department of State.

U.S. Energy Information Administration. 2011. *U.S. Energy Related Carbon Dioxide Emissions 2011.* Washington, DC: U.S. Department of Energy. **www .eia.gov**

U.S. Energy Information Administration. 2012. Washington, DC: U.S. Department of Energy. **www.eia.gov**

U.S. State Department. 2012. *Trafficking in Persons Report, June 2012.* Washington, DC: U.S. State Department. **www .state.gov**

Vail, D. Angus. 1999. "Tattoos Are Like Potato Chips . . . You Can't Have Just One: The Process of Becoming and Being a Collector." *Deviant Behavior* 20: 253–273.

Valentine, Kathyrn, Mary Prentice, Monica F. Torres, and Eduardo Arellano. 2012. "The Importance of Cross-Racial Interactions as Part of College Education: Perceptions of Faculty." *Journal of Diversity in Higher Education* 5 (December): 191–206.

Van Ausdale, and Joe R. Feagin. 2010. *The First R: How Children Learn Race and Racism.* Lanham, MD: Rowman and Littlefield.

Van Ausdale, Debra, and Joe R. Feagin. 1996. "The Use of Racial and Ethnic Concepts by Very Young Children." *American Sociological Review* 61 (October): 779–793.

Vanneman, Reeve, and Lynn Weber Cannon. 1987. *The American Perception of Class.* Philadelphia: Temple University Press.

VanRoosmalen, Erica, and Susan A. McDaniel. 1998. "Sexual Harassment in Academia: A Hazard to Women's Health." *Women and Health* 28: 33–54.

Varelas, Nicole, and Linda A. Foley. 1998. "Blacks' and Whites' Perceptions of Interracial and Intraracial Date Rape." *Journal of Social Psychology* 138 (June): 392–400.

Vaughan, Diane. 1996. *The Challenger Launch Decision: Risky Technology, Culture, and Deviance at NASA.* Chicago, IL: The University of Chicago Press.

Veblen, Thorstein. 1994 [1899]. *Theory of the Leisure Class.* New York: Penguin.

Veblen, Thorstein. 1953 [1899]. *The Theory of the Leisure Class: An Economic Study of Institutions.* New York: The New American Library.

Vespa, Jonathan. 2009. "Gender Ideology Construction." *Gender & Society* 23 (June): 363–387.

Vidmar, Neil, and Valerie P. Hans. 2007. *American Juries: The Verdict.* Amherst, NY: Prometheus Books.

Villareal, Andres. 2010. "Stratification by Skin Color in Contemporary Mexico." *American Sociological Review* 75 (6): 652–678.

Volscho, Thomas W., and Nathan J. Kelly. 2012. "The Rise of the Super-Rich: Power Resources, Taxes, Financial Markets, and the Dynamics of the Top 1 Percent, 1949–2008." *American Sociological Review* 77 (October): 679–699.

Waldfogel, Jane, Wen-Jui Han, and Jeanne Brooks-Gunn. 2002. "The Effects of Early Maternal Employment in Child Cognitive Development." *Demography* 39: 369–392.

Walker-Barnes, Chanequa J., and Craig A. Mason. 2001. "Perceptions of Risk Factors for Female Gang Involvement among African American and Hispanic Women." *Youth and Society* 32 (March): 303–336.

Wallace, Walter L. 1971. *The Logic of Science in Sociology.* Chicago: Aldine-Atherton.

Wallace, Walter L. 1983. *Principles of Scientific Sociology.* New York: Aldine Publishing Co.

Wallerstein, Immanuel M. 1974. *The Modern World System: Capitalist Agriculture and the Origins of the European World Economy in the Sixteenth Century.* New York: Academic Press.

Wallerstein, Immanuel M. 1980. *The Modern World-System II.* New York: Academic Press.

Walters, Suzanna Danuta. 1999. "Sex, Text, and Context: (In) Between Feminism and Cultural Studies." Pp. 222–260 in *Revisioning Gender,* edited by Myra Marx Ferree, Judith Lorber, and Beth B. Hess. Thousand Oaks, CA: Sage.

Wang, Wendy. 2012. "The Rise of Intermarriage: Rates, Characteristics Vary by Race and Gender." Washington, DC: Pew Research Center. **www .pewresearch.org**

Washington, Scott. 2006. "Racial Taxonomy." Unpublished manuscript, Princeton University, Princeton, NJ.

Washington, Scott. 2011. "Who Isn't Black? The History of the One-Drop Rule." PhD dissertation, Department of Sociology, Princeton University, Princeton, NJ.

Wasserman, Stanley, and Katherine Faust (eds.). 1994. *Social Network Analysis: Methods and Applications.* Cambridge, MA: Cambridge University Press.

Waters, Mary C. 1990. *Ethnic Options: Choosing Identities in America.* Berkeley, CA: University of California Press.

Waters, Mary C., and Peggy Levitt (eds.). 2002. *The Changing Face of Home: The Transnational Lives of the Second Generation.* New York: Russell Sage Foundation.

Watkins, Susan. 1987. "The Fertility Transition: Europe and the Third World Compared." *Sociological Forum* 2 (Fall): 645–673.

Watkins, S. Craig. 1999. *Representing: Hip Hop Culture and the Production of Black Cinema.* Berkeley: University of California Press.

Watt, Toni Terling, and Susan F. Sharp. 2001. "Gender Differences in Strains Associated with Suicidal Behavior among Adolescents." *Journal of Youth and Adolescence* 30 (June): 333–348.

Watts, Duncan J. 1999. "Networks, Dynamics, and the Small-World Phenomenon." *American Journal of Sociology* 105 (September): 493–527.

Watts, Duncan J., and Stephen H. Strogatz. 1998. "Collective Dynamics of 'Small World' Networks." *Nature* 393: 440–442. **www.sentencingproject.org**

Weber, Max. 1947 [1925]. *The Theory of Social and Economic Organization.* New York: Free Press.

Weber, Max. 1958 [1904]. *The Protestant Ethic and the Spirit of Capitalism.* New York: Scribner's.

Weber, Max. 1962 [1913]. *Basic Concepts in Sociology.* New York: Greenwood.

Weber, Max. 1978 [1921]. *Economy and Society: An Outline of Interpretive Sociology,* edited by Guenther Roth and Claus Wittich. Berkeley, CA: University of California Press.

Webster, Murray, Jr., and Stuart J. Hysom. 1998. "Creating Status Characteristics." *American Sociological Review* 63 (June): 351–378.

Weeks, John R. 2011. *Population: An Introduction to Concepts and Issues,* 11th ed. Belmont, CA: Wadsworth Publishing Co.

Weisburd, David, Cynthia M. Lum, and Anthony Petrosino. 2001. "Does Research Design Affect Study Outcomes in Criminal Justice?" *The Annals of the American Academy of Political and Social Science* 578 (November): 50–70.

Weisburd, David, Stanton Wheeler, Elin Waring, and Nancy Bode. 1991. *Crimes of the Middle Class: White Collar Defenders in the Courts.* New Haven, CT: Yale University Press.

West, Candace, and Sarah Fenstermaker. 1995. "Doing Difference." *Gender & Society* 9 (February): 8–37.

West, Candace, and Don Zimmerman. 1987. "Doing Gender." *Gender & Society* 1 (June): 125–151.

West, Carolyn M. 1998. "Leaving a Second Closet: Outing Partner Violence in Same-Sex Couples." Pp. 163–183 in *Partner Violence: A Comprehensive*

Review of 20 Years of Research, edited by Jana L. Jasinksi and Linda M. Williams. Thousand Oaks, CA: Sage.

West, Cornel. 1993. *Race Matters*. Boston, MA: Beacon Press.

West, Heather C. and William J. Sabol. 2010. *Prisoners in 2010*. Washington, DC: Bureau of Justice Statistics. **www.bjs.gov**

Western, Bruce. 2007. *Punishment and Inequality in America*. New York: Russell Sage Foundation

White, Deborah Gray. 1985. *Ar'n't I a Woman?: Female Slaves in the Plantation South*. New York: W. W. Norton.

White, Jonathan R. 2002. *Terrorism: An Introduction*. Belmont, CA: Wadsworth.

Whitehead, Andrew L., and Joseph O. Baker. 2012. "Homosexuality, Religion, and Science: Moral Authority and the Persistence of Negative Attitudes." *Sociological Inquiry* 82: 487–509.

White House. 2012. *Critical Issues Facing Asian Americans and Pacific Islanders*. Washington DC: Initiative on Asian Americans and Pacific Islanders. **www.whitehouse.gov**

Whiten, A., J. Goodall, W. C. McGrew, T. Nishida, V. Reynolds, Y. Sugiyama, C. E. G. Tutin, R. W. Wrangham, and C. Boesch. 1999. "Cultures in Chimpanzees." *Nature* 399 (June 17): 682–684.

Whorf, Benjamin. 1956. *Language, Thought, and Reality: Selected Writings*. Cambridge, MA: Technology Press, MIT.

Whyte, William F. 1943. *Street Corner Society*. Chicago, IL: University of Chicago Press.

Wickelgren, Ingrid. 1999. "Discovery of 'Gay Gene' Questioned." *Science* 284 (April 23): 571.

Wilder, Esther I., and Toni Terling Watt. 2002. "Risky Parental Behavior and Adolescent Sexual Activity at First Coitus." *The Milbank Quarterly* 80 (September): 481–524.

Wilkins, Amy. 2012. "Stigma and Status: Interracial Intimacy and Intersectional Identities among Black College Men." *Gender & Society* 26 (April): 165–189.

Williams, Christine L. 1992. "The Glass Escalator: Hidden Advantages for Men in the 'Female' Professions." *Social Problems* 39 (August): 253–267.

Williams, Christine L. 1995. *Still a Man's World: Men Who Do Women's Work*. Berkeley, CA: University of California Press.

Williams, Christine L., and Arlene Stein. 2002. *Sexuality and Gender*. Hoboken, NJ: Wiley-Blackwell.

Williams, Timothy. 2012. "Suicides Outpacing War Deaths for Troops." *The New York Times*, July 8, p. A10.

Willie, Charles Vert. 1979. *The Caste and Class Controversy*. Bayside, NY: General Hall.

Wilson, Barbara J., Stacy L. Smith, W. James Potter, Dale Kunkel, Daniel Linz, Carolyn M. Colvin, and Edward Donnerstein. 2002. "Violence in Children's Television Programming: Assessing the Risks." *Journal of Communication* 52 (March): 5–35.

Wilson, John F. 1978. *Religion in American Society: The Effective Presence*. Englewood Cliffs, NJ: Prentice Hall.

Wilson, William Julius. 1978. *The Declining Significance of Race: Blacks and Changing American Institutions*. Chicago, IL: University of Chicago Press.

Wilson, William Julius. 1987. *The Truly Disadvantaged: The Inner City, the Underclass and Public Policy*. Chicago, IL: University of Chicago Press.

Wilson, William Julius. 1996. *When Work Disappears: The World of the New Urban Poor*. New York: Knopf.

Wilson, William Julius. 2009. *More Than Just Race: Being Black and Poor in the Inner City*. New York: W. W. Norton.

Winnick, Louis. 1990. "America's Model Minority." *Commentary* 90 (August): 222–229.

Wirth, Louis. 1928. *The Ghetto*. Chicago: University of Chicago Press.

Wolcott, Harry F. 1996. "Peripheral Participation and the Kwakiutl Potlatch." *Anthropology and Education Quarterly* 27 (December): 467–492.

Woo, Deborah. 1998. "The Gap between Striving and Achieving." Pp. 247–256 in *Race, Class, and Gender: An Anthology*, 3rd ed., edited by Margaret L. Andersen and Patricia Hill Collins. Belmont, CA: Wadsworth.

Wood, Julia T. 1994. *Gendered Lives: Communication, Gender and Culture*. Belmont, CA: Wadsworth.

Wood, Richard L. 2002. *Faith in Action: Religion, Race, and Democratic Organizing in America*. Chicago, IL: University of Chicago Press.

Woodhams, Jessica, Claire Cooke, Leigh Harkins, and Teresa da Silva. 2012. "Leadership in Multiple Perpetrator Stranger Rape." *Journal of Interpersonal Violence* 27: 728–752.

Worchel, Stephen, Joel Cooper, George R. Goethals, and James L. Olsen. 2000. *Social Psychology*. Belmont, CA: Wadsworth.

World Bank. 2010. *Independent Evaluation Group, An Evaluation of World Bank Support 1997–2007: Water and Development*. New York: World Bank. **www.worldbank.org**

World Bank. 2012. *GNI Per Capita, Atlas Index*. New York: World Bank. **www.worldbank.org**

World Health Organization. 2010. *AIDS*. New York: World Health Organization. **www.who.org**

World Hunger Education Service. 2011. "2011 World Hunger and Poverty Facts and Statistics." Washington, DC: World Hunger Education Service. **www.worldhunger.org**

World Hunger Organization. 2012. *2012 World Hunger and Poverty Facts and Statistics*. New York: World Hunger Organization. **www.worldhunger.org**

Wright, Erik Olin. 1979. *Class Structure and Income Determination*. New York: Academic Press.

Wright, Erik Olin. 1985. *Classes*. London: Verso.

Wright, Erik Olin, Karen Shire, Shu-Ling Hwang, Maureen Dolan, and Janeen Baxter. 1992. "The Non-Effects of Class on the Gender Division of Labor in the Home: A Comparative Study of Sweden and the United States." *Gender & Society* 6 (June): 252–282.

Wright, R. 2000. *Nonzero: The Logic of Human Destiny*. New York: Pantheon.

Wuthnow, Robert (ed.). 1994. *I Come Away Stronger: How Small Groups Are Shaping American Religion*. Grand Rapids, MI: Eerdmans.

Wuthnow, Robert. 1998. *After Heaven: Spirituality in America since the 1950s*. Berkeley, CA: University of California Press.

Wuthnow, Robert. 2010. *Boundless Faith: The Outreach of American Churches*. Berkeley: University of California Press.

Wysocki, Diane Kholos. 1998. "Let Your Fingers Do the Talking: Sex on an Adult Chat-Line." *Sexualities* 1 (November): 425–452.

Xu, Wu, and Ann Leffler. 1992. "Gender and Race Effects on Occupational Prestige, Segregation, and Earnings." *Gender & Society* 6 (September): 376–392.

Xuewen, Sheng, Norman Stockman, and Norman Bonney. 1992. "The Dual Burden: East and West (Women's Working Lives in China, Japan, and Great Britain)." *International Sociology* 7 (June): 209–223.

Zajonc, Robert B. 1968. "Attitudinal Effects of Mere Exposure." *Journal of Personality and Social Psychology*. Monograph Supplement, Part 2: 1–29.

Zald, Mayer, and John McCarthy. 1975. "Organizational Intellectuals and the Criticism of Society." *Social Service Research* 49: 344–362.

Zhou, Min, and Carl L. Bankston. 2000. "Immigrant and Native Minority Groups, School Performance, and the Problem of Self-Esteem." Paper presented at the Southern Sociological Society, April, New Orleans, LA.

Zimbardo, Phillip G., Ebbe B. Ebbesen, and Christina Maslach. 1977. *Influencing Attitudes and Changing Behavior*. Reading, MA: Addison-Wesley.

Zola, Irving Kenneth. 1989. "Toward the Necessary Universalizing of a Disability Policy." *Milbank Quarterly* 67 (Suppl. 2): 401–428.

Zola, Irving Kenneth. 1993. "Self, Identity and the Naming Question: Reflections on the Language of Disability." *Social Science and Medicine* 36 (January): 167–173.

Zoroya, Gregg. 2012. "Homeless, At-Risk Vets Double." *USA Today* (December 27).

Zorza, J. 1991. "Woman Battering: A New Cause of Homelessness." *Clearinghouse Review* 25 (4).

Zuckerman, Phil. 2002. "The Sociology of Religion of W. E. B. DuBois." *Sociology of Religion* 63: 239–253.

Zweigenhaft, Richard L., and G. William Domhoff. 2006. *Diversity in the Power Elite: How It Happened, Why It Matters*. Lanham, MD: Rowman and Littlefield.

Zwick, Rebecca, (ed.). 2004. *Rethinking the SAT: The Future of Standardized Testing in University Admissions*. New York: Routledge Falmer.

Name Index

Page numbers in *italics* refer to boxed material or illustrations.

Subject Index

Page numbers in **bold** refer to definitions/explanations of key terms.
Page numbers in *italics* refer to boxed material or illustrations.

Employment. *See also* Unemployment; Workplace
  disability and, 372-3
  of immigrants, 366-7, *367*
  precarious work, *366*
  recidivism and, 156
  social network and, *129, 130*
  of women, 269-70, 271-2, *277*
  work-home balance, 272, 322
End-of-life care, *351*
Energy use and global stratification, *213,* 214, *214*
Engels, Friedrich, 15
Enlightenment, **13-4**
Enron Corporation, 163-4
Environment
  global stratification and, 213
  inequality and, 397-8
  natural disasters, 396
  pollution, *392,* 392-5, *394, 395*
  population studies, 398-404
Environmental justice movement, **397-8**
Environmental racism, **397,** *397, 398*
Environmental sociology, **392**
Equal Pay Act of 1963, 255-6, **278**
Equal Rights Amendment, **278**
Erikson, Erik, **88**
Estate system, **174**
Ethics
  human genome project, 73
  research, 68-70
Ethnic group/ethnicity, **228-9.** *See also* African
    Americans; Latinos; Race
  Asian Americans, 245-6
  assimilation perspective, 239, 248-9
  body image, 113
  census classifications, *400*
  foreclosures by, *176*
  gender identity, 265
  Middle Easterners, 246-7
  minority population, *12*
  Native Americans, 241
  organizational leadership, 140
  policing and, 166-7
  religious belief, 330-1
  SAT scores by, *345*
  segregation, 249-50
  socialization of young adults, *89*
  teenage childbearing by, *299*
  television usage by, *44*
  theories of, *240*
  voters by, *381*
  wealth distribution, 181
  White, 247-8
Ethnic identity, **228**
Ethnicity. *See* Ethnic group/ethnicity
Ethnic prejudice, **235**
Ethnic quotas, **248**
Ethnocentrism, **39, 235**
Ethnomethodology, **33,** 35, *114,* **115-6**
Eufunction, **142**
Eugenics, **295**
Evaluation research, *66,* **68**
Evil, as banal, 133

Evolution, **16**
Experimental group, **63**
Experimental randomization, **64**
Expressive need, **137**
Extended family, **308,** *323*
Extreme conversion, **95**
Extreme poverty, **217**
Eye contact, 112

Fair trade, *408*
Faking data, *65*
False consciousness, **188, 329**
Family, **307.** *See also* Children; Marriage
  changing character of, *306*
  diversity among, 312-6
  divorce, 317-9
  extended, 308, *323*
  father-absent, *313*
  gender roles, 316-7, *318*
  gender socialization, 260-1
  ideal of, 305, *313*
  interracial, *75,* 307-8, *308, 309*
  kinship systems, 307-8
  nuclear, 308-10
  social change and, 321-2
  socialization, 78-9, 87, *311*
  social policy and, 322-4
  theories of, 18, *310,* 310-2
  transnational, *207,* 321-2
  violence within, 319-21
  work-home balance, 272, 322
Family and Medical Leave Act of 1993, **322**
Family-wage economy, **106**
Family wage system, **308**
Farkas, Lee, 164
Father-absent family, *313*
Female-headed household, 312-3
Feminism, **273**
Feminist theory, **21-2, 273**
  family, *310,* 311-2
  gender, 273-5, *274*
  power, *377*
  sexuality, *289,* 290-1
  state, 379
Feminization of poverty, **193,** 313
Feral children, 74
Fertility rate, **212-3,** 219
Field experiment, **65**
Field research, **63**
Filipinos, 246
Folkways, **33**
Foraging society, *104,* 104-5
Formal deviance, **146**
Formal organization, **107-8, 135-6.** *See also* Bureaucracy
Freud, Sigmund, **83,** 86
Functionalism/functionalist theory, **14, 18,** 22
  age stratification, 92, *93*
  crime, *160*
  deviance, *149,* 149-54
  economy and work, *373,* 373-4
  education, 339-40, *343*
  emphasis of, 21

Informal structure of bureaucracy, **137–8**
Informant, **63**
Informed consent, **70**
In-group, **128–9, 235**
Initiation into group, *128*
Innate trait, **76**
Institutional racism, **238–9**
Instrumental need, **127**, 137
Intelligence, 346–7, *347*
Interest group, **377**
Interlocking directorates, **378**
Internalization, **74**
International division of labor, **208**, 211
Internet
    cybersex, 302
    cyberspace revolution, 407–8
    cyberterrorism, 168
    dating, 113
    demographics of users, *120*
    email scams, *9*
    in research, 55
    social class, *42*
    social interaction, *80,* 107, 119–20
Interpersonal attraction, **112–4**
Intersection perspective, **240**
Intersexed, **257–8**
Intervening variable, **57**
Islamic views of West, *39*
Issues, social, **6**

Japan
    culture of, *28, 39, 94*
    people of, *46,* 245–6
Jewish immigrants, *247,* 248
Jihad, **39**
Job displacement, **365**
Job network, *130*
Jobs. *See* Employment; Workplace
Johnson and Johnson Co., 163
Juries, 124, 227–8

Kinship system, **307–8**
Koreans, 246
Kozlowski, Dennis, 164
Kurdish people, 46
Kwakiutl society, 36

Label, **156**
Labeling theory, **146, 156–9**
Labor, division of, **103,** 208, 211, **367–8**
Labor, emotional, **367**
Labor force participation rate, **268**
Labor unions, *184*
Laddered model of class, 181–4
Laissez-faire approach to social change, **16**
Laissez-faire racism, **238**
Language, **31**
    "Baby Einstein" early intervention program, *67*
    diversity of, *32, 37*
    as element of culture, *31,* 31–2

    socialization and, 78
    social meaning of, *34*
Lanza, Adam, 150, *152,* 153
Latent function, **18, 340**
Latinos
    ethnic identity, socialization of, *89*
    history and groups of, 243–5
    religious beliefs of, 331
    social class and, 185–6
Laws, **33,** 375
Learned, culture as, 28–9
Leisure time by gender, *317*
Lesbians. *See* Gays and lesbians
Liberal feminism, **273**
Life chances, **174**
Life course perspective, **87,** 87–94
Lindh, John Walker, 95
Literature review, **55**
Living wage campaign, **195**
Looking-glass self, *80, **84***
Lorenz, Konrad, *112*
Loughner, Jared, 151
Lower class, **184**
Lower-middle class, **184**
Lumpenproletariat, **15, 189**

Macedonia, culture of, *94*
Macroanalysis, **100,** 125, 135
Macrochange, **406**
Macrosociology, **18**
Macrostructural approach, **154**
Madoff, Bernard L., 162–3, *163*
Mafia, 164
Majority group, *403*
Mali, culture of, *28*
Malthus, Thomas R., **404**
Malthusian theory, **404–5**
Manifest function, **18**
Marriage
    culture and, *94,* 285
    gender roles, 316–7, *318*
    infidelity, 286
    interracial, 307–8, *309*
    monogamy in, 307
    race and, *316*
    same-sex, 284, 285, 314, *315*
    teen pregnancy and, 298–9
Married-couple families, 313–4
Martineau, Harriet, **14**
Marx, Karl, **15,** 18, 19, *20,* 185, 188, **189–90,** 210, 329, 332, **378.** *See also* Conflict theory
Mascots, Native American names for, *29*
Masculinity, perceptions of, *294*
Mass media, **41**
    age and news consumption, *48*
    class stereotypes, *188*
    death of celebrities, *49*
    gender socialization, 262
    images of family, *313*
    images of violent crime, *162*
    living without, *43*

Mass media (*continued*)
  organization of, 41–2
  popular culture and, 41, 42–3
  research reports, *58–9*
  sexuality, 302–3
  as socialization agent, 79
  theoretical perspectives on culture, 45–8, *46*
  violence, *66*, 79, *162*
Master status, **108–9**, 157
Material culture, **26–7**
Maternity leave, *323*
Matriarchal religion, **327**
Matriarchy, **268**
Matrilineal kinship system, **307**
Matrilocal, **307**
Matrix of domination, **250**
McDonaldization of society, *139*, **139–40**, 202
Mead, George Herbert, 17, *80*, **84–6**
Mean, *64*
Meaning and religion, 325–6
Means of production, **189**
Mechanical solidarity, **102**, **411**
Media. *See* Mass media
Median, *64*
Median income, **181**
Medicaid, **351**, 353
Medicalization of deviance, **148–9**
Medicare, **351**, 354
Megachurch, **331**
Membership group, **127**
Mental illness, as deviance, 159
Mere exposure effect, **113**
Meritocracy, **187**
Merton, Robert, **18**, 73, 152
Mexican Americans, 243–4
Microanalysis, **100–1**, 124–5, 135
Microchange, **406**
Microsociological approach, **155**
Microsociology, **18**
Middle class, *180*, **184**, 185, 189–90
Middle Easterners, 246–7
Migration, *207*, 212, 402. *See also* Immigration
Milgram, Stanley, 68–9
Milgram Obedience Studies, *132*, 132–3
Military, **375**
  gays and lesbians, 387
  race and, 386
  resocialization and, 95
  as social institution, 385–6
  veterans, *15*, 387
  women, 386–7
Mills, C. Wright, **5–7**, 13, 87, 92, 202, 321, 378
Minority group, **233**, **294**, 384–5, ***403***. *See also* Ethnic group/ethnicity; Race
Mode, *64*
Modernization, **406**, **412**, 412–4
Modernization theory, **208–10**, *209*
Mommy tax, *267*
Monogamy, **307**
Monotheism, **327**
Mores, **33**
Mortality rate, **352**, *401*, **401–2**

Mortgage, subprime, **176–7**
Motherhood, Black teenage, *9*
Multidimensional analysis of society, **16**
Multidimensional poverty index, **217–8**, *218*
Multinational corporation, **210**, **363**
Multiracial, as ascribed status, 108
Multiracial feminism, **274–5**
Muslims, 331
Myth of model minority, **186**

National Crime Victimization Surveys, 161
Nationalism, **375**
Native Americans, 241
Natural resources, waste of, 395
Nature-nurture controversy, 75–6
Negative reinforcement, **83**
Neocolonialism, **210**
Net worth, **177**
  distribution of, 179
  by income quintile, *183*
  by race, *181*
Newly industrializing country, **223**
News consumption and age, *48*
Nigeria, culture of, *94*
Nike, 201, 202, 211
9/11 terrorist attack, 168
No Child Left Behind Act of 2003, 339, 346, 348–9
Nonmaterial culture, **27**, 45
Non-participant observation, **62**
Nonverbal communication, **110**
Non-zero sum game, **118–9**
Normative organization, **136**
Norm-restoration behavior, 116
Norms, **33**. *See also* Socialization
  as element of culture, *31*, 33, 35, 45
  informal, in bureaucracy, 137–8
  religion and, 325
Not-me syndrome, **131**
Nouveau riche, **183**
Nuclear family, **308–10**

Obama, Barack, 228, *231*, 308
Obedience experiment, *132*, 132–3
Obesity, *350*, **350–1**
Observation
  empirical, 7
  nonparticipant, 62
  participant, 53–4, 62–3, *66*, *69*
Occupational prestige, **182**, **369–70**
Occupational segregation, **270–1**, **369**
Occupational system, **368**
Occupational training, **340**
Occupy America movement, *16*, *374*
Old-fashioned racism, **237**
Online community, *80*, 107
Open class system, **181**, **187**
Open-ended question, **62**
Organic food, *408*
Organic metaphor, **16**
Organic solidarity, **103**, **411**
Organizational culture, **135**

Spatial mismatch, **365**
Sports
    social change in, *409*
    social class and, *173*
    as socialization agent, 81–2
    social stratification and, 172–3
    women in, *258*, 278
Spurious correlation, *64*
State, **374–5**. *See also* Government
    global interdependence and, 375–6
    power, authority, and bureaucracy, 376–7
    social order and, 375
    war, revolution, and social change, 410–1
    women heads of, *385*
Statistical concepts, *64–5*
Status, **108–9**, 159, **172**, **190**
    ascribed, 92, 108, 174
Status attainment, **181**
Status inconsistency, **108**
Status symbol, **175**
Stepfamilies, 314
Stereotype, **233**
    age, 90–1, 93
    class, *188*
    gender, conformity to, 262–5
    race, gender, class, and, 233, 234
    salience and, 233–4
Stereotype threat, **261**, **342**
Stigma, **157**, 197–8, **353**
Stigmatization, 156–7
Stockholm syndrome, **96**
Stranger, role of in group, 9–10
Streaking, *131*
Street drug trade, *69*
Structural strain theory, **152–3**, *153*
Student debt crisis, *179*
Subculture, **37–8**, 148
Subprime mortgage, **176–7**
Substance use/abuse
    alcohol, 7, 148–9
    conflict theories, *155*
    as deviant behavior, 148, 159–60
    by high school seniors, *160*
Suicide
    among veterans, *15*
    as deviant behavior, 146
    Durkheim and, 14–5
    rampage shootings as, 150–2
    rates by state, *151*
    types of, 149–50
Sui generis society, **14**, **100**
Support group, *126*
Surveys, 62, *66*
Sustainability, **391–2**, *408*
Sutherland, Edwin, 156
Symbolic, culture as, 29
Symbolic interaction theory, **19**, *22*
    age stratification, 93, *93*
    crime, *160*
    culture, 47–8
    cyberspace interaction, 120
    deviance, *149*, 155–9

    economy and work, *373*, 374
    education, 341–2, *343*
    emphasis of, 21
    family, *310*, 311
    gender, 273, *274*
    health care, *355*, 355–6
    online communities and, *80*
    organizations, *141*, 142
    prejudice and racism, 239–40
    race and ethnicity, *240*
    religion, *328*, 329
    sexuality, *289*
    social change, 412, *413*
    socialization, 84–5

Taboos, **33**
Tactile communication, **110**, *111*
Tattoos, *30*, 148, 157, *157*
Tax rates, 177, 185
Taylor, Howard, *11*
Tchikrin people, 26, *26*
Teacher expectancy effect, **341–2**
Tea Party, *383*
Technological innovation
    cyberspace revolution, 407–8
    economy and, 365–6
    social change and, 50
Teens. *See* Adolescents
Temporary Assistance for Needy Families, **196**
Tenure, **137**
Terrorism, 168, 215–7, **216**
Tertius gaudens, **125**
Testing, educational, 344–5, *345*, 349
Theoretical frameworks
    on culture and media, 45–8, *46*
    described, 18–9, *20*, 21–2, *22*
    economic recession, *20*
    education, 339–42, *343*
    family, *310*, 310–2
    gender, 272–5, *274*
    global stratification, 208–12, *209*
    health care, 354–5, *355*
    organizations, *141*, 141–2
    population growth, 404–6
    power, *377*, 377–9
    prejudice and racism, 239–40, *240*
    religion, 327–9, *328*
    sexuality, 288–91, *289*
    social change, 411–2, *413*
    social inequality, 189–92, *191*
    social interaction, *114*, 114–9
    socialization, *83*, 83–6
Third shift of housework, **316**
Thomas, W. I., **128**, **155**, 156
Time
    concepts of, 32
    culture as varying across, 29–30
Title IX, **278**
Tocqueville, Alexis de, **14**
Tönnies, Ferdinand, **413**
Torture of Iraqi prisoners of war, 133–4, 135